STRUCTURAL TIMBER DESIGN

to Eurocode 5

To Margaret and John
Dawn and Romi

STRUCTURAL TIMBER DESIGN
to Eurocode 5

2nd Edition

Jack Porteous BSc, MSc, DIC, PhD, CEng, MIStructE, FICE
Director
Jack Porteous Consultancy

and

Abdy Kermani BSc, MSc, PhD, CEng, FIStructE, FIWSc
Professor and Director of Centre for Timber Engineering
Edinburgh Napier University

⊛WILEY-BLACKWELL
A John Wiley & Sons, Ltd., Publication

Registered Office
John Wiley & Sons, Ltd, The Atrium, Southern Gate, Chichester, West Sussex, PO19 8SQ, United Kingdom.

Editorial Offices
9600 Garsington Road, Oxford, OX4 2DQ, United Kingdom.
The Atrium, Southern Gate, Chichester, West Sussex, PO19 8SQ, United Kingdom.

For details of our global editorial offices, for customer services and for information about how
to apply for permission to reuse the copyright material in this book please see our website at
www.wiley.com/wiley-blackwell.

Library of Congress Cataloging-in-Publication Data

Porteous, Jack.
　Structural timber design to Eurocode 5 / Jack Porteous and Abdy Kermani. – 2nd edition.
　　pages　cm
　Includes bibliographical references and index.
　ISBN 978-0-470-67500-7 (pbk.)
1. Building, Wooden–Standards–Europe.　2. Structural frames–Design and construction–Standards–
Europe.　3. Timber–Standards–Europe.　I. Kermani, Abdy.　II. Title.
　TA666.P66 2013
　694.02′184–dc23
　　　　　　　　　　　　　　　　　　　　　　2012051136

A catalogue record for this book is available from the British Library.

Cover design by Sandra Heath

Set in 10/12pt Times by SPi Publisher Services, Pondicherry, India

1　2013

Contents

Preface to the Second Edition

As a natural material, timber is unique, innovative and easy to handle. It is sustainable, environmentally friendly, can be readily recycled and as sawn sections or quality controlled engineered products, timber has a large potential market for use as a structural and building material. However, the existing civil and structural engineering curricula neglect, to a large extent, the importance of timber as a viable engineering material and as a consequence relatively few textbooks provide information on the design of timber structures. Also, most books have tended to concentrate on designs in accordance with BS 5268, a permissible stress based design, with limited information on designs to Eurocode 5, which is based on a limit states design philosophy.

However, on 31 March 2010 the structural design code used for timber, BS 5268, was withdrawn and is now replaced by Eurocode 5 and from 2015 all timber designs in the UK will have to be carried out in accordance with this code. This book is based solely on the use of *Eurocode 5: Design of structures – Part 1-1: General – Common Rules and Rules for Building*, referred to as EC5 in the book, and incorporates the requirements of the associated UK National Annex. There is a pressing need for practising engineers as well as specialist contractors, postgraduate and undergraduate students of civil and structural engineering courses to become familiar with the design rules in EC5 and this book offers a detailed explanation and guide to the use of the code. It provides comprehensive information and a step-by-step approach to the design of elements, connections and structures using numerous worked examples and encourages the use of computers to carry out design calculations.

The version of EC5 covered in the first edition of the book was revised in 2008 (BS EN: 1995-1-1:2004+A1:2008) and any changes that are relevant to the book content have been taken into account in this edition. The UK NA to BS EN 1995-1-1 was also revised in 2012 and this book makes reference to the current issue, *NA to BS EN 1995-1-1:2004+A1:2008, Incorporating National Amendment No. 2*. The first issue of non-contradictory complementary information relating to EC5 was finalised in 2012 and published in PD6693-1 and where the content of this document is considered to be of relevance to an issue covered in the book, reference has been made to it in the relevant chapter. Further, a corrigendum relating to the above version of EC5 is about to be issued to address errors in some of the code clauses and also provide some clarification on the implementation of some of the existing clause requirements. In anticipation of the publication of the corrigendum, those matters of relevance to the content of the book have also been taken into account in this edition. New content has been added to the text, covering cases where composite sections are subjected to combined axial and bending stresses, a new chapter has been added covering racking design of wall diaphragms and the opportunity has been taken to correct any areas of possible misinterpretation of the code, minor presentation and typographical errors in the previous edition.

Chapter One introduces the nature and inherent characteristics of timber and gives an overview of timber and its engineered products as structural and building materials, and includes design related information on the strength and stiffness properties required for design in accordance with the requirements of EC5. In Chapter Two the design philosophy used in Eurocodes is explained. It includes information on the relevance of the requirements of Eurocode 0 to EC5 as well as the significance of the effects of moisture content, load duration, creep behaviour and size factors etc., in the design process.

Chapter Three gives an overview of Mathcad®, a computer software used to carry out mathematical calculations, and details its simplicity and the advantages that it provides when used for design calculations. The software is commonly used in design offices and universities and the aim is to encourage readers to use computing as a tool to increase their understanding of how design solutions vary in response to a change in one or more of the variables and how alternative design options can be easily investigated. The design of basic elements is explained and illustrated in Chapters Four and Five, whilst the design of more specialised elements such as glued-laminated straight, tapered and curved beams and columns, thin webbed and thin flanged beams and built-up columns is covered in Chapters Six, Seven and Eight using numerous worked examples.

In Chapter Nine, the lateral stability requirements of timber structures are addressed, and the design of stability bracing as well as the racking resistance of floor diaphragms together with a reference to wall diaphragms, using the rules in EC5, are explained.

The design of connections using metal dowel-type fasteners is covered in Chapter Ten. It includes an overview of the theory used for connection design together with a comprehensive coverage of the lateral and axial strength requirements of nailed, screwed and bolted joints. The lateral stiffness behaviour of these types of connections is also covered in Chapter Ten as well as the design of connections with multiple shear planes. Several step-by-step worked examples are provided to illustrate the design methods explained in this chapter.

Chapter Eleven covers the strength and stiffness behaviour of connected joints such as toothed-plates, split-rings and shear-plates. In Chapter Twelve, the design of rigid and semi-rigid connections subjected to combined moment and lateral forces is addressed with examples showing the significant effect on joint and member behaviour when semi-rigid behaviour is included in the design process. The final chapter, Chapter Thirteen, is a new chapter covering the simplified analysis of wall diaphragms using Method A, referred to in EC5, together with the new simplified analysis method developed for use in the UK, referred to in PD6693-1, which replaces the UK requirement to use Method B.

As stated earlier, a corrigendum to BS EN: 1995-1-1:2004+A1:2008 is about to be issued and an outline of the proposed changes being considered is given in Appendix C and where an amendment will result in a change to the design procedure described in the book, reference is made to the draft proposal in the text.

All design examples given in this book are produced in the form of worksheet files and are available on a CD to run under Mathcad Version 11, or higher. Details are given at the back of the book. The examples are fully self-explanatory and well annotated and the authors are confident that the readers, whether practising design engineers, course instructors or students, will find them extremely useful to produce design solutions or prepare course handouts. In particular, the worksheets will allow design engineers to undertake sensitivity analyses to arrive at the most suitable/economic solution(s) very quickly.

To prevent any confusion between the numbering system used in the book and that used in the Eurocodes, where reference is made in the text to a specific section, item number, or table in a Eurocode and/or its accompanying UKNA, it is given in italics. For example *6.4.2* refers to Item *6.4.2* of EC5 whereas 6.4.2 refers to Section 6.4.2 in Chapter 6 of the book.

Permission to reproduce extracts from British Standards is granted by BSI. British Standards can be obtained from BSI Customer Services, 389 Chiswick High Road, London W4 4AL. Tel: +44 (0)20 8996 9001. email: cservices@bsigroup.com. Web: www.bsigroup.com/en/standards-and-publications/.

Chapter 1

Timber as a Structural Material

1.1 INTRODUCTION

Timber from well-managed forests is one of the most sustainable resources available and it is one of the oldest known materials used in construction. It has a very high strength to weight ratio, is capable of transferring both tension and compression forces and is naturally suitable as a flexural member. Timber is a material that is used for a variety of structural forms such as beams, columns, trusses, girders, and is also used in building systems such as piles, deck members, railway sleepers and in formwork for concrete.

There are a number of inherent characteristics that make timber an ideal construction material. These include its high strength to weight ratio, its impressive record for durability and performance and good insulating properties against heat and sound. Timber also benefits from its natural growth characteristics such as grain patterns, colours and its availability in many species, sizes and shapes that make it a remarkably versatile and an aesthetically pleasing material. Timber can easily be shaped and connected using nails, screws, bolts and dowels or adhesively bonded together.

The limitations in maximum cross-sectional dimensions and lengths of solid sawn timbers, due to available log sizes and natural defects, are overcome by the recent developments in composite and engineered wood products. Finger jointing and various lamination techniques have enabled timbers (elements and systems) of uniform and high quality in any shape, form and size to be constructed; being only limited by the manufacturing and/or transportation boundaries.

Timber structures can be highly durable when properly treated, detailed and built. Examples of this are seen in many historic buildings all around the world. Timber structures can easily be reshaped or altered, and if damaged they can be repaired. Extensive research over the past few decades has resulted in comprehensive information on material properties of timber and its reconstituted and engineered products and their effects on structural design and service performance. Centuries of experience of use of timber in buildings has shown us the safe methods of construction, connection details and design limitations.

This chapter provides a brief description of the engineering properties of timber that are of interest to design engineers and architects, and it highlights that, unlike some structural materials such as steel or concrete, the properties of timber are very

Structural Timber Design to Eurocode 5, Second Edition. Jack Porteous and Abdy Kermani.
© Jack Porteous and Abdy Kermani 2013. Published 2013 by Blackwell Publishing Ltd.

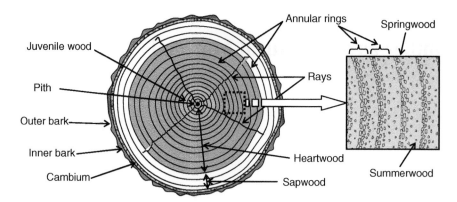

Fig. 1.1. Cross-section of tree trunk.

sensitive to environmental conditions; for example moisture content, which has a direct effect on the strength and stiffness, swelling or shrinkage of timber. A proper understanding of the physical characteristics of timber enables the building of safe and durable timber structures.

1.2 THE STRUCTURE OF TIMBER

Structural timber is sawn (milled) from the trunk of the tree, which provides rigidity, mechanical strength and height to maintain the crown. *Trunk* resists loads due to gravity and wind acting on the tree and also provides for the transport of water and minerals from the tree roots to the crown. *Roots*, by spreading through the soil and acting as a foundation, absorb moisture-containing minerals from the soil and transfer them via the trunk to the crown. *Crown*, comprising branches and twigs to support leaves, provides a catchment area producing chemical reactions that form sugar and cellulose that allow the growth of the tree.

As engineers we are mainly concerned with the trunk of the tree. A typical cross-section of a tree trunk, shown in Figure 1.1, illustrates its main features such as *bark*, the outer part of which is a rather dry and corky layer and the inner living part. The *cambium*, a very thin layer of cells underside the inner bark, is the growth centre of the tree. New wood cells are formed on the inside of the cambium (over the old wood) and new bark cells are formed on the outside and as such increase the diameter of the trunk. Although tree trunks can grow to a large size, in excess of 2 m in diameter, commercially available timbers are more often around 0.5 m in diameter.

Wood, in general, is composed of long thin tubular cells. The cell walls are made up of cellulose and the cells are bound together by a substance known as lignin. Most cells are oriented in the direction of the axis of the trunk except for cells known as *rays*, which run radially across the trunk. The rays connect various layers from the pith to the bark for storage and transfer of food. Rays are present in all trees but are more pronounced in some species such as oak. In countries with a temperate climate, a tree produces a new layer of wood just under the cambium in the early part of every growing season. This growth ceases at the end of the growing season or during winter months. This process results in clearly visible concentric rings known as *annular rings*, *annual rings* or *growth rings*. In tropical countries, where trees grow through-out the year, a tree produces wood cells that are essentially uniform. The age of a tree may be determined by counting its growth rings [1, 2].

The annular band of the cross-section nearest to the bark is called *sapwood*. The central core of the wood, which is inside the sapwood, is *heartwood*. The sapwood is lighter in colour compared to heartwood and is 25–170 mm wide depending on the species. It contains both living and dead cells and acts as a medium for transportation of sap from the roots to the leaves, whereas the heartwood, which consists of inactive cells, functions mainly to give mechanical support or stiffness to the trunk. As sap-wood changes to heartwood, the size, shape and the number of cells remain unchanged. In general, in hardwoods the difference in moisture content of sapwood and heart-wood depends on the species but in softwoods the moisture content of sapwood is usually greater than that of heartwood. The strength and weights of the two are nearly equal. Sapwood has a lower natural resistance to attacks by fungi and insects and accepts preservatives more easily than heartwood.

In many trees and particularly in temperate climates, where a definite growing sea-son exists, each annular ring is visibly subdivided into two layers: an inner layer made up of relatively large hollow cells called *springwood* or *earlywood* (due to the fast growth), and an outer layer of thick walls and small cavities called *summerwood* or *latewood* (due to a slower growth). Since summerwood is relatively heavy, the amount of summerwood in any section is a measure of the density of the wood; see Figure 1.1.

1.3   TYPES OF TIMBER

Trees and commercial timbers are divided into two types: *softwoods* and *hardwoods*. This terminology refers to the botanical origin of timber and has no direct bearing on the actual softness or hardness of the wood as it is possible to have some physically softer hardwoods like balsa from South America and wawa from Africa, and some physically hard softwoods like the pitchpines.

1.3.1   Softwoods

Softwoods, characterised by having naked seeds or as cone-bearing trees, are gener-ally evergreen with needle-like leaves (such as conifers) comprising single cells called *tracheids*, which are like straws in plan, and they fulfil the functions of conduction and support. Rays, present in softwoods, run in a radial direction perpendicular to the growth rings. Their function is to store food and allow the convection of liquids to where they are needed. Examples of the UK grown softwoods include spruce (white-wood), larch, Scots pine (redwood) and Douglas fir.

1.3.1.1   Softwood characteristics

- Quick growth rate (trees can be felled after 30 years) resulting in low-density timber with relatively low strength.
- Generally poor durability qualities, unless treated with preservatives.
- Due to the speed of felling they are readily available and comparatively cheaper.

1.3.2   Hardwoods

Hardwoods are generally broad-leaved (deciduous) trees, which often lose their leaves at the end of each growing season. The cell structure of hardwoods is more complex than that of softwoods with thick-walled cells, called fibres, providing the structural support and thin-walled cells, called *vessels*, providing the medium for food conduction. Due to the necessity of growing new leaves every year the demand for sap is high and in some instances larger vessels may be formed in the springwood, these are referred to as 'ring-porous' woods such as oak and ash. When there is no definite growing period the pores tend to be more evenly distributed, resulting in 'diffuse-porous' woods such as poplar and beech. Examples of the UK grown hardwoods include oak, beech, ash, alder, birch, maple, poplar and willow.

1.3.2.1   *Hardwood characteristics*

- Hardwoods grow at a slower rate than softwoods, which generally results in a timber of high density and strength, which takes time to mature, over 100 years in some instances.
- There is less dependence on preservatives for durability qualities.
- Due to the time taken to mature and the transportation costs of hardwoods, as most are tropical, they tend to be expensive in comparison with softwoods.

British Standard BS 7359:1991 [3] provides a list of some 500 timbers of economic interest in the United Kingdom and tabulates softwoods and hardwoods including their standard names, botanical names/species type and also, where relevant, their alternative commercial names with sources of supply and average densities.

1.4   NATURAL CHARACTERISTICS OF TIMBER

Wood as a natural material is highly varied in its structure and has many natural characteristics or defects which are introduced during the growing period and during the conversion and seasoning process. Often such characteristics or defects can cause problems in timber in use either by reducing its strength or impairing its appearance.

1.4.1   Knots

These are common features of the structure of wood. A knot is a portion of a branch enclosed by the natural growth of the tree, normally originating at the centre of the trunk or a branch. The influence of knots depends on their size, shape, frequency and location in the structural member. The presence of knots has adverse effects on most mechanical properties of timber as they distort the fibres around them, causing fibre discontinuity and stress concentrations or non-uniform stress distributions. Their effects are further magnified in members subjected to tensile stress either due to direct or bending stresses. For example, the presence of a knot on the lower side of a flexural member, being subjected to tensile stresses due to bending, has a greater effect on the load capacity of the member than a similar knot on the upper side being subjected to compressive stresses.

Table 1.1 Effect of grain deviation on strength properties of timber

Slope of grain	Bending strength (%)	Compression parallel to grain (%)	Impact loading (%)
Straight grain	100	100	100
1 in 20 (3°)	93	100	95
1 in 10 (6°)	81	99	62
1 in 5 (11.5°)	55	93	36

The presence of knots in round timber has much less effect on its strength properties than those in a sawn timber. When a log is sawn, the knots and fibres surrounding it will no longer be continuous – thus adversely affecting the strength properties; whereas in the round timber there are no discontinuities in the wood fibres and often the angle of grain to the longitudinal axis is smaller than that in the sawn timber.

In general, the size, shape, frequency and location of knots influence the quality and hence the grade of softwood timbers for structural use, with better grades having fewer and smaller knots.

1.4.2   Slope of grain

Wood grain refers to the general direction of the arrangement of fibres in wood and is expressed with respect to the longitudinal axis of the sawn timber or the round timber (log or pole). In general, the direction of the fibres does not lie truly parallel to the longitudinal axis of the sawn or round timbers. In softwoods, the deviation with respect to the log (longitudinal) axis is often constant, resulting in the production of *spiral* grain. *Interlocked* grains are often produced in tropical hardwoods where the grain direction changes routinely from one direction to another.

A *cross* grain occurs when the grain direction is at an angle to the longitudinal axis of the sawn section. A cross grain occurs during conversion (sawing process) as a result of conversion of a bent or heavily tapered log or a log with spiral or interlocked grain.

Grain deviation can severely impair the strength properties of timber. Visual grading rules limit the grain deviation; in general, a grain deviation of 1 in 10 is accepted for high-grade timber whereas 1 in 5 often relates to a low-grade one. The effect of grain deviation on some properties of timber is shown in Table 1.1.

1.4.3   Reaction wood

Reaction wood refers to abnormal wood tissues produced in tree trunks subjected to strong wind pressures. Horizontal branches and leaning branches are believed to form reaction wood in an attempt to prevent them from excessive bending and cracking under their own weight. There are two types of reaction wood: in softwoods it is referred to as *compression wood* and in hardwoods as *tension wood*. Compression wood, Figure 1.2, forms on the underside of branches of leaning softwoods and contains more lignin than normal wood. Tension wood forms on the upper sides of leaning hardwoods and contains more cellulose than normal wood.

Fig. 1.2. Compression wood (dark patch).

Reaction wood is much denser than normal wood with the specific gravity of around 35% greater in compression wood and 7% greater in tension wood. Longitudinal shrinkage is also greater, 10 times more than normal for compression wood and 5 times for tension wood. Timber containing compression wood is liable to excessive distortion during drying and tends to fail in a brittle manner. It is harder to drive a nail in compression wood, there is a greater chance of it splitting, and compression wood may take a strain differently than normal wood. Most visual strength grading rules limit the amount of compression wood in high quality grades.

1.4.4   Juvenile wood

This is a wood that is produced early in the first 5–20 rings of any trunk cross-section (Figure 1.1) and, in general, exhibits lower strength and stiffness than the outer parts of the trunk and much greater longitudinal shrinkage than mature, normal wood. Juvenile wood is mainly contained within the heartwood. In this regard, in young, fast grown trees with a high proportion of juvenile wood, heartwood may be inferior to sapwood, but is not normally considered a problem.

1.4.5   Density and annual ring widths

Density is an important physical characteristic of timber affecting its strength properties. *Annual ring width* is also critical in respect of strength in that excessive width of such rings can reduce the density of the timber. Density can be a good indicator of the mechanical properties provided that the timber section is straight grained, free from knots and defects. The value of density as an indicator of mechanical properties can also be reduced by the presence of gums, resins and extractives, which may adversely affect the mechanical properties. In this regard, the prediction of strength on the basis of density alone is not always satisfactory. Research studies show a coefficient of determination, R^2, ranging between 0.16 and 0.4 for density and 0.2 and 0.44 for the annual ring width [4].

Specific gravity or relative density is a measure of timber's solid substance. It is generally expressed as the ratio of the oven-dry weight of the timber to the weight of an equal volume of water. Because water volume varies with the moisture content of the timber, the specific gravity of timber is normally expressed at a certain moisture content. Basic oven-dry specific gravity of commercial timber ranges from 0.29 to 0.81, most falling between 0.35 and 0.60.

1.4.6   Conversion of timber

Once the tree is felled in the forest, the crown is removed and often it is also debarked in the forest. Logs are then classed and stockpiled under water sprays to prevent them from drying out. Some of the better quality ones are sent to peeling plants for the manufacture of veneers but the majority (depending on the quality) are sent to saw-millers to convert round logs to sawn timber. There are many cutting patterns used to produce timber, but the first step in most sawmill operations is to scan the log for the best alignment and cutting pattern for optimum return; then remove one or two wings (slabs) from the logs to give some flat surfaces to work from. The log, referred to as a *cant*, is turned on a flat face and sawn through and through to give boards (sections) of the required thickness.

Each sawmill establishes its own cutting patterns for different sized logs; maximising the number of pieces cut in the most popular sizes. *Through conversion* produces mostly tangentially sawn timber and some quarter sawn sections. Tangential timber is economical to produce because of the relatively fewer repetitive production methods. Boxing the heart (Figure 1.3) eliminates the heartwood from the boards that would otherwise produce shakes, juvenile wood or may even be rotten.

The *quarter sawn* techniques are more expensive processes, with more wastage, because of the need to double (or more) handle the log. They are, however, more decorative and less prone to cupping or distortion.

There are several alternative variations of tangential and radial cuts to obtain the best or most economical boards for the end use. Examples of methods of log break-down and different cutting patterns are shown in Figure 1.3.

In growing trees, all cell walls including their voids, in both heartwood and sap-wood, are saturated with water (moisture content in excess of 100%). When a tree is cut and its moisture content falls to around 27%, the only moisture left is the bound water, which is the moisture that is part of the cell wall. This state is referred to as *fibre saturation point*. Wood, in general, is dimensionally stable when its moisture content is greater than the fibre saturation point. The process of drying (seasoning) timber should ideally remove over a third of the moisture from the cell walls. Timber at this stage is referred to as *seasoned* with a moisture content of between 12 and 25% (depending on the method and duration of drying, i.e. air, kiln, solar, microwave, etc.). Wood changes dimensionally with change in moisture below its fibre saturation point: it shrinks when it loses moisture and swells as it gains moisture. These dimensional changes are mostly in the direction of the annual

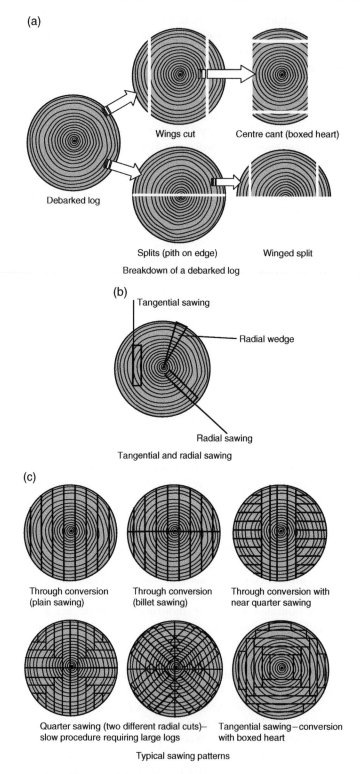

(a)

Wings cut   Centre cant (boxed heart)

Debarked log

Splits (pith on edge)   Winged split

Breakdown of a debarked log

(b)

Tangential sawing

Radial wedge

Radial sawing

Tangential and radial sawing

(c)

Through conversion   Through conversion   Through conversion with
(plain sawing)   (billet sawing)   near quarter sawing

Quarter sawing (two different radial cuts)–   Tangential sawing–conversion
slow procedure requiring large logs   with boxed heart

Typical sawing patterns

Fig. 1.3. Examples of log breakdown and cutting pattern.

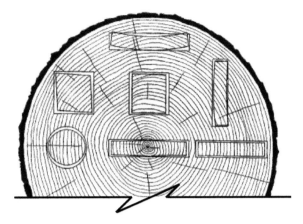

Fig. 1.4. Distortion of various cross-sections [5].

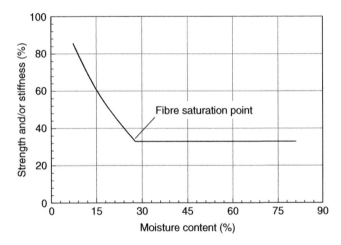

Fig. 1.5. General relationship between strength and/or stiffness and moisture content.

growth rings (tangentially), with about half as much across the rings (radially) and as such mainly affect cross-sectional dimensions (perpendicular to the grain) and can result in warping, checking or splitting of wood. Longitudinal shrinkage of wood (shrinkage parallel to the grain) for most species is generally very small. The combined effects of radial and tangential shrinkage (differential shrinkage) can distort the sawn timber. The major types of distortion as a result of these effects after drying for various cross-sections cut from different locations in a log are shown in Figure 1.4.

The change in moisture content of timber also affects its strength, stiffness and resistance to decay. Most timber in the United Kingdom is air-dried to a moisture content of between 17 and 23% (which is generally below the fibre saturation point) at which the cell walls are still saturated but moisture is removed from the cell cavities. Figure 1.5 highlights a general relationship between strength and/or stiffness characteristics of timber and its moisture content. The figure shows that there is an almost linear loss in strength and stiffness as moisture content increases to about

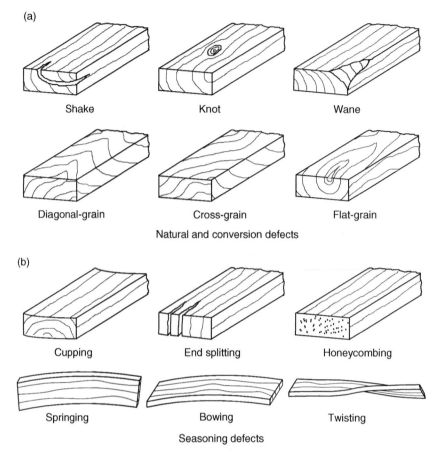

(a)

Shake   Knot   Wane

Diagonal-grain   Cross-grain   Flat-grain

Natural and conversion defects

(b)

Cupping   End splitting   Honeycombing

Springing   Bowing   Twisting

Seasoning defects

Fig. 1.6. Defects in timber.

27%, corresponding to the fibre saturation point. Further increase in moisture content has no influence on either strength or stiffness. It should be noted that although for most mechanical properties the pattern of change in strength and stiffness characteristics with respect to change in moisture content is similar, the magnitude of change is different from one property to another. It is also to be noted that as the moisture content decreases shrinkage increases. Timber is described as being hygroscopic, which means that it attempts to attain an equilibrium moisture content with its surrounding environment, resulting in a variable moisture content. This should always be considered when using timber, particularly softwoods, which are more susceptible to shrinkage than hardwoods.

As logs vary in cross-section along their length, usually tapering to one end, a board that is rectangular at one end of its length might not be so at the other end. The rectangular cross-section may intersect with the outside of the log, the *wane* of the log, and consequently have a rounded edge. The effect of a wane is a reduction in the cross-sectional area resulting in reduced strength properties. A wane is an example of a conversion defect and this, as well as other examples of conversion or natural defects, is shown in Figure 1.6a.

1.4.7   Seasoning

Seasoning is the controlled process of reducing the moisture content of the timber so that it is suitable for the environment and intended use. There are two main methods of seasoning timber in the United Kingdom, air-drying and kiln-drying; other less common methods include solar and microwave techniques. All methods require the timber to be stacked uniformly, separated by spacers of around 25 mm to allow the full circulation of air etc. around the stack. Often, ends of boards are sealed by a suitable sealer or cover to prevent rapid drying out through the end grains. However, with air-drying it is not possible to obtain less than 16–17% moisture content in the United Kingdom. Further seasoning would need to be carried out inside a heated and ventilated building.

The kiln-drying method relies on a controlled environment that uses forced air circulation through large fans or blowers, heating of some form provided by piped steam together with a humidity control system to dry the timber. The amount and duration of air, heat and humidity depend on species, size, quantity, etc.

1.4.8   Seasoning defects

Seasoning defects are directly related to the movements which occur in timber due to changes in moisture content. Excessive or uneven drying, as well as the presence of compression wood, juvenile wood or even knots, exposure to wind and rain, and poor stacking and spacing during seasoning can all produce defects or distortions in timber. Examples of seasoning defects such as cupping (in tangential cuts), end splitting, springing, bowing, twisting, etc. are illustrated in Figure 1.6. All such defects have an effect on structural strength as well as on fixing, stability, durability and finished appearance.

1.4.9   Cracks and fissures

These are caused by separation of the fibres along the grain forming fissures and cracks that appear on one face or at the end grain but do not necessarily continue through to the other side. Their presence may indicate decay or the beginnings of decay.

1.4.10   Fungal decay

This may occur in growing mature timber or even in recently converted timber, and in general it is good practice to reject such timber.

1.5   STRENGTH GRADING OF TIMBER

The strength capability of timber is difficult to assess as often there is no control over its quality and growth. The strength of timber is a function of several parameters including the species type, density, size and form of members, moisture content, duration of the applied load and presence of various strength reducing characteristics such as slope of grain, knots, fissures and wane. To overcome this difficulty, the strength grading method of strength classification has been devised. Several design properties are associated with a strength grade; these include modulus of elasticity and bending strength parallel to the grain, strength properties in tension and compression parallel and perpendicular to the grain, shear strength parallel to the grain and density. The design properties of timber are determined non-destructively through *visual strength grading* criteria or by *machine strength grading* via measurements such as the following: flatwise bending stiffness, using a three-point or four-point loading system; density, using x-rays or gamma rays techniques; and modulus of elasticity, by means of resonant vibrations (dynamic response) using one or a combination of these methods.

The requirements for strength grading of timber are detailed in the following standards:

- BS EN 14081-1:2005+A1:2011 [6]
- BS EN 14081-2:2010 [7].

Most European Union countries have their own long-established visual grading rules and as such guidance for visual strength grading of softwoods and hardwoods is provided in the following British Standards:

- BS 4978:2007+A1:2011 [8]
- BS 5756:2007 [9].

1.5.1   Visual grading

Visual grading is a manual process carried out by an approved grader. The grader examines each piece of timber to check the size and frequency of specific physical characteristics or defects, e.g. knots, slope of grains, rate of growth, wane, resin pockets and distortion.

The required specifications are given in BS 4978 and BS 5756 to determine if a piece of timber is accepted into one of the two visual stress grades or rejected. These are general structural (GS) and special structural (SS) grades. *Table 2* of BS 5268-2:2002 [10] (reproduced here as Table 1.2) refers to main softwood combinations of species (available in the United Kingdom) visually graded in accordance with BS 4978.

1.5.2   Machine grading

Machine grading of timber sections is carried out on the principle that stiffness is related to strength; where the relationship between the modulus of elasticity, E, and the modulus of rupture of a species of timber from a certain geographical location is determined from a statistical population, based on a substantial number of laboratory controlled tests. There are a number of ways of determining the modulus of elasticity, including resonant vibration (dynamic response), but the most common methods are either load- or deflection-controlled bending tests. The machine exerts pressure and bending is induced at increments along the timber length. The resulting deflection (or the load to induce a known deflection) is then automatically measured and compared

Table 1.2 Softwood combinations of species and visual grades that satisfy the requirements for various strength classes[*]

Timber species	Grade and related strength classes
British grown timber	
Douglas fir	GS (C14), SS (C18)
Larch	GS (C16), SS (C24)
British pine	GS (C14), SS (C22)
British spruce	GS (C14), SS (C18)
Imported timber	
Parana pine	GS (C16), SS (C24)
Caribbean pitch pine	GS (C18), SS (C27)
Redwood	GS (C16), SS (C24)
Whitewood	GS (C16), SS (C24)
Western red cedar	GS (C14), SS (C18)
Douglas fir-larch (Canada and USA)	GS (C16), SS (C24)
Hem-fir (Canada and USA)	GS (C16), SS (C24)
Spruce-pine-fir (Canada and USA)	GS (C16), SS (C24)
Sitka spruce (Canada)	GS (C14), SS (C18)
Western white woods (USA)	GS (C14), SS (C18)
Southern pine (USA)	GS (C18), SS (C24)

[*]Timber graded in accordance with BS 4978:1996; based on Table 1.2, BS 5268-2:2002.

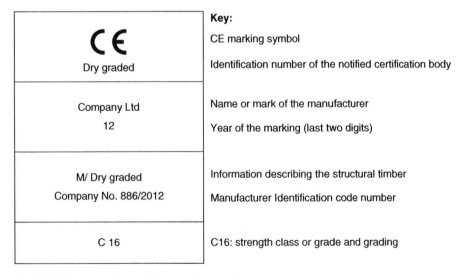

Fig. 1.7. Example of simplified grading marking.

with pre-programmed criteria, which leads to the direct grading of the timber section and marking with the appropriate strength class. An example of the grading marking, based on the requirements of BS EN 14081-1:2005+A1:2011, is shown in Figure 1.7.

In general less material is rejected if machine graded; however, timber is also visually inspected during machine grading to ensure that major, strength-reducing, defects do not exist.

Table 1.3 Strength and stiffness properties and density values for structural timber strength classes, (in accordance with **Table 1**, of BS EN 338: 2009)

Bending parallel to grain: f_m and E_0
Shear: f_v and G

Tension or compression parallel to grain: $f_{t,0}$, $f_{c,0}$ and E_0

Tension or compression perpendicular to grain: $f_{t,90}$, $f_{c,90}$ and E_{90}

Strength class	Characteristic strength properties (N/mm²)						Stiffness properties (kN/mm²)				Density (kg/m³)	
	Bending	Tension 0	Tension 90	Compression 0	Compression 90	Shear	Mean modulus of elasticity 0	5% modulus of elasticity 0	Mean modulus of elasticity 90	Mean shear modulus	Density	Mean density
	$(f_{m,k})$	$(f_{t,0,k})$	$(f_{t,90,k})$	$(f_{c,0,k})$	$(f_{c,90,k})$	$(f_{v,k})$	$(E_{0,mean})$	$(E_{0,05})$	$(E_{90,mean})$	(G_{mean})	(ρ_k)	(ρ_{mean})
Softwood and poplar species												
C14	14	8	0.4	16	2.0	3.0	7.0	4.7	0.23	0.44	290	350
C16	16	10	0.4	17	2.2	3.2	8.0	5.4	0.27	0.50	310	370
C18	18	11	0.4	18	2.2	3.4	9.0	6.0	0.30	0.56	320	380
C20	20	12	0.4	19	2.3	3.6	9.5	6.4	0.32	0.59	330	390
C22	22	13	0.4	20	2.4	3.8	10.0	6.7	0.33	0.63	340	410
C24	24	14	0.4	21	2.5	4.0	11.0	7.4	0.37	0.69	350	420
C27	27	16	0.4	22	2.6	4.0	11.5	7.7	0.38	0.72	370	450
C30	30	18	0.4	23	2.7	4.0	12.0	8.0	0.40	0.75	380	460
C35	35	21	0.4	25	2.8	4.0	13.0	8.7	0.43	0.81	400	480
C40	40	24	0.4	26	2.9	4.0	14.0	9.4	0.47	0.88	420	500
C45	45	27	0.4	27	3.1	4.0	15.0	10.0	0.50	0.94	440	520
C50	50	30	0.4	29	3.2	4.0	16.0	10.7	0.53	1.00	460	550
Hardwood species												
D18	18	11	0.6	18	7.5	3.4	9.5	8.0	0.63	0.59	475	570
D24	24	14	0.6	21	7.8	4.0	10.0	8.5	0.67	0.62	485	580
D30	30	18	0.6	23	8.0	4.0	11.0	9.2	0.73	0.69	530	640
D35	35	21	0.6	25	8.1	4.0	12.0	10.1	0.80	0.75	540	650
D40	40	24	0.6	26	8.3	4.0	13.0	10.9	0.86	0.81	550	660
D50	50	30	0.6	29	9.3	4.0	14.0	11.8	0.93	0.88	620	750
D60	60	36	0.6	32	10.5	4.5	17.0	14.3	1.13	1.06	700	840
D70	70	42	0.6	34	13.5	5.0	20.0	16.8	1.33	1.25	900	1080

Subscripts used are: 0, direction parallel to grain; 90, direction perpendicular to grain; m, bending; t, tension; c, compression; v, shear; k, characteristic.

1.5.3   Strength classes

The concept of grouping timber into strength classes was introduced into the United Kingdom with BS 5268-2 in 1984. Strength classes offer a number of advantages both to the designer and the supplier of timber. The designer can undertake the design without the need to check on the availability and price of a large number of species and grades that might be used. Suppliers can supply any of the species/grade combinations that meet the strength class called for in a specification. The concept also allows new species to be introduced to the market without affecting existing specifications for timber.

BS EN 338:2009 [11] defines a total of 20 strength classes: 12 for softwoods – C14, C16, C18, C20, C22, C24, C27, C30, C35, C40, C45 and C50; and 8 for hardwoods – D18, D24, D30, D35, D40, D50, D60 and D70. The letters C and D refer to coniferous species (C classes) or deciduous species (D classes), and the number in each strength class refers to its 'characteristic bending strength' in N/mm² units; for example, C40 timber has a characteristic bending strength of 40 N/mm². It ranges from the weakest grade of softwood, C14, to the highest grade of hardwood, D70, often used in Europe.

1.5.3.1   Material properties

Section 3 of BS EN 1995-1-1:2004 + A1:2008 (referred to in the text as EC5) [12] deals with the material properties and defines the strength and stiffness parameters, stress–strain relations and gives values for modification factors for strength and deformation under various service classes and/or load duration classes. EC5, in common with other Eurocodes, does not contain the material property values and this information is given in a supporting standard, i.e. in *Table 1* of BS EN 338:2009, reproduced here as Table 1.3.

The characteristic values are defined as the population 5th-percentile values obtained from the results of tests with a duration of approximately 5 min at the equilibrium moisture content of the test pieces relating to a temperature of 20 °C and a relative humidity of 65%.

In addition to providing characteristic strength and stiffness properties and density values for each strength class (and the rules for allocation of timber populations, i.e. combinations of species, source and grade, to the classes), BS EN 338:2009 lists the equations that form the relations between some of the characteristic values given in Table 1.3 for properties other than bending strength, mean modulus of elasticity in bending and density.

The relationships between the characteristic strength and stiffness properties are given as follows:

- Tensile strength parallel (0) to grain, $f_{t,0,k} = 0.6 f_{m,k}$
- Compression strength parallel (0) to grain, $f_{c,0,k} = 5(f_{m,k})^{0.45}$
- Shear strength, $f_{v,k}$ shall be taken from Table 1.3 (Table 1, BS EN 338:2009)
- Tensile strength perpendicular (90) to grain
  $$f_{t,90,k} = 0.4 \text{ N/mm}^2 \text{ for softwoods}$$
  $$f_{t,90,k} = 0.6 \text{ N/mm}^2 \text{ for hardwoods}$$
- Compression strength perpendicular (90) to grain,
  $$f_{c,90,k} = 0.007\rho_k \text{ for softwoods}$$
  $$f_{c,90,k} = 0.015\rho_k \text{ for hardwoods}$$

- Modulus of elasticity parallel (0) to grain,

$$E_{0.05} = 0.67 \, E_{0,\,mean} \text{ for softwoods}$$
$$E_{0.05} = 0.84 \, E_{0,\,mean} \text{ for hardwoods}$$

- Mean modulus of elasticity perpendicular (90) to grain,

$$E_{90,\,mean} = E_{0,\,mean}/30 \text{ for softwoods}$$
$$E_{90,\,mean} = E_{0,\,mean}/15 \text{ for hardwoods}$$

- Mean shear modulus, $G_{mean} = E_{0,\,mean}/16$
- Mean density, $\rho_{mean} = 1.2 \, \rho_k$.

1.6 SECTION SIZES

In general, it is possible to design timber structures using any size of timber. However, since the specific use is normally not known at the time of conversion, sawmills tend to produce a range of standard sizes known as 'common target' sizes. Specifying such common target sizes will often result in greater availability and savings in cost.

There are a number of alternative sizes and finishes of cross-sections. BS EN 1313:2010 [13] specifies permitted deviations for thickness and width at reference moisture content of 20% and adjustments for changes in section sizes due to change in moisture content. The deviation in *sawn* sections at a moisture content of 20% are as follows: for thicknesses and widths up to 100 mm, −1 mm and +3 mm, and for over 100 mm sizes, −2 mm and +4 mm. Sawn sections should only be used in situations where dimensional tolerances are of no significance. Planing two parallel edges to a specified dimension is referred to as *regularising* and if all four edges are planed to specified sizes, the process is referred to as *planed all round*. The requirements of EC5 for timber target sizes (i.e. specified sizes) are those given in BS EN 336:2003 [14] and in its National Annex. This standard specifies two tolerance classes: tolerance class 1 (T1) is applicable to sawn surfaces, and tolerance class 2 (T2) applicable to planed timber. Regularised timber can be achieved by specifying T1 for the thickness and T2 for the width. For T1, dimensions of up to 100 mm are limited to −1/+3 mm and dimensions of over 100 mm to −2/+4 mm. For T2, dimensions of up to 100 mm are limited to −1/+1 mm and those over 100 mm to −1.5/+1.5 mm.

The commonly available lengths and cross-section sizes are also listed in the UK National Annex of BS EN 336, and are referred to as target sizes. The 'target size' is defined as the specified timber section size at a reference moisture content of 20%, and to which the deviations, which would ideally be zero, are to be related. The target sizes can be used, without further modification, in design calculations.

The common target sizes, whose sizes and tolerances comply with BS EN 336, for sawn softwood structural timber, for structural timber machined on the width and for structural timber machined on all four sides are given in Table 1.4. In Table 1.5 the range of lengths of sawn softwood structural timber are detailed.

1.7 ENGINEERED WOOD PRODUCTS (EWPs)

The readily available sawn sections of softwood are limited in size and quality. The largest section sizes available are 75 mm thick×225 mm wide and at most 5 m in length. Any larger section sizes would suffer from both conversion and seasoning

Table 1.4 Common target sizes of structural timber (softwoods)*

Sawn thickness (to tolerance class 1) (mm)	Machined thickness (to tolerance class 2) (mm)	Sawn width (to tolerance class 1) (mm) / Machined width (to tolerance class 2) (mm)									
		75 / 72	100 / 97	125 / 120	150 / 145	175 / 170	200 / 195	225 / 220	250 / 245	275 / 270	300 / 295
22		√									
38	35	√			√	√	√	√			
47	44	√	√	√	√	√	√	√	√	×	
63	60				√	√	√	√			
75	72	√			√	√	√	√	√	√	√
100	97	√			√		√	√	√		√
150	145	√			√						√
300											×

Certain sizes may not be obtainable in the customary range of species and grades that are generally available.
BS EN 336 has a lower limit of 35 mm for machined thicknesses.
×applies only to sections with sawn width or thickness.
* In accordance with Tables NA.2, NA.3 and NA.4 of BS EN 336:2003; for (i) sawn to tolerance class 1,
(ii) machined on the width to tolerance class 2, (iii) machined on all four sides to tolerance class 2.

Table 1.5 Commonly available lengths of structural softwood timber*

Length (m)
2.40
3.00, 3.30, 3.60 or 3.90
4.20, 4.50 or 4.80
5.10 or 5.40

Lengths of 5.40 m and over may not be readily available without finger jointing.
* In accordance with Table NA.1, BS EN 336:2003.

defects. EWPs are developed to overcome the limitations of sawn timber and are produced, in combination with adhesives, in a variety of forms:

- dried thin planks of wood are glued together to form glued-laminated timber or glulam; or
- dried thin planks of wood are bonded together in different layouts, consisting of several layers of timber planks stacked crosswise and glued together either on wide faces only, or on both wide and narrow faces to form multi-layered cross-laminated timber (CLT) panels;
- veneered, by peeling logs, and bonded together in different layouts to produce plywood or laminated veneer lumber (LVL);
- chipped, to different sizes, to produce fibreboards, chipboards or oriented strand board (OSB); and
- sliced in different forms to produce parallel strand lumber (PSL) known as Parallam® or laminated strand lumber (LSL) known as TimberStrand®.

These products are engineered and tested to predetermined design specifications to meet national or international standards.

EWPs may also include products that are made by bonding or mechanically fixing together two or more of the above products to form structurally efficient composite members or systems such as I-beams and box beams or in combination with other materials to make a range of value-added systems such as structural insulated panels (SIPs).

EWPs may be selected over solid sawn timber in many applications due to certain comparative advantages:

- They can be manufactured to meet application-specific performance requirements.
- Large sections or panels in long lengths can be manufactured from small logs with defects being removed or dispersed.
- They are often stronger and less prone to humidity-induced warping than equivalent solid timbers, although most particle- and fibre-based boards readily soak up water unless they are treated with sealant or painted.

EWPs are more expensive to produce than solid timber, but offer advantages, including economic ones, when manufactured in large sizes due to the rarity of trees suitable for cutting large sections.

1.7.1   Glued-laminated timber (glulam)

Glued-laminated timber, glulam, is fabricated from small sections of timber boards (called laminates) bonded together with adhesives and laid up so that the grain of all laminates is essentially parallel to the longitudinal axis. Individual laminates are typically 19–50 mm in thickness, 1.5–5 m in length, end-jointed by the process of finger jointing as shown in Figure 1.8a and then placed randomly throughout the glulam component. Normally, the laminates are dried to around 12–18% moisture content before being machined and assembled. Edge-gluing permits beams wider and larger sections than those commercially available to be manufactured after finger jointing. Assembly is commonly carried out by applying a carefully controlled adhesive mix to the faces of the laminates. They are then placed in mechanical or hydraulic jigs of the appropriate shape and size, and pressurised at right angles to the glue lines and held until curing of the adhesive is complete. Glulam is then cut, shaped and any specified preservative and finishing treatments are applied.

Timber sections with a thickness of around 33 mm to a maximum of 50 mm are used to laminate *straight* or *slightly curved* members, whereas much thinner sections (12 or 19 mm, up to about 33 mm) are used to laminate *curved* members. Glued-laminated members can also be constructed with variable sections to produce tapering beams, columns, arches and portals (Figure 1.8).

The laminated lay-up of glulam makes it possible to match the lamination quality to the level of design stresses. Beams can be manufactured with the higher grade laminates at the outer highly stressed regions and the lower grade of laminates in the inner parts. Such combined concepts permit the timber resource to be used more efficiently.

Design of glued-laminated timber members is covered in Chapter 6 where the strength, stiffness and density properties of homogeneous (single grade) and combined (having outer laminations of higher grade) glued-laminated members are detailed.

(d) Truss system (Scottish Parliament)

(a) Finger joint

(b) Post & beam

(c) Curved portal

Fig. 1.8. Glued-laminated structures. (Part (b) photo courtesy of APA, The Engineered Wood Association. (c) photo courtesy of Axis Timber Limited, a member of the Glued Laminated Timber Association, UK.)

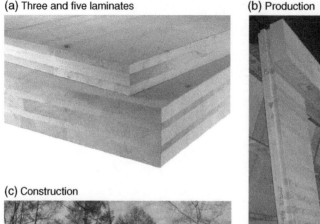

Fig. 1.9. Cross-laminated timber.

1.7.2   Cross-laminated timber (CLT or X-Lam)

Cross-laminated timber, known as CLT or X-Lam, is a prefabricated solid timber panel, formed with a minimum of three orthogonally bonded layers of solid timber boards (laminates). For improved performance, long continuous lengths of timber boards can be produced by the finger jointing process. Cross-laminated timber panels can have three, five, seven or more layers in odd numbers, symmetrically formed around the middle layer (Figure 1.9). The layers are stacked perpendicular to one another and are glued together either on their wide faces only or on both wide and narrow faces and then pressed together over their entire surface area mechanically, or by means of a vacuum bag.

The European standard prEN 16351:2011 [15] deals with the performance requirements and minimum requirements for the production of the cross-laminated timber products for use in buildings and bridges. CLT laminations comprise timber boards that are strength graded according to EN 14081-1 or wood-based panels such as LVL. The common panel thicknesses range between 50 to 300 mm, but panels as thick as 500 mm are also produced. Various panel sizes from 0.6 m wide up to 3 m wide by 16 m long, 4.8 m wide by 20 m long, or even 1.2 m wide by 24 m long are possible and are produced by a number of manufacturers. However, often the ability to transport, shipping or crane lifting of the panels, is the limiting factor governing their size. In the UK, CLT is currently imported from mainland Europe (e.g. Austria, Germany and Switzerland) and Scandinavia, see Table 1.6. But the situation is likely

Table 1.6 Examples of European suppliers or producers of cross-laminated timber (CLT)

Supplier/Producer	Width (mm)	Length (mm)	Thickness (mm)	Species used	Country of origin	Product
Eurban	3400	13 500	60–500	Spruce, Larch, Douglas fir	Austria	Crosslam panels
Binderholz	1250	24 000	66–34	Spruce, Larch, Pine, Douglas fir	Austria	BBS
Metsa Wood (Finnforest Merk)	4800	14 800	51–300	Spruce	Germany	Leno
KLH	2950	16 500	57–500	Spruce, Pine, Fir	Austria	KLH solid timber panels
Stora Enso	2950	2950	57–296	Spruce, Larch, Pine	Austria	CLT
Kaufmann	3000	16500	78–278	Spruce	Austria	BSP Crossplan

to change as the UK market for CLT develops; it is likely that a number of factories will be established using UK grown timber.

The CLT panels have improved dimensional stability compared to that of solid timber and provide relatively high strength and stiffness properties in both longitudinal and transverse directions, i.e. enabling two-way spanning capability. As panels can be manufactured with their outer layers orientated in either direction relative to the production length, to minimise waste and offcuts the design and manufacturing should be coordinated such that for walls the outer layers of CLT panels are oriented in the vertical direction and for floors and roofs in the direction of their major span.

CLT based structures also provide a number of other benefits, including: enhanced connector strength and splitting resistance, increased dead weight and robustness, high axial load capacity for walls due to large bearing areas, as well as offering high thermal, acoustic and fire performance and having a very low carbon footprint.

For structural design the characteristic strength and stiffness values of CLT products with CE certification (marking) should be obtained from the manufacturers or suppliers. Often such information is available from manufacturers' websites. CLT elements and systems can be designed using the rules in EC5 and they are also available as proprietary systems.

1.7.3 Plywood

Plywood is a flat panel made by bonding together, under pressure, a number of thin layers of veneer, often referred to as plies (or laminates). Plywood was the first type of EWP to be invented. Logs are debarked and steamed or heated in hot water for about 24 hours. They are then rotary-peeled into veneers of 2–4 mm in thickness and clipped into sheets of some 2 m wide. After kiln-drying and gluing, the veneers are laid up with the grain perpendicular to one another and bonded under pressure in an odd number of laminates (at least three), as shown in Figure 1.10a. The outside plies, always made of veneer, are referred to as faces (face ply or back ply) and the inner laminates, which could be made of either veneers or sliced/sawn

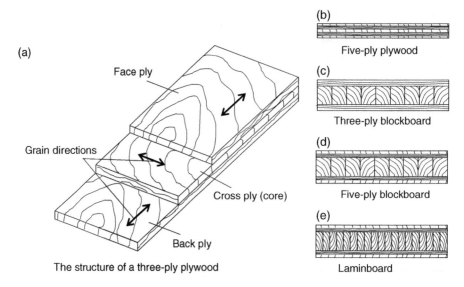

(a)

Face ply

Grain directions

Cross ply (core)

Back ply

The structure of a three-ply plywood

(b)

Five-ply plywood

(c)

Three-ply blockboard

(d)

Five-ply blockboard

(e)

Laminboard

Fig. 1.10. Examples of plywood and wood core plywood.

wood, are called core. Examples of wood core plywood include blockboards and laminboards, as shown in Figures 1.10c–1.10e.

Plywood is produced in many countries from either softwood or hardwood or a combination of both. The structural grade plywoods that are commonly used in the United Kingdom are as follows:

- American construction and industrial plywood
- Canadian softwood plywood and Douglas fir plywood
- Finnish birch-faced (combi) plywood, Finnish birch plywood and Finnish conifer plywood
- Swedish softwood plywood.

The plywood sheet sizes available sizes are 1200 mm × 2400 mm or 1220 mm × 2440 mm. The face veneer is generally oriented with the longer side of the sheet except for Finnish made plywoods in which face veneers run parallel to the shorter side. Structural plywood and plywood for exterior use are generally made with waterproof adhesive that is suitable for severe exposure conditions.

The structural properties and strength of plywood depend mainly on the number and thickness of each ply, the species and grade and the arrangement of the individual plies. As with timber, the structural properties of plywood are functions of the type of applied stresses, their direction with respect to grain direction of face ply and the duration of load.

Plywood may be subjected to bending in two different planes, depending on its intended use, and the direction of the applied stress and, therefore, it is important to differentiate between them:

(i)   Bending about either of the axes (i.e. x–x or y–y) in the plane of the board, as shown in Figure 1.11a; for example, in situations where it is used as shelving or as floor board.

(a)

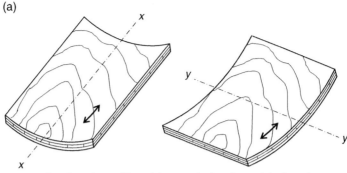

Bending about either of the axes in the plane of the board

(b)

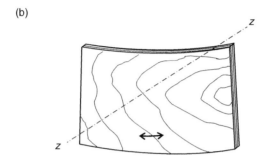

Bending about an axis perpendicular to the plane of the board

Fig. 1.11. Plywood – axes of bending.

(ii)   Bending about an axis perpendicular to the plane of the panel (i.e. z–z axis as shown in Figure 1.11b); for example, when it is acting as a web of a flexural member such as in ply-webbed beams.

BS EN 636:2003 [16] details the requirements for plywood for general purposes and for structural application in dry, humid or exterior conditions. It also gives a classification system based on the bending properties. The information on how to utilise the classification system of BS EN 636 in order to determine the characteristic values for plywood panels for use in structural design, in accordance with EN 1995-1-1 (EC5), is given in BS EN 12369-2:2011 [17].

BS EN 12369-2 includes the characteristic values of the mechanical properties for load-bearing plywood panels, complying with BS EN 636, under service class 1 conditions and provides the class designations of F3, F5, F10, F15, F20, F25, F30, F40, F50, F60, F70 and F80 with corresponding characteristic strength values in bending of 3 N/mm² to 80 N/mm² respectively as set out in Table 2 of the Standard together with tensile and compressive strength values parallel and perpendicular to the face grain of the panels. Similarly, *Table 3* of the Standard provides the classification for modulus of elasticity in bending, tension and compression with designations of classes E5 to E140 corresponding to mean modulus of elasticity in bending of 500 N/mm² to 14 000 N/mm² respectively as well as their corresponding values in tension and compression parallel and perpendicular to the face grain of the panels. Shear properties

Table 1.7 Details of the commonly used structural grade plywoods in the United Kingdom

American plywood grades			Canadian plywood grades			Finnish plywood grades			Swedish plywood grades		
Grade	American standard	Quality control agency	Grade	Canadian standard	Quality control agency	Grade	Finnish and CEN standards	Quality control agency	Grade	Swedish standard	Quality control agency
C-D Exposure 1 (CDX)	PS1-95	APA and TECO	CSP Select Tight Face Exterior	CSA 0151-M 1978		Birch (Finply all birch)	SFS 2413 EN 635-2 EN 636-2&3	VTT	P30	SBN 1975.5	The National Swedish Testing Institute (Statens Provningsanstalt)
C-C Exterior (CCX)	PS1-95	APA and TECO	CSP Select Exterior	CSA 0151-M 1978		Birch-faced (Finply combi)	SFS 2413 EN635-2 EN636-2&3	VTT			
A-C Exterior (ACX)	PS1-95	APA and TECO	CSP Sheathing Grade Exterior		CANPLY (Formerly COFI)	Conifer plywood (Finply conifer)	EN635-3 EN 636-3	VTT			
B-C Exterior (BCX)	PS1-95	APA and TECO	DFP Select Tight Face Exterior	CSA 0121-M 1978		Birch-faced (Finply combi mirror)	SFS 2413 EN635-2 EN636-2&3	VTT			
Sturd-I-Floor Exposure 1 and Exterior	PS1-95	APA	DFP Select Exterior	CSA 0121-M 1978		Birch-faced (Finply twin)	SFS 2413 EN 635-2 EN636-2&3	VTT			
Floor span Exposure 1 and Exterior	PS1-95	TECO	DFP Sheathing Grade Exterior								
C-D Plugged Exposure 1	PS1-95	APA and TECO									
C-C Plugged Exterior	PS1-95	APA and TECO									

Quality control agencies: APA, The Engineered Wood Association; Canadian Plywood Association (CANPLY); Technical Research Centre of Finland (VTT); The National Swedish Testing Institute (Statens Provningsanstalt); TECO Corporation (TECO).

are detailed in *Table 4* of the Standard. Plywood of these classes can also be used under service classes 2 and 3 in accordance with the requirements of EC5.

However, the Standard also recommends that, where optimised values are required, the characteristic values are determined directly by testing in accordance with BS EN 789:2004 [18] and BS EN 1058:2009 [19] or by a combination of testing to these two standards and calculation to BS EN 14272:2011 [20]. In this regard, the characteristic strength and stiffness values of products with CE certification (marking) should be obtained from the manufacturers or suppliers. Often such information is available from manufacturers' websites.

The relevant grades, national standards and the quality control agencies relating to the structural grade plywoods that are commonly used in the United Kingdom are detailed in Table 1.7.

Indicative strength, stiffness and density values for the American plywood grade: C-D exposure 1 (CDX) and Swedish plywood grade P30 are given in Table 1.8.

In Tables 1.9, 1.10, 1.11 and 1.12 characteristic values for a range of Finnish plywoods that are used in the United Kingdom are given, based on the *Handbook of Finnish Plywood* [21].

In Tables 1.13 and 1.14 strength, stiffness and density values for unsanded CANPLY Canadian Douglas fir plywood and Canadian softwood plywood are given, respectively, based on data published by CANPLY Canadian Plywood Association [22].

1.7.4 Laminated Veneer Lumber (LVL)

LVL, shown in Figure 1.12, is an engineered timber composite manufactured by laminating wood veneers using exterior-type adhesives. In production, LVL is made with thin veneers similar to those in most plywoods. Veneers, 3–4 mm in thickness, are peeled off good quality logs and vertically laminated, but unlike plywood, successive veneers are generally oriented in a common grain direction, which gives orthotropic properties similar to those in sawn timber. Certain grades of LVL also include a few sheets of veneer in its lay-up in the direction perpendicular to the long direction of the member to enhance the strength properties. LVL was first produced some 40 years ago and currently it is being manufactured by a number of companies in the United States, Finland, Australia, New Zealand and Japan.

In the USA, LVL is manufactured from species such as southern yellow pine or Douglas fir by Weyerhaeuser under the name Microllam® LVL; and in Finland LVL is manufactured from Spruce by Metsa Wood (Finnforest) under the name Kerto. Kerto is produced as a standard product when all veneers are parallel (Kerto-S®) and also as Kerto-Q® in which approximately every fifth veneer is in the perpendicular direction. Kerto-T, a new product by Metsa Wood, is similar to Kerto-S but is made from lighter veneers and is produced for use as a stud in both load-bearing and non load-bearing walls.

Standard dimensions of cross-section for Kerto-LVLs are shown in Table 1.15 and the characteristic values for their strength and stiffness properties are given in Table 1.16.

1.7.5 Laminated Strand Lumber (LSL), TimberStrand®

LSL, shown in Figure 1.13, is manufactured in the USA by Weyerhaeuser under the registered name TimberStrand®. LSL is produced from strands of wood species (often

Table 1.8 Strength and stiffness properties and density values of selected American and Swedish structural plywoods

Section properties		Characteristic strength (N/mm²)							Density (kg/m³)		Mean modulus of rigidity (N/mm²)	Mean modulus of elasticity (N/mm²)			
		Bending	Compression	Tension		Panel shear	Planar (rolling) shear		Characteristic	Mean	Panel shear	Bending		Tension and compression	
Plywood type	Nominal thickness (mm)	$f_{m,0,k}$ $f_{m,90,k}$	$f_{c,0,k}$ $f_{c,90,k}$	$f_{t,0,k}$ $f_{t,90,k}$		$f_{v,k}$	$f_{r,k}$		ρ_k	ρ_{mean}	$G_{v,mean}$	$E_{m,0,mean}$ $E_{m,90,mean}$		$E_{t/c,0,mean}$ $E_{t/c,90,mean}$	
American plywood Grade: C-D Exposure 1 (CDX)	12.5	23.5 12.2	13.9 8.1	13.6 7.2		3.2	0.9		410	460	500	10 300 2500		6800 4600	
	21	14.8 10.1	10.6 7.7	10.5 6.9		3.2	0.9		410	460	500	7800 2500		5200 3900	
Swedish plywood Grade: P30	12	23.0 11.4	15.0 12.0	15.0 12.0		2.9	0.9		410	460	500	9200 4600		7200 4800	
	24	21.6 12.4	15.4 11.4	15.4 11.4		2.9	0.9		410	460	500	8700 5000		7400 4600	

Note: 1. Characteristic value of modulus of elasticity, $E_{i,k}=0.8\times E_{i,mean}$.
2. Number of plies ≥5.

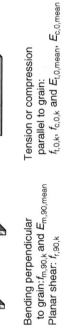

Bending parallel to grain: $f_{m,0,k}$ and $E_{m,0,mean}$
Planar shear: $f_{r,0,k}$

Bending perpendicular to grain: $f_{m,90,k}$ and $E_{m,90,mean}$
Planar shear: $f_{r,90,k}$

Tension or compression parallel to grain: $f_{t,0,k}$, $f_{c,0,k}$ and $E_{t,0,mean}$, $E_{c,0,mean}$

Tension or compression perpendicular to grain: $f_{t,90,k}$, $f_{c,90,k}$ and $E_{t,90,mean}$, $E_{c,90,mean}$

Panel shear: $f_{v,k}$

Table 1.9 Finnish plywood: density values

	Mean density (kg/m³)	Characteristic density (kg/m³)
Plywood	ρ_{mean}	ρ_k
Birch (1.4 mm plies)	680	630
Birch-faced (1.4 mm plies)	620	560
Conifer (1.4 mm (thin) plies)	520	460
Conifer (thick plies)	460	400

aspen), up to 300 mm in length and 30 mm in width, or species combinations blended with a polyurethane-based adhesive. The strands are oriented in a parallel direction and formed into mats 2.44 m wide by up to 14.63 m long, of various thicknesses of up to 140 mm. The mats are then pressed by steam injection to the required thickness. TimberStrands are available in dimensions of up to 140 mm thick × 1220 mm deep × 14.63 m long. Design values for the strength and stiffness properties of TimberStrand are given in Table 1.17.

1.7.6   Parallel Strand Lumber (PSL), Parallam®

PSL, shown in Figure 1.14, is manufactured in the USA by Weyerhaeuser under the registered name Parallam®. The manufacturing process involves peeling small-diameter logs into veneer sheets. The veneers are then dried to a moisture content of 2–3% and then cut into thin long strands oriented parallel to one another.

The process of stranding reduces many of the timber's natural growth and strength-reducing characteristics such as knots, pitch pockets and slope of grain. This results in a dimensionally stable material that is more uniform in strength and stiffness characteristics and also in density than its parent timbers. For bonding strands, waterproof structural adhesive, mixed with a waxed component, is used and redried under pressure in a microwave process to dimensions measuring 275 × 475 mm² in section by up to 20 m in length.

1.7.7   Oriented Strand Board (OSB)

OSB is an engineered structural board manufactured from thin wood strands, flakes or wafers sliced from small-diameter round timber logs and bonded with an exterior-type adhesive (comprising 95% wood, 5% resin and wax) under heat and pressure; see Figure 1.15.

OSB panels comprise exterior or surface layers that are composed of strands oriented in the long panel direction, with inner layers comprising randomly oriented strands. Their strength is mainly due to their multi-layered make-up and the cross-orientation of the strands. The use of water and boil-proof resins/adhesives provide strength, stiffness and moisture resistance.

In the United Kingdom, OSB is often referred to as Sterling board or Sterling OSB. OSB has many applications and often is used in preference to plywood as a more cost-effective, environmentally friendly and dimensionally stable panel. It is available

Table 1.10 Finnish birch plywood: Strength and stiffness properties

Section properties			Characteristic strength (N/mm²)									Mean modulus of rigidity (N/mm²)			Mean modulus of elasticity (N/mm²)			
			Bending		Compression		Tension		Panel shear	Planar (rolling) shear		Panel shear	Planar shear		Bending		Tension and compression	
Nominal thickness (mm)	Number of plies	Mean thickness (mm)	$f_{m,0,k}$	$f_{m,90,k}$	$f_{c,0,k}$	$f_{c,90,k}$	$f_{t,0,k}$	$f_{t,90,k}$	$f_{v,k}$	$f_{r,0,k}$	$f_{r,90,k}$	$G_{v,mean}$	$G_{r,0,mean}$	$G_{r,90,mean}$	$E_{m,0,mean}$	$E_{m,90,mean}$	$E_{t/c,0,mean}$	$E_{t/c,90,mean}$
4	3	3.6	65.9	10.6	31.8	20.2	45.8	29.2	9.5	2.77	–	620	169	–	16 471	1029	10 694	6806
6.5	5	6.4	50.9	29.0	29.3	22.8	42.2	32.8	9.5	3.20	1.78	620	169	123	12 737	4763	9844	7656
9	7	9.2	45.6	32.1	28.3	23.7	40.8	34.2	9.5	2.68	2.35	620	206	155	11 395	6105	9511	7989
12	9	12.0	42.9	33.2	27.7	24.3	40.0	35.0	9.5	2.78	2.22	620	207	170	10 719	6781	9333	8167
15	11	14.8	41.3	33.8	27.4	24.6	39.5	35.5	9.5	2.62	2.39	620	207	178	10 316	7184	9223	8277
18	13	17.6	40.2	34.1	27.2	24.8	39.2	35.8	9.5	2.67	2.34	620	206	183	10 048	7452	9148	8352
21	15	20.4	39.4	34.3	27.0	25.0	39.0	36.0	9.5	2.59	2.41	620	206	186	9858	7642	9093	8407
24	17	23.2	38.9	34.4	26.9	25.1	38.8	36.2	9.5	2.62	2.39	620	206	189	9717	7783	9052	8448
27	19	26.0	38.4	34.5	26.8	25.2	38.7	36.3	9.5	2.57	2.43	620	205	190	9607	7893	9019	8481
30	21	28.8	38.1	34.6	26.7	25.3	38.5	36.5	9.5	2.59	2.41	620	205	192	9519	7981	8993	8507
35	25	34.4	37.6	34.7	26.6	25.4	38.4	36.6	9.5	2.57	2.43	620	204	193	9389	8111	8953	8547
40	29	40.0	37.2	34.7	26.5	25.5	38.3	36.8	9.5	2.56	2.44	620	204	195	9296	8204	8925	8575
45	32	44.2	37.0	34.7	26.5	25.5	38.2	36.8	9.5	2.55	2.46	620	203	195	9259	8241	8914	8586
50	35	48.4	36.8	34.8	26.4	25.6	38.1	36.9	9.5	2.54	2.46	620	203	196	9198	8302	8895	8605

Bending parallel to grain: $f_{m,0,k}$ and $E_{m,0,mean}$
Planar shear: $f_{r,0,k}$ and $G_{r,0,mean}$

Bending perpendicular to grain: $f_{m,90,k}$ and $E_{m,90,mean}$
Planar shear: $f_{r,90,k}$ and $G_{r,90,mean}$

Tension or compression parallel to grain: $f_{t,0,k}$, $f_{c,0,k}$ and $E_{t,0,mean}$, $E_{c,0,mean}$

Tension or compression perpendicular to grain: $f_{t,90,k}$, $f_{c,90,k}$ and $E_{t,90,mean}$, $E_{c,90,mean}$

Panel shear: $f_{v,k}$ and $G_{v,mean}$

Table 1.11 Finnish combi plywood: Strength and stiffness properties

Section properties			Characteristic strength (N/mm²)									Mean modulus of rigidity (N/mm²)			Mean modulus of elasticity (N/mm²)			
			Bending		Compression		Tension		Panel shear	Planar (rolling) shear		Panel shear	Planar shear		Bending		Tension and compression	
Nominal thickness (mm)	Number of plies	Mean thickness (mm)	$f_{m,0,k}$	$f_{m,90,k}$	$f_{c,0,k}$	$f_{c,90,k}$	$f_{t,0,k}$	$f_{t,90,k}$	$f_{v,k}$	$f_{r,0,k}$	$f_{r,90,k}$	$G_{v,mean}$	$G_{r,0,mean}$	$G_{r,90,mean}$	$E_{m,0,mean}$	$E_{m,90,mean}$	$E_{t/c,0,mean}$	$E_{t/c,90,mean}$
6.5	5	6.4	50.8	29.0	24.5	22.8	19.1	32.8	7.0	3.20	1.14	600	169	41	12 690	4763	8859	7656
9	7	9.2	43.9	32.1	22.5	23.7	17.5	34.2	7.0	2.68	1.51	593	206	52	10 983	6105	8141	7989
12	9	12.0	40.0	33.2	21.5	24.3	16.7	35.0	7.0	2.78	1.42	589	207	57	10 012	6781	7758	8167
15	11	14.8	37.5	33.8	20.8	24.6	16.2	35.5	7.0	2.62	1.53	586	207	59	9386	7184	7520	8277
18	13	17.6	35.8	34.1	20.4	24.8	15.8	35.8	7.0	2.67	1.50	584	206	61	8950	7452	7358	8352
21	15	20.4	34.5	34.3	20.0	25.0	15.6	36.0	7.0	2.59	1.55	583	206	62	8628	7642	7240	8407
24	17	23.2	32.9	34.4	19.8	25.1	15.4	36.2	7.0	2.62	1.53	582	206	63	8381	7783	7151	8448
27	19	26.0	31.2	34.5	19.6	25.2	15.3	36.3	7.0	2.57	1.56	581	205	63	8185	7893	7081	8481
30	21	28.8	29.9	34.6	19.5	25.3	15.1	36.5	7.0	2.59	1.54	581	205	64	8026	7981	7024	8507

Bending parallel to grain: $f_{m,0,k}$ and $E_{m,0,mean}$
Planar shear: $f_{r,0,k}$ and $G_{r,0,mean}$

Bending perpendicular to grain: $f_{m,90,k}$ and $E_{m,90,mean}$
Planar shear: $f_{r,90,k}$ and $G_{r,90,mean}$

Tension or compression parallel to grain: $f_{t,0,k}$, $f_{c,0,k}$ and $E_{t,0,mean}$, $E_{c,0,mean}$

Tension or compression perpendicular to grain: $f_{t,90,k}$, $f_{c,90,k}$ and $E_{t,90,mean}$, $E_{c,90,mean}$

Panel shear: $f_{v,k}$ and $G_{v,mean}$

Table 1.12 Finnish conifer plywood with thin veneers: strength and stiffness properties

Section properties			Characteristic strength (N/mm²)										Mean modulus of rigidity (N/mm²)			Mean modulus of elasticity (N/mm²)			
			Bending		Compression		Tension		Panel shear	Planar (rolling) shear		Panel shear	Planar shear		Bending		Tension and compression	
Nominal thickness (mm)	Number of plies	Mean thickness (mm)	$f_{m,0,k}$	$f_{m,90,k}$	$f_{c,0,k}$	$f_{c,90,k}$	$f_{t,0,k}$	$f_{t,90,k}$	$f_{v,k}$	$f_{r,0,k}$	$f_{r,90,k}$	$G_{v,mean}$	$G_{r,0,mean}$	$G_{r,90,mean}$	$E_{m,0,mean}$	$E_{m,90,mean}$	$E_{t/c,0,mean}$	$E_{t/c,90,mean}$
4	3	3.6	37.6	6.0	22.0	14.0	17.1	10.9	7.0	1.77	–	530	56	–	12 235	765	7944	5056
6.5	5	6.4	29.1	16.6	20.3	15.8	15.8	12.3	7.0	2.05	1.14	530	66	41	9462	3538	7313	5688
9	7	9.2	26.0	18.3	19.6	16.4	15.2	12.8	7.0	1.72	1.51	530	69	52	8465	4535	7065	5935
12	9	12.0	24.5	19.0	19.2	16.8	14.9	13.1	7.0	1.78	1.42	530	69	57	7963	5037	6933	6067
15	11	14.8	23.6	19.3	19.0	17.0	14.8	13.2	7.0	1.68	1.53	530	69	59	7663	5337	6851	6149
18	13	17.6	23.0	19.5	18.8	17.2	14.6	13.4	7.0	1.71	1.50	530	69	61	7464	5536	6795	6205
21	15	20.4	22.5	19.6	18.7	17.3	14.5	13.5	7.0	1.66	1.55	530	69	62	7323	5677	6755	6245
24	17	23.2	22.2	19.7	18.6	17.4	14.5	13.5	7.0	1.68	1.53	530	69	63	7218	5782	6724	6276
27	19	26.0	22.0	19.7	18.6	17.4	14.4	13.6	7.0	1.65	1.56	530	68	63	7137	5863	6700	6300
30	21	28.8	21.8	19.8	18.5	17.5	14.4	13.6	7.0	1.66	1.54	530	68	64	7072	5928	6681	6319

Bending parallel to grain: $f_{m,0,k}$ and $E_{m,0,mean}$
Planar shear: $f_{r,0,k}$ and $G_{r,0,mean}$

Bending perpendicular to grain: $f_{m,90,k}$ and $E_{m,90,mean}$
Planar shear: $f_{r,90,k}$ and $G_{r,90,mean}$

Tension or compression parallel to grain: $f_{t,0,k}$, $f_{c,0,k}$ and $E_{t,0,mean}$, $E_{c,0,mean}$

Tension or compression perpendicular to grain: $f_{t,90,k}$, $f_{c,90,k}$ and $E_{t,90,mean}$, $E_{c,90,mean}$

Panel shear: $f_{v,k}$ and $G_{v,mean}$

Table 1.13 Canadian Douglas fir plywood (unsanded CANPLY): strength and stiffness properties and density values

Section properties		Mean density (kg/m²)	Characteristic strength (N/mm²)									Mean modulus of rigidity (N/mm²)	Mean modulus of elasticity (N/mm²)			
			Bending		Compression		Tension		Panel shear	Planar (rolling) shear		Panel shear	Bending		Tension and compression	
Nominal thickness (mm)	Number of plies	ρ_{mean}	$f_{m,0,k}$	$f_{m,90,k}$	$f_{c,0,k}$	$f_{c,90,k}$	$f_{t,0,k}$	$f_{t,90,k}$	$f_{v,0,k}$	$f_{r,0,k}$	$f_{r,90,k}$	$G_{v,mean}$	$E_{m,0,mean}$	$E_{m,90,mean}$	$E_{t/c,0,mean}$	$E_{t/c,90,mean}$
7.5	3	460	26.4	5.5	25.4	8.1	16.8	4.4	3.5	1.07	0.33	500	12 950	510	9730	3300
9.5	3	460	24.9	5.4	20.1	8.0	13.3	4.3	3.5	0.89	0.33	500	12 290	490	7680	3250
12.5	4	460	22.1	7.0	15.2	11.7	10.1	6.4	3.5	0.95	0.48	500	10 980	1230	5840	4780
12.5	5	460	29.5	10.4	20.4	9.7	13.5	7.4	3.5	1.25	0.64	500	11 050	2270	7810	3960
15.5	4	460	25.5	7.8	19.7	12.4	13.1	6.7	3.5	0.91	0.51	500	12 830	1460	7550	5050
15.5	5	460	26.2	9.6	16.5	7.8	10.9	5.9	3.5	1.31	0.68	500	9930	2110	6300	3190
18.5	5	460	31.0	11.1	21.1	10.1	14.0	7.7	3.5	1.27	0.66	500	11 840	2510	8080	4120
18.5	6	460	23.8	10.7	17.3	6.5	11.4	5.0	3.5	1.07	0.63	500	9100	2640	6620	2670
18.5	7	460	25.2	10.8	17.3	9.8	11.4	7.5	3.5	1.13	0.83	500	9620	2670	6620	4010
20.5	5	460	24.0	14.4	17.0	11.1	11.3	8.4	3.5	1.06	0.59	500	9170	3760	6520	4520
20.5	6	460	22.2	10.8	16.6	5.9	11.0	4.5	3.5	1.09	0.68	500	8490	2820	6370	2410
20.5	7	460	23.4	10.9	15.6	8.9	10.3	6.7	3.5	1.14	0.89	500	8930	2840	5970	3620
22.5	7	460	25.3	10.3	16.3	8.5	10.8	6.5	3.5	1.16	0.94	500	9650	2800	6250	3480
22.5	8	460	26.2	10.3	16.3	10.8	10.8	8.2	3.5	0.89	0.98	500	10 010	2790	6250	4390
25.5	7	460	24.0	12.6	16.6	10.6	1.10	8.1	3.5	1.12	0.97	500	9210	3610	6360	4320
25.5	8	460	24.1	10.8	14.4	10.6	9.5	8.1	3.5	0.90	1.04	500	9260	3070	5520	4320
25.5	9	460	24.7	10.8	16.9	9.5	11.2	7.2	3.5	1.18	0.85	500	9490	3090	6480	3880
25.5	10	460	25.5	11.6	19.5	9.5	12.9	7.2	3.5	1.19	0.66	500	9800	3320	7450	3880
28.5	8	460	22.8	12.7	13.8	12.2	9.2	9.3	3.5	0.90	1.07	500	8790	3760	5300	4990
28.5	9	460	22.8	10.4	15.2	8.5	10.0	6.5	3.5	1.20	0.90	500	8800	3100	5800	3470
28.5	10	460	23.5	11.2	17.4	8.5	11.5	6.5	3.5	1.21	0.69	500	9050	3320	6670	3470
28.5	11	460	24.4	11.7	17.4	10.6	11.5	8.1	3.5	1.12	0.90	500	9410	3490	6670	4340
31.5	8	460	23.2	14.1	15.2	10.6	10.1	10.2	3.5	0.86	1.10	500	8930	3490	5830	5490
31.5	9	460	21.8	12.3	15.1	10.1	10.0	7.7	3.5	1.18	0.91	500	8400	3750	5770	4120
31.5	10	460	21.6	10.7	15.8	7.7	10.4	5.9	3.5	1.22	0.72	500	8330	3280	6030	3140
31.5	11	460	22.6	11.2	15.8	9.6	10.4	7.3	3.5	1.13	0.94	500	8690	3420	6030	3920
31.5	12	460	23.3	11.8	15.8	11.5	10.4	8.8	3.5	0.87	0.96	500	8990	3600	6030	4710

Bending parallel to grain: $f_{m,0,k}$ and $E_{m,0,mean}$
Planar shear: $f_{r,0,k}$ and $G_{r,0,mean}$

Bending perpendicular to grain: $f_{m,90,k}$ and $E_{m,90,mean}$
Planar shear: $f_{r,90,k}$ and $G_{r,90,mean}$

Tension or compression parallel to grain: $f_{t,0,k}$, $f_{c,0,k}$ and $E_{t,0,mean}$, $E_{c,0,mean}$

Tension or compression perpendicular to grain: $f_{t,90,k}$, $f_{c,90,k}$ and $E_{t,90,mean}$, $E_{c,90,mean}$

Panel shear: $f_{v,k}$ and $G_{v,mean}$

Table 1.14 Canadian Softwood plywood (unsanded CANPLY): strength and stiffness properties and density values

Nominal thickness (mm)	Number of plies	Mean density (kg/m³) ρ_{mean}	Bending $f_{m,0,k}$	Bending $f_{m,90,k}$	Compression $f_{c,0,k}$	Compression $f_{c,90,k}$	Tension $f_{t,0,k}$	Tension $f_{t,90,k}$	Panel shear $f_{v,0,k}$	Planar (rolling) shear $f_{r,0,k}$	Planar (rolling) shear $f_{r,90,k}$	Panel shear $G_{v,mean}$	Bending $E_{m,0,mean}$	Bending $E_{m,90,mean}$	Tension and compression $E_{t/c,0,mean}$	Tension and compression $E_{t/c,90,mean}$
7.5	3	420	24.0	5.5	16.1	8.1	12.3	4.4	3.5	1.07	0.33	430	8780	510	6590	3300
9.5	3	420	22.6	5.4	12.7	8.0	9.7	4.3	3.5	0.89	0.33	430	8330	490	5200	3250
12.5	4	420	22.3	7.0	12.1	11.7	9.2	6.4	3.5	0.91	0.48	430	8320	1230	4940	4780
12.5	5	420	20.2	10.4	14.5	9.7	11.1	7.4	3.5	1.25	0.64	430	7510	2270	5930	3960
15.5	4	420	23.1	7.8	12.5	12.4	9.5	6.7	3.5	0.91	0.51	430	8690	1460	5120	5050
15.5	5	420	17.9	9.6	11.7	7.8	8.9	5.9	3.5	1.31	0.68	430	6740	2110	4780	3190
18.5	5	420	21.2	11.1	15.0	10.1	11.4	7,7	3.5	1.27	0.66	430	8040	2510	6120	4120
18.5	6	420	19.0	10.7	14.7	6.5	11.2	5.0	3.5	1.34	0.63	430	7210	2640	6010	2670
18.5	7	420	17.8	10.8	13.1	9.8	10.0	7.5	3.5	1.13	0.83	430	6740	2670	5340	4010
20.5	5	420	16.5	14.4	12.8	11.1	9.7	8.4	3.5	1.06	0.59	430	6260	3760	5230	4520
20.5	6	420	15.3	10.8	12.8	5.9	9.7	4.5	3.5	1.09	0.68	430	5820	2820	5230	2410
20.5	7	420	16.5	10.9	11.8	8.9	9.0	6.7	3.5	1.14	0.89	430	6260	2840	4820	3620
22.5	7	420	17.7	10.3	12.1	8.5	9.2	6.5	3.5	1.16	0.94	430	6720	2800	4940	3480
22.5	8	420	18.6	10.3	12.1	10.8	9.2	8.2	3.5	0.89	0.98	430	7080	2790	4940	4390
25.5	7	420	17.0	12.6	12.7	10.6	9.7	8.1	3.5	1.12	0.97	430	6500	3610	5210	4320
25.5	8	420	17.2	10.8	10.7	10.6	8.1	8.1	3.5	0.90	1.04	430	6550	3070	4360	4320
25.5	9	420	17.8	10.8	13.0	9.5	9.9	7.2	3.5	1.18	0.85	430	6780	3090	5330	3880
25.5	10	420	18.6	11.6	15.4	9.5	11.7	7.2	3.5	1.19	0.66	430	7090	3320	6300	3880
28.5	8	420	16.4	12.7	10.4	12.2	7.9	9.3	3.5	0.90	1.07	430	6280	3760	4260	4990
28.5	9	420	16.4	10.4	11.7	8.5	8.9	6.5	3.5	1.20	0.90	430	6290	3100	4770	3470
28.5	10	420	17.1	11.2	13.8	8.5	10.5	6.5	3.5	1.21	0.69	430	6540	3320	5640	3470
28.5	11	420	18.0	11.7	13.8	10.6	10.5	8.1	3.5	1.12	0.90	430	6900	3490	5640	4340
31.5	8	420	17.0	14.1	11.8	10.6	9.0	10.2	3.5	0.86	1.10	430	6490	3490	4840	5490
31.5	9	420	15.9	12.3	11.8	10.1	9.0	7.7	3.5	1.18	0.91	430	6080	3750	4840	4120
31.5	10	420	15.7	10.7	12.5	7.7	9.5	5.9	3.5	1.22	0.72	430	6010	3280	5100	3140
31.5	11	420	16.7	11.2	12.5	9.6	9.5	7.3	3.5	1.13	0.94	430	6380	3420	5100	3920

Bending parallel to grain: $f_{m,0,k}$ and $E_{m,0,mean}$
Planar shear: $f_{r,0,k}$ and $G_{r,0,mean}$

Bending perpendicular to grain: $f_{m,90,k}$ and $E_{m,90,mean}$
Planar shear: $f_{r,90,k}$ and $G_{r,90,mean}$

Tension or compression parallel to grain: $f_{t,0,k}$, $f_{c,0,k}$ and $E_{t,0,mean}$, $E_{c,0,mean}$

Tension or compression perpendicular to grain: $f_{t,90,k}$, $f_{c,90,k}$ and $E_{t,90,mean}$, $E_{c,90,mean}$

Panel shear: $f_{v,k}$ and $G_{v,mean}$

(a)

(b)

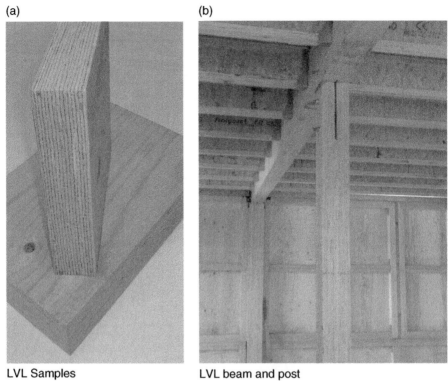

LVL Samples

LVL beam and post

(c)

LVL truss members

Fig. 1.12. Laminated veneer lumber (LVL).

Table 1.15 Standard dimensions of cross-section for Kerto-LVL

Kerto-LVL type	Thickness (mm)	Width or depth (mm)								
		200	225	260	300	360	400	450	500	600
S/Q	27	√	√							
S/Q	33	√	√	√						
S/Q	39	√	√	√	√					
S/Q	45	√	√	√	√	√				
S/Q	51	√	√	√	√	√	√			
S/Q	57	√	√	√	√	√	√	√		
S/Q	63	√	√	√	√	√	√	√	√	
S/Q	69	√	√	√	√	√	√	√	√	√
S	75	√	√	√	√	√	√	√	√	√
S	90	√	√	√	√	√	√	√	√	√

Note: Kerto (LVL) may also be supplied in widths up to 2500 mm; for availability contact Metsa Wood.

Table 1.16 Kerto (LVL): strength and stiffness properties and density values

	Symbol	Units	Kerto-S® 21–90 mm	Kerto-Q® 27–69 mm	Kerto-T®	
Characteristic values						
Bending						
Edgewise	$f_{m,0,edge,k}$	N/mm²	44.0	32.0	$27(300/h)^s$	
Size effect parameter*	s		0.12	0.12	0.15	
Flatwise, parallel to grain	$f_{m,0,flat,k}$	N/mm²	50.0	36.0	32.0	
Flatwise, perpendicular to grain	$f_{m,90,flat,k}$	N/mm²	–	8.0	–	
Tension						
Parallel to grain	$f_{t,0,k}$	N/mm²	35.0	26.0	$24(3000/L)^{s/2}$	
Perpendicular to grain	$f_{t,90,k}$	N/mm²	0.8	6.0	–	
Compression						
Parallel to grain	$f_{c,0,k}$	N/mm²	35.0	26.0	26.0	
Perpendicular to grain edgewise	$f_{c,90,edge,k}$	N/mm²	6.0	9.0	4.0	
Perpendicular to grain flatwise	$f_{c,90,flat,k}$	N/mm²	1.8	2.2	1.0	
Shear						
Edgewise	$f_{v,0,edge,k}$	N/mm²	4.1	4.5	2.4	
Flatwise	$f_{v,0,flat,k}$	N/mm²	2.3	1.3	1.3	
Modulus of elasticity						
Parallel to grain	$E_{0,k}$	N/mm²	11600	8800	8400	
Shear modulus						
Edgewise	$G_{0,k}$	N/mm²	400	400	270	
Density	ρ_k	kg/m³	480	480	410	
Mean values						
Modulus of elasticity						
Parallel to grain	$E_{0,mean}$	N/mm²	13800	10500	10000	
Shear modulus						
Edgewise	$G_{0,mean}$	N/mm²	600	600	400	
Density	ρ_{mean}	kg/m³	510	510	440	

Bending edgewise: $f_{m,0,edge,k}$ and $E_{0,mean/k}$
Shear edgewise: $f_{v,edge,k}$ and $G_{0,mean}$

Bending flatwise: $f_{m,0,flat,k}$ and $E_{0,mean/k}$
Shear flatwise: $f_{v,flat,k}$

Tension or compression parallel to grain. $f_{t,0,k}$, $f_{c,0,k}$ and $E_{0,mean/k}$

Tension or compression perpendicular to grain edgewise: $f_{t,90,k}$ and $f_{c,90,edge,k}$

Compression perpendicular to grain flatwise: $f_{c,90,flat,k}$

*s is the size effect exponent referred to in Clause 3.4 of EC5 [12].

Fig. 1.13. TimberStrand (LSL), courtesy of Weyerhaeuser.

Table 1.17 TimberStrand® (LSL): Strength and stiffness properties and density values

	Symbol	Units	Grade: 1.5E	Grade: 1.7E
Characteristic values				
Bending				
Edgewise	$f_{m,0,edge,k}$	N/mm²	32.4	37.6
Flatwise	$f_{m,0,flat,k}$	N/mm²	36.3	42.0
Tension				
Parallel to grain	$f_{t,0,k}$	N/mm²	24.4	28.9
Compression				
Parallel to grain	$f_{c,0,k}$	N/mm²	25.4	31.0
Perpendicular to grain edgewise	$f_{c,90,edge,k}$	N/mm²	8.9	10.1
Perpendicular to grain flatwise	$f_{c,90,flat,k}$	N/mm²	5.4	5.9
Shear				
Edgewise	$f_{v,0,edge,k}$	N/mm²	8.6	8.6
Flatwise	$f_{v,0,flat,k}$	N/mm²	3.2	3.2
Density	ρ_k	kg/m³	420	420
Mean values				
Modulus of elasticity				
Parallel to grain	$E_{0,mean}$	N/mm²	10300	11700
Shear modulus				
Edgewise	$G_{0,mean}$	N/mm²	645	730
Density	ρ_{mean}	kg/m³	650	690

Bending edgewise: $f_{m,0,edge,k}$ and $E_{0,mean}$
Shear edgewise: $f_{v,edge,k}$ and $G_{0,mean}$

Bending flatwise: $f_{m,0,flat,k}$ and $E_{0,mean}$
Shear flatwise: $f_{v,flat,k}$

Compression or tension parallel to grain: $f_{c,0,k}$ and $f_{t,0,k}$

Compression perpendicular to grain
flatwise: $f_{t,90,flat,k}$
edgewise: $f_{t,90,edge,k}$

Fig. 1.14. Parallam (PSL) – manufacturing process and in construction; photos (a) and (b) courtesy of Weyerhaeuser.

(a)
(b)
(c)
(d)

Fig. 1.15. OSB: (a) production before the press (photo courtesy of Wikipedia Foundation); (b) and (c) OSB boards and wall panels; (d) OSB webs in I-joists.

in various thicknesses of 8–25 mm with panel sizes of up to 2.4 m wide × 4.8 m long, which makes it an attractive product for floor decking, roof cladding, wall sheathing and for composite constructions such as SIPs, etc.

BS EN 12369-1:2001 [23] provides information on the characteristic values for the three grades of OSB complying with BS EN 300:2006 [24] for use in designing structures to EC5:

- OSB/2 is a general purpose (unconditioned) load-bearing panel for use in dry conditions only (service class 1).
- OSB/3 is a load-bearing structural panel for use in humid conditions (service classes 1 or 2).
- OSB/4 is a heavy-duty load-bearing structural panel for use in humid conditions (service classes 1 and 2).

OSB/3 and OSB/4 grades are intended for use in design and construction of load-bearing or stiffening building elements such as walls, flooring, roofing and I-beams. BS EN 12369-1 gives the minimum characteristic values for OSB complying with BS EN 300, which are summarised in Table 1.18.

Table 1.18 Strength and stiffness properties and density values for OSB boards complying with EN 300: 2006 (based on BS EN 12369-1:2001)

Section properties	Characteristic strength (N/mm²)								Characteristic Density (kg/m³)	Mean modulus of rigidity (N/mm²)		Mean modulus of elasticity (N/mm²)					
	Bending		Compression		Tension		Panel shear	Planar (rolling) shear		Panel shear	Planar shear	Bending		Tension		Compression	
Thickness (mm)	$f_{m,0,k}$	$f_{m,90,k}$	$f_{c,0,k}$	$f_{c,90,k}$	$f_{t,0,k}$	$f_{t,90,k}$	$f_{v,k}$	$f_{r,k}$	ρ_k	$G_{v,mean}$	$G_{r,mean}$	$E_{m,0,mean}$	$E_{m,90,mean}$	$E_{t,0,mean}$	$E_{t,90,mean}$	$E_{c,0,mean}$	$E_{c,90,mean}$
OSB/2: load-bearing boards for use in dry conditions; OSB/3: load-bearing boards for use in humid conditions																	
>6–10	18.0	9.0	15.9	12.9	9.9	7.2	6.8	1.0	550	1080	50	4930	1980	3800	3000	3800	3000
>10–18	16.4	8.2	15.4	12.7	9.4	7.0	6.8	1.0	550	1080	50	4930	1980	3800	3000	3800	3000
>18–25	14.8	7.4	14.8	12.4	9.0	6.8	6.8	1.0	550	1080	50	4930	1980	3800	3000	3800	3000
OSB/4: heavy-duty load-bearing boards for use in humid conditions																	
>6–10	24.5	13.0	18.1	14.3	11.9	8.5	6.9	1.1	550	1090	60	6780	2680	4300	3200	4300	3200
>10–18	23.0	12.2	17.6	14.0	11.4	8.2	6.9	1.1	550	1090	60	6780	2680	4300	3200	4300	3200
>18–25	21.0	11.4	17.0	13.7	10.9	8.8	6.9	1.1	550	1090	60	6780	2680	4300	3200	4300	3200

The 5% characteristic values for stiffness (i.e. G_k and E_k) should be taken as 0.85 times the mean values given in this table. Other properties not given in this table shall comply with the requirements given in EN 300 for the grades OSB/2, OSB3, OSB/4

Bending parallel to grain: $f_{m,0,k}$ and $E_{m,0,mean}$
Planar shear: $f_{r,k}$ and $G_{r,mean}$

Bending perpendicular to grain: $f_{m,90,k}$ and $E_{m,90,mean}$
Planar shear: $f_{r,k}$ and $G_{r,mean}$

Tension or compression parallel to grain: $f_{t,0,k}$, $f_{c,0,k}$ and $E_{t,0,mean}$, $E_{c,0,mean}$

Tension or compression perpendicular to grain: $f_{t,90,k}$, $f_{c,90,k}$ and $E_{t,90,mean}$, $E_{c,90,mean}$

Panel shear: $f_{v,k}$ and $G_{v,mean}$

1.7.8   Particleboards and fibre composites

Particle and fibre composites are usually available in panel form and are widely used in housing construction and furniture manufacture. There are several products in this category and all are processed in a similar way. Examples include high-density fibreboard, medium-density fibreboard, tempered hardboard, cement-bonded particleboard, etc. For fibreboards, chips are refined to wood fibres by the aid of steam and then dried and adhesive is added to form a mat of wood particles and pressed until the adhesive is cured. After cooling the boards are cut to the required sizes. Wood chipboard is a particular derivative of this product family, and is made from small particles of wood and binder.

In BS EN 312:2010 [25] seven types of particleboards (chipboards) are classified and are distinguished as follows:

- P1 general purpose boards for use in dry conditions
- P2 boards for interior fitments (including furniture) for use in dry conditions
- P3 non-load-bearing boards for use in humid conditions
- P4 load-bearing boards for use in dry conditions
- P5 load-bearing boards for use in humid conditions
- P6 heavy-duty load-bearing boards for use in dry conditions
- P7 heavy-duty load-bearing boards for use in humid conditions.

Note that dry conditions refer to service class 1 only and humid conditions refer to service classes 1 and 2.

The P1, P2 and P3 grade particleboards (chipboards) are for general applications including furniture manufacturing and kitchen worktops. Boards of type P4–P7 are intended for use in design and construction of load-bearing or stiffening building elements such as walls, flooring, roofing and I-beams. For dry internal applications grade P4 can be used. The P4 grade is adequate where no moisture will be encountered during or after construction. The moisture-resistant grade P5 is the most commonly specified flooring in the United Kingdom. It is used extensively in new build house building and refurbishment projects. Durability is achieved by using advanced moisture-resistant resins. A green identification dye is added to the surface of the P5 grade to visually differentiate it on-site.

BS EN 12369-1 gives the minimum characteristic values for particleboards complying with BS EN 312, which are summarised in Table 1.19.

1.7.9   Thin webbed joists (I-joists)

I-joists are structurally engineered timber joists comprising flanges made from solid timber or LVL and a web made from OSB, plywood or particleboard. The flanges and web are bonded together to form an I-section member, a structurally efficient alternative to conventional solid timber. I-joists are economical and versatile structural elements in which the geometry permits efficient use of material by concentrating the timber in the outermost areas of the cross-section where it is required to resist the stresses. The flanges are commonly designed to provide the moment capacity of the beam and the web to predominantly carry the shear force; see Figure 1.16.

Table 1.19 Minimum strength, stiffness and density values for particleboards P4 and P5 complying with EN 312: 2003 (based on BS EN 12369-1:2001)

Section properties	Characteristic strength (N/mm²)					Characteristic density (kg/m³)	Mean modulus of rigidity (N/mm²)	Mean modulus of elasticity (N/mm²)		
	Bending	Compression	Tension	Panel shear	Planar shear		Panel shear	Bending	Tension	Compression
Thickness (mm)	$f_{m,k}$	$f_{c,k}$	$f_{t,k}$	$f_{v,k}$	$f_{r,k}$	ρ_k	$G_{v,mean}$	$E_{m,mean}$	$E_{t,mean}$	$E_{c,mean}$
Particleboard P4: load-bearing boards for use in dry conditions (service class 1 only)										
> 6–13	14.2	12.0	8.9	6.6	1.8	650	860	3200	1800	1800
> 13–20	12.5	11.1	7.9	6.1	1.6	600	830	2900	1700	1700
> 20–25	10.8	9.6	6.9	5.5	1.4	550	770	2700	1600	1600
> 25–32	9.2	9.0	6.1	4.8	1.2	550	680	2400	1400	1400
> 32–40	7.5	7.6	5.0	4.4	1.1	500	600	2100	1200	1200
> 40	5.8	6.1	4.4	4.2	1.0	500	550	1800	1100	1100
Particleboard P5: load-bearing boards for use in humid conditions (service classes 1 and 2)										
> 6–13	15.0	12.7	9.4	7.0	1.9	650	960	3500	2000	2000
> 13–20	13.3	11.8	8.5	6.5	1.7	600	930	3300	1900	1900
> 20–25	11.7	10.3	7.4	5.9	1.5	550	860	3000	1800	1800
> 25–32	10.0	9.8	6.6	5.2	1.3	550	750	2600	1500	1500
> 32–40	8.3	8.5	5.6	4.8	1.2	500	690	2400	1400	1400
> 40	7.5	7.8	5.5	4.4	1.0	500	660	2100	1300	1300

The 5% characteristic values for stiffness (i.e. G_k and E_k) should be taken as 0.8 times the mean values given in this table. Other properties not given in this table should comply with the requirements given in EN 312:2003

Bending: $f_{m,k}$ and $E_{m,mean}$
Planar shear: $f_{r,k}$

Tension: $f_{t,k}$, $E_{t,mean}$
Compression: $f_{c,k}$ and $E_{c,mean}$

Panel shear: $f_{v,k}$ and $G_{v,mean}$

(a)   (b)   (c)

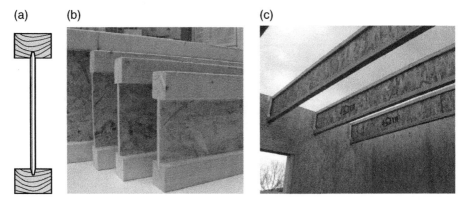

Fig. 1.16. Typical I-joists and their application.

I-joists are lightweight and can easily be handled by one or two persons, they generally possess higher strength and stiffness than comparable-sized solid timber, resist shrinkage, warping, splitting and checking, and are more efficient than solid timber for large spans and loads. They can be used as structural framing in floors, walls and in flat and pitched roofs. They are susceptible to shear buckling and are unstable until braced laterally. Compression flanges should be supported to prevent lateral deflection and buckling. Web stiffeners may be required at the bearing supports and positions of concentrated loads; service holes in the web should only be located in areas where shear loads are low.

I-joists can be designed using the rules in EC5 (see Chapter 7) and they are also available as proprietary systems. Manufacturers produce design guidance literature such as load span tables, permitted web hole requirements, joist hanger details, stiffener requirements, etc. In the United Kingdom, two main manufacturers or suppliers of I-joists are James Jones & Sons Ltd and Metsa Wood:

- James Jones JJI-joists® manufactured at the company's Timber Systems Division in Forres, Scotland, are available in a range of sizes familiar to the UK construction industry. Flanges are made with solid timber grade C24 and webs with 9-mm-thick OSB/3. Section depths range from 145 to 450 mm with flange widths from 47 to 97 mm all 45 mm deep.
- Metsa Wood Finnjoists or FJI-joists® manufactured at the company's factory in King's Lynn, England, have flanges produced from LVL, and a web of 10 mm thick OSB/3. Section depths range from 200 mm to 400 mm with flange widths from 38 mm to 89 mm all 39 mm deep.

1.7.10   Thin webbed beams (box beams)

Box beams comprise solid timber, LVL or glulam flanges with plywood or OSB webs. The webs are generally glued to the flanges on each side to form a box shape. Machine driven nails/staples can be used to aid fabrication.

Similar to I-joists, the larger parts of the cross-section (flanges) of box beams are at the top and bottom where the flexural stresses are highest. Plywood box beam showing veneer on its webs can be used as part of the aesthetic finish as well as the structure.

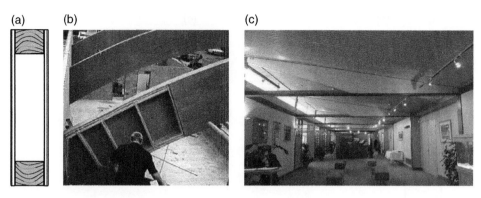

Fig. 1.17. Box beam: (a) cross-section, (b) curved ply-box beam during construction (photo courtesy of Peter Yttrup and Fred Bosveld), and (c) tapered ply-box beam (photo courtesy of Catriona McLeod and Port of Brisbane Corporation, Australia).

The hollow cross-section of the box profile also permits services to be run in the void inside the member giving a cleaner finish. It also gives the member torsional rigidity, which makes it more able to resist lateral torsional buckling or stresses due to eccentric loads.

Box beams are manufactured in depths up to 1.2 m. Web stiffeners are used to help control shear buckling of the web and provide convenient locations for web butt joints; see Figure 1.17. They are also located at positions of point loads to counter localised web buckling. In box beams, the web joint locations are ideally alternated from side to side and away from the areas of highest shear.

Unlike I-beams, which are factory produced in their final sections, it is not currently possible to buy box beams 'off the shelf'. Box beams are normally designed specifically for each contract requirement and assembled on-site.

The design of thin webbed joists and beams (I-joists and box beams) is covered in Chapter 7.

1.7.11   Structural Insulated Panels (SIPs)

SIPs are factory produced, pre-fabricated building panels that can be used as wall, floor and roof components in all types of residential and commercial buildings. They were developed in North America and have experienced wide-scale utilisation around the world. The biggest benefit with the system is that the structural support and the insulation are incorporated into a single system during manufacture. This enables high quality, more accurate thermal efficiency and a greater level of structural support to be achieved.

They are composed of a core of rigid foam insulation, which is laminated between two layers of structural timber panels (boards) by industrial adhesives. This process produces a single solid building element that provides both structural and insulation qualities. These panels are produced in varying sizes and thicknesses depending on the application and thermal/structural requirements.

The materials used to produce these building components can vary greatly in both the structural sheathing and the inner insulation core. Materials commonly used in the United Kingdom for the panels are OSB grade 3, or plywood combined with a variety of plastic foams including expanded polystyrene, extruded polystyrene,

Fig. 1.18. SIPs during construction.

urethane and other similar insulation cores. Typical SIPs can be seen in Figure 1.18. Further information on the structural performance of SIPs may be obtained from the Structural Insulated Panel Association (SIPA) website and relevant research publications [26, 27].

Table 1.20 Summary of the current engineered wood products and their structural applications

Product	Category	Application	Common sizes
Glulam	Laminate	• Beams, columns, trusses, bridges, portal frames, post and beam systems • Industrial, commercial, recreational, residential and institutional	No theoretical limits to size, length or shape
CLT (X-Lam)	Laminate	• Floors, walls, roofs, lift shafts, stairwells and bridge decks • Industrial, commercial, recreational, residential and institutional	Length: up to 24 m Width: up to 4.8 m Thickness: 50–500 mm available
LVL	Laminate	• Beams, columns, vehicle decking, door and window frame, formwork system, flanges of I-joists • Industrial, commercial, recreational, residential and institutional	Length: up to 24 m Width: 19–90 mm Depth: 200–600 mm up to 2.5 m available
TimberStrand	Composite	• Beams, columns, truss members, headers, portal frames, post and beam systems • Industrial, commercial, recreational, residential and institutional	Length: up to 14.6 m Width: 45–140 mm Depth: 1220 mm
Parallam	Composite	• Beams, columns, truss members, headers, portal frames, post and beam systems • Industrial, commercial, recreational, residential and institutional	Length: up to 20 m Width: 45–275 mm Depth: 200–475 mm
I-joists	System	• Floor and roof joists, formwork, ceiling ties, load-bearing stud wall units, available as complete systems (cassettes). • Industrial, commercial, recreational, residential and institutional.	Length: up to 15 m Width: 38–97 mm Depth: 0.2–0.6 m
Box beams	System	• Beams and columns • Industrial and residential buildings	Spans of 30–40 m are possible with portal frames

In Table 1.20 the available range of EWPs is summarised and their applications are outlined.

1.8   SUSPENDED TIMBER FLOORING

A suspended flooring system generally comprises a series of joists closely spaced, being either simply supported at their ends or continuous over load-bearing partition walls. The floor boarding or decking is applied on the top of the joists and underneath ceiling linings are fixed. A typical suspended floor arrangement is shown in Figure 1.19a.

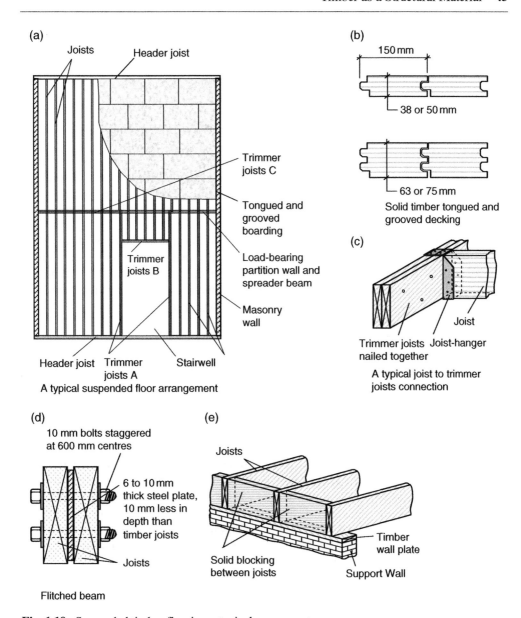

Fig. 1.19. Suspended timber flooring – typical components.

The distance between the centres of the joists is normally governed by the size of the decking and ceiling boards, which are normally available in dimensions of 1200 mm wide × 2400 mm long. The size of the decking and ceiling boards allows convenient joist spacings of 300 mm, 400 mm or 600 mm centre to centre. In addition, the choice of joist spacing may also be affected by the spanning capacity of the flooring material, joist span and other geometrical constraints such as an opening for a stairwell.

The most common floor decking in domestic dwellings and timber-framed buildings uses some form of wood-based panel products, for example chipboard,

OSB or plywood. Solid timber decking such as softwood tongued and grooved (T&G) decking is often used in roof constructions, in conjunction with glued-laminated members, to produce a pleasant, natural timber ceiling with clear spans between the main structural members. The solid timber T&G boards are normally machined from 150-mm-wide sections with 38 to 75 mm basic thicknesses; see Figure 1.19b.

The supports for joists are provided in various forms depending on the type of construction. Timber wall plates are normally used to support joists on top of masonry walls and foundations; see Figure 1.19e. In situations where joists are to be supported on load-bearing timber-framed walls or internal partitions, header beams or spreader members are provided to evenly distribute the vertical loads. Joist hangers are often used to attach and support joists onto the main timber beams, trimmer members or masonry walls; see Figure 1.19c.

Timber trimmer joists are frequently used within timber floors of all types of domestic buildings; see Figure 1.19a. There are two main reasons for which trimmer joists may be provided. First is to trim around an opening such as a stairwell or loft access (trimmer joists A), and to support incoming joists (trimmer joists B), and second is to reduce the span of floor joists over long open spans (trimmer joists C), as shown in Figure 1.19a.

Trimming around openings can usually be achieved by using two or more joists nailed together to form a trimmer beam, as shown in Figure 1.19c, or by using a single but larger timber section, if construction geometry permits. Alternatively, trimmers can be of hardwood or glued-laminated timber, boxed ply-webbed beams, or as shown in Figure 1.19d, composite timber and steel flitched beams.

All flooring systems are required to have fire resistance from the floor below and this is achieved by the ceiling linings, the joists and the floor boarding acting together as a composite construction. For example, floors in two-storey domestic buildings require modified 30-min fire resistance (30-min load-bearing, 15-min integrity and 15-min insulation). In general, a conventional suspended timber flooring system comprising 12.5 mm plasterboard taped and filled, T&G floor boarding with at least 16 mm thickness directly nailed/screwed to floor joists, meets this requirement provided that where joist hangers are used they are formed from at least 1-mm-thick steel of strap or shoe type. Further information on fire safety and resistance is given in 1.11.

1.9   ADHESIVE BONDING OF TIMBER

In recent years there has been a significant advance in adhesive technology achieving high strength, stiffness and durability. These adhesives are now being used with timber, in the production of EWPs and also in timber construction to manufacture adhesively bonded components and connections (glued joints).

Connections bonded by adhesives can result in a better appearance and are often stiffer, requiring less timber; if formed by thermosetting resins they can perform better in fire than mechanical connectors. Their main disadvantage is the high level of quality control that is required in their manufacture and they can also degrade in conditions of fluctuating moisture content, in particular where dissimilar materials are involved. Examples of uses of adhesives in structural timber connections include finger joints, scarf joints, splice and gusset plates (using high-quality structural plywood plates), in the manufacture of I-beams, box beams, stress skin panels and in composite (sandwich) constructions where OSB or plywood side panels are bonded to a core of polystyrene such as in the manufacture of SIPs.

Structural adhesives should be weather and boil proof to BS EN 301:2006 [28]. Acceptable strength and durability can be achieved by using phenolic and aminoplastic-type adhesives as defined in BS EN 301. The adhesives should meet the requirements for adhesives type I or II as follows:

- Type I adhesives, which will withstand full outdoor exposure and temperatures up to and above 50 °C.
- Type II adhesives, which may be used in heated and ventilated buildings and exterior protected from weather. They may not be able to withstand prolonged exposure to weather or to temperatures above 50 °C.

The following adhesives may be considered:

- Resorcinol formaldehyde and phenol resorcinol formaldehyde
    Type: phenolic thermoset resin – for exterior use.
    Uses: finger jointing, laminating, timber jointing, etc.
- Phenol-formaldehyde (PF) hot setting
    Type: phenolic thermoset resin – for exterior use.
    Uses: plywood, laminating, particleboard, etc.
- Melamine urea formaldehyde
    Type: aminoplastic thermoset resin – for semi-exterior and humid interior use.
    Uses: plywood, laminating, particleboard, timber jointing etc.
- Urea formaldehyde
    Type: aminoplastic thermoset resin – for interior use.
    Uses: plywood, laminating, particleboard, timber jointing etc.
- Casein adhesives
    Type: milk product – for interior use only.
    Uses: general purpose timber jointing.

It should be noted that not all adhesives are classified in accordance with BS EN 301. It is therefore important for the designer to ensure that the adhesives selected are suitable for the specified service class and comply with the relevant building regulations. It is also important that timber is conditioned to a moisture content corresponding to the average moisture content likely to be attained in service and that surfaces are properly prepared prior to gluing.

1.10 PRESERVATIVE TREATMENT FOR TIMBER

Under ideal conditions timber should not deteriorate, but when timber is used in exposed (outdoor) conditions, it becomes susceptible to degradation due to a variety of natural causes. It will suffer rot and insect attack unless it is naturally durable or is protected by a preservative. In general, timber with a moisture content of over 20% is susceptible to fungal decay; timber of any species kept in dry conditions will remain sound; however, dry timber may be subjected to insect attack. Timber can be protected from the attacks by fungi, harmful insects or marine borers by applying chemical preservatives. The degree of protection achieved depends on the preservative used and the proper penetration and retention of the chemicals, as treatability varies among the species and also between their heartwood and sapwood. Some preservatives are more effective than others, and some are more adaptable to specific use requirements.

There are a number of widely used methods of application of preservative treatments. Pressure impregnation with a water-borne agent is appropriate for timber in ground contact or high hazard conditions. Double vacuum impregnation with a solvent-based organic preservative is a preferred method for treating joinery timbers. Micro-emulsion treatments, which are water borne, with new more environmentally acceptable products, are now available in the market. Preservatives should be applied under the controlled conditions of an authorised wood treatment plant.

British Standard BS 8417:2011 [29] provides guidance on the treatment of timber for use in the United Kingdom, and includes the requirements of key parts of other relevant BS and EN standards. Issues related to the requirement for preservative treatment include service life, in-service environment, species type and its natural durability as well as the type and form of the preservative treatment.

BS EN 335-1:2006 [30] provides a description of use classes and lists the potential biological organisms and insects that may challenge the timber in a particular service condition. A summary of use classes is given in Table 1.21.

Recommendations for the treatment of softwood timbers are given in BS 8417. Further information on the protection of timber and timber products may be found in the following:

- TRADA publication: Wood Information Sheet WIS 2/3-33 Wood preservation – Chemical and processes, 2005.
- Wood Protection Association (WPA): Industrial Wood Preservation, Specification and Practice, Derby, 2006. (www.wood-protection.org).

1.11   FIRE SAFETY AND RESISTANCE

Fire safety involves prevention, detection, containment and evacuation; requiring prevention of the ignition of combustible materials by controlling either the source of heat, reducing the combustibility of the materials or providing protective barriers. This involves proper design and detailing, insulation or construction and maintenance of the building and its components.

Timber and wood-based materials comprise mainly cellulose and lignin, which are combustible and will burn if exposed to an ignition source under suitable conditions. But this does not mean that due to its combustibility timber is an unacceptable material for construction use. Often the opposite is true. Due to its good thermal insulation properties, when timber burns a layer of char is created, which helps to protect and maintain the strength and structural integrity of the wood inside. This is why timber in

Table 1.21 Use classes and possible biological organisms*

Use classes	Definition of service situation (location of timber component)	Exposure to wetting during service life	Biological organisms			
			Fungi	Beetle[†]	Termite	Marine borers
1	Above ground, covered (dry)	None	No	Yes	Possible	No
2	Above ground, covered (risk of wetting)	Occasionally	Yes	Yes	Possible	No
3	(i) Above ground, exterior, protected	Occasionally	Yes	Yes	Possible	No
	(ii) Above ground, exterior, not covered	Frequently				
4	(i) In contact with ground, exterior and/ or fresh water	Predominantly or permanently	Yes	Yes	Possible	No
	(ii) In contact with ground, severe exterior and/or fresh water	Permanently				
5	In salt water	Permanently	Yes	Yes	Possible	Yes

*In accordance with BS EN 335-1:2006.
[†]The risk of attack can be insignificant depending on specific service situations.

large sections can often be used in unprotected situations where non-combustible materials such as steel would require special fire protection.

The fire protection of timber depends on many factors including size, species type and moisture content. Smaller section sizes, low-density species and sections with cracks and fissures are more likely to ignite and burn more easily than larger and denser ones, and as such may require treatment with flame-retardant chemicals. This may also be a design requirement for situations that require the use of materials with better fire-resistance properties. The treatments used are based on formulations of water-soluble inorganic salts such as ammonium phosphate or water-soluble humidity-resistance formulations and organic resins.

The choice of fire-retardant treatment depends upon many different factors, including the standard of performance required and the conditions in which the treated timber or panel products are to be used. There is much literature available on the choice of fire-retardant treatment and information is also available from specialist organisations including the following:

- TRADA publication: Wood Information Sheet WIS 2/3-3, Flame Retardant Treatments for Timber, 2003.
- Wood Protection Association (WPA): Flame Retardant Specification Manual, 2nd Edition, Derby, 2011. (www.wood-protection.org).

The fire performance of all materials to be used in buildings (of various use), including wood and wood-based products, is given in the relevant Building Regulations operating in Scotland, England and Wales, and Northern Ireland.

The design of timber structures for the accidental situations of fire exposure should be carried out in accordance with the requirements of Eurocode BS EN 1995-1-2:2004 [31] in conjunction with BS EN 1995-1-1:2004+A1:2008 [12] and EN 1991-1-2:2002 [32]. This standard describes the principles, requirements and rules for the structural design of buildings exposed to fire, so that

- fire risks are limited with respect to the individual, neighbouring property, society, and where required, directly exposed property, in the case of fire, and
- a detailed structural fire design is carried out covering the behaviour of the structural system at elevated temperatures, the potential heat exposure and the beneficial effects of active fire protection systems, together with the uncertainties associated with these three features and the consequences of failure.

1.12   REFERENCES

1 Somayaji, S. *Structural Wood Design.* West Publishing Company, St Paul, MN, 1990.
2 Illston, J.M. *Construction Materials – Their Nature and Behaviour.* E&FN Spon, London, 1994.
3 BS 7359:1991. *Nomenclature of Commercial Timbers Including Sources of Supply.* British Standards Institution.
4 Johansson, C.J. Grading of timber with respect to mechanical properties. In: Thelandersson, S., Larsen, H.J. (eds), *Timber Engineering.* Wiley, London, 2003.
5 Hoffmeyer, P. *Wood as a Building Material, 'Timber Engineering' STEP 1,* Blass, H.J., Aune, P., Choo, B.S., et al. (eds). Centrum Hout, Almere, 1995.
6 BS EN 14081-1:2005+A1:2011. *Timber Structures – Strength Graded Structural Timber with Rectangular Cross Section. Part 1: General Requirements,* British Standards Institution.
7 BS EN 14081-2:2010. *Timber Structures – Strength Graded Structural Timber with Rectangular Cross Section. Part 2: Machine Grading, Additional Requirements for Initial Type Testing,* British Standards Institution.
8 BS 4978:2007+A1:2011. *Specification for Visual Strength Grading of Softwood,* British Standards Institution.
9 BS 5756:2007. *Visual Strength Grading of Hardwood,* British Standards Institution.
10 BS 5268-2:2002. *Structural Use of Timber. Part 2: Code of Practice for Permissible Stress Design, Materials and Workmanship,* British Standards Institution.
11 BS EN 338:2009. *Structural Timber – Strength Classes,* British Standards Institution.
12 BS EN 1995-1-1:2004+A1:2008. *Design of Timber Structures: Common Rules and Rules for Buildings,* British Standards Institution.
13 BS EN 1313-1:2010. *Round and Sawn Timber – Permitted Deviations and Preferred Sizes. Part 1: Softwood Sawn Timber,* British Standards Institution.
14 BS EN 336:2003. *Structural Timber – Sizes, Permitted Deviations,* British Standards Institution.

15 prEN 16351:2011. Draft BS EN 16351. *Timber Structures – Cross Laminated Timber – Requirements*, British Standards Institution.

16 BS EN 636:2003. *Plywood – Specifications*, British Standards Institution.

17 BS EN 12369-2:2011. *Wood-Based Panels – Characteristic Values for Structural Design. Part 2: Plywood*, British Standards Institution.

18 BS EN 789:2004. *Timber Structures. Test Methods. Determination of Mechanical Properties of Wood Based Panels*, British Standards Institution.

19 BS EN 1058:2009. *Wood-Based Panels. Determination of Characteristic 5-Percentile Values and Characteristic Mean Values*, British Standards Institution.

20 BS EN 14272:2011. *Plywood. Calculation Method for some Mechanical Properties*, British Standards Institution.

21 Finnish Forest Industries Federation. *Handbook of Finnish Plywood*. Kirjapaino Markprint oy, Lahti, Finland, 2002, ISBN 952-9506-63-5.

22 Canadian Plywood Association (CANPLY). *CE Engineering Values*, www.canply.org.

23 BS EN 12369-1:2001. *Wood-Based Panels – Characteristic Values for Structural Design. Part 1: OSB, Particleboards and Fibreboards*, British Standards Institution.

24 BS EN 300:2006. *Oriented Strand Boards (OSB) – Definitions, Classification and Specifications*, British Standards Institution.

25 BS EN 312:2010. *Particleboards – Specifications*, British Standards Institution.

26 Kermani, A. Performance of structural insulated panels. *Proceedings of the Institution of Civil Engineers. Journal of Buildings & Structures*, Vol. 159, Issue SB1, 2006, pp. 13–19.

27 Kermani, A., Hairstans, R. Racking resistance of structural insulated panels. American Society of Civil Engineers (ASCE). *Journal of Structural Engineering*, Vol. 132, No. 11, 2006, pp. 1806–1812.

28 BS EN 301:2006. *Adhesives Phenolic and Aminoplastic, for Load-Bearing Timber Structures: Classification and Performance Requirements*, British Standards Institution.

29 BS 8417:2011. *Preservation of Wood – Code of Practice*, British Standards Institution.

30 BS EN 335-1:2006. *Durability of Wood and Wood-Based Products. Definitions of Use Classes. General*, British Standards Institution.

31 BS EN 1995-1-2:2004. *Design of Timber Structures. Part 1-2: General – Structural Fire Design*, British Standards Institution.

32 BS EN 1991-1-2:2002. *Eurocode 1: Action on Structures. Part 1-1: General Actions – Actions on Structures Exposed to Fire*, British Standards Institution.

Chapter 2

Introduction to Relevant Eurocodes

2.1 EUROCODES: GENERAL STRUCTURE

Eurocodes form a set of documents that enable building and civil engineering structures to be designed to common standards across the European Union (EU) using different structural materials. The documents are structured on a hierarchical basis, led by EN 1990, Eurocode – Basis of structural design, defining the basis of structural design, followed by EN 1991, Eurocode 1: Actions on structures, which comprises ten parts, defining the actions that have to be withstood. These documents are supported by a number of Eurocodes detailing the particular methods of design to be followed for the structural materials being used, i.e. structural timber, steel, concrete, etc.

The Eurocode for the design of timber structures is EN 1995, Eurocode 5: Design of timber structures. It comprises three parts:

- EN 1995-1-1   Design of timber structures – Part 1-1: General – Common rules and rules for buildings
- EN 1995-1-2   Design of timber structures – Part 1-2: General – Structural fire design
- EN 1995-2   Design of timber structures – Part 2: Bridges.

EN 1995 covers the requirements for strength, serviceability, durability and fire resistance, with matters such as thermal or sound insulation etc., having to be obtained from other standards.

The content of this book relates to the design of timber and wood-related products for buildings in accordance with the requirements of EN 1995-1-1. As stated in Chapter 1, the design of timber structures for the accidental situation of fire exposure should be carried out in accordance with the requirements of EN 1995-1-2, and this design condition has not been addressed in the book.

On 31 March 2010 the structural design code used for timber, BS 5268-2 [1], which is based on a permissible stress design philosophy, was withdrawn and BS EN 1995-1-1 became the British Standards Institution (BSI) structural timber design code. EN 1995-1-1 is a limit states design philosophy code in which the requirements concerning structural reliability are related to limit states, i.e. states beyond which the

Structural Timber Design to Eurocode 5, Second Edition. Jack Porteous and Abdy Kermani.
© Jack Porteous and Abdy Kermani 2013. Published 2013 by Blackwell Publishing Ltd.

structure or its elements will no longer satisfy performance criteria. This approach provides a more realistic representation of the overall behaviour of the structure than that used in BS 5268-2 and is the philosophy that has been adopted for the Eurocode design suite.

In every Eurocode each item is defined as being either a Principle or an Application rule. A Principle is a statement or requirement that must be fully complied with unless an alternative is given in the document and an Application rule is a rule that will satisfy the Principle. Alternative design rules can be used by the designer provided it can be demonstrated that these will fully comply with the Principles and will produce an alternative design equivalent in regard to serviceability, structural integrity and durability. An important point to note, however, is that in such a situation the design cannot be claimed to be fully compliant with the Eurocode and this may prove to be a problem if a CE marking is required for the design or substantiation of a product. Where an item in a Eurocode is prefixed by a number in brackets followed by the letter P it is a Principle and where it is only prefixed by a number in brackets it is an Application rule.

Where it is considered that a national choice is appropriate for certain design rules or values of functions in a Eurocode, these items can be varied and are defined as Nationally Determined Parameters (NDPs). This information is given in a National Annex, which may also incorporate what is termed 'non-contradictory complimentary information' (NCCI), giving additional guidance on the interpretation or implementation of the design rules in the Eurocode. If not included in the National Annex the NCCI should be published in a separate document and this has been the format used to publish this type of information in the UK for structural timber design.

For application in the United Kingdom, the Eurocodes are published by the BSI incorporating the prefix BS before the Eurocode reference and when implemented nationally, the full text of each Eurocode will be preceded by the associated United Kingdom National Annex (UKNA). When designing to the Eurocode rules the NDP given in the UKNA must be used rather than the equivalent requirement in the Eurocode and because of the significance of the NDP in timber design, the authors consider it important that attention is drawn to these requirements when discussing the design rules in EN 1995-1-1:2004 [2]. This has been included for in the book. The NCCI is published in Published Document, PD6693-1:2012, and is entitled 'Recommendations for the design of timber structures to Eurocode 5 Design of timber structures – Part 1-1: General – Common rules and rules for buildings'.

Currently UKNAs are published as separate documents and the versions associated with BS EN 1990:2002 [3] and BS EN 1995-1-1:2004 referenced within the text and the examples given in the book, are as follows:

- UK National Annex for Eurocode 0 – Basis of structural design [4]
- UK National Annex to Eurocode 5: Design of timber structures – Part 1-1: General – Common rules and rules for buildings [5].

Since publication of the first edition of the book, EN 1995-1-1:2004 has been revised and republished and the latest versions of the particular Eurocode documents that are referred to in this edition of the book together with the abbreviation used in the text for the relevant document are as follows:

British Standard title	Abbreviation used in text
BS EN 1990:2002+A1:2005 'Eurocode – Basis of structural design' [3]	EC0
BS EN 1991-1-1:2002 'Eurocode 1 – Part 1-1: General Actions – Densities, self-weight and imposed loads for Buildings' [6]	EC1
BS EN 1995-1-1:2004+A1:2008 'Design of timber structures – Part 1-1: General – Common rules and rules for Buildings' [2]	EC5
NA to BS EN 1995-1-1:2004+A2:2012 'UK National Annex to Eurocode 5: Design of timber structures – Part 1-1: General – Common rules and rules for buildings' [5].	UKNA to EC5

From consideration of the feedback from users of EC5 in several countries in the EU, it was concluded that clarification is required on the implementation of some requirements of EC5 and errors were also identified in some of the code clauses. The response to these matters has still to be fully agreed by the relevant European Committee for Standardisation responsible for this code (CEN/TC 250) and the intention is to publish a revision to the code in the form of a 'corrigendum' statement, expected to be issued in 2013. A summary of the likely changes being considered by the Committee for inclusion in the corrigendum and taken to be relevant to the content of this book are given in APPENDIX C and where appropriate have been referenced in the text.

Because of the importance of the design framework set by EC0, those matters that have a significant relevance to designs carried out in accordance with the requirements of EC5 are briefly reviewed in 2.2.

2.2   EUROCODE 0: BASIS OF STRUCTURAL DESIGN (EC0)

EC0 provides the framework within which design must be carried out and, as stated in 2.1, uses the limit states design method.

In this section, the methodology to be used and its application to the design of structures made from timber and wood-related products is discussed, drawing on the content of EC5 to show how the requirements have been addressed and interpreted.

2.2.1   Terms and definitions (EC0, *1.5*)

Some of the terms and definitions used in EC0 are slightly different to those normally used in UK timber design practice and the following, including some terms for those not familiar with limit states design, are to be noted:

Action. This is the term used for a load or force applied to the structure (i.e. a direct action). This term is also used for imposed displacements, e.g. settlement (i.e. an indirect action).

Effect of action. This is the term used for the internal stress resultants or displacements in the structure arising from the effect of the action.

Permanent action. This is the term used for an action that will always act in the same direction (i.e. is monotonic) over a given reference period with negligible variation in magnitude, e.g. self-weight.

Variable action. This is the term used for an action that is not monotonic and can vary with time, e.g. live loading.

Limit states. States beyond which the structure will not comply with the design requirements that have been set.

Ultimate limit states (ULS). Limit states associated with collapse or equivalent forms of failure.

Serviceability limit states (SLS). Limit states beyond which defined service criteria will not be met.

Irreversible SLS. SLS where some effects of actions having exceeded the SLS criteria will remain after the SLS actions have been removed.

Reversible SLS. SLS where no effect of actions exceeding the SLS criteria will remain after the SLS have been removed.

Serviceability criterion. A design requirement for a SLS.

Resistance. This is the capacity of a structural element to withstand actions without failing, e.g. shear resistance, bearing resistance.

Strength. This is the withstand capacity of the material at a failure condition, e.g. shear strength, bearing strength.

Reliability. This is the ability of a structure or structural element to fulfil its design requirements over the design working life and is normally expressed in probabilistic terms.

2.2.2   Basic requirements (EC0, *2.1*)

The fundamental Principles that must be satisfied by any structure are given in EC0, *2.1* and can be summarised as follows:

- During its intended life it must sustain all actions likely to occur and meet the specified requirements.
- It must have adequate structural resistance, serviceability and durability.
- The fire-resistance requirements must be fully met.
- It must not be susceptible to damage disproportionate to the original cause.

The adequacy of the design for structural resistance, serviceability and durability will be satisfied by compliance with EC5; fire resistance will be met by designing in accordance with the requirements of BS EN 1995-1-2:2004 [7] and robustness will be achieved by:

(1) minimising hazards to which the structure can be exposed,
(2) choosing a structural form that will be least affected by the types of hazard to be designed for,
(3) selecting a structure that can survive localised damage including the removal of an individual member or a limited part of the structure,
(4) avoiding as far as possible structural systems that can collapse without warning,
(5) tying the structural members together.

Table 2.1 Recommended minimum values for the reliability index β*

Reliability class (RC) associated with CC2: RC2	Minimum values for β	
	ULS	SLS (irreversible)
1-year reference period	4.7	2.9
50-year reference period	3.8	1.5

*Based on *Tables B2 and C2* in EC0.

General guidance on the approaches that can be used to satisfy robustness requirements is given in the Designers' Guide to EN 1990 [8] and the Designers' Guide to
Eurocode 1: Actions on Buildings; EN 1991-1-1 and -1-3 to -1-7. [9].

2.2.3   Reliability management (EC0, *2.2*)

The design of the structure must satisfy reliability criteria and the conceptual requirements to be met are given in EC0, *2.2*. Consequence classes, categorised as high
(CC1), medium (CC2) and low (CC3), dependent on the consequences of the loss of
human life as well as economic, social or environmental consequences in the event of
failure, have been set. The consequence category for most facilities in which timber or
timber-related materials are used for structure or structural elements will be CC2.
Each consequence class is linked to a reliability class (RC), with classes CC1, CC2
and CC3 being linked to reliability classes RC1, RC2 and RC3 respectively. A reliability class has an associated reliability index β, which can be considered as the safety
index to be achieved for that class.

For those facilities that come within the CC2 consequence category, the recommended minimum values of the reliability indices are given in Table 2.1 and, as stated
in EC0, structures designed in accordance with the requirements of EC0, BS EN 1991,
Eurocode 1: Actions on structures, and EC5 will generally result in a structure having
a reliability index greater than 3.8 for a 50-year reference period. For a 50-year reference period, the recommended reliability indices will result in a probability of failure
of the structure between 10^{-4} and 10^{-5} at the ULS and 10^{-1} and 10^{-2} at the SLS.

It should also be noted that achievement of the above reliability levels will depend
on the checking standard used for drawings, calculations and specifications and for
compliance with the RC2 class, the minimum standard will require checking by
different persons to those originally responsible for the design and in accordance with
the quality management system of the organisation (EC0, *Annex B, Table B4*).

2.2.4   Design working life (EC0, *2.3*)

At the start of the design process the 'design working life' of the facility to be designed
must be specified and is defined in EC0 as the

'assumed period for which a structure or part of it is to be used for its intended
purpose with anticipated maintenance but without major repair being necessary'.

In EC0, *Table 2.1*, five categories of design working life are specified and the one most typically associated with facilities supported by timber structures designed in accordance with the requirements of EC5 will be Category 4. The indicative design working life of this category is 50 years. Where necessary or beneficial, it is acceptable to have a Category 4 design working life for the structure and incorporate structural elements with a shorter design working life provided these can readily be replaced without any adverse effect on the facility.

It is to be noted that the design working life may or may not coincide with the reference period used to determine the design values for environmental factors, i.e. wind speed and temperature extremes etc. However, it provides a guide for selection.

Provided the client implements a sound maintenance inspection policy and properly undertakes the maintenance requirements of the building and structure, the facility will fulfil its specified requirements for the design working life. When dealing with timber structures, it is of particular importance that the maintenance policy also ensures that the environment within which the structure functions complies with the service class (see 2.2.20) for which it has been designed.

2.2.5   Durability (EC0, *2.4*)

Durability is the ability of the structure and its elements to remain fit for use when properly maintained during the design working life and must include the effects of deteriorating factors that can arise during this period.

The particular factors highlighted in EC0, *2.4(2)* that are relevant to timber structures are as follows:

- *The intended or foreseeable use of the structure.* For example, a factor to be included for in timber or wood product flooring will be an allowance for wear and tear during the design working life.
- *The required design criteria.* The design intent has to be defined at the outset, identifying where elements of the structure will require replacing during the working life (if this is needed) and ensuring that the design is such that replacement can be undertaken while also retaining a facility that will comply with its specified requirements.
- *The expected environmental conditions.* This is particularly relevant to timber structures. The strength properties of timber and wood product structures are affected by changes in environmental conditions and it is essential that during the working life they function under the service class conditions for which they have been designed.
- *The composition, properties and performance of the materials and products.* It is advisable to consider the use of materials that will enhance durability, and the use of preservative treatments for timber structures in appropriate conditions should be an option. Where preservative treatments are used and they are critical in achieving the durability requirement, the design must be such that re-application during the design working life can be carried out while ensuring the facility will meet its specified requirements.
- *The choice of the structural system.* Where possible a robust structural system should be used, able to safely withstand the known design hazards with, where

economically possible, inbuilt redundancy in excess of the robustness criteria required by the code.

- *The quality of workmanship and the level of control.* The fabrication and assembly must be fully in accordance with the specification requirements, otherwise problems can arise, reducing the durability of the elements of the structure, e.g. preparation and application of adhesives.
- *The intended maintenance during the design working life.* A maintenance strategy should be prepared at the outset of the design process and the building designed such that the structure can be accessed to allow the strategy to be implemented during the design working life.

2.2.6   Quality management (EC0, *2.5*)

A requirement of EC0 is that a quality system must be in place and one in accordance with the requirements of the International System Standard, ISO 9000 [10], will be acceptable. Unless such a system is operated, the design will not be accepted as compliant with the requirements of EC5.

2.2.7   Principles of limit state design: General (EC0, *3.1*)

The concept of limit states design is that for defined states a structure may be classified as either satisfactory or unsatisfactory. The limit state being the state beyond which the structure will no longer satisfy the design criteria and will be classed as unsatisfactory.

It is possible to establish limit states to define limits of satisfaction for numerous issues, and to simplify the design process EC0 has defined two such states:

ultimate limit states – associated with forms of structural failure/collapse,
serviceability limit – associated with normal service conditions, e.g. deflection
states   and vibration conditions.

For the ULS one is dealing with extreme safety conditions and for the SLS it is the level of comfort and appearance that is being addressed and, as shown in Table 2.1, the level of reliability used in the design process will be different for each of the states.

2.2.8   Design situations (EC0, *3.2*)

The structure must be designed for the effect of actions and environmental factors that will occur during the design working life and EC0 defines four design situations that must be considered:

Persistent design situations: conditions of normal use, i.e. self-weight and imposed loading, including wind, snow, etc.
Transient design situations: refers to temporary conditions, e.g. during construction or repair.
Accidental design situations: exceptional conditions, e.g. explosion or impact.
Seismic: conditions arising from a seismic event.

For normal structures there will be no requirement to design for seismic conditions in the United Kingdom and this design situation is not addressed in the book. With regard to the other design situations, i.e. persistent, transient and accidental, the designer must establish the loading conditions to which the structure can reasonably be expected to be subjected over the design working life.

2.2.9   Ultimate limit states (EC0, *3.3*)

These are the limit states that concern the safety of people and/or of the structure. Also protection of the contents supported by the structure can be included for in these states provided this requirement is agreed with the designer.

Specific attention is drawn to the following ULS that are relevant to timber structures and must be considered:

- Loss of equilibrium of part or all of the structure
- Failure by excessive deformation
- Failure as a mechanism
- Failure due to rupture
- Failure due to loss of stability.

2.2.10   Serviceability limit states (EC0, *3.4*)

These are the limit states that concern the functioning and appearance (excessive displacement and cracking/distress) of the structure as well as the comfort of the users. The SLS should be agreed at the outset of the project and a distinction made between those states that will be irreversible and those that will be reversible.

In the irreversible condition, the SLS will be permanently infringed even when the actions that cause the exceedance are removed. Such conditions are treated in the same manner as those for ULS, i.e. at the first exceedance of the SLS the design will be non-compliant.

In the reversible condition, when the actions causing the infringement are removed the exceedance is also removed. In such instances, an agreement can be established with the client on those situations that will fall into this category and the frequency of occurrence that will be acceptable.

EC0 has been structured to accommodate the following three types of SLS:

(a) No exceedance will be permitted.
(b) The frequency and duration of exceedance events will be agreed with the client.
(c) Long-term exceedance events will be agreed with the client.

The combinations of actions associated with these types of SLS are as follows:

- No exceedance permitted – the *characteristic combination* is to be used.
- Frequency and duration of exceedance events agreed – the *frequent combination* is to be used.
- Long-term exceedance events agreed – the *quasi-permanent combination* is to be used.

The action combinations referred to above are discussed in EC0, *4.1.3* and *6.5* and in *2.2.25*.

It is stated in EC5, *2.2.3(2)*, that calculations for instantaneous deformation should be based on the application of the characteristic combination of actions; however, as this is not a Principle in EC5, where an economic case can be substantiated and agreement reached with the client to accept a reversible condition, the design can be based on the frequent rather than the characteristic combination.

The criteria used for the verification of the SLS should be associated with the following matters:

(i) Deformations that affect appearance; that cause damage to finishes or non-structural members; that affect user comfort and the functioning of the structure.
(ii) Vibrations that cause discomfort to users or limit the functionality of the structure.
(iii) Damage that adversely affects appearance, durability or the functionality of the structure.

2.2.11   Limit states design (EC0, *3.5*)

In limit states design, structural and load models are set up for each limit state and the design is verified by demonstrating that none of the states will be exceeded when design values of actions, material or product properties and geometry are used in the models.

Although it is possible to use probabilistic methods of analysis for verification, the deterministic partial factor method referred to in *Section 6* of EC0 (see also 2.2.17) is the practical design method recommended for use.

2.2.12   Classification of actions (EC0, *4.1.1*)

The following symbols and terminology apply when dealing with actions:

(a) *Permanent actions (G).* These are the actions that remain monotonic and will vary by a negligible amount with time, e.g. self-weight, fixed equipment, fixed partitions, finishes and indirect actions caused by shrinkage and/or settlement.
(b) *Variable actions (Q).* These are the actions that do not remain monotonic and may vary with time, e.g. imposed loading, wind, snow and thermal loading.
(c) *Accidental actions (A).* For example, explosion or impact loading.

2.2.13   Characteristic values of actions (EC0, *4.1.2*)

Actions are defined by 'representative' values and the main representative value used in design is the characteristic value. The characteristic value should, where possible, be derived from the statistical data associated with the action and, depending on the design condition, it will be a mean value, an upper or lower value or a nominal value (this being used when it cannot be derived from statistical data).

Characteristic values:

- *Permanent actions* (G_k)
  Where this relates to the self-weight of the material, because the variability of the action is small (i.e., the coefficient of variation of the action during the design working life is less than 0.05–0.1), when dealing with timber or wood-related products, G_k, is generally derived using the mean density of the material.
  Characteristic values for self-weight, G_k, are obtained from the standards that give mechanical properties (e.g., BS EN 338:2009 [11] for timber) or from EC1.
- *Variable actions* (Q_k)
  The characteristic values of the variable actions referred to in Eurocode EN 1991: 'Actions on structures', are given in the relevant parts of that code.
  When dealing with climatic actions (e.g. wind, temperature etc.), EC0 states that the characteristic value is based on the probability of 0.02 of its time-varying part being exceeded for a reference period of 1 year. The probability of exceedance (p) and the reference period (r) are linked by the approximate relationship $T \cong \dfrac{r}{p}$, (where T is the return period) and represents the likely time between two successive occurrences when the characteristic value will be exceeded. On this basis the return period will be 50 years, which equates to a probability of 0.64 that the characteristic value will be exceeded during this period. This criterion also applies to imposed loading on the floors of buildings.
  It should also be noted that where a building carries more than one floor, a reduction in loading is possible and guidance on this is given in EC1.
- *Accidental actions* (A_d)
  Because of the lack of statistical data relating to this condition, the design value, A_d, should be specified and agreed for each project.

2.2.14 Other representative values of variable actions (EC0, *4.1.3*)

Other representative values of variable actions are specified in addition to the characteristic values, Q_k, referred to in 2.2.13. These are as follows:

- The *combination value* $(\psi_0 Q_k)$: used for verification of ULS and for the characteristic combinations of irreversible SLS.
- The *frequent value* $(\psi_1 Q_k)$: used for verification of ULS involving accidental actions and for the verification of reversible SLS.
- The *quasi-permanent value* $(\psi_2 Q_k)$: used for the assessment of long-term effects; for the representation of variable actions in accidental (and seismic) combinations at the ULS, and for the verification of frequent and long-term effects of SLS.

The factors ψ_0, ψ_1 and ψ_2 are reduction factors. Factor ψ_0 takes into account the reduced probability of simultaneous occurrence of the most unfavourable values of several independent variable actions. Factor ψ_1 is a time-related function and sets an upper limit for the value of the variable action to which it applies. For buildings, it is set such that the proportion of time it is exceeded is 1% of the reference period. Factor ψ_2 is also

Table 2.2 Values for ψ factors*

Variable action	ψ_0	ψ_1	ψ_2
Category of imposed loads on buildings (see EC1)			
Category A: domestic and residential areas	0.7	0.5	0.3
Category B: office areas	0.7	0.5	0.3
Category C: areas where people congregate	0.7	0.7	0.6
Category H: roofs (noting that imposed load should not be applied with snow load or wind action (see clause 3.3.2(1) in EC0)	0.7	0	0
Snow loads on buildings for sites with an altitude no greater than 1000 m above sea level	0.5	0.2	0
Wind loads on buildings	0.5	0.2	0

*Based on *Table NA.A1.1*, UKNA to EC0 [4].

a time-related function and in timber engineering its primary role is to convert variable actions to equivalent permanent actions (referred to as quasi-permanent actions) in order to derive the creep loading on the structure. For floor loading on buildings, the value of ψ_2 is set such that the proportion of time it is exceeded is 50% of the reference period and in the case of wind loading and temperature loading (as well as snow loading for sites at an altitude up to 1000 m above sea level) the value will be zero.

Values for ψ_0, ψ_1 and ψ_2 are given in *Table NA.A1.1* of the UKNA to EC0 and some of the values relevant to loading conditions associated with timber buildings are given in Table 2.2.

2.2.15 Material and product properties (EC0, *4.2*)

As in the case of actions, the properties of materials or products are also represented by characteristic values.

When dealing with timber or wood-related properties the characteristic value will be either the 5th-percentile value or the mean value. The 5th-percentile value will apply to strength-related properties and the mean value will normally be used for stiffness-related properties. The exception to this rule is when stiffness-related functions are used in the derivation of a strength property; for example when used for the evaluation of the critical bending strength of a timber beam, in which case the 5th-percentile value rather than the mean value is used.

Characteristic values of the properties of timber and some of the commonly used wood-related products in timber design are given in Chapter 1.

2.2.16 Structural analysis (EC0, *5.1*)

2.2.16.1 *General*

EC0 gives no specific guidance on the method(s) of structural analysis to be used in design other than to require that the structural models be appropriate for the limit state

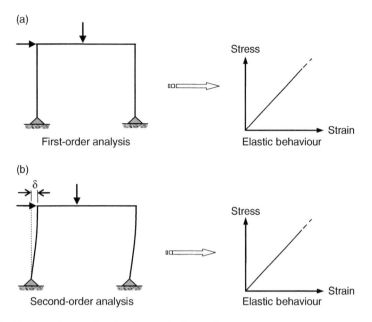

Fig. 2.1. First- and second-order linear elastic analysis.

considered, be able to predict behaviour with an acceptable degree of accuracy, and be based on established engineering theory and practice.

When analysing a structure, several alternative methods of analysis are possible and these are outlined in 2.2.16.1.1 and 2.2.16.1.2.

2.2.16.1.1 *Linear elastic analysis*

This is based on linear stress/strain and moment curvature laws:

(a) *First-order analysis without redistribution*. This is performed on the initially defined geometry of the structure and its elements without any adjustment of internal forces/moments due to redistribution. It is the basis of most first-order linear elastic analysis computer programs (Figure 2.1a).

(b) *First-order analysis with redistribution*. This is performed on the initially defined geometry of the structure and its elements but internal forces/moments are adjusted without further calculation to also adjust rotations and check rotation capacity.

(c) *Second-order analysis*. This is performed on the geometry of the deformed structure (Figure 2.1b).

2.2.16.1.2 *Non-linear analysis*

It is based on a non-linear stress–strain relationship as shown in Figure 2.2:

(a) *First-order analysis*. This is performed on the initially defined geometry of the structure.

(b) *Second-order analysis*. This is performed on the geometry of the deformed structure.

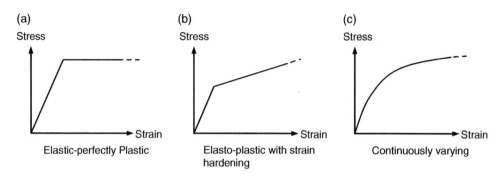

Fig. 2.2. Alternative stress–strain relationships commonly used in non-linear analysis.

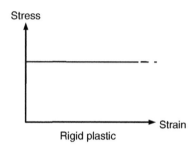

Fig. 2.3. Rigid plastic behaviour.

The following descriptions are commonly associated with non-linear analyses incorporating plastic behaviour:

- *Elastic-perfectly plastic* (Figure 2.2a) – linear elastic followed by pure plastic behaviour.
- *Elasto-plastic* (Figure 2.2b) – linear elastic followed by plastic behaviour with strain hardening.
- *Rigid plastic* (Figure 2.3) – pure plastic behaviour, using limit states analysis for the assessment of ultimate loading (e.g. for the behaviour of joints formed with metal dowel type fasteners in timber structures).

2.2.16.2   *Structural analysis requirements for timber and wood-based structures*

In regard to timber and wood-based structures, the structural analysis requirements are addressed in EC5, *Section 5*.

Because of the brittle nature of timber under tension-induced stress configurations, plastic analysis should not be used and EC5 requires that the forces in the elements of the structure be determined using a linear elastic analysis. The effects of deviation from straightness of members have to be taken into account and this will be achieved by validation of the element strength using the design rules in EC5. Where, however, it is considered that a second-order linear analysis is necessary when dealing with plane frames or arches, this should be carried out in accordance with the requirements of EC5, *5.4.4*.

Although all timber connections will exhibit semi-rigid behaviour to varying degrees, where their rotational deformation will have a negligible effect on the force distribution in the structure, EC5 states that the connections may be considered to be rigid and, where this is not the case, they may generally be assumed to be rotationally pinned (EC5, *5.4.2(7)*). As this is not a Principle in EC5, where it is felt that the effect of semi-rigid behaviour of connections should be taken into account in the analysis, provided it can be considered to be effectively linear and have adequate ductility, by incorporating the stiffness behaviour of the connections into the structural model a linear elastic analysis of the semi-rigid structure can still be undertaken.

When designing joints formed with metal dowel type fasteners, excluding those situations where brittle failure can arise, the strength equations in EC5 assume that failure at the joint will be in accordance with the principles of plastic theory, as shown in Figure 2.3. In such situations, the joint forces will be derived from an elastic analysis of the structure at the ULS and the associated joint strengths determined from the application of the relevant EC5 strength equations, which are derived primarily from the assumption that rigid plastic behaviour, will apply. This is one of the apparent anomalies between modelling to determine the action effect (i.e. the global analysis model) and modelling for strength verification where a different model can be assumed.

2.2.17 Verification by the partial factor method: General (EC0, *6.1*)

For normal designs, the partial factor design method should be used for the design of the structure and its elements. In this method, the effects of actions are multiplied by partial factors to obtain the design value, Ef_d, and resistances, which are generally derived from material strengths, are divided by partial factors to obtain the design resistance, R_d, at the ULS and SLS. Verification is undertaken at the relevant state to demonstrate that Ef_d is less than or equal to the design resistance R_d, i.e.

$$Ef_{d_{ULS}} \leq R_{d_{ULS}} \text{ and } Ef_{d_{SLS}} \leq R_{d_{SLS}}$$

The values used for actions and material properties are the characteristic or other representative values and the values used for the partial factors vary depending on the limit state being considered and must be such that the level of reliability referred to in 2.2.3 for the structure at that limit state will be achieved.

2.2.18 Design values of actions (EC0, *6.3.1*)

In general terms, the design value F_d of an action can be written as

$$F_d = \gamma_f F_{rep} \tag{2.1}$$

where

$$F_{rep} = \psi F_k \tag{2.2}$$

In the above, F_{rep} is the value to be taken into account in the relevant combination of actions. It can be the main representative value (i.e. the characteristic value, F_k), the combination value, $\psi_0 F_k$, the frequent value, $\psi_1 F_k$, or the quasi-permanent value, $\psi_2 F_k$; F_k is the characteristic value of the action; ψ is either 1.00 or ψ_0, ψ_1 or ψ_2.

2.2.19   Design values of the effects of actions (EC0, 6.3.2)

The 'effects' of actions, Ef_d, are the response of the structure to the imposed actions and cover the internal stress resultants (e.g. moments, shear forces, axial forces, stress or strain) and the structural deformations (e.g. deflections and rotations).

Based on the content of EC0, 6.3.2(1) the design value of the effects of actions can be written in general terms as

$$Ef_d = \gamma_{Sd} Ef \left\{ \gamma_{f,i} F_{rep,i}; a_d \right\} \qquad i \geq 1 \tag{2.3}$$

where γ_{Sd} is a partial factor taking account of uncertainties in modelling the effects of actions, $\gamma_{f,i}$ is a partial factor for action i that takes account of the possibility of unfavourable deviations of the action values from the representative values, a_d is the design value of the geometrical data (discussed in 2.2.22), and i is the number of representative actions.

For the design of timber and wood product structures in accordance with EC5, partial factors γ_f and γ_{Sd} are combined into one factor γ_F (i.e. $\gamma_F = \gamma_f \gamma_{Sd}$) simplifying equation (2.3) to

$$Ef_d = Ef \left\{ \gamma_{F,i} F_{rep,i}; a_d \right\} \qquad i \geq 1 \tag{2.4}$$

In the Eurocodes, when determining the design value of a permanent action, γ_F is defined as γ_G and when determining the design value of a variable action, it is defined as γ_Q.

The values of γ_G and γ_Q are dependent on the limit states being considered and this is addressed in 2.2.24 for ULS and 2.2.25 for SLS.

2.2.20   Design values of material or product properties (EC0, 6.3.3)

The design value X_d of a material or product property can be derived for the ULS and the SLS and reference to this value in timber design will invariably mean to the value used at the ULS. It is obtained from

$$X_d = \eta \frac{X_k}{\gamma_m} \qquad\qquad (EC0, \text{ equation } (6.3)) \tag{2.5}$$

where X_k is the characteristic value of the property, η is the mean value of a conversion factor that takes into account volume and scale effects, the effects of moisture and temperature and any other relevant parameters, and γ_m is a partial factor that takes into

Table 2.3 Load-duration class definitions*

Class	Period of time	Examples given in *NA.2.1* of the UKNA to EC5
Permanent	>10 years	Self-weight
Long term	6 months to 10 years	Storage loading (including in lofts) Water tanks
Medium term	1 week to 6 months	Imposed floor loading
Short term	<1 week	Snow Maintenance or man loading on roofs Residual structure after an accidental event
Instantaneous	Instantaneous	Wind Impact loading Explosion

*Based on the content of *NA.2.1* in the UKNA to EC5.

account the possibility of the characteristic value of a material or product property (e.g. strength or stiffness) being less than the specified value and also the effect of scatter around the mean value of the conversion factor.

In EC5, η covers the effects of duration of load and variation in moisture content on the properties of timber and wood products and is referred to as the modification factor, k_{mod}. Factors covering scale and volume effects are considered separately in EC5 and are discussed in 2.3.6.

The modification factor is extremely important in timber design and a brief overview of how load duration and moisture content effects are taken into account is given in the following sub-sections.

2.2.20.1 Load duration classes

When subjected to loading, the strength properties of members reduce and the longer the duration of the load the greater the reduction will be. In order to establish a common basis for design, load duration classes (see EC5, *2.3.1.2*) have been defined to cover the range of durations likely to arise in practice and the duration associated with each class is given in Table 2.3, based on the content of *NA.2.1* in the UKNA to EC5.

The load duration class for self-weight is classed as 'permanent', and 'variable' actions come into one of the remaining classes defined in the table, and is determined by duration.

2.2.20.2 Service classes

Because the strength (and creep behaviour) of timber and wood-related products is affected by the moisture content of the material, these properties are dependent on the temperature and relative humidity conditions the materials are subjected to over the design life of the structure. A typical relationship between strength adjustment and moisture content derived from tests is shown for Douglas fir in Figure 2.4. When the moisture content is low, the strength property will be at its maximum and as the moisture content increases the strength is reduced and will reach a minimum value at the fibre saturation point.

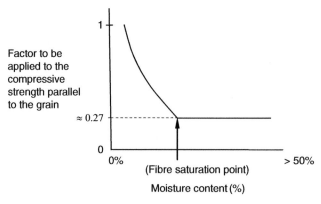

Fig. 2.4. The effect of moisture content on the compressive strength of Douglas fir when loaded parallel to the grain.

To take this effect into account in design, three service classes have been defined in EC5, *2.3.1.3*, covering the typical environmental conditions that timber structures will function under. These are as follows:

- *Service class 1 – where the average moisture content in most softwoods will not exceed 12%.*
  This corresponds to a temperature of 20 °C and a relative humidity of the surrounding air only exceeding 65% for a few weeks per year.
- *Service class 2 – where the average moisture content in most softwoods will not exceed 20%.*
  This corresponds to a temperature of 20 °C and a relative humidity of the surrounding air only exceeding 85% for a few weeks per year.
- *Service class 3 – where the average moisture content in most softwoods exceeds 20%.*
  This corresponds to climatic conditions leading to higher moisture contents than service class 2.

The highest values of timber strength will be obtained when structures function in service class 1 conditions and the lowest when they function in service class 3 conditions. The level of service class to be used in the United Kingdom for the type of element to be constructed is given in *NA.2.2* in the UKNA to EC5.

Values for k_{mod} based on the load duration referred to in 2.2.20.1 and the above service classes are given in EC5, *Table 3.1*, and those for use with timber and some wood-related products are summarised in Table 2.4.

With timber, glued-laminated timber and laminated veneer lumber (LVL), certain characteristic strengths are affected by member size and where the appropriate size is less than the reference size given in EC5, the relevant characteristic strength can be increased by multiplying by a factor k_h or k_l. These factors are referred to in Table 2.5 and discussed in 2.3.6.

In EC5 the partial factor for material γ_m is enhanced to γ_M, taking into account uncertainty in the resistance model used for design together with the adverse effects of geometric deviations in addition to the effect of unfavourable deviation of material

Table 2.4 Values of k_{mod} *

Material – standard	Service class		Load-duration class of action				
		Permanent	Long term	Medium term	Short term	Instantaneous
Solid timber – EN 14081-1 [12]	1	0.60	0.70	0.80	0.90	1.10
Glued-laminated timber – EN 14080 [13]	2	0.60	0.70	0.80	0.90	1.10
Laminated veneered lumber (LVL) – BS EN 14374 [14] or EN 14279 [15]	3	0.50	0.55	0.65	0.70	0.90
Plywood – BS EN 636						
Parts: 1 [16], 2 [17] and 3 [18]	1	0.60	0.70	0.80	0.90	1.10
Parts: 2 [17] and 3 [18]	2	0.60	0.70	0.80	0.90	1.10
Part: 3 [18]	3	0.50	0.55	0.65	0.70	0.90
OSB – BS EN 300 [19], Board type:						
OSB/2	1	0.30	0.45	0.65	0.85	1.10
OSB/3 and/4	1	0.40	0.50	0.70	0.90	1.10
OSB/3 and/4	2	0.30	0.40	0.55	0.70	0.90

* Based on values given in *Table 3.1* of EC5.

Table 2.5 Factors used in EC5 to take account of the effect of member size on strength properties

Size factor symbol	Function of factor	Characteristic property affected	Relevant item in EC5
k_h	Adjusts the characteristic bending strength and/or the characteristic tensile strength parallel to the grain to take into account the reduced effect of defects when the member size is less than the reference size given in EC5	$f_{m,k}$	Solid timber, *3.2(3)*; glued-laminated timber, *3.3(3)*; LVL, *3.4(3)*
		$f_{t,0,k}$	Solid timber, *3.2(3)*; glued-laminated timber, *3.3(3)*
k_ℓ	Adjusts the characteristic tensile strength parallel to the grain of LVL to take into account the reduced effect of defects when the member length is less than the reference length given in EC5	$f_{t,0,k}$	LVL, *3.4(4)*

or product property. The value of γ_M for the ULS combinations of actions for persistent or transient design situations (referred to in EC0, *6.4.3.2* as the *fundamental combinations*) as well as any accidental combination is given in *Table NA.3* in the UKNA to EC5. A summary of the values to be used to derive the design value at the ULS, including for accidental combinations, is given in Table 2.6.

Table 2.6 Partial factors for material properties and resistances, γ_M*

States/combinations	γ_M
Ultimate limit states (fundamental combinations)	
Solid timber, grade stamp individually marked	1.3
Solid timber, grade stamp package marked	2.0
Glued-laminated timber	1.25
LVL, plywood and OSB	1.2
Particleboard	1.3
Fibreboards – hard, medium, MDF, soft	1.3
Punched metal plate fasteners	
Anchorage strength	1.3
Plate (steel) strength	1.15
Connections – excluding punched metal plate fasteners	1.3
Ultimate limit states – accidental combinations	
Any material and connection	1.0
Serviceability limit states – all combinations	
Any material and connection	1.0

*Incorporating the requirements of *NA.2.3* of the UKNA to EC5 and EC0.

Stiffness properties (e.g. $G_{0,mean}$ and $E_{0,mean}$) are also functions of the moisture content in the material and for structural timber when drying from the fibre saturation point (m.c. approximately 27%) to 12% (the service class 1 condition) the increase in stiffness will be approximately 12.5% [20]. In other words the greatest stiffness will be at service class 1 and the value will reduce by approximately 0.75% for each 1% increase in moisture content. In the interest of design simplification, this behaviour is ignored in EC5 meaning that for service classes 2 and 3 conditions, deformations will be slightly underestimated when using the stiffness values given in the relevant harmonised standards.

No guidance is given in EC0 as to how the design value of a stiffness related property shall be derived and the values to be used in deformation and strength related analyses are given in EC5. With timber and wood-related products the value is dependent on the design state being considered; i.e. whether it is a deflection analysis at the SLS or a stress analysis at the ULS, whether or not the distribution of internal forces in the structure (or a member) is affected by the stiffness distribution, and also whether a first- or second-order linear analysis is to be undertaken.

In EC5, *2.4.1(2)P*, it is stated that the design member stiffness property, E_d or G_d, is obtained from the respective mean property value as follows:

$$E_d = \frac{E_{mean}}{\gamma_M} \qquad (2.6)$$

$$G_d = \frac{G_{mean}}{\gamma_M} \qquad (2.7)$$

where E_{mean} is the mean value of the modulus of elasticity of the timber or wood product, and G_{mean} is the mean value of the shear modulus of the timber or wood product.

However, equations (2.6) and (2.7) will only apply when undertaking a first-order linear elastic analysis at the instantaneous loading condition (i.e. when the load is immediately applied) at the SLS, at which state $\gamma_M = 1$, or when undertaking a second-order linear elastic analysis at the ULS, for which condition γ_M will be obtained from Table 2.6. The value of the member stiffness property to be used for other conditions will be different and the values to be used for the alternative situations that will arise in timber design are discussed in 2.3.4.

When dealing with connections, at the instantaneous loading condition the design stiffness at the SLS is obtained using the slip modulus, K_{ser}, given in EC5, *Table 7.1*, for the type of fastener being used and at the ULS the design stiffness is K_u, which, as stated in EC5, *2.2.2(2)*, is equal to $\frac{2}{3} K_{ser}$. As above, the value of connection stiffness at other conditions will be different and the requirements are also discussed in 2.3.4.

At the failure condition, the design value of the modulus of elasticity and the shear modulus used in strength calculations are the characteristic values $E_{0,05}$ and $G_{0,05}$ respectively.

2.2.21   Factors applied to a design strength at the ULS

Having derived the design value of a property, X_d, in accordance with the requirements outlined in 2.2.20, further adjustment may be necessary to take account of other factors that can affect the property, e.g. the effect of instability, system strength, etc. This is taken into account by the application of modification factors and a summary of the factors referred to in EC5 that are most likely to arise in design; the functions they fulfil and the property or strength ratio to which they will apply are given in Table 2.7.

2.2.22   Design values of geometrical data (EC0, *6.3.4*)

The design values of geometrical data are the sizes used in the design of the structure and its elements, including, where relevant, the maximum deviation from straightness allowed in *Section 10* of EC5. The design value is defined as

$$a_d = a_{nom} \tag{2.8}$$

where a_d is the nominal reference dimension, and a_{nom} is the dimension used in the design drawings and documentation.

For timber and wood-related products the design values may be taken to be the nominal values from product standards or drawings.

Where relevant, the design equations in EC5 include for the effect of the occurrence of the maximum permitted deviations specified in EC5, *Section 10*.

2.2.23   Design resistance (EC0, *6.3.5*)

When dealing with timber and wood-related product structures, in line with the requirements of EC0, the design value of a resistance is expressed in EC5, *2.4.3* as

Table 2.7 Commonly used modification factors in EC5 that affect design values

Factor symbol	Function of factor	Property or stress ratio affected	Relevant item(s) in EC5
k_m	Allows for stress redistribution when a section is subjected to bending about both the y–y and z–z axes and is stressed beyond the elastic limit. It also accounts for variation in material strength across the member section.	$\dfrac{\sigma_{m,d}}{f_{m,d}}$	6.1.6(2)
k_{vol}	Adjusts the design tensile strength perpendicular to the grain for glued-laminated timber and LVL when the stressed volume in the apex zone of double tapered, curved or pitched cambered beams exceeds the reference volume stated in EC5.	$f_{t,90,d}$	3.3(5), 3.4(7), 6.4.3(6), 6.4.3(7)
k_{dis}	Adjusts the design tensile stress strength perpendicular to the grain for glued-laminated timber and LVL in the apex zone of double tapered, curved or pitched cambered beams.	$f_{t,90,d}$	6.4.3(6), 6.4.3(7)
k_{crit}	Takes account of the effect of lateral torsional instability, reducing the design bending strength of a member bent about the strong y–y axis when the associated relative slenderness ratio for bending is >0.75.	$f_{m,y,d}$	6.3.3(3)
k_{cy}, k_{cz}	Takes account of the effect of axial instability, reducing the design compression strength of a member when subjected to axial compression and the relative slenderness ratio about the y–y (for k_{cy}) and/or z–z (for k_{cz} axis is >0.3.	$f_{c,0,d}$	6.3.2
$k_{c,90}$	Takes account of loading and geometry conditions to permit the compressive strength of the timber or wood-based structural product perpendicular to the grain to be increased whilst not exceeding the strain limit set for the design strength.	$f_{c,90,d}$	6.1.5
k_v	Takes account of the effect of a notch on the shear strength of a member.	$f_{v,d}$	6.5.2
k_{sys}	Increases member strength properties when several similar members, assemblies or components are equally spaced and connected by a continuous load distribution system that is capable of transferring load between the member and those adjacent.	All of the member strength properties that benefit from the load sharing system	6.6

$$R_{\mathrm{d}} = k_{\mathrm{mod}} \frac{R_{\mathrm{k}}}{\gamma_{\mathrm{M}}} \qquad (2.9)$$

where k_{mod} is a modification factor that takes into account the effect of load duration and moisture content (see 2.2.20), γ_{M} is the partial factor for a material property at the ULS (see 2.2.20), and R_{k} is the characteristic value of the load-carrying capacity at the ULS.

In general, however, the resistance properties are defined in EC5 as F functions and the more representative expression for the design resistance for a timber or wood-related product is

$$F_{\mathrm{Rd}} = k_{\mathrm{mod}} \frac{F_{\mathrm{Rk}}}{\gamma_{\mathrm{M}}} \qquad (2.10)$$

2.2.24 Ultimate limit states (EC0, *6.4.1–6.4.5*)

Where they are relevant, the following ULS must be verified:

(a) *Equilibrium (EQU)*. To confirm that the structure or any part of it is not unstable.
(b) *Strength (STR)*. To confirm that the structure and its elements will not fail under stress, by element instability or at connections. Where displacements will affect the behaviour of the structure, their effect must be taken into account.
(c) *Geotechnical (GEO)*. To confirm that the foundations of the facility provide the strength and stiffness required by the structure.
(d) *Fatigue (FAT)*. To confirm that the elements of the structure will not fail under fatigue.

For timber or wood product structures, ULS (a), (b) and (c) will generally be relevant and any condition where fatigue could apply is taken into account in EC5 by the Strength (STR) requirements.

Load combinations are applied at each relevant ULS and by the application of the partial factor method (see 2.2.17) it must be verified that the design value of the effect of the design actions is less than or equal to the design value of the equivalent resistance. For example, considering the strength verification of the structure and its elements at the STR ULS, the requirement will be

$$Ef_{\mathrm{d}} \leq F_{\mathrm{Rd}} \qquad (2.11)$$

and for a material property,

$$Ef_{\mathrm{d}} \leq (\Pi k) X_{\mathrm{d}} \qquad (2.12)$$

where Ef_{d} is the design value of the effect of actions (e.g. internal moment, internal stress, etc.) (see 2.2.20), F_{Rd} is the design value of the corresponding resistance (see 2.2.23), X_d is the design value of the timber or wood product material property (see 2.2.20), and Πk is the product of those modification factors that will affect the design value. (The principal modification factors in EC5 are summarised in Table 2.7 and discussed in the appropriate chapters in the book.)

For each relevant limit state the design value of the effect of actions must be derived. To achieve this, those actions that are considered to be able to occur simultaneously are combined, and, where more than one variable action exists, each combination will include one of the variable actions in turn as the leading variable action.

To derive the combination of actions for persistent or transient design situations (referred to in EC0 as the *fundamental combinations*) and ignoring pre-stressing actions as they are not generally relevant to timber design, the combination to be satisfied is given in *equation (6.10)* in EC0 as follows:

$$\sum_{j\geq1}\gamma_{G,j}G_{k,j}+\gamma_{Q,1}Q_{k,1}+\sum_{i>1}\gamma_{Q,i}\psi_{0,i}Q_{k,i} \qquad (EC0,\ equation\ (6.10)) \quad (2.13)$$

The less favourable of the following combination expressions may be considered as an alternative to equation (2.13) for STR (and GEO) limit states,

$$\sum_{j\geq1}\gamma_{G,j}G_{k,j}+\gamma_{Q,1}\psi_{0,1}Q_{k,1}+\sum_{i>1}\gamma_{Q,i}\psi_{0,i}Q_{k,i} \qquad (EC0,\ equation\ (6.10a)) \quad (2.14)$$

$$\sum_{j\geq1}\xi_j\gamma_{G,j}G_{k,j}+\gamma_{Q,1}Q_{k,1}+\sum_{i>1}\gamma_{Q,i}\psi_{0,i}Q_{k,i} \qquad (EC0,\ equation\ (6.10b)) \quad (2.15)$$

where γ_G is the partial factor for permanent loading, γ_Q is the partial factor for variable loading, ψ_0 is the factor that converts a variable action into its combination value, ξ is a reduction factor for unfavourable permanent actions, G_k is the permanent action, $Q_{k,1}$ is the leading variable action and Q_k is an accompanying variable action.

With accidental design situations, one combination of actions applies to all limit states and is given in *equation (6.11b)* in EC0,

$$\sum_{j\geq1}G_{k,j}+A_d+\left(\psi_{1,1}\ or\ \psi_{2,1}\right)Q_{k,1}+\sum_{i>1}\psi_{2,i}Q_{k,i} \qquad (EC0,\ equation\ (6.11b)) \quad (2.16)$$

where A_d is the design value for a specific accidental event (e.g. the action due to an impact or the indirect thermal action due to a fire) or relates to the situation after an accidental event, in which case $A_d=0$; ψ_1 and ψ_2 convert the variable action into the frequent and quasi-permanent value, respectively, and are referred to in 2.2.14.

Numerical values of the γ and ξ factors to be used to derive the design values of actions for the EQU and STR (not involving geotechnical actions) states when subjected to persistent and transient design situations are given in Table 2.8, and the values applicable to all ULS when subjected to accidental design situations are given in Table 2.9. Based on the content of *Table NA.A1.1* in the UKNA to EC0, values for the ψ factors are summarised in Table 2.2.

For the STR limit state, when subjected to combinations of actions under persistent and transient design situations, as stated previously and shown in Table 2.8, alternative combinations of actions can be used. From calibration work undertaken by Gulvanessian and Holicky [21], the use of *equation (6.10)* in EC5 will result in the

Table 2.8 Design values of actions for equilibrium (EQU) and strength (STR) limit states[*]

Ultimate limit state (under persistent and transient design situations – fundamental combinations)	Relevant equation in EC0	Permanent actions		Leading variable action	Accompanying variable actions	
		Unfavourable‖	Favourable‖		Main	Others
EQU (a)†	(6.10)	$1.10G_{k,j,sup}$	$0.90G_{k,j,inf}$	$1.5Q_{k,1}$ (0 when favourable)	–	$1.5\psi_{0,i}\,Q_{k,i}$ (0 when favourable)
(b)‡ The highest design value from combination (i) or (ii) (i) (6.10)		$1.35G_{k,j,sup}$	$1.15G_{k,j,inf}$	$1.5Q_{k,1}$ (0 when favourable)	–	$1.5\psi_{0,i}\,Q_{k,i}$ (0 when favourable)
(ii) (6.10)		$1.0G_{k,j,sup}$	$1.0G_{k,j,inf}$	$1.5Q_{k,1}$ (0 when favourable)	–	$1.5\psi_{0,i}\,Q_{k,i}$ (0 when favourable)
STR§ (not involving geotechnical actions)} (c)	(i) (6.10)	$1.35G_{k,j,sup}$	$1.0G_{k,j,inf}$	$1.5Q_{k,1}$ (0 when favourable)	–	$1.5\psi_{0,i}\,Q_{k,i}$ (0 when favourable)
(d)	(ii) (6.10a)	$1.35G_{k,j,sup}$	$1.0G_{k,j,inf}$	–	$1.5\psi_{0,1}\,Q_{k,1}$	$1.5\psi_{0,i}\,Q_{k,i}$
	(iii) (6.10b)	$0.925\times1.35G_{k,j,sup}$	$1.0G_{k,j,inf}$	$1.5Q_{k,1}$	–	$1.5\psi_{0,i}\,Q_{k,i}$

* Based on *Tables NA.A1.2(A)* and *NA.A1.2(B)* of the UKNA to EC0.

† The values apply when static equilibrium does not also involve the resistance of structural members. If the static equilibrium involves the resistance of structural members, to assess the integrity of the structural element(s) the design values for STR (c). (i) must be considered in addition to EQU (a).

‡ Where static equilibrium involves the resistance of structural members, as an alternative to the option referred to in the note above (†), a combined verification based on the maximum design value derived from combinations EQU (b), (i) and EQU (b), (ii) should be adopted.

§ At the STR limit state, the designer is given a choice of using the design values given in STR (c), (i) or the less favourable of STR (d), (ii) and STR (d), (iii). The examples given in the book are based on STR (c), (i).

‖ For timber structures, where the variability of G_k is small, $G_{k,sup}$ and $G_{k,j,inf}$ shall be replaced by a single value, $G_{k,j}$, based on the mean value of density.

Table 2.9 Design values of actions for accidental combinations[*]

Ultimate limit states (under an accidental combination of actions)	Relevant equation in EC0	Permanent actions		Leading variable action	Accompanying variable actions	
		Unfavourable[†]	Favourable[†]		Main	Others
Any ultimate limit state	(6.11a/b)	$1.0G_{k,j,sup}$	$1.0G_{k,j,inf}$	A_d	$1.0\psi_{1,1}Q_{k,1}$	$1.0\psi_{2,i}Q_{k,i}$

[*]Based on *Table NA.A1.3* of the UKNA to EC0.
[†]For timber structures, where the variability of G_k is small, $G_{k,j,sup}$ and $G_{k,j,inf}$ shall be replaced by a single value, $G_{k,j}$, based on the mean value of density.

highest reliability index, and closely approximates that achieved by design in accordance with current BS requirements, generally well exceeding the minimum reliability index given in Table 2.1. The alternative use of the less favourable of *equations (6.10a)* and *(6.10b)* will, on the other hand, achieve a more uniform reliability that is better aligned with the minimum level set in the code, and is likely to be more economical, but will be below the reliability index achieved by design in accordance with BSI codes. Unless otherwise stated, *equation (6.10)* has been used in the examples given in the book to determine the design values of actions at the ULS.

To determine the load case producing the greatest design effect (i.e. the maximum bending moment, shear force, etc.), the load combination equation(s) must be applied in turn with each variable action acting as the leading variable. Also, where the variable loads are not related, all possible combinations must be considered. For example, consider the STR limit state for a simply supported beam loaded by its own weight, $G_{k,1}$, a permanent load, $G_{k,2}$, a medium-term duration variable load, $Q_{k,1}$, and an unrelated short-term variable load, $Q_{k,2}$. Adopting *equation (6.10)*, the alternative loading conditions that have to be considered to determine an effect, Ef, e.g. a bending moment, are

$$1.35\left(G_{k,1}+G_{k,2}\right) \rightarrow Ef_1 \tag{2.17}$$

$$1.35\left(G_{k,1}+G_{k,2}\right)+1.5Q_{k,1} \rightarrow Ef_2 \tag{2.18}$$

$$1.35\left(G_{k,1}+G_{k,2}\right)+1.5Q_{k,2} \rightarrow Ef_3 \tag{2.19}$$

$$1.35\left(G_{k,1}+G_{k,2}\right)+1.5Q_{k,1}+1.5\psi_{0,2}Q_{k,2} \rightarrow Ef_4 \tag{2.20}$$

$$1.35\left(G_{k,1}+G_{k,2}\right)+1.5Q_{k,2}+1.5\psi_{0,1}Q_{k,1} \rightarrow Ef_5 \tag{2.21}$$

Where a load combination consists of actions belonging to different load duration classes, the effect of load duration on the property of the timber or wood product has to be taken into account by the use of the k_{mod} modification factor discussed in 2.2.20. The effects of combinations of permanent and variable actions have a less degrading effect on strength properties than permanent action alone, and where a combination of permanent and several variable actions is applied, the design condition will be dictated by the variable action having the shortest duration. On this basis, and as required by EC5, *3.1.3(2)*, the modification factor corresponding to the action having the shortest duration used in the combined load case is applied to the strength property being considered. Where there is a linear relationship between actions and effects, the design condition will be that having the largest value after division by the associated k_{mod} factor. For the example given in equations (2.17)–(2.21), taking $k_{mod,perm}$, $k_{mod,med}$ and $k_{mod,short}$ as the modification factors for the permanent, medium-term and short-term actions, respectively, and with a linear relationship between action and corresponding stress, the design value Ef_d of effect Ef will be the largest value given in equation (2.22):

$$Ef_d = \max \begin{pmatrix} Ef_1 / k_{mod,perm} \\ Ef_2 / k_{mod,med} \\ Ef_3 / k_{mod,short} \\ Ef_4 / k_{mod,short} \\ Ef_5 / k_{mod,short} \end{pmatrix} \tag{2.22}$$

It should be noted that when a favourable value of the variable action is to be applied (i.e. $Q_k = 0$), this means that the variable action is not being applied in that particular load case and the k_{mod} to be used will be the one associated with the shortest duration of the variable actions that are being applied.

2.2.25 Serviceability limit states: General (EC0, *6.5*)

For timber structures the following SLS shall be verified:

- Vibration
- Deformation.

At these states it has to be demonstrated that

$$Ef_d \leq C_d \qquad\qquad (EC0, \text{ equation } (6.13)) \quad (2.23)$$

where Ef_d is the design value of the effect of actions at the SLS and C_d is the limiting design value of the relevant serviceability criterion (i.e. vibration or deflection criterion), given in EC5 and the UKNA to EC5.

At the SLS the partial factors γ_G and γ_Q used to derive the design value of the effects of actions are set equal to 1 and the particular loading conditions used for these states are summarised below.

2.2.25.1 *Vibration*

Vibration criteria in EC5 are limited to the vibration behaviour of residential floors, and the design loading conditions producing the design value at the SLS are as follows:

(a) The mass of the floor – to be used to determine the lowest natural frequency of the floor structure.

(b) The application of a 1 kN vertical force at the point on the floor that will result in the maximum vertical deflection – to be used to determine the maximum instantaneous vertical deflection due to foot force effect.

(c) The application of a 1 N s impulse at the point of maximum instantaneous vertical deflection referred to in item (b) – to be used to determine the maximum initial value of the vertical floor vibration velocity due to heel impact effect.

2.2.25.2   *Deflection*

The combinations of actions associated with the deflection states at the SLS that will result in the design value of the displacement are given in EC0 under the following headings:

- Characteristic combination
- Frequent combination
- Quasi-permanent combination.

Using the symbols defined in 2.2.24, these combinations are determined as follows:

(a) *Characteristic combination*

$$\sum_{j\geq1}G_{k,j}+Q_{k,1}+\sum_{i>1}\psi_{0,i}Q_{k,i} \qquad (EC0,\ equation\ (6.14b))\quad(2.24)$$

This combination is normally used for irreversible limit states (i.e. states where the SLS will be permanently infringed even when the actions that caused the exceedance are removed) and is the combination used in EC5, 2.2.3. It will be noted that equation (2.24) equates to equation (2.13) when the partial factors (γ_G and γ_Q) in equation (2.13) are set equal to unity.

(b) *Frequent combination*

$$\sum_{j\geq1}G_{k,j}+\psi_{1,1}Q_{k,1}+\sum_{i>1}\psi_{2,i}Q_{k,i} \qquad (EC0,\ equation\ (6.15b))\quad(2.25)$$

This combination is normally used for reversible limit states (i.e. where an infringement of a state disappears when the action causing the exceedance is removed). Although this option is not referred to in EC5, provided an agreement can be established with the client on those situations that will fall into this category together with a frequency of occurrence that will be acceptable, the combination can be used.

(c) *Quasi-permanent combination*

$$\sum_{j\geq1}G_{k} + \sum_{i\geq1}^{n}\psi_{2,i}Q_{k,i} \qquad (EC0,\ equation\ (6.16b))\quad(2.26)$$

This is the combination that is used for the assessment of long-term (creep) effects.

To determine the design value, the characteristic combination is applied with each variable action acting as the leading variable, from which the maximum loading condition will be obtained. If a reversible limit state condition has been accepted for the deformation criteria, the frequent rather than the characteristic combination will be used.

The design displacement, Ef_d, arising from the application of the design load case will be an instantaneous deformation, i.e. the displacement that will immediately arise due to elastic deformation of the structure as well as any deformation arising from joint movements in the structure.

Assuming all of the members, components and connections have the same creep behaviour, and on the assumption that there is a linear relation between actions and their associated displacements, to obtain the displacement caused solely by creep the quasi-permanent load combination is used.

The final deformation is obtained by combining the instantaneous and the creep displacement.

Where the structure comprises members with different time-dependent properties, the final deformation requirements are discussed in 2.3.2(b).

2.3   EUROCODE 5: DESIGN OF TIMBER STRUCTURES – PART 1-1: GENERAL – COMMON RULES AND RULES FOR BUILDINGS (EC5)

In this section, those matters relating to the content of EC5 that have not been addressed in 2.2, and are considered to require an explanation or some further clarification, are discussed.

2.3.1   General matters

In the Eurocodes, the procedure for displaying a decimal point is to use a comma, e.g. 4,5, in accordance with the requirements of ISO 3898 [22]. The United Kingdom, however, uses a full stop for the decimal point, i.e. 4.5, and the UK practice has been used in the book.

The traditional practice in the United Kingdom is to show the z–z axis as the longitudinal axis of a member and the x–x and y–y axes to denote the respective major and minor axes of its cross-section. In the Eurocode suite the longitudinal axis is referred to as the x–x axis and the y–y and z–z axes are the respective major and minor axes of the cross-section. The Eurocode convention, which is shown in Figure 2.5, is used in the book.

Where it is relevant to show the direction of the grain of the timber it is defined by the symbol used in Figure 2.5.

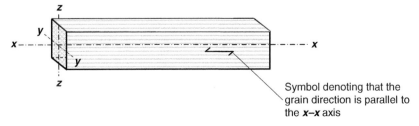

Fig. 2.5. Member axes.

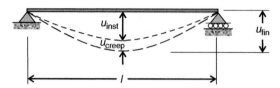

Fig. 2.6. Deformation.

2.3.2   Serviceability limit states (EC5, *2.2.3*)

In EC5 the deformation of a member or structure is required at two stages:

(i) When the loading is immediately applied; this is called the instantaneous deformation: u_{inst}.

(ii) After all time-dependent displacement (i.e. creep deformation, u_{creep}) has taken place; this is called the final deformation: u_{fin}.

These deformations are shown diagrammatically in Figure 2.6 in relation to a simply supported beam without any pre-camber.

Deformation is calculated in two different ways, depending on the creep behaviour of the structure:

(a) *Structures comprising members, components and connections having the same creep behaviour*

Creep behaviour in timber and wood-related products is a function of several factors, and to simplify the design process the assumption is made in EC5 that when subjected to a permanent load over the lifetime of a building, the instantaneous deflection (u_{inst}) and the creep deflection (u_{creep}) are related as follows,

$$u_{creap} = k_{def} u_{inst} \qquad\qquad (2.27)$$

where k_{def} is a deformation factor whose value is dependent on the type of material being stressed as well as its moisture content. Values for the factor have been derived for timber and wood-based materials at defined environmental conditions when subjected to constant loading at the SLS over the design life, and are given in EC5, *Table 3.2*. The environmental conditions are referred to as service class 1, 2 or 3 (discussed in 2.2.20) and values for k_{def} for timber and some wood-related products at these conditions are given in Table 2.10.

For structures or members complying with the above conditions the final deformation, u_{fin}, can then be written as

$$u_{fin} = u_{inst} + u_{creep} = u_{inst}\left(1 + k_{def}\right) \qquad\qquad (2.28)$$

Where an action is not permanent, to obtain the creep effect it must be converted to an equivalent permanent action and this is done by using the quasi-permanent value of the action, referred to in 2.2.25.2(c).

Table 2.10 Values of k_{def} at service class 1, 2 or 3*

	Value of k_{def}		
Material – standard	Service class 1	Service class 2	Service class 3
Solid timber			
EN 14081-1	0.60	0.80	2.00
Glued-laminated timber			
EN 14080	0.60	0.80	2.00
LVL			
EN 14374, EN 14279	0.60	0.80	2.00
Plywood			
EN 636-1	0.80	–	–
EN 636-2	0.80	1.00	–
EN 636-3	0.80	1.00	2.50
OSB			
EN 300 – type OSB/2	2.25	–	–
EN 300 – types OSB/3, OSB/4	1.50	2.25	–

*Based on data given in *Table 3.2*, EC5.

EC5 uses the characteristic combination of actions to derive the instantaneous deformation and the quasi-permanent load combination to derive the creep deformation. In *3.4(2)P* of EC0, which is a Principle, a distinction is made between reversible and irreversible SLS and where it is acceptable to the client to adopt a reversible SLS condition, to obtain an economical design the frequent combination given in equation (2.25) should be used rather than the characteristic combination to derive the displacement.

Assuming irreversible SLS conditions will apply, the final deformation under permanent and variable loading will be as follows:

(i) For permanent actions, G, on a member or connection:

$$u_{fin,G} = u_{inst,G} + u_{creep,G} = u_{inst,G}\left(1 + k_{def}\right) \quad \text{(EC5, equation (2.3))} \quad (2.29)$$

(ii) For the leading variable action, Q_1, on the member or connection:

$$u_{fin,Q,1} = u_{inst,Q,1} + u_{creep,Q,1} = u_{inst,Q,1}\left(1 + \psi_2 k_{def}\right)$$
$$\text{(EC5, equation (2.4))} \quad (2.30)$$

(iii) For the accompanying variable action(s), Q_i, on member i or connection i:

$$u_{fin,Q,i} = u_{inst,Q,i} + u_{creep,Q,i} = u_{inst,Q,i}\left(\psi_{0,i} + \psi_{2,i} k_{def}\right)$$
$$\text{(EC5, equation (2.5))} \quad (2.31)$$

(iv) The final condition for n variable actions will be

$$u_{fin} = u_{fin,G} + u_{fin,Q,1} + \sum_{i=2}^{n} u_{fin,Q,i}$$
$$\text{(based on } EC5\text{, equation (2.2))} \quad (2.32)$$

The ψ values are obtained from *Table NA.A1.1* in the UKNA to EC0 (reproduced in part in Table 2.2) and where there is only one variable action, equation (2.31) is not relevant.

(b) *Structures comprising members, components and connections having different creep behaviour*

In these situations, the creep behaviour will affect the stiffness and stress distribution and the linear relationship between the instantaneous and the creep deflection referred to in 2.3.2(a) will not apply.

The final deformation will be obtained by adding the instantaneous and the creep deformation but as the requirements of *2.2.3(3), 2.2.3(4)* and *2.3.2.2(1)*, for calculating the combined instantaneous and the creep deformation for these types of structure have been accepted as being in conflict and open to misinterpretation they are to be revised as outlined in Appendix C. Under the proposed revision, for these types of structure the instantaneous deformation (u_{inst}) will be calculated using the characteristic combination of actions and mean stiffness properties as described in 2.2.3(a), and for the creep deformation the quasi-permanent combination of actions is used, as also referred to in 2.3.2(a), but the stiffness properties in the calculation are to be the final mean values of the appropriate moduli of elasticity, shear moduli and slip moduli, in accordance with the requirements of *2.3.2.2(1)*, (as referred to in 2.3.4.1). If no modification were to be made to this loading condition, when calculating the creep deformation the instantaneous deformation associated with the quasi-permanent load combination would be included in the result. This can be simply demonstrated by considering the bending deformation of a simply supported composite beam designed in accordance with the *equivalent section method* referred to in Chapter 7 and a solid beam having the same cross-section and made from the material selected for the equivalent section. To again keep the problem simple, consider the sections only being subjected to the same permanent actions (G_{sls}). Using this approach and taking the mean value of the material forming the transformed section as E_{mean}, with an associated deformation factor of k_{def} and applying the same properties to the solid section beam, adopting the above procedure for the composite section and the procedure referred to in 2.3.2(a) for the solid section, the respective creep deformations will be a function of the loading and stiffness properties as follows:

For the composite section;

$$u_{creep1} = \text{fn1}\left[G_{sls}/\left(E_{mean}/(1+k_{def})\right)\right] = \text{fn1}\left[G_{sls}/E_{mean} + k_{def}G_{sls}/E_{mean}\right]$$

For the solid section:

$$u_{creep2} = \text{fn1}\left[G_{sls}/\left(E_{mean}/k_{def}\right)\right] = \text{fn1}\left[k_{def}G_{sls}/E_{mean}\right]$$

In both cases the instantaneous deformation will be:

$$u_{inst} = \text{fn1}\left(G_{sls}/E_{mean}\right)$$

The net final bending deformation will therefore be:
For the composite section:

$$u_{fin1} = u_{inst} + u_{creep1} = fn1\left[G_{sls}/E_{mean}\right] + fn1\left[G_{sls}/E_{mean} + k_{def}G_{sls}/E_{mean}\right]$$
$$= fn1\left[2G_{sls}/E_{mean} + k_{def}G_{sls}/E_{mean}\right]$$

For the solid section:

$$u_{fin2} = u_{inst} + u_{creep2} = fn1\left[G_{sls}/E_{mean}\right] + fn1\left[k_{def}G_{sls}/E_{mean}\right]$$
$$= fn1\left[G_{sls}/E_{mean} + k_{def}G_{sls}/E_{mean}\right]$$

As will be seen from this exercise, although these sections should have the same final deformation, applying the quasi-permanent loading condition without modification will mean the instantaneous deformation associated with the permanent actions in the combination will be double counted. The same exercise based on the variable loading associated with the quasi-permanent combination of actions will show that the instantaneous deformation associated with the variable loading in that combination will also be double counted.

To take this into account, in structures comprising members, components and connections having different creep behaviour, the procedure referred to in Appendix A requires the final deformation (u_{fin}) to be obtained from a combination of the creep deformation (u_{creep}) and the instantaneous deformation (u_{inst}) minus the instantaneous deformation arising from the quasi-permanent combination of actions $(u_{inst,QP})$ as follows:

$$u_{fin} = u_{inst} + u_{creep} - u_{inst,QP}$$

Calculation of the final deformation using the above procedure will involve different loading combinations and stiffness properties and in practical analyses an alternative, simpler, but more conservative approach to determine the final deformation, is the procedure given in the previous edition of this book. In such an approach the instantaneous deformation is calculated as in 2.3.2(a) but the creep deformation is derived using only the instantaneous loading condition (i.e. the quasi-permanent load combination is not used) and reduced stiffness properties for the elements of the structure referred to above. For structures or members complying with the above conditions, the final deformation, u_{fin}, will be obtained from a single analysis and will be:

$$u_{fin} = u_{(inst+creep)}$$

where $u_{(inst+creep)}$ is the deformation derived from a linear elastic analysis of the structure subjected to the instantaneous loading condition and based on the reduced stiffness properties in *2.3.2.2(1)*.

For both situations, i.e. 2.3.2(a) and 2.3.2(b), when timber is being used and installed at or near its fibre saturation point, but is to function in an environment where it is likely

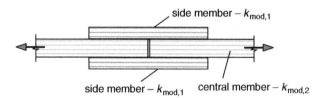

Fig. 2.7. A connection comprising two timber elements with different time-dependent behaviour.

to dry out under load, as required by EC5, *3.2(4)*, the value of k_{def} given in EC5, *Table 3.2* (reproduced in part in Table 2.3), used in the analysis must be increased by 1.0.

Further, as required by EC5, *2.3.2.2*, when dealing with a connection, if it is constituted of timber elements with the same creep behaviour, the value of k_{def} used in the analysis is to be taken as double the value given in Table 2.10, and if the connection comprises two wood-based elements with different creep behaviour, $k_{def,1}$ and $k_{def,2}$, the value to be used will be $k_{def} = 2\sqrt{k_{def,1}k_{def,2}}$.

The application of the factor 2 to derive k_{def} values for connections may be inappropriate for certain conditions and sizes of fixing; however, as no guidance is given in the UKNA to EC5 on this matter, the code requirements should be complied with.

2.3.3   Load duration and moisture influences on strength (EC5, *2.3.2.1*)

The adjustment of strength and resistance properties due to the effect of load duration and moisture content is taken into account in the design process by the modification factor, k_{mod}. The factor is obtained from *Table 3.1* in EC5, which has been reproduced in part in Table 2.4, and is discussed in 2.2.20.

When the strength of a connection is being considered and it comprises two timber elements, each having a different time-dependent behaviour, $k_{mod,1}$ and $k_{mod,2}$ (e.g. as shown in Figure 2.7), the requirement of EC5, *2.3.2.1(2)* is that the modification factor k_{mod} to be used to calculate the design load-carrying capacity of the connection be obtained from

$$k_{mod} = \sqrt{k_{mod,1}k_{mod,2}} \qquad\qquad \text{(EC5, \emph{equation} (2.6))} \quad (2.33)$$

2.3.4   Load duration and moisture influences on deformations (EC5, *2.3.2.2*)

2.3.4.1   *SLS analyses*

At the SLS, in order to demonstrate compliance with SLS criteria at the instantaneous and final displacement conditions, displacement analyses must be undertaken at each condition and, where a structure comprises members, components or connections having different time-dependent properties, the effect of creep on stiffness properties must be taken into account:

(a) At the instantaneous condition, the analysis is undertaken using the design value of the combination of actions for the SLS, i.e. either equation (2.24) or (2.25) depending on whether the characteristic or the frequent combination of actions will apply. As the creep behaviour of the member is not relevant at this condition, in accordance with the requirements of EC5, *2.2.3(2)*, the mean value of the appropriate modulus of elasticity, shear modulus and slip modulus should be used to derive the stiffness properties.

(b) For analysis at the final deformation condition, where the structure has a linear elastic behaviour and consists of members, components and connections having the same creep behaviour, the design value of the combination of actions will be the summation of the actions used in the instantaneous condition analysis referred to in (a) plus the quasi-permanent combination, i.e. equation (2.26). In this situation, creep behaviour will not influence stress behaviour, and consequently the stiffness properties remain the same as those used for the instantaneous condition analysis.

(c) Where the structure comprises members, components and connections that have different creep behaviour, this will influence displacement behaviour. For this condition, the loading used for the final deformation analysis will be the same as for (a) and the creep effect on displacement behaviour will be achieved by using a reduced stiffness property for the structural elements. In accordance with the requirements of EC5, *2.3.2.2(1)*, reduced stiffness properties must be derived using the final mean values of the modulus of elasticity, shear modulus and slip modulus as given in the following equations,

$$E_{\text{mean,fin}} = \frac{E_{\text{mean}}}{\left(1 + k_{\text{def}}\right)} \qquad \text{(EC5, } equation \text{ (2.7))} \quad (2.34)$$

$$G_{\text{mean,fin}} = \frac{G_{\text{mean}}}{\left(1 + k_{\text{def}}\right)} \qquad \text{(EC5, } equation \text{ (2.8))} \quad (2.35)$$

$$K_{\text{ser,fin}} = \frac{K_{\text{ser}}}{\left(1 + k_{\text{def}}\right)} \qquad \text{(EC5, } equation \text{ (2.9))} \quad (2.36)$$

where $E_{\text{mean,fin}}$ is the final mean value of the modulus of elasticity, E_{mean} is the mean value of elasticity, $G_{\text{mean,fin}}$ is the final mean value of the shear modulus, G_{mean} is the mean value of the shear modulus, $K_{\text{ser,fin}}$ is the final slip modulus, K_{ser} is the slip modulus, defined in EC5, *7.1*, and k_{def} is the deformation factor for timber and wood-based products and, for connections, it will be as defined in 2.3.2. Values for k_{def} for timber and some wood-related products are given in Table 2.10.

The design value of stiffness properties used in the analysis will therefore be

$$E_{\text{d,SLS}} = \frac{E_{\text{mean}}}{\left(1 + k_{\text{def}}\right)}, \quad G_{\text{d,SLS}} = \frac{G_{\text{mean}}}{\left(1 + k_{\text{def}}\right)}, \quad K_{\text{d,SLS}} = \frac{K_{\text{ser}}}{\left(1 + k_{\text{def}}\right)} \qquad (2.37)$$

where $E_{\text{d,SLS}}$ is the design value of the final mean value of the modulus of elasticity at the SLS, $G_{\text{d,SLS}}$ is the design value of the final mean value of the shear modulus at the SLS, and $K_{\text{d,SLS}}$ is the design value of the final slip modulus at the SLS.

2.3.4.2   ULS analyses

At the ULS, analyses are undertaken to validate strength and stability behaviour, and the loading to be used will be that producing the greatest design effect, selected from the action combinations referred to in 2.2.24

(a) When undertaking a first-order linear elastic analysis (see 2.2.16.1) on a structure and the stiffness distribution within the structure does not affect the distribution of internal stress resultants, in accordance with the requirements of EC5, 2.2.2(1)P, the mean values of the appropriate modulus of elasticity, shear modulus and slip modulus shall be used to derive stiffness properties. This condition will apply where all members have the same time-dependent properties and the relevant stiffness related properties will be E_{mean}, G_{mean} and K_{ser}.

(b) When undertaking a first-order linear elastic analysis of a structure and the stiffness distribution within the structure does affect the distribution of internal stress resultants, for the instantaneous condition, the stiffness related properties given in 2.3.4.2(a) will still apply. This will be the case for structures where the members, including connections, have different time-dependent properties, or with composite members where the materials being used have different time-dependent properties. For the final condition, in accordance with the requirements of EC5, 2.2.2(1)P, the properties must be derived using the final mean value of the appropriate modulus of elasticity, shear modulus and slip modulus given in EC5, 2.3.2.2(2). The final mean values are adjusted to the load component causing the largest stress in relation to strength and are as follows:

$$E_{\text{mean,fin}} = \frac{E_{\text{mean}}}{\left(1 + \psi_2 k_{\text{def}}\right)} \qquad \text{(EC5, equation (2.10))} \quad (2.38)$$

$$G_{\text{mean,fin}} = \frac{G_{\text{mean}}}{\left(1 + \psi_2 k_{\text{def}}\right)} \qquad \text{(EC5, equation (2.11))} \quad (2.39)$$

$$K_{\text{ser,fin}} = \frac{K_{\text{ser}}}{\left(1 + \psi_2 k_{\text{def}}\right)} \qquad \text{(EC5, equation (2.12))} \quad (2.40)$$

where the functions are as previously defined and ψ_2 is the factor for the quasi-permanent value of the action causing the largest stress in relation to the strength. If this is a permanent action, a value of 1 should be used. If the determination of ψ_2 is assessed to be a complicated or difficult exercise, a safe result will be obtained by adopting a value of 1 for the factor (see Table 2.2 for the value of ψ_2).

(c) When undertaking a second-order linear elastic analysis of a structure (see 2.2.16.1.1(c)), in accordance with the requirements of EC5, 2.2.2(1)P, stiffness-related properties must be derived using the design values of the appropriate modulus of elasticity and/or shear modulus as defined in EC5, *2.4.1(2)P*. For connections, the value used for the slip modulus will be K_u, as given in EC5, *2.2.2(2)*.

After derivation of the stiffness-related properties in accordance with the above requirements, the design value used in the ULS analysis will be

Case (a)

$$E_{d,\mathrm{ULS}} = E_{\mathrm{mean}}, \quad G_{d,\mathrm{ULS}} = G_{\mathrm{mean}}, \quad K_{d,\mathrm{ULS}} = K_{\mathrm{ser}} \tag{2.41}$$

Case (b) (final condition)

$$E_{d,\mathrm{ULS}} = \frac{E_{\mathrm{mean}}}{\left(1 + \psi_2 k_{\mathrm{def}}\right)}, \quad G_{d,\mathrm{ULS}} = \frac{G_{\mathrm{mean}}}{\left(1 + \psi_2 k_{\mathrm{def}}\right)}, \quad K_{d,\mathrm{ULS}} = \frac{K_{\mathrm{ser}}}{\left(1 + \psi_2 k_{\mathrm{def}}\right)} \tag{2.42}$$

Case (c)

$$E_{d,\mathrm{ULS}} = \frac{E_{\mathrm{mean}}}{\gamma_{\mathrm{M}}}, \quad G_{d,\mathrm{ULS}} = \frac{G_{\mathrm{mean}}}{\gamma_{\mathrm{M}}}, \quad K_{d,\mathrm{ULS}} = K_u \tag{2.43}$$

where the functions are as previously defined and $E_{d,\mathrm{ULS}}$ is the design value of the modulus of elasticity at the ULS, $G_{d,\mathrm{ULS}}$ is the design value of the shear modulus at the ULS, $K_{d,\mathrm{ULS}}$ is the design value of the slip modulus at the ULS, and γ_{M} is the partial factor for a material property (or connection) given in *NA.2.3* of the UKNA to EC5.

In the above equations, it has to be remembered that the value of k_{def} used for conditions must be as explained in 2.3.2.

2.3.5   Stress–strain relations (EC5, *3.1.2*)

Although the actual stress–strain relationship for timber and wood-related products when loaded to failure is generally non-linear, the characteristic strengths of structural timbers and wood products are derived assuming that a linear relationship exists. Consequently, when calculating the design stress in a section, e.g. flexural, axial, shear, etc., it is to be assumed that elastic behaviour will apply up to the failure condition. Where EC5 considers that plastic behaviour can be taken into account to enhance member strength, this is incorporated into the relevant strength validation rules given in the code.

2.3.6   Size and stress distribution effects (EC5, *3.2, 3.3, 3.4* and *6.4.3*)

Timber is not a homogeneous material and due to the presence of defects, variability in strength across and along member lengths, as well as the loading configuration being used, the strength between and within members will vary. Although there has been considerable research and theoretical investigation into member size, length, volume, load configuration and stress distribution effects, there is not yet full agreement on the effect of these factors and how they should be incorporated into the design process.

Because timber and wood-related products are brittle materials, the most widely adopted theoretical approach used in investigations has been based on the application of weakest link theory, using a Weibull distribution. Although not valid for some timber species, it is assumed that the strength-degrading defects in timber and wood-related products are randomly distributed throughout the sample volume and are of random size, enabling the sample to be considered as a chain element comprising several small volumes of different strength when subjected to tension. In this condition, the sample strength will be dictated by the strength of the weakest volume. If there are two members of differing volumes, V_1 and V_2, and the strength distribution fits the Weibull distribution, it has readily been demonstrated (e.g. [23, 24]) that the theory will conclude that the ratio of the respective failure strength of the members (σ_1 and σ_2) can be written as

$$\left(\frac{\sigma_2}{\sigma_1}\right) = \left(\frac{V_1}{V_2}\right)^{1/k} \tag{2.44}$$

where k is the shape factor of the Weibull distribution.

For bending stresses, equation (2.44) should be further adjusted to take account of the effect of the variation in stress distribution over the length of the member arising from the loading configuration being used; however, this effect has not been included for in EC5.

For timber sections, the volumes V_1 and V_2 can be represented by the member breadth, depth and length, i.e. $b_1 \times h_1 \times L_1$ and $b_2 \times h_2 \times L_2$, respectively, and equation (2.44) can be written as

$$\left(\frac{\sigma_2}{\sigma_1}\right) = \left(\frac{b_1}{b_2}\right)^{1/k_b} \left(\frac{h_1}{h_2}\right)^{1/k_h} \left(\frac{L_1}{L_2}\right)^{1/k_L}$$

where the exponents $1/k_b$, $1/k_h$ and $1/k_L$ are the factors associated with each dimension.

Because the width of timber members does not vary significantly, the width effect is generally ignored and if members are strength tested at a constant span to depth ratio, the depth and length exponents can be combined to give a single factor, i.e.

$$\left(\frac{\sigma_2}{\sigma_1}\right) = \left(\frac{h_1}{h_2}\right)^{1/k_{hL}}$$

This is the approach used in EC5 for bending, where bending strengths obtained in accordance with BS EN408:2010 [25] require all tests to be carried out using a two-point bending configuration on a beam having a span to depth ratio of 18 (plus a tolerance allowance).

The Weibull theory (as it is commonly referred to) has also been used to investigate volume effects as well as the effect of varying the types of loading configuration applied to members. However, opinions vary among researchers on the application of these effects, and in particular loading configuration effects. Also, there is not full agreement on the values to be used for the relevant exponents. In such circumstances,

EC5 has adopted a simplified approach and ignored certain effects. Factors have been included for bending and tensile strengths in solid timber, glued-laminated timber and LVL as well as for stress distribution and volume effects in double tapered, curved and pitched camber beams. The effects are not applicable to wood-based panel products, e.g. plywood, OSB, fibreboard, etc.

The consequence of the above is that unlike designs in structural steel and reinforced concrete, when using timber from the same strength class the effect of defects can result in members of different sizes having different characteristic strengths.

To take these effects into account, the characteristic values of strength properties that are influenced by the effect are derived using reference sizes (which for depth, width and length effects are the sizes above which the effect can be ignored, unless stated otherwise in the code) and the characteristic strength of the property used for design is obtained by multiplying the characteristic strength given in the relevant British Standard by a factor, k, derived from the member size, the reference size and the factor $1/k_{hL}$ or $1/k_L$, as appropriate. The characteristic strengths in bending and tension given in BS EN 338 have been derived using a reference depth of 150 mm for solid timber, 600 mm for glulam (EN 14080) and in BS EN 14374:2004, the reference depth for LVL bent edgewise is 300 mm. No size factor is applicable for LVL when bent flatwise. The tensile strength of LVL is also affected by the length of the member and the reference length used in EN 14374 is 3000 mm.

In EC5 factor k_h, which relates to depth, and k_ℓ, which relates to length, are derived from exponential functions of the reference size divided by the member size, where the exponent is the value derived for $1/k_{hL}$ or $1/k_L$, as appropriate. For sizes greater than the reference size the factor is less than unity, but, as the reduction is relatively small, it is ignored in EC5 for solid timber and glulam and taken to equal 1. When sizes are less than the reference value the factor will be greater than unity, resulting in an increase in the property strength. Also, as the member size decreases, an upper limit of 1.3, 1.1 and 1.2 has been set for k_h for solid timber, glulam and LVL, respectively, to derive the characteristic bending strength and 1.1 for k_ℓ when deriving the characteristic tensile strength of LVL along the grain direction. The factor k_h will apply to bending about the strong or the weak axis when dealing with solid timber but for horizontally laminated glued-laminated timber, it will only apply to the beam depth where the section is loaded perpendicular to the plane of the wide faces of the laminations. For LVL, k_h will only apply when a section is subjected to edgewise bending.

As an example, the theoretical value of factor k_h to be used to determine the characteristic bending and tensile strength of glulam beams is $(600/h)^{0.1}$, where h is the depth for a member in bending or the width (the maximum size of the cross-section) for a member in tension and $1/k_{hL} = 0.1$. The relationship is given in EC5, *equation (3.2)* and a comparison with the theoretical value is shown in Figure 2.8. Above the 600 mm reference size k_h is taken to be 1 and as the size of the beam decreases it follows the theoretical function until it reaches a maximum value of 1.1, which, for glulam members, occurs around a depth of 230 mm.

Relationships for the size effects used in EC5 for timber, glued-laminated timber and LVL as well as the volume and stress distribution effects in the apex zone of double tapered, curved and pitched cambered glued-laminated timber and LVL beams are summarised in Table 2.11 together with the associated reference criteria.

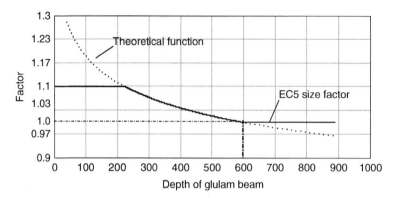

Fig. 2.8. EC5 factor (k_h) for glulam beams in bending and tension compared with the theoretical value.

2.3.7   System strength (EC5, 6.6)

Where a continuous load distribution system laterally connects a series of equally spaced similar members, the distribution system will enable load sharing to take place between the members. This allows the member strength properties to be increased in value and is achieved by multiplying the relevant properties by a system strength factor, k_{sys}. It takes advantage of the fact that stiffer members will take a greater share of the applied load than weaker/less stiff members and that the reliability index set for the EC5 design rules is such that the level of probability that adjacent members in the system will all have minimum strength and minimum stiffness characteristics will not be achieved. On this basis one member in the system can be considered to have characteristic strength/stiffness properties and will deform by a larger amount than adjacent stronger/stiffer members and when reaching its failure strength the adjacent members will be withstanding higher value stresses. In other words the strength of the structure functioning as a system will be greater than the strength of the equivalent structure in which the beams act individually.

For this to work the continuous load distribution system must have adequate stiffness and strength to transfer loading from one member to neighbouring members and for this condition k_{sys} shall be taken to equal 1.1. This can be taken to apply where the load distribution system is as follows:

(a) Structural flooring connected to floor beams where the flooring is continuous over at least two spans, necessitating at least four members, and any joints in the flooring are staggered.

(b) Stud walling connected by sheathing fixed to the studs in accordance with the fixing manufacturers recommendations or as required by the design. The maximum spacing of studs should be taken to be 610 mm c/c.

(c) Tiling battens, purlins or structural panels connected to roof trusses where the load distribution members are continuous over at least two spans, necessitating at least four trusses, with any joints being staggered. The spacing of the trusses must not be greater than 1.2 m.

Table 2.11 Values for k_h, k_ℓ, k_{vol} and k_{dis}*

Material	Factor	Definitions/conditions	Characteristic or design value
Solid timber	For bending and tension: $k_h = \min\left\{\begin{array}{l}\left(\dfrac{150}{h}\right)^{0.2}\\ \text{or } 1.3\end{array}\right\}$	Characteristic density ≤ 700 kg/m^3 (i) Bending: reference depth $h = 150$ mm. (ii) Tension: reference width (maximum cross sectional dimension) $h = 150$ mm.	(i) Bending strength: $= k_h f_{m,k}$ (ii) Tensile strength parallel to the grain: $= k_h f_{t,0,k}$
Glued-laminated timber	For bending and tension and stress distribution: $k_h = \min\left\{\begin{array}{l}\left(\dfrac{600}{h}\right)^{0.1}\\ \text{or } 1.1\end{array}\right\}$ In the apex zone of a double tapered, curved and pitched cambered beam with all veneers parallel to the beam axis: $k_{vol} = \left(\dfrac{V_0}{V}\right)^{0.2}$ In the apex zone of double tapered and curved beams: $k_{dis} = 1.4$ In the apex zone of pitched cambered beams: $k_{dis} = 1.7$	*For the evaluation of k_h* (i) Bending: reference depth $h = 600$ mm. (ii) Tension: reference width (maximum cross sectional dimension) $h = 600$ mm. *For the evaluation of volume factor k_{vol}* (i) Tension: reference volume $V_0 = 0.01$ m^3. The stressed volume of the apex zone (in m^3) as defined in EC5, *Figure 6.9, Figure 6.7.* (NB: the value used for V should not be greater than $2V_b/3$ where V_b is the volume of the beam.)	(i) Bending strength: $= k_h f_{m,k}$ (ii) Tensile strength parallel to the grain: $= k_h f_{t,0,k}$ (i) Tensile strength perpendicular to the grain: $= k_{dis}\ k_{vol} f_{t,90,d}$

(continued)

Table 2.11 (*continued*)

Material	Factor	Definitions/conditions	Characteristic or design value
LVL	For bending edgewise; tension and stress distribution: $$k_h = \min\left\{ \left(\dfrac{300}{h}\right)^s \text{ or } 1.2 \right\}$$ For length: $k_\ell = \min\{(3000/\ell)^{s/2} \text{ or } 1.1\}$ In the apex zone of a double tapered, curved and pitched cambered beam with all veneers parallel to the beam axis: $$k_{vol} = \left(\dfrac{V_0}{V}\right)^{0.2}$$ In the apex zone of double tapered, curved beams and pitched cambered beams, the values for k_{dis} are as for glued-laminated beams.	*For the evaluation of k_h* (i) The size effect exponent s is obtained from EN 14374: $s = 2(cov) - 0.05$, where cov is the coefficient of variation of the test results. (ii) Bending: reference depth $h = 300$ mm. *For the evaluation of k_ℓ* (i) Tension: reference length $\ell = 3000$ mm and s is as defined as for k_h. *For the evaluation of k_{vol}* (i) Tension: reference volume $V_0 = 0.01$ m³. The stressed volume of the apex zone (in m³) as defined in EC5, *Figure 6.9, Figure 6.7*). (NB: the value used for V should not be greater than $2V_b/3$ where V_b is the volume of the beam.)	(i) Bending strength: $= k_h f_{m,k}$ (i) Tensile strength along the grain: $= k_\ell f_{t,0,k}$ (i) Tensile strength perpendicular to the grain: $= k_{dis}\, k_{vol}\, f_{t,90,d}$

*Based on EC5, 3.2, 3.3, 3.4 and 6.4.3.

Where there is a requirement to have to verify that the load distribution system is strong enough to transfer the loading between the members, the loading shall be assumed to be of short-term duration.

If laminated timber flooring is to be used in the structure, the value of k_{sys} will be dependent on the number of loaded laminations being used and whether the floor will be nailed/screwed or pre-stressed/glued. Values for the system strength factor for laminated floor plates made from solid timber or glued-laminated timber are given in EC5, *Figure 6.12*.

2.4 SYMBOLS

The symbols and subscripts used in the text are generally the same as those used in EC0 and EC5.

Latin upper case letters

A	Accidental action
A_d	Design value of an accidental action
A_f	Cross-sectional area of a flange
A_w	Cross-sectional area of a web
C	A function of certain design properties of a material
E	Modulus of elasticity of a material
$E_{0.05}$	Fifth percentile (characteristic value) of modulus of elasticity
$E_{0,mean}$	Mean value of modulus of elasticity parallel to the grain
$E_{90,mean}$	Mean value of modulus of elasticity perpendicular to the grain
E_{mean}	Mean value of modulus of elasticity
$E_{mean,fin}$	Final mean value of modulus of elasticity
$E_{p,osb,0,mean}$	Mean value of modulus of elasticity parallel to the surface grain of plywood or OSB panel length
$E_{p,osb,90,mean}$	Mean value of modulus of elasticity perpendicular to the surface grain of plywood or OSB panel length
Ef	The effect of actions
Ef_d	Design value of the effect of actions
F	Action or force
$F_{ax,Ed}$	Design axial force on a fastener
$F_{ax,Rd}$	Design value of the axial withdrawal capacity of the fastener
$F_{ax,Rk}$	Characteristic axial withdrawal capacity of the fastener
F_c	Compressive action or force
F_d	Design value of a force
$F_{d,ser}$	Design force at the serviceability limit state
$F_{f,Rd}$	Design load capacity per fastener in wall diaphragm
$F_{i,c,Ed}$	Design compressive reaction force at the end of shear wall i
$F_{i,t,Ed}$	Design tensile reaction force at the end of shear wall i
$F_{i,vert,Ed}$	Design vertical load on wall i
$F_{i,v,Rd}$	Design racking resistance of panel i (when using method A) or wall diaphragm i when using the procedure in PD6693-1
F_k	Characteristic value of an action or force
F_{rep}	Representative value of an action
F_t	Tensile force
$F_{v,0,Rk}$	Characteristic load capacity of a connector along the grain

$F_{v,Ed}$	Design shear force per shear plane of fastener; horizontal design effect on wall diaphragm
$F_{v,Rd}$	Design load capacity per shear plane per fastener; design racking load capacity
$F_{v,Rk}$	Characteristic load capacity per shear plane per fastener
$F_{v,w,Ed}$	Design shear force on web
$F_{x,Ed}$	Design value of a force in the x-direction
$F_{y,Ed}$	Design value of a force in the y-direction
G	Permanent action
$G_{0,05}$	Fifth-percentile value of shear modulus
G_d	Design value of shear modulus
$G_{d,inf}$	Lower design value of a permanent action
$G_{d,sup}$	Upper design value of a permanent action
G_k	Characteristic value of a permanent action
$G_{k,j}$	Characteristic value of permanent action j
$G_{k,j,inf}$	Lower characteristic value of permanent action j
$G_{k,j,sup}$	Upper characteristic value of permanent action j
$G_{k,j}$	Characteristic value of permanent action j
G_{mean}	Mean value of shear modulus
I	Second moment of area of a section
I_f	Second moment of area of a flange
I_{tor}	Torsional moment of inertia
I_y	Second moment of area about the strong axis
I_z	Second moment of area about the weak axis
K_{ser}	Slip modulus of a fastener or connector for the serviceability limit state at the instantaneous condition
$K_{ser,fin}$	Slip modulus of a fastener or connector at the final condition
K_u	Slip modulus of a fastener or connector for the ultimate limit state at the instantaneous condition
M_d,	Design moment
$M_{y,d}; M_{z,d}$	Design moment about the principal y–y axis and the z–z axis
$M_{y,Rk}$	Characteristic yield moment of fastener
N	Axial force
Q	Variable action
Q_k	Characteristic variable action
$Q_{k,1}$	Leading variable action
$Q_{k,i}$	Characteristic value of accompanying variable action i
$R_{90,d}$	Design splitting capacity
$R_{90,k}$	Characteristic splitting capacity
$R_{ax,d}$	Design load capacity of an axially loaded connection
$R_{ax,k}$	Characteristic load capacity of an axially loaded connection
$R_{ax,\alpha,k}$	Characteristic load capacity at an angle α to the grain
R_d	Design value of load capacity of the resistance
$R_{ef,k}$	Effective characteristic load capacity of a connection
$R_{i,v,d}$	Design racking capacity of a wall
R_k	Characteristic value of the resistance
$R_{sp,k}$	Characteristic splitting capacity
$R_{v,d}$	Design racking capacity of a wall diaphragm
S	First moment of area
V	Shear force; volume
W_y	Section modulus about axis y

W_z	Section modulus about axis z
X_d	Design value of a material strength property
X_k	Characteristic value of a material strength property

Latin lower case letters

a	Distance
a_d	Design value of geometrical data
a_{nom}	Nominal value of geometrical data
a_1	Spacing, parallel to the grain, of fasteners within one row
a_2	Spacing, perpendicular to the grain, between rows of fasteners
$a_{3,c}$	Distance between fasteners at an unloaded end
$a_{3,t}$	Distance between fasteners at a loaded end
$a_{4,c}$	Distance between fasteners at an unloaded edge
$a_{4,t}$	Distance between fasteners at a loaded edge
$a_{4,c}$	Distance between fasteners at an unloaded edge
b	Width
b_i	Width of panel i (Method A or the PD6693-1 method)
b_{net}	Clear distance between studs
b_w	Web width
d	Diameter (the nominal diameter)
d_c	Connector diameter
d_{ef}	Effective diameter of a screw
$f_{h,i,k}$	Characteristic embedment strength of timber member i
$f_{ax,k}$	Characteristic withdrawal parameter for nails (or for screws)
$f_{c,0,d}$	Design compressive strength along the grain
$f_{c,w,d}$	Design compressive strength of a web
$f_{c,f,d}$	Design compressive strength of a flange
$f_{c,90,k}$	Characteristic compressive strength perpendicular to the grain
$f_{t,f,d}$	Design tensile strength of a flange
$f_{h,k}$	Characteristic embedment strength
$f_{head,k}$	Characteristic pull-through parameter for smooth nails
f_1	Fundamental frequency
$f_{m,k}$	Characteristic bending strength
$f_{m,y,d}$	Design strength about the principal y-axis
$f_{m,z,d}$	Design strength about the principal z-axis
$f_{m,\alpha,d}$	Design strength at an angle α to the grain
$f_{p,osb,c,0,d}$	Design compressive strength of plywood or OSB along the surface grain
$f_{p,osb,c,90,k}$	Characteristic compressive strength of plywood or OSB perpendicular to the surface grain
$f_{p,osb,m,0,k}$	Characteristic bending strength parallel to the surface grain of plywood or OSB
$f_{p,osb,m,90,k}$	Characteristic bending strength perpendicular to the surface grain of plywood or OSB
$f_{p,osb,t,0,d}$	Design tensile strength of plywood or OSB parallel to the surface grain
$f_{p,osb,t,0,k}$	Characteristic tensile strength of plywood or OSB parallel to the surface grain
$f_{p,osb,v,0,d}$	Design transverse shear strength in bending parallel to the surface grain of plywood or OSB
$f_{p,osb,v,0,k}$	Characteristic transverse shear strength in bending parallel to the surface grain of plywood or OSB

$f_{p,osb,v,90,k}$	Characteristic transverse shear strength in bending perpendicular to the surface grain of plywood or OSB
$f_{t,0,d}$	Design tensile strength along the grain
$f_{t,0,k}$	Characteristic tensile strength along the grain
$f_{t,90,d}$	Design tensile strength at right angles to the grain
$f_{t,w,d}$	Design tensile strength of the web
f_u	Tensile strength of wire used for nails
$f_{u,k}$	Characteristic tensile strength of bolts
$f_{v,0,k}$	Design panel shear strength
$f_{v,ax,\alpha,k}$	Characteristic withdrawal strength at an angle α to the grain
$f_{v,ax,90,k}$	Characteristic withdrawal strength perpendicular to the grain
$f_{v,d}$	Design shear strength
$f_{v,k}$	Characteristic shear strength
$f_{v,r,k}$	Characteristic rolling shear strength of plywood
h	Depth of member; height of wall
h_{ap}	Depth of the apex zone
h_e	Embedment depth; distance between centre of the most distant fastener and the loaded edge
h_{ef}	Effective depth
$h_{f,c}$	Depth of compression flange
$h_{f,t}$	Depth of tension flange
h_w	Web depth
i	Notch inclination
i_y or i_z	Radius of gyration
$k_{c,90}$	Bearing strength modification factor
$k_{c,y}$ or $k_{c,z}$	Instability factor
k_{crit}	Factor used for lateral buckling
k_{ef}	Exponent factor to derive the effective number of fasteners in a row
k_d	Dimension factor for a panel
k_{def}	Deformation factor
k_{dis}	Factor for taking account of the stress distribution in an apex zone
$k_{f,1}, k_{f,2}, k_{f,3}$	Modification factors for bracing resistance
k_h	Depth factor
$k_{i,q}$	Uniformly distributed load factor
k_ℓ	Factor for limiting lateral deflection in bracing design; factor for length effect in LVL
k_m	Factor for the redistribution of bending stresses in a cross-section
k_{mod}	Modification factor for duration of load and moisture content
k_n	Sheathing material factor; notch factor
k_s	Fastener spacing factor; modification factor for spring stiffness
k_{shape}	Factor depending on the shape of the cross-section
k_{shear}	Amplification factor for shear deflections
k_{sys}	System strength factor
k_v	Reduction factor for notched beams
k_{vol}	Volume factor in apex zone
k_y or k_z	Instability factor
ℓ	Span; contact length
ℓ_{ef}	Effective length; design span of a beam
m	Mass per unit area
n_{40}	Number of frequencies below 40 Hz
n_{ef}	Effective number of fasteners

q_d	Design value of a distributed load; the design value of the internal stability load per unit length provided by a bracing system
q_i	Equivalent uniformly distributed vertical load acting on a wall
r	Radius of curvature
s	Spacing
s_0	Basic fastener spacing
r_{in}	Inner radius of a curved or pitched cambered beam at the apex zone
t	Thickness
t_{pen}	Penetration depth
u	Deformation; horizontal displacement of a structure or structural element
u_{creep}	Creep deformation
u_{fin}	Final deformation
$u_{fin,G}$	Final deformation for a permanent action
$u_{fin,Q,1}$	Final deformation for a leading variable action
$u_{fin,Q,i}$	Final deformation for accompanying variable action i
u_{inst}	Instantaneous deformation
$u_{inst'G}$	Instantaneous deformation for a permanent action, G
$u_{inst,Q,1}$	Instantaneous deformation for a leading variable action, Q_1
$u_{inst,Q,i}$	Instantaneous deformation for accompanying variable action Q_i
v	Unit impulse velocity response
w	Vertical deflection limit of a structural member
w_c	Pre-camber
w_{creep}	Creep deflection limit
w_{fin}	Final deflection limit
w_{inst}	Instantaneous deflection limit
$w_{net,fin}$	Net final deflection limit

Greek upper case letters

Πk	Product of the k factors that affect the design value

Greek lower case letters

α and β	Angles
β_c	Straightness factor
γ_F	Partial factor for actions also accounting for model uncertainties and dimensional variations
$\gamma_{f,i}$	Partial factor for action i that takes account of the possibility of unfavourable deviations of the action values from the representative values
γ_G	Partial factor for permanent actions
$\gamma_{G,j}$	Partial factor for permanent action j
γ_Q	Partial factor for variable actions
$\gamma_{Q,i}$	Partial factor for variable action i
γ_m	Partial factor for material properties
γ_M	Partial factor for material properties also accounting for model uncertainties and dimensional variations
γ_{Sd}	Partial factor taking account of uncertainties in modelling the effects of actions
ζ	Modal damping ratio
η	Conversion factor
λ_y	Slenderness ratio corresponding to bending about the y-axis
λ_z	Slenderness ratio corresponding to bending about the z-axis

$\lambda_{rel,y}$	Relative slenderness ratio corresponding to bending about the y-axis
$\lambda_{rel,z}$	Relative slenderness ratio corresponding to bending about the z-axis
ρ_k	Characteristic density
ρ_m	Mean density
$\sigma_{c,0,d}$	Design compressive stress along the grain
$\sigma_{c,\alpha,d}$	Design compressive stress at an angle α to the grain
$\sigma_{f,c,d}$	Mean design compressive stress in a flange
$\sigma_{f,c,max,d}$	Design compressive stress of the extreme fibres in a flange
$\sigma_{f,t,d}$	Mean design tensile stress in a flange
$\sigma_{f,t,max,d}$	Design tensile strength of the extreme fibres in a flange
$\sigma_{m,crit}$	Critical bending stress
$\sigma_{m,y,d}$	Design bending stress about the principal y-axis
$\sigma_{m,z,d}$	Design bending stress about the principal z-axis
$\sigma_{m,\alpha,d}$	Design bending stress at an angle α to the grain
σ_N	Axial stress
$\sigma_{t,0,d}$	Design tensile stress along the grain
$\sigma_{t,90,d}$	Design tensile stress perpendicular to the grain
$\sigma_{w,c,d}$	Design compressive stress in a web
$\sigma_{w,t,d}$	Design tensile stress in a web
τ_v and $\tau_{v,d}$	Shear stress and design shear stress
$\tau_{tor,d}$	Design torsional shear stress
ψ_0	Factor for the combination value of a variable action
ψ_1	Factor for the frequent value of a variable action
ψ_2	Factor for the quasi-permanent value of a variable action

2.5   REFERENCES

1 BS 5268-2-2002. *Structural Use of Timber. Part 2: Code of Practice for Permissible Stress Design, Materials and Workmanship*, British Standards Institution.

2 BS EN 1995-1-1:2004+A1:2008. *Eurocode 5: Design of Timber Structures. Part 1-1: General – Common Rules and Rules for Buildings*, British Standards Institution.

3 BS EN 1990:2002+A1:2005 *Incorporating Corrigendum December 2008. Eurocode: Basis of Structural Design*, British Standards Institution.

4 NA to BS EN 1990:2002+A1:2005 *Incorporating National Amendment No1. UK National Annex for Eurocode 0 – Basis of Structural Design*, British Standards Institution.

5 NA to BS EN 1995-1-1:2004+A1:2008, Incorporating National Amendment No 2 *UK National Annex to Eurocode 5: Design of Timber Structures. Part 1-1: General – Common Rules and Rules for Buildings*, British Standards Institution.

6 BS EN 1991-1-1:2002. *Eurocode 1: Actions on Structures. Part 1-1: General Actions – Densities, Self-Weight and Imposed Loads for Buildings*, British Standards Institution.

7 BS EN 1995-1-2:2004. *Design of Timber Structures. Part 1-2: Structural Fire Design*, British Standards Institution.

8 Gulvanessian, H., Calgaro, J.-A., Holicky, M. *Designers' Guide To EN 1990 Eurocode: Basis of Structural Design*, Thomas Telford, London.

9 Gulvanessian, H., Formichi, P., Calgaro, J.-A. *Designers' Guide to Eurocode 1: Actions on Buildings: EN 1991-1-1 and -1-3 to -1-7*, Thomas Telford, London.

10 BS EN ISO 9000:2005. *Quality Management Systems, Fundamentals and Vocabulary*, British Standards Institution.

11 BS EN 338:2009. *Structural Timber – Strength Classes*, British Standards Institution.

12 EN 14081-1:2005+A1:2011. *Timber Structures – Strength Graded Structural Timber with Rectangular Cross-Section. Part 1: General Requirements*.

13 EN 14080:2005. *Timber Structures – Glued Laminated Timber – Requirements*.

14 BS EN 14374:2004. *Timber Structures – Structural Laminated Veneered Lumber – Requirements*, British Standards Institution.

15 EN 14279:2004+A1:2009. *Laminated Veneer Lumber (LVL) – Specifications, Definitions, Classification and Requirements*.

16 BS EN 636-1:1997. *Plywood – Specifications. Part 1: Requirements for Plywood for Use in Dry Conditions*, British Standards Institution.

17 BS EN 636-2:1997. *Plywood – Specifications. Part 2: Requirements for Plywood for Use in Humid Conditions*, British Standards Institution.

18 BS EN 636-3:1997. *Plywood – Specifications. Part 3: Requirements for Plywood for Use in Exterior Conditions*, British Standards Institution.

19 BS EN 300:2006. *Oriented Strand Board (OSB) – Definition, Classification and Specifications*, British Standards Institution.

20 Larsen, H.J., 'Properties affecting reliability design of timber structures', An overview prepared for COST E24 Seminar on reliability of timber structures, Coimbra, Portugal, 4-5 May 2001.

21 Gulvanessian, H., Holicky, M. Eurocodes: using reliability analysis to combine action effects. In: *Proceedings of the I.C.E.*, August, 2005, ISSN 0965 0911.

22 International Organisation for Standardisation, *ISO 3898, Basis of Design for Structures – Notation – General Symbols*.

23 Rouger, F., Volume and stress distribution effects. In: Blass, H.J., Aune, P., Choo, B.S., et al. (eds), *Timber Engineering STEP 1*, 1st edn. Centrum Hout, Almere, 1995.

24 Isaksson, T., Structural timber – variability and statistical modelling. In: Thelandersson, S., Larsen, H.J. (eds), *Timber Engineering*, Wiley, London, 2003.

25 BS EN 408:2010. *Timber Structures – Structural Timber and Glued Laminated Timber – Determination of Some Physical and Mechanical Properties*, British Standards Institution.

Chapter 3

Using Mathcad® for Design Calculations

3.1 INTRODUCTION

Mathcad is a powerful and easy to use computational tool that is used by most academic institutions and many design offices. The aim of this chapter is to demonstrate how the analysis and design calculations for structural timber can be incorporated into simple-to-use electronic notepads or worksheets using this software. Access to a personal computer (PC) and the associated software 'Mathcad' is not a prerequisite for understanding the design calculations in the examples given in the book. All of the design examples in the book are fully self-explanatory and well annotated. They have been produced in the form of worksheet files and are available on a CD to run under Mathcad 11 or higher. The later versions of Mathcad include more user friendly features, have more functions, are easier to integrate with other software packages and can solve more complicated problems, however, to suit readers familiar with the first edition of the book, version 11 has been retained for the examples given in this edition. The use of a higher version of Mathcad will automatically permit its enhanced features and functions. Mathcad details are given at the end of the book.

The design worksheets given are intended as a source of study, practice and further development by the reader. They should not be seen as complete and comprehensive but rather as the foundations of a design system that can be developed further. The aim is to encourage readers to use computing as a tool to increase their understanding of how design solutions vary in response to a change in one or more of the variables and how alternative design options can be readily obtained. This will allow design engineers to arrive at the most suitable and economic solution quickly.

It is important to note that this chapter is not intended to teach Mathcad. It aims only to familiarise the readers with the Mathcad worksheet formats that are used to produce the design examples given in the book.

3.2 WHAT IS Mathcad?

Mathcad (produced by the Parametric Technology Corporation (PTC)) [1] is an electronic notepad (live worksheet) that allows mathematical calculations to be performed on a computer screen in a format similar to the way they would be carried out for hand

Structural Timber Design to Eurocode 5, Second Edition. Jack Porteous and Abdy Kermani.
© Jack Porteous and Abdy Kermani 2013. Published 2013 by Blackwell Publishing Ltd.

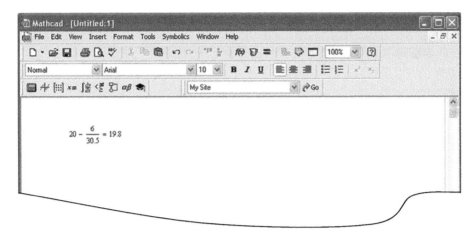

Fig. 3.1. A simple calculation.

calculations. While Mathcad employs the usual mathematical symbols (i.e. $+$, $-$, $/$, $=$), for algebraic operations it also uses the conventional symbols of calculus for differentiation and integration to perform these operations. It preserves the conventional symbolic form for subscribing, special mathematical and trigonometric functions, series operations and matrix algebra. When expository text is added, Mathcad's symbolic format leads to reports that are easily understood by others. Data can be presented in both tabular and graphical forms.

Mathcad can also be used to answer, amongst many others, the 'what-if' questions in engineering problems. With a well-structured worksheet, design calculations can be performed whereby parameters can be changed and the results viewed almost immediately on the computer display and/or printed. Through this procedure designs can be optimised with minimum effort.

3.3   WHAT DOES Mathcad DO?

Mathcad combines the live document interface of a spreadsheet with the WYSIWYG interface of a word processor [2]. With Mathcad, functions can be represented graphically and equations can be typeset on the screen, exactly the way they are presented in textbooks but with the advantage that the calculations can also be done.

Mathcad comes with multiple fonts and has the ability to print what you see on the screen through any Windows supported printer. This, combined with Mathcad's live document interface, makes it easy to produce up-to-date, publication-quality engineering reports and/or design solution sheets.

The following sub-sections demonstrate how some simple operations are carried out in Mathcad. This is to illustrate the format and meaning of the operations used to produce the examples in this text.

3.3.1   A simple calculation

Although Mathcad can perform sophisticated mathematics, it can just as easily be used as a simple calculator [2]. For example, click anywhere in the worksheet; you

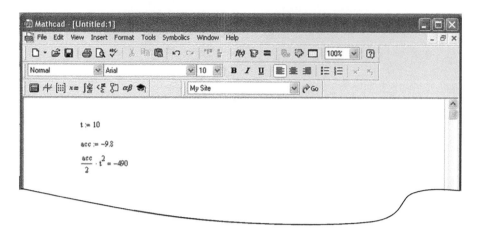

Fig. 3.2. Calculating with variables and functions.

will see a small crosshair. Type 20 – 6/30.5=. As soon as the 'equal to' key is pressed, Mathcad computes and shows the result; see Figure 3.1.

3.3.2   Definitions and variables

Mathcad's power and versatility quickly becomes apparent when variables and functions are being used [2]. By defining variables and functions, equations can be linked together and intermediate results can be used in further calculations.

For example, to define a value of say 10 to a variable, say t, click anywhere in the worksheet and type t: (the letter t followed by a colon).

Mathcad will show the colon as the definition symbol : = and will create an empty placeholder to its right. Then type 10 in the empty placeholder to complete the definition for t.

To enter another definition, press return [↵] to move the crosshair below the first equation. For example, to define acc as –9.8 type acc:–9.8. Then press [↵] again.

Now that the variables acc and t are defined, they can be used in other expressions. For example, to calculate the magnitude of $\frac{acc}{2}t^2$, type acc/2shift*t^2. The reference to 'shift' in the expression means to press the spacebar key; the caret symbol ^ represents raising to a power, the asterisk * is multiplication, and the slash / is division.

To obtain the result, type=for Mathcad to return the result as shown in Figure 3.2.

3.3.3   Entering text

Mathcad handles text as easily as it does equations [2]. To begin typing text, click in an empty space and choose **Text Region** from the **Insert** menu or simply type ". Mathcad will then create a text box in which you can type, change font, format and so on as you would when using a simple Windows based word processor. The text box will grow as the text is entered.

Now type, say, 'Equation of motion'; see Figure 3.3. To exit text mode simply click outside the text box.

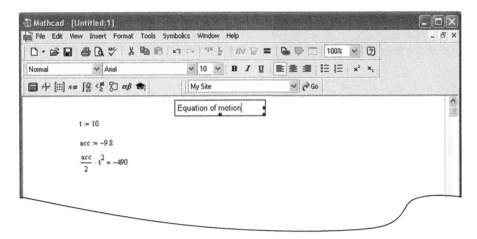

Fig. 3.3. Entering text.

3.3.4   Working with units

Mathcad's unit capabilities take care of many of the usual chores associated with using units and dimensions in engineering analysis and design calculations [2]. Once the appropriate definitions are entered, Mathcad automatically performs unit conversions and flags up incorrect and inconsistent dimensional calculations.

The SI system is used and, although Mathcad version 11 or greater recognises most of the common units used in practice, by default a result will be displayed using its fundamental units. If the result is to be expressed in other units compliant with the SI system, this will be achieved by clicking the unit placeholder at the end of the result and inserting the units to be used. You can also define your own units if you so wish, and to assign units to a number, simply multiply the number by the name or letter(s) that defines the unit.

To illustrate the above, calculate the magnitude of the bending moment M at the built-in end of a cantilever of length $L=2\,$m induced by a force of $P=10\,$kN acting at its free end. To do this, click anywhere in a Mathcad worksheet and type the following:

$$L := 2*m$$
$$P := 10*kN$$
$$M := P*L$$

Then type M=. As soon as the=sign is typed, Mathcad will compute the result and also display the units of M (as shown in Figure 3.4). With version Mathcad 14 or higher, just type=after P*L and the answer is displayed. The default unit for force times distance is joule and to display the answer in kN m, click the placeholder to the right of 10 and insert kN*m.

Several of the equations given in EC5 [3] are empirical and dimensionally incorrect. In such circumstances, to obtain the correct result from the Mathcad equation, each symbol within the equation must be made dimensionless by dividing by its associated

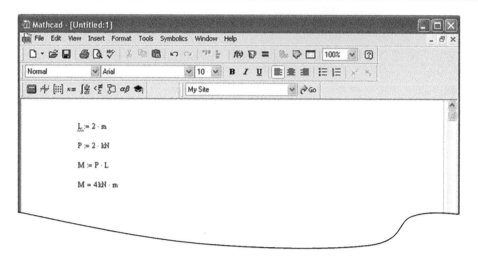

Fig. 3.4. Equations using units.

units and the units used for the symbol to be imported into the equation must use the units defined in EC5.

For example, to obtain the yield moment of a bolt, $M_{y,Rk}$, EC5, *equation (8.30)* must be used. The equation is structured to give the answer in N mm units providing the tensile strength of the bolt, $f_{u,k}$ is in N/mm² and the bolt diameter, d, is in mm units. The equation is

$$M_{y,Rk} = 0.3 f_{u,k} d^{2.6}$$

(EC5, *equation (8.30)*)

To obtain the answer, the units for $f_{u,k}$ must be N/mm² and d must be in mm. The equation is then set up with the dimensions removed from each function, and at the end of the equation, type *N*mm. To obtain the solution, type $M_{y,Rk}=$ and the answer will be given in N mm units as shown in the following example:

$$f_{u,k} := 400\,\text{N} \cdot \text{mm}^{-2} \quad d := 3 \cdot \text{mm}$$

$$M_{y,R,k} := 0.3 \cdot \left(f_{u,k} \cdot \frac{\text{mm}^2}{\text{N}} \right) \cdot \left(\frac{d}{\text{mm}} \right)^{2.6} \cdot \text{N mm}$$

$$M_{y,Rk} = 2.09 \times 10^3 \,\text{N mm}$$

3.3.5   Commonly used Mathcad functions

In the examples in the book, there are some Mathcad functions that are commonly used and an explanation of how two of these functions are set up for use in calculations is given.

3.3.5.1   *Maximum and minimum values*

The Mathcad function will determine the maximum or minimum value where a function in a problem has more than one value.

Example. The effective number of bolts in a row loaded parallel to the grain will be the minimum of value n (the number of bolts in the row) and function $n^{0.9}\sqrt[4]{a_1/(13d)}$ (where a_1 is the bolt spacing along the row and d is the bolt diameter). When these functions have been defined within the calculation the minimum value, say $n_{ef.r}$, will be obtained by typing *n.ef.r:min(n,n^0.9shift*(a.1/13*d)^0.25)*. To obtain the answer on a new line press return [↵] and type *n.ef.r=*.

For example, where the number of bolts (n) in a row in a connection equals 3, each bolt diameter (d) is 12 mm, the bolt spacing parallel to the grain (a_1) is 90 mm, and the connection is loaded parallel to the grain, the effective number of bolts in the row, n_{ef}, will be

$$n_{ef} := \min\left[n_b, n_b^{0.9} \cdot \left(\frac{90 \cdot mm}{13 \cdot d} \right)^{0.25} \right]$$

$$n_{ef} = 2.34$$

There can be any number of functions within the brackets and if a maximum value is required the procedure is as explained but type max rather than min.

3.3.5.2   *Conditional statement*

This function will enable a solution to be obtained when the evaluation process involves compliance with particular conditions.

Example. The determination of the lateral torsional instability function, k_{crit} is dependent on the value of the relative slenderness ratio for bending, $\lambda_{rel,m}$. When $k_{crit} \leq 0.75$ the value of $\lambda_{rel,m}$ is 1, when $0.75 < k_{crit} \leq 1.54$ the value is $1.56 - 0.75\,\lambda_{rel,m}$ and when $k_{crit} > 1.54$ it is $1/\lambda_{rel,m}^2$. To obtain the value within a calculation in which the function $\lambda_{rel,m}$ has been determined, the following procedure will apply:

(1) Open the programming toolbar from the **View** menu.
(2) On a new line in the calculation, type *k.crit:* and press].
(3) Click the top placeholder and type 1. Click the 'if' statement on the programming toolbar and it will appear after 1 in the calculation.
(4) In the placeholder after the 'if' type $\lambda.rel.m \leq 0.75$.
(5) Move to the placeholder on the next line and press] and another placeholder will appear on the line below.
(6) In the first placeholder, type $1.56 - 0.75* \lambda.rel.m$ and then click the 'if' statement on the programming toolbar.
(7) In the placeholder after 'if' type $0.75 < \lambda.rel.m < 1.54$.
(8) Move to the placeholder on the bottom line and type $1/\lambda.rel.m^2$ then press the shift key two times and click the 'otherwise' statement on the programming toolbar.
(9) To obtain the value for k_{crit} press return [↵] and type *k.crit=*.

For example, if the relative slenderness ratio for bending of a member is 1.33, the value of k_{crit} will be

$$k_{crit} := \begin{vmatrix} 1 & \text{if} & \lambda_{rel,m} \leq 0.75 \\ 1.56 - 0.75 \cdot \lambda_{rel,m} & \text{if} & 0.75 < \lambda_{rel,m} \leq 1.4 \\ \dfrac{1}{\lambda_{rel,m}^2} & & \text{otherwise} \end{vmatrix}$$

3.4 SUMMARY

The above examples aim to demonstrate the simplicity of using Mathcad in producing the design examples given in the following chapters of this book. To learn more about Mathcad, readers are referred to [1–2].

3.5 REFERENCES

1 Wieder, S., *Introduction to Mathcad for Scientists and Engineers*. McGraw-Hill, Hightstown, 1992.
2 *Mathcad 11 User's Guide*, MathSoft Engineering and Education, Inc., Bagshot, Surrey, UK.
3 BS EN 1995-1-1:2004:+A1:2008. *Eurocode 5: Design of Timber Structures. Part 1-1: General – Common Rules and Rules for Buildings*, British Standard Institution.

Chapter 4
Design of Members Subjected to Flexure

4.1 INTRODUCTION

Flexural members are those elements in a structure that are subjected to bending, and several types and forms of such members are used in timber construction. Typical examples are solid section rectangular beams, floor joists, rafters and purlins. Other examples include glulam beams and composites (thin webbed beams and thin flanged beams), and the design requirements of these types of members are covered in Chapters 6 and 7 respectively. Typical examples of timber beams, floor joists and purlins are shown in Figure 4.1.

Although the design principles used for the design of timber members in bending are essentially the same as those used for members constructed from other materials, e.g. concrete or steel, material characteristics peculiar to timber and wood-based structural products dictate that additional design checks are undertaken. With timber, the material properties are different in the two main directions, i.e. parallel and perpendicular to the grain and, unlike steel and concrete, they are affected by changes in moisture content and duration of load.

This chapter deals in detail with the general requirements that are necessary for the design of flexural members made from straight solid timber or wood-based structural products of uniform cross-section in which the grain runs essentially parallel to the member lengths. The design of tapered, curved and pitched cambered beams is covered in Chapter 6.

4.2 DESIGN CONSIDERATIONS

The main design requirements for flexural members are listed in Table 4.1.

Flexural members have to satisfy the relevant design rules and requirements of EC5 [1], and the limit states associated with the main design effects are given in Table 4.1. The equilibrium states and strength conditions (i.e. bending, shear and bearing) relate to failure situations and are therefore ultimate limit states, whereas the displacement and vibration conditions relate to normal usage situations and are serviceability limit states. If, however, a design condition is able to arise where a displacement could result in structural collapse, ultimate limit states would have to be checked for the

Structural Timber Design to Eurocode 5, Second Edition. Jack Porteous and Abdy Kermani.
© Jack Porteous and Abdy Kermani 2013. Published 2013 by Blackwell Publishing Ltd.

(a) (b)

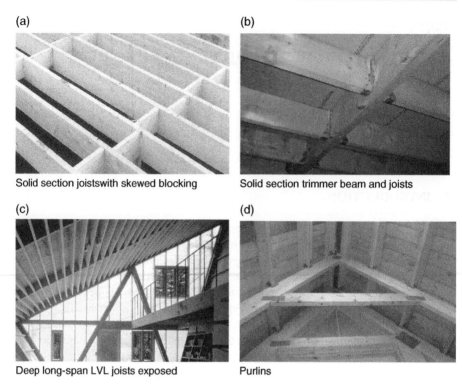

Solid section joistswith skewed blocking   Solid section trimmer beam and joists

(c) (d)

Deep long-span LVL joists exposed   Purlins

Fig. 4.1. Examples of flexural members.

Table 4.1 Main design requirements for flexural members and the associated EC5 limit states

Design or displacement effect	EC5 limit states
Retention of static equilibrium (sliding, uplift)	ULS
Bending stress and the prevention of lateral torsional instability	ULS
Shear stress	ULS
Bearing stress	ULS
Torsion stress (where relevant)	ULS
Deflection	SLS
Vibration	SLS

largest design effects arising from the application of the fundamental combination of actions referred to in 2.2.24.

Small deflection bending theory is taken to apply and limitations on permissible deviations from straightness must comply with the criteria given in *Section 10* of EC5. In general, bending is the most critical criterion for medium-span beams, deflection and vibration for long-span beams, and shear for heavily loaded short-span beams, but in practice, however, design checks have to be carried out for all of the design and displacement conditions.

For strength-related conditions, the design value of each stress (i.e. the design stress) is calculated and compared with the design value of its equivalent strength (i.e. the design strength modified where appropriate by strength factors) and, to meet the code reliability requirements, when using the partial factor method the following condition must be met:

Design stress ≤ (Strength factors) × Design strength

4.3   DESIGN VALUE OF THE EFFECT OF ACTIONS

To verify the ultimate and serviceability limit states, each design effect has to be checked and for each effect the largest value caused by the relevant combination of actions must be used. In the case of the ultimate limit states, the largest values will be derived from the application of the fundamental combination of actions referred to in 2.2.24 and given in Table 2.8. Where accidental situations have to be designed for, the combination of actions given in Table 2.9 must also be used. For the serviceability limit states, the combination of actions discussed in 2.2.25 will apply.

For the strength-related states, the design effect will also be a function of the strength modification factor, k_{mod}, referred to in 2.2.20. Where a load combination comprises actions having different load duration classes, as explained in 2.2.24, the modification factor corresponding to the action with the shortest duration used in the combined load case is applied to the strength property being considered, and where there is a linear relationship between action and effect, the design condition will be that giving the largest value after division by the associated k_{mod} factor.

With the equilibrium related states the design effects will apply solely to matters associated with static instability, and k_{mod} will not be relevant.

For the SLS, the k_{mod} modification factor is again not applicable as the design effects being considered are displacement and vibration under normal usage. The loading conditions to be used for these states are defined in 2.2.25.

An indication of the work involved in determining the critical load cases that will result in the greatest design effects at the ULS and the SLS is given in Example 4.8.1. The example covers the basic case of a simply supported beam subjected to permanent and variable actions, anchored at its ends to supporting structure, with all possible load cases being considered. Although the determination of the critical load cases is not difficult, it is time consuming, particularly when dealing with redundant structures. To be able to validate that the critical design effect of actions is being used, the design effects arising from all possible load combinations should be covered. For the examples given in the book, however, to ensure that attention is primarily focused on the EC5 design rules for the timber or wood-based product being used, only the design load case producing the largest design effect has generally been given or evaluated in the calculations.

4.4   MEMBER SPAN

In EC5 the bearing stress at the end of a beam is taken to be uniformly distributed over the bearing area, and for simply supported beams where the bearing area is much greater than is required for strength reasons, the beam need only be designed to span onto sufficient area to ensure that the design bearing strength is not exceeded.

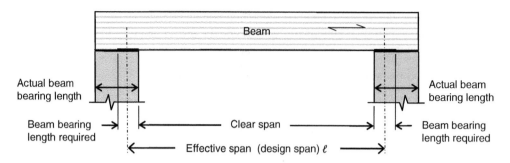

Fig. 4.2. Beam span.

In such instances the design span, i.e. the effective span of the beam, will be the clear span plus half the bearing length at each end, as shown in Figure 4.2. For solid timber beams and joists as well as built-up flooring beams it is usually acceptable to assume an additional length of 50 mm to be added to the clear span, and for built-up beams with spans up to 12 m an allowance of 100 mm should be made. For cases where the beam loading is high or longer span beams are being used, initial checks should be made on the bearing requirement to determine the allowance to be made.

4.5   DESIGN FOR ULTIMATE LIMIT STATES (ULS)

At the ULS, beam design is validated against failure state conditions and the design procedures for members of solid timber or wood-based structural products at these states are given in *Section 6* of EC5.

4.5.1   Bending

The elastic theory of bending states that when, for example, a solid rectangular member as shown in Figure 4.3 is subjected to a bending moment M about the y–y axis (the strong axis), the design stress at any distance z from this axis will be

$$\sigma = \frac{M \cdot z}{I_y} \tag{4.1}$$

where I_y is the second moment of area of the cross-section about the y–y axis.

The term I_y/z is referred to as the section modulus about the strong axis and is denoted by the symbol W_y. The value of W_y is dependent on the distance from the y–y axis and the values used in design for z are the distances to the extreme tension and compression fibre position on the section. Adopting the symbols defined in Chapter 2, the design bending stress, $\sigma_{m,y,d}$, at the extreme fibre position in the rectangular section shown in Figure 4.3 when bent about its strong axis by a design moment, M_d, will be

$$\sigma_{m,y,d} = \frac{M_d}{W_y} \tag{4.2}$$

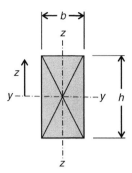

Fig. 4.3. Cross-section of a rectangular beam.

where

$$W_y = \frac{I_y}{z} = \frac{bh^3}{12} \times \frac{2}{h} = \frac{bh^2}{6} \tag{4.3}$$

Similar expressions may be derived for the design bending stress about the z–z axis (the weak axis) at the extreme fibre positions based on the design moment about that axis and the W_z section modulus.

When a member is subjected to bending, providing there is no reduction in strength due to lateral torsional instability of the section, its design strength will be based on the bending strength of the member material. Lateral torsional instability affects a member bent about the y–y axis when the compression face of the member is not fully restrained against lateral movement and the relative slenderness for bending, $\lambda_{rel,m}$, about this axis is ≥ 0.75. The design procedure for members that are not affected by lateral torsional instability is addressed in 4.5.1.1.

Where the relative slenderness for bending of a member about the y–y axis exceeds 0.75, the design procedure for taking the effect of lateral torsional instability into account is given in 4.5.1.2.

The procedure for determining the relative slenderness ratio for bending of a member about its strong axis is given in 4.5.1.2.

4.5.1.1   *Bending (where the relative slenderness ratio for bending about the major axis is ≤ 0.75)*

When a member is subjected to uni-axis bending about its y–y axis and the relative slenderness ratio for bending about this axis is ≤ 0.75, the design condition for bending is that the maximum bending stress in the section shall not exceed the design bending strength of the timber or wood-based structural material used for the member. This is achieved by compliance with equation (4.4a):

$$\frac{\sigma_{m,y,d}}{f_{m,y,d}} \leq 1 \tag{4.4a}$$

For a member subjected to uni-axis bending about its z–z axis, lateral torsional instability will not occur and the design condition is again that the maximum bending stress in the section must not exceed the design bending strength of the timber or wood-based structural product material used for the member. For bending about this axis, equation (4.4b) must be satisfied:

$$\frac{\sigma_{m,z,d}}{f_{m,z,d}} \leq 1 \tag{4.4b}$$

If a member is subjected to bending about both axes and the relative slenderness ratio for bending about the y–y axis is ≤ 0.75, the design conditions to be met are that equations (4.4c) and (4.4d) must be satisfied,

$$\frac{\sigma_{m,y,d}}{f_{m,y,d}} + k_m \frac{\sigma_{m,z,d}}{f_{m,z,d}} \leq 1 \qquad\qquad (EC5,\ equation\ (6.11)) \quad (4.4c)$$

$$k_m \frac{\sigma_{m,y,d}}{f_{m,y,d}} + \frac{\sigma_{m,z,d}}{f_{m,z,d}} \leq 1 \qquad\qquad (EC5,\ equation\ (6.12)) \quad (4.4d)$$

where $\sigma_{m,y,d}$ and $\sigma_{m,z,d}$ are the design bending stresses about the strong and weak axes as shown in Figure 4.3, and, for a rectangular section of width b (mm) and depth h (mm) as shown:

$$\sigma_{m,y,d} = \frac{M_{y,d}}{W_y}$$

where $M_{y,d}$ is the design bending moment about the y–y axis and $W_y = (bh^2)/6$ is the associated section modulus;

$$\sigma_{m,z,d} = \frac{M_{z,d}}{W_z}$$

where $M_{z,d}$ is the design bending moment about the z–z axis and $W_z = (hb^2)/6$ is the associated section modulus. $f_{m,y,d}$ and $f_{m,z,d}$ are the design bending strengths about the strong and the weak axes, respectively, and

$$f_{m,y/z,d} = \frac{k_{mod} \cdot k_{sys} \cdot k_h \cdot f_{m,k}}{\gamma_M} \tag{4.5}$$

Here

- k_{mod} is the modification factor for load duration and service classes discussed in 2.2.20 and given in Table 2.4.
- k_{sys} is the system strength factor discussed in 2.3.7.
- k_h is the modification factor for member size effect, referred to in Tables 2.5 and 2.11 and discussed in 2.3.6. As the effect only applies to solid timber and LVL (when bent flatwise), as well as glulam, for design using other wood-based structural products, the factor should be taken to equal 1. Because the factor is dependent on the member size in the direction of bending, the value for bending about the y–y axis can differ from the value for bending about the x–x axis.
- $f_{m,k}$ is the characteristic bending strength of the timber or, in the case of a wood-based structural product, the characteristic bending strength relating to the axis of bending being considered. Strength information for timber and the commonly used wood-based structural products is given in Chapter 1.

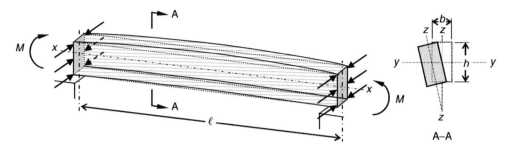

Fig. 4.4. Lateral torsional buckling of a beam subjected to uniform end moments M applied about the major axis (lateral buckled position shown solid).

- γ_M is the partial coefficient for material properties, given in Table 2.6.
- k_m is a modification factor that allows for an element of redistribution of stress (yield behaviour) in the stress block and also takes into account the effect of variation in material properties. Values for the factor are given in *6.1.6(2)*, EC5:
  - For solid timber and LVL (and glued-laminated timber), $k_m = 0.7$ for a rectangular section, and $k_m = 1.0$ for other cross-sections.
  - For other wood-based structural products, $k_m = 1.0$ for any cross-section.

For sections in which the second moment of area about the y–y and z–z axes has the same value (e.g. square or circular solid timber members) lateral torsional instability will not occur and $f_{m,y,d}$ will equal $f_{m,z,d}$. When using a circular cross-section, the section modulus will be $(\pi d^3)/32$, where d is the diameter of the member section.

4.5.1.2 *Bending (where the relative slenderness ratio for bending about the major axis is > 0.75)*

When a perfectly straight beam is subjected to bending about its y–y axis, it can be shown by elastic buckling theory that there is an elastic critical load at which the beam will become unstable, failing suddenly by deflecting sideways and twisting about its longitudinal x–x axis. This mode of failure is termed *lateral torsional buckling* and is shown in section A–A in Figure 4.4 for a rectangular member subjected to pure bending about the y–y axis. It is to be noted that in every case a beam must be prevented from moving out of plane laterally at its ends otherwise it cannot support load, and this will be achieved by blocking at these positions (examples of blocking along the beam length are shown on Figure 4.7) or fixing to other elements of the structure (e.g. ring beams).

The bending moment at which elastic buckling will occur is termed the elastic critical moment and is a function of the nature of the loading on the beam; the length of the beam and its support conditions; the position of the beam loading relative to its shear centre; the shear modulus and modulus of elasticity of the beam as well as its section properties. For a member with a design span, ℓ, restrained against torsional movement at its ends but free to rotate laterally in plan, subjected to a pure moment applied at its ends about the y–y axis, as indicated in Figure 4.4, it can be shown that the elastic critical moment of the beam, $M_{y,crit}$, will be:

$$M_{y,crit} = \frac{\pi}{\ell}\sqrt{\frac{E_{0.05}I_z G_{0.05}I_{tor}}{\left(1-(I_z/I_y)\right)}} \tag{4.6a}$$

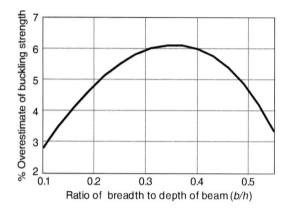

Fig. 4.5. Per cent overestimate in EC5 of the elastic critical moment of a rectangular beam when ignoring factor *a*.

For a rectangular section of breadth b, depth h, and $I_z = (1/12)hb^3$, I_{tor} can be taken to approximately equal $(1/3)(b)^3 h(1 - 0.63(b/h))$, and inserting these relationships in equation (4.6a) the elastic critical moment can be written as:

$$M_{y,crit} = \frac{\pi(b^3 h)\sqrt{E_{0,05}G_{0,05}\left((1 - 0.63(b/h))/(1 - (b/h)^2)\right)}}{6\ell} \tag{4.6b}$$

where $E_{0,05}$ is the 5th-percentile value of the modulus of elasticity parallel to the grain and $G_{0,05}$ is the 5th-percentile value of the shear modulus.

For all practical sizes of solid timber beam, the factor $((1 - 0.63(b/h))/(1 - (b/h)^2))^{0.5}$ (referred to as factor *a*) is less than unity and will only influence the buckling strength by a small percentage. This is shown in Figure 4.5 where the effect of deleting the function, expressed as a percentage of the buckling strength, is plotted over a practical range of beam breadth to depth ratios. It is seen that the elastic buckling strength will be overestimated by 3–6% and the maximum value will occur when the breadth to depth ratio is between 0.3 and 0.4.

In EC5 factor *a* is ignored for solid rectangular timber softwood beams and the elastic critical moment becomes:

$$M_{y,crit} = \frac{\pi(b^3 h)\sqrt{E_{0,05}G_{0,05}}}{6\ell} \tag{4.7a}$$

For solid rectangular sections made with hardwood, LVL (or glued-laminated timber), in EC5 the full torsional rigidity is retained but function $(1 - (b/h)^2)$ in factor *a* is ignored so that the elastic critical moment relationship for these materials is:

$$M_{y,crit} = \frac{\pi(b^3 h)\sqrt{E_{0,05}G_{0,05}\left(1 - 0.63\dfrac{b}{h}\right)}}{6\ell} \tag{4.7b}$$

When the section modulus about the strong y–y axis is W_y, from equation (4.7b) the stress in the section will be:

$$\sigma_{m,crit} = \frac{M_y}{W_y} = \frac{\pi b^2}{h\ell}\sqrt{E_{0,05}G_{0,05}\left(1-0.63\frac{b}{h}\right)} \qquad (4.7c)$$

Equation (4.7c) relates to a hardwood, LVL (or glued-laminated timber) rectangular section beam subjected to a uniform moment at each end and will equate to the solution derived from *equation (6.31)* in EC5. With softwood rectangular sections, the ratio of $E_{0,05}/G_{0,05}$ for timber is taken to be approximately 16, and by applying this to equation (4.7a), the critical bending stress (i.e. the buckling strength), $\sigma_{m,crit}$, of a rectangular softwood beam bent about its strong axis can be written as:

$$\sigma_{m,crit} = \frac{M_{y,crit}}{W_y} = \frac{0.78b^2}{h\ell}E_{0.05} \qquad (4.8)$$

where $E_{0.05}$ is the 5th-percentile value of the modulus of elasticity parallel to the grain, b is the breadth of the beam, h is the depth of the beam, ℓ is the design span of the simply supported beam between lateral supports at the ends of the beam, W_y is the section modulus of the beam about the y–y axis.

Equations (4.7a) and (4.7b) are only valid for a pure moment condition applied to a simply supported beam where the beam ends are prevented from moving laterally, are free to rotate in plan, and are fully restrained against torsional rotation. For situations where different end fixing conditions exist and moment is induced by other types of loading, as well as the cases where load is applied at the compression (or tension) face rather than the centroidal axis of the beam, the elastic critical moment can be obtained by using the same expression but replacing the design span ℓ by what is termed the 'effective length', ℓ_{ef}, of the beam. The effective length is obtained by adjusting the design span to take account of the effect of the change in loading and end fixing conditions and values for commonly used cases in timber design are given in Table 4.2.

As it is extremely difficult to achieve full restraint against lateral rotation in plan at the ends of a single span beam, with the exception of cantilever beams (where full restraint is assumed to exist), all of the other cases given in Table 4.2 assume that the beam is fully restrained against torsional rotation but able to rotate in plan at its ends.

When designing solid softwood rectangular beams, based on the use of the effective length of the beam, equation (4.8) can be written as:

$$\sigma_{m,crit} = \frac{0.78b^2}{h\ell_{ef}}E_{0.05} \qquad \text{(EC5, equation (6.32))} \quad (4.9a)$$

and when designing for hardwood, LVL (or glued-laminated) rectangular beams, equation (4.7c) will be as given in (4.9b), which is equivalent to *equation (6.31)* in EC5:

$$\sigma_{m,crit} = \frac{M_y}{W_y} = \frac{\pi b^2}{h\ell_{ef}}\sqrt{E_{0,05}G_{0,05}\left(1-0.63\frac{b}{h}\right)} \qquad (4.9b)$$

where ℓ_{ef} is the effective length obtained from Table 4.2 for the loading configuration being used.

Table 4.2 The ratio of the 'effective length' to the design span of the beam (incorporating the cases in *Table 6.1*, EC5)

Beam end condition: restrained in position laterally; restrained torsionally; free to rotate in plan	Applied loading	ℓ_{ef}/ℓ^* – EC5	ℓ_{ef}/ℓ^* – other cases
Simply supported	Constant moment	1.0	
	Uniformly distributed load	0.9	
	Concentrated load at mid-span	0.8	
	Point loads at quarter and three quarter points		0.96
	Moment (M) at one end and $M/2$ in opposite direction at the other end		0.76
	Moment (M) at one end and zero moment at the other end		0.53
Fully fixed at both ends	Uniformly distributed load		0.78
	Concentrated load at mid-span		0.64
Simply supported and restrained from lateral torsional movement at mid-span	Concentrated load at mid-span		0.28
Cantilever[†]	Uniformly distributed load	0.5	
	Concentrated load at the free end	0.8	

*The ratio between the effective length ℓ_{ef} and the design span ℓ is valid for a beam loaded at its centre of gravity. If the load is applied at the compression face of the beam, ℓ_{ef} should be increased by $2h$ (where h is the depth of the beam) and, for a load at the tension face of the beam, may be decreased by $0.5h$.
[†]For this case, at the fixed end the cantilever is restrained laterally in position, restrained torsionally and prevented from rotating in plan while free to move laterally and rotate at the other end.

To link the buckling strength of a beam, $\sigma_{m,crit}$, to its bending strength, $f_{m,k}$, the following relationship is used in EC5:

$$(\lambda_{rel,m})^2 \sigma_{m,crit} = f_{m,k}$$

and is defined in EC5, *equation (6.30)*, as follows:

$$\lambda_{rel,m} = \sqrt{\frac{f_{m,k}}{\sigma_{m,crit}}} \tag{4.10}$$

where $\lambda_{rel,m}$ is defined as the relative slenderness for bending.

Equation (4.10) is only valid when the critical bending stress is less than or equal to the elastic limit of the material. Beyond this limit, the relationship has to be modified to take account of the effect of inelastic behaviour of the material and to set a minimum value for the relative slenderness below which lateral torsional buckling will not arise.

It should be noted in equation (4.10) the characteristic strength is the value obtained directly from the relevant design standards, e.g. BS EN 338:2009 [2] for timber. However, as the strength can be increased by k_h and k_{sys} when these factors are greater

Table 4.3 Value of k_{crit} for beams with an initial deviation from straightness no greater than the limits set in EC5, *10.2*

Value of k_{crit}	Value of relative slenderness $\lambda_{rel,m}$
1	$\lambda_{rel,m} \leq 0.75$
$1.56 - 0.75\,\lambda_{rel,m}$	$0.75 < \lambda_{rel,m} \leq 1.4$
$1/\lambda_{rel,m}^2$	$1.4 < \lambda_{rel,m}$

than 1 it can be argued in such circumstances that the enhanced value should be used in the equation as it will produce the more critical condition. It is generally accepted for design purposes that the value used within equation (4.10) is the value obtained directly from the relevant design standards and is the procedure followed in this book.

In EC5 a value of $\lambda_{rel,m} = 0.75$ has been adopted as the limit below which the beam will be stiff enough not to buckle laterally and for all values less than 0.75 the design condition to be met is that the maximum bending stress in the section will not exceed the bending strength of the member. For this condition, the beam must comply with the requirements of 4.5.1.1.

When $\lambda_{rel,m} \geq 1.4$ (1.4 is the value at which the ratio of $f_{m,k}/\sigma_{m,crit}$ is approximately equal to 2 and is taken to be the elastic limit of the beam material), the beam is considered to fail solely by elastic buckling and the elastic critical bending strength will become the design condition. For relative slenderness values between 0.75 and 1.4 the section will fail in an inelastic manner and EC5 adopts the approximation of a linear relationship between relative slenderness and member strength between these limits.

To take the effect of the reduction in the strength of the beam into account as $\lambda_{rel,m}$ increases, a lateral buckling reduction factor, k_{crit}, is applied to the design bending strength of the beam such that at the ULS the beam strength, i.e. its design buckling strength, will be obtained from $k_{crit} \times$ the design bending strength.

The value of k_{crit} to be used is given in Table 4.3.

Where lateral instability effects can occur in beams, the maximum initial mid-length deviation from straightness in the beam permitted in *Section 10*, EC5, is limited to $\ell/300$ for solid timber and $\ell/500$ for LVL (and glued-laminated timber), where ℓ is the beam length in mm. The design rules in EC5 for strength validation are considered to take the effect of this imperfection into account.

When $\lambda_{rel,m} \leq 0.75$, the design strength will equate to the bending strength of the member, resulting in $k_{crit} = 1$. For behaviour within the linear elastic range of the beam, i.e. where $\sigma_{m,crit}/f_{m,k} < 0.51$, failure is solely by buckling and the design strength will equal the elastic buckling stress. For this condition, k_{crit} relates to the relative slenderness ratio as follows:

$$k_{crit} = \frac{1}{\lambda_{rel,m}^2} = \frac{\sigma_{m,crit}}{f_{m,k}} \qquad (4.11)$$

At the elastic limit, $\sigma_{m,crit}/f_{m,k} = 0.51$ and when applying this to equation (4.10), $\lambda_{rel,m} = 1.4$. The relationship in equation (4.11) applies for all values of $\lambda_{rel,m} \geq 1.4$.

Equation (4.10) will give the buckling strength of a beam that is perfectly straight; however, any deviation from straightness will result in a strength that is lower than the value obtained from this equation. By limiting the out of alignment in a beam to the

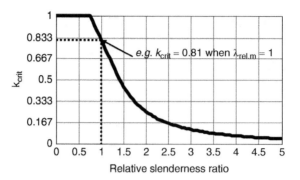

Fig. 4.6. The value of k_{crit} for different values of $\lambda_{\text{rel,m}}$.

maximum values given in EC5, it is to be understood that EC5 considers that the buckling strength obtained from equation (4.11) will still be acceptable for design purposes.

When $\lambda_{\text{rel,m}}$ is between 0.75 and 1.4, a linear transition relationship is adopted, reducing from the material bending strength when $\lambda_{\text{rel,m}} = 0.75$, at which $k_{\text{crit}} = 1$, to the elastic critical strength when $\lambda_{\text{rel,m}} = 1.4$, at which $k_{\text{crit}} = 0.51$, as follows:

$$k_{\text{crit}} = 1.56 - 0.75\lambda_{\text{rel,m}} \tag{4.12}$$

The effect of deviation from straightness is also relevant within this range of relative slenderness however the adoption of linear behaviour as defined in equation (4.12) is a conservative approach and will offset the strength reduction effect arising from this imperfection.

A graphical representation of k_{crit} plotted against the relative slenderness ratio for bending, $\lambda_{\text{rel,m}}$, is shown in Figure 4.6.

To demonstrate that lateral buckling of the beam will not occur, the bending stress in the beam must be less than or equal to the reduced bending strength, i.e.

$$\sigma_{m,d} \leq k_{\text{crit}} f_{m,d} \qquad \text{(EC5, equation (6.33))} \quad (4.13)$$

where $\sigma_{m,d}$ is the design bending stress (i.e. the moment on the section divided by the section modulus), $f_{m,d}$ is the design bending strength, and k_{crit} is a factor taking account of the reduced bending strength due to lateral torsional buckling.

The bending strength is derived from the characteristic bending strength in accordance with equation (4.5), i.e.

$$f_{m,y/z,d} = \frac{k_{\text{mod}} \cdot k_{\text{sys}} \cdot k_{h} \cdot f_{m,k}}{\gamma_{M}} \tag{4.5}$$

Consequently, for bending about the y–y axis the reduced bending strength can be written as:

$$k_{\text{crit}} \cdot f_{m,y,d} = \frac{k_{\text{mod}} \cdot k_{\text{sys}} \cdot k_{h}}{\gamma_{M}} \left(k_{\text{crit}} \cdot f_{m,k} \right) \tag{4.14}$$

Knowing the relative slenderness ratio, the value of k_{crit} × characteristic bending strength can be derived by calculation or, for the softwood species of timber listed in BS EN338:2009 [2], can be obtained from Table 4.4.

Table 4.4 The value of $k_{crit} \times$ characteristic bending strength tabulated against the relative slenderness ratio for bending, $\lambda_{rel,m}$, for all of the strength classes of softwood given in BS EN 338:2003 [E]

Strength class Characteristic bending strength (N/mm²)		C14 14	C16 16	C18 18	C20 20	C22 22	C24 24
Relative slenderness ratio for bending, $\lambda_{rel,m}$	Value of k_{crit}	$k_{crit} \times f_{m,k}$ (N/mm²)	$k_{crit} \times f_{m,k}$ (N/mm²)	$k_{crit} \times f_{m,k}$ (N/mm²)	$k_{crit} \times f_{m,k}$ (N/mm²)	$k_{crit} \times f_{m,k}$ (N/mm²)	$k_{crit} \times f_{m,k}$ (N/mm²)
Up to 0.75	1.0	14	16	18	20	22	24
0.8	0.960	13.440	15.360	17.280	19.200	21.120	23.040
0.85	0.923	12.915	14.760	16.605	18.45	20.295	22.140
0.9	0.885	12.390	14.160	15.930	17.700	19.470	21.240
0.95	0.848	11.865	13.560	15.255	16.950	18.645	20.340
1.0	0.810	11.340	12.960	14.580	16.200	17.820	19.440
1.1	0.735	10.290	11.760	13.230	14.700	16.170	17.640
1.2	0.660	9.240	10.560	11.880	13.200	14.520	15.840
1.3	0.585	8.190	9.360	10.530	11.700	12.870	14.040
1.4	0.510	7.140	8.160	9.180	10.200	11.220	12.240
1.5	0.444	6.222	7.111	8.000	8.889	9.778	10.667
1.6	0.391	5.469	6.250	7.031	7.813	8.594	9.375
1.7	0.346	4.844	5.536	6.228	6.920	7.612	8.304
1.8	0.309	4.321	4.938	5.556	6.173	6.790	7.407
1.9	0.277	3.878	4.432	4.986	5.540	6.094	6.648
2.0	0.250	3.500	4.000	4.500	5.000	5.500	6.000
2.1	0.227	3.175	3.628	4.082	4.535	4.989	5.442
2.2	0.207	2.893	3.306	3.719	4.132	4.545	4.959
2.3	0.189	2.647	3.025	3.403	3.781	4.159	4.537
2.4	0.174	2.431	2.778	3.125	3.472	3.819	4.167
2.5	0.160	2.240	2.560	2.880	3.200	3.520	3.840
2.6	0.148	2.071	2.367	2.663	2.959	3.254	3.550
2.7	0.137	1.920	2.195	2.469	2.743	3.018	3.292
2.8	0.128	1.786	2.041	2.296	2.551	2.806	3.061
2.9	0.119	1.665	1.902	2.140	2.378	2.616	2.854
3.0	0.111	1.556	1.778	2.000	2.222	2.444	2.667

Strength class Characteristic bending strength (N/mm²)		C27 27	C30 30	C35 35	C40 40	C45 45	C50 50
Relative slenderness ratio for bending, $\lambda_{rel,m}$	Value of k_{crit}	$k_{crit} \times f_{m,k}$ (N/mm²)	$k_{crit} \times f_{m,k}$ (N/mm²)	$k_{crit} \times f_{m,k}$ (N/mm²)	$k_{crit} \times f_{m,k}$ (N/mm²)	$k_{crit} \times f_{m,k}$ (N/mm²)	$k_{crit} \times f_{m,k}$ (N/mm²)
Up to 0.75	1.0	27	30	35	40	45	50
0.8	0.960	25.920	28.800	33.600	38.400	43.200	48.000
0.85	0.923	24.908	27.675	32.288	36.900	41.513	46.125
0.9	0.885	23.895	26.550	30.975	35.400	39.825	44.250
0.95	0.848	22.883	25.425	29.663	33.900	38.138	42.375
1.0	0.810	21.870	24.300	28.350	32.400	36.450	40.500
1.1	0.735	19.845	22.050	25.725	29.400	33.075	36.750
1.2	0.660	17.820	19.800	23.100	26.400	29.700	33.000
1.3	0.585	15.795	17.550	20.475	23.400	26.325	29.250
1.4	0.510	13.770	15.300	17.850	20.400	22.950	25.500

(continued)

Table 4.4 (*continued*)

Strength class Characteristic bending strength (N/mm²)		C27 27	C30 30	C35 35	C40 40	C45 45	C50 50
Relative slenderness ratio for bending, $\lambda_{rel,m}$	Value of k_{crit}	$k_{crit} \times f_{m,k}$ (N/mm²)	$k_{crit} \times f_{m,k}$ (N/mm²)	$k_{crit} \times f_{m,k}$ (N/mm²)	$k_{crit} \times f_{m,k}$ (N/mm²)	$k_{crit} \times f_{m,k}$ (N/mm²)	$k_{crit} \times f_{m,k}$ (N/mm²)
1.5	0.444	12.000	13.333	15.556	17.778	20.000	22.222
1.6	0.391	10.547	11.719	13.672	15.625	17.578	19.531
1.7	0.346	9.343	10.381	12.111	13.841	15.571	17.301
1.8	0.309	8.333	9.259	10.802	12.346	13.889	15.432
1.9	0.277	7.479	8.310	9.695	11.080	12.465	13.850
2.0	0.250	6.750	7.500	8.750	10.000	11.250	12.500
2.1	0.227	6.122	6.803	7.937	9.070	10.204	11.338
2.2	0.207	5.579	6.198	7.231	8.264	9.298	10.331
2.3	0.189	5.104	5.671	6.616	7.561	8.507	9.452
2.4	0.174	4.688	5.208	6.076	6.944	7.813	8.681
2.5	0.160	4.320	4.800	5.600	6.400	7.200	8.000
2.6	0.148	3.994	4.438	5.178	5.917	6.657	7.396
2.7	0.137	3.704	4.115	4.801	5.487	6.173	6.859
2.8	0.128	3.444	3.827	4.464	5.102	5.740	6.378
2.9	0.119	3.210	3.567	4.162	4.756	5.351	5.945
3.0	0.111	3.000	3.333	3.889	4.444	5.000	5.556

In Table 4.4, values of $k_{crit} \times$ characteristic bending strength are given for beams with relative slenderness ratios between 0.75 and 3.0, which should cover the range likely to be used in practical designs. For a strength class of timber, the reduced bending strength at a particular relative slenderness ratio is the value given in the column containing the material strength class, e.g. for a strength class C18 rectangular timber beam with a relative slenderness ratio of 2, the value of $k_{crit} \times$ characteristic bending strength is 4.5 N/mm². At intermediate values of relative slenderness ratio, the result is obtained by linearly interpolating between the values in the table immediately above and below the required value.

Lateral restraint can be provided by blocking or strutting the beam at positions along its length, as illustrated in Figure 4.7, and for these situations the effective length will be based on the beam length between adjacent blocking/strutting positions. Where the beam is supported laterally along the length of its compression flange, e.g. due to flooring structure secured to the compression flange, and the beam is relatively rigid across its depth (e.g. a solid timber beam), it can be considered to be fully restrained and k_{crit} will be unity. The lateral restraint must provide adequate strength and stiffness to the beam and guidance on these requirements is given in Chapter 9.

4.5.1.3   Bending of notched members

When members are notched, stress concentrations can arise at the notch position and, depending on the stress condition, the effects have to be taken into account in the design.

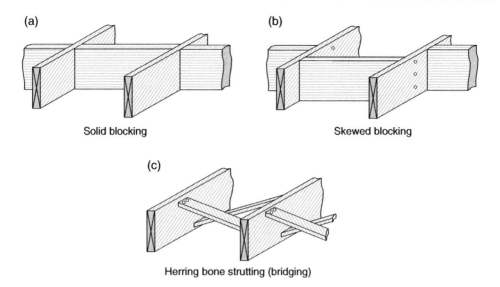

(a)

Solid blocking

(b)

Skewed blocking

(c)

Herring bone strutting (bridging)

Fig. 4.7. Examples of the provision of lateral support.

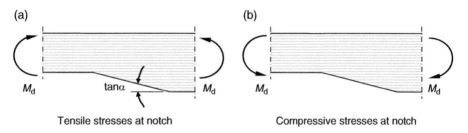

(a)

M_d $\tan\alpha$ M_d

Tensile stresses at notch

(b)

M_d M_d

Compressive stresses at notch

Fig. 4.8. Notch subjected to bending.

For members having a notch and subjected to bending, as stated in EC5, *6.5.1(2)*, the stress concentration effect can be ignored when:

    (i)   bending will result in tensile stresses at the notch and the slope at the notch is less than 1:10, i.e. tan $\alpha \leq 0.1$ (see Figure 4.8a);
   (ii) bending will result in compressive stresses at the notch (see Figure 4.8b).

Where a beam has a rectangular cross-section, its grain runs essentially parallel to the member length and there is a notch at the support, the effect of stress concentrations has to be taken into account and this is considered in 4.5.2.1.

See Examples 4.8.2 and 4.8.3.

4.5.2   Shear

When a beam is loaded laterally and subjected to bending, shear stresses will also arise. In accordance with elastic bending theory, shear stresses will be generated parallel to the longitudinal axis of the beam and, to achieve equilibrium, equal

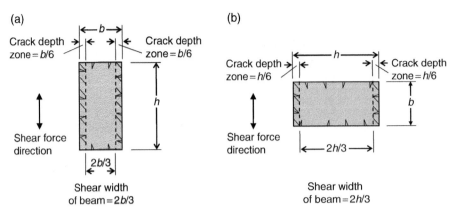

Fig. 4.9. Shear width of a solid (or glued-laminated) timber beam when subjected to a shear force in the direction shown.

value shear stresses will be generated in the beam perpendicular to the longitudinal axis as shown in Figure 4.9a.

The value of the shear stress at any level in the cross-section of a beam, as derived from elastic theory, is

$$\tau = \frac{VS}{Ib} \tag{4.15}$$

where τ is the shear stress at the required level, V is the shear force at the position being considered, S is the first moment of the area above the shear stress level about the neutral axis, I is the second moment of area of the cross-section about the neutral axis, and b is the width of the cross-section at the shear stress level.

When dealing with solid timber or glued-laminated timber, cracks may occur across the section reducing the shear resistance of the member and this is taken into account in EC5 by the use of an effective width for the section. The effective width, b_{ef}, is obtained by multiplying the member width, b, by a modification factor, k_{cr}, as follows:

$$b_{ef} = k_{cr}b \qquad\qquad \text{(EC5, equation (6.13a))} \quad (4.15a)$$

The derivation of *equation (6.13a)* is based on the premise that cracks having a depth up to 1/6th of the width of the solid timber or glued-laminated section may occur on each external face, effectively reducing the width that can withstand shear to the section, as shown on Figure 4.9. In Figure 4.9a the shear force acts across the beam width b and the effective shear width able to resist shear stresses will be the uncracked width $k_{cr}b=2b/3$, with the beam depth h taken to be fully effective in shear. In Figure 4.9b the shear force acts across the beam width h and the effective shear width able to resist shear stresses will be the uncracked width $k_{cr}h=2\,h/3$ and the beam depth b is taken to be fully effective in shear.

The introduction of the modification factor followed from investigations of failures of timber structures in Germany caused by severe snow falls in 2006 which concluded that the incidence of timber splits in the external surface of members had contributed to shear failure. This was found to occur in solid timber and glued-laminated timber

Table 4.5 The value of modification factor k_{cr} (Based on *Table NA.4* in the UKNA to EC5)

Material	Value for k_{cr}
Solid timber	0.67
Glued-laminated timber	0.67
Laminated veneer lumber	1.0
Wood-based panels	1.0

sections but was not a problem in wood-based panels or LVL sections. It is to be noted the finding made no reference to the service class the beam sections had been designed to function within and is applicable to any service class condition.

The use of the factor in design is a matter where national choice is permitted and not all countries have agreed to its introduction in their National Annex. For UK design the modification factor has been accepted and the recommended values are given in *Table NA.4* in the UK NA and summarised in Table 4.5. For wood-based panels and LVL, as stated above, cracking across the section is not a problem and for these materials the value of k_{cr} is 1.0.

At any position along the beam the shear stress at the top and bottom faces of the cross-section will be zero and the maximum shear stress will arise at the neutral axis position. For a rectangular section of width b and depth h, as shown on Figure 4.10(a), the maximum shear stress will occur at mid-depth and will be:

$$\tau = \frac{3V}{2b_{ef}h} \tag{4.16}$$

The limit state for shear design is the ULS and the requirement in EC5 is that for shear with a stress component parallel to the grain or where both shear stress components are perpendicular to the grain (e.g. rolling shear), as shown in Figure 4.10b,

$$\tau_d \le f_{v,d} \qquad\qquad (EC5,\ equation\ (6.13)) \quad (4.17)$$

where $f_{v,d}$ is the design shear strength for the condition being investigated, i.e. Figures 4.10a or 4.10b.

The introduction of k_{cr} will reduce the shear capacity of solid timber and glulam beams. However, as the normal design shear stress condition using these materials will be that shown in Figure 4.10a, this reduction will generally be offset, particularly in beams of medium to low strength class timber, by the increased shear strengths in BS EN 338:2009 over the values given in the 2003 revision of this standard.

4.5.2.1   *Shear stress with a stress component parallel to the grain*

For a rectangular cross-section beam of breadth b and depth h with a shear stress component parallel to the grain, as shown in Figure 4.10a, the validation requirement for the design shear stress, τ_d, will be as follows:

(a) For a beam without a notch (as shown in Figure 4.11a):

$$\tau = \frac{3V}{2b_{ef}h}$$ (4.16)

$$\tau_d \leq f_{v,d}$$ (EC5, *equation (6.13)*)   (4.17)

(b) For a beam notched on the opposite side to the support (as shown in Figure 4.11b):

$$\tau = \frac{3V}{2b_{ef}h_{ef}}$$ (4.16b)

$$\tau_d = k_v f_{v,d}$$ (EC5, *equation (6.60)*)   (4.17a)

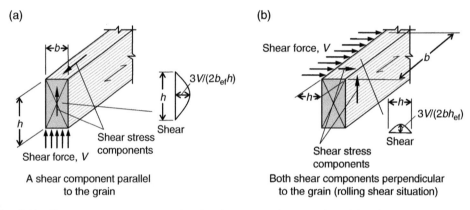

(a)

$3V/(2b_{ef}h)$

Shear

Shear stress
components

Shear force, V

A shear component parallel
to the grain

(b)

Shear force, V

$3V/(2bh_{ef})$

Shear

Shear stress
components

Both shear components perpendicular
to the grain (rolling shear situation)

Fig. 4.10. Shear stress components plus the shear stress in a member: (a) a shear component parallel to the grain; (b) both components perpendicular to the grain (rolling shear).

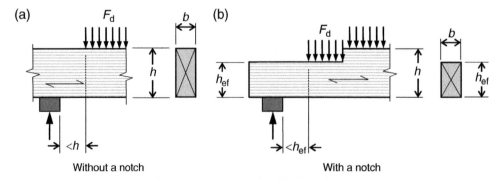

(a) F_d b

h

$<h$

Without a notch

(b) F_d b

h_{ef}

h

h_{ef}

$<h_{ef}$

With a notch

Fig. 4.11. Rectangular beams without a notch at a support (a) and with a notch on the opposite side to the support (b).

(c) For a beam with a notch on the same side as the support (as shown in Figure 4.12):

$$\tau = \frac{3V}{2b_{ef}h_{ef}}$$

(4.16a)

$$\tau_d = k_v f_{v,d}$$

(EC5, *equation (6.60)*)   (4.17a)

In the above cases,

- V_d is the design shear force acting on the beam. In determining the design shear force, the contribution made by any loading acting on the top face of the beam without a notch and within a distance of h from the edge of the support may be ignored due to the effect of the bearing stress (Figure 4.11a). This also applies to beams with a notch on the opposite side to the support where the contribution within a distance of h_{ef} from the edge may be ignored (Figure 4.11b). No reduction is permitted when the notch is on the same side as the support (Figure 4.12).
- $f_{v,d}$ is the design shear strength and is defined as:

$$f_{v,d} = \frac{k_{mod}k_{sys}f_{v,k}}{\gamma_M}$$

(4.18)

where k_{mod}, k_{sys}, γ_M are as described in 4.5.1.1 and $f_{v,k}$ is the characteristic shear strength. The strength is based on the shear strength parallel to the grain (as this is smaller than the shear strength perpendicular to the grain) and for softwood is derived from tests. Values for the shear strength of timber and wood-based structural products are given in Chapter 1.

k_v is a factor that takes account of the effect of a notch in the beam. Stress concentrations are generated by a notch, and from the application of linear elastic fracture mechanics combined with experimental testing, values for the factor have been derived to remove the risk of beam failure due to the effect of crack propagation. Although the equations in EC5 are written in a format that suggests shear is the critical factor, where a notch exists the critical failure criteria will change from shear to tension perpendicular to the grain and this is taken into account by the use of the k_v factor [3]. For beams with a notch, EC5, 6.5.2, requires the following: where the notch is on the opposite side to the

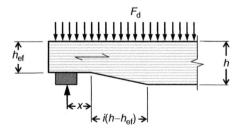

Fig. 4.12. Beams with an end notch.

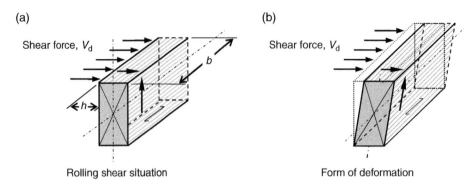

(a)

Shear force, V_d

b

$\leftarrow h \rightarrow$

Rolling shear situation

(b)

Shear force, V_d

i

Form of deformation

Fig. 4.13. Shear stress with both stress components perpendicular to the grain (rolling shear condition).

support (see Figure 4.11b), $k_v = 1.0$; where the notch on the same side as the support (see Figure 4.12), k_v is the lesser of:

$$k_v = \frac{k_n \left(1 + 1.1 i^{1.5} / \sqrt{h}\right)}{\sqrt{h}\left(\sqrt{\alpha(1-\alpha)} + 0.8 \frac{x}{h}\sqrt{(1-\alpha)-\alpha^2}\right)} \quad \text{and } 1 \qquad (4.19)$$

where:
i is the inclination of the notch and is 0 for a right angle notch,
h is the beam depth in mm, x is the distance from the centroid of the support reaction to the corner of the notch in mm,
k_n is a factor that equals 4.5 for LVL, 5.0 for solid timber (6.5 for glued-laminated timber):

$$\alpha = \frac{h_{ef}}{h}$$

4.5.2.2   Shear stress with both stress components perpendicular to the grain (rolling shear)

For a rectangular section with both shear stress components perpendicular to the grain, where the design shear force, V_d, is applied over length b as shown in Figure 4.13, this type of loading condition will in general be applied to wood-based panel materials, for which $k_{cr} = 1$, and the shear stress, τ_d, over a loaded length of b will be:

$$\tau_d = \frac{3V_d}{2bh} \qquad (4.20)$$

$$\tau_d \le f_{v,d} \qquad \text{(EC5, equation (6.13))} \quad (4.17)$$

where $f_{v,d}$ is the design shear strength of the member and is defined as

$$f_{v,d} = \frac{k_{mod} k_{sys} f_{v,k}}{\gamma_M} \qquad (4.21)$$

where k_{mod}, k_{sys}, γ_M are as described in 4.5.1.1 and $f_{v,k}$ is the characteristic shear strength based on the shear strength across the grain (the rolling shear strength).

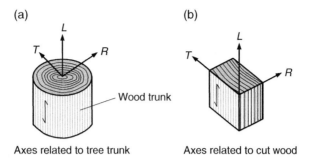

Fig. 4.14. Principal axes of wood.

If timber is being loaded this way (which will not be a common occurrence in structural timber design), because of the assumption of splits in the timber penetrating to 1/6th of its depth (see Figure 4.9), the shear resistance within the crack depth zone will have to be ignored and the design detail adopted will have to be able to transfer the shear force across this zone to the solid core section. No guidance is given in the code on the length of the crack zone to be used (*b*) but it is not unreasonable to consider it will apply near the ends of beams. The k_{cr} factor will be applied to the depth *h* when determining the shear stress across the section and for a design shear force V_d the design shear stress will be $3V_d/(2bh_{ef})$, as shown in Figure 14.10b. As stated in *6.1.7(1)P*, for this condition the rolling shear strength, also referred to as the planar shear strength, is to be taken as approximately equal to twice the tension strength perpendicular to the grain. Values for timber and wood-based structural products are given in Chapter 1.

See Examples 4.8.2 and 4.8.3.

4.5.3   Bearing (compression perpendicular to the grain)

Wood is an anisotropic material (i.e. its physical properties depend upon grain direction), and when used for structural purposes it is assumed to be orthotropic. This means that it is presumed to have directional properties in three mutually perpendicular (orthogonal) axes. The orthogonal axes are aligned with the grain direction (*L*), the radial direction (*R*) and the tangential direction (*T*) as shown in Figures 4.14a and 4.14b.

The properties of the wood along the direction of the *L* axis are referred to as properties parallel to the grain, and because of the direction of its cellular structure, the properties associated with this axis are considerably stronger and stiffer than equivalent properties associated with the *R* and *T* axes. When loaded in the *R* or *T* directions, the cellular structure is compressed in its weakest directions and although the values of the respective properties in each direction are not the same, the differences are small and for practical design purposes these properties are treated as one group. Properties in the *R* and *T* directions are referred to as properties perpendicular to the grain.

The symbols used in EC5 for compressive strength properties parallel and perpendicular to the grain are as follows:

$\sigma_{c,0,k}$:   characteristic compressive strength parallel to the grain.
$\sigma_{c,90,k}$:   characteristic compressive strength perpendicular to the grain.

The subscripts:

c (or t) refers to the type of stress (c – for compressive; t – for tensile).

0 (or 90) refers to the direction of the applied stress relative to the grain direction (0 – parallel to the grain direction; 90 – perpendicular to the grain direction).

k (or d) refers to the nature of the stress (k – characteristic strength; d – design strength).

When timber is compressed perpendicular to the grain the wood fibres, which can in principle be likened to a bundle of narrow thin-walled tubes loaded laterally, withstand increased loading as they are squeezed together and as they start to collapse the rate of load increase reduces. This behaviour continues until the fibres are fully squashed and as the wood is strained beyond this stage the sustained load will continue to rise until eventually failure occurs, usually by shearing across the grain. The strain in the wood can exceed 30% and failure may still not arise. To control deformations, BS EN 408 [4] requires the compression strength of timber perpendicular to the grain to be determined using a 1% strain offset, which equates to a strain of approximately 3% at the ultimate limit state and approximately 1% at the serviceability limit state.

The design approach in EC5 is significantly different to the method given in the 2004 code publication and has been developed from a proposal by Blass and Gorlacher [5] based on theoretical analysis and tests results using solid softwood timber and glued-laminated softwood timber members. In this approach the serviceability limit state is taken to be the design condition with the strain limit set at approximately 1%. Depending on the support condition, the distance between adjacent point loads/supports and the beam depth, the strain in a member may be less than this value and in such a circumstance is able to be increased to approximate this limit by the application of a modification factor, $k_{c,90}$, which is defined below. This will also result in an increase in strength. Examples of bearing effects (failures) are illustrated in Figure 4.15.

For compression perpendicular to the grain the condition to be met is:

$$\sigma_{c,90,d} \le k_{c,90} f_{c,90,d} \qquad\qquad \text{(EC5, equation (6.3))} \quad (4.22)$$

Here $\sigma_{c,90,d}$ is the design compressive stress perpendicular to the grain, and is calculated from:

$$\sigma_{c,90,d} = \frac{F_{c,90,d}}{A_{ef}} \qquad\qquad \text{(EC5, equation (6.4))} \quad (4.23)$$

where:

$F_{c,90,d}$ is the design compressive load on the member, perpendicular to the grain,

A_{ef} is the effective contact area in compression perpendicular to the grain. It is obtained by multiplying the member width, b, by an effective contact length parallel to the grain of the member. The effective contact length is derived by adding 30 mm, but no more than a, ℓ or $\ell_1/2$, to each side of the loaded length ℓ. The symbols referred to are as shown on Figure 4.16, where ℓ is the contact length.

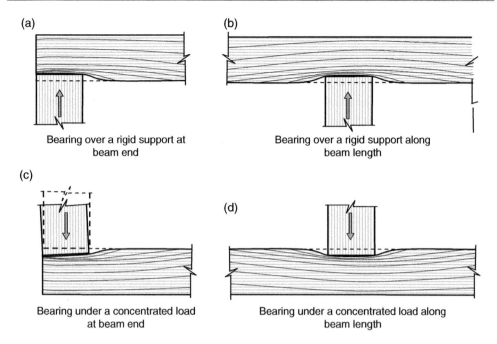

Fig. 4.15. Bearing (compression) failure perpendicular to the grain.

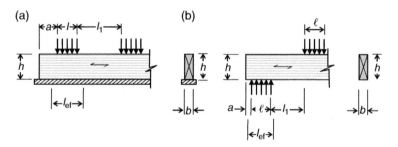

Fig. 4.16. Bearing stress in a beam on (a) a continuous support, (b) a discrete support (based on *Figure (6.2)* in EC5)

The function $f_{c,90,d}$ is the design compressive strength perpendicular to the grain and is defined as:

$$f_{c,90,d} = \frac{k_{mod} \cdot k_{sys} \cdot f_{c,90,k}}{\gamma_M} \qquad (4.24)$$

where k_{mod}, k_{sys}, γ_M are as defined in 4.5.1.1 and:

$f_{c,90,k}$ is the characteristic compressive strength perpendicular to the timber grain and for solid softwood timber and glued-laminated timber is derived in accordance with the requirements of BS EN 408. Because of its grain structure, large strains arise when timber is loaded in this direction and the strength is calculated so that the maximum strain occurring under serviceability limit state loading will not have an adverse effect on the building functionality. As stated previously, based on the test procedure the maximum strain at this condition will be approximately 1%. Values for the compressive strength of timber and wood-based structural products are given in Chapter 1.

$k_{c,90}$ is a factor that can be considered to take account of the strain distribution across the member depth under the loading condition being applied. For loading conditions where the induced strain across the depth is uniform (which effectively equates to the BS EN 408 test set-up) the value will be 1.0, as referred to in *6.1.5(2)*. Where the applied load can spread beyond the loaded length within the depth of the member the strain across the depth will reduce and by increasing the value of $k_{c,90}$ it can be re-set to the maximum limit, effectively allowing greater loads to be supported. Those cases where the factor can be increased are given in *6.1.5(3)* and *(4)*, and are referred to in 4.5.3.1 and 4.5.3.2. The maximum value permitted is associated with glulam beams at supports where $k_{c,90} = 1.75$ and was derived from tests [5].

Values for the compressive strength of timber and wood-based structural products are given in Chapter 1.

4.5.3.1   Bearing stress when the member is resting on supports
Consider a solid softwood or a glued-laminated softwood member supported as shown in Figure 4.16b:

(i)   When $l_1 \geq 2h$ there will be no overlapping of stresses from the bearing and applied loading condition and the strain level within the bearing zone below the bearing load will be lower than the limit associated with the BS EN 408 compression test. Under this condition, for a solid softwood glued member and a glued-laminated softwood member, $k_{c,90}$ can be taken to equal 1.5 and 1.75 respectively without breaching the serviceability limit state criteria. Because of limitations in the test results used to validate/derive the code theory, with glued laminated softwood members $k_{c,90} = 1.75$ can only be used when the bearing length, l, does not exceed 400 mm.

(ii)   When $l_1 < 2h$ stresses from the bearing and the beam loading will overlap with increased strain within the bearing zone and to ensure the strain limit will not be exceeded, $k_{c,90} = 1$ for both materials.

(iii)   Although not clearly stated in EC5 cases (i) and (ii) will also apply when the member carries a uniformly distributed load in addition to the localised point load(s) and this is to be clarified as stated in Appendix C. When the member only supports a uniformly distributed load, the $k_{c,90}$ values will be as stated in (i).

(iv)   When other materials are used e.g. LVL, hardwood members etc, irrespective of the ratio of l_1/h, the value for $k_{c,90}$ will be 1.

Other conditions not referred to in EC5 are given in PD6693-1 [6].

4.5.3.2   Bearing stress when the member is resting on a continuous support
Consider a solid softwood or a glued-laminated softwood member on a continuous support along its length and across its full width as shown in Figure 4.16a:

(i)   When $l_1 \geq 2h$ as in 4.5.3.1(i) there will be no overlapping of stresses within the bearing zone below the bearing load and for this condition the value for $k_{c,90}$ for a solid softwood glued member and a glued-laminated softwood member can be increased to 1.25 and 1.5 respectively without breaching the serviceability limit state criteria. The limits are lower than 4.5.3.1 because of the effect of the continuous bearing support.

(ii) When $l_1 < 2h$ stresses from the bearing and the beam loading will overlap with increased strain within the bearing zone and to ensure the strain limit will not be exceeded, $k_{c,90} = 1$ for both materials.

(iii) When other materials are used, e.g. LVL, hardwood members etc., irrespective of the ratio of l_1/h, the value for $k_{c,90}$ will be 1.

See Examples 4.8.2 and 4.8.3.

4.5.4 Torsion

When a member is subjected to torsion, shear stresses are generated across its section and from elastic theory, for solid sections, it can be shown that when the member is subjected to a torsional moment, T, the maximum torsional stress, τ_{tor}, and the angle of twist per unit length of the member, ϑ / ℓ, will be as follows:

For a circular section:

$$\tau_{tor} = \frac{2T}{\pi \cdot r^3} \tag{4.25}$$

$$\frac{\vartheta}{\ell} = \frac{2T}{\pi \cdot r^4 G} \tag{4.26}$$

where r is the radius of the cross-section of the member and G is the shear modulus of the material.

For a rectangular section:

$$\tau_{tor} = \frac{T}{k_2 hb^2} \tag{4.27}$$

$$\frac{\vartheta}{\ell} = \frac{T}{k_1 hb^3 G} \tag{4.28}$$

where b and h are the cross-sectional sizes of the member and h is the larger dimension. The constants k_1 and k_2 have been determined by Timoshenko and Goodier [7] and are given in Table 4.6 for varying ratios of h/b.

For solid rectangular sections, where $h/b \geq 1.5$, k_1 can be approximated to:

$$k_1 = \frac{1}{3}\left(1 - 0.63\left(\frac{b}{h}\right)\right) \tag{4.29}$$

and when $h/b \geq 2.5$, factor k, which equals k_1/k_2 can be taken to be unity. It is to be noted that at this ratio the shear stress will be overestimated by approximately 3.5% and when the ratio is 5 the overestimate will have reduced to effectively zero.

Table 4.6 Stress (k_2) and rotation (k_1) factors based on equations from Timoshenko and Goodier

h/b	k	k_1	$k_2 = \dfrac{k_1}{k}$
1.0	0.675	0.1406	0.208
1.2	0.759	0.166	0.219
1.3	0.793	0.177	0.223
1.5	0.848	0.196	0.231
1.7	0.888	0.211	0.237
2.0	0.930	0.229	0.246
2.5	0.968	0.249	0.258
3.0	0.985	0.263	0.267
4.0	0.997	0.281	0.282
5.0	0.999	0.291	0.291
6.0	0.9999	0.298	0.298
8.0	1.000	0.307	0.307
10.0	1.000	0.312	0.312
∞	1.000	0.333	0.333

The value of k and k_1 is obtained from the relationships given in equation (4.30), derived from functions in [7]:

$$k = 1 - \frac{8}{\pi^2} \sum_{n=1,3,5...}^{\infty} \frac{1}{n^2 \cosh\left(nh\pi / 2b\right)};$$

$$k_1 = \frac{1}{3}\left(1 - \frac{192}{\pi^5}\cdot\frac{b}{h} \sum_{n=1,3,5...}^{\infty} \frac{1}{n^5}\cdot\tanh\left(\frac{n\pi h}{2b}\right)\right) \qquad (4.30)$$

Equation (4.28) applies where the member is free to rotate under the applied torsional moment, as will be the case with statically determinate structures. When dealing with statically indeterminate structures, however, depending on the configuration of the member within the structure, the detailing of the end joint(s), the effect of semi-rigid behaviour at the connections etc., the rotation of the member may be less than that derived from equation (4.28), and in such cases the torsion stress will be smaller than the stress obtained from equation (4.27).

The limit state for torsion is the ULS and the EC5 design requirement, given in EC5, *6.1.8*, is:

$$\tau_{\text{tor,d}} \leq k_{\text{shape}} f_{\text{v,d}} \qquad \text{(EC5, } equation\ (6.14)) \quad (4.31)$$

Here $\tau_{\text{tor,d}}$ is the design torsional stress derived from equations (4.25) or (4.27) for circular and rectangular sections, respectively, when subjected to a design torsional moment T_{d}.

f_{vd} is the design shear strength as defined in 4.5.2.1, i.e.

$$f_{\text{v,d}} = \frac{k_{\text{mod}} k_{\text{sys}} f_{\text{v,k}}}{\gamma_{\text{M}}} \qquad (4.18)$$

where the factors and functions are as described in 4.5.2.1:

k_{shape} is a factor that takes account of the effect on strength arising from the cross-sectional shape of the member by applying this factor to the design shear strength. The value of the factor is obtained from *equation (6.15)* in EC5 however, from more recent research it has been concluded that the current rules for circular cross-sections can be retained but those for rectangular sections are potentially unsafe and the values will be revised as defined in Appendix C. The new proposal will require the factor to be determined as follows:

$$k_{\text{shape}} = 1.2 \text{ for a circular cross-section;}$$

$$k_{\text{shape}} = \min\left[(1+0.05\,h/b),\ 1.3\right] \text{ for a rectangular cross-section.} \qquad (4.32)$$

h being the larger cross-sectional dimension. b the smaller cross-sectional dimension.

4.5.5   Combined shear and torsion

EC5 only addresses members subjected to shear or to torsion but not to a combination of shear and torsion.

When a member is subjected to combined torsion and shear, the respective torsional and direct shear stresses will combine and the section must be designed for the resulting maximum shear stress condition. Limited research has been carried out on this combined stress condition and using the torsional strength relationship in equation (4.31) it is proposed in STEP 1 [8] that the following failure criterion may be taken to apply:

$$\frac{\tau_{\text{tor,d}}}{k_{\text{shape}}f_{v,d}} + \left(\frac{\tau_{v,d}}{f_{v,d}}\right)^2 \leq 1 \qquad (4.33)$$

where the functions are as described in 4.5.2.1 and 4.5.4.

Equation (4.33) assumes an element of stress redistribution, increasing the combined shear resistance of the section. As an alternative, a more conservative approach can be adopted in which the respective shear stresses are added linearly and the failure criterion will be:

$$\frac{\tau_{\text{tor,d}}}{k_{\text{shape}}f_{v,d}} + \frac{\tau_{v,d}}{f_{v,d}} \leq 1 \qquad (4.34)$$

Torsional shear stresses will not interact with bending, compression, tension or bearing stresses and when acting in members also subjected to such stresses the torsional shear stress condition need only be checked for compliance with equation (4.31).

4.6   DESIGN FOR SERVICEABILITY LIMIT STATES (SLS)

At the SLS, members must be shown to behave satisfactorily when in normal use and the primary requirements to be checked are deflection and vibration behaviour.

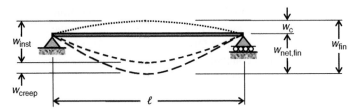

Fig. 4.17. Components of deflection (based on *Figure 7.1* in EC5).

For these states, as stated in 2.2.20.2 and 2.2.25, the partial factors for actions, γ_G and γ_Q, and the partial material factor, γ_M, are to be taken as unity.

4.6.1   Deformation

4.6.1.1   *Deformation due to bending and shear*

EC5 requires that the deformation of a beam must be such that the facility it supports will be able to function as designed, that there will be no adverse visual effects, no structural implications (e.g. there will be no damage to non-structural elements due to deformations) and services must be able to function satisfactorily.

The deformation of a timber or wood-based structural beam is made up of several components and when subjected to the SLS design loading the limiting values of these components are shown in Figure 4.17, where the symbols used in EC5 are as follows:

w_c is the precamber, where used.

w_{inst} is the instantaneous deformation; i.e. the deformation that is permitted immediately under the action of the design load.

w_{creep} is the creep deformation; i.e. the deformation that is permitted to arise with time under the combination of loading causing the creep behaviour.

w_{fin} is the final deformation; i.e. the combination of the instantaneous and the creep deformation.

$w_{net,fin}$ is the net final deformation; i.e. the deformation below the straight line joining the supports.

The net final deformation can be written as

$$w_{net,fin} = w_{inst} + w_{creep} - w_c \qquad (4.35)$$

or

$$w_{net,fin} = w_{fin} - w_c \qquad (4.36)$$

To prevent the occurrence of unacceptable damage arising due to excessive deflections as well as to meet functional and visual requirements, deflections have to be specified for the project and agreed with the client. Recommended ranges of limiting values for the deflection of simply supported beams and cantilevers are given in *7.2(2)* of EC5 and specific values are given in *NA.2.6* of the UKNA to EC5 [9] for $w_{net,fin}$. A summary of these requirements is given in Table 4.7.

Table 4.7 Guidance on limiting values for vertical deflection given in *Table 7.2* of EC5 and *NA.2.5* of the UKNA to EC5

Document	Element	w_{inst}	w_{fin}	$w_{net,fin}$
EC5 *Table 7.2*	Beam on two supports	$\ell/300$ to $\ell/500$	$\ell/150$ to $\ell/300$	$\ell/250$ to $\ell/350$
	Cantilever beam	$\ell/150$ to $\ell/250$	$\ell/75$ to $\ell/150$	$\ell/125$ to $\ell/175$
UKNA to EC5, *NA.2.5* Roof or floor members with a plastered or plasterboard ceiling	Beam between two supports			$\ell/250$
	Cantilever beam			$\ell/125$
Roof or floor members without a plastered or plasterboard ceiling	Beam between two supports			$\ell/150$
	Cantilever beam			$\ell/75$

Note: ℓ is the beam span or the length of the cantilever.

In EC5, the design loading used to determine displacements at the SLS is based on the characteristic combination of actions and is discussed in 2.2.25.2. With the characteristic combination of actions, the design is based on an irreversible SLS, which means that no consequences of actions exceeding the specified SLS requirements can remain when the actions are removed.

The terminology used in EC5 to describe the actual displacement in a structure is the same as that used in Figure 4.17 for the limiting displacement but with the letter w replaced by u. The components of deformation are described in 2.3.2, and adopting the EC5 approach the instantaneous deflection (u_{inst}) and the final deflection (u_{fin}) of a structural member will be obtained as follows:

(i) The instantaneous deflection u_{inst} of a solid member acting alone should be calculated using the characteristic combination of actions (equation (2.24)), the appropriate mean value modulus of elasticity parallel to the grain, $E_{0,mean}$ and/or the mean value shear modulus, G_{mean}. The relationship between G_{mean} and $E_{0,mean}$ is taken to be $G_{mean} = E_{0,mean}/16$ for timber and values of these properties for timber and other wood-based structural products are given in Chapter 1.

(ii) The final deformation u_{fin} is obtained by combining the instantaneous deflection derived from (i) and the creep deflection and the method used will be in accordance with the requirements of 2.3.4.1(b) where the structure has a linear elastic behaviour and consists of members, components and connections having the same creep behaviour or 2.3.4.1(c) where this is not the case. For the case where 2.3.4.1(b) applies, and assuming that a pre-camber is not used, as stated in 2.3.2, the final displacement will be as follows.

For permanent actions, G:

$$u_{fin,G} = u_{inst,G} + u_{creep,G} = u_{inst,G}\left(1 + k_{def}\right) \qquad (4.37)$$

For the leading variable action Q_1:

$$u_{\text{fin},Q,1} = u_{\text{inst},Q,1} + u_{\text{creep},Q,1} = u_{\text{inst},Q,1}\left(1 + \psi_2 k_{\text{def}}\right) \tag{4.38}$$

For the accompanying variable actions, Q_i, where $i > 1$:

$$u_{\text{fin},Q,i} = u_{\text{inst},Q,i} + u_{\text{creep},Q,i} = u_{\text{inst},Q,i}\left(\psi_{0,i} + \psi_{2,i} k_{\text{def}}\right) \tag{4.39}$$

And the final condition will be:

$$u_{\text{fin}} = u_{\text{fin},G} + u_{\text{fin},Q,1} + \sum_{i=2}^{n} u_{\text{fin},Q,i} \tag{4.40}$$

where the functions and symbols are as described in 2.2.14, 2.2.24 and 2.3.2.

Creep in wood arises due to the combined effects of load duration, moisture content, temperature and stress level. Provided the temperature does not exceed 50 °C the influence of temperature on creep can be ignored and when the stress level in the wood is at the SLS condition it has also been found that the rate of creep due to this effect will stabilise. For these reasons, the creep deformation caused by moisture content and load duration effects derived using the factors in EC5 is only relevant to the SLS loading condition.

As the duration of the load extends and also as the moisture content rises, the creep deformation of structural elements will increase and from the results of creep tests, values of a deformation factor, k_{def} (which is a factor that is used in the determination of creep deformation), have been derived for timber and wood-based materials under service class 1, 2 and 3 conditions. Values of the factor are given for timber and some wood-based products in Table 2.10 and the methodology used for the calculation of the final deformation of a structure (or a structural element) is explained in 2.3.2.

When a member is subjected to bending by shear forces, in addition to deformation due to the effect of the bending moment it will also deform due to the effect of the shear forces, and the significance of the shear deformation will primarily be a function of the ratio of the modulus of elasticity $E_{0,\text{mean}}$ of the member to its shear modulus G_{mean}. Consider, for example, a simply supported rectangular beam of depth h and design span ℓ carrying a point load at mid-span. The ratio of the instantaneous deflection at mid-span caused by the shear forces, $u_{\text{inst},v}$, to the instantaneous deflection at mid-span caused by the bending moment, $u_{\text{inst},m}$, will be:

$$\frac{u_{\text{inst},v}}{u_{\text{inst},m}} = 1.2 \frac{E_{0,\text{mean}}}{G_{\text{mean}}}\left(\frac{h}{\ell}\right)^2 \tag{4.41}$$

For structural steel, the ratio $E_{0,\text{mean}}/G_{\text{mean}}$ is approximately 2 and consequently in steel design when using normal sections, the shear deformation effect is generally ignored. With timber, however, $E_{0,\text{mean}}/G_{\text{mean}}$ is approximately 16 and for practical beam design, h/ℓ will range between 0.1 and 0.05 resulting in a shear deformation between 5 and 20% of the flexural value. As this is a significant percentage, the effect of shear deformation must be taken into account when designing timber structures.

Table 4.8 Bending deflection and shear amplification factors for standard load cases on simply supported or cantilevered beams of rectangular cross-section

Load case	Bending deflection (mm)	Shear amplification factor
Uniformly distributed load (udl) equal to a total load Q (kN) along the length of a simply supported beam	At mid-span $= \dfrac{5Q\ell^3}{32E_{0,\text{mean}}b(h)^3}$	$\left(1+0.96\left(\dfrac{E_{0,\text{mean}}}{G_{0,\text{mean}}}\right)\left(\dfrac{h}{\ell}\right)^2\right)$
Point load P (kN) at mid-span of a simply supported beam	At mid-span $= \dfrac{P}{4E_{0,\text{mean}}b}\left(\dfrac{\ell}{h}\right)^3$	$\left(1+1.20\left(\dfrac{E_{0,\text{mean}}}{G_{0,\text{mean}}}\right)\left(\dfrac{h}{\ell}\right)^2\right)$
Point load P (kN) at the end of a cantilever	At the end of the cantilever $= \dfrac{4P}{E_{0,\text{mean}}b}\left(\dfrac{\ell}{h}\right)^3$	$\left(1+0.3\left(\dfrac{E_{0,\text{mean}}}{G_{0,\text{mean}}}\right)\left(\dfrac{h}{\ell}\right)^2\right)$
Point load P (kN) at the quarter and three quarter points of a simply supported beam	At mid-span $= \dfrac{11P}{32E_{0,\text{mean}}b}\left(\dfrac{\ell}{h}\right)^3$	$\left(1+0.873\left(\dfrac{E_{0,\text{mean}}}{G_{0,\text{mean}}}\right)\left(\dfrac{h}{\ell}\right)^2\right)$
Point load P (kN) at the quarter, mid-span and three quarter points of a simply supported beam	At mid-span $= \dfrac{19P}{32E_{0,\text{mean}}b}\left(\dfrac{\ell}{h}\right)^3$	$\left(1+1.011\left(\dfrac{E_{0,\text{mean}}}{G_{0,\text{mean}}}\right)\left(\dfrac{h}{\ell}\right)^2\right)$

Note: $E_{0,\text{mean}}$ is the mean modulus of the beam material parallel to the grain (in kN/mm ^{2}); $G_{0,\text{mean}}$ is the mean shear modulus (in kN/mm ^{2}); b is the member breadth (in mm); h is the member depth (in mm); ℓ is the design span (in mm).

Shear deformation can be expressed in terms of the flexural deflection multiplied by a shear amplification factor, and the value of the factor associated with a simply supported rectangular beam of width b, depth h and design span ℓ for some standard load cases is given in Table 4.8. The combined shear and flexural deflection in the beam for the selected load case is obtained by multiplying the bending deflection by the accompanying shear amplification factor. Approximate values for shear amplification factors for other beam configurations are also given in *NA.2.7.2* of the UKNA to EC5.

See Examples 4.8.2 and 4.8.3.

4.6.1.2 Deformation due to compression over supports

With the revised approach in EC5 for the design of members subjected to compression perpendicular to the grain, discussed in 4.5.3, the strain in such members at the serviceability limit state condition will be limited to a value that should not result in adverse functionality effects. Consequently there will be no requirement to consider compression deformation effects at supports at these states.

4.6.2 Vibration

The human body is extremely sensitive to vibration, and compliance with the criteria given in EC5, which relates to the design requirements of wood-based structural floors in residential property, will ensure that the vibration behaviour of the structure is kept within an acceptable zone when subjected to the SLS loading condition.

Human sensitivity to vibration is a complex subject and from research into human discomfort it has been concluded that the major factors influencing a person's view on the acceptability, or otherwise, of vibration are as follows:

- Discomfort due to machine-enforced vibrations
- Discomfort due to footstep-enforced vibrations
- Proximity to and transfer of the vibration.

The effect of these factors can be reduced to acceptable levels by design and by appropriate detailing of the structure as well as non-structural elements. In regard to the specific design requirements, design criteria are given in EC5, and in particular, in the UKNA to EC5, and these are discussed in the following sub-sections.

4.6.2.1 Machine-enforced vibrations
Structural vibrations arising from machinery can affect human sensitivity, and acceptable levels of continuous vibration due to vibrating machinery will be obtained from ISO 2631:2 [10]. Where the machine vibration exceeds the acceptable level, the most common way to deal with the problem is to either isolate the machine foundations from the structure or to install anti-vibration mountings between the machine and the structure, detuning the structural response. This will normally involve an analysis of the dynamic behaviour of the structure, and because of the specialist nature of the problem, it is not addressed in EC5.

4.6.2.2 Footstep (footfall) induced vibrations
This is a matter that has been the subject of investigation for some considerable time and over this period the criteria set for the boundaries of the problem have tended to change. The design requirements in EC5 relate solely to residential floors having a fundamental frequency greater than 8 Hz. Floors with a fundamental frequency less than 8 Hz require special investigation, and are not covered in the code.

EC5 requires that the fundamental frequency of the residential floor be greater than 8 Hz and for a rectangular floor with overall dimensions $\ell \times b$, simply supported along four sides and with timber beams spanning in the ℓ direction, the approximate value of the fundamental frequency, f_1, can be calculated from equation (4.42). The relationship is also applicable to timber floors designed to span in one direction having an effective span $= l$ and to simply supported beams. The functions in equation (4.42) are:

$$f_1 = \frac{\pi}{2\ell^2}\sqrt{\frac{(EI)_\ell}{m}} \qquad\qquad (EC5,\ equation\ (7.5)) \quad (4.42)$$

Where:

- ℓ is the design span of the floor beams (in metres).
- $(EI)_\ell$ is the equivalent flexural rigidity of the floor supporting structure about an axis perpendicular to the direction of the beam span, in $N\,m^2/m$. Unless the floor decking is designed to act with the floor beams as a composite structure in the direction of the beam span (e.g. in the case of thin flanged composite beams), $(EI)_\ell$ should only be based on the flexural rigidity of the floor beams. Composite

action between the floor decking and the timber joists can only be assumed to occur where the floor decking is glued to the joists and designed in accordance with *9.1.2*, EC5 and noting that the adhesives must comply with the requirements of *3.6* and *10.3* in EC5.

- *m* is the mass per unit area of the floor, in kg/m², and is based on permanent actions only without including partition loads.

It is common in residential floors for some members to support point loads in addition to the floor loading (e.g. trimmer joists as referred to in Figure 1.19 in Chapter 1) and for such a condition, using the Rayleigh energy approach, an approximate expression for the fundamental frequency, f_1, of such a beam can be developed as follows:

$$f_1 = \frac{\pi}{2L^2} \sqrt{\frac{gEI}{w + \frac{2}{L}\sum_{i=1}^{n} P_i \sin^2\left(\frac{\pi a_i}{L}\right)}} \qquad (4.43)$$

Where the symbols are as defined above and:

- *L* is the design span of the beams (in metres).
- *EI* is the flexural rigidity of the beam about an axis perpendicular to the direction of the beam span, in kN m².
- *w* is the floor loading on the beam in kN/m (from tributary width supported by the beam) based on permanent actions only without including partition loads.
- P_i is the point load in kN at position a_i, (which is the distance of P_i to the nearest support in m), from permanent actions.

For such beams, it has been decided in the UK that vibration behaviour must stay within the behaviour achieved by complying with the deflection limits set in BS 5268-2 [11] and to try and ensure this, the above expression has been adjusted and the design requirement to be met is given in the NA as follows:

$$f_1 = \frac{50\pi}{L^2}\left[(EI)_{joist} \Big/ \left[w_t s_j + \sum_{i \geq 1}^{n}(2P_i/L)\sin^2(\pi a_i/L)\right]\right]^{0.5} \qquad (equation\ NA.\ 1b)$$

$$w_t = \max(w, 0.75) \qquad (equation\ NA.\ 1c)$$

Where the symbols remain as given above unless otherwise noted below:

- *L* is the design span of the beam (in mm).
- *EI* is the flexural rigidity of the beam in Nmm².
- P_i is the point load in N at position a_i from the nearest support, in mm, calculated where appropriate using w_t.
- w_t is the uniformly distributed load calculated using expression (*NA.1c*) in kN/m².
- s_j is the tributary width for the uniformly distributed loading, in m.
- *w* is the weight of the floor, being permanent actions only without including partition loads, in kN/m².

As noted in *NA. 1c*, where the self weight of the floor structure is less than 0.75 kN/m², a value of 0.75 kN/m² must still be used in *equation (NA. 1b)*. This is to ensure the flexural rigidity of the member will be comparable with that obtained from following the design requirements of BS 5268-2.

For residential floors having a fundamental frequency greater than 8 Hz, human sensitivity relates to the effects of vibration amplitude and velocity caused by dynamic footfall forces. To ensure compliance with the SLS criteria set for these issues, design criteria for residential wood-based plate type floors are given in EC5 and the UK requirements are given in the UKNA to EC5. The UKNA guidance is based on the use of joist-type floors, which is the most common type of floor structure used in the United Kingdom, and it is this category of floor that is considered in the following sub-sections. Where the modulus of elasticity is used in equations, unless otherwise stated, E_{mean} will apply. The requirements to be satisfied in EC5 are as follows:

(a) *Low-frequency effects (step frequency effect).*

The enforcing frequency on the floor from this action will be less than 8 Hz and consequently the effect of the step action can be considered to be the same as that caused by a static load. The static load simulating the foot force effect is 1 kN applied at the centre of the floor and the deflection of the floor at this point, a, must be no greater than the limit given in *Table NA.6* in the UKNA to EC5, i.e.

$$a \leq 1.8 \, (\text{mm}) \text{ for floor spans} \leq 4000 \, \text{mm} \tag{4.44}$$

$$a \leq \frac{16500}{\ell^{1.1}} (\text{mm}) \text{ for floor spans} > 4000 \, \text{mm} \tag{4.45}$$

where ℓ is the span of the floor joist in mm.

In *NA.2.7.2* of the UKNA to EC5 the deflection a of the floor under a 1 kN point load at mid-span on a floor having a design span ℓ mm may be evaluated using the following equation:

$$a = \frac{1000 k_{\text{dist}} \, \ell_{\text{eq}}^{3} \, k_{\text{amp}}}{48 \, (EI)_{\text{joist}}} \qquad (\textit{equation NA. 1a}) \quad (4.46)$$

Where:

- k_{dist} gives the proportion of the 1 kN load supported by a single joist. The value to be used is:

$$k_{\text{dist}} = \max \left\{ k_{\text{strut}} \left[0.38 - 0.08 \ln \left[\frac{14 (EI)_{\text{b}}}{s^{4}} \right]; 0.30 \right] \right\} \tag{4.47}$$

where $k_{\text{strut}} = 1$ or, in the case of solid timber joists which have transverse stiffness provided by single or multiple lines of herringbone strutting, or blocking with a depth of at least 75% of the joist depth, in addition to the stiffness provided by the decking or ceiling $= 0.97$, $(EI)_{\text{b}}$ is the flexural rigidity of the floor perpendicular to the joists in $N\,mm^2/m$ (see *Notes relating to* $(EI)_b$ (box below) for the factors to be taken into account when evaluating this rigidity); s is the joist spacing in mm.

- ℓ_{eq} is the equivalent span of the floor joists (mm); for simply supported joists, ℓ_{eq} should be taken to be the span ℓ in mm; for end spans of continuous joists, ℓ_{eq} should be taken to be 0.875ℓ in mm; for internal spans of continuous joists, ℓ_{eq} should be taken to be 0.8ℓ in mm.

- k_{amp} is an amplification factor that takes into account the effect of shear deflections in the case of solid timber and glued thin webbed joists or joint slip in the case of mechanically jointed floor trusses, and, after a review of the factor for simply supported and for continuous span conditions is to be taken as:

  1.05 for solid timber joists,

  1.20 for glued thin-webbed joists,

  1.30 for mechanically-jointed floor trusses,

- EI_{joist} is the bending stiffness of a joist in $N\,mm^2$, calculated using the mean value of the modulus of elasticity of the joist.

Notes relating to $(EI)_b$

1. The flexural stiffness is calculated as $(EI)_b$ using the mean value of the modulus of elasticity of the floor decking and discontinuities at the ends of floor boards or at the edges of floor panels may be ignored.

2. Where plasterboard ceilings are fixed directly to the soffit of the floor joists, the flexural rigidity of the plasterboard can be added. It is to be assumed that $E_{plasterboard} = 2000\,N/mm^2$.

3. Where the floor comprises open web joists fitted with a continuous transverse member secured to all joists within 0.1ℓ from mid-span, $(EI)_b$ may be increased by adding the bending stiffness of the transverse member (in $N\,mm^2$) divided by the span ℓ (in metres). Also, $(EI)_b$ may be increased for open web joists where a transverse bracing member is secured to all joists within a distance of $0.05l$ from each of the one third span points, by adding the bending stiffness of **one** of the transverse members in Nmm^2 divided by the span l in metres.

(b) *High-frequency effects (heel impact effect).*

Under the action of a unit impulse force of $1.0\,Ns$ at the centre of the floor simulating heel contact, the maximum initial value of the vertical floor vibration velocity v (in m/s) must comply with *equation (7.4)* in EC5:

$$v \leq b^{(f_1 \cdot \zeta - 1)}\, m\,/(Ns^2)\qquad\qquad (ECS,\ equation\ (7.4))\quad (4.48)$$

Where:

- v is the unit impulse velocity response in $m/(Ns^2)$ units; i.e. the maximum value of the floor vibration velocity (in m/s) in a vertical direction caused by an impulse of $1.0\,Ns$, simulating the heel impact condition applied at the point on the floor giving maximum displacement.

- b is a constant for the control of unit impulse response and is related to the floor deflection a as shown in EC5, *Figure 7.2*. It can be expressed in equation format as follows,

$$b = 150 - \left(30(a - 0.5)/0.5\right) = 180 - 60a \quad \text{when } a \le 1 \text{ mm} \qquad (4.49)$$

$$b = 120 - \left(40(a - 1)\right) = 160 - 40a \quad \text{when } a > 1 \text{ mm} \qquad (4.50)$$

where a is in mm and is obtained from equation (4.46).
- f_1 is the natural frequency of the floor obtained from equation (4.42) (or (*NA. 1b*) for trimmer joists) (in Hz).
- ζ is the modal damping ratio of the floor and for typical UK floors *NA.2.7* of the UKNA to EC5 states that a value of 0.02 is appropriate.

To derive the actual unit impulse velocity response of the structure, the EC5 requirement is as follows:

For a rectangular floor with overall dimensions $b \times \ell$ and simply supported on four sides, the approximate value for v can be obtained from *equation (7.6)* in EC5 as follows,

$$v = \frac{4(0.4 + 0.6n_{40})}{mb\ell + 200} \qquad (EC5, \ equation \ (7.6)) \quad (4.51)$$

where b is the floor width (in metres), ℓ is the design span of the floor (in metres), m is as defined in equation (4.42) (in kg/m²), and n_{40} is the number of first-order vibration modes with natural frequencies up to 40 Hz.

The value of n_{40} can be calculated from the approximate expression given in *equation (7.7)* of EC5,

$$n_{40} = \left(\left(\left(\frac{40}{f_1} \right)^2 - 1 \right) \left(\frac{b}{\ell} \right)^4 \frac{(EI)_\ell}{(EI)_b} \right)^{0.25} \qquad (EC5, \ equation \ (7.7)) \quad (4.52)$$

where $(EI)_b$ is as defined in equation (4.47), but the units to be used for this equation are $\mathrm{N\,m^2/m}$, and $(EI)_\ell$ is the equivalent plate bending stiffness of the floor in the direction of the span of the joists as defined in equation (4.42) (in $\mathrm{N\,m^2/m}$) and $(EI)_b < (EI)_\ell$.

It is to be noted that when using $\zeta = 0.02$ for the modal damping ratio, the UKNA to EC5 states that the unit impulse velocity will not normally govern the size of the joists used in residential timber floors.

See Example 4.8.4.

4.7   REFERENCES

1 BS EN 1995-1-1:2004+A1:2008. *Eurocode 5: Design of Timber Structures. Part 1-1: General – Common Rules and Rules for Buildings*, British Standard Institution.
2 BS EN 338:2009. *Structural Timber – Strength Classes*, British Standards Institution.
3 Aicher, S., Hofflin, L. 'New design model for round holes in glulam beams', 8th World Conference on Timber Engineering, 2004, Lahti, Finland.
4 BS EN 408:2010. *Timber Structures — Structural Timber and Glued Laminated Timber – Determination of some Physical and Mechanical Properties*, British Standards Institution

5 Blass, H.J, Gorlacher, R. 'Compression perpendicular to the grain', Proceedings of the W.C.T.E. 2004, Volume II, Lahti, Finland, 2004.

6 PD6693-1:2012, Incorporating Corrigendum No1, *PUBLISHED DOCUMENT – Recommendations for the design of timber structures in Eurocode 5: Design of Timber Structures – Part 1-1: General – Common Rules and Rules for Buildings*, British Standards Institution

7 Timoshenko, S.P., Goodier, J.N. *Theory of Elasticity*, 2nd edn. McGraw-Hill Book Company, New York, 1951.

8 Aune, P., Shear and torsion. In: Blass, H.J., Aune, P., Choo, B.S., Gorlacher, R., Griffiths, R., Hilson, B.O., Racher, P., Steck, G. (eds), *Timber Engineering STEP 1*, 1st edn. Centrum Hout, Almere, 1995.

9 NA to BS EN 1995-1-1:2004+A1:2008, Incorporating National Amendment No 2; *UK National Annex to Eurocode 5: Design of Timber Structures. Part 1-1: General – Common Rules and Rules for Buildings*, British Standards Institution.

10 International Organisation for Standardisation ISO 2631-2 *Mechanical Vibration and Shock – Evaluation of Human Exposure to Whole-Body Vibration – Part 2: Vibrations in Buildings*.

11 BS 5268-2-2002. *Structural Use of Timber. Part 2: Code of Practice for Permissible Stress Design, Materials and Workmanship*, British Standards Institution.

12 NA to BS EN 1990:2002+A1:2005 *Incorporating National Amendment No1. UK National Annex for Eurocode 0: Basis of Structural Design*, British Standards Institution.

13 BS EN 12369-1:2001. *Wood-Based Panels – Characteristic Values for Structural Design. Part 1: OSB, Particleboards and Fibreboards*, British Standards Institution.

4.8   EXAMPLES

To be able to validate that the critical design effect of actions is being used, the design effects arising from all possible load combinations have to be investigated as part of the design process. Example 4.8.1 is given to show the load cases that have to be investigated in the case of a simply supported beam subjected to a permanent and a variable action.

In the remaining examples, unless otherwise stated, only the load combination(s) giving the greatest design effects have been considered, and these have been based on the use of the fundamental combination (equation (2.13)) at the ULS and the characteristic combination (equation (2.24)) at the SLS.

Example 4.8.1 A simply supported rectangular beam in an office area is secured to a supporting structure at A and B as shown in Figure E4.8.1, and functions under service class 1 conditions. It is subjected to characteristic permanent loading, $G_k = 0.6 \,\text{kN/m}$, a characteristic medium duration variable load $Q_{k,1} = 1.2 \,\text{kN/M}$ acting downwards and a characteristic instantaneous duration variable load (wind action) $Q_{k,2} = 0.6 \,\text{kN/m}$ acting upwards. The associated strength modification factors for the beam material are $k_{\text{mod,perm}} = 0.60$, $k_{\text{mod,med}} = 0.80$ and $k_{\text{mod,inst}} = 1.1$ respectively.

Determine the design load cases that will produce the design effect for

(a) the equilibrium ultimate limit states (ULS);
(b) for the strength ULS, i.e. bending, shear and bearing;
(c) the design effect for the instantaneous deflection of the beam at the serviceability limit state (SLS).

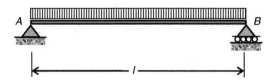

Fig. E4.8.1.

1. Actions on the beam

Permanent loading on the beam, G_k	$G_k = 0.6$ kN/m
Medium duration variable loading acting downwards on the beam, $Q_{k.1}$	$Q_{k.1} = 1.2$ kN/m
Instantaneous duration variable loading acting upwards on the beam, $Q_{k.2}$	$Q_{k.2} = 0.6$ kN/m

(In this example, as an alternative to setting the values of $Q_k = 0$ for unfavourable conditions, $\gamma_{Q.f}$ has been set at 0 for such a condition)

2. Partial safety factors

Table 2.2, ψ values (UKNA to EN 1990:2002+A1:2005 *Table NA.A1.1*):

Factor for the combination value of medium duration variable action $Q_{k.1}$, $\psi_{0.1}$	$\psi_{0.1} = 0.7$
Factor for the combination value of wind action $Q_{k.2}$, $\psi_{0.2}$	$\psi_{0.2} = 0.5$

(a) Partial factors for equilibrium ULS:

Table 2.8, equation EQU (a) (UKNA to EN 1990:2002+A1:2005 *Table NA.A1.2(A)*)

For permanent favourable actions, $\gamma_{G.e}$	$\gamma_{G.e} = 0.9$
For variable unfavourable actions, $\gamma_{Q.e.u}$	$\gamma_{Q.e.u} = 1.5$
For variable favourable actions, $\gamma_{Q.e.f}$	$\gamma_{Q.e.f} = 0$

(b) Partial factors for strength ULS:

Table 2.8, equation STR (c), (i) (UKNA to EN 1990:2002+A1:2005 *Table NA.A1.2(B)*)

For permanent unfavourable actions, $\gamma_{G.u}$	$\gamma_{G.u} = 1.35$
For permanent favourable actions, $\gamma_{G.f}$	$\gamma_{G.f} = 1.00$
Unfavourable variable actions, $\gamma_{Q.u}$	$\gamma_{Q.u} = 1.5$
Favourable variable actions, $\gamma_{Q.f}$	$\gamma_{Q.f} = 0$

(In this example, as stated earlier, $\gamma_{Q.f} = 0$ has been used as an alternative to changing the values of Q_k to $= 0$ for this condition.)

(c) Partial factors for serviceability deflection limit states:

Section 2.2.25

Partial factors for permanent and variable actions = 1

3. Modification factors

Table 2.4 (EC5, *Table 3.1*)

Load duration factor for permanent duration loading at service class 1, $k_{mod.perm}$	$k_{mod.perm} = 0.6$
Load duration factor for medium duration loading at service class 1, $k_{mod.med}$	$k_{mod.med} = 0.8$
Load duration factor for instantaneous duration loading at service class 1, $k_{mod.inst}$	$k_{mod.inst} = 1.1$

4. Critical loading conditions for design effects

(a) Equilibrium states

(i) Table 2.8, equation EQU (a):

Ignoring all load cases where the permanent action is unfavourable, eight alternative load cases have to be considered and the critical load case will be:

Critical load case: Permanent (favourable) G_k + leading variable (unfavourable) $Q_{k,2}$ + accompanying variable (favourable) $Q_{k,1}$:

$\text{Equi}1_d = \gamma_{G.e} \cdot G_k - \gamma_{Q.e.u} \cdot Q_{k.2} + \gamma_{Q.e.f} \cdot \psi_{0,1} \cdot Q_{k.1}$

$\text{Equi}1_d = -0.36$ kN/m i.e. loading is acting upwards on the beam

(ii) Table 2.8, equation STR (c),(i):

Ignoring all load cases where the permanent action is unfavourable, eight alternative load cases have to be considered and the critical load case will be:

Critical load case: Permanent (favourable) G_k + leading variable (unfavourable) $Q_{k,2}$ + accompanying variable (favourable) $Q_{k,1}$:

$\text{Equi}2_d = \gamma_{G.f} \cdot G_k - \gamma_{Q.u} \cdot Q_{k.2} + \gamma_{Q.f} \cdot \psi_{0,1} \cdot Q_{k.1}$

$\text{Equi}2_d = -0.3$ kN/m loading is acting upwards on the beam

The design load case for equilibrium will be $\text{Equi}1_d$.

It must also be noted that the supporting structure to which the beam is anchored must have sufficient static stability to withstand the uplift force generated at each end of the beam.

(b) Strength states

Table 2.8, equation STR (c),(i):

There will be eight load cases where the permanent action is unfavourable and eight where the permanent action is favourable. The design load case will be the load case generating the largest value when divided by the associated k_{mod}, and must include for any reversal of stress condition. The design load case/k_{mod} will be as follows:

(i) For bending compression on the top face, for shear, for bearing:

(Permanent G_k (unfavourable) + leading variable (unfavourable) $Q_{k.1}$ + accompanying variable (favourable) $Q_{k.2}$)/$k_{mod.med}$

NB: content of the last paragraph in 2.2.24:

$$\text{STR}1_d = \frac{\gamma_{G.u} \cdot G_k + \gamma_{Q.u} \cdot Q_{k.1} - \gamma_{Q.f} \cdot \psi_{0.2} \cdot Q_{k.2}}{k_{mod.med}} \qquad \text{STR}1_d = 3.26 \text{ KN/m}$$

The design loading will therefore be:

$\text{DL}1_d = \gamma_{G.u} \cdot G_k + \gamma_{Q.u} \cdot Q_{k.1} - \gamma_{Q.f} \cdot \psi_{0.2} \cdot Q_{k.2} \qquad \text{DL}1_d = 2.61$ kN/m

(ii) For bending compression on the bottom face (e.g. to check lateral torsional instability under stress reversal):

(Permanent G_k (favourable) + leading variable $Q_{k.2}$ + accompanying variable (favourable) $Q_{k.1}$)/$k_{mod,inst}$

$$\text{STR}2_d = \frac{\gamma_{G.f} \cdot G_k - \gamma_{Q.u} \cdot Q_{k.2} + \gamma_{Q.f} \cdot \psi_{0.1} \cdot Q_{k.1}}{k_{mod.inst}} \qquad \text{STR}2_d = -0.27 \text{ kN/m}$$

The design loading will therefore be

$$DL2_d = \gamma_{G.f} \cdot G_k - \gamma_{Q.u} \cdot Q_{k.2} + \gamma_{Q.f} \cdot \psi_{0.1} \cdot Q_{k.1} \qquad DL2_d = -0.3 \text{ kN/m}$$

i.e the loading is acting upwards on the beam

(c) Deflection states – adopting the irreversible limit state condition

Load case for determining the instantaneous downward deflection:

Using the characteristic combination, i.e. equation (2.24) in 2.2.25.2c. (*equation (6.14b) in EC0*), four load cases have to be considered and the design condition will be:

Permanent + variable $Q_{k.1}$:

$$DL3_d = G_k + Q_{k.1} \qquad DL3_d = 1.8 \text{ kN/m}$$

Example 4.8.2 A 63 mm by 225 mm deep sawn timber beam in a domestic residence supports the characteristic loading shown in Figure E4.8.2. The beam has a clear span of 3.10 m, the bearing length has been restricted to 85 mm at each end, is of strength class C24 to BS EN 338:2009, and functions in service class 2 conditions. The beam is laterally restrained against lateral buckling along its length and is notched at its ends by 15 mm, extending 150 mm into the beam from the centre of each support position. The notches are on the same side as the beam supports.

Given that

$G_{k.udl} = 1.0$ kN/m (characteristic uniformly distributed permanent action)
$Q_{k.udl} = 2.5$ kN/m (characteristic uniformly distributed medium-term action)
$G_{k.p} = 1.00$ kN characteristic point load at mid-span

carry out a design check to confirm that

(a) the beam will meet the ULS design requirements of EC5;
(b) the deflection criteria will also be acceptable, assuming that the structure is exposed without applied finishes.

1. Beam geometric properties

Breadth of the beam, b	$b = 63$ mm
Depth of the beam, h	$h = 225$ mm
Clear span of the beam, ℓ_c	$\ell_c = 3100$ mm

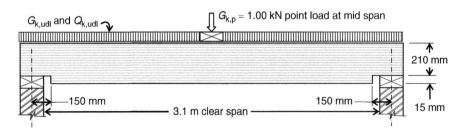

Fig. E4.8.2.

Bearing length of the beam at each end, ℓ_b	$\ell_b = 85\,\text{mm}$	
Design span of the beam, ℓ	$\ell = (\ell_c + \ell_b)$	$\ell = 3.19\,\text{m}$
Notch depth at each end of beam, h_n	$h_n = 15\,\text{mm}$	
Notch inclination, i (including dimension correction factor required by Mathcad)	$i = 0\,\text{mm}^{1/3}$	
Beam depth at each end, h_{ef}	$h_{ef} = h - h_n$	$h_{ef} = 210\,\text{mm}$
Ratio of h_{ef}/h, α	$\alpha = \dfrac{h_{ef}}{h}$	$\alpha = 0.93$
Length of the notch from the centre line of the end support, x	$x = 150\,\text{mm}$	
Section modulus of the beam about the y–y axis, W_y	$W_y = \dfrac{b \cdot h^2}{6}$	$W_y = 5.32 \times 10^5\,\text{mm}^3$

2. Timber properties

Table 1.3, strength class C24 (BS EN 338:2009, *Table 1*):

Characteristic bending strength, $f_{m.k}$	$f_{m.k} = 24\,\text{N/mm}^2$	
Characteristic shear strength, $f_{v.k}$	$f_{v.k} = 2.5\,\text{N/mm}^2$	
Characteristic bearing strength, $f_{c.90.k}$	$f_{c.90.k} = 2.5\,\text{N/mm}^2$	
Fifth-percentile modulus of elasticity parallel to the grain, $E_{0.05}$	$E_{0.05} = 7.4\,\text{kN/mm}^2$	
Mean modulus of elasticity parallel to the grain, $E_{0.mean}$	$E_{0.mean} = 11\,\text{kN/mm}^2$	
Mean shear modulus, $G_{0.mean}$	$G_{0.mean} = 0.69\,\text{k N/mm}^2$	
Mean density of the beam timber, ρ_m	$\rho_m = 420\,\text{kg/m}^3$	

3. Partial safety factors

Table 2.8 (UKNA to BS EN 1990:2002 + A1:2005, *Table NA.A1.2(B)*)) for the ULS

Permanent actions, $\gamma_{G.ULS}$	$\gamma_{G.ULS} = 1.35$
Variable actions, $\gamma_{Q.ULS}$	$\gamma_{Q.ULS} = 1.5$

Table 2.2 (UKNA to BS EN 1990:2002 + A1:2005, *Table NA.A1.1* – Category A)

Factor for the quasi-permanent value of the variable action, ψ_2	$\psi_2 = 0.3$

Table 2.6 (UKNA to EC5, *Table NA.3*)

Material factor for solid timber at the ULS, γ_M	$\gamma_M = 1.3$

4. Actions

(i) ULS

Characteristic self-weight of the beam, $G_{k.swt}$	$G_{k.swt} = b \cdot h \cdot g \cdot \rho_m$	$G_{k.swt} = 0.06\,\text{kN/m}$
(1) Design action from the self-weight of the beam, $F_{d.swt}$	$F_{d.swt} = \gamma_{G.ULS} \cdot G_{k.swt}$	$F_{d.swt} = 0.08\,\text{kN/m}$

Characteristic permanent action due to the point load, $G_{k.p}$ $G_{k.p} = 1.00$ kN

(2) Design permanent action due to the point load, $F_{d.p}$ $F_{d.p} = \gamma_{G.ULS} \cdot G_{k.p}$ $F_{d.p} = 1.35$ kN

Characteristic permanent action due to the udl, $G_{k.udl}$ $G_{k.udl} = 1.0$ kN/m

Characteristic medium-term action due to the udl, $Q_{k.udl}$ $Q_{k.udl} = 2.5$ kN/m

(3) Design action due to the permanent action udl, $F_{d.p.udl}$ $F_{d.p.udl} = \gamma_{G.ULS} \cdot G_{k.udl}$ $F_{d.p.udl} = 1.35$ kN/m

(4) Design action due to the variable action udl, $F_{d.q.udl}$ $F_{d.q.udl} = \gamma_{Q.ULS} \cdot Q_{k.udl}$ $F_{d.q.udl} = 3.75$ kN/m

5. Modification factors

Factor for medium-duration loading and service class 2, $k_{mod.med}$, (Table 2.4 (EC5, *Table 3.1*)) $k_{mod.med} = 0.8$

Size factor for depth greater than 150 mm, k_h (Table 2.11; EC5, *equation (3.1)*) $k_h = 1.0$

Lateral stability of the beam: k_{crit}, (Table 4.3 (EC5, *6.3.3*)) $k_{crit} = 1$

Reduction factor for the notch, k_v, (equation (4.19) (EC5, *Clause 6.5.2*))

    Shear factor for solid timber, k_n (incorporating the dimension factor for Mathcad) $k_n = 5$ mm$^{0.5}$

    Reduction factor for notched beams, k_v

$$k_v = \min\left[1, \frac{k_n \cdot \left(1 + \dfrac{1.1 \cdot i^{1.5}}{\sqrt{h}}\right)}{\sqrt{h}\cdot\left[\sqrt{\alpha\cdot(1-\alpha)} + 0.8\cdot\dfrac{x}{h}\sqrt{\dfrac{1}{\alpha} - \alpha^2}\right]}\right]$$ $k_v = 0.68$

Deformation factor for service class 2, k_{def}, (Table 2.10 (EC5, *Table 3.2*)) $k_{def} = 0.8$

Modification factor for the influence of cracks, k_{cr} (Table NA.4 in UKNA to EC5) $k_{cr} = 0.67$

Compression factor for strain effects, $k_{c.90}$ (where $l_1 > 2h$) (see EC5, *equation 6.3*) $k_{c.90} = 1.5$

Load sharing factor, k_{sys} is not relevant $k_{sys} = 1$

6. Bending strength

Note 1: As explained in 2.2.24, the greatest value of design effect will be associated with the largest value of design action divided by the associated k_{mod} for the characteristic combinations of load cases that can exist. The critical load case will be the combination of design loads $(1)+(2)+(3)+(4)$ and only this case has been considered.

Design bending moment, M_d,

$$M_d = \frac{F_{d.p} \cdot \ell}{4} + \frac{F_{d.swt} \cdot \ell^2}{8} + \frac{F_{d.p.udl} \cdot \ell^2}{8} + \frac{F_{d.p.udl} \cdot \ell^2}{8}$$
$M_d = 7.64$ kN m

Design bending stress, $\sigma_{m.y.d}$,
$$\sigma_{m.y.d} = \frac{M_d}{W_y}$$
$\sigma_{m.y.d} = 14.38$ N/mm²

Design bending strength, $f_{m.y.d}$, (equation (4.5))
$$f_{m.y.d} = \frac{k_{mod.med} \cdot k_{sys} \cdot k_h \cdot f_{m.k}}{\gamma_M}$$
$f_{m.y.d} = 14.77$ N/mm²

Bending strength is satisfactory

7. Shear strength

As with Note 1 above, design value for end shear force, V_d:

$$V_d = \frac{F_{d.p}}{2} + F_{d.p.udl} \cdot \frac{\ell}{2} + F_{d.p.udl} \cdot \frac{\ell}{2} + F_{d.swt} \cdot \frac{\ell}{2}$$
$V_d = 8.92$ kN

Effective shear width of the beam, b_{ef} (equation 4.15(a) (EC5, *equation 6.13a)*)
$b_{ef} = k_{cr} \cdot b$
$b_{ef} = 42.21$ mm

Design shear stress, $\tau_{v.d}$ (equation (4.16) (EC5, *equation (6.60)))*)
$$\tau_{v.d} = \frac{3}{2} \cdot \frac{V_d}{b \cdot h_{ef}}$$
$\tau_{v.d} = 1.51$ N/mm²

Design shear strength, $f_{v.d}$, (equation (4.18))
$$f_{v.d} = \frac{k_{mod.med} \cdot k_{sys} \cdot f_{v.k}}{\gamma_M}$$
$f_{v.d} = 2.46$ N/mm²

Design shear strength, taking the notch effect into account,
$k_v \cdot f_{v.d}$ (equation (4.17a)) (EC5, *equation (6.60)))*
$k_v \cdot f_{v.d} = 1.68$ N/mm²

Shear strength is satisfactory

8. Bearing strength

With reference to Note 1 above, the design bearing force will equal the design shear force in the beam, V_d:

Effective contact area, A_{ef} (equation 4.23 (EC5, *equation 6.4))*)
$A_{ef} = b \cdot (l_b) + 30\,\text{mm})$
$A_{ef} = 7.25 \times 10^3$ mm²

Design bearing stress, $\sigma_{c.90.d}$ (equation 4.23 (EC5, *equation 6.4))*)
$$\sigma_{c.90.d} = \frac{V_d}{A_{ef}}$$
$\sigma_{c,90,d} = 1.23$ N/mm⁻²

Design bearing strength, $f_{c,90,d}$, (equations 4.22, 4.23 (EC5, *equation 6.3 and 6.4)))*
$$f_{c.90.d} = \frac{k_{mod.med} \cdot k_{sys} \cdot f_{c.90.k}}{\gamma_M}$$

$f_{c,90,d} = 1.54$ N/mm⁻²

Factored design bearing strength, $k_{c,90} f_{c,90,d}$, (Equation 4.22)
$k_{c,90} f_{c,90,d} = 2.31$ N/mm⁻²

Bearing strength is OK.

9. Beam deflection

The greatest value of design effect at the SLS will be associated with the largest value of bending moment and shear force obtained from the combinations of load cases that can exist, i.e. for this example: (i) load case (1)+(2)+(3); (ii) load case (1)+(2)+(3)+(4) but with the partial factors set equal to 1. The critical combination will be load case (ii), and it is used to determine the deflection. The effect of the notch at the ends of the beam on deflection is negligible and has been ignored in the calculation.

Deflection due to bending and shear effects:

At the SLS the partial safety factor is 1.

As the member is made from one material, the mean value of stiffness will be used to derive the instantaneous and the creep deflection of the beam.

Instantaneous deflection under permanent actions, $u_{inst.G}$ (Table 4.8)

$$u_{inst.point.G} = \frac{1}{4} \cdot \frac{G_{k.p}}{E_{0.mean} \cdot b \cdot h^3} \cdot \left(\ell^3\right) \cdot \left[1 + 1.2 \cdot \frac{E_{0.mean}}{G_{0.mean}} \cdot \left(\frac{h}{\ell}\right)^2\right]$$

$$u_{inst.udl.G} = \frac{5}{32} \cdot \frac{G_{k.swt} + G_{k.udl}}{E_{0.mean} \cdot b \cdot h^3} \cdot \left(\ell^4\right) \cdot \left[1 + 0.96 \cdot \frac{E_{0.mean}}{G_{0.mean}} \cdot \left(\frac{h}{\ell}\right)^2\right]$$

$$u_{inst.G} = u_{inst.point.G} + u_{inst.udl.G} \qquad\qquad u_{inst.G} = 3.44\,mm$$

Instantaneous deflection due to variable action, $u_{inst.Q}$ (Table 4.8)

$$u_{inst.Q} = \frac{5}{32} \cdot \frac{Q_{k.udl}}{E_{0.mean} \cdot b \cdot h^3} \cdot \left(\ell^4\right) \cdot \left[1 + 0.96 \cdot \frac{E_{0.mean}}{G_{0.mean}} \cdot \left(\frac{h}{\ell}\right)^2\right]$$

$$u_{inst.Q} = 5.48\,mm$$

Combined permanent and variable instantaneous deflection, u_{inst}, $\qquad u_{inst} = u_{inst.G} + u_{inst.Q} \qquad u_{inst} = 8.92\,mm$

EC5 limitation on deflection – use span/300, w_{inst} $\qquad\qquad w_{inst} = \dfrac{\ell}{300} \qquad\qquad w_{inst} = 10.62\,mm$

(Table 4.7 (EC5, *Table 7.2*)) $\qquad\qquad$ i.e. OK

Final deflection due to the permanent actions, $u_{fin.G}$ (equation 4.37 EC5, *equation (2.3)*) $\qquad u_{fin.G} = u_{inst.G} \cdot (1 + k_{def}) \qquad u_{fin.G} = 6.19\,mm$

Final deflection due to the variable and quasi-permanent actions, $u_{fin.Q}$ (equation 4.38 EC5, *equation (2.4)*) $\qquad u_{fin.Q} = u_{inst.Q} \cdot (1 + \psi_2 \cdot k_{def}) \qquad u_{fin.Q} = 6.8\,mm$

Final deflection due to the permanent and quasi-permanent actions, $u_{net.fin}$ $\qquad u_{net.fin} = u_{fin.G} + u_{fin.Q} \qquad u_{net.fin} = 12.99\,mm$

UKNA to EC5 limitation on deflection – use span/150, $w_{net.fin}$ (Table 4.7 (UKNA to EC5, *Table NA.5*)) $\qquad w_{net.fin} = \dfrac{\ell}{150} \qquad\qquad w_{net.fin} = 21.23\,mm$

$\qquad\qquad\qquad\qquad$ i.e. OK

The deflection of the beam is satisfactory

Example 4.8.3 A 75 mm wide by 300 mm deep simply supported LVL (Kerto-S) beam, AB, supports another beam CD on its compression face at mid-span as shown in Figure E4.8.3. Beam CD provides lateral restraint to beam AB at mid-span and applies a characteristic permanent load of 5.25 kN and a characteristic variable medium duration load of 7.35 kN. Beam AB has a clear span of 5.00 m, a bearing length of 100 mm at each end, and functions in service class 1 conditions.
    Carry out a design check to confirm that the beam will satisfy

(a) the ULS design requirements of EC5, and
(b) the instantaneous deflection requirement at the SLS.

Loading from central point load:
$G_{k.p} = 5.25$ kN characteristic permanent action.
$Q_{k.p} = 7.35$ kN characteristic variable action.

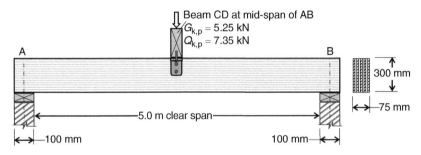

Fig. E4.8.3.

1. Beam AB geometric properties

Breadth of the beam, b	$b = 75$ mm	
Depth of the beam, h	$h = 300$ mm	
Clear span of the beam, ℓ_c	$\ell_c = 5000$ mm	
Bearing length, ℓ_b	$\ell_b = 100$ mm	
Beam effective span, ℓ	$\ell = (\ell_c + \ell_b)$	$\ell = 5100$ mm
Section modulus of the beam about the y–y axis, W_y	$W_y = \dfrac{b \cdot h^2}{6}$	$W_y = 1.12 \times 10^6$ mm³

2. LVL properties

Table 1.16 – Kerto S

Characteristic bending strength, $f_{m.k}$	$f_{m.k} = 44$ N/mm²
Characteristic shear strength, $f_{v.k}$	$f_{v.k} = 4.1$ N/mm²
Characteristic bearing strength, $f_{c.90.k}$	$f_{c.90.k} = 6.0$ N/mm²
Fifth-percentile modulus of elasticity parallel to the grain, $E_{0.05}$	$E_{0.05} = 11.6$ kN/mm²
Fifth-percentile shear modulus, $G_{0.05}$	$G_{0.05} = 0.4$ kN/mm²
Mean modulus of elasticity parallel to the grain, $E_{0.mean}$	$E_{0.mean} = 13.8$ kN/mm²
Mean shear modulus, $G_{0.mean}$	$G_{0.mean} = 0.6$ k N/mm²
Mean density of the beam,	$\rho_m = 510$ kg/m³

3. Partial safety factors

Table 2.8 (UKNA to BS EN 1990:2002+A1:2005, *Table NA.A1.2(B)*) for the ULS:

Permanent actions, γ_G $\gamma_G = 1.35$

Variable actions, γ_Q $\gamma_Q = 1.5$

Table 2.6 (UKNA to EC5, *Table NA.3*):

Material factor for LVL, γ_M $\gamma_M = 1.2$

4. Actions

Self-weight of the beam, $G_{k.selfwt}$	$G_{k.selfwt} = b \cdot h \cdot g \cdot \rho_m$	$G_{k.selfwt} = 0.11 \text{ kN/m}$
Design action from the self-weight of the beam, $F_{d.selfwt}$	$F_{d.selfwt} = \gamma_G \cdot G_{k.selfwt}$	$F_{d.selfwt} = 0.15 \text{ kN/m}$

Characteristic permanent action from the point load, $G_{k.p}$ $G_{k.p} = 5.25 \text{ kN}$

Characteristic variable (imposed) action from the point load, $Q_{k.p}$ $Q_{k.p} = 7.35 \text{ kN}$

Design action from the point load for the critical load case at the ULS, $F_{d.p}$

(Table 2.8, equation c) using the unfavourable condition variable action) $F_{d.p} = \gamma_G \cdot G_{k.p} + \gamma_Q \cdot Q_{k.p}$ $F_{d.p} = 18.11 \text{ kN}$

5. Modification factors

Factor for medium-duration loading and service class 1, $k_{mod.med}$ (Table 2.4 (EC5, *Table 3.1*)) $k_{mod.med} = 0.8$

Size factor for depth = 300 mm, k_h (Table 2.11 (EC5, *3.4*)); $k_h = 1.0$

Lateral stability of the beam, k_{crit}, (4.5.1.2 (EC5, *6.3.3*))

Effective length of the beam – adopt case for the most critical condition – the point load at mid-span, ℓ_{ef} (Table 4.2) $\ell_{ef} = 0.28\ell + 2h$ $\ell_{ef} = 2.03 \text{ m}$

Critical bending stress, $\sigma_{m.crit}$, (equation (4.9b); EC5, *equation (6.31)*)

$$\sigma_{m.crit} = \frac{\pi \cdot b^2 \left[E_{0.05} \cdot G_{0.05} \cdot \left(1 - 0.63 \cdot \dfrac{b}{h}\right) \right]^{0.5}}{h \cdot \ell_{ef}}$$ $\sigma_{m.crit} = 57.43 \text{ N/mm}^2$

Relative slenderness for bending, $\lambda_{rel.m}$ (equation (4.10); EC5, *equation (6.30)*) $\lambda_{rel.m} = \sqrt{\dfrac{f_{m.k}}{\sigma_{m.crit}}}$ $\lambda_{rel.m} = 0.88$

Lateral stability factor, k_{crit}
Table 4.3 (EC5, *equation (6.34)*)

$$k_{crit} = \begin{vmatrix} 1 & if \quad \lambda_{rel.m} \leq 0.75 \\ 1.56 - 0.75 \cdot \lambda_{rel.m} & if \quad 0.75 < \lambda_{rel.m} \leq 1.4 \\ \dfrac{1}{\lambda_{rel.m}^2} & otherwise \end{vmatrix}$$

$$k_{crit} = 0.9$$

Modification factor for the influence of cracks, k_{cr} (Table NA.4 in UKNA to EC5) $k_{cr}=1.0$

Compression factor for stain effects, $k_{c,90}$ (for LVL, the value will be 1.0) (see EC5, *6.1.5(2)*) $k_{c,90}=1.0$

Load sharing factor, k_{sys} is not relevant $k_{sys}=1$

6. Bending strength

The design load case at the ULS will be due to a combination of self-weight of the beam plus the combined permanent and variable point loads at mid-span:

Design bending moment, M_d, $M_d = \dfrac{F_{d.p} \cdot \ell}{4} + \dfrac{F_{d.selfwt} \cdot \ell^2}{8}$ $M_d=23.59\,\text{kN m}$

Design bending stress, $\sigma_{m.y.d}$, $\sigma_{m.y.d} = \dfrac{M_d}{W_y}$ $\sigma_{m.y.d}=20.97\,\text{N/mm}^2$

Design bending strength, $f_{m.y.d}$, $f_{m.y.d} = \dfrac{k_{mod.med} \cdot k_{sys} \cdot k_h \cdot k_{m.k}}{\gamma_M}$

$f_{m.y.d}=29.33\,\text{N/mm}^2$

Design bending strength taking lateral torsional buckling effect into account, $f_{mr.y.d}$, equation (4.13) (EC5, *equation (6.33)*) $f_{mr.y.d}=k_{crit} \cdot f_{m.y.d}$ $f_{mr.y.d}=26.5\,\text{N/mm}^2$

Bending strength is satisfactory

7. Shear strength

The design load case will be due to a combination of self-weight of the beam plus the combined permanent and variable point loads at mid-span:

Design value for end shear force, (ignoring shear reduction due to self-weight at the support), V_d $V_d = \dfrac{F_{d.p}}{2} + F_{d.selfwt} \cdot \dfrac{\ell}{2}$ $V_d=9.44\,\text{kN}$

Effective shear width of the beam, b_{ef} (equation 4.15(a) (EC5, *equation 6.13a)*) $b_{ef}=k_{cr} \cdot b$ $b_{ef}=75\,\text{mm}$

Design shear stress, $\tau_{v.d}$ (equation 4.16 (EC5, *equation 6.60)*) $\tau_{v.d} = \dfrac{3}{2} \cdot \dfrac{V_d}{b_{ef} \cdot h}$ $\tau_{v.d}=0.63\,\text{N/mm}^2$

Design shear strength, $f_{v.d}$ (equation (4.18)) $f_{v.d} = \dfrac{k_{mod.med} \cdot k_{sys} \cdot f_{v.k}}{\gamma_M}$ $f_{v.d}=2.73\,\text{N/mm}^2$

Shear strength is satisfactory

8. Bearing strength

The design load case will be due to a combination of the self-weight of the beam plus the combined permanent and variable point loads at mid-span:

Design value of the end reaction, $Reac_d$ $Reac_d=V_d$ $Reac_d=9.44\,\text{kN}$

Effective contact area, A_{ef} (equation 4.23 (*EC5, equation 6.4)*) $A_{ef}=b \cdot (lb + 30\,\text{mm})$ $A_{ef}=9.75 \times 10^3\,\text{mm}^2$

Design bearing stress, $\sigma_{c.90.d}$ (equation 4.23 (EC5, *equation 6.4)*) $\sigma_{c.90.d} = \dfrac{Reac_d}{b \cdot \ell_b}$ $\sigma_{c.90.d}=0.97\,\text{N/mm}^{-2}$

Design bearing strength, $f_{c.90.d}$ (equation (4.23))

$$f_{c.90.d} = \frac{k_{mod.med} \cdot k_{sys} \cdot f_{c.90.k}}{\gamma_M}$$

$f_{c.90.d} = 4 \text{ N/mm}^{-2}$

Factored design bearing strength, $k_{c.90} \cdot f_{c.0.d}$ (equation (4.22))

$k_{c.90} \cdot f_{c.90.d} = 4 \text{ N/mm}^{-2}$

Bearing strength is OK

9. Beam deflection

Deflection due to bending and shear effects:

At the SLS the partial safety factor is 1.

As the member is made from one material, the mean value of stiffness will be used to derive the instantaneous and the creep deflection of the beam.

The greatest value of instantaneous deflection at the SLS will be associated with the largest value of bending moment at that state and will be due to the characteristic combination of the combined self-weight of the beam plus the combined dead and variable point loads at mid-span.

Instantaneous deflection due to the self-weight of the beam $u_{inst,selfwt}$ (Table 4.8):

$$u_{inst.selfwt} = \frac{5 \cdot G_{k.selfwt} \cdot \ell^4}{32 \cdot E_{0.mean} \cdot b \cdot h^3} \cdot \left[1 + 0.96 \cdot \frac{E_{0.mean}}{G_{0.mean}} \cdot \left(\frac{h}{\ell}\right)^2\right]$$

$u_{inst.selfwt} = 0.46 \text{ mm}$

Instantaneous deflection due to the point loads at mid-span $u_{inst.point loads}$ (Table 4.8):

$$u_{inst.point.loads} = \frac{(G_{k.p} + Q_{k.p}) \cdot \ell^3}{4 \cdot E_{0.mean} \cdot b \cdot h^3} \cdot \left[1 + 1.2 \cdot \frac{E_{0.mean}}{G_{0.mean}} \cdot \left(\frac{h}{\ell}\right)^2\right]$$

$u_{inst.point loads} = 16.38 \text{ mm}$

Instantaneous deflection under combined self-weight and point loads, u_{inst},

$u_{inst} = u_{inst.selfwt} + u_{inst.point loads}$

$u_{inst} = 16.84 \text{ mm}$

EC5 limitation on deflection – use span/300, w_{inst} (Table 4.7 (EC5, *Table 7.2*))

$$w_{inst} = \frac{\ell}{300}$$

$w_{inst} = 17 \text{ mm}$

The deflection is satisfactory

Example 4.8.4 A timber floor in a domestic property has a clear span of 3.70 m between supports and a bearing length of 50 mm at each end, as shown on Figure E4.8.4. The structure comprises 47 mm by 200 mm deep sawn timber joists at 400 mm c/c, strength class C18 to BS EN 338:2009, and functions in service class 1 conditions. The flooring is 18 mm thick OSB/3 boarding to BS EN 12369-1:2001 and is nailed to the joists. Although the floor structure is finished on its underside with plasterboard, no increase in the flexural stiffness of the floor will be allowed for this. The floor width is 4.0 m and the floor mass, based on permanent loading only, is 35 kg/m².

Carry out a design check to confirm that the vibration behaviour of the floor will be acceptable.

1. Joist geometric properties

Breadth of joist, b $b = 47 \text{ mm}$

Depth of joist, h $h = 200 \text{ mm}$

Joist spacing, J_s $J_s = 400 \text{ mm}$

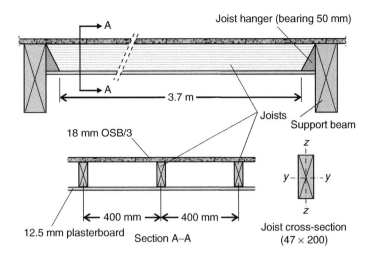

Fig. E4.8.4.

Clear span of joists between inside edge of hanger supports, ℓ_c	$\ell_c = 3700\,\text{mm}$	
Bearing length, ℓ_b	$\ell_b = 50\,\text{mm}$	
Flooring thickness, t	$t = 18\,\text{mm}$	
Width of flooring per metre, b_t	$b_t = 1000\,\text{mm}$	
Effective span of a joist, ℓ	$\ell = (\ell_c + \ell_b)$	$\ell = 3750\,\text{mm}$
Second moment of area of a joist, I_{joist}	$I_{joist} = \dfrac{b \cdot h^3}{12}$	$I_{joist} = 3.13 \times 10^7\,\text{mm}^4$
Second moment of area of flooring/metre width – ignoring discontinuities at the edges of the floor panels, I_f	$I_f = \dfrac{b_t \cdot t^3}{12}$	$I_f = 4.86 \times 10^5\,\text{mm}^4$
Width of floor structure, w_f	$w_f = 4.0\,\text{m}$	

2. Timber and OSB stiffness properties

Table 1.3, strength class C18 (BS EN 338:2009, *Table 1*)

Mean modulus of elasticity of the joists parallel to the grain, $E_{0.\text{mean}}$ $E_{0.\text{mean}} = 9\,\text{kN/mm}^2$

Table 1.17 (BS EN 12639-1:2001):

Mean modulus of elasticity of flooring parallel to the grain, $E_{f.0.\text{mean}}$ $E_{f.0.\text{mean}} = 4.93\,\text{kN/mm}^2$

3. Actions at the SLS

Mass of the floor (kg/m²), m_f $m_f = 35\,\text{kg/m}^2$

4. Modification factors

4.6.2.2 (UKNA to EC5, *NA.2.7*)

Distribution factor, k_{dist}, equation (4.47) (UKNA to EC5, *NA.2.7.2*)

Evaluation of ln function, fa
(Mathcad adjusted to make it dimensionally correct)

$$fa = \left[14 \cdot \left[\frac{Ef_{0.mean} \cdot I_f}{Js^4} \cdot (N^{-1} \cdot mm^2)\right]\right]$$

$$fa = 1.31$$

Evaluation of function value for, k_{strut}

$$k_{strut} = 1$$

Evaluation of inequality function for k_{dist} (equation (4.47))

$$k_{dist} = \max\lfloor k_{strut} \cdot (0.38 - 0.08 \cdot \ln(fa)), 0.3 \rfloor$$
$$k_{dist} = 0.36$$

Evaluation of function value for, k_{amp}

$$k_{amp} = 1.05$$

Equivalent floor span, in mm, ℓ_{eq}

$$\ell_{eq} = \ell \qquad\qquad \ell_{eq} = 3.75 \times 10^3 \text{ mm}$$

The modal damping ratio, ζ (equation (4.48)) (UKNA to EC5, *NA.2.7*))

$$\zeta = 0.02$$

5. Vibration check

(a) Check fundamental frequency of the floor (equation (4.42); EC5, *equation (7.5)*)

Approximate natural frequency of the floor $-f_1$
(ignoring the floor stiffness as the flooring is not glued to the joists)
(equation (4.42); EC5, *equation (7.5)*)

(Mathcad adjusted to make it dimensionally correct)

$$f_1 = \frac{\pi}{2\ell^2} \cdot \sqrt{\frac{E_{0.mean} \cdot \dfrac{I_{joist}}{Js}}{m_f}}$$

$$f_1 = 15.85 \text{ Hz}$$

The fundamental frequency of the floor is acceptable as it is greater than 8 Hz.

(b) Check the static deflection of the floor:

Maximum allowable deflection under the 1 kN point load, a_p
(equation (4.44))
(UKNA to EC5, *Table NA.6*))
(Mathcad adjusted to make it dimensionally correct)

$$a_p = 1.8 \text{ mm}$$

1000 N point load at the centre of the floor, P kN

$$P = 1000 \text{ N}$$

Actual deflection of floor under 1 kN, a (equation (4.46))
(UKNA to EC5, *NA,2,7,2*))

$$a = \frac{P \cdot k_{dist} \cdot \ell_{eq}^3 \cdot k_{amp}}{48 \cdot E_{0.mean} \cdot I_{joist}} \qquad a = 1.47 \text{ mm}$$

The deflection of the floor is less than the allowable value and this requirement is acceptable

(c) Check unit impulse velocity response of the floor:

The constant for the control of unit impulse response, b_v
(equation (4.50)) (UKNA to EC5, *Table NA.6*))
(Mathcad adjusted to make it dimensionally correct)

$$b_v = \left(160 - 40 \cdot \frac{a}{mm}\right) \qquad b_v = 101.36$$

Maximum allowable unit impulse velocity, v_p (equation (4.48); EC5, *equation (7.4)*)

$$v_p = b_v^{f1 \cdot \zeta \cdot Hz^{-1} - 1} \cdot m \cdot (N \cdot s^2)^{-1}$$

$$v_p = 0.04 \text{ m/(N} \cdot \text{s}^2)$$

The number of first-order modes with natural frequencies up to 40 Hz, n_{40} (equation (4.52); EC5, *equation 7.7*))

(Mathcad adjusted to make it dimensionally correct)

$$n_{40} = \left[\left[\left(\frac{40 \cdot s^{-1}}{f_1} \right)^2 - 1 \right] \cdot \left(\frac{w_f}{\ell} \right)^4 \cdot \frac{E_{0.mean} \cdot \left(\frac{I_{joist}}{J_s} \right)}{Ef_{0.mean} \cdot I_f m^{-1}} \right]^{0.25}$$

$$n_{40} = 6.72$$

The approximate actual unit impulse velocity, v, (equation (4.51); EC5, *equation (7.6)*) (Mathcad adjusted to make it dimensionally correct)

$$v = \frac{4 \cdot (0.4 + 0.6 \cdot n_{40})}{(m_f \cdot w_f \cdot \ell \cdot kg^{-1}) + 200} \cdot m \cdot (N \cdot s^2)^{-1}$$

$$v = 0.02 \text{ m/(N s}^2)$$

As the unit impulse velocity response of the floor the allowable value, v_p, is acceptable.

The floor has complied with all of the vibration requirements in EC5 and is acceptable.

Chapter 5

Design of Members and Walls Subjected to Axial or Combined Axial and Flexural Actions

5.1 INTRODUCTION

Timber sections are commonly used in construction as axially loaded members or as members subjected to combined axial and bending actions. Members of a truss, posts or columns, vertical wall studs and bracing elements are typical examples. Some examples are shown in Figure 5.1.

This chapter covers the design of straight solid timber or wood-based structural products of uniform cross-section in which the grain runs essentially parallel to the member lengths and the members are subjected to the effects of axial compression or tension or combined axial and flexural actions. The design of glued-laminated section columns and of built-up columns is covered in Chapters 6 and 8 respectively.

The general information in 4.3 is relevant to the content of this chapter.

5.2 DESIGN CONSIDERATIONS

Axially loaded members have to satisfy the relevant design rules and requirements of EC5 [1], and the limit states associated with the main design effects for members loaded in this manner are given in Table 5.1. The equilibrium states and strength conditions relate to failure situations and are therefore ultimate limit states. The displacement condition relates to normal usage situations, however no guidance is given in EC5 regarding limiting criteria for this state. Where lateral instability of a member can occur, a limitation is set for the maximum deviation from straightness allowed and this is given in *Section 10*, EC5.

For strength-related conditions, the design stress is calculated and compared with the design strength modified where appropriate by strength factors and, to meet the code reliability requirements, when using the partial factor method the design stress must not exceed the design strength.

When members or structures are subjected to combined stresses, e.g. due to the combined effects of axial and bending actions, additional design effects will arise and the design requirements for such conditions are also covered in this chapter.

Structural Timber Design to Eurocode 5, Second Edition. Jack Porteous and Abdy Kermani.
© Jack Porteous and Abdy Kermani 2013. Published 2013 by Blackwell Publishing Ltd.

(a)

(b)

Simple post

Eccentrically loaded column

(c)

(d)

Truss system

Beam and post construction

(e)

(f)

Stud wall – multiple studs to carry higher
applied loads

Tree truss column

Fig. 5.1. Examples of columns, posts and a stud wall.

Table 5.1 Main design requirements for axially loaded members and the associated EC5 limit

Design or displacement effect	EC5 limit states
Retention of static equilibrium (sliding, overturning, uplift)	ULS
Axial stress, including the effect of lateral instability	ULS
Deflection	SLS

5.3   DESIGN OF MEMBERS SUBJECTED TO AXIAL ACTIONS

5.3.1   Members subjected to axial compression

These are members that are subjected to a compressive action acting parallel to the grain and along the centroidal x–x axis of the member, as shown in Figure 5.2. Such members function as columns, posts, stud members in walls or struts in pin jointed trusses.

When a member is subjected to axial compression its failure strength is dependent on several factors:

- strength/stiffness – compressive strength and modulus of elasticity of the timber;
- geometry of the member – cross-sectional sizes and length;
- support condition – the amount of lateral support and fixity at its ends;
- geometric imperfections – deviations from nominal sizes, initial curvature and inclination;
- material variations and imperfections – density, effect of knots, effect of compression wood and moisture content.

The rules in EC5 take these factors into account.

When subjected to an axial load, because of imperfections in the geometry of the member or variations in its properties, or a combination of both, as the slenderness ratio, λ, of the member increases there is a tendency for it to displace laterally and to eventually fail by buckling as shown in Figure 5.3.

The slenderness ratio is defined as the effective length of the member, L_e, divided by its radius of gyration, i,

$$\lambda = \frac{L_e}{i} \tag{5.1}$$

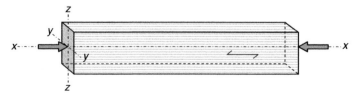

Fig. 5.2. Axial compression.

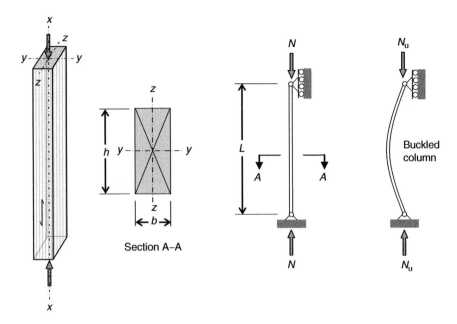

Fig. 5.3. Buckling of a column.

where the radius of gyration about an axis $i = \sqrt{I/A}$, I is the second moment of area about the axis, and A is the cross-sectional area of the member.

For any member there will be a slenderness ratio, λ_y, about the y–y axis and, λ_z, about the z–z axis and when using a rectangular section, as shown in section A–A in Figure 5.3, the respective slenderness ratios are

$$\lambda_y = \frac{L_{e,y}}{i_y} = \frac{L_{e,y}}{h/\sqrt{12}} \quad \text{and} \quad \lambda_z = \frac{L_{e,z}}{i_z} = \frac{L_{e,z}}{b/\sqrt{12}}$$

where $L_{e,y}$ and $L_{e,z}$ are the effective length about the y–y axis and the z–z axis respectively. Buckling will occur about the axis with the highest slenderness ratio.

The effective length L_e (or buckling length) of a compression member is the distance along its length between adjacent points of contra-flexure. These are adjacent points at which the bending moment in the member is zero. Although EC5 gives no information on how to determine the effective length of a compression member, provided the end connections of the member ensure full positional and directional control where required, the guidance given in Table 5.2 can be used. When dealing with small span trusses used in domestic dwellings and assembled in one plane from same thickness material and connected by punched metal plate fasteners or nailed/screwed metal or plywood gusset plates (commonly referred to in the UK as trussed rafters), guidance on the effective length to be used for compression members is provided in PD6693-1 [2].

The content of Table 5.2 covers the cases shown in Figure 5.4, where L_e is the effective length and L is the actual column length. Where full positional and directional control cannot be assured and the stiffness properties of the end connections are known, approximate solutions can be determined using second-order analysis methods.

Table 5.2 Effective length of compression members*

Support condition at the ends of the member	L_e/L
Held effectively in position and direction at both ends	0.7
Held effectively in position at both ends and in direction at one end	0.85
Held effectively in position at both ends but not in direction	1
Held effectively in position and direction at one end and in direction but not position at the other end	1.5
Held effectively in position and direction at one end and completely free at the other end	2

*Based on *Table 21*, BS 5268-2:2002 [3].

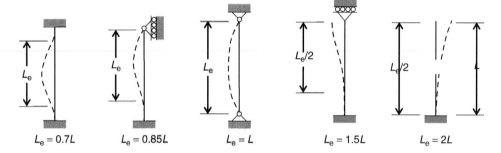

$L_e = 0.7L$ $L_e = 0.85L$ $L_e = L$ $L_e = 1.5L$ $L_e = 2L$

Fig. 5.4. Effective length and end conditions.

For an idealised perfectly straight column of length L having uniform properties and pin jointed at both ends, the theoretical axial load at which buckling will occur about the $y-y$ or the $z-z$ axes within the elastic limit of the column material will be the Euler buckling load for the respective axis. Expressing the Euler buckling loads in terms of the slenderness ratios of the member, they can be written as

$$P_{E,y} = \frac{\pi^2 E_{0.05} A}{\lambda_y^2} \quad \text{and} \quad P_{E,z} = \frac{\pi^2 E_{0.05} A}{\lambda_z^2} \tag{5.2}$$

where $P_{E,y}$ is the Euler buckling load about the $y-y$ axis, $P_{E,z}$ is the Euler buckling load about the $z-z$ axis, $E_{0.05}$ is the characteristic modulus of elasticity of the member, A is the cross-sectional area of the member, λ_y is the slenderness ratio about the $y-y$ axis $=(1.0 \times L)/i_y$, λ_z is the slenderness ratio about the $z-z$ axis $=(1.0 \times L)/i_z$.

Dividing the respective Euler buckling loads by the cross-sectional area of the member, A, the buckling strength of the member about the $y-y$ axis, $\sigma_{E,y}$, and about the $z-z$ axis, $\sigma_{E,z}$, is obtained as follows:

$$\sigma_{E,y} = \frac{\pi^2 E_{0.05}}{\lambda_y^2} \quad \text{and} \quad \sigma_{E,z} = \frac{\pi^2 E_{0.05}}{\lambda_z^2} \tag{5.2a}$$

In EC5 the square root of the ratio of the characteristic compressive strength of the timber parallel to the grain, $f_{c,0,k}$, to its buckling strength is defined as the relative slenderness ratio, λ_{rel}, giving the following relationships for $\lambda_{rel,y}$ and $\lambda_{rel,z}$:

$$\lambda_{\text{rel},y} = \frac{\lambda_y}{\pi}\sqrt{\frac{f_{c,0,k}}{E_{0.05}}} \quad \text{and} \quad \lambda_{\text{rel},z} = \frac{\lambda_z}{\pi}\sqrt{\frac{f_{c,0,k}}{E_{0.05}}} \tag{5.3}$$

where, for the axes as shown in Figure 5.3, $\lambda_{\text{rel},y}$ is the relative slenderness ratio corresponding to bending about the y–y axis (i.e. the member will deflect in the z-direction), and $\lambda_{\text{rel},z}$ is the relative slenderness ratio corresponding to bending about the z–z axis (i.e. the member will deflect in the y-direction).

These relationships are given in *equations (6.21)* and *(6.22)* of *6.3.2* in EC5.

When the member is short and stocky, buckling will not occur and failure will be due to the timber failing under stress and in EC5 this applies when both $\lambda_{\text{rel},y}$ and $\lambda_{\text{rel},z}$ are ≤ 0.3. Inserting the characteristic values for $f_{c,0,k}$ and $E_{0.05}$ given in BS EN 338:2009 [4] for softwood species into equation (5.3), when $\lambda_{\text{rel}} = 0.3$, the maximum slenderness ratio of the member will be between 16.2 and 18.1, the range being slightly greater for hardwood species. For a member having a rectangular cross-section with a least lateral dimension, b, the effective length will be obtained from $L_E = b/\sqrt{12}\ \lambda_{\text{rel}}$ resulting in column lengths of $4.66b$ and $5.23b$ respectively. It is clear from these results that for most practical situations the design condition is likely to be based on a buckling failure rather than a stress failure.

When $\lambda_{\text{rel},y}$ and/or $\lambda_{\text{rel},z}$ exceed 0.3, the effect of member buckling has to be taken into account. From equation (5.3), at $\lambda_{\text{rel}} = 0.3$, the factor of safety between failure by buckling under the Euler buckling load and failure by direct compression is approximately 11. This is based on a theoretically idealised condition assuming no defects or out of alignment imperfections, and pure elastic behaviour applies with no upper limit on the elastic strength of the material.

According to Blass [5], under axial load the stress–strain curve for timber will be as indicated in Figure 5.5 and using these relationships, taking into account the effects of increase in slenderness ratio and member imperfections, the buckling strength of compression members has been modelled. The evaluation was based on a second-order plastic iterative analysis that incorporated the yield behaviour of timber and the consequent change in member stiffness, and the design guidance in EC5 on members subjected to axial compression has been developed from this work.

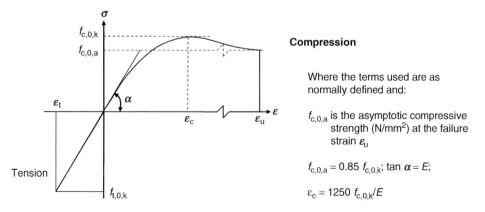

Fig. 5.5. Stress–strain curve; based on the stress–strain relationship developed by Blass.

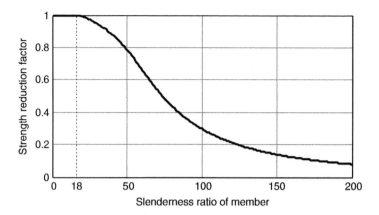

Fig. 5.6. Typical graph showing the strength reduction in a solid timber compression member as the slenderness ratio is increased.

From the analyses, for different strengths of timber a plot of the characteristic buckling strength against slenderness ratio was obtained, and a typical plot showing the strength reduction as the slenderness ratio increased is shown in Figure 5.6.

By covering the range of timber strength class properties and associated geometric imperfections, the buckling strengths at varying slenderness ratios were obtained for timber (and glued-laminated members). From these results, an approximate curve has been derived in terms of $\lambda_{\text{rel},y}$ (or $\lambda_{\text{rel},z}$) from which a buckling strength reduction factor, $k_{c,y}$ (or $k_{c,z}$), called the instability factor in EC5, is obtained for solid timber, LVL (and glulam) members subjected to axial compression. The relevant equations are as follows,

$$k_{c,y} = \frac{1}{k_y + \sqrt{k_y^2 - \lambda_{\text{rel},y}^2}}$$
(EC5, equation (6.25))   (5.4a)

$$k_{c,z} = \frac{1}{k_z + \sqrt{k_z^2 - \lambda_{\text{rel},z}^2}}$$
(EC5, equation (6.26))   (5.4b)

where

$$k_y = 0.5\left(1 + \beta_c\left(\lambda_{\text{rel},y} - 0.3\right) + \lambda_{\text{rel},y}^2\right)$$
(EC5, equation (6.27))   (5.5a)

$$k_z = 0.5\left(1 + \beta_c\left(\lambda_{\text{rel},z} - 0.3\right) + \lambda_{\text{rel},z}^2\right)$$
(EC5, equation (6.28))   (5.5b)

and β_c has been derived from the analyses and applies to solid rectangular members compliant with the straightness limits given in *Section 10* of EC5 (i.e. the deviation from straightness measured midway between supports should be less than or equal to member length/300 for timber members and less than or equal to member length/500 for LVL (and glulam) members) as well as the tolerance limits for member sizes. For these cases, *equation (6.29)* in EC5 states

$$\beta_c = 0.2 \text{ for solid timber}$$
$$= 0.1 \text{ for glued-laminated timber and LVL}$$
(5.6)

From the above, the strength of the member at the design condition will be as follows,

Buckling strength about the y–y axis $= k_{c,y} f_{c,0,d}$ (5.7a)
Buckling strength about the z–z axis $= k_{c,z} f_{c,0,d}$ (5.7b)

where $k_{c,y}$ and $k_{c,z}$ are as defined in equation (5.4) and $f_{c,0,d}$ is the design compressive strength of the member parallel to the grain.

An alternative to evaluating $k_{c,y}$ and $k_{c,z}$ using equations (5.4)–(5.6) is to use tables set out in terms of the strength class of the timber and the slenderness ratio of the member. Slenderness ratio, λ, has been used rather than relative slenderness ratio, λ_{rel}, as it is the ratio UK designers are currently more familiar with and it also has a more direct relevance to the design process. The value of the instability factor for solid timber softwood members of strength classes C14–C50 and solid timber hardwood members classes D18 to D70 in accordance with BS EN 338:2009 for a range of slenderness ratios up to 180 is given in Table 5.3.

In Table 5.3 the lowest value of slenderness ratio given for each strength class is the value below which buckling will not arise and at which $k_{c,y}$ and $k_{c,z}$ will always be taken as unity.

Using equations (5.3) to evaluate the relative slenderness ratio, λ_{rel}, of the member about the y–y and z–z axes and taking account of the above, the EC5 procedure for the design of axially loaded elements is as follows.

(a) *Where both $\lambda_{rel,y}$ and $\lambda_{rel,z}$ are ≤ 0.3.*
Under this condition, the member will not buckle and the design requirement will be that the design stress does not exceed the design strength, i.e.

$$\sigma_{c,0,d} \leq f_{c,0,d}$$ (EC5, *equation (62)*)   (5.8)

Where

- $\sigma_{c,0,d}$ is the design compressive stress parallel to the grain and:

$$\sigma_{c,0,d} = \frac{N_d}{A}$$ (5.9)

  where N_d is the design axial load and A is the cross-sectional area of the member.

- $f_{c,0,d}$ is the design compressive strength parallel to the grain and:

$$f_{c,0,d} = \frac{k_{mod} \cdot k_{sys} \cdot f_{c,0,k}}{\gamma_M}$$ (5.10)

  where k_{mod} is the modification factor for load duration and service classes as given in Table 2.4; k_{sys} is a strength factor for load-sharing systems. When dealing with single column members this factor is not relevant and is taken to be unity; γ_M is the partial coefficient for material properties, given in Table 2.6; $f_{c,0,k}$ is the characteristic compressive strength of the timber or wood-based product parallel to the grain. Strength information for timber and the commonly used wood-based structural products is given in Chapter 1.

Table 5.3 Value of $k_{c,y}$ or $k_{c,z}$ based on equations (6.25)–(6.29) inclusive in EC5 with $\beta_c = 0.2$ – for softwoods

Strength class C14			Strength class C16			Strength class C18			Strength class C22		
$f_{c,0,k}$ 16 N/mm²			$f_{c,0,k}$ 17 N/mm²			$f_{c,0,k}$ 18 N/mm²			$f_{c,0,k}$ 20 N/mm²		
$E_{0.05}$ 4700 N/mm²			$E_{0.05}$ 5400 N/mm²			$E_{0.05}$ 6000 N/mm²			$E_{0.05}$ 6700 N/mm²		
λ	λ_{rel}	$k_{c,y}(k_{c,z})$	λ	λ_{rel}	$k_{c,y}(k_{c,z})$	λ	λ_{rel}	$k_{c,y}(k_{c,z})$	λ	λ_{rel}	$k_{c,y}(k_{c,z})$
16.153	0.300	1.000	16.797	0.300	1.000	17.207	0.300	1.000	17.250	0.300	1.000
20	0.371	0.984	20	0.357	0.987	20	0.349	0.989	20	0.348	0.989
25	0.464	0.960	25	0.446	0.965	25	0.436	0.968	25	0.435	0.968
30	0.557	0.932	30	0.536	0.939	30	0.523	0.943	30	0.522	0.944
35	0.650	0.899	35	0.625	0.908	35	0.610	0.914	35	0.609	0.915
40	0.743	0.856	40	0.714	0.870	40	0.697	0.878	40	0.696	0.879
45	0.836	0.804	45	0.804	0.823	45	0.785	0.834	45	0.783	0.835
50	0.929	0.741	50	0.893	0.766	50	0.872	0.781	50	0.870	0.782
55	1.021	0.673	55	0.982	0.702	55	0.959	0.720	55	0.957	0.721
60	1.114	0.605	60	1.072	0.636	60	1.046	0.655	60	1.043	0.657
65	1.207	0.540	65	1.161	0.572	65	1.133	0.591	65	1.130	0.593
70	1.300	0.482	70	1.250	0.512	70	1.220	0.531	70	1.217	0.533
75	1.393	0.430	75	1.339	0.459	75	1.308	0.477	75	1.304	0.479
80	1.484	0.387	80	1.429	0.412	80	1.395	0.429	80	1.391	0.431
85	1.579	0.347	85	1.518	0.371	85	1.482	0.387	85	1.478	0.389
90	1.671	0.313	90	1.607	0.336	90	1.569	0.351	90	1.565	0.352
95	1.764	0.284	95	1.697	0.305	95	1.656	0.318	95	1.652	0.320
100	1.857	0.258	100	1.786	0.278	100	1.743	0.290	100	1.739	0.291
105	1.950	0.236	105	1.875	0.254	105	1.831	0.265	105	1.826	0.267
110	2.043	0.217	110	1.965	0.233	110	1.918	0.244	110	1.913	0.245
115	2.136	0.199	115	2.054	0.214	115	2.005	0.224	115	2.000	0.225
120	2.229	0.184	120	2.143	0.198	120	2.092	0.207	120	2.087	0.208
125	2.322	0.170	125	2.232	0.183	125	2.179	0.192	125	2.174	0.193
130	2.414	0.158	130	2.322	0.170	130	2.266	0.178	130	2.261	0.179
135	2.507	0.147	135	2.411	0.158	135	2.354	0.166	135	2.348	0.167
140	2.600	0.137	140	2.500	0.148	140	2.441	0.155	140	2.435	0.155
145	2.693	0.128	145	2.590	0.138	145	2.528	0.145	145	2.522	0.145
150	2.786	0.120	150	2.679	0.129	150	2.615	0.136	150	2.609	0.136
155	2.879	0.113	155	2.768	0.122	155	2.702	0.127	155	2.696	0.128
160	2.972	0.106	160	2.858	0.114	160	2.789	0.120	160	2.783	0.120
165	3.064	0.100	165	2.947	0.108	165	2.877	0.113	165	2.870	0.113
170	3.157	0.094	170	3.036	0.102	170	2.964	0.107	170	2.956	0.107
175	3.250	0.089	175	3.125	0.096	175	3.051	0.101	175	3.043	0.101
180	3.343	0.084	180	3.215	0.091	180	3.138	0.095	180	3.130	0.096

Note: The relative slenderness ratio, λ_{rel}, has been included for information.

(b) *Where $\lambda_{rel,y}$ or $\lambda_{rel,z}$ (or both) > 0.3.*

Where $\lambda_{rel,y}$ or $\lambda_{rel,z}$ (or both) exceed 0.3, the member can buckle about the associated *y–y* and *z–z* axis, respectively, and for such a condition it must be demonstrated that when taking into account the effect of the related instability factors, the compressive stress in the member is less than the compressive strength of the member. This requires:

Table 5.3 (*continued*)

Strength class C24			Strength class C27			Strength class C30			Strength class C35		
$f_{c,0,k}$ 21 N/mm²			$f_{c,0,k}$ 22 N/mm²			$f_{c,0,k}$ 23 N/mm²			$f_{c,0,k}$ 25 N/mm²		
$E_{0.05}$ 7400 N/mm²			$E_{0.05}$ 7700 N/mm²			$E_{0.05}$ 8000 N/mm²			$E_{0.05}$ 8700 N/mm²		
λ	λ_{rel}	$k_{c,y}(k_{c,z})$	λ	λ_{rel}	$k_{c,y}(k_{c,z})$	λ	λ_{rel}	$k_{c,y}(k_{c,z})$	λ	λ_{rel}	$k_{c,y}(k_{c,z})$
17.692	0.300	1.000	17.632	0.300	1.000	17.577	0.300	1.000	17.582	0.300	1.000
20	0.339	0.991	20	0.340	0.991	20	0.341	0.991	20	0.341	0.991
25	0.424	0.971	25	0.425	0.970	25	0.427	0.970	25	0.427	0.970
30	0.509	0.948	30	0.510	0.947	30	0.512	0.947	30	0.512	0.947
35	0.593	0.920	35	0.596	0.919	35	0.597	0.919	35	0.597	0.919
40	0.678	0.887	40	0.681	0.886	40	0.683	0.885	40	0.683	0.885
45	0.763	0.846	45	0.766	0.844	45	0.768	0.843	45	0.768	0.843
50	0.848	0.796	50	0.851	0.794	50	0.853	0.793	50	0.853	0.793
55	0.933	0.739	55	0.936	0.736	55	0.939	0.734	55	0.938	0.734
60	1.017	0.676	60	1.021	0.674	60	1.024	0.671	60	1.024	0.672
65	1.102	0.614	65	1.106	0.611	65	1.109	0.608	65	1.109	0.608
70	1.187	0.554	70	1.191	0.551	70	1.195	0.548	70	1.194	0.549
75	1.272	0.499	75	1.276	0.496	75	1.280	0.494	75	1.280	0.494
80	1.357	0.450	80	1.361	0.447	80	1.365	0.445	80	1.365	0.445
85	1.441	0.406	85	1.446	0.404	85	1.451	0.402	85	1.450	0.402
90	1.526	0.368	90	1.531	0.366	90	1.536	0.364	90	1.536	0.364
95	1.611	0.335	95	1.616	0.333	95	1.621	0.331	95	1.621	0.331
100	1.696	0.305	100	1.701	0.303	100	1.707	0.302	100	1.706	0.302
105	1.780	0.279	105	1.787	0.278	105	1.792	0.276	105	1.792	0.276
110	1.865	0.256	110	1.872	0.255	110	1.877	0.253	110	1.877	0.253
115	1.950	0.236	115	1.957	0.235	115	1.963	0.233	115	1.962	0.233
120	2.035	0.218	120	2.042	0.217	120	2.048	0.216	120	2.048	0.216
125	2.120	0.202	125	2.127	0.201	125	2.133	0.200	125	2.133	0.200
130	2.204	0.188	130	2.212	0.186	130	2.219	0.185	130	2.218	0.185
135	2.289	0.175	135	2.297	0.174	135	2.304	0.173	135	2.304	0.173
140	2.374	0.163	140	2.382	0.162	140	2.389	0.161	140	2.389	0.161
145	2.459	0.153	145	2.467	0.152	145	2.475	0.151	145	2.474	0.151
150	2.544	0.143	150	2.552	0.142	150	2.560	0.141	150	2.559	0.141
155	2.628	0.134	155	2.637	0.133	155	2.645	0.133	155	2.645	0.133
160	2.713	0.126	160	2.722	0.126	160	2.731	0.125	160	2.730	0.125
165	2.798	0.119	165	2.807	0.118	165	2.816	0.118	165	2.815	0.118
170	2.883	0.112	170	2.892	0.112	170	2.901	0.111	170	2.901	0.111
175	2.967	0.106	175	2.978	0.106	175	2.987	0.105	175	2.986	0.105
180	3.052	0.101	180	3.063	0.100	180	3.072	0.099	180	3.071	0.100

Note: The relative slenderness ratio, λ_{rel}, has been included for information.

$$\text{when} \quad \lambda_{rel,y} > 0.3 \quad \frac{\sigma_{c,0,d}}{k_{c,y}f_{c,0,d}} \leq 1 \tag{5.11a}$$

$$\text{when} \quad \lambda_{rel,z} > 0.3 \quad \frac{\sigma_{c,0,d}}{k_{c,z}f_{c,0,d}} \leq 1 \tag{5.11b}$$

Table 5.3 (*continued*)

Strength class C40			Strength class C45			Strength class C50		
$f_{c,0,k}$ 26 N/mm²			$f_{c,0,k}$ 27 N/mm²			$f_{c,0,k}$ 29 N/mm²		
$E_{0.05}$ 9400 N/mm²			$E_{0.05}$ 10 000 N/mm²			$E_{0.05}$ 10 700 N/mm²		
λ	λ_{rel}	$k_{c,y}(k_{c,z})$	λ	λ_{rel}	$k_{c,y}(k_{c,z})$	λ	λ_{rel}	$k_{c,y}(k_{c,z})$
17.920	0.300	1.000	17.920	0.300	1.000	17.920	0.300	1.000
20	0.335	0.992	20	0.331	0.993	20	0.331	0.993
25	0.419	0.972	25	0.413	0.973	25	0.414	0.973
30	0.502	0.950	30	0.496	0.951	30	0.497	0.951
35	0.586	0.923	35	0.579	0.925	35	0.580	0.925
40	0.670	0.890	40	0.662	0.894	40	0.663	0.893
45	0.753	0.851	45	0.744	0.855	45	0.746	0.855
50	0.837	0.803	50	0.827	0.809	50	0.829	0.808
55	0.921	0.747	55	0.910	0.755	55	0.911	0.754
60	1.004	0.686	60	0.992	0.695	60	0.994	0.694
65	1.088	0.624	65	1.075	0.633	65	1.077	0.632
70	1.172	0.564	70	1.158	0.574	70	1.160	0.572
75	1.256	0.509	75	1.240	0.518	75	1.243	0.517
80	1.339	0.459	80	1.323	0.468	80	1.326	0.467
85	1.423	0.415	85	1.406	0.424	85	1.409	0.422
90	1.507	0.376	90	1.489	0.384	90	1.491	0.383
95	1.590	0.342	95	1.571	0.350	95	1.574	0.348
100	1.674	0.312	100	1.654	0.319	100	1.657	0.318
105	1.758	0.286	105	1.737	0.292	105	1.740	0.291
110	1.841	0.263	110	1.819	0.268	110	1.823	0.267
115	1.925	0.242	115	1.902	0.247	115	1.906	0.246
120	2.009	0.223	120	1.985	0.229	120	1.989	0.228
125	2.093	0.207	125	2.067	0.212	125	2.071	0.211
130	2.176	0.192	130	2.150	0.197	130	2.154	0.196
135	2.260	0.179	135	2.233	0.183	135	2.237	0.183
140	2.344	0.167	140	2.316	0.171	140	2.320	0.170
145	2.427	0.156	145	2.398	0.160	145	2.403	0.159
150	2.511	0.147	150	2.481	0.150	150	2.486	0.149
155	2.595	0.138	155	2.564	0.141	155	2.569	0.140
160	2.267	0.130	160	2.646	0.133	160	2.651	0.132
165	2.762	0.122	165	2.729	0.125	165	2.734	0.124
170	2.846	0.115	170	2.812	0.118	170	2.817	0.118
175	2.930	0.109	175	2.894	0.112	175	2.900	0.111
180	3.013	0.103	180	2.977	0.106	180	2.983	0.105

Note: The relative slenderness ratio, λ_{rel}, has been included for information.

i.e.

$$\sigma_{c,0,d} \le k_{c,y} \cdot f_{c,0,d} \tag{5.12}$$

and

$$\sigma_{c,0,d} \le k_{c,z} \cdot f_{c,0,d} \tag{5.13}$$

Table 5.3 Value of $k_{c,y}$ or $k_{c,z}$ based on equations (6.25)–(6.29) inclusive in EC5 with $\beta_c = 0.2$ – for hardwoods

Strength class D18			Strength class D24			Strength class D30			Strength class D35		
$f_{c,0,k}$ 18 N/mm² $E_{0.05}$ 8000 N/mm²			$f_{c,0,k}$ 21 N/mm² $E_{0.05}$ 8500 N/mm²			$f_{c,0,k}$ 23 N/mm² $E_{0.05}$ 9200 N/mm²			$f_{c,0,k}$ 25 N/mm² $E_{0.05}$ 10100 N/mm²		
λ	λ_{rel}	$k_{c,y}$ $(k_{c,z})$	λ	λ_{rel}	$k_{c,y}$ $(k_{c,z})$	λ	λ_{rel}	$k_{c,y}$ $(k_{c,z})$	λ	λ_{rel}	$k_{c,y}$ $(k_{c,z})$
19.869	0.300	1.000	18.961	0.300	1.000	18.850	0.300	1.000	18.944	0.300	1.000
20	0.302	1.000	20	0.316	0.998	20	0.318	0.998	20	0.317	0.998
25	0.377	0.991	25	0.396	0.989	25	0.398	0.989	25	0.396	0.989
30	0.453	0.981	30	0.475	0.978	30	0.477	0.978	30	0.475	0.978
35	0.528	0.970	35	0.554	0.965	35	0.557	0.965	35	0.554	0.965
40	0.604	0.955	40	0.633	0.949	40	0.637	0.948	40	0.633	0.949
45	0.679	0.937	45	0.712	0.928	45	0.716	0.927	45	0.713	0.928
50	0.755	0.913	50	0.791	0.899	50	0.796	0.897	50	0.792	0.899
55	0.830	0.881	55	0.870	0.860	55	0.875	0.857	55	0.871	0.859
60	0.906	0.838	60	0.949	0.807	60	0.955	0.803	60	0.950	0.807
65	0.981	0.783	65	1.028	0.745	65	1.035	0.740	65	1.029	0.744
70	1.057	0.721	70	1.108	0.677	70	1.114	0.672	70	1.109	0.676
75	1.132	0.656	75	1.187	0.611	75	1.194	0.606	75	1.188	0.610
80	1.206	0.596	80	1.266	0.550	80	1.273	0.545	80	1.267	0.550
85	1.283	0.538	85	1.345	0.496	85	1.353	0.491	85	1.346	0.495
90	1.359	0.487	90	1.424	0.448	90	1.432	0.443	90	1.425	0.447
95	1.434	0.442	95	1.503	0.406	95	1.512	0.402	95	1.504	0.406
100	1.510	0.403	100	1.582	0.369	100	1.592	0.365	100	1.584	0.369
105	1.585	0.368	105	1.661	0.337	105	1.671	0.334	105	1.663	0.337
110	1.661	0.337	110	1.740	0.309	110	1.751	0.305	110	1.742	0.308
115	1.736	0.310	115	1.819	0.284	115	1.830	0.281	115	1.821	0.283
120	1.812	0.286	120	1.899	0.262	120	1.910	0.259	120	1.900	0.261
125	1.887	0.265	125	1.978	0.242	125	1.989	0.239	125	1.980	0.242
130	1.963	0.246	130	2.057	0.224	130	2.069	0.222	130	2.059	0.224
135	2.038	0.228	135	2.136	0.209	135	2.149	0.206	135	2.138	0.208
140	2.114	0.213	140	2.215	0.194	140	2.228	0.192	140	2.217	0.194
145	2.189	0.199	145	2.294	0.182	145	2.308	0.180	145	2.296	0.181
150	2.265	0.186	150	2.373	0.170	150	2.387	0.168	150	2.375	0.170
155	2.340	0.175	155	2.452	0.159	155	2.467	0.158	155	2.455	0.159
160	2.416	0.164	160	2.531	0.150	160	2.546	0.148	160	2.534	0.150
165	3.491	0.155	165	2.611	0.141	165	2.626	0.140	165	2.613	0.141
170	3.567	0.146	170	2.690	0.133	170	2.706	0.132	170	2.692	0.133
175	3.642	0.133	175	2.769	0.126	175	2.785	0.124	175	2.771	0.126
180	3.718	0.126	180	2.848	0.119	180	2.865	0.118	180	2.851	0.119

Note: The relative slenderness ratio, λ_{rel}, has been included for information.

where $\sigma_{c,0,d}$ and $f_{c,0,d}$ are as previously defined, and $k_{c,y}$ and $k_{c,z}$ are the instability factors defined in equations (5.4)–(5.6).

The critical design condition arising from equations (5.12) and (5.13) will be the one with the lower value of instability factor, which will also be associated with the member having the highest slenderness ratio.

See Example 5.7.1.

Table 5.3 (*continued*)

Strength class D40 $f_{c,0,k}$ 26 N/mm² $E_{0.05}$ 10 900 N/mm²			Strength class D50 $f_{c,0,k}$ 29 N/mm² $E_{0.05}$ 11 800 N/mm²			Strength class D60 $f_{c,0,k}$ 32 N/mm² $E_{0.05}$ 14 300 N/mm²			Strength class D70 $f_{c,0,k}$ 34 N/mm² $E_{0.05}$ 16 800 N/mm²		
λ	λ_{rel}	$k_{c,y}(k_{c,z})$	λ	λ_{rel}	$k_{c,y}(k_{c,z})$	λ	λ_{rel}	$k_{c,y}(k_{c,z})$	λ	λ_{rel}	$k_{c,y}(k_{c,z})$
19.297	0.300	1.000	19.011	0.300	1.000	19.923	0.300	1.000	20.950	0.300	1.000
20	0.311	0.999	20	0.316	0.998	20	0.301	1.000	-	-	-
25	0.389	0.990	25	0.395	0.989	25	0.376	0.991	25	0.358	0.993
30	0.466	0.979	30	0.473	0.978	30	0.452	0.981	30	0.430	0.984
35	0.544	0.967	35	0.552	0.965	35	0.527	0.970	35	0.501	0.974
40	0.622	0.952	40	0.631	0.949	40	0.602	0.956	40	0.573	0.962
45	0.700	0.932	45	0.710	0.928	45	0.678	0.938	45	0.644	0.946
50	0.777	0.905	50	0.789	0.900	50	0.753	0.914	50	0.716	0.927
55	0.855	0.868	55	0.868	0.861	55	0.828	0.882	55	0.788	0.901
60	0.933	0.819	60	0.947	0.809	60	0.903	0.839	60	0.859	0.866
65	1.010	0.760	65	1.026	0.747	65	0.979	0.785	65	0.931	0.821
70	1.088	0.694	70	1.105	0.680	70	1.054	0.723	70	1.002	0.766
75	1.166	0.628	75	1.184	0.614	75	1.129	0.659	75	1.074	0.706
80	1.244	0.567	80	1.262	0.553	80	1.205	0.597	80	1.146	0.645
85	1.321	0.511	85	1.341	0.498	85	1.280	0.540	85	1.217	0.587
90	1.399	0.463	90	1.420	0.450	90	1.355	0.489	90	1.289	0.534
95	1.477	0.419	95	1.499	0.408	95	1.430	0.444	95	1.360	0.486
100	1.555	0.382	100	1.578	0.371	100	1.506	0.405	100	1.432	0.444
105	1.632	0.349	105	1.657	0.339	105	1.581	0.370	105	1.504	0.406
110	1.710	0.319	110	1.736	0.310	110	1.656	0.339	110	1.575	0.373
115	1.788	0.294	115	1.815	0.285	115	1.732	0.312	115	1.647	0.343
120	1.866	0.271	120	1.894	0.263	120	1.807	0.288	120	1.718	0.316
125	1.943	0.250	125	1.973	0.243	125	1.882	0.266	125	1.790	0.293
130	2.021	0.232	130	2.051	0.226	130	1.957	0.247	130	1.862	0.272
135	2.099	0.216	135	2.130	0.210	135	2.033	0.230	135	1.933	0.253
140	2.176	0.201	140	2.209	0.195	140	2.108	0.214	140	2.005	0.236
145	2.254	0.188	145	2.288	0.183	145	2.183	0.200	145	2.076	0.220
150	2.332	0.176	150	2.367	0.171	150	2.259	0.187	150	2.148	0.206
155	2.410	0.165	155	2.446	0.160	155	2.334	0.176	155	2.220	0.194
160	2.487	0.155	160	2.525	0.151	160	2.409	0.165	160	2.291	0.182
165	2.565	0.146	165	2.604	0.142	165	2.485	0.155	165	2.363	0.171
170	2.643	0.138	170	2.683	0.134	170	2.560	0.147	170	2.434	0.162
175	2.721	0.130	175	2.762	0.126	175	2.635	0.139	175	2.506	0.153
180	2.798	0.123	180	2.840	0.120	180	2.710	0.131	180	2.578	0.145

Note: The relative slenderness ratio, λ_{rel}, has been included for information.

5.3.2   Members subjected to compression at an angle to the grain

This is an ultimate limit state condition and using the empirical relationship developed by Hankinson [6] the failure strength of a wood-related product when subjected to a resultant compressive action at an angle α to the grain, $f_{c,\alpha}$, as shown in Figure 5.7, can

Table 5.3 cont'd Value of $k_{c,y}$ or $k_{c,z}$ based on *equations (6.25)–(6.29)* inclusive in EN 1995-1-1 with $\beta_c = 0.1$ – for LVL

Kerto-S®		Kerto-Q®	
$f_{c,0,k}$ 35 N/mm² $E_{0,05}$ 11.6k N/mm²		$f_{c,0,k}$ 26 N/mm² $E_{0,05}$ 8.8k N/mm²	
λ	$k_{cy} (k_{cz})$	λ	$k_{cy} (k_{cz})$
17.158	1	17.339	1
20	0.994	20	0.995
25	0.983	25	0.984
30	0.970	30	0.971
35	0.954	35	0.955
40	0.932	40	0.934
45	0.901	45	0.904
50	0.857	50	0.863
55	0.798	55	0.806
60	0.727	60	0.736
65	0.653	65	0.663
70	0.582	70	0.592
75	0.518	75	0.528
80	0.463	80	0.471
85	0.415	85	0.423
90	0.373	90	0.381
95	0.337	95	0.344
100	0.306	100	0.312
105	0.279	105	0.285
110	0.255	110	0.260
115	0.234	115	0.239
120	0.216	120	0.220
125	0.200	125	0.204
130	0.185	130	0.189
135	0.172	135	0.175
140	0.160	140	0.163
145	0.149	145	0.153
150	0.140	150	0.143
155	0.131	155	0.134
160	0.123	160	0.126
165	0.116	165	0.118
170	0.109	170	0.112
175	0.103	175	0.106
180	0.098	180	0.100

Note: The relative slenderness ratio, λ_{rel}, has been included for information.

be written in terms of its strength parallel to the grain ($f_{c,0}$) and perpendicular to the grain ($f_{c,90}$) as follows:

$$f_{c,\alpha} = \frac{f_{c,0} f_{c,90}}{f_{c,0} \sin^n \alpha + f_{c,90} \cos^n \alpha}$$

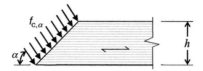

Fig. 5.7. Compressive strength of a member loaded at an angle α to the grain.

The value of the exponent n for wood-related products is generally taken to be 2 and from the relationship given in equation (4.22) for members loaded in compression perpendicular to the grain (i.e. $k_{c,90} f_{c,90,d}$) the compressive strength of the material at this angle, $f_{c,\alpha,d}$, can be written as:

$$f_{c,\alpha,d} = \frac{f_{c,0,d}}{\dfrac{f_{c,0,d}}{k_{c,90} f_{c,90,d}} \sin^2 \alpha + \cos^2 \alpha} \tag{5.14}$$

where the function $f_{c,0,d}$ is as described in 5.3.1, and $f_{c,90,d}$ is the design compressive strength perpendicular to the timber grain and is obtained as described in equation (4.23). The factor $k_{c,90}$ is as described against equation (4.24), but, because of the complexity of the various support conditions that can arise, it is recommended that a value of $k_{c,90} = 1$ is used.

This is the strength criteria used in EC5 and where the member in Figure 5.7 is b mm wide and subjected to a design compressive stress, $\sigma_{c,\alpha,d}$, at an angle α to the grain, from equation (5.14) the design condition to be satisfied will be

$$\sigma_{c,\alpha,d} \leq \frac{f_{c,0,d}}{\dfrac{f_{c,0,d}}{k_{c,90} f_{c,90,d}} \sin^2 \alpha + \cos^2 \alpha} \qquad \text{(EC5, equation (6.16))} \tag{5.15}$$

where the functions are as previously described and $\sigma_{c,\alpha,d}$ is the design compressive stress, and if it is generated by a design load N_d acting over the loaded area ($bh/\cos\alpha$), then:

$$\sigma_{c,\alpha,d} = \frac{N_d \cos \alpha}{bh}$$

5.3.3   Members subjected to axial tension

These are members that are subjected to a tensile action acting parallel to the grain and through the centroidal x–x axis of the member as shown in Figure 5.8. Such members function as ties in pin jointed trusses and provide tensile resistance to overturning forces in stud walls.

Although the tensile strength, $f_{t,0,k}$, of clear wood samples is greater than the compression strength, $f_{c,0,k}$, because tension failure occurs in a brittle rather than a ductile mode and also because of its sensitivity to the effects of grain slope, knots and other defects, the tensile strength of structural timber is generally less than the compression strength. This is particularly the case at the lower strength classes.

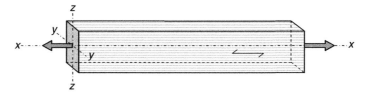

Fig. 5.8. Axial tension.

With tension members buckling will not arise and so slenderness considerations are only relevant if there is a risk of a stress reversal condition, for example due to the effect of wind loading on members in truss frameworks or in wall studs.

A tension member should be checked at the weakest point and this will normally be at connections. As connection design will follow after the member has been selected, the design of tension members normally involves a trial and error process. Assumptions are made as to the type and size of connections to be used and the adequacy of the tension member is confirmed at the connection location allowing for a loss of area due to the assumed connection. After the connections have been designed, the member is rechecked taking into account the actual net cross-sectional area. It is to be noted that in EC5, *5.2 (3)*, the effect of loss of cross-sectional area may be ignored where nails and screws with a diameter of 6 mm or less driven without pre-drilling are used. However, as required by EC5, *5.2 (4)*, all holes within a distance of half the minimum fastener spacing measured parallel to the grain from a given cross-section should be taken as occurring at that cross-section.

The EC5 procedure for the design of axially loaded members is as follows:

The design tensile stress in the member must be less than the design tensile strength:

$$\sigma_{t,0,d} \leq f_{t,0,d} \qquad\qquad (EC5,\ equation\ (6.1)) \qquad (5.16)$$

where:

- $\sigma_{t,0,d}$ is the design tensile stress parallel to the grain, and

$$\sigma_{t,0,d} = \frac{N_d}{A_{net}} \qquad\qquad (5.17)$$

  where N_d is the design axial load and A_{net} is the net cross-sectional area after allowing for the effect of the connection. If the connections are different at each end of the member, the end with the minimum net cross-sectional area must be used.
- $f_{t,0,d}$ is the design tensile strength parallel to grain, and

$$f_{t,0,d} = \frac{k_{mod} \cdot k_{sys} \cdot k_h \cdot f_{t,0,k}}{\gamma_M} \qquad\qquad (5.18)$$

  where k_{mod} and k_{sys} are as defined in equation (5.10), and k_h is the size effect modification factor for members under tension, discussed in Chapter 2 and given in Table 2.11. The largest cross-sectional dimension of the member should be used to evaluate the factor. When dealing with LVL, the factor is defined as k_ℓ, and is associated with member length. $f_{t,0,k}$ is the characteristic tensile strength of the timber or wood-based product parallel to the grain. Strength information for timber and the commonly used wood-based structural products is given in Chapter 1.

See Example 5.7.2.

5.4   MEMBERS SUBJECTED TO COMBINED BENDING AND AXIAL LOADING

Where the effects of instability have to be taken into account in EC5, the strength validation requirements for a column subjected to combined compression and bending are different to the strength validation requirements for a beam subjected to combined compression and bending.

It is accepted that there are cases where different rules will apply, however it is considered that a general categorisation may be unsafe and the design procedure given in the following sub-sections applies to any member subjected to combined axial compression and bending with limitations given for the applicability of the procedure in line with the criteria used in EC5.

5.4.1   Where lateral torsional instability due to bending about the major axis will not occur

These are members that are subjected to a combination of direct compression and bending, and where lateral torsional instability of the member cannot occur. This implies that the relative slenderness for bending, $\lambda_{rel,m}$, about the major axis (described in 4.5.1.2), will be ≤ 0.75. Members subjected to combined axial and lateral loading or, as shown in Figure 5.9, to eccentric compressive loading acting along the direction of the x–x axis, come into this category.

The behaviour of a member under pure axial compression is discussed in 5.3.1 and, where lateral torsional instability cannot occur, two failure modes can arise:

(a) When both $\lambda_{rel,y}$ and $\lambda_{rel,z}$ are ≤ 0.3, buckling behaviour is not relevant and failure will be based on the compressive strength of the member.

(b) If either $\lambda_{rel,y}$ or $\lambda_{rel,z}$ is > 0.3, buckling can arise and failure will be based on the compression strength of the member multiplied by the associated instability factor, i.e. $k_{c,y}$ (or $k_{c,z}$).

The design requirements in EC5 for the above conditions are as follows:

(a) *Both $\lambda_{rel,y}$ and $\lambda_{rel,z}$ are ≤ 0.3.*
As there is no strength reduction due to buckling under this condition, EC5 takes advantage of the strength benefits associated with the plastic behaviour of timber when subjected to compression stresses.

Figure 5.10 shows interaction diagrams for a member subjected to combined bending moment and axial compression based on the application of elastic and plastic theory. Under elastic theory, the failure condition will occur when the combined compression stress in the member reaches the compressive strength of the material. When plastic theory applies, the material yields when it reaches the compressive strength allowing the stress in the section to extend over the surface and enhance its strength.

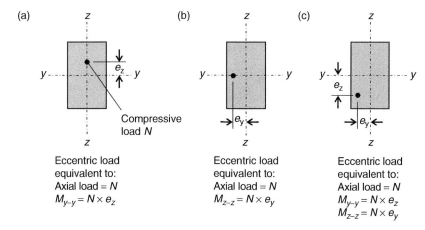

Fig. 5.9. Eccentric loading.

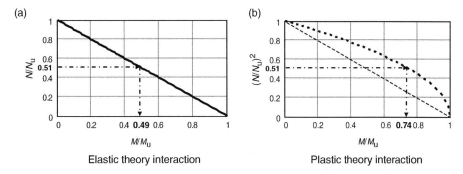

Fig. 5.10. Interaction diagrams for combined bending and axial compression of a member about an axis.

With the elastic theory approach, the sum of the combined axial and bending stress to strength ratios cannot exceed unity. With the plastic theory solution, however, the axial stress to axial strength ratio becomes a squared function enabling the member to take a higher value of bending stress to bending strength ratio for the same value of axial stress to strength ratio. For example, with an axial stress to strength ratio of 0.51, the strength increase when using plastic theory rather than elastic theory is approximately 50%.

Because of the plastic behaviour of timber under compression, EC5 adopts the plastic theory approach when both $\lambda_{rel,y}$ and $\lambda_{rel,z}$ are ≤ 0.3, and the interaction diagram for the design of members subjected to combined bending and axial compression with factor k_m applied to the ratio of moments about the z–z axis will be as shown in Figure 5.11. The design equations for this condition and for the other condition where the k_m factor is applied to the ratio of moments about the y–y axis are:

$$\left(\frac{\sigma_{c,0,d}}{f_{c,0,d}}\right)^2 + \frac{\sigma_{m,y,d}}{f_{m,y,d}} + k_m \frac{\sigma_{m,z,d}}{f_{m,z,d}} \leq 1 \qquad \text{(EC5, equation (6.19))} \qquad (5.19)$$

$$\left(\frac{\sigma_{c,0,d}}{f_{c,0,d}}\right)^2 + k_m \frac{\sigma_{m,y,d}}{f_{m,y,d}} + \frac{\sigma_{m,z,d}}{f_{m,z,d}} \leq 1 \qquad \text{(EC5, equation (6.20))} \qquad (5.20)$$

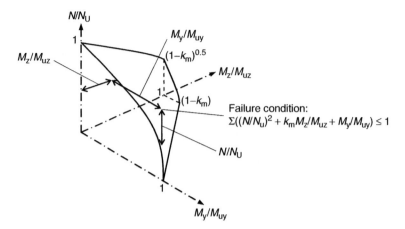

Fig. 5.11. Axial force–moment interaction curve for bi-axial bending when both $\lambda_{\text{rel},y}$ and $\lambda_{\text{rel},z} \leq 0.3$ and with factor k_m applied to the ratio of moments about the z–z axis.

where the bending stress functions are as defined in Chapter 4 and, because of size factors, the bending strengths can differ about the y–y and z–z axes.

As equations (5.19) and (5.20) do not include for the effects of lateral torsional instability, they are only valid for the cases where lateral torsional buckling of the member cannot occur or is prevented (e.g. members whose cross-section is square or circular, or the relative slenderness ratio for bending is ≤ 0.75). If lateral torsional instability can occur, the equations are valid when the member can only be subjected to bending about the weak axis.

(b) If $\lambda_{\text{rel},y}$ and/or $\lambda_{\text{rel},z} > 0.3$.
Under this condition, because axial load buckling effects have to be taken into account, no benefit is taken of any plastic behaviour in the member and the ultimate load is achieved when the material reaches its failure strength in the extreme fibre. This is in line with the elastic theory interaction approach shown in Figure 5.10a.

With this approach, the EC5 procedure for the design of members subjected to combined bending and axial compression when $\lambda_{\text{rel},y}$ and/or $\lambda_{\text{rel},z}$ exceed 0.3 is as shown in Figure 5.12 and requires:

$$\frac{\sigma_{c,0,d}}{k_{c,y} f_{c,0,d}} + \frac{\sigma_{m,y,d}}{f_{m,y,d}} + k_m \frac{\sigma_{m,z,d}}{f_{m,z,d}} \leq 1 \qquad \text{(EC5, equation (6.23))} \qquad (5.21)$$

$$\frac{\sigma_{c,0,d}}{k_{c,z} f_{c,0,d}} + k_m \frac{\sigma_{m,y,d}}{f_{m,y,d}} + \frac{\sigma_{m,z,d}}{f_{m,z,d}} \leq 1 \qquad \text{(EC5, equation (6.24))} \qquad (5.22)$$

where the functions remain as previously defined and, as with equations (5.19) and (5.20), where size factors are relevant, the bending strengths about the y–y and z–z axes can differ.

As in 5.4.1(a), equations (5.21) and (5.22) are only valid for situations where lateral torsional buckling of the member will not or cannot occur, otherwise the member can only be subjected to bending about the weak axis.

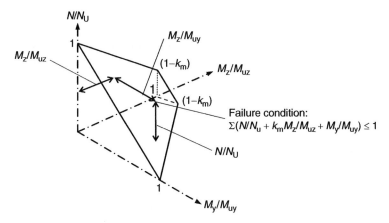

Fig. 5.12. Axial force–moment interaction curve for bi-axial bending when either $\lambda_{rel,y}$ or $\lambda_{rel,z} > 0.3$, and with factor k_m applied to the ratio of moments about the z–z axis.

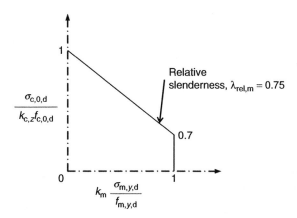

Fig. 5.13. Interaction diagram for combined axial loading and bending about the y–y axis.

In the case where there is axial loading and bending only about the major y–y axis, the strength validation equations will reduce to equation (5.23):

$$\frac{\sigma_{c,0,d}}{k_{c,z}f_{c,0,d}} + k_m\frac{\sigma_{m,y,d}}{f_{m,y,d}} \leq 1 \qquad\qquad (5.23)$$

Equation (5.23) is valid for the condition where lateral torsional buckling will not occur, i.e. $\lambda_{rel,m} \leq 0.75$, at which $k_{crit} = 1$. For this situation the boundary condition for equation (5.23), based on the use of solid timber, LVL (or glulam) rectangular sections, will be as defined by the solid line in Figure 5.13.

Where the relative slenderness ratio for bending exceeds 0.75, the EC5 strength criteria will be as given in 5.4.2.

See Example 5.7.3.

5.4.2   Lateral torsional instability under the effect of bending about the major axis

This situation will apply to members that are subjected to a combination of direct compression and bending about the major axis *only*, and where lateral torsional instability of the member can occur. This will apply to members in which the relative slenderness ratio for bending about the major axis, $\lambda_{rel,m}$, is greater than 0.75. No condition is given for a member subjected to axial compression with bending about the y–y and the x–x axes and $\lambda_{rel,m}$ is greater than 0.75.

For the condition where a member is subjected to direct compression and bending about the major axis and lateral torsional instability of the member can occur, no plastic behaviour is allowed to occur under the effects of the axial load but is permitted under the effect of the moment. The interaction between axial load and moment at failure is based on a solution involving plastic behaviour similar to that shown in Figure 5.10b and the design requirement is:

$$\left(\frac{\sigma_{m,d}}{k_{crit}f_{m,d}}\right)^2 + \frac{\sigma_{c,d}}{k_{c,z}f_{c,0,d}} \leq 1 \qquad\qquad (EC5,\ equation\ (6.35)) \quad (5.24)$$

where the terms are as previously described, and $\sigma_{m,d}$ is the design bending stress about the strong axes y–y, and $\sigma_{m,d} = M_{y,d}/W_y$, where $M_{y,d}$ is the design bending moment about the y–y axis and W_y is the associated section modulus; $\sigma_{c,d}$ is the design compressive stress and equates to $\sigma_{c,0,d}$ as defined in equation (5.10); k_{crit} is the factor that takes into account the reduced bending strength due to lateral buckling. It is discussed in Chapter 4 and defined in Table 4.3.

In applying equation (5.24) it is to be noted that if the relative slenderness ratio for bending of the member, $\lambda_{rel,m}$, is close to 0.75, because there is no bending about the z–z axis, the member state can also be considered to approximate the same condition as addressed by equation (5.23).

A comparison of equations (5.23) and (5.24) is shown in Figure 5.14.

For such a condition, when $\sigma_{m,y,d}/f_{m,y,d} \leq 0.7$, equation (5.23) will dictate the limiting design condition and when $\sigma_{m,y,d}/f_{m,y,d} > 0.7$, equation (5.24) should be complied with.

See Example 5.7.4.

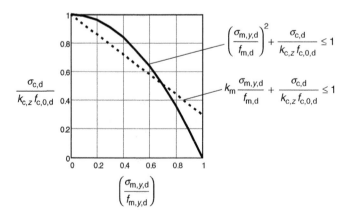

Fig. 5.14. Comparison between equation (5.24) (when $k_{crit} = 1$) and equation (5.23).

5.4.3   Members subjected to combined bending and axial tension

Although an element of plastic behaviour due to bending is permitted, because members in tension fail in a brittle mode, EC5 takes the approach that the ultimate load will be achieved when the material reaches its failure strength in the extreme fibre. This is in line with the elastic theory solution shown in Figure 5.10a.

From this the design requirement for members subjected to combined bending and axial tension, given in EC5, *6.2.3(1)P*, is as follows:

$$\frac{\sigma_{t,0,d}}{f_{t,0,d}} + \frac{\sigma_{m,y,d}}{f_{m,y,d}} + k_m \frac{\sigma_{m,z,d}}{f_{m,z,d}} \leq 1 \qquad \text{(EC5, } equation\ (6.17)\text{)} \qquad (5.25)$$

$$\frac{\sigma_{t,0,d}}{f_{t,0,d}} + k_m \frac{\sigma_{m,y,d}}{f_{m,y,d}} + \frac{\sigma_{m,z,d}}{f_{m,z,d}} \leq 1 \qquad \text{(EC5, } equation\ (6.18)\text{)} \qquad (5.26)$$

where the functions remain as previously defined.

Equations (5.25) and (5.26) assume that lateral torsional buckling of the member when bent about its major axis (y–y) is prevented. Where this is not the case, as indicated in Appendix C, the code is to be amended such that the method given in *6.3* will apply but the tensile stress should be taken to be 0, i.e. the section can be designed taking into account instability requirements but ignoring any stress due to tensile forces. This is a fairly conservative approach where the design bending (compression) stress is critical, and it has still to be remembered that the combined tension condition may be the critical condition and must still be checked. For the common design condition where there is only bending about the strong axis together with axial tension on a member, it is suggested in this book that the conditions to be checked will be:

(a) where the design bending compression stress ($\sigma_{m(c),y,d}$) exceeds $k_{crit} f_{m,y,d}$:

$$\frac{\sigma_{t,0,d}}{f_{t,0,d}} + \frac{\sigma_{m,(c),y,d}}{k_{crit} f_{m,y,d}} \leq 1 \qquad (5.26a)$$

   noting the axial tension fraction will be subtracted from the bending compression fraction as they will be of opposite signs.

(b) where the design bending compression stress ($\sigma_{m(c),y,d}$) is less than or equal to $k_{crit} f_{m,y,d}$, using the design bending tensile stress (which must have the same absolute value as the design bending compression stress, but will be of opposite sign), equation (5.25) with the z–z axis bending stress=0, will apply.

In equation (5.26a) k_{crit} is the factor that takes into account the reduced bending strength due to lateral buckling, discussed in Chapter 4 (also defined in Table 4.3) and the other functions are as previously defined.

See Example 5.7.5.

5.5 DESIGN OF STUD WALLS

In timber frame construction, the main functions of walls are to provide vertical support for floor and roof structures and strength and stability against the effects of lateral loading, generally caused by wind actions.

Fig. 5.15. Stud walls during construction.

The design of walls subjected to vertical loading and out of plane lateral actions, and where the sheathing is not designed to function compositely with the wall studs, is addressed in this chapter. The strength of stud walling in which the sheathing and the studs are designed to function as a composite section is covered in Chapter 8 and the in-plane racking strength of a wall is addressed in Chapters 9 and 13. Design guidance on how to calculate the lateral deflection of walls under combined out of plane bending and axial load is given in this chapter. Examples of stud walls during construction are shown in Figure 5.15.

5.5.1   Design of load-bearing walls

This covers the design of walls subjected to axial stresses as well as a combination of axial stress and bending stress due to the effect of out of plane actions (e.g. wind loading).

In general, load-bearing walls in timber-framed buildings are constructed using vertical timber members spaced at regular intervals and secured at their ends to continuous timber header and sole plates, as shown in Figure 5.16.

The vertical timbers are generally called studs and the walls are commonly referred to as stud walls. The studs are aligned so that the stronger axis (y–y) is parallel to the face of the wall and are secured in position by the header and sole plates, as shown in Figure 5.17. In-plane restraint is provided by battens that are prevented from moving laterally by diagonal or equivalent bracing members and that function during construction as well as for persistent design situations. If the wall sheathing cannot provide adequate lateral resistance, provided diagonal or equivalent bracing is used, the effective length of the stud about the z–z axis will be based on the greatest length of stud between the plate and the batten support. Where the sheathing material is able to provide adequate lateral restraint, the risk of buckling of the studs about the z–z axis can be ignored. Adequate lateral resistance will be provided by the sheathing material, provided it is secured to the studs and plates in accordance with the manufacturer's fixing recommendations or as required by the design. If sheathing is only fixed to one side of the wall, the studs will not be fully restrained laterally and it can be argued that a reduced effective length should be used.

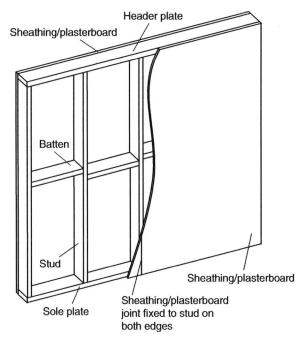

Fig. 5.16. Details of a typical stud wall (insulation, breather membrane, etc. not shown).

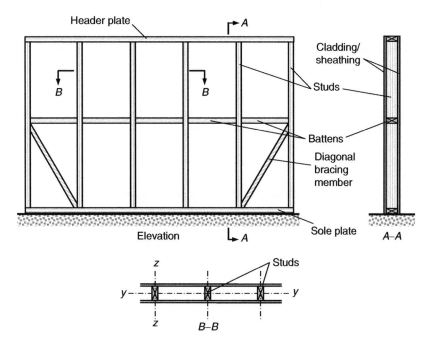

Fig. 5.17. Typical stud wall construction.

It is unlikely that stud walls will have any degree of fixity at their ends and, for out of plane buckling about the y–y axis, the studs are considered to be effectively held in position and torsionally restrained by the fixings to the header and sole plates but are free to rotate laterally at these positions. The effective length of the stud about this axis is taken to equal the height of the stud wall. Where the stud wall is sheathed on one or both sides the guidance in PD6693-1 is that the effective length for out of plane buckling should be taken to be 0.85 times the length of the stud, to take into account the stiffening effect of the sheathing. It is however common practice in domestic construction for stud walls to be loaded with permanent actions and construction loads during the construction process before all of the sheathing has been fixed, and in accordance with the requirements of BS EN 1991-1-6 [7] the design has to be checked for that condition. In that condition, where no intermediate battens/bracing are fitted to the studs, the effective length will be the length of the stud and the design condition for buckling instability will be about the axis having the largest slenderness ratio. In the examples given in the book only the final in-service design condition has been considered.

The strength of the wall is primarily derived from the studs and all concentrated loading should ideally be located directly over the studs and not in the span area of the header plate. For any loading over the span area, the plate must be checked to confirm whether it can function satisfactorily as a beam or alternatively additional studs can be inserted to carry the load.

When several equally spaced similar members are laterally connected by a continuous load distribution system, EC5 permits the member strength properties to be multiplied by a system strength factor, k_{sys}, as discussed in 2.3.7. In the case of stud walls, provided the sheathing is secured to the studs and plates in accordance with the manufacturer's fixing recommendations, it will function as a continuous load distribution system, allowing the k_{sys} factor to be used. Because the header and sole plates are single members the factor will not apply to these elements, but it will be relevant to the studs, being applied to the compression strength, $f_{c,0,k}$, the bending strength, $f_{m,k}$, and the shear strength, $f_{v,k}$, (where relevant). The factor is defined in EC5, *6.7*, and where the stud spacing is no greater than 610 mm centre to centre and the sheathing is fixed to the studs in accordance with the fixing manufacturer's recommendations, or in accordance with the design requirements, based on experience in the United Kingdom with such structures where there are four or more wall studs, k_{sys} can be taken to equal 1.1. For stud spacing greater than 610 mm centre to centre or when the sheathing is not properly secured or contains openings, unless it can be demonstrated by calculation, the factor should be taken to equal 1.

5.5.1.1   *Design of stud walls subjected to axial compression*
5.5.1.1.1   *Stud design*
(a) *Where the relative slenderness ratios* $\lambda_{rel,y}$ *and* $\lambda_{rel,z}$ *are* > 0.3.

This condition will normally apply when there is no sheathing or the sheathing cannot provide adequate lateral stiffness or it is not properly secured to the studs and plates and the design requirement will be:

$$\frac{\sigma_{c,0,d}}{k_{c,y}f_{c,0,d}} \leq 1 \text{ and } \frac{\sigma_{c,0,d}}{k_{c,z}f_{c,0,d}} \leq 1 \qquad (5.27)$$

i.e.

$$\sigma_{c,0,d} \le k_{c,y} \cdot f_{c,0,d} \tag{5.28a}$$

and

$$\sigma_{c,0,d} \le k_{c,z} \cdot f_{c,0,d} \tag{5.28b}$$

where $\sigma_{c,0,d}$ is the design compressive stress parallel to the grain
and

$$\sigma_{c,0,d} = \frac{N_d}{A} \tag{5.29}$$

where N_d is the design axial load on the stud and A is its cross-sectional area; $f_{c,0,d}$ is the design compressive strength of the stud parallel to the grain and

$$f_{c,0,d} = \frac{k_{mod} \cdot k_{sys} \cdot f_{c,0,k}}{\gamma_M} \tag{5.30}$$

where the functions are as previously defined and $k_{sys} = 1$ for the cases where the sheathing is not adequately secured to the studs.

The design procedure is as described in 5.3.1 for the design of a compression member under axial load. The critical design condition arising from equations (5.28a) and (5.28b) will be the one with the lowest value of instability factor and will be associated with the highest slenderness ratio of the stud.

(b) *Where the relative slenderness ratio $\lambda_{rel,y} > 0.3$ and $\lambda_{rel,z} \le 0.3$.*
If the sheathing material is properly secured to the studs, $k_{c,z}$ will always exceed $k_{c,y}$ and the design condition will be:

$$\sigma_{c,0,d} \le k_{c,y} \cdot f_{c,0,d} \tag{5.31}$$

(c) *Where both $\lambda_{rel,y}$ and $\lambda_{rel,z}$ are ≤ 0.3.*
This situation is unlikely to arise in practice, but if it does the values of $k_{c,y}$ and $k_{c,z}$ should be taken to be unity and the design condition will be:

$$\frac{\sigma_{c,0,d}}{f_{c,0,d}} \le 1 \tag{5.32}$$

where the functions are as previously defined but the k_{sys} factor used to derive $f_{c,0,d}$ in cases (b) and (c) can be taken to be 1.1 where there are at least four studs and the stud spacing is not greater than 610 mm c/c.

5.5.1.1.2 *Plate design*

The header and sole plates provide lateral support to the ends of the studs and also function as bearing members at these positions. Normal fixings between the studs and the plates and the structure will provide adequate lateral restraint and the design condition will generally relate to a design check of the strength of the plates under compression perpendicular to the grain. If there is any loading directly onto the plates between the stud positions, they will also need to be designed for bending and shear forces, as for beams. As stud walls are generally not meant to be loaded in this manner, this design condition has not been considered.

For compression perpendicular to the grain the condition to be satisfied is described in 4.5.3 and will be:

$$\frac{\sigma_{c,90,d}}{k_{c,90} f_{c,90,d}} \leq 1 \tag{5.33}$$

where $\sigma_{c,90,d}$ is the design compressive stress perpendicular to the grain and $\sigma_{c,0,d} = N_d / A_{ef}$ where N_d is the design axial load in the stud and A_{ef} is the effective contact area in compression perpendicular to the grain. As defined in 6.1.5(1)P and stated in 4.5.3, A_{ef} is derived using an effective contact length parallel to the grain, which is obtained by adding 30 mm, but no more than a, ℓ or $\ell_1/2$, to each side of the loaded length ℓ, where the symbols are as shown on Figure 4.16.

$f_{c,90,d}$ is the design compressive strength perpendicular to the grain and is defined as:

$$f_{c,90,d} = \frac{k_{mod} \cdot f_{c,90,k}}{\gamma_M} \tag{5.34}$$

where the functions are as previously defined and $f_{c,90,k}$ is the characteristic compressive strength of the timber or wood-based product perpendicular to the grain. Strength information for timber and the commonly used wood-based structural products is given in Chapter 1.

$k_{c,90}$ is a factor that takes the strain distribution across the compression member into account and is described in 4.5.3. When dealing with this type of problem the header plate as well as the sole plate condition must be checked as the value of the $k_{c,90}$ factor will commonly be different for each case, as shown in the example given in Figure 5.18.

See Example 5.7.6.

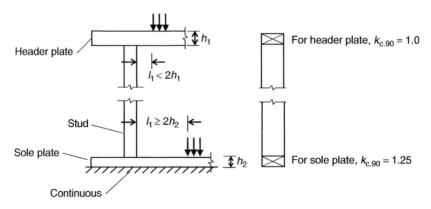

Fig. 5.18. Value of $k_{c,90}$ for a timber stud fitted with timber header and sole plates.

5.5.1.2   *Design of stud walls subjected to combined out of plane bending and axial compression*

This situation most commonly arises when stud walls are subjected to the effect of out of plane wind loading in addition to the axial load being carried. The design procedure for the studs follows the method given in 5.4 for the design of compression members also subjected to bending moment.

5.5.1.2.1   *Stud design*

(a) *Where $\lambda_{rel,y}$ and $\lambda_{rel,z}$ are >0.3.*
This will apply when the cladding cannot provide full buckling restraint about the z–z axis.

    (i)   For the case where the relative slenderness ratio for bending of each stud is ≤ 0.75 (i.e. lateral torsional buckling of the stud will not arise), the design condition will be:

$$\frac{\sigma_{c,0,d}}{k_{c,y}f_{c,0,d}} + \frac{\sigma_{m,y,d}}{f_{m,y,d}} \leq 1 \qquad (\text{EC5, equation (6.23), with } \sigma_{m,z,d} = 0) \qquad (5.35)$$

$$\frac{\sigma_{c,0,d}}{k_{c,z}f_{c,0,d}} + k_m \frac{\sigma_{m,y,d}}{f_{m,y,d}} \leq 1 \qquad (\text{EC5, equation (6.24), with } \sigma_{m,z,d} = 0) \quad (5.36)$$

where the functions are as previously defined and $\sigma_{m,y,d}$ are the design bending stresses about the y–y axis of the stud and $\sigma_{m,y,d} = M_{y,d}/W_y$ where $M_{y,d}$ is the design bending moment about the y–y axis and $W_y = bh^2/6$ is the associated section modulus of the stud; $f_{m,y,d}$ is the design bending strength about y–y axis and:

$$f_{m,y/z,d} = \frac{k_{mod} \cdot k_h \cdot k_{sys} \cdot f_{m,k}}{\gamma_M} \qquad (5.37)$$

where the functions are as previously defined and $k_{sys} = 1$ for the cases where the sheathing is not adequately secured to the studs.

    (ii) For the condition where the relative slenderness ratio for bending of each stud is >0.75 (i.e. lateral torsional buckling of the stud can arise), the requirements of EC5, *6.3.3(6)*, must be checked:

$$\left(\frac{\sigma_{m,y,d}}{k_{crit}f_{m,y,d}}\right)^2 + \frac{\sigma_{c,0,d}}{k_{c,z}f_{c,0,d}} \leq 1 \qquad (5.38)$$

where the functions are as previously defined.

For the condition where a stud wall is subjected to combined bending and axial compression, where $\lambda_{rel,z} > 0.3$ and $\lambda_{rel,m}$ approximates 0.75, as discussed in 5.4.2, equations (5.35), (5.36) and (5.38) should be complied with.

(b) *Where $\lambda_{rel,y}$ is >0.3 and $\lambda_{rel,z} \leq 0.3$.*
For this situation, $k_{c,z}$ will be unity and on the basis that lateral torsional buckling cannot arise the design condition will be:

$$\frac{\sigma_{c,0,d}}{k_{c,y}f_{c,0,d}} + \frac{\sigma_{m,y,d}}{f_{m,y,d}} \leq 1 \qquad (5.39)$$

where the functions are as previously defined and the k_{sys} factor can be taken to be 1.1 where there are at least four studs and the stud spacing is not greater than 610 mm c/c.

(c) *Where $\lambda_{rel,y}$ and $\lambda_{rel,z}$ are ≤ 0.3*.
This situation is unlikely to arise in practice, but if it does the values of $k_{c,y}$ and $k_{c,z}$ will be unity. With the understanding that for this condition the relative slenderness ratio for bending of each stud will also be ≤ 0.75, the design condition will be:

$$\left(\frac{\sigma_{c,0,d}}{f_{c,0,d}}\right)^2 + \frac{\sigma_{m,y,d}}{f_{m,y,d}} \leq 1 \qquad \text{(EC5, equation (6.19)}, \text{ with } \sigma_{m,z,d} = 0) \quad (5.40)$$

where the functions are as previously defined and the k_{sys} factor can be taken to be 1.1 where there are at least four studs and the stud spacing is not greater than 610 mm c/c.

It is to be noted that when subjected to lateral loading the wall studs will also be subjected to shear forces. However, for normal design conditions the shear stress will be small and for the examples given no shear check has been carried out.

Also, it should be noted that when considering combined stress problems (e.g. axial stress plus bending stress states) the largest stress to strength condition will not be a linear relationship. For such conditions the largest stress/k_{mod} approach cannot be used. All possible combinations states have to be investigated to demonstrate they will not exceed the design limiting condition.

5.5.1.2.2 Plate design
The procedure remains as described in 5.5.1.1.2.

See Example 5.7.7.

5.5.2 Out of plane deflection of load-bearing stud walls (and columns)

The behaviour of stud walls (and columns) with an initial deviation and subjected to axial loading is addressed in EC5 as a strength problem. Out of plane deflection is not considered in the code. There may, however, be a design situation where the out of plane deflection is required and the following methodology is given for calculating this deformation.

When a member with an initial out of plane displacement is subjected to axial loading, due to the additional moment induced in the member by the axial load, the displacement will be amplified. In the case of a stud wall the displacement will comprise an initial deviation from straightness, δ_0, and, if subjected to out of plane loading, an additional displacement, δ_q, as shown in Figures 5.19a and 5.19b respectively.

From EC5, *10.2*, it is to be noted that the maximum initial deviation from straightness measured halfway along the member, δ_0, cannot exceed $L/300$ for solid timber and $L/500$ for glued-laminated timber or LVL, where L is the length of the member.

Maximum out of plane deflection will occur at mid-height and from classical elastic stability theory it can be shown that if a stud wall comprises members having an initial out of plane deflection δ_0 (i.e. from the principal y–y axis position shown in Figure 5.19)

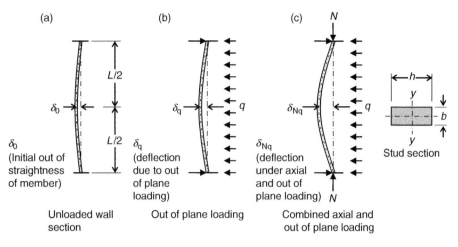

Fig. 5.19. Deflection of wall under vertical and out of plane loading.

and each stud is subjected to an axial design load N_d, the out of plane deflection of the wall (or a column) will increase to δ_{N0}, where:

$$\delta_{N0} = \alpha \delta_0 \tag{5.41}$$

In equation (5.41) α is an amplification factor derived from:

$$\alpha = \left(\frac{1}{1 - N_d/P_E} \right) \tag{5.42}$$

where:

$$P_E = \frac{\pi^2 E_{0.05} (bh^3/12)}{L^2}$$

and P_E is the Euler buckling load of each stud about its y–y axis, L is the effective length of the stud, b is the width of the stud, h is the depth of the stud, and $E_{0.05}$ is the fifth percentile modulus of elasticity of the stud material parallel to the grain.

Although this is an approximate solution, it is accurate to within 2% for values of N/P_E less than 0.6, which will be the case when deriving the displacement at the SLS. For the above condition the increase in wall deflection, δ_{N0}, can therefore be written as:

$$\delta_{N0} = \alpha \delta_0 - \delta_0 = \delta_0 (\alpha - 1) \tag{5.43}$$

Under the action of out of plane loading only on the wall, and taking shear deformation into account, from the data given in Table 4.7, the deflection δ_q per stud at the mid-height of the wall caused by a load of q kN/m² will be:

$$\delta q = \frac{5qL^4}{32E_{0,m}b(h)^3} \left[1 + 0.96 \left(\frac{E_{0,\text{mean}}}{G_{0,\text{mean}}} \right) \left(\frac{h}{L} \right)^2 \right] (S_p) \tag{5.44}$$

where δ_q is the out of plane deflection of the stud (in mm), L is the wall height (in metres), $E_{0,m}$ is the mean modulus of elasticity of the stud parallel to the grain (in kN/m²), b is the breadth of the stud (in metres), h is the depth of the stud (in metres), and S_p is the lateral spacing of the studs (in mm).

From equation (5.42), when each wall stud is subjected to an axial design load N_d, δ_q will be increased to $\alpha\delta_q$ and when added to the out of straightness deflection of the wall, the total deflection δ_{Nq} will be:

$$\delta_{Nq} = \alpha\delta_0 + \alpha\delta_q \qquad (5.45)$$

and the net increase in deflection due to the combined axial load and lateral load effect, δ_{net}, will be:

$$\delta_{net} = \left(\alpha\delta_0 + \alpha\delta_q\right) - \delta_0$$

which reduces to

$$\delta_{net} = \frac{(N_d/P_E)\delta_0 + \delta_q}{(1 - N_d/P_E)} \qquad (5.46)$$

It will be noted from equation (5.46) that the out of plane deflection is a function of the initial out of straightness of the wall (or column) as well as the axial load. If the wall (or column) is perfectly straight, δ_0 will equal zero and if there is no axial load, the additional deflection δ_{net} will equate to (5.44).

See Example 5.7.8.

5.6   REFERENCES

1 BS EN 1995-1-1:2004:+A1:2008. *Eurocode 5: Design of Timber Structures. Part 1-1: General – Common Rules and Rules for Buildings*, British Standard Institution.

2 PD6693-1:2012, Incorporating Corrigendum No1; PUBLISHED DOCUMENT - *Recommendations for the design of timber structures to Eurocode 5: Design of Timber Structures – Part 1: General – Common Rules and Rules for Buildings*, British Standard Institution.

3 BS 5268-2:2002. *Structural Use of Timber. Part 2: Code of Practice for Permissible Stress Design, Materials and Workmanship*, British Standards Institution.

4 BS EN 338:2009. *Structural Timber – Strength Classes*, British Standards Institution.

5 Blass, H.J. *International Council for Building Research Studies and Documentation, Working Commission W18A – Timber Structures, Design of Timber Columns. Meeting 20*, Dublin, Ireland, 1987.

6 Hankinson, R.L. Investigation of crushing strength of spruce at varying angles to the grain. *Air Service Information Circular*, Vol. 3, No. 259 (Material Section Paper No. 130), 1921.

7 BS EN 1991-1-6:2005, incorporating corrigendum July 2008. *Eurocode 1: Actions on Structures – Part 1-6: General Actions – Actions during Execution*, British Standard Institution.

8 NA to BS EN 1990:2002+A1:2005. *Incorporating National Amendment No1. UK National Annex for Eurocode 0 – Basis of Structural Design*, British Standards Institution.

9 NA to BS EN 1995-1-1:2004+A1:2008, Incorporating National Amendment No 2; *UK National Annex to Eurocode 5: Design of Timber Structures. Part 1-1: General – Common Rules and Rules for Buildings*, British Standards Institution.

5.7 EXAMPLES

As stated in 4.3, to be able to verify the ultimate and serviceability limit states, each design effect has to be checked and for each effect the largest value caused by the relevant combination of actions must be used.

However, to ensure that attention is primarily focused on the EC5 design rules for the timber or wood product being used, only the design load case producing the largest design effect has generally been given or evaluated in the following examples.

Example 5.7.1 The column shown in Figure E5.7.1 has a cross-section 150 mm×200 mm, is of strength class C18 to BS EN 338:2009, and functions under service class 2 conditions. It supports a characteristic permanent compressive axial action (including its self-weight) of 30 kN and a characteristic variable medium-term compressive axial action of 50 kN. The column is 3.75 m high and at each end is effectively held in position but not in direction about the z–z and the y–y axes.

Check that the column will meet the ultimate limit state (ULS) requirements of EC5.

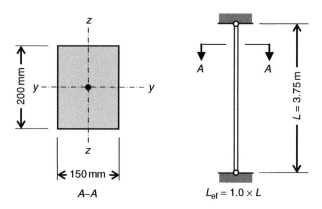

Fig. E5.7.1.

1. Column geometric properties

Column length, L	$L=3.7$ m	
Effective length about the y–y axis, $L_{e.y}$ Table 5.2	$L_{e.y}=1.0 \cdot L$	i.e. $L_{e.y}=3.75$ m
Effective length about the z–z axis, $L_{e.z}$ Table 5.2	$L_{e.z}=1.0 \cdot L$	i.e. $L_{e.z}=3.75$ m
Width of the member, b	$b=150$ mm	
Depth of the member, h	$h=200$ mm	

Cross-sectional area of the column, A	$A = b \cdot h$	$A = 3 \times 10^4 \text{ mm}^2$
Second moment of area about the y–y axes, I_y	$I_y = \dfrac{b \cdot h^3}{12}$	$I_y = 1 \times 10^8 \text{ mm}^4$
Radius of gyration about the y–y axis, i_y	$i_y = \sqrt{\dfrac{I_y}{A}}$	$i_y = 57.74 \text{ mm}$
Slenderness ratio about the y–y axis, λ_y	$\lambda_y = \dfrac{L_{e.y}}{i_y}$	$\lambda_y = 64.95$
Second moment of area about the z–z axis, I_z	$I_z = \dfrac{h \cdot b^3}{12}$	$I_z = 5.63 \times 10^7 \text{ mm}^4$
Radius of gyration about the z–z axis, i_z	$i_{I_z} = \sqrt{\dfrac{I_z}{A}}$	$i_z = 43.3 \text{ mm}$
Slenderness ratio about the z–z axis, λ_z	$\lambda_z = \dfrac{L_{e.z}}{i_z}$	$\lambda_z = 86.6$

2. Timber properties
Table 1.3, strength class C18 (BS EN 338:2009(E), *Table 1*)

Characteristic compression strength parallel to the grain, $f_{c.0.k}$	$f_{c.0.k} = 18 \text{ N/mm}^2$	
Fifth-percentile modulus of elasticity parallel to the grain, $E_{0.05}$	$E_{0.05} = 6.0 \text{ kN/mm}^2$	

3. Partial safety factors
Table 2.8 (UKNA to BS EN 1990:2002+A1:2005, *Table NA.A1.2(B)*) for the ULS

Permanent actions, γ_G	$\gamma_G = 1.35$
Variable actions, γ_Q	$\gamma_Q = 1.5$

Table 2.6 (UKNA to EC5, *Table NA.3*)

Material factor for solid timber, γ_M	$\gamma_M = 1.3$

4. Actions

Characteristic permanent compressive action, G_k	$G_k = 30 \text{ kN}$	
Characteristic medium-term compressive variable action, Q_k	$Q_k = 50 \text{ kN}$	
Design compressive action for the critical load combination, N_d (Table 2.8, equation (c) using the unfavourable condition)	$N_d = \gamma_G \cdot G_k + \gamma_Q \cdot Q_k$	$N_d = 1.16 \times 10^5 \text{ N}$

5. Modification factors

Factor for medium duration loading and service class 2, $k_{mod.med}$ (Table 2.4 (EC5, *Table 3.1*))	$k_{mod.med} = 0.8$
System strength factor, k_{sys} not relevant	$k_{sys} = 1.0$

6. Compression strength of column

The critical design load case at the ULS will be due to the combination of permanent and unfavourable medium-duration variable action:

Design compression stress, $\sigma_{c.0.d}$
$$\sigma_{c.0.d} = \frac{N_d}{A}$$
$\sigma_{c.0.d} = 3.85$ N/mm²

Design compression strength, $f_{c.0.d}$
$$f_{c.0.d} = \frac{k_{mod.med} \cdot k_{sys} \cdot f_{c.0.k}}{\gamma_M}$$
$f_{c.0.d} = 11.08$ N/mm²

Buckling resistance condition (5.3.1 (EC5, *6.3.2*)):

Relative slenderness about the y–y axis, $\lambda_{rel.y}$ (equation (5.3); EC5, *equation (6.21)*)
$$\lambda_{rel.y} = \frac{\lambda_y}{\pi} \sqrt{\frac{f_{c.0.k}}{E_{0.05}}}$$
$\lambda_{rel.y} = 1.13$

Relative slenderness about the z–z axis, $\lambda_{rel.z}$ (equation (5.3); EC5, *equation (6.22)*)
$$\lambda_{rel.z} = \frac{\lambda_z}{\pi} \sqrt{\frac{f_{c.0.k}}{E_{0.05}}}$$
$\lambda_{rel.z} = 1.51$

As both relative slenderness ratios are greater than 0.3, the conditions in 5.3.1.(b) apply. (EC5, *6.3.2(3)*):

Maximum relative slenderness ratio of the column
$\lambda_{rel.z} = 1.51$

Factor β_c for solid timber (equation (5.6); (EC5, *equation (6.29)*))
$\beta_c = 0.2$

Factor k_z (equation (5.5b); EC5, *equation (6.28)*)
$$k_z = 0.5 \cdot \left[1 + \beta_c \cdot \left(\lambda_{rel.z} - 0.3 \right) + \lambda_{rel.z}^2 \right]$$
$k_z = 1.76$

Instability factor, $k_{c.z}$ (equation (5.4b); (EC5, *equation (6.26)*))
$$k_{c.z} = \frac{1}{k_z + \sqrt{k_z^2 - \lambda_{rel.z}^2}}$$
$k_{c.z} = 0.37$

Design buckling strength, $k_{c.z} f_{c.0.d}$ (equation(5.7b))
$k_{c.z} \cdot f_{c.0.d} = 4.15$ N/mm²

Design stress/design buckling strength ratio (equation (5.11b))
$$\frac{\sigma_{c.0.d}}{k_{c.z} \cdot f_{c.0.d}} = 0.93$$

The design stress is less than the design buckling strength; therefore the 150 mm by 200 mm timber section in strength class C18 meets the ULS requirements of EC5.

Example 5.7.2 A 47 mm × 125 mm sawn timber member of strength class C18 to BS EN 338:2009 functioning under service class 2 conditions in a timber truss is subjected to a characteristic permanent tensile axial action of 2.5 kN and a characteristic variable medium-term tensile axial action of 3.0 kN. The member is effectively pin jointed at each end to accommodate end fixings, 8% of the cross-sectional area is removed at one end and 10% at the other end. The effect of bending of the member due to its self-weight can be ignored.

Check that the member complies with the requirements of EC5 at the ULS.

1. **Geometric properties**

   Thickness of the member, t_1 $t_1 = 47\,\text{mm}$

   Width of the member, h $h = 125\,\text{mm}$

   Cross-sectional area of the member, A $A = t_1 \cdot h$ $A = 5.88 \times 10^3\,\text{mm}^2$

   Net area of the member – based on the minimum area – $(100\% - 10\%)$, A_{net} $A_{\text{net}} = 0.9 \cdot A$ $A_{\text{net}} = 5.29 \times 10^3\,\text{mm}^2$

2. **Timber properties**

   Table 1.3, strength class C18 (BS EN 338:2009(E), Table 1)

   Characteristic tensile strength parallel to the grain, $f_{t.0.k}$ $f_{t.0.k} = 11\,\text{N/mm}^2$

3. **Partial safety factors**

   Table 2.8 (UKNA to BS EN 1990:2002+A1:2005, *Table NA.A1.2(B)*) for the ULS

   Permanent actions, γ_G $\gamma_G = 1.35$

   Variable actions, γ_Q $\gamma_Q = 1.5$

   Table 2.6 (UKNA to EC5, *Table NA.3*)

   Material factor for solid timber, γ_M $\gamma_M = 1.3$

4. **Actions**

   Characteristic permanent tensile action, G_k $G_k = 2.5\,\text{kN}$

   Characteristic variable medium term action, Q_k $Q_k = 3\,\text{kN}$

   Design tensile action for the critical load combination, N_d
   (Table 2.8, equation (c) using the unfavourable condition)

   $N_d = \gamma_G \cdot G_k + \gamma_Q \cdot Q_k$

   $N_d = 7.88 \times 10^3\,\text{N}$

5. **Modification factors**

   Factor for medium-duration loading and service class 2, $k_{\text{mod.med}}$
   (Table 4.1 (EC5, *Table 3.1*))

   $k_{\text{mod.med}} = 0.8$

   Size factor, k_h (Table 2.11; EC5, *equation (3.1)*) (the equation incorporates a dimensional correction factor for Mathcad)

   $$k_h = \begin{vmatrix} 1.0 & \text{if} \quad h \geq 150\text{mm} \\ \left(\dfrac{150\text{mm}}{h}\right)^{0.2} & \text{if} \quad 1.3 > \left(\dfrac{150\text{mm}}{h}\right)^{0.2} \\ 1.3 & \text{otherwise} \end{vmatrix}$$

   $$k_h = 1.04$$

   System strength factor, k_{sys} – not relevant $k_{\text{sys}} = 1.0$

6. **Tensile strength of timber**

   The critical design load case at the ULS will be due to the combination of permanent and unfavourable medium-duration actions:

   Design tension stress parallel to the grain, $\sigma_{t.0.d}$ $\sigma_{t.0.d} = \dfrac{N_d}{A_{\text{net}}}$ $\sigma_{t.0.d} = 1.49\,\text{N/mm}^2$

Design tension strength parallel to the grain, $f_{t.0.d}$ (equation (5.18))

$$f_{t.0.d} = \frac{k_{mod.med} \cdot k_{sys} \cdot k_h \cdot f_{t.0.k}}{\gamma_M} \qquad f_{t.0.d} = 7.02 \text{ N/mm}^{-2}$$

The design tension stress is less than the design tensile strength; therefore the 47 mm by 125 mm sawn section in strength class C18 meets the ULS requirements of EC5.

Example 5.7.3 The column shown in Figure E5.7.3 has a cross-section 100 mm × 200 mm, is of strength class C24 to BS EN 338:2009, and functions under service class 2 conditions. It supports a characteristic permanent compressive action of 10 kN and a characteristic variable medium-term compressive action of 17.5 kN. The loading is applied 25 mm eccentric from the y–y axis and 10 mm from the z–z axis as shown in the figure and the permanent action includes an allowance to cover for the effect of the self-weight of the column. The column is 3.75 m high and at each end is effectively held in position but not in direction about the z–z and about the y–y axes.

Check that the column will meet the ULS requirements of EC5.

Note: the relative slenderness ratio for bending about the y–y axis does not exceed 0.75.

Note: Because the relative bending slenderness is less than 0.75, there is no need to investigate lateral torsional instability effects and Section 5.4.1 will apply.

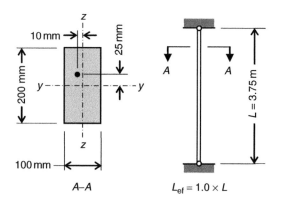

Fig. E5.7.3.

1. Geometric properties

Actual column length, L	$L = 3.75$ m	
Effective length for buckling about the y–y axis, $L_{e.y}$ (Table 5.2)	$L_{e.y} = 1.0 \cdot L$	i.e. $L_{e.y} = 3.75$ m
Effective length for buckling about the z–z axis, $L_{e.z}$ (Table 5.2)	$L_{e.z} = 1.0 \cdot L$	i.e. $L_{e.z} = 3.75$ m
Width of the member, b	$b = 100$ mm	
Depth of the member, h	$h = 200$ mm	
Cross-sectional area, A	$A = b \cdot h$	$A = 2 \times 10^4$ mm^2
Second moment of area about the y–y axes, I_y	$I_y = \dfrac{b \cdot h^3}{12}$	$I_y = 6.67 \times 10^7$ mm^4

Section modulus about the y–y axes, W_y $\quad W_y = \dfrac{2I_y}{h}$ $\qquad\qquad W_y = 6.67 \times 10^5 \text{ mm}^3$

Radius of gyration about the y–y axis, i_y $\quad i_y = \sqrt{\dfrac{I_y}{A}}$ $\qquad\qquad i_y = 57.74 \text{ mm}$

Slenderness ratio about the y–y axis, λ_y $\quad \lambda_y = \dfrac{L_{e.y}}{i_y}$ $\qquad\qquad \lambda_y = 64.95$

Second moment of area about the z–z axis, I_z $\quad I_z = \dfrac{h \cdot b^3}{12}$ $\qquad\qquad I_z = 1.67 \times 10^7 \text{ mm}^4$

Section modulus about the z–z axis, W_z $\quad W_z = \dfrac{2I_z}{b}$ $\qquad\qquad W_z = 3.33 \times 10^5 \text{ mm}^3$

Radius of gyration about the z–z axis, i_z $\quad i_z = \sqrt{\dfrac{I_z}{A}}$ $\qquad\qquad i_z = 28.87 \text{ mm}$

Slenderness ratio about the z–z axis, λ_z $\quad \lambda_z = \dfrac{L_{e.z}}{i_z}$ $\qquad\qquad \lambda_z = 129.9$

2. Timber properties

Table 1.3, strength class C24 (BS EN 338:2009(E), *Table 1*)

Characteristic bending strength about the y–y axis, $f_{m.y.k}$ $\qquad f_{m.y.k} = 24 \text{ N/mm}^2$

Characteristic bending strength about the z–z axis, $f_{m.z.k}$ $\qquad f_{m.z.k} = 24 \text{ N/mm}^2$

Characteristic compression strength parallel to the grain, $f_{c.0.k}$ $\qquad f_{c.0.k} = 21 \text{ N/mm}^2$

Fifth-percentile modulus of elasticity parallel to the grain, $E_{0.05}$ $\qquad E_{0.05} = 7.4 \text{ kN/mm}^2$

3. Partial safety factors

Table 2.8 (UKNA to BS EN 1990:2002+A1:2005, *Table NA.A1.2(B)*) for the ULS

Permanent actions, γ_G $\qquad\qquad \gamma_G = 1.35$

Variable actions, γ_Q $\qquad\qquad \gamma_Q = 1.5$

Table 2.6 (UKNA to EC5, *Table NA.3*)

Material factor for solid timber, γ_M $\qquad\qquad \gamma_M = 1.3$

4. Actions

Characteristic permanent compressive action, G_k $\qquad\qquad G_k = 10 \text{ kN}$

Characteristic short-term compressive variable action, Q_k $\qquad\qquad Q_k = 17.5 \text{ kN}$

Design compressive action for a critical load combination, N_d (Table 2.8, equation (c) using the unfavourable condition variable action)

$\qquad\qquad N_d = \gamma_G \cdot G_k + \gamma_Q \cdot Q_k$

$\qquad\qquad N_d = 3.98 \times 10^4 \text{ N}$

Moment about the y–y axis:

    Eccentricity, e_z $e_z = 25$ mm

    Design moment about the y–y axis, $M_{y.d}$ $M_{y.d} = N_d \cdot e_z$ $M_{y.d} = 0.99$ kN m

Moment about the z–z axis:

    Eccentricity, e_y $e_y = 10$ mm

    Design moment about the z–z axis, $M_{z.d}$ $M_{z.d}\ N_d \cdot e_y$ $M_{z.d} = 0.4$ kN m

5. Modification factors

Factor for medium-duration loading and service class 2, $k_{mod.med}$
(Table 2.4 (EC5, *Table 3.1*)) $k_{mod.med} = 0.8$

System strength factor, k_{sys}
– not relevant $k_{sys} = 1.0$

Depth factor for solid timber – bending about the y–y axis, $k1_h$
(Table 2.11 (EC5, *3.2*)) $k1_h = 1.0$

Depth factor for solid timber – bending about the z–z axis, $k2_h$
(Table 2.11 (EC5, *3.2*)) (Mathcad adjusted to make it dimensionally correct)

$$k2_h = \begin{vmatrix} 1.0 & \text{if} \quad b \geq 150\,\text{mm} \\ \left(\dfrac{150\,\text{mm}}{b} \right)^{0.2} & \text{if} \quad 1.3 > \left(\dfrac{150\,\text{mm}}{b} \right)^{0.2} \\ 1.3 & \text{otherwise} \end{vmatrix}$$

$$k2_h = 1.08$$

6. Strength of column

The critical design load case at the ULS will be due to the combination of permanent and unfavourable medium-duration variable actions:

Moment condition

Design bending moment about the y–y axis, $M_{y.d}$ $M_{y.d} = 0.99$ kN m

Design bending stress about the y–y axis, $\sigma_{m.y.d}$ $\sigma_{m.y.d} = \dfrac{M_{y.d}}{W_y}$ $\sigma_{m.y.d} = 1.49$ N/mm^2

Design bending moment about the z–z axis, $M_{z.d}$ $M_{z.d} = 0.4$ kN m

Design bending stress about the z–z axis, $\sigma_{m.z.d}$ $\sigma_{m.z.d} = \dfrac{M_{z.d}}{W_z}$ $\sigma_{m.z.d} = 1.19$ N/mm^2

Design bending strength about the y–y axis, $f_{m.y.d}$

$$f_{m.y.d} = \frac{k_{mod.med} \cdot k_{sys} \cdot k1_h \cdot f_{m.y.k}}{\gamma_M}$$ $f_{m.y.d} = 14.77$ N/mm^2

Design bending strength about the z–z axis, $f_{m.z.d}$

$$f_{\text{m.z.d}} = \frac{k_{\text{mod.med}} \cdot k_{\text{sys}} \cdot k2_h \cdot f_{\text{m.y.k}}}{\gamma_M} \qquad\qquad f_{\text{m.z.d}} = 16.02 \ \text{N/mm}^2$$

Axial compression condition

Design compression stress, $\sigma_{\text{c.0.d}}$ $\qquad\qquad \sigma_{\text{c.0.d}} = \dfrac{N_d}{A} \qquad\qquad \sigma_{\text{c.0.d}} = 1.99 \ \text{N/mm}^2$

Design compression strength, $f_{\text{c.0.d}}$
(equation (5.10)) $\qquad\qquad f_{\text{c.0.d}} = \dfrac{k_{\text{mod.med}} \cdot k_{\text{sys}} \cdot f_{\text{c.0.k}}}{\gamma_M} \qquad f_{\text{c.0.d}} = 12.92 \ \text{N/mm}^2$

Buckling resistance condition (5.3.1 (EC5, 6.3.2)):

Relative slenderness about the y–y axis,
$\lambda_{\text{rel.y}}$ (equation (5.3); EC5, *equation (6.21)*) $\qquad \lambda_{\text{rel.y}} = \dfrac{\lambda_y}{\pi} \cdot \sqrt{\dfrac{f_{\text{c.0.k}}}{E_{0.05}}} \qquad \lambda_{\text{rel.y}} = 1.1$

Relative slenderness about the z–z axis,
$\lambda_{\text{rel.z}}$ (equation (5.3); EC5, *equation (6.22)*) $\qquad \lambda_{\text{rel.z}} = \dfrac{\lambda_z}{\pi} \cdot \sqrt{\dfrac{f_{\text{c.0.k}}}{E_{0.05}}} \qquad \lambda_{\text{rel.z}} = 2.2$

As both relative slenderness ratios are greater than 0.3, conditions in EC5, *6.3.2(3)*, apply:

Buckling about z–z axes

Factor β_c for solid timber (equation
(5.6); EC5, *equation (6.29)*) $\qquad\qquad \beta_c = 0.2$

Factor k_z (equation (5.5b); *EC5,
equation (6.28)*) $\qquad\qquad k_z = 0.5 \cdot \left[1 + \beta_c \cdot (\lambda_{\text{rel.z}} - 0.3) + \lambda_{\text{rel.z}}^2 \right]$

$\qquad\qquad\qquad\qquad\qquad\qquad\qquad\qquad\qquad k_z = 3.12$

Instability factor about the z–z axis, $k_{\text{c.z}}$
(equation (5.4b); EC5, *equation (6.26)*) $\qquad k_{\text{c.z}} = \dfrac{1}{k_z + \sqrt{k_z^2 - \lambda_{\text{rel.z}}^2}} \qquad k_{\text{c.z}} = 0.19$

Buckling about y–y axes

Factor k_y (equation (5.5a); EC5,
equation 6.27)) $\qquad\qquad k_y = 0.5 \left[1 + \beta_c \cdot (\lambda_{\text{rel.y}} - 0.3) + \lambda_{\text{rel.y}}^2 \right]$

$\qquad\qquad\qquad\qquad\qquad\qquad\qquad\qquad\qquad k_y = 1.19$

Instability factor about the y–y axis, $k_{\text{c.y}}$
(equation (5.4a); EC5, *equation (6.25)*) $\qquad k_{\text{c.y}} = \dfrac{1}{k_y + \sqrt{k_y^2 - \lambda_{\text{rel.y}}^2}} \qquad k_{\text{c.y}} = 0.61$

Redistribution factor for a rectangular
section, k_m
(equation (4.4c) (EC5, *6.1.6)*) $\qquad\qquad k_m = 0.7$

Combined stress condition
Equations (5.21) and (5.22) (EC5, *equations (6.23) and (6.24)*):

$$\frac{\sigma_{\text{c.0.d}}}{k_{\text{c.y}} \cdot f_{\text{c.0.d}}} + \frac{\sigma_{\text{m.y.d}}}{f_{\text{m.y.d}}} + k_m \frac{\sigma_{\text{m.z.d}}}{f_{\text{m.z.d}}} = 0.4$$

$$\frac{\sigma_{\text{c.0.d}}}{k_{\text{c.z}} \cdot f_{\text{c.0.d}}} + k_m \frac{\sigma_{\text{m.y.d}}}{f_{\text{m.y.d}}} + \frac{\sigma_{\text{m.z.d}}}{f_{\text{m.z.d}}} = 0.96$$

Relationships less than unity; therefore the 100 mm by 200 mm sawn section in strength class C24 will meet the ULS requirements of EC5.

Example 5.7.4 The LVL (Kerto-S) column shown in Figure E5.7.4 has a cross-section 90 mm×200 mm, and functions under service class 2 conditions. It supports a characteristic permanent compressive action of 8 kN and a characteristic variable short-term compressive action of 19.5 kN. The loading is applied 65 mm eccentric from the $y–y$ axis as shown in the figure and the permanent action includes an allowance to cover for the effect of the self-weight of the column. The column is 4.15 m high and at each end is effectively held in position but not in direction about the $z–z$ axis and the $y–y$ axis.

Check that the column will meet the ULS requirements of EC5.

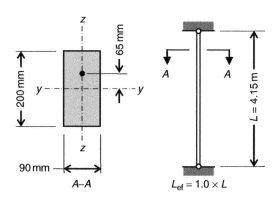

Fig. E5.7.4.

1. **Column geometric properties**

Actual column length, L	$L=4.15$ m	
Effective length for buckling about the $y–y$ axis, $L_{e.y}$ (Table 5.2)	$L_{e.y}=1.0 \cdot L$	i.e. $L_{e.y}=4.15$ m
Effective length for buckling about $z–z$ axis, $L_{e.z}$ (Table 5.2)	$L_{e.z}=1.0 \cdot L$	i.e. $L_{e.z}=4.15$ m
Effective length of member acting as a beam with a constant moment along the length, ℓ_{ef} (Table 4.2 (EC5, *Table 6.1*))	$\ell_{ef} = L$	$\ell_{ef} = 4.15$ m
Width of member, b	$b=90$ mm	
Depth of member, h	$h=200$ mm	
Cross-sectional area, A	$A=b \cdot h$	$A=1.8 \times 10^4$ mm^2
Second moment of area about the $y–y$ axes, I_y	$I_y = \dfrac{b \cdot h^3}{12}$	$I_y=6 \times 10^7$ mm^4
Section modulus about the $y–y$ axes, W_y	$W_y = \dfrac{2I_y}{h}$	$W_y=6 \times 10^5$ mm^3

Radius of gyration about the y–y axis, i_y	$i_y = \sqrt{\dfrac{I_y}{A}}$	$i_y = 57.74$ mm
Slenderness ratio about the y–y axis, λ_y	$\lambda_y = \dfrac{L_{e.y}}{i_y}$	$\lambda_y = 71.88$
Second moment of area about the z–z axis, I_z	$I_z = \dfrac{h \cdot b^3}{12}$	$I_z = 1.22 \times 10^7$ mm^4
Radius of gyration about the z–z axis, i_z	$i_z = \sqrt{\dfrac{I_z}{A}}$	$i_z = 25.98$ mm
Slenderness ratio about the z–z axis, λ_z	$\lambda_z = \dfrac{L_{e.z}}{i_z}$	$\lambda_z = 159.73$

2. LVL properties
Table 1.16 – Kerto S

Characteristic bending strength about the y–y axis, $f_{m.y.k}$	$f_{m.y.k} = 44$ N/mm^2
Characteristic compression strength parallel to the grain, $f_{c.0.k}$	$f_{c.0.k} = 35$ N/mm^2
Fifth-percentile modulus of elasticity parallel to the grain, $E_{0.05}$	$E_{0.05} = 11.6$ kN/mm^2
Fifth-percentile shear modulus, $G_{0.05}$	$G_{0.05} = 0.4$ kN/mm^2

3. Partial safety factors
Table 2.8 (UKNA to BS EN 1990:2002+A1:2005, *Table NA.A1.2(B)*) for the ULS

Permanent actions, γ_G	$\gamma_G = 1.35$
Variable actions, γ_Q	$\gamma_Q = 1.5$

Table 2.6 (UKNA to EC5, *Table NA.3*):

Material factor for LVL, γ_M	$\gamma_M = 1.2$

4. Actions

Characteristic permanent compressive action, G_k	$G_k = 8$ kN	
Characteristic short-term compressive variable action, Q_k	$Q_k = 19.5$ kN	
Design compressive action for the critical load combination, N_d (Table 2.8, equation (c) using the unfavourable condition variable action)	$N_d = \gamma_G \cdot G_k + \gamma_Q \cdot Q_k$	$N_d = 4 \times 10^4$ N

Moment about the y–y axis:

Eccentricity, e_z	$e_z = 65$ mm	
Design moment about the y–y axis, $M_{y.d}$	$M_{y.d} = N_d \cdot e_z$	$M_{y.d} = 2.6$ kN m

5. Modification factors

Factor for short-duration loading and service class 2, $k_{mod.short}$ (Table 2.4 (EC5, *Table 3.1*))

$$k_{mod.short} = 0.9$$

System strength factor, k_{sys} – not relevant

$$k_{sys} = 1$$

Size effect exponent, s (Table 1.15)

$$s = 0.12$$

Depth factor for LVL – for bending edgewise, k_h (Table 2.11 (EC5, *3.4*; BS EN 14374-2004))

$$k_h = \begin{vmatrix} 1.0 & if \quad h \geq 300 \text{ mm} \\ \left(\dfrac{300 \text{ mm}}{h}\right)^s & if \quad 1.2 > \left(\dfrac{300 \text{ mm}}{h}\right)^s \\ 1.2 & otherwise \end{vmatrix}$$

$$k_h = 1.05$$

6. Strength of the column

The design load case at the ULS will be due to the combination of permanent and unfavourable short-duration variable action:

Moment condition

Design bending moment about the y–y axis, $M_{y.d}$

$$M_{y.d} = 2.6 \text{ kN m}$$

Design bending stress about the y–y axis, $\sigma_{m.y.d}$

$$\sigma_{m.y.d} = \frac{M_{y.d}}{W_y} \qquad\qquad \sigma_{m.y.d} = 4.34 \text{ N/mm}^2$$

Design bending strength about the y–y, $f_{m.y.d}$

$$f_{m.y.d} = \frac{k_{mod.short} \cdot k_{sys} \cdot k_h \cdot f_{m.y.k}}{\gamma_M} \qquad f_{m.y.d} = 34.65 \text{ N/mm}^2$$

Redistribution factor for a rectangular section, k_m (Equation (4.4c) (EC5, *6.1.6*))

$$k_m = 0.7$$

Buckling resistance condition – lateral torsional buckling under major axis bending (4.5.1.2 (EC5, *6.3.3*)):

Lateral stability factor, k_{crit} (4.5.1.2 (EC5, *6.3.3*)):

Critical bending stress, $\sigma_{m.crit}$ (equation (4.7c); EC5, *equation (6.31)*)

$$\sigma_{m.crit} = \frac{\pi \cdot b^2 \left[E_{0.05} \cdot G_{0.05} \cdot \left(1 - 0.63 \cdot \dfrac{b}{h}\right) \right]^{0.5}}{h \cdot \ell_{ef}}$$

$$\sigma_{m.crit} = 55.9 \text{ N/mm}^2$$

Relative slenderness for bending, $\lambda_{rel.m}$ (equation (4.10); EC5, *equation (6.30)*)

$$\lambda_{rel.m} = \sqrt{\frac{f_{m.y.k}}{\sigma_{m.crit}}} \qquad\qquad \lambda_{rel.m} = 0.89$$

Lateral stability factor,
k_{crit} (Table 4.3 (EC5,
equation (6.34))

$$k_{crit} = \begin{vmatrix} 1 & if \quad \lambda_{rel.m} \leq 0.75 \\ 1.56 - 0.75 . \lambda_{rel.m} & if \quad 0.75 < \lambda_{rel.m} \leq 1.4 \\ \dfrac{1}{\lambda_{rel.m}^2} & otherwise \end{vmatrix}$$

$$k_{crit} = 0.89$$

Axial compression condition

Design compression stress, $\sigma_{c.0.d}$

$$\sigma_{c.0.d} = \frac{N_d}{A}$$

$\sigma_{c.0.d} = 2.23 \text{ N/mm}^2$

Design compression strength, $f_{c.0.d}$

$$f_{c.0.d} = \frac{k_{mod.short} \cdot k_{sys} \cdot f_{c.0.k}}{\gamma_M}$$

$f_{c.0.d} = 26.25 \text{ N/mm}^2$

Buckling resistance condition (5.4.1 (EC5, 6.3.2)):

Relative slenderness about the y–y
axis, $\lambda_{rel.y}$
(equation (5.3); EC5, *equation (6.21)*)

$$\lambda_{rel.y} = \frac{\lambda_y}{\pi} \cdot \sqrt{\frac{f_{c.0.k}}{E_{0.05}}}$$

$\lambda_{rel.y} = 1.26$

Relative slenderness about the z–z
axis, $\lambda_{rel.z}$
(equation (5.3); EC5, *equation (6.22)*)

$$\lambda_{rel.z} = \frac{\lambda_z}{\pi} \cdot \sqrt{\frac{f_{c.0.k}}{E_{0.05}}}$$

$\lambda_{rel.z} = 2.79$

Factor β_c for LVL (equation (5.6);
EC5, *equation (6.29)*)

$\beta_c = 0.1$

Factor k_y (equation (5.5a); EC5,
equation (6.27))

$$k_y = 0.5 \cdot \left[1 + \beta_c \cdot (\lambda_{rel.y} - 0.3) + \lambda_{rel.y}^2 \right]$$

$k_y = 1.34$

Instability factor about the y–y
axis, $k_{c.y}$ (equation (5.4a); EC5,
equation (6.25))

$$k_{c.y} = \frac{1}{k_y + \sqrt{k_y^2 - \lambda_{rel.y}^2}}$$

$k_{c.y} = 0.56$

Factor k_z (equation (5.5b); EC5,
equation (6.28))

$$k_z = 0.5 \cdot \left[1 + \beta_c \cdot (\lambda_{rel.z} - 0.3) + \lambda_{rel.z}^2 \right]$$

$k_z = 4.52$

Instability factor about the z–z
axis, $k_{c.z}$ (equation (5.4b); EC5,
equation (6.26))

$$k_{c.z} = \frac{1}{k_z + \sqrt{k_z^2 - \lambda_{rel.z}^2}}$$

$k_{c.z} = 0.12$

Combined stress conditions
Including for equations (5.21) and (5.23) in addition to equation (5.24) in the strength check:

Compression stress condition about
the y–y axis
(equation (5.21) (EC5, 6.3.2(3)))

$$\frac{\sigma_{c.0.d}}{k_{c.y} \cdot f_{c.0.d}} + \frac{\sigma_{m.y.d}}{f_{m.y.d}} = 0.28$$

Compression stress condition about the z–z axis (equation (5.23) (EC5, 6.3.2(3)))	$\dfrac{\sigma_{c.0.d}}{k_{c.z} \cdot f_{c.0.d}} + k_m \dfrac{\sigma_{m.y.d}}{f_{m.y.d}} = 0.77$
Combined stress condition (equation (5.24); EC5, *equation (6.35)*)	$\left(\dfrac{\sigma_{m.y.d}}{k_{crit} \cdot f_{m.y.d}}\right)^2 + \dfrac{\sigma_{c.0.d}}{k_{c.z} \cdot f_{c.0.d}} = 0.7$

As all relationships are less than unity, the 90 mm by 200 mm LVL member will meet the ULS requirements of EC5.

Example 5.7.5 A 63 mm by 125 mm sawn timber section of strength class C24 to BS EN 338:2009 functioning under service class 2 conditions is shown in Figure E5.7.5 and is subjected to a characteristic permanent tensile action of 1.0 kN and a characteristic variable tensile medium-term action of 4 kN along the direction of the x–x axis of the member. The variable tensile action also induces a variable medium-term moment of 1.0 kNm about the y–y axis and 0.10 kNm about the z–z axis. There is no loss of area in the member at each end connection.

Check that the member will meet the ULS requirements of EC5.

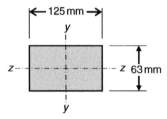

Fig. E5.7.5. Section of member.

1. Geometric properties

Thickness of the member, t	$t=63$ mm	
Width of the member, h	$h=125$ mm	
Length of the member, l	$l=3000$ mm	
Cross-sectional area of the member, A	$A=t \cdot h$	$A=7.88 \times 10^3$ mm^2
Second moment of area about the y–y axes, I_y	$I_y = \dfrac{t \cdot h^3}{12}$	$I_y = 1.03 \times 10^7$ mm^4
Section modulus about the y–y axes, W_y	$W_y = \dfrac{2I_y}{h}$	$W_y = 1.64 \times 10^5$ mm^3
Second moment of area about z–z axis, I_z	$I_z = \dfrac{h \cdot t^3}{12}$	$I_z = 2.6 \times 10^6$ mm^4
Section modulus about the z–z axes, W_z	$W_z = \dfrac{2I_z}{t}$	$W_z = 8.27 \times 10^4$ mm^3

2. Timber properties

Table 1.3, strength class C24 (BS EN 338:2009(E), *Table 1*)

Characteristic tensile strength parallel to the grain, $f_{t.0.k}$	$f_{t.0.k} = 14 \text{ N/mm}^2$
Characteristic bending strength about the y–y axis, $f_{m.y.k}$	$f_{m.y.k} = 24 \text{ N/mm}^2$
Characteristic bending strength about the z–z axis, $f_{m.z.k}$	$f_{m.z.k} = 24 \text{ N/mm}^2$
5% tile modulus of elasticity parallel to the grain, $E_{0.05}$	$E_{0.05} = 7.4 \text{ kN/mm}^2$
5% tile shear modulus, $G_{0.05}$	$G_{0.05} = \dfrac{E_{0.05}}{16}$ $G_{0.05} = 0.46 \text{ kN/mm}^2$

3. Partial safety factors

Table 2.8 (UKNA to BS EN 1990:2002+A1:2005, *Table NA.A1.2(B)*) for the ULS

Permanent actions, γ_G	$\gamma_G = 1.35$
Variable actions, γ_Q	$\gamma_Q = 1.5$

Table 2.6 (UKNA to EC5, *Table NA.3*)

Material factor for solid timber, γ_M	$\gamma_M = 1.3$

4. Actions

Characteristic permanent compressive action, G_k 　　　　　 $G_k = 1.0 \text{ kN}$

Characteristic medium-term compressive variable action, Q_k 　　　　　 $Q_k = 4.0 \text{ kN}$

Design tensile action for the critical load combination, N_d (Table 2.8, equation (c) using the unfavourable condition variable action) 　　　　　 $N_d = \gamma_G \cdot G_k + \gamma_Q \cdot Q_k$ 　　　　　 $N_d = 7.35 \times 10^3 \text{ N}$

Characteristic moment due to the variable action about the y–y axis, $M_{Q.y.k}$ 　　　　　 $M_{Q.y.k} = 1.0 \text{ kN m}$

Design moment about the y–y axis due to the variable actions, $M_{y.d}$ 　　　　　 $M_{y.d} = \gamma_Q \cdot M_{Q.y.k}$ 　　　　　 $M_{y.d} = 1.5 \text{ kN m}$

Characteristic moment due to the variable action about the z–z axis, $M_{Q.z.k}$ 　　　　　 $M_{Q.z.k} = 0.1 \text{ kN m}$

Design moment about the z–z axis due to the variable actions, $M_{z.d}$ 　　　　　 $M_{z.d} = \gamma_Q \cdot M_{Q.z.k}$ 　　　　　 $M_{z.d} = 0.15 \text{ kN m}$

5. Modification factors

Factor for medium-duration loading
and service class 2, $k_{\mathrm{mod.med}}$
(Table 2.4 (EC5, *Table 3.1*))

$$k_{\mathrm{mod.med}} = 0.8$$

System strength factor, k_{sys}
– not relevant

$$k_{\mathrm{sys}} = 1$$

Size factor for bending:

Depth factor for solid timber –
bending about y–y axis, k_{h}
(Table 2.11 (EC5, *3.2*))

$$k_{\mathrm{h}} = \begin{vmatrix} 1.0 & if \quad h \geq 150 \text{ mm} \\ \left(\dfrac{150 \text{ mm}}{h}\right)^{0.2} & if \quad 1.3 > \left(\dfrac{150 \text{ mm}}{h}\right)^{0.2} \\ 1.3 & otherwise \end{vmatrix}$$

$$k_{\mathrm{h}} = 1.04$$

Depth factor for solid timber –
bending about z–z axis, k_{t}
(Table 2.11 (EC5, *3.2*))

$$k_{\mathrm{t}} = \begin{vmatrix} 1.0 & if \quad t \geq 150 \text{ mm} \\ \left(\dfrac{150 \text{ mm}}{t}\right)^{0.2} & if \quad 1.3 > \left(\dfrac{150 \text{ mm}}{t}\right)^{0.2} \\ 1.3 & otherwise \end{vmatrix}$$

$$k_{\mathrm{t}} = 1.19$$

Size factor for tension:

Depth factor for solid timber in
tension, $k1_{\mathrm{h}}$
(Table 2.11 (EC5, *3.2*))

$$k1_{\mathrm{h}} = k_{\mathrm{h}}$$

$$k1_{\mathrm{h}} = 1.04$$

Lateral torsional stability of the
column acting as a beam, k_{crit},
(4.5.1.2 (EC5, *6.3.3*))

Effective length of the column acting
as a beam - taking moment about the
NA as the design condition, l_{ef}
(Table 4.2)

$$l_{\mathrm{ef}} = l$$

$$l_{\mathrm{ef}} = 3 \text{ m}$$

Critical bending stress, $\sigma_{\mathrm{m.crit}}$,
(Equation 4.9b (EC5,
equation 6.31))

$$\sigma_{\mathrm{m.crit}} = \dfrac{\pi \cdot t^2 \left[E_{0.05} \cdot G_{0.05} \cdot \left(1 - 0.63 \cdot \dfrac{t}{h}\right) \right]^{0.5}}{h \cdot \ell_{\mathrm{ef}}}$$

$$\sigma_{\mathrm{m,\,crit}} = 50.82 \text{ N/mm}^2$$

Relative slenderness for bending,
$\lambda_{\mathrm{rel,m}}$
(Equation 4.10 (EC5,
equation 6.30))

$$\lambda_{\mathrm{rel.m}} = \sqrt{\dfrac{f_{\mathrm{m.y.k}}}{\sigma_{\mathrm{m.crit}}}}$$

$$\lambda_{\mathrm{rel,m}} = 0.69$$

Lateral stability factor, k_{crit}
Table 4.3 (EC5, *equation 6.34*)

$$k_{crit} = \begin{vmatrix} 1 & \text{if} & \lambda_{rel.m} \le 0.75 \\ 1.56 - 0.75 \lambda_{rel.m} & \text{if} & 0.75 < \lambda_{rel.m} \le 1.4 \\ \dfrac{1}{\lambda_{rel.m}^2} & & \text{otherwise} \end{vmatrix}$$

$$k_{crit} = 1$$

i.e. lateral torsional instability of the column under bending will not arise

6. Strength of member

The critical design load case at the ULS will be due to the combination of permanent and unfavourable medium-duration variable actions:

Moment condition

Design bending moment about the y–y axis, $M_{y.d}$

$$M_{y.d} = 1.5 \, \text{kN m}$$

Design bending stress about the y–y axis, $\sigma_{m.y.d}$

$$\sigma_{m.y.d} = \frac{M_{y.d}}{W_y} \qquad\qquad \sigma_{m.y.d} = 9.14 \, \text{N/mm}^2$$

Design bending moment about the z–z axis, $M_{z.d}$

$$M_{z.d} = 0.15 \, \text{kN m}$$

Design bending stress about the z–z axis, $\sigma_{m.z.d}$

$$\sigma_{m.z.d} = \frac{M_{z.d}}{W_z} \qquad\qquad \sigma_{m.z.d} = 1.81 \, \text{N/mm}^2$$

Design bending strength about the y–y axis, $f_{m.y.d}$

$$f_{m.y.d} = \frac{k_{mod.med} \cdot k_{sys} \cdot k_h \cdot f_{m.y.k}}{\gamma_M} \qquad f_{m.y.d} = 15.32 \, \text{N/mm}^2$$

Design bending strength about the z–z axis, $f_{m.z.d}$

$$f_{m.z.d} = \frac{k_{mod.med} \cdot k_{sys} \cdot k_t \cdot f_{m.z.k}}{\gamma_M} \qquad f_{m.z.d} = 17.57 \, \text{N/mm}^2$$

Axial tensile condition

Design tension stress parallel to the grain, $\sigma_{t.0.d}$

$$\sigma_{t.0.d} = \frac{N_d}{A} \qquad\qquad \sigma_{t.0.d} = 0.93 \, \text{N/mm}^2$$

Design tension strength parallel to the grain, $f_{t.0.d}$

$$f_{t.0.d} = \frac{k_{mod.med} \cdot k_{sys} \cdot k1_h \cdot f_{t.0.k}}{\gamma_M} \qquad f_{t.0.d} = 8.94 \, \text{N/mm}^2$$

Redistribution factor for a rectangular section, k_m
(equation (4.4c) (EC5, *6.1.6*))

$$k_m = 0.7$$

Combined stress condition

Equations (5.25) and (5.26); EC5, *equations (6.17) and (6.18)*:

$$\frac{\sigma_{t.0.d}}{f_{t.0.d}} + \frac{\sigma_{m.y.d}}{f_{m.y.d}} + k_m \cdot \frac{\sigma_{m.z.d}}{f_{m.z.d}} = 0.77$$

$$\frac{\sigma_{t.0.d}}{f_{t.0.d}} + k_m \frac{\sigma_{m.y.d}}{f_{m.y.d}} + \frac{\sigma_{m.z.d}}{f_{m.z.d}} = 0.63$$

Relationships less than unity; therefore the 63 mm by 125 mm sawn section in strength class C24 will meet the ULS requirements of EC5.

(Had the value of k_{crit} been greater than 0.75, the procedure referred to in 5.43 can be used.)

Example 5.7.6 The stud wall shown in Figure E5.7.6 has an overall height of 3.75 m and the studs are spaced at 600 mm centre to centre with braced battens at mid-height. Sawn timber of 44 mm by 100 mm is used for the studs and the header and sole plates are 50 mm by 100 mm sawn timber, all class C16 to BS EN 338:2009. The wall functions in service class 2 conditions and supports a characteristic permanent action of 0.6 kN (inclusive of the panel self-weight) and a characteristic variable long-term action of 2.4 kN per stud. There is wall sheathing on both faces and the fixings provide lateral support to the studs about the z–z axis.

Check that the wall will meet the ULS requirements of EC5.

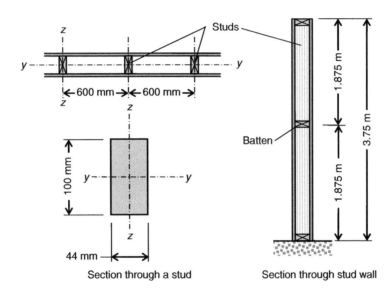

Fig. E5.7.6.

1. Geometric properties

Stud length, L	$L = 3.75$ m	
Effective length of a stud buckling about the y–y axis, $L_{e.y}$ (based on PD6693-1 and Table 5.2)	$L_{e.y} = 0.85 \cdot 1.0 \cdot L$	i.e. $L_{e.y} = 3.19$ m
Width of each stud, b	$b = 44$ mm	
Depth of each stud, h	$h = 100$ mm	
Bearing area of each stud on the sole plate, A_b	$A_b = b \cdot h$	
Cross-sectional area of each stud, A	$A = b \cdot h$	$A = 4.4 \times 10^3 \, \text{mm}^2$
Second moment of area of a stud about the y–y axes, I_y	$I_y = \dfrac{b \cdot h^3}{12}$	$I_y = 3.67 \times 10^6 \, \text{mm}^4$

Radius of gyration of a stud about the y–y axis, i_y	$i_y = \sqrt{\dfrac{I_y}{A}}$	$i_y = 28.87\,\text{mm}$
Slenderness ratio of a stud about the y–y axis, λ_y	$\lambda_y = \dfrac{L_{e.y}}{i_y}$	$\lambda_y = 110.42$

2. **Timber strength properties**

Table 1.3, strength class C16 (BS EN 338:2009(E), *Table 1*)

Characteristic compression strength parallel to the grain, $f_{c.0.k}$	$f_{c.0.k} = 17/\text{mm}^2$
Characteristic compressive strength perpendicular to the grain, $f_{c.90.k}$	$f_{c.90.k} = 2.2\,\text{N/mm}^2$
Fifth-percentile modulus of elasticity parallel to the grain, $E_{0.05}$	$E_{0.05} = 5.4\,\text{kN/mm}^2$

3. **Partial safety factors**

Table 2.8 (UKNA to BS EN 1990:2002+A1:2005, *Table NA.A1.2(B)*) for the ULS

Permanent actions, γ_G	$\gamma_G = 1.35$
Variable actions, γ_Q	$\gamma_Q = 1.5$

Table 2.6 (UKNA to EC5, *Table NA.3*)

Material factor for solid timber, γ_M	$\gamma_M = 1.3$

4. **Actions per stud**

Characteristic permanent compressive action, G_k	$G_k = 0.6\,\text{kN}$
Characteristic long-term compressive variable action, Q_k	$Q_k = 2.4\,\text{kN}$
Design compressive action for the critical load combination, N_d (Table 2.8, equation (c) using the unfavourable condition variable action)	$N_d = \gamma_G \cdot G_k + \gamma_Q \cdot Q_k$ $\qquad N_d = 4.41 \times 10^3\,\text{N}$

5. **Modification factors**

Factor for long-duration loading and service class 2, $k_{mod.long}$ Table 2.4 (EC5, *Table 3.1*))	$k_{mod.long} = 0.7$
System strength factor, k_{sys} (2.3.7 (EC5, *6.6*))	$k_{sys} = 1.1$
Adopt bearing factor, $k_{c.90} = 1$ (For header plate condition) (equation (4.22) (EC5, *6.1.5 (2)*)))	$k_{c.90} = 1$

6. Compression strength of a stud

The critical design load case at the ULS will be due to the combination of permanent and unfavourable long-duration variable action:

Design compression stress in a stud, $\sigma_{c.0.d}$

$$\sigma_{c.0.d} = \frac{N_d}{A}$$

$\sigma_{c.0.d} = 1 \text{ N/mm}^2$

Design compression strength of a stud, $f_{c.0.d}$

$$f_{c.0.d} = \frac{k_{mod.med} \cdot k_{sys} \cdot f_{c.0.k}}{\gamma_M}$$

$f_{c.0.d} = 10.07 \text{ N/mm}^2$

Buckling resistance condition (5.3.1 (EC5, 6.3.2)):

Relative slenderness about the y–y axis, $\lambda_{rel.y}$ (Equation (5.3); EC5, equation (6.21))

$$\lambda_{rel.y} = \frac{\lambda_y}{\pi} \cdot \sqrt{\frac{f_{c.0.k}}{E_{0.05}}}$$

$\lambda_{rel.y} = 1.97$

As the relative slenderness ratio is greater than 0.3, conditions in 5.3.1 apply (EC5, Clause 6.3.2(3)):

Factor β_c for solid timber (equation (5.6); EC5, equation (6.29))

$\beta_c = 0.2$

Factor k_y

(equation (5.5a); EC5, equation (6.27))

$$k_y = 0.5 \cdot \left[1 + \beta_c \cdot \left(\lambda_{rel.y} - 0.3 \right) + \lambda_{rel.y}^2 \right]$$

$k_y = 2.61$

Instability factor about the y–y axis (equation (5.4a); EC5, equation (6.25))

$$k_{c.y} = \frac{1}{k_y + \sqrt{k_y^2 - \lambda_{rel.y}^2}}$$

$k_{c.y} = 0.23$

Strength condition (equation 5.12)

$$\frac{\sigma_{c.0.d}}{k_{c.y} \cdot f_{c.0.d}} = 0.43$$

The relationship is less than unity; therefore the 44 mm by 100 mm studs, strength class C16, will meet the ULS compression strength requirement of EC5.

7. Bearing strength of sole plates – header plate will be the critical design condition

Design bearing load on the stud, N_d

$N_d = 4.41 \times 10^3 \text{ N}$

Effective contact length, l_{ef}

$l_{ef} = b + 60 \cdot \text{mm}$

Effective contact area perpendicular to the grain, A_{ef}

$A_{ef} = h \cdot l_{ef}$

$A_{ef} = 1.04 \times 10^4 \text{ mm}^2$

Design bearing stress, $\sigma_{c,90,d}$

$$\sigma_{c.90.d} = \frac{N_d}{A_{ef}}$$

$\sigma_{c.90.d} = 0.42 \text{ N/mm}^2$

Design bearing strength, $f_{c.90.d}$

$$f_{c.90.d} = \frac{k_{mod.long} \cdot f_{c.90.k}}{\gamma_M}$$

$f_{c.90.d} = 1.18 \text{ N/mm}^2$

Design bearing stress/strength ratio:

$$\frac{\sigma_{c.90.d}}{k_{c.90} \cdot f_{c.90.d}} = 0.36$$

The bearing stress to strength ratio is less than unity, so the bearing strength of the plates will be satisfactory.

Example 5.7.7 A stud wall in a domestic property, of the same layout as shown in Figure E5.7.6, has an overall height of 3.20 m. The studs are spaced at 600 mm centre to centre and braced battens are at mid-height. The studs, header and sole plates are 38 mm by 125 mm section, all strength class C18 to BS EN 338:2009. The wall functions in service class 2 conditions and each stud supports a characteristic vertical permanent action of 0.5 kN, a characteristic vertical variable medium-term action of 2.0 kN and a characteristic lateral wind action of 0.75 kN/m². Sheathing material is fixed on both faces but, because of openings, it does not provide full buckling restraint about the z–z axis of the studs.

Check that the member will meet the ULS requirements of EC5.

1. Geometric properties

Stud length, L	$L = 3.2\,\text{m}$	
Effective length of a stud buckling about the y–y axis, $L_{e.y}$ (Table 5.2)	$L_{e.y} = 1.0 \cdot L$	$L_{e.y} = 3.2\,\text{m}$
Effective length of a stud buckling about the z–z axis, $L_{e.z}$ (Table 5.2)	$L_{e.z} = 0.5 \cdot L$	$L_{e.z} = 1.6\,\text{m}$
Width of each stud, b	$b = 38\,\text{mm}$	
Depth of each stud, h	$h = 125\,\text{mm}$	
Lateral spacing of each stud, s_{stud}	$s_{\text{stud}} = 600\,\text{mm}$	
Bearing area of each stud on the sole plate, A_b	$A_b = b \cdot h$	
Cross-sectional area of each stud, A	$A = b \cdot h$	$A = 4.75 \times 10^3\,\text{mm}^2$
Second moment of area of a stud about the y–y axes, I_y	$I_y = \dfrac{b \cdot h^3}{12}$	$I_y = 6.18 \times 10^6\,\text{mm}^4$
Radius of gyration of a stud about the y–y axis, i_y	$i_y = \sqrt{\dfrac{I_y}{A}}$	$i_y = 36.08\,\text{mm}$
Slenderness ratio of a stud about the y–y axis, λ_y	$\lambda_y = \dfrac{L_{e.y}}{i_y}$	$\lambda_y = 88.68$
Section modulus of a stud about the y–y axes, W_y	$W_y = \dfrac{2 I_y}{h}$	$W_y = 9.9 \times 10^4\,\text{mm}^3$
Second moment of area of a stud about the z–z axis, I_z	$I_z = \dfrac{h \cdot b^3}{12}$	$I_z = 5.72 \times 10^5\,\text{mm}^4$
Radius of gyration of a stud about the z–z axis, i_z	$i_z = \sqrt{\dfrac{I_z}{A}}$	$i_z = 10.97\,\text{mm}$
Slenderness ratio of a stud about the z–z axis, λ_z	$\lambda_z = \dfrac{L_{e.z}}{i_z}$	$\lambda_z = 145.86$

2. **Timber strength properties**
   Table 1.3, strength class C18 (BS EN 338:2009(E), *Table 1*)

   | Characteristic compression strength parallel to the grain, $f_{c.0.k}$ | $f_{c.0.k}=18 \text{ N/mm}^2$ |

   Characteristic compression strength parallel to the grain, $f_{c.0.k}$ $f_{c.0.k}=18 \text{ N/mm}^2$

   Characteristic bending strength, $f_{m.y.k}$ $f_{m.y.k}=18 \text{ N/mm}^2$

   Characteristic compressive strength perpendicular to the grain, $f_{c.90.k}$ $f_{c.90.k}=2.2 \text{ N/mm}^2$

   Fifth-percentile modulus of elasticity parallel to the grain, $E_{0.05}$ $E_{0.05}=6.0 \text{ kN/mm}^2$

3. **Partial safety factors**
   Table 2.8 (UKNA to BS EN 1990:2002+A1:2005, *Table NA.A1.2(B)*) for the ULS

   Permanent actions, γ_G $\gamma_G=1.35$

   Variable actions, γ_Q $\gamma_Q=1.5$

   Table 2.2 (UKNA to EN 1990:2002(E)+A1:2005, *TableNA.A1.1*)

   Combination factor for a variable action, ψ_0 $\psi_0=0.7$

   Combination factor for a wind action, $\psi_{0.w}$ $\psi_{0.w}=0.5$

   Table 2.6 (UKNA to EC5, *Table NA.3*)

   Material factor for solid timber, γ_M $\gamma_M=1.3$

4. **Actions**

   Characteristic permanent compressive action, G_k $G_k=0.5 \text{ kN}$

   Characteristic medium-term compressive variable action, Q_k $Q_k=2.0 \text{ kN}$

   Characteristic variable lateral (wind) action, W_k $W_k=0.75 \text{ kN/m}^2$

   Combined load cases to obtain the design condition for the stud design

   (i) Vertical loading – associated with critical design load cases

   Design axial loading with permanent action only on the wall, $N1_d$ $N1_d=\gamma_G \cdot G_k$ $N1_d=0.68 \text{ kN}$

Design axial loading with permanent action; unfavourable medium-term variable and wind (dominant) actions, $N2_d$
$$N2_d = \gamma_G \cdot G_k + \psi_0 \cdot \gamma_Q \cdot Q_k$$
$N2_d = 2.77$ kN

(ii) Associated design lateral loading

Design lateral loading with permanent action only on the wall, $W1_d$
$$W1_d = 0 \text{ kN/m}^2$$

Design lateral loading with permanent action; unfavourable medium-term variable and wind (dominant) actions, $W2_d$
$$W2_d = \gamma_Q \cdot W_k$$
$W2_d = 1.13 \times 10^{-3}$ N/mm^2

Design moment per stud, $M2_{y.d}$
$$M2_{y.d} = W2_d \cdot s_{stud} \cdot \frac{L^2}{8}$$
$M2_{y.d} = 864$ N m

(iii) Combined load cases for plate design

Design axial loading due to permanent and medium-term variable action, $N3_d$
$$N3_d = \gamma_G \cdot G_k + \gamma_Q \cdot Q_k$$
$N3_d = 3.67$ kN

5. Modification factors

Factor for permanent-duration loading and service class 2, $k_{mod.perm}$ (Table 2.4 (EC5, *Table 3.1*))
$$k_{mod.perm} = 0.6$$

Factor for medium-duration loading and service class 2, $k_{mod.med}$ (Table 2.4 (EC5, *Table 3.1*))
$$k_{mod.med} = 0.8$$

Factor for instantaneous-duration loading and service class 2, $k_{mod.inst}$ (Table 2.4 (EC5, *Table 3.1*))
$$k_{mod.inst} = 1.1$$

System strength factor, k_{sys} – 2.3.7 (EC5, *6.7*)
$$k_{sys} = 1.1$$

Adopt bearing factor, $k_{c.90} = 1$ (For header plate condition) (equation (4.22) (EC5, *6.1.5 (2)*))
$$k_{c.90} = 1$$

Depth factor for solid timber – bending about y–y axis, k_h (Table 2.11 (EC5, *3.2*))
$$k_h = \min\left[\left(\frac{150}{h \cdot \text{mm}^{-1}}\right)^{0.2}, 1.3\right]$$
$$k_h = 1.04$$

6. Strength of the studs

The critical design load case at the ULS will be due to the combination of permanent, unfavourable wind (dominant) and unfavourable medium-term variable action:

Axial compression condition

Design compression stress due to permanent and combined medium-term and wind (dominant) actions, $\sigma2_{c.0.d}$

$$\sigma2_{c.0.d} = \frac{N2_d}{A}$$

$\sigma2_{c.0.d} = 0.58 \text{ N/mm}^2$

Design compression strength of a stud under permanent action only, $f1_{c.0.d}$

$$f1_{c.0.d} = \frac{k_{mod.perm} \cdot k_{sys} \cdot f_{c.0.k}}{\gamma_M}$$

$f1_{c.0.d} = 9.14 \text{ N/mm}^2$

Design compression strength of a stud under permanent action, medium-term variable and wind (dominant) action, $f3_{c.0.d}$

$$f3_{c.0.d} = \frac{k_{mod.inst} \cdot k_{sys} \cdot f_{c.0.k}}{\gamma_M}$$

$f3_{c.0.d} = 16.75 \text{ N/mm}^2$

Buckling resistance condition (5.3.1 (EC5, 6.3.2)):

Relative slenderness about the y–y axis, $\lambda_{rel.y}$ (equation (5.3); EC5, *equation (6.21)*)

$$\lambda_{rel.y} = \frac{\lambda_y}{\pi} \cdot \sqrt{\frac{f_{c.0.k}}{E_{0.05}}}$$

$\lambda_{rel.y} = 1.55$

Relative slenderness about the z–z axis, $\lambda_{rel.z}$ (equation (5.3); EC5, *equation (6.22)*)

$$\lambda_{rel.z} = \frac{\lambda_z}{\pi} \cdot \sqrt{\frac{f_{c.0.k}}{E_{0.05}}}$$

$\lambda_{rel.z} = 2.54$

As both relative slenderness ratios are greater than 0.3, conditions in 5.3.1 b) apply (EC5, *6.3.2(3)*):

Redistribution factor for a rectangular section, k_m (see equation (4.4) (EC5, *6.1.6*))

$k_m = 0.7$

Factor β_c for solid timber (equation (5.6); EC5, *equation (6.29)*)

$\beta_c = 0.2$

Factor k_z (equation (5.5b); EC5, *equation (6.28)*)

$$k_z = 0.5 \cdot \left[1 + \beta_c \cdot \left(\lambda_{rel.z} - 0.3 \right) + \lambda_{rel.z}^2 \right]$$

$k_z = 3.96$

Instability factor about the z–z axis, $k_{c.z}$

$$k_{c.z} = \frac{1}{k_z + \sqrt{k_z^2 - \lambda_{rel.z}^2}}$$

$k_{c.z} = 0.14$

(equation 5.4b (EC5, *equation 6.26*))

Factor, k_y
(equation (5.5a); EC5,
equation (6.27))

$$k_y = 0.5 \cdot \left[1 + \beta_c \cdot \left(\lambda_{\text{rel.}y} - 0.3 \right) + \lambda_{\text{rel.}y}^2 \right]$$

$k_y = 1.82$

Instability factor about the
y–y axis, $k_{c.y}$
(equation (5.4a); EC5,
equation (6.25))

$$k_{c.y} = \frac{1}{k_y + \sqrt{k_y^2 - \lambda_{\text{rel.}y}^2}}$$

$k_{c.y} = 0.36$

Moment condition

Design bending stress
about the y–y axis due to
permanent and combined
medium-term vertical and
wind (dominant) action,
$\sigma2_{y.d}$

$$\sigma2_{\text{m.}y.d} = \frac{M2_{y.d}}{W_y}$$

$\sigma2_{\text{m.}y.d} = 8.73 \, \text{N/mm}^2$

Design bending strength
about the y–y axis, $f_{\text{m.}y.d}$

$$f_{\text{m.}y.d} = \frac{k_{\text{mod.inst}} \cdot k_{\text{sys}} \cdot k_h \cdot f_{\text{m.}y.k}}{\gamma_M}$$

$f_{\text{m.}y.d} = 17.38 \, \text{N/mm}^2$

Lateral torsional stability of
stud functioning as a beam,
k_{crit} (5.3.1 (EC5, *6.3.3*))

Effective length of a stud,
$L_{\text{ef.b}}$: With braced battens, the
distance between lateral
restraint $= L/2$. Due to udl
loading a factor of 0.9 has
been used (approximate
solution) and because
loading is applied at the
compression edge, $2h$ has
been added.
(Table 4.2 (EC5, *Table 6.1*))

$$L_{\text{ef.b}} = 0.9 \cdot \left(\frac{L}{2} \right) + 2 \cdot h$$

$L_{\text{ef.b}} = 1.69 \, \text{m}$

Critical bending stress, $\sigma_{\text{m.crit}}$
(equation (4.8); EC5,
equation (6.32))

$$\sigma_{\text{m.crit}} = \frac{0.78 \cdot b^2}{h \cdot L_{\text{ef.b}}} \cdot E_{0.05}$$

$\sigma_{\text{m.cirt}} = 31.99 \, \text{N/mm}^2$

Lateral torsional buckling
condition (4.5.1.2 (EC5,
6.3.3)):

| Relative slenderness for bending, $\lambda_{\text{rel.m}}$ (equation (4.10); EC5, equation (6.30)) | $\lambda_{\text{rel.m}} = \sqrt{\dfrac{f_{\text{m.y.k}}}{\sigma_{\text{m.crit}}}}$ | $\lambda_{\text{rel.m}} = 0.75$ |

| Lateral stability factor, k_{crit} (Table 4.3 (EC5, equation (6.34)) | $k_{\text{cirt}} = \begin{vmatrix} 1 & if & \lambda_{\text{rel.m}} \leq 0.75 \\ 1.56 - 0.75 \cdot \lambda_{\text{rel.m}} & if & 0.75 < \lambda_{\text{rel.m}} \leq 1.4 \\ \dfrac{1}{\lambda_{\text{rel.m}}^2} & & otherwise \end{vmatrix}$ | $k_{\text{cirt}} = 1$ |

Combined stress conditions

Including for equations (5.35) and (5.36) in addition to equation (5.38) in the strength check:

| Combined axial and bending stress condition about the y–y axis – functioning as a column (equation (5.35) (EC5, 6.3.2(3))) | $\dfrac{\sigma 2_{\text{c.0.d}}}{k_{\text{c.y}} \cdot f3_{\text{c.0.d}}} + \dfrac{\sigma 2_{\text{m.y.d}}}{f_{\text{m.y.d}}} = 0.599$ |

| Compression stress condition about the z–z axis (equation (5.36) (EC5 6.3.2(3))) | $\dfrac{\sigma 2_{\text{c.0.d}}}{k_{\text{c.z}} \cdot f3_{\text{c.0.d}}} + k_{\text{m}} \cdot \dfrac{\sigma 2_{\text{m.y.d}}}{f_{\text{m.y.d}}} = 0.595$ |

| Combined bending and axial stress condition functioning as a beam (equation (5.38); EC5, equation (6.35)): | $\left(\dfrac{\sigma 2_{\text{m.y.d}}}{k_{\text{crit}} \cdot f_{\text{m.y.d}}} \right)^2 + \dfrac{\sigma 2_{\text{c.0.d}}}{k_{\text{c.z}} \cdot f3_{\text{c.0.d}}} = 0.498$ |

The critical condition is due to combined axial stress and bending about the y–y axis, functioning as a column and with wind as the dominant variable action. The studs are OK as the combined compression and bending ratio is less than unity.

7. Bearing strength of the sole plates – header plate will be the critical design condition

For this condition, the greatest vertical load arises with the vertical variable load as the only variable load case:

| Design bearing load per stud, $N3_{\text{d}}$ | $N3_{\text{d}} = 3.67 \times 10^3$ N | |

| Effective contact length, | $l_{\text{ef}} = b + 60 \cdot$ mm
$l_{\text{ef}} = 98$ mm | |

| Effective contact area perpendicular to the grain, A_{ef} | $A_{\text{ef}} = h \cdot l_{\text{ef}}$ | $A_{\text{ef}} = 1.23 \times 10^4$ mm^2 |

| Design bearing stress, $\sigma_{\text{c.90.d}}$ | $\sigma_{\text{c.90.d}} = \dfrac{N3_{\text{d}}}{A_{\text{ef}}}$ | $\sigma_{\text{c.90.d}} = 0.3$ N/mm^2 |

Design bearing strength, $f_{c.90.d}$ $\qquad f_{c.90.d} = \dfrac{k_{mod.med} \cdot f_{c.90.k}}{\gamma_M}$ $\qquad f_{c.90.d} = 1.35 \text{ N/mm}^2$

Design bearing stress/ $\qquad\qquad \dfrac{\sigma_{c.90.d}}{k_{c.90} \cdot f_{c.90.d}} = 0.22$
strength ratio:

The bearing stress to strength ratio is less than unity, so the bearing strength of the plates will be satisfactory.

Example 5.7.8 A stud wall in a domestic property with the same layout as shown in Figure E5.7.6 has an overall height of 2.4 m. The studs are spaced at 600 mm centre to centre and no battens are fitted at mid-height. Studs are 44 mm by 100 mm section, class C18 to BS EN 338:2009, and the wall functions in service class 2 conditions. Each wall stud supports a characteristic permanent action of 0.4 kN, a characteristic medium-duration variable action of 1.8 kN and a characteristic lateral wind action of 0.85 kN/m². Sheathing material is fixed on both faces and provides full buckling restraint about the z–z axis of the studs.

Determine the increase in the instantaneous lateral deflection of the wall at the serviceability limit states (SLS) under the characteristic combination of actions, taking wind loading as the dominant variable action.

1. Geometric properties

Stud length, L	$L = 2.4$ m	
Effective length of a stud for buckling about the y–y axis, $L_{e.y}$ (Table 5.2)	$L_{e.y} = 0.85 \cdot L$	i.e. $L_{e.y} = 2.04$ m
Width of a stud, b	$b = 44$ mm	
Depth of a stud, h	$h = 100$ mm	
Lateral spacing of each stud, s_{stud}	$s_{stud} = 600$ mm	

2. Timber strength properties

Table 1.3, strength class C18 (BS EN 338:2009(E), *Table 1*)

Fifth-percentile modulus of elasticity parallel to the grain, $E_{0.05}$	$E_{0.05} = 6.0 \text{ kN/mm}^2$
Mean modulus of elasticity parallel to the grain, $E_{0,mean}$	$E_{0,mean} = 9.0 \text{ kN/mm}^2$
Mean shear modulus, $G_{0,mean}$	$G_{0,mean} = 0.56 \text{ kN/mm}^2$

3. Partial safety factors

Table 2.2 (UKNA to EN 1990:2002(E)+A1:2005, *TableNA.A1.1*) – Category A

Combination factor for a variable action ψ_0	$\psi_0 = 0.7$
Combination factor for a wind action, $\psi_{0,w}$	$\psi_{0,w} = 0.5$

4. Actions

Characteristic permanent compressive action, G_k	$G_k = 0.4$ kN	
Characteristic medium-term compressive variable action, Q_k	$Q_k = 1.8$ kN	
Characteristic variable (wind) action, W_k	$W_k = 0.85$ kN·mm^{-2}	
Variable lateral action, F_{ser}	$F_{ser} = W_k$	$F_{ser} = 0.85$ kN/m^2
Permanent vertical action, N_{serdl}	$N_{serdl} = G_k$	$N_{serdl} = 0.4$ kN
Variable vertical action, N_{serll}	$N_{serll} = Q_k$	$N_{serll} = 1.8$ kN
Characteristic combination for vertical load with wind as the dominant action, G_{VLw}	$G_{VLw} = N_{serdl} + \psi_0 \cdot N_{serll}$	$G_{VLw} = 1.66$ kN
Characteristic lateral wind loading with wind as the dominant action, G_{HLW}	$G_{HLw} = F_{ser}$	

5. Deflection of the stud wall at the SLS

The critical design load case at the SLS will be due to the combination of permanent, unfavourable wind (dominant) and unfavourable medium-term variable action:

Maximum permitted out of straightness of a stud, δ_0 (EC5, *Section 10*)

$$\delta_0 = \frac{L}{300} \qquad\qquad \delta_0 = 8 \text{ mm}$$

Euler load of a strut about the y–y axis, P_E (equation (5.42))

$$P_E = \pi^2 \cdot \frac{E_{0.05}}{L_{e.y}^2} \cdot \frac{b \cdot h^3}{12} \qquad\qquad P_E = 52.17 \text{ kN}$$

Lateral displacement due to the horizontal loading, $\delta_{\text{inst. Gw}}$

(Table 4.7)

$$\delta_{\text{inst.Gw}} = \frac{5}{32} \cdot \frac{G_{HLw}}{E_{0,mean} \cdot b \cdot h^3} \cdot (L^4) \cdot \left[1 + 0.96 \cdot \left(\frac{E_{0,mean}}{G_{0,mean}}\right)\left(\frac{h}{L}\right)^2\right] \cdot (s_{stud}) \qquad \delta_{\text{inst. Gw}} = 6.86 \text{ mm}$$

Increase in the instantaneous value of the lateral deflection of the wall under the critical characteristic combination of loading, $\delta_{char,instw}$
(equation (5.46))

$$\delta_{char.instw} = \frac{1}{1 - \dfrac{G_{VLw}}{P_E}} \cdot \left(\frac{G_{VLw}}{P_E} \cdot \delta_0 + \delta_{inst.G_w}\right) \qquad\qquad \delta_{char.\,instw} = 7.34 \text{ mm}$$

The increase in the instantaneous lateral deflection of the wall at the SLS under the critical characteristic combination of loading will be 7.34 mm.

Chapter 6

Design of Glued-Laminated Members

6.1 INTRODUCTION

The use of glued-laminated members (glulam) for structural purposes offers the advantages of excellent strength and stiffness to weight ratio, can be designed to have a substantial fire resistance, and can achieve a high standard of architectural finish. Some examples of glued-laminated structures are shown in Figure 6.1. A summary of some of the main advantages of glulam members is as follows:

(a) As glulam sections are built up from thin members it is possible to manufacture complicated shapes. They can be produced in any size, length and shape. Manufacturing and transportation facilities remain as the only practical limiting factors affecting dimensions.

(b) The use of a number of laminates results in a more even distribution of defects throughout the section, reducing the variability and generally increasing the strength of the timber.

(c) Pre-cambers can easily be incorporated into the section during the manufacturing process.

In this chapter, the design procedure is given for glued-laminated timber members designed in accordance with the requirements of EC5 [1]. The particular requirements of curved members, tapered members and pitched cambered members functioning as beams are addressed, noting that the methodology used will also apply to members manufactured from laminated veneer lumber (LVL).

It should be noted that EN standards for the production and design of glued-laminated timber are under major revision with EN 14080 [2], when approved, becoming the design standard which will replace current key standards, including BS EN 386:2001 [3] and BS EN 1194 [4]. In the new standard there will be clarification between glued-laminated timber and glued solid timber with associated design procedures, however, because the content of the standard is still under discussion the design procedures covered in this edition of the book continue to be based on existing standards.

The general information in 4.3 is relevant to the content of this chapter.

Structural Timber Design to Eurocode 5, Second Edition. Jack Porteous and Abdy Kermani.
© Jack Porteous and Abdy Kermani 2013. Published 2013 by Blackwell Publishing Ltd.

(a) A straight beam

(b) Curved structure (photo courtesy of Constructional Timber Limited, a member of Glued Laminated Timber Association)

(c) Sloped columns (photo courtesy of APA, The Engineered Wood Association)

(d) Curved portals (photo courtesy of APA, The Engineered Wood Association)

(e) Portal frame (photo courtesy of Constructional Timber Ltd., a member of Glued Laminated Timber Association)

(f) A roof structure (photo courtesy of Lilleheden ltd., a member of Glued Laminated Timber Association)

(g) Footbridge (photo courtesy of Engineered Wood Products Association of Australia (EWPAA))

(h) Curved columns

Fig. 6.1. Examples of glued-laminated structures.

Table 6.1 Main design requirements for glued-laminated members and the associated EC5 limit states

Element	Design or displacement effect	EC5 limit states
Beam members	As Table 4.1	As Table 4.1
Columns	As Table 5.1	As Table 5.1

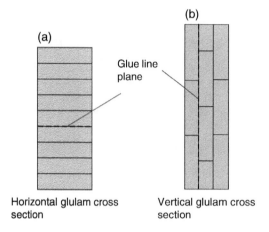

Horizontal glulam cross section

Vertical glulam cross section

Fig. 6.2. Horizontal and vertical glulam.

6.2 DESIGN CONSIDERATIONS

Glued-laminated members can function as beams or columns or members subjected to combined bending and axial loading, and the limit states associated with the main design effects are given in Table 6.1. The general comments in 4.2 and 5.2 are also applicable to glued-laminated elements.

The production requirements of glulam must comply with the requirements of BS EN 386 and the design procedure used is dependent on whether the member is constructed using laminations of the same grade and species (homogeneous glulam) or formed with inner and outer laminations of different grades and species (combined glulam). The requirements for each type are discussed in 6.3.

6.3 GENERAL

6.3.1 Horizontal and vertical glued-laminated timber

Glulam is formed by bonding timber laminations with the grain running essentially parallel and can be produced as horizontal glued-laminated timber or vertical glued-laminated timber, as defined in BS EN 386.

Horizontal glulam is glued-laminated timber in which the glue line planes are perpendicular to the long length of the cross-section. Vertical glulam is glued-laminated timber where the glue line planes are perpendicular to the short length of the cross-section. An example of each is shown in Figure 6.2.

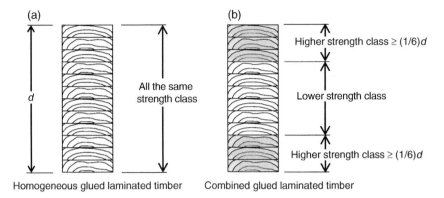

Fig. 6.3. Example of a homogeneous and a combined glued-laminated timber member.

6.3.2   Design methodology

Glued-laminated timber must comply with the requirements of BS EN386 and BS EN1194. BS EN1194 defines the strength classes associated with glulam and also states how the characteristic values of strength related and stiffness related properties are to be obtained. It enables grades and species as well as combinations of grades and species to be classified using a glulam strength class system in a similar manner to that used in BS 338:2009 [5] for structural timber.

The classification system only applies to horizontally laminated glulam made with at least four laminations, all of which must be softwood. Strength classes are given for homogeneous and for combined glued-laminated timber, and an example of each is shown in Figure 6.3.

The strength classes are defined as follows:

- *Homogeneous glued-laminated timber*: where all of the laminations in the section are of the same grade (strength class) and species, or species combinations.
- *Combined glued-laminated timber*: where the section is made from inner and outer laminations of different grades (strength classes) and species, or species combinations. The outer laminations will have the higher strength class and must equate to at least one-sixth of the member depth on both sides.

In BS EN 1194 the characteristic strength, stiffness and density values of glued-laminated members are given for four strength classes, defined as GL 24, GL 28, GL 32 and GL 38, each followed by the suffix h, where the glulam is homogeneous, or c, if it is combined.

As in the case of the reference system used for timber in BS EN 338, the numbers used in the glulam system represent the characteristic bending strength of the glulam in N/mm^2 and the properties of some of the strength classes given in BS EN 1194 are reproduced in Table 6.2.

Mechanical properties of glued-laminated timber can be derived from the equivalent properties of the softwood laminations being used. The relationships between the materials are given in *Table A.1* of BS EN 1194 and summarised in Table 6.3.

Table 6.2 The strength, stiffness and density properties of homogeneous glulam strength classes (GL 24h, GL 28h and GL 32h) and combined glulam strength classes (GL 24c, GL 28c and GL 32c)*

Glulam strength class	Homogeneous glulam			Combined glulam		
	GL 24h	GL 28h	GL 32h	GL 24c	GL 28c	GL 32c
Bending strength $f_{m,g,k}$ (N/mm²)	24	28	32	24	28	32
Tension strength						
$f_{t,0,g,k}$ (N/mm²)	16.5	19.5	22.5	14	16.5	19.5
$f_{t,90,g,k}$ (N/mm²)	0.4	0.45	0.5	0.35	0.4	0.45
Compression strength						
$f_{c,0,g,k}$ (N/mm²)	24	26.5	29	21	24	26.5
$f_{c,90,g,k}$ (N/mm²)	2.7	3.0	3.3	2.4	2.7	3.0
Shear strength						
$f_{v,g,k}$ (N/mm²)	2.7	3.2	3.8	2.2	2.7	3.2
Modulus of elasticity						
$E_{0,g,mean}$ (kN/mm²)	11.6	12.6	13.7	11.6	12.6	13.7
$E_{0,g,05}$ (kN/mm²)	9.4	10.2	11.1	9.4	10.2	11.1
$E_{90,g,mean}$ (kN/mm²)	0.39	0.42	0.46	0.32	0.39	0.42
Shear modulus						
$G_{g,mean}$ (kN/mm²)	0.72	0.78	0.85	0.59	0.72	0.78
Density $\rho_{g,k}$ (kg/m³)	380	410	430	350	380	410

Note: For strength-related calculations, take $G_{g,05} = E_{0,g,05}/16$.
Mean density is taken to be the average of the mean density of the inner and outer laminates, based on BS EN 338.
* Based on the properties given in BS EN 1194:1999.

The properties of the laminations are derived from the tension characteristic strength, the mean tension modulus and the characteristic density and values can be obtained from the equivalent properties given in BS EN 338:2009 for the softwood species to be used, or they can be derived as described in BS EN 1194. Particular strength requirements must also be met for the end joints in all laminations and alternative options for achieving this are given in BS EN 1194, *Clause 6.3.2.*

Horizontally glued-laminated timber is also able to be formed from some UK sourced temperate hardwoods (oak and sweet chestnut) in accordance with the rules in BS EN 1194, and details of the strength class permitted for laminations of such hardwoods can be obtained from PD6693-1 [6]. A list of other hardwood species suitable for horizontal glued-laminated timber is also given in that document.

The equations in Table 6.3 are valid for homogeneous glulam made from softwood laminations or laminations from hardwoods listed in PD6693-1 and can be used to calculate the properties of glued-laminated timber that does not conform to any of the four strength classes given in the standard.

Where the mechanical properties of a combined glued-laminated member are required, on the basis that the zones of different laminations equate to at least one-sixth

Table 6.3 The characteristic strength, stiffness and density properties of homogeneous glued-laminated timber derived from the properties of the softwood laminations being used[†]

Characteristic property of homogeneous glued-laminated timber made from softwood laminations	BS EN 1194:1999 relationship (based on the properties of the softwood laminations[*])
Bending strength $f_{m,g,k}$ (N/mm^2)	$=7+1.15f_{t,0,l,k}$
Tension strength $f_{t,0,g,k}$ (N/mm^2) $f_{t,90,g,k}$ (N/mm^2)	$=5+0.8f_{t,0,l,k}$ $=0.2+0.015f_{t,0,l,k}$
Compression strength $f_{c,0,g,k}$ (N/mm^2) $f_{c,90,g,k}$ (N/mm^2)	$=7.2f_{t,0,l,k}^{0.45}$ $=0.7f_{t,0,l,k}^{0.5}$
Shear strength $f_{v,k}$ (N/mm^2)	$=0.32f_{t,0,l,k}^{0.8}$ **
Modulus of elasticity $E_{0,g,mean}$ (N/mm^2) $E_{0,g,05}$ (N/mm^2) $E_{90,g,mean}$ (N/mm^2)	$=1.05E_{0,l,mean}$ $=0.85E_{0,l,mean}$ $=0.035E_{0,l,mean}$
Shear modulus $G_{g,mean}$ (N/mm^2)	$=0.065E_{0,l,mean}$
Density $\rho_{g,k}$ (Kg/m^3)	$=1.10\rho_{\ell,k}$

* The lamination properties are as follows: $f_{t,0,l,k}$ – the characteristic tensile strength of the softwood lamination in N/mm^2; $E_{0,l,mean}$ – the mean value of the modulus of elasticity of the lamination parallel to the grain in N/mm^2; $\rho_{\ell,k}$ – the characteristic density of the softwood lamination in kg/m^3.
† Based on *Table A.1* in BS EN 1194:1999.
** The values used for shear strength are based on BS EN 338:2003. The shear strength values in the currents standard BS EN 338:2009, are different.

of the beam depth (as shown in Figure 6.3b) or two laminations, whichever is greater, the relationships given in Table 6.3 will apply to the properties of the individual parts of the cross-section of the member.

In BS EN 1194 examples are given of beam lay-ups complying with the requirements of Table 6.2, where the properties of the laminations in the section have been derived using the characteristic values in BS EN 338:2003 and these are summarised in Table 6.4. It should be noted that in BS EN 338:2009 there has been a reduction in the characteristic tensile strength and an increase in the characteristic shear strength of softwood laminations from the 2003 version and these changes have not resulted in amendments to the strength properties of the strength classes given in *Tables 1* and *2* in BS EN 1194.

The characteristic bending strength, $f_{m,g,k}$, in Table 6.2 relates to members with a minimum depth of 600 mm and a minimum thickness of 150 mm, and the tensile strength parallel to the grain, $f_{t,0,g,k}$, relates to members with a minimum width of 600 mm and a minimum thickness of 150 mm. These strengths can be increased when glulam sections with smaller sizes are used by the application of a size effect factor,

Table 6.4 Examples of beam lay-ups compliant with Table 6.2, in which the characteristic properties are derived from the equations given in Table 6.3 using properties obtained from BS EN 338:2003 for the strength class being used for the laminations

Glulam strength classes	GL 24	GL 28	GL 32
Homogeneous glulam – lamination strength class	C24	C30	C40
Combined glulam: outer/inner – lamination strength classes	C24/C18	C30/C24	C40/C30

referred to in 2.3.6 and Table 2.11. EC5 only permits the factor to be applied to the depth for bending and the width for tension and makes no adjustment for thickness. The value of the factor, k_h, is:

$$k_h = \min \left\{ \begin{array}{c} (600/h)^{0.1} \\ 1.1 \end{array} \right\} \qquad \text{(EC5, equation (3.2))} \quad (6.1)$$

where h is the depth of the glulam member (in mm) when subjected to bending and the width of the glulam member (in mm) when subjected to tension.

When using homogeneous glued-laminated timber or combined glued-laminated timber compliant with the strength classes given in BS EN1194:1999, *Table 2* (i.e. the combined glulam strength classes given in Table 6.2 above), conventional bending theory will apply. The design procedure for uniform section straight members subjected to bending, shear, torsion and axial loading will be as described in Chapters 4 and 5, respectively, and the relevant strength property given in Table 6.2 will be used in the strength condition being validated.

If combined glued-laminated timber with properties that are not compliant with the GL 24c to GL 38c strength classes is to be formed, bending strength verification must be carried out at all relevant points in the cross-section, and because this will involve the analysis of members with different values of modulus of elasticity, the *equivalent section method* [7, 8] (commonly referred to as the modular ratio approach or the transformed–section method) can be used for the analysis. In this method, the material used in the inner or the outer laminations is selected for use and the other material is replaced by an equivalent area of the selected material such that the force in the replaced material at any distance from the neutral axis caused by bending of the section will be the same as that to be taken by the original material at the same position. By this method, a section using only one material is formed and the theory of bending can be applied enabling the stress in the material selected for the section to be found directly. For the converted laminations, the stress is obtained by multiplying the calculated stress in these members by the ratio of the E value of the original material to the E value of the selected material (called the modular ratio). The mean value of the glulam stress class, $E_{0,g,mean}$, should be used in the analysis. Further information on the method is given in Chapter 7.

Design values of glued-laminated members are derived in the same way as for solid timber sections, using those factors in EC5 that are applicable to the material. For example, to derive the design bending strength of a glulam member compliant with BS EN 1194,

$$f_{m,d} = \frac{k_{mod} \cdot k_h \cdot k_{sys} \cdot f_{m,g,k}}{\gamma_M} \qquad (6.2)$$

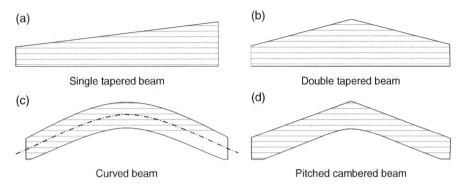

(a)

Single tapered beam

(b)

Double tapered beam

(c)

Curved beam

(d)

Pitched cambered beam

Fig. 6.4. Single and double tapered, curved and pitched cambered beams.

where k_{mod} is the modification factor for load duration and service classes as given in Table 2.4, k_h is the modification factor for member size effect, referred to in Table 2.5, k_{sys} is the system strength factor discussed in 2.3.7, $f_{m,g,k}$ is the characteristic bending strength of the glulam strength class being used, and γ_M is the partial factor for material properties, given in Table 2.6.

See Examples 6.7.1 and 6.7.2.

6.4   DESIGN OF GLUED-LAMINATED MEMBERS WITH TAPERED, CURVED OR PITCHED CURVED PROFILES (ALSO APPLICABLE TO LVL MEMBERS)

Glued-laminated (and LVL) beams can be tapered or curved in order to meet architectural requirements, to provide pitched roofs, to obtain maximum interior clearance and/or to reduce wall height requirements at end supports. The most common forms used in timber structures are double tapered beams and curved beams having a rectangular cross-section, as shown in Figure 6.4.

Because of the sloping surface, with these members the distribution of bending stress is non-linear across any section and, in the apex zone of the beam types shown in Figures 6.4b–6.4d, radial stresses perpendicular to the grain are also induced. If the bending moment tends to increase the radius of curvature, the radial stresses will be in tension perpendicular to the grain and if it tends to decrease the radius of curvature, the radial stresses will be in compression perpendicular to the grain.

With these types of beams it is recommended that the laminations are set parallel with the tension edge such that the tapered sides will be on the compression face when subjected to normal loading conditions.

6.4.1   Design of single tapered beams

These beams are rectangular in section and slope linearly from one end to the other as shown in Figure 6.5. No upper limit is set in EC5 for the angle of slope, α, but in practice it would normally be within the range 0–10°.

These types of beams are used in roof construction and the critical design checks will relate to the maximum shear stress and the maximum bending stress condition at the ULS and the deflection behaviour at the SLS.

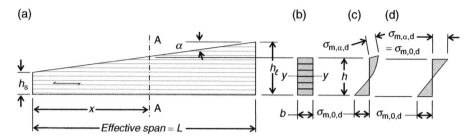

Fig. 6.5. A single tapered beam: (a) elevation; (b) section A–A; (c) bending stress; and (d) bending stress distribution used in EC5 at section A–A.

With regard to the maximum shear stress, because of the taper in the beam, the distribution of shear stress will vary across the depth of any section and along the beam length. When subjected to uniformly distributed loading or to a point load at mid-span, the maximum shear stress will occur at mid-height of the beam at the end where the beam depth is h_s and will be:

$$\tau_{v,d} = \frac{1.5 V_d}{b h_s} \tag{6.3}$$

where V_d is the design shear force at the end of the beam.

If there is a requirement to determine the shear stress at any other position along the beam, a reasonably accurate approximation can be obtained using the approach adopted by Maki and Keunzi [9].

The design shear strength of the tapered rectangular beam is derived in the same way as described in 4.5.2 for a uniform rectangular timber section, and is:

$$f_{v,g,d} = \frac{k_{mod} \cdot k_{sys} \cdot f_{v,g,k}}{\gamma_M} \tag{6.4}$$

where the factors are as previously defined and $f_{v,g,k}$ is the characteristic shear strength of the glulam strength class being used. Where beam lay-ups not compliant with Table 6.2 are used, the shear strength should be based on the characteristic shear strength of the inner laminations.

Also, because of the taper, when subjected to a bending moment, M, the bending stress distribution across the section will be non-linear as indicated in Figure 6.5c and, on the basis of analysis by Riberholt [10], at any cross-section along the tapered beam the maximum stress parallel to the tapered face at the tapered face and the maximum horizontal tensile stress on the horizontal face can be approximated to:

$$\text{Tapered face bending stress} = \left(1 - 3.7 \tan^2 \alpha\right) M/W_y \tag{6.5a}$$

$$\text{Parallel face bending stress} = \left(1 + 3.7 \tan^2 \alpha\right) M/W_y \tag{6.5b}$$

where W_y is the section modulus about the y–y axis.

In EC5, to simplify the design process, stresses in the section are derived using conventional elastic bending theory ignoring the taper effect. The bending stress at the tapered face acting parallel to the tapered face and the bending stress at the horizontal surface on the bottom face of the beam, as shown in Figure 6.5c, are taken to equal the bending stresses derived assuming there is no taper, as shown in Figure 6.5d. As is seen from equations (6.5) this will be a safe approximation for the stress at the tapered face but will slightly underestimate the stress at the bottom face. For beam tapers up to about 10% the stress on the bottom parallel face will be underestimated by a maximum of around 11%. When the taper increases beyond 10% the underestimate in the value of the stress on the bottom face increases relatively rapidly and it is recommended that in such circumstances equation (6.5b) be used to derive the maximum bending stress at this position and that the stress be validated against the bending strength of the section.

On the basis of the EC5 approximation, when subjected to a design moment M_d at a position x measured from the end where the beam depth is h_s, as shown in Figure 6.5, the maximum bending stress at the tapered face of the beam at an angle α to the grain and at the outermost fibre on the beam face parallel to the grain will be:

$$\sigma_{m,\alpha,d} = \sigma_{m,0,d} = \frac{M_d}{W} \qquad \text{(EC5, equation (6.37))} \quad (6.6)$$

where $\sigma_{m,\alpha,d}$ is the design bending stress at an angle α to the grain as shown in Figure 6.5d, $\sigma_{m,0,d}$ is the design bending stress parallel to the grain as shown in Figure 6.5d, and W is the section modulus at the position of the applied moment $= bh^2/6$, with b the member width and h the member depth at position x.

These stresses will be a function of the type of loading and the beam geometry, and where the beam is simply supported with a profile as shown in Figure 6.5, the maximum stress condition will occur when x has the value given in Table 6.5. The values of the maximum stresses in the section at position x are also given in the table.

Table 6.5 The position and value of the maximum bending stress in a single tapered beam when subjected to different loading conditions

Loading condition on a simply supported tapered beam	Position of maximum bending stress measured from the end where the depth is h_s, x	Section modulus at the position of maximum stress, W	Maximum bending stresses in the section at x, $\sigma_{m,\alpha,d}$, $\sigma_{m,0,d}$
Uniformly distributed load, q_d	$x = \dfrac{L}{\left(1 + \dfrac{h_\ell}{h_s}\right)}$	$W_x = \dfrac{2b}{3}\left(\dfrac{h_\ell}{\left(1 + \dfrac{h_\ell}{h_s}\right)}\right)^2$	$\dfrac{q_d x (L - x)}{2 W_x}$
Point load at mid span, Q_d	when $\dfrac{h_\ell}{h_s} \leq 3$: $x = \dfrac{L}{2}$	$W1_x = \dfrac{b\left(h_\ell + h_s\right)^2}{24}$	$\dfrac{Q_d x}{2\left(W1_x\right)}$
	when $\dfrac{h_\ell}{h_s} > 3$ $x = \dfrac{L}{\left(\dfrac{h_\ell}{h_s} - 1\right)}$	$W2_x = \dfrac{2}{3} b h_s^2$	$\dfrac{Q_d x}{2\left(W2_x\right)}$

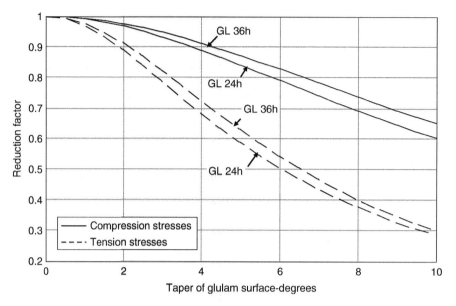

Fig. 6.6. Strength reduction factor $k_{m,\alpha}$ for tapered glulam members (dashed lines for tension stresses and solid lines for compression stresses parallel to the tapered face).

The design requirement in EC5 is that at the outermost fibre of the tapered edge the design stress, $\sigma_{m,\alpha,d}$, must satisfy the following expression:

$$\sigma_{m,\alpha,d} \leq k_{m,\alpha} \cdot f_{m,d} \qquad \text{(EC5, equation (6.38))} \quad (6.7)$$

where $f_{m,d}$ is the design bending strength of the glulam member as defined in equation (6.2) and $k_{m,\alpha}$ is as follows:

(a) For tensile stresses parallel to the tapered face of the beam:

$$k_{m,\alpha} = \frac{1}{\sqrt{1 + \left(\left(f_{m,d}/\left(0.75 \cdot f_{v,d}\right)\right)\tan\alpha\right)^2 + \left(\left(f_{m,d}/f_{t,90,d}\right)\tan^2\alpha\right)^2}} \qquad (6.8)$$

$$\text{(EC5, equation (6.39))}$$

where $f_{v,d}$ is the design shear strength of the beam and for a glulam beam:

$$f_{v,d} = \frac{k_{mod} \cdot k_{sys} \cdot f_{v,g,k}}{\gamma_M}$$

$f_{t,90,d}$ is the design tensile strength of the beam and for a glulam beam:

$$f_{t,90,d} = \frac{k_{mod} \cdot k_{sys} \cdot f_{t,90,g,k}}{\gamma_M}$$

(b) For compressive stresses parallel to the tapered face of the beam:

$$k_{m,\alpha} = \frac{1}{\sqrt{1 + \left(\left(f_{m,d} / \left(1.5 \cdot f_{v,d} \right) \right) \tan \alpha \right)^2 + \left(\left(f_{m,d} / f_{c,90,d} \right) \tan^2 \alpha \right)^2}} \qquad (6.9)$$

(EC5, *equation (6.40)*)

where $f_{c,90,d}$ is the design compressive strength perpendicular to the grain and for a glulam beam:

$$f_{c,90,d} = \frac{k_{mod} \cdot k_{sys} \cdot f_{c,90,g,k}}{\gamma_M}$$

In the above relationships the functions k_{mod}, k_{sys} and γ_M are as previously defined and $f_{v,g,k}$, $f_{t,90,g,k}$ and $f_{c,90,g,k}$ are the characteristic shear strength, the characteristic tension strength perpendicular to the grain and the characteristic compression strength perpendicular to the grain respectively. Values of the strength properties for GL 24, GL 28 and GL 32 are given in Table 6.2.

Figure 6.6 shows the effect of the reduction factor, $k_{m,\alpha}$ when applied to glulam beams of homogeneous glulam compliant with the strength classes given in *Table 1* of BS EN 1194. For tensile stresses parallel to the tapered edge, the factor is represented by a dashed line and for compression stresses parallel to the tapered edge, by a solid line. It can be seen that the value of $k_{m,\alpha}$ is primarily dependent on the angle of taper of the beam and the largest strength reduction will always be associated with the condition where the tapered face is subjected to tension. For example, when using a beam with a taper of 10°, irrespective of the strength grade of the glulam, the bending strength will only be approximately 30% of the non-tapered beam of the same depth when subjected to tension bending stresses parallel to the tapered edge, compared to approximately 60 to 65% when it is subjected to compression bending stresses on the same face.

Where the relevant stress being considered is a compression bending stress, the effect of lateral instability will also have to be taken into account and for such conditions:

$$\sigma_{m,0,d} \leq k_{crit} f_{m,d}$$

and/or

$$\sigma_{m,\alpha,d} \leq k_{crit} k_{m,\alpha} f_{m,d} \qquad (6.10)$$

In equation (6.10) the functions are as previously defined and k_{crit} is the lateral torsional instability factor referred to in 4.5.1.2. An approximate conservative value of k_{crit} will be obtained by assuming a uniform beam depth of h_ℓ.

If the benefit of the size factor is to be taken into account or the design condition arises from some loading arrangement other than those referred to in Table 6.5, the position and value of the maximum stress condition will have to be obtained by trial and error, calculating the stresses at intervals along the beam length.

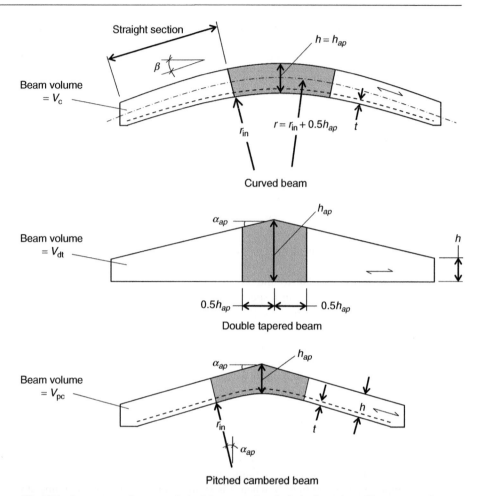

Fig. 6.7. Apex zones for curved, double tapered and pitched cambered beams are shown shaded.

Because of the tapered profile, deflection calculations are more involved than those required for beams with a uniform profile and equations to calculate the flexural and shear deflection for this type of beam when subjected to a uniformly distributed load or a point load at mid-span are given in Annex 6.1. They are based on the deflection equations for tapered beams given in the *Timber Designers' Manual* [11]. The maximum deflection is taken to occur at the centre of the beam span, which is acceptable for design purposes.

See Example 6.7.3.

6.4.2   Design of double tapered beams, curved and pitched cambered beams

These types of beams are shown in Figure 6.7 and are rectangular in cross-section.

The critical design checks for these beams are the same as those referred to in 6.4.1 for single tapered beams. At the ULS the maximum shear stress and the maximum

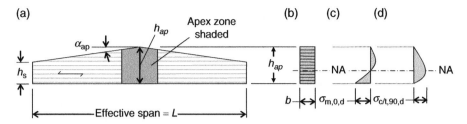

Fig. 6.8. Bending and radial stresses in the apex zone of a double tapered beam: (a) elevation of a double tapered beam; (b) section at apex; (c) bending stress at apex; (d) radial stress at apex.

bending stress condition must be validated, and at the SLS the deflection behaviour must be shown to be acceptable. With these profiles, however, in addition to design checks in the tapered area of the beam, the stress condition in the apex zone must also be validated, taking into account the effect on material strength arising from:

- residual stresses caused by the production process,
- stress distribution and volume effects,
- the combination of shear stresses in the zone and radial tension stresses perpendicular to the laminations caused by bending.

The stressed volume in the area of the apex affected by the radial stresses referred to above is illustrated in Figure 6.7, and in EC5 it is called the apex zone. The zone must be limited to a maximum of 2/3 V_b, where V_b is the total volume of the beam.

6.4.2.1 Bending and radial stresses in the apex zone – for double tapered and pitched cambered beams

In the apex zone of double tapered beams and pitched cambered beams, the bending stress distribution is complex and non-linear and is shown in Figure 6.8 for a double pitched beam. The bending stress at the apex will be zero and the bending stress distribution will be as shown in Figure 6.8c. The radial stress induced in the section will be a maximum at the neutral axis position and the distribution will be as indicated in Figure 6.8d.

6.4.2.1.1 Bending stress in the apex zone

Under the action of a design moment $M_{ap,d}$ at the apex, the maximum bending stress, $\sigma_{m,0,d}$, shown in Figure 6.8c is defined in EC5 as:

$$\sigma_{m,0,d} = k_\ell \frac{6M_{ap,d}}{bh_{ap}^2} \qquad \text{(EC5, equation (6.42))} \quad (6.11)$$

where b and h_{ap} are the width and depth, respectively, of the section at the apex as shown in Figure 6.8b and:

k_ℓ is obtained from

$$k_\ell = k_1 + k_2\left(\frac{h_{ap}}{r}\right) + k_3\left(\frac{h_{ap}}{r}\right)^2 + k_4\left(\frac{h_{ap}}{r}\right)^3 \qquad \text{(EC5, equation (6.43))} \quad (6.12)$$

$$k_1 = 1 + 1.4\tan(\alpha_{ap}) + 5.4\tan^2(\alpha_{ap})$$ (EC5, *equation (6.44)*) (6.13)

$$k_2 = 0.35 - 8\tan(\alpha_{ap})$$ (EC5, *equation (6.45)*) (6.14)

$$k_3 = 0.6 + 8.3\tan(\alpha_{ap}) - 7.8\tan^2(\alpha_{ap})$$ (EC5, *equation (6.46)*) (6.15)

$$k_4 = 6\tan^2(\alpha_{ap})$$ (EC5, *equation (6.47)*) (6.16)

$$r = r_{in} + 0.5h_{ap}$$ (EC5, *equation (6.48)*) (6.17)

where r_{in} is the inner radius of the beam, as shown in Figure 6.7c for a pitched cambered beam.

As an alternative to calculation, the value of k_ℓ can be obtained from Figure A6.2.1 in ANNEX 6.2 where the function is plotted for various angles of α_{ap} over a range of h_{ap}/r_{in} from 0 to 0.5.

6.4.2.1.2   *Radial stress in the apex zone*

The radial stress in a double tapered and a pitched cambered beam can be taken to be a maximum at the neutral axis position, as shown in the case of a double tapered beam in Figure 6.8d, and reduces to zero at the top and bottom faces.

Under the action of a design moment, $M_{ap,d}$, at the apex the critical radial stress condition will be the maximum tensile stress perpendicular to the grain, $\sigma_{t,90,d}$, and in accordance with the requirement of *NA.2.4* of the UKNA to EC5 [12] the stress will be derived from

$$\sigma_{t,90,d} = k_p \frac{6M_{ap,d}}{bh_{ap}^2}$$ (EC5, *equation (6.54)*) (6.18)

where b and h_{ap} are the width and depth, respectively, of the section at the apex as shown in Figure 6.8b and for a double tapered beam $k_p = 0.2\tan(\alpha_{ap})$ and for a pitched cambered beam it is obtained from:

$$k_p = k_5 + k_6\left(\frac{h_{ap}}{r}\right) + k_7\left(\frac{h_{ap}}{r}\right)^2$$ (EC5, *equation (6.56)*) (6.19)

$$k_5 = 0.2\tan(\alpha_{ap})$$ (EC5, *equation (6.57)*) (6.20)

$$k_6 = 0.25 - 1.5\tan(\alpha_{ap}) + 2.6\tan^2(\alpha_{ap})$$ (EC5, *equation (6.58)*) (6.21)

$$k_7 = 2.1\tan(\alpha_{ap}) - 4\tan^2(\alpha_{ap})$$ (EC5, *equation (6.59)*) (6.22)

$$r = r_{in} + 0.5h_{ap}$$ as defined in equation (6.17).

As an alternative to calculation, the value of k_p can be obtained from Figure A6.4, where the function is plotted for various angles of α_{ap} over a range of h_{ap}/r_{in} from 0 to 0.5.

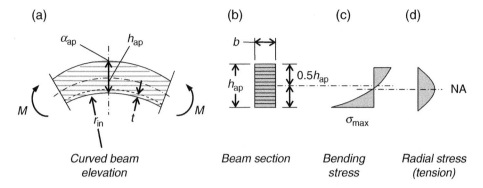

Fig. 6.9. Bending and radial stresses in a curved beam.

6.4.2.2 *Bending and radial stresses in the apex zone – of a curved beam*

Unlike the case of a uniform straight beam, under the action of a pure moment on a uniform section curved beam, as shown in Figure 6.9, the bending stress distribution across any section along the beam will not be linear. The position of the neutral axis will be below the centroidal axis and the bending stress distribution will be as shown in Figure 6.9c, with the maximum bending stress occurring at the inner radius position.

The radial tension stress induced in the section will be a maximum at the neutral axis position reducing in a non-linear manner to zero at the outer faces as indicated in Figure 6.9d.

6.4.2.2.1 *Bending stress in the apex zone*

The bending stress distribution in a curved beam can be derived using the Airy stress function written in terms of polar coordinates, and the maximum bending stress will be shown to occur at the inner radius position as indicated in Figure 6.9c. In EC5 a close approximation to the value of the maximum stress at the inner radius position in the apex zone of a curved beam of breadth b and depth h_{ap} when subjected to a design moment $M_{ap,d}$ is obtained from EC5, *equation (6.42)*, by setting the angle of taper, α_{ap}, equal to 0 in EC5, *equation (6.43)*, as follows,

$$\sigma_{m,0,d} = k_{curve,b}\,\frac{6M_d}{bh_{ap}^2} \qquad (6.23)$$

where

$$k_{curve,b} = 1 + 0.35\left(\frac{h_{ap}}{r}\right) + 0.6\left(\frac{h_{ap}}{r}\right)^2 \qquad (6.24)$$

and $r = r_{in} + 0.5h_{ap}$ as defined in equation (6.17).

As an alternative to calculation, the value of $k_{curve,b}$ can be obtained from Figure A6.2.1 where the value of the function will equal the value of k_ℓ when taking $\alpha_{ap} = 0°$.

6.4.2.2.2 *Radial stress in the apex zone*

As with double tapered and pitched cambered beams, the radial stress can be taken to be a maximum at the neutral axis position, and will reduce to zero at the top and bottom faces as shown in Figure 6.9d.

Under the action of a design moment, $M_{ap,d}$, at the apex, in EC5 the maximum tensile stress perpendicular to the grain, $\sigma_{t,90,d}$, at the neutral axis position in the apex zone of a curved beam of breadth b and depth h_{ap} is obtained from EC5, *equation (6.54)*, by setting α_{ap} (the angle of taper) equal to zero in EC5, *equation (6.56)*. In accordance with the requirement of *NA.2.4* in the UKNA to EC5 the stress will be:

$$\sigma_{t,90,d} = k_{curve,t}\frac{6M_{ap,d}}{bh_{ap}^2} \qquad \text{(EC5, equation (6.54)) \quad (6.25)}$$

where $k_{curve,t} = 0.25(h_{ap}/r)$, and, as defined in equation (6.17), $r = r_{in}+0.5h_{ap}$.

6.4.2.3   *Bending strength in the apex zone – for double tapered beams, curved beams and pitched cambered beams*

When a curved or pitched cambered glulam beam is being formed, the laminates are bent to the required curvature for the beam and as a consequence are subjected to a bending stress. This effect is not relevant to double tapered beams as they are produced from conventional glulam beams formed using straight laminations.

To take the effect of this bending stress into account, a stress reduction factor derived from tests is applied to the bending strength of the glulam beam and the reduced design bending strength, $f_{r,m,0,d}$, is given in EC5, *equation (6.41)*, as follows:

$$f_{r,m,0,d} = k_r f_{m,0,d} \qquad (6.26)$$

where $f_{m,0,d}$ is defined in equation (6.2) and k_r is a reduction factor. As there is no stress reduction in the case of double tapered beams, for this type the factor is taken to be 1. For curved and pitched cambered beams, the factor is based on the ratio of r_{in}/t, and the EC5 requirement is

- when $r_{in}/t \geq 240$ no strength reduction is required,
- when r_{in}/t is less than 240 the modification factor is $0.76 + 0.001(r_{in}/t)$.

Here the inner radius of the curved beam, r_{in}, and the lamination thickness, t, are shown in Figure 6.7.

6.4.2.4   *Radial tensile strength in the apex zone – for double tapered beams, curved beams and pitched cambered beams*

Larsen [13] and Colling [14] have shown that with these beam types the tensile strength in the apex zone of the section will be affected by stress and volume effects. Factors for these effects have been developed, and in EC5 the design tensile strength of the beam member at right angles to the grain, $f_{r,t,90,d}$, is obtained from

$$f_{r,t,90,d} = k_{dis}k_{vol}f_{t,90,d} \qquad (6.27)$$

where $f_{t,90,d}$ is defined in equation (6.8) and k_{dis} is a stress distribution factor in the apex zone. It is

- 1.4 for double tapered and curved beams, and
- 1.7 for pitched cambered beams.

k_{vol} is a volume factor, which for solid timber is equal to 1, and for glued-laminated timber (and LVL),

$$k_{vol} = \left(\frac{V_0}{V}\right)^{0.2}$$

Table 6.6 The stressed volume of the apex zone

Figure reference	Beam type	Stressed volume† (V)	Maximum allowable value of the stressed volume*
Figure 6.7a	Curved beam	$\dfrac{\beta\pi}{180} b\,(h_{ap}^2 + 2h_{ap}r_{in})$	$\dfrac{2}{3}V_c$
Figure 6.7b	Double tapered beam	$bh_{ap}^2\,(1 - 0.25\tan(\alpha_{ap}))$	$\dfrac{2}{3}V_{dt}$
Figure 6.7c	Pitched cambered beam	$b\left(\sin(\alpha_{ap})\cos(\alpha_{ap})(r_{in} + h_{ap})^2 - r_{in}^2\,\dfrac{\alpha_{ap}\pi}{180}\right)$	$\dfrac{2}{3}V_{pc}$

* The beam volume reference is shown in Figure 6.7.
† Angles β and α are in degrees.

where all veneers are parallel to the beam axis and V_0 is the reference volume against which the function exponent has been derived, and equals $0.01\,\text{m}^3$; V is the stressed volume (in m^3) of the apex zone with a maximum value of $2V_{beam}/3$, where V_{beam} is the volume of the beam.

From the geometry of the apex zone, for the beam types shown in Figure 6.7, all having a thickness b, the stressed volume, V, is given in Table 6.6.

6.4.2.5 Criterion for bending stress – for double tapered beams, curved beams and pitched cambered beams

To achieve an acceptable design, the design bending stress in the apex zone of the beam section must be no greater than its reduced design bending strength, i.e.:

$$\sigma_{m,0,d} \le k_r f_{m,0,d} \qquad\qquad \text{(EC5, equation (6.41))} \quad (6.28)$$

where $\sigma_{m,0,d}$ is the design tensile bending stress in the apex zone and is obtained from 6.4.2.1.1 for double tapered beams, pitched cambered beams and curved beams; $k_r f_{m,0,d}$ is the reduced design bending strength of the apex zone of the beam and is obtained from 6.4.2.3 for double tapered beams, pitched cambered beams and curved beams.

6.4.2.6 Criterion for radial tension stress – for double tapered beams, curved beams and pitched cambered beams

To achieve an acceptable design, the design tensile stress perpendicular to the grain in the apex zone of the beam section for the relevant beam type must be no greater than its reduced design tension strength, i.e.:

$$\sigma_{t,90,d} \le k_{dis}k_{vol} f_{t,90,d} \qquad\qquad \text{(EC5, equation (6.50))} \quad (6.29)$$

where $\sigma_{t,90,d}$ is the design tension stress in the apex zone and is obtained from 6.4.2.1.2 for double tapered beams and pitched cambered beams, and 6.4.2.2.2 for curved beams; $k_{dis}k_{vol}f_{t,90,d}$ is the reduced design tensile strength of the apex zone of the beam and is obtained from 6.4.2.4 for double tapered beams, pitched cambered beams and curved beams.

See Example 6.7.4.

6.4.3 Design of double tapered beams, curved and pitched cambered beams subjected to combined shear and tension perpendicular to the grain

When double tapered beams, curved beams and pitched cambered beams of breadth b and depth h_{ap} at the apex are subjected to combined shear stresses and tension stresses perpendicular to the grain, these stresses will interact and in EC5 the criterion to be met at the design condition is:

$$\frac{\tau_d}{f_{v,d}} + \frac{\sigma_{t,90,d}}{k_{dis}k_{vol}f_{t,90,d}} \leq 1 \qquad\qquad (EC5, \text{ equation } (6.53)) \quad (6.30)$$

Where

- τ_d is the design shear stress at the apex section. When the design shear force at the apex is F_d, the design shear stress can be assumed to be $\tau_d = 3F_d/2bh_{ap}$;
- $f_{v,d}$ is the design shear strength of the beam and for a glulam beam it will be $k_{mod}k_{sys}f_{v,g,k}/\gamma_M$, where the functions are as previously defined;
- $\sigma_{t,90,d}$ is the design tension stress on the beam at the apex and, for the relevant beam type, is obtained from 6.4.2.1.2 and 6.4.2.2.2;
- $k_{dis}k_{vol}f_{t,90,d}$ is the reduced tensile strength of the beam perpendicular to the grain and is obtained from 6.4.2.4.

6.5 FINGER JOINTS

If large finger joints complying with the requirements of BS EN 387:2001 [15] are required to be incorporated into a design where the direction of grain changes at the joint, this is only permitted for products to be installed in service class 1 or 2 conditions.

ANNEX 6.1

Deflection formulae for simply supported tapered and double tapered beams subjected to a point load at mid-span or to a uniformly distributed load.

The equations for deflection due to bending moment and to shear at mid-span in simply supported tapered and double tapered beams when subjected to a mid-span point load or a uniformly distributed load along the length of each beam type are based on the equations in the *Timber Designers' Manual* and are given in Table A6.1.1. Although the maximum deflection in a tapered beam when subjected to a point load at mid-span will be marginally off centre, the use of the central deflection value is good enough for design purposes.

A graphical representation of the equations in Table A6.1.1 is given in Figures A6.1.1, A6.1.2 and A6.1.3.

Table A6.1.1 Deflection formulae for simply supported tapered and double tapered beams subjected to a point load at mid-span or a uniformly distributed load along the beam length

Beam type and loading condition	Bending deflection at mid-span	Shear deflection at mid-span
Tapered beam with a point load F_d at mid-span, and $M_d = \dfrac{F_d L}{4}$	$\dfrac{5 M_d L^2 k1_{\delta b}}{96 E_{0,\text{g,mean}} I_{h_s}}$ $k1_{\delta b} = 19.2 \left(\dfrac{1}{a-1}\right)^3 \left(\ln a - 2\left(\dfrac{a-1}{a+1}\right)\right)$	$\dfrac{1.2 M_d k1_{\delta s}}{G_{\text{g,mean}} A_{h_s}}$ $k1_{\delta s} = \dfrac{1}{(a-1)} \ln a$
Tapered beam with a uniformly distributed load q_d/unit length along the span, and $M_d = \dfrac{q_d L^2}{8}$	$\dfrac{5 M_d L^2 k2_{\delta b}}{48 E_{0,\text{g,mean}} I_{h_s}}$ $k2_{\delta b} = 19.2 \left(\dfrac{1}{a-1}\right)^4 \left(3(a+1)\ln\left(\dfrac{a+1}{2}\right) - (2a+1)\ln a - \left(\dfrac{(a-1)^2}{2(a+1)}\right)\right)$	$\dfrac{1.2 M_d k2_{\delta s}}{G_{\text{g,mean}} A_{h_s}}$ $k2_{\delta s} = \dfrac{2(a+1)}{(a-1)^2} \ln\dfrac{(a+1)^2}{4a}$
Double tapered beam with a point load F_d at mid-span, and $M_d = \dfrac{F_d L}{4}$	$\dfrac{5 M_d L^2 k3_{\delta b}}{96 E_{0,\text{g,mean}} I_{h_s}}$ $k3_{\delta b} = 38.4 \left(\dfrac{1}{a-1}\right)^3 \left(\ln\left(\dfrac{a+1}{2}\right) + \dfrac{4}{(a+1)} - \dfrac{2}{(a+1)^2} - 1.5\right)$	$\dfrac{1.2 M_d k3_{\delta s}}{G_{\text{g,mean}} A_{h_s}}$ $k3_{\delta s} = \dfrac{2}{(a-1)} \ln\left(\dfrac{a+1}{2}\right)$
Double tapered beam with a uniformly distributed load q_d/unit length along the span, and $M_d = \dfrac{q_d L^2}{8}$	$\dfrac{5 M_d L^2 k4_{\delta b}}{48 E_{0,\text{g,mean}} I_{h_s}}$ $k4_{\delta b} = 19.2 \left(\dfrac{1}{a-1}\right)^3 \left(2\left(\dfrac{a+2}{a-1}\right)\ln\left(\dfrac{a+1}{2}\right) + \dfrac{3}{(a+1)} - \dfrac{2}{(a+1)^2} - 4\right)$	$\dfrac{1.2 M_d k4_{\delta s}}{G_{\text{g,mean}} A_{h_s}}$ $k4_{\delta s} = \dfrac{4}{(a-1)} \left(\left(\dfrac{a+1}{a-1}\right)\ln\left(\dfrac{a+1}{2}\right) - 1\right)$

$E_{0,\text{g,mean}}$ is the mean modulus of the glulam parallel to the grain; $G_{\text{g,mean}}$ is the shear modulus of the glulam; I_{h_s} is the second moment of area of the glulam beam with a depth $= h_s$, i.e.

$I_{h_s} = b h_s^3 / 12$

A_{h_s} is the cross-sectional area of the glulam beam with a depth $= h_s$, i.e.

$A_s = b h_s$

a is the ratio h_ℓ / h_s for tapered beams and h_{ap} / h_s for double tapered beams.

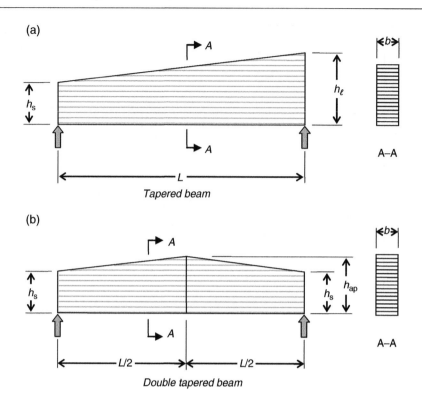

(a)

(b)

Fig. A6.1.1. Tapered and double tapered beams.

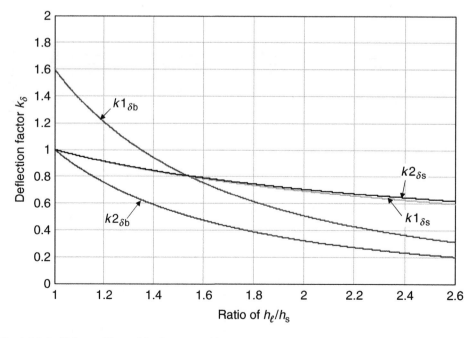

Fig. A6.1.2. Values of $k_{\delta b}$ and $k_{\delta s}$ for tapered beams.

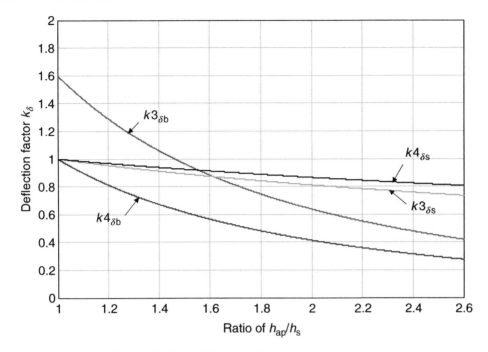

Fig. A6.1.3. Values of $k_{\delta b}$ and $k_{\delta s}$ for double tapered beams.

ANNEX 6.2

Graphical representation of factors k_ℓ and k_p used in the derivation of the bending and radial stresses in the apex zone of double tapered curved and pitched cambered beams.

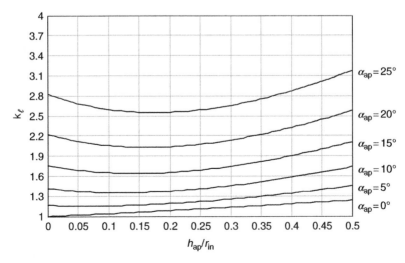

Fig. A6.2.1. Plot of k_ℓ against the ratio h_{ap}/r_{in} for various angles of slope α_{ap}.

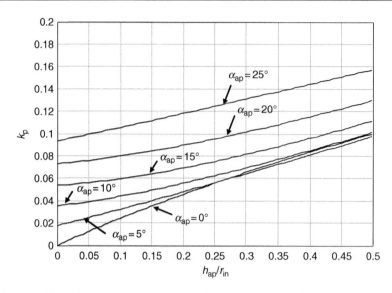

Fig. A6.2.2. Plot of k_p against the ratio h_{ap}/r_{in} for various angles of slope α_{ap}.

6.6   REFERENCES

1   BS EN 1995-1-1:2004+A1:2008. *Eurocode 5: Design of Timber Structures. Part 1-1: General – Common Rules and Rules for Buildings, British Standards Institution.*

2   prEN 14080. *Timber Structures – Glued-Laminated Timber and Glued Solid Timber.* CEN/TC 124.

3   BS EN 386:2001. *Glued-Laminated Timber – Performance Requirements and Minimum Production Requirements,* British Standards Institution.

4   BS EN 1194:1999. *Timber Structures – Glued-Laminated Timber – Strength Classes and Determination of Characteristic Values,* British Standards Institution.

5   BS EN 338:2009. *Structural Timber – Strength Classes,* British Standards Institution.

6   PD6693-1-1:2012, Incorporating Corrigendum No1: *PUBLISHED DOCUMENT – Recommendations for the design of timber structures to Eurocode 5: Design of Timber Structures – Part 1-1: General – Common Rules and Rules for Buildings,* British Standards Institution.

7   Hearn, E. J. *Mechanics of Materials, Volume 1,* 2nd edn. International Series on Materials Science and Technology, Volume 19. BPCC Wheatons Ltd, Exeter, 1985.

8   Gere, J.M., Timoshenko, S.P. *Mechanics of Materials,* 3rd SI edn. Chapman and Hall, London, 1991.

9   Maki, A.C., Kuenzi, E.W. Deflection and stresses of tapered wood beams. *Research Paper* FPL 34, US Forest Service, 1965.

10   Riberholt, H., 'Tapered timber beams'. In: *The Proceedings of the CIB W18 Meeting,* 1979. (Paper W18/11-10-1)

11   Ozelton, E.C., Baird, J.A. *Timber Designers' Manual,* 3rd edn. Blackwell Science, New York, 2002.

12 NA to BS EN 1995-1-1:2004+A1:2008, Incorporating National Amendment No. 2, *UK National Annex to Eurocode 5: Design of Timber Structures. Part 1-1: General – Common Rules and Rules for Buildings*, British Standards Institution.

13 Larsen, H.J. 'Eurocode 5 and CIB structural timber design code'. In: *The Proceedings of the CIB W18 Meeting*, Florence, Italy, 1986. (Paper 19-102-2)

14 Colling, F. 'Influence of volume and stress distribution on the shear strength and tensile strength perpendicular to grain'. In: *The Proceedings of the CIB W18 Meeting*, Florence, Italy, 1986. (Paper 19-12-3)

15 BE EN 387:2001. *Glued-Laminated Timber – Production Requirements for Large Finger Joints. Performance Requirements and Minimum Production Requirements*, British Standards Institution.

16 NA to BS EN 1990:2002 + A1:2005 *Incorporating National Amendment No1. UK National Annex for Eurocode 0 – Basis of Structural Design*, British Standards Institution.

6.7 EXAMPLES

As stated in 4.3, Chapter 4, in order to verify the ultimate and serviceability limit states, each design effect has to be checked and for each effect the largest value caused by the relevant combination of actions must be used.

However, to ensure attention is primarily focussed on the EC5 design rules for the timber or wood product being used, only the design load case producing the largest design effect has generally been given or evaluated in the following examples.

Example 6.7.1 A series of glulam beams 115 mm wide by 560 mm deep (a typical beam section is shown in Figure E6.7.1) with an effective span of 10.5 m is to be used in the construction of the roof of an exhibition hall. The roof comprises exterior tongued and grooved solid

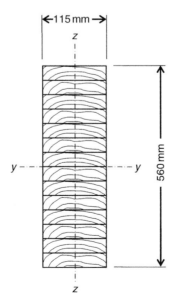

Fig. E6.7.1.

softwood decking exposed on the underside and covered on the top with insulation and a weather protective roof covering. The decking will provide full lateral support to the beam but load sharing between the beams is assumed not to apply. The beams are glulam strength class GL 24h to BS 1194:1999 and will function in service class 1 conditions. The bearing length at the end of each beam is 155 mm. Assume that the limiting value for vertical deflection at the instantaneous condition is span/300 and at the net final condition is span/250, for the loading condition given below:

(a) determine the required pre-camber in each beam;
(b) confirm that the beams will comply with the design rules in EC5.

Beam loading:

Characteristic permanent vertical load on each beam	1.4 kN/m
Characteristic short-term variable vertical load on each beam	2.5 kN/m

1. **Glulam beam geometric properties**

Breadth of each beam, b	$b = 115$ mm
Depth of each beam, h	$h = 560$ mm
Effective span of each beam, ℓ	$\ell = 10.5$ m
Bearing length at each end of a beam, ℓ_b	$\ell_b = 155$ mm

   Section modulus of each beam about the y–y axis, W_y
   $$W_y = \frac{b \cdot h^2}{6} \qquad W_y = 6.01 \times 10^6 \text{ mm}^3$$

2. **Glulam properties**

   Table 6.2, homogeneous grade GL 24h

Characteristic bending strength, $f_{m.g.k}$	$f_{m.g.k} = 24$ N/mm²
Characteristic shear strength, $f_{v.g.k}$	$f_{v.g.k} = 2.7$ N/mm²
Characteristic bearing strength, $f_{c.90.g.k}$	$f_{c.90.g.k} = 2.7$ N/mm²
Mean modulus of elasticity parallel to grain, $E_{0.g.mean}$	$E_{0.g.mean} = 11.6$ kN/mm²
Mean shear modulus, $G_{0.g.mean}$	$G_{0.g.mean} = 0.72$ kN/mm³
Mean density of each beam	$\rho_m = 1.2 \times 380$ kg/m³
(based on the ratio of ρ_m/ρ_k obtained from BS EN 338:2009)	$\rho_m = 456$ kg/m³

3. **Partial safety factors**

   Table 2.8 (UKNA to BS EN 1990:2002+A1:2005, *Table NA.A1.2(B)*)) for the ultimate limit states (ULS)

Permanent actions, γ_G	$\gamma_G = 1.35$
Variable actions, γ_Q	$\gamma_Q = 1.5$

   Table 2.2 (UKNA to BS EN 1990:2002+A1:2005, *Table NA.A1.1*)

Factor for quasi-permanent value of variable action, ψ_2	$\psi_2 = 0.0$

   Table 2.6 (UKNA to EC5, *Table NA.3*)

Material factor for glulam at ULS, γ_M	$\gamma_M = 1.25$

4. Actions

Self-weight of a beam, $G_{k.selfwt}$

$$G_{k.selfwt} = b \cdot h \cdot g \cdot \rho_m$$

$$G_{k.selfwt} = 0.29 \text{ kN/m}$$

Design action from the self-weight of a beam, $F_{d.selfwt}$

$$F_{d.selfwt} = \gamma_G \cdot G_{k.selfwt}$$

$$F_{d.selfwt} = 0.39 \text{ kN/m}$$

Characteristic permanent action on a beam, $G_{k.p}$

$$G_{k.p} = 1.4 \text{ kN/m}$$

Characteristic variable (imposed) short-term action on a beam, $Q_{k.p}$

$$Q_{k.p} = 2.75 \text{ kN/m}$$

Design action from permanent action, short-term action and self-weight for the critical load case at the ULS, $F_{d.p}$
(Table 2.8, equation (c) using the unfavourable condition variable action)

$$F_{d.p} = \gamma_G \cdot G_{k.p} + \gamma_Q \cdot Q_{k.p} + \gamma_G \cdot G_{k.selfwt}$$

$$F_{d.p} = 6.4 \text{ kN/m}$$

5. Modification factors

Factor for short-duration loading and service class 1, $k_{mod.short}$
(Table 2.4 (EC5, *Table 3.1*))

$$k_{mod.short} = 0.9$$

Size factor for depth less than 600 mm, k_h,
(Table 2.11 (EC5, *3.3*))

$$k_h = \begin{vmatrix} 1.0 & \textit{if } h \geq 600 \text{ mm} \\ \left(\dfrac{600 \text{ mm}}{h} \right)^{0.1} & \textit{if } 1.1 > \left(\dfrac{600 \text{ mm}}{h} \right)^{0.1} \\ 1.1 & \textit{otherwise} \end{vmatrix}$$

$$k_h = 1.01$$

Lateral stability of a beam, k_{crit}
(4.5.1.2 (EC5, *6.3.3*))

$$k_{crit} = 1$$

Modification factor for the influence of cracks, k_{cr}
(Table NA.4 in UKNA to EC5)

$$k_{cr} = 0.67$$

Compression factor for strain effects, $k_{c.90}$
(Equation 4.22 (EC5, 6.1.5,(1) and (4)))

$$k_{c.90} = 1.75$$

Bearing factor, $k_{c.90}$
(equation (4.22) (EC5, *6.1.5 (1)*))

$$k_{c.90} = 1$$

Deformation factor for service class 1, k_{def},
Table 2.10 (EC5, *Table 3.2*)

$$k_{def} = 0.6$$

Load sharing factor, k_{sys} – not relevant

$$k_{sys} = 1$$

6. Bending strength

The design load case will be due to a combination of the self-weight of the beam plus the permanent and variable loading:

Design bending moment, M_d,

$$M_d = \frac{F_{d.p} \cdot \ell^2}{8}$$

$$M_d = 88.25 \text{ kNm}$$

Design bending stress, $\sigma_{m.y.d}$,

$$\sigma_{m.y.d} = \frac{M_d}{W_y}$$

$$\sigma_{m.y.d} = 14.68 \text{ N/mm}^2$$

Design bending strength, $f_{m.y.d}$,

$$f_{m.y.d} = \frac{k_{mod.short} \cdot k_{sys} \cdot k_h \cdot f_{m.g.k}}{\gamma_M}$$

$$f_{m.y.d} = 17.4 \text{ N/mm}^2$$

| Design bending strength taking lateral torsional buckling effect into account, $f_{mr.y.d}$ (equation (4.13); EC5, *equation (6.33)*) | $f_{mr.y.d} = k_{crit} \cdot f_{m.y.d}$ |
| | $f_{mr.y.d} = 19.01$ N/mm² |

Bending strength of a glulam beam greater than the bending stress and is satisfactory for this loading condition

7. Shear strength

The design load case will be due to a combination of self-weight of the beam plus the permanent and variable loading:

Design value of the end shear force, V_d	$V_d = \dfrac{F_{d.p} \cdot \ell}{2}$	$V_d = 33.62$ kN
Effective shear width of the beam, b_{ef} (equation 4.15(a) (EC5, *equation 6.13a*))	$b_{ef} = k_{cr} \cdot b$	$b_{ef} = 77.05$ mm
Design shear stress, $\tau_{v.d}$ (equation 4.16 (EC5, *equation 6.60*))	$\tau_{v.d} = \dfrac{3}{2} \cdot \dfrac{V_d}{b_{ef} \cdot h}$	$\tau_{v.d} = 1.17$ N/mm²
Design shear strength, $f_{v.d}$ (equation (6.4))	$f_{v.d} = \dfrac{k_{mod.short} \cdot k_{sys} \cdot f_{v.g.k}}{\gamma_M}$	
		$f_{v.d} = 1.94$ N/mm²

Shear strength is satisfactory

8. Bearing strength

The design load case will be due to a combination of self-weight of the beam plus the permanent and variable loading:

Design value of the end reaction, $Reac_d$	$Reac_d = V_d$	$Reac_d = 33.62$ kN
Effective contact area, A_{ef} (equation 4.23 (EC5, equation 6.4))	$A_{ef} = b \cdot (lb + 30 \cdot mm)$	$A_{ef} = 2.13 \times 10^4$ mm²
Design bearing stress, $\sigma_{c.90.d}$ (equation (4.23 (EC5, *equation 6.4*))	$\sigma_{c.90.d} = \dfrac{Reac_d}{\gamma_M}$	$\sigma_{c.90.d} = 1.58$ N/mm²
Design bearing stress, $\sigma_{c.90.d}$ (equation (4.23) (EC5, *equation 6.4*))	$f_{c.90.d} = \dfrac{k_{mod.short} \cdot k_{sys} \cdot f_{c.90.g.k}}{\gamma_M}$	
		$f_{c.90.d} = 1.94$ N/mm²
Design bearing strength, $f_{c.0.d}$, (Equation 4.22, 4.23 (EC5, *equations 6.3 and 6.4*))	$k_{c.90} \cdot f_{c.90.d} = 3.4$ N/mm²	

Bearing strength is OK

9. Beam deflection

At the SLS the partial safety factor is 1.

As the member is material having the same creep properties, the mean value of stiffness will be used to derive the instantaneous and the final deflection of the beam.

The greatest value of instantaneous deflection at the SLS will be associated with the largest value of bending moment and will be due to the characteristic combination of the self-weight of the beam plus the permanent and variable loading.

Deflection due to bending and shear effects
Instantaneous deflection due to loading on the beam u_{inst} (Table 4.8)

$$u_{inst.d\ell} = \frac{5\cdot\left(G_{k.selfwt}+G_{k.p}\right)\cdot\ell^4}{32\cdot E_{0.g.mean}\cdot b\cdot h^3}\cdot\left[1+0.96\cdot\frac{E_{0.g.mean}}{G_{0.g.mean}}\cdot\left(\frac{h}{\ell}\right)^2\right] \qquad u_{inst.d\ell}=14.29\,mm$$

$$u_{inst.Q} = \frac{5\cdot\left(Q_{k.p}\right)\cdot\ell^4}{32\cdot E_{0.g.mean}\cdot b\cdot h^3}\cdot\left[1+0.96\cdot\frac{E_{0.g.mean}}{G_{0.g.mean}}\cdot\left(\frac{h}{\ell}\right)^2\right] \qquad u_{inst.Q}=23.27\,mm$$

Instantaneous deflection at the mid-span of a beam, u_{inst} $\qquad u_{inst}=u_{inst.d\ell}+u_{inst.Q}$ $\qquad u_{inst}=37.56\,mm$

Limitation on deflection at the instantaneous state – span/300, w_{inst} $\qquad w_{inst}=\dfrac{\ell}{300}$ $\qquad w_{inst}=35\,mm$

(Table 4.7 (EC5, *Table 7.2*)) $\qquad$ i.e. beam deflection exceeds the limit

Final deflection due to permanent actions, $u_{fin.G}$ (equation (4.37); EC5, *equation (2.3)*) $\qquad u_{fin.G}=u_{inst.d\ell}\cdot\left(1+k_{def}\right)$ $\qquad u_{fin.G}=22.86\,mm$

Final deflection due to variable and quasi-permanent actions, $u_{fin.Q}$ (equation (4.42); EC5, *equation (2.4)*) $\qquad u_{fin.Q}=u_{inst.Q}\cdot\left(1+\psi_2\cdot k_{def}\right)$ $\qquad u_{fin.Q}=23.27\,mm$

Final deflection due to permanent and quasi-permanent actions, $u_{net.fin}$ $\qquad u_{net.fin}=u_{fin.G}+u_{fin.Q}$ $\qquad u_{net.fin}=46.13\,mm$

Adopting EC5 limitation on deflection – use span/250, $w_{net.fin}$ (Table 4.7) $\qquad w_{net.fin}=\dfrac{\ell}{250}$ $\qquad w_{net.fin}=42\,mm$

Precamber to be provided
precamber$=max[(u_{inst}-w_{inst}),\,(u_{net.fin}-w_{net.fin})]$ $\qquad$ precamber$=4.13\,mm$
Provide a precamber of 10 mm

Example 6.7.2 The glulam column shown in Figure E6.7.2 is made from combined glulam grade GL 24c to BS EN 1194:1999, has a cross-section of 125 mm × 450 mm, and functions under service class 2 conditions. It supports a characteristic permanent compressive action (including an allowance for the effect of its self-weight) of 40 kN and a characteristic variable medium-term compressive action of 75 kN. The column is 4.0 m high and at each end it is effectively held in position but not in direction about the z–z axis and the y–y axis. The axial load is offset from the centroid of the column as shown in the figure.
Check that the column will comply with the ULS requirements of EC5.

1. **Glulam column geometric properties**
   Actual column length, L $\qquad$ L=4.0 m
   Effective length of the column buckling about the y–y axis, $L_{e.y}$ (Table 5.2) $\qquad$ $L_{e.y}$=1.0 L $\quad$ i.e. $L_{e.y}$=4 m
   Effective length of the column buckling about the z–z axis, $L_{e.z}$ (Table 5.2) $\qquad$ $L_{e.z}$=1.0 L $\quad$ i.e. $L_{e.z}$=4 m
   Effective length of the member acting as a beam with a constant moment along its length, ℓ_{ef} (Table 4.2 (EC5, *Table 6.1*)) $\qquad$ $\ell_{ef}=L$ $\qquad$ $\ell_{ef}=4m$

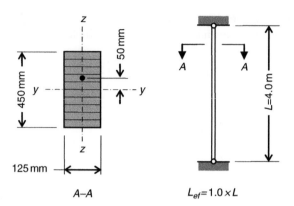

Fig. E6.7.2.

Width of the column, b	$b = 125\,\text{mm}$	
Depth of the column, h	$h = 450\,\text{mm}$	
Cross-sectional area of the column, A	$A = b \cdot h$	$A = 5.63 \times 10^4\,\text{mm}^2$
Second moment of area of the column about the y–y axes, I_y	$I_y = \dfrac{b \cdot h^3}{12}$	$I_y = 9.49 \times 10^8\,\text{mm}^4$
Section modulus about the y–y axes, W_y	$W_y = \dfrac{2 I_y}{h}$	$W_y = 4.22 \times 10^6\,\text{mm}^3$
Radius of gyration about the y–y axis, i_y	$i_y = \sqrt{\dfrac{I_y}{A}}$	$i_y = 129.9\,\text{mm}$
Slenderness ratio about the y–y axis, λ_y	$\lambda_y = \dfrac{L_{e.y}}{i_y}$	$L_y = 30.79$
Second moment of area of the column about the z–z axis, I_z	$I_z = \dfrac{h \cdot b^3}{12}$	$I_z = 7.32 \times 10^7\,\text{mm}^4$
Radius of gyration of the column about the z–z axis, i_z	$i_z = \sqrt{\dfrac{I_z}{A}}$	$i_z = 36.08\,\text{mm}$
Slenderness ratio about the z–z axis, λ_z	$\lambda_z = L_{e.z}/I_z$	$\lambda_z = 110.85$

2. **Glulam properties**

Table 6.2, combined grade GL 24c

Characteristic bending strength about y–y axis, $f_{m.y.g.k}$	$f_{m.y.g.k} = 24\,\text{N/mm}^2$	
Characteristic compression strength parallel to the grain, $f_{c.0.g.k}$	$f_{c.0.g.k} = 21\,\text{N/mm}^2$	

Fifth-percentile modulus of elasticity parallel to the grain, $E_{0.05.g}$

$E_{0.05.g} = 9.4\,\text{kN/mm}^2$

Fifth-percentile shear modulus, $G_{0.05.g}$

$G_{0.05.g} = \dfrac{E_{0.05.g}}{16}$

$G_{0.05.g} = 0.59\,\text{kN/mm}^2$

3. Partial safety factors

Table 2.8 (UKNA to BS EN 1990:2002+A1:2005, *Table NA.A1.2(B)*)) for the ULS

Permanent actions, γ_G	$\gamma_G = 1.35$
Variable actions, γ_Q	$\gamma_Q = 1.5$

Table 2.6 (UKNA to EC5, *Table NA.3*)

Material factor for glulam, γ_M $\gamma_M = 1.25$

4. Actions

Characteristic permanent compressive action, G_k

$G_k = 40\,\text{kN}$

Characteristic medium-term compressive variable action, Q_k

$Q_k = 75\,\text{kN}$

Design compressive action for the critical load combination, N_d
(Table 2.8, equation (c) using the unfavourable condition variable action)

$N_d = \gamma_G \cdot G_k + \gamma_Q \cdot Q_k$ $N_d = 1.67 \times 10^5\,\text{N}$

Moment about the *y–y* axis:

Eccentricity, e_z $e_z = 50\,\text{mm}$

Design moment about the *y–y* axis, $M_{y.d}$ $M_{y.d} = N_d \cdot e_z$ $M_{y.d} = 8.32\,\text{kN\,m}$

5. Modification factors

Factor for medium-duration loading and service class 2, $k_{mod,med}$
(Table 2.4 (EC5, *Table 3.1*))

$k_{mod,med} = 0.8$

System strength factor, k_{sys} – not relevant

$k_{sys} = 1$

Depth factor for glulam – for bending, k_h
(Table 2.11 (EC5, *3.3*))

$$k_h = \begin{vmatrix} 1.0 & \text{if} & h \geq 600\,\text{mm} \\ \left(\dfrac{600\,\text{mm}}{h}\right)^{0.1} & \text{if} & 1.1 \leq \left(\dfrac{600\,\text{mm}}{h}\right)^{0.1} \\ 1.1 & & \text{otherwise} \end{vmatrix}$$

$k_h = 1.03$

6. Strength of column

The critical design load case at the ULS will be due to the combination of permanent and unfavourable medium-term duration variable action:

Moment condition

Design bending moment about the *y–y* axis, $M_{y.d}$

$M_{y.d} = 8.32\,\text{kN\,m}$

Design bending stress about the y–y axis, $\sigma_{m.y.d}$

$$\sigma_{m.y.d} = \frac{M_{y.d}}{W_y}$$

$\sigma_{m.y.d} = 1.97\,\text{N/mm}^2$

Design bending strength about the y–y axis, $f_{m.y.d}$

$$f_{m.y.d} = \frac{k_{mod,med} \cdot k_{sys} \cdot k_h \cdot f_{m.y.g.k}}{\gamma_M}$$

$f_{m.y.d} = 15.81\,\text{N/mm}^2$

Redistribution factor for a rectangular section, k_m
(4.5.1 (EC5, 6.1.6))

$k_m = 0.7$

Buckling resistance condition – lateral torsional buckling under major axis bending
(4.5.1.2 (EC5, 6.3.3))

Lateral stability factor, k_{crit} (4.5.1.2 (EC5, 6.3.3))

Critical bending stress, $\sigma_{m.crit}$
(equation (4.7c); EC5, *equation (6.31)*)

$$\sigma_{m.crit} = \frac{\pi \cdot b^2 \left[E_{0.05.g} \cdot G_{0.05.g} \cdot \left(1 - 0.63 \cdot \dfrac{b}{h}\right) \right]^{0.5}}{h \cdot \ell_{ef}}$$

$\sigma_{m.crit} = 58.21\,\text{N/mm}^2$

Relative slenderness for bending, $\lambda_{rel.m}$
(equation (4.10); EC5, *equation (6.30)*)

$$\lambda_{rel.m} = \sqrt{\frac{f_{m.y.g.k}}{\sigma_{m.crit}}}$$

$\lambda_{rel.m} = 0.64$

Lateral stability factor, k_{crit}
(Table 4.3 (EC5, *equation (6.34)*))

$k_{crit} = 1$

Axial compression condition

Design compression stress, $\sigma_{c.0.d}$

$$\sigma_{c.0.d} = \frac{N_d}{A}$$

$\sigma_{c.0.d} = 2.96\,\text{N/mm}^2$

Design compression strength, $f_{c.0.d}$

$$f_{c.0.d} = \frac{k_{mod,med} \cdot k_{sys} \cdot f_{c.0.g.k}}{\gamma_M}$$

$f_{c.0.d} = 13.44\,\text{N/mm}^2$

Buckling resistance condition (5.4.1 (EC5, 6.3.2)):

Relative slenderness about the y–y axis, $\lambda_{rel.y}$
(equation (5.3);
EC5, *equation (6.21)*)

$$\lambda_{rel.y} = \frac{\lambda_y}{\pi} \cdot \sqrt{\frac{f_{c.0.g.k}}{E_{0.05.g}}}$$

$\lambda_{rel.y} = 0.46$

Relative slenderness about the z–z axis, $\lambda_{rel.z}$
(equation (5.3);
EC5, *equation (6.22)*)

$$\lambda_{rel.z} = \frac{\lambda_z}{\pi} \cdot \sqrt{\frac{f_{c.0.g.k}}{E_{0.05.g}}}$$

$\lambda_{rel.z} = 1.67$

Factor β_c for glulam
(equation (5.6); EC5, *equation (6.29)*)

$\beta_c = 0.1$

Factor k_y
(equation (5.5a); EC5, *equation (6.27)*)

$$k_y = 0.5 \cdot \left[1 + \beta_c \cdot \left(\lambda_{rel.y} - 0.3\right) + \lambda^2_{rel.y}\right] \qquad\qquad k_y = 0.62$$

Instability factor about the y–y axis
(equation (5.4a); EC5, *equation (6.25)*)

$$k_{c.y} = \frac{1}{k_y + \sqrt{k_y^2 - \lambda^2_{rel.y}}} \qquad\qquad k_{c.y} = 0.98$$

Factor k_z
(equation (5.5b); EC5, *equation (6.28)*)

$$k_z = 0.5 \left[1 + \beta_c \cdot \left(\lambda_{rel.z} - 0.3\right) + \lambda^2_{rel.z}\right] \qquad\qquad k_z = 1.96$$

Instability factor about the z–z axis
(equation (5.4b); EC5, *equation (6.26)*)

$$k_{c.z} = \frac{1}{k_z + \sqrt{k_z^2 - \lambda^2_{rel.z}}} \qquad\qquad k_{c.z} = 0.33$$

Combined stress conditions

Compression stress condition about the y–y axis
(equation (5.35) (EC5, *6.3.2(3)*))

$$\frac{\sigma_{c.0.d}}{k_{c.y} \cdot f_{c.0.d}} + \frac{\sigma_{m.y.d}}{f_{m.y.d}} = 0.35$$

Compression stress condition about the z–z axis
(equation (5.36) (EC5, *6.3.2(3)*))

$$\frac{\sigma_{c.0.d}}{k_{c.z} \cdot f_{c.0.d}} + k_m \cdot \frac{\sigma_{m.y.d}}{f_{m.y.d}} = 0.75$$

Combined stress condition
(equation (5.38); EC5, *equation (6.35)*)

$$\left(\frac{\sigma_{m.y.d}}{k_{crit} \cdot f_{m.y.d}}\right)^2 + \frac{\sigma_{c.0.d}}{k_{c.z} \cdot f_{c.0.d}} = 0.67$$

As all relationships are less than unity, the glulam member will meet the ULS requirements of EC5.

Example 6.7.3 Single tapered glulam beams 150 mm wide, having a profile as shown Figure E6.7.3, and with an effective span of 9.0 m are to be used in the construction of the roof of an exhibition hall. The roof is braced laterally at 3.0 m centres along its top face to provide full lateral support at these positions and load sharing between glulam beams will not apply. The beams are glulam strength class GL 28h in accordance with BS EN 1194:1999 and will function in service class 1 conditions and are to be subjected to the characteristic loading given below. Confirm the beam will comply with the design rules in EC5.
    Beam loading:

Characteristic permanent vertical load on each beam   1.65 kN/m
Characteristic short-term variable vertical load on each beam   2.25 kN/m

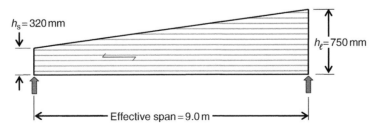

$h_s = 320$ mm

$h_\ell = 750$ mm

Effective span = 9.0 m

Fig. E6.7.3.

1. Glulam beam geometric properties

Breadth of each beam, b	$b = 150\,mm$
Minimum depth at the tapered end, h_s	$h_s = 320\,mm$
Maximum depth of each beam, h_ℓ	$h_\ell = 750\,mm$
Effective span of each beam, ℓ	$\ell = 9.0\,m$
Distance between bracing positions along the tapered face, ℓ_c	$\ell_c = 3.0\,m$
Bearing length at each end of the beam, ℓ_b	$\ell_b = 150\,mm$

Angle of slope of the tapered face, α $\tan \alpha = \dfrac{h_\ell - h_s}{\ell}$ $\tan \alpha = 0.05$

2. Glulam properties

Table 6.2, homogeneous grade GL 28h

Characteristic bending strength, $f_{m.g.k}$	$f_{m.g.k} = 28\,N/mm^2$
Characteristic shear strength, $f_{v.g.k}$	$f_{v.g.k} = 3.2\,N/mm^2$
Characteristic bearing strength, $f_{c.90.g.k}$	$f_{c.90.g.k} = 3.0\,N/mm^2$
Mean modulus of elasticity parallel to the grain, $E_{0.g.mean}$	$E_{0.g.mean} = 12.6\,kN/mm^2$
Characeristic modulus of elasticity parallel to the grain, $E_{0.05.g}$	$E_{0.05.g} = 10.2\,kN/mm^2$
Mean shear modulus, $G_{0.g.mean}$	$G_{0.g.mean} = 0.78\,kN/mm^2$

Characteristic shear modulus, $G_{0.05.g}$ $G_{0.05.g} = \dfrac{E_{0.05.g}}{16}$ $G_{0.05.g} = 0.64\,kN/mm^2$

Mean density of each beam (based on the ratio of ρ_m/ρ_k obtained from BS EN 338:2009) $\rho_m = 1.2 \cdot 410\,kg/m^3$ $\rho_m = 492\,kg/m^3$

3. Partial safety factors

Table 2.8 (UKNA to BS EN 1990:2002: A1:2005, *Table NA.A1.2(B)*)) for the ULS

Permanent actions, γ_G	$\gamma_G = 1.35$
Variable actions, γ_Q	$\gamma_Q = 1.5$

Table 2.2 (UKNA to BS EN 1990:2002+A1:2005, *Table NA.A1.1*)

Factor for quasi-permanent value of variable action, ψ_2 $\psi_2 = 0.0$

Table 2.6 (UKNA to EC5, *Table NA.3*)

Material factor for glulam at ULS, γ_M $\gamma_M = 1.25$

4. Actions

Self-weight of the beam, $G_{k.selfwt}$ – as a safe approximation assume that it is based on its greatest depth and is uniform along the beam length $G_{k.selfwt} = (b \cdot h_\ell) \cdot g \cdot \rho_m$ $G_{k.selfwt} = 0.54\,kN/m$

Design action from the self-weight of the beam, $F_{d.selfwt}$	$F_{d.selfwt} = \gamma_G \cdot G_{k.selfwt}$	$F_{d.selfwt} = 0.73 \, kN/m$
Characteristic permanent action on the beam, $G_{k.p}$	$G_{k.p} = 1.65 \, kN/m$	
Characteristic variable (imposed) short-term action on the beam, $Q_{k.p}$	$Q_{k.p} = 2.25 \, kN/m$	
Design action from permanent action, short-term action and self-weight for the critical load case at the ULS, $F_{d.p}$	$F_{d.p} = \gamma_G \cdot G_{k.p} + \gamma_Q \cdot Q_{k.p} + \gamma_G \cdot G_{k.selfwt}$	
(Table 2.8, equation (c) using the unfavourable condition variable action)		$F_{d.p} = 6.34 \, kN/m$

5. Modification factors

Factor for short-duration loading and service class 1, $k_{mod.short}$ (Table 2.4 (EC5, *Table 3.1*))	$k_{mod.short} = 0.9$	
Adopt size factor based on maximum depth = 750 mm, k_h (Table 2.11 (EC5, *3.3*))	$h = 750 \, mm$	$k_h = 1$
Modification factor for the influence of cracks, k_{cr} (Table NA.4 in UKNA to EC5)	$k_{cr} = 0.67$	
Compression factor for strain effects, $k_{c,90}$ (where l_1 is not applicable and l is $< 400 mm$) (see EC5, *equation 6.3 and 6.1.5(4)*)	$k_{c,90} = 1.75$	
Bearing factor, $k_{c.90}$ (equation (4.22) (EC5, *6.1.5,(1)*))	$k_{c.90} = 1$	
Deformation factor for service class 1, k_{def} (Table 2.10 (EC5, *Table 3.2*))	$k_{def} = 0.6$	
Load sharing factor, k_{sys} (2.3.7 (EC5, *6.6*))	$k_{sys} = 1.0$	

6. Bending strength

The design load case will be due to a combination of the self-weight of the beam plus the permanent and variable loading:

Position of the maximum bending stress from the 320 mm deep end under the design loading, x (Table 6.5)	$x = \dfrac{\ell}{1 + \dfrac{h_\ell}{h_s}}$	$x = 2.69 \, m$

Section modulus about the y–y axis at position x, W_y (Table 6.5)

$$W_y = \frac{2}{3} \cdot b \cdot \left(\frac{h_\ell}{1 + \dfrac{h_\ell}{h_s}} \right)^2 \qquad W_y = 5.030 \times 10^6 \, mm^3$$

Design loading on the beam, $F_{d.p}$ $F_{d.p} = 6.34 \, \text{kN/m}$

Design bending stress, $\sigma_{m.y.d}$
(Table 6.5)

$$\sigma_{m.y.d} = \frac{F_{d.p} \cdot x \cdot (\ell - x)}{2 \cdot W_y}$$ $\sigma_{m.y.d} = 10.69 \, \text{N/mm}^2$

Design bending strength, $f_{m.g.d}$
(equation (6.2))

$$f_{m.g.d} = \frac{k_{mod.short} \cdot k_{sys} \cdot f_{m.g.k}}{\gamma_M}$$ $f_{m.g.d} = 20.16 \, \text{N/mm}^2$

Design shear strength, $f_{v.g.d}$
(equation (6.4))

$$f_{v.g.d} = \frac{k_{mod.short} \cdot k_{sys} \cdot f_{v.g.k}}{\gamma_M}$$ $f_{v.g.d} = 2.3 \, \text{N/mm}^2$

Design compression strength, $f_{c.90.g.d}$,

$$f_{c.90.g.d} = \frac{k_{mod.short} \cdot k_{sys} \cdot f_{c.90.g.k}}{\gamma_M}$$ $f_{c.90.g.d} = 2.16 \, \text{N/mm}^2$

The strength reduction factor for a tapered
beam with compressive stresses parallel to the tapered edge, $k_{m,\alpha}$
(equation (6.9); EC5, *equation (6.40)*)

$$k_{m,\alpha} = \frac{1}{\left[1 + \left(\dfrac{f_{m.g.d}}{1.5 f_{v.g.d}} \cdot \tan \alpha \right)^2 + \left(\dfrac{f_{m.g.d}}{f_{c.90.g.d}} \cdot \tan \alpha^2 \right)^2 \right]^{0.5}}$$ $k_{m,\alpha} = 0.96$

Lateral stability of the beam - assume the beam is of uniform depth $= h_\ell$ (safe
approximation), k_{crit} (4.5.1.2 (EC5, *6.3.3*))

Effective length of beam, ℓ_{ef} $\ell_{ef} = \ell_c$ $\ell_{ef} = 3 \, \text{m}$
(Table 4.2 (EC5, *Table 6.1*))

Critical bending stress, $\sigma_{m.g.crit}$
(as an approximation use equation
(4.9b); EC5, *equation (6.31)*)

$$\sigma_{m.g.crit} = \frac{\pi \cdot b^2 \left[E_{0.05.g} \cdot G_{0.05.g} \cdot \left(1 - 0.63 \cdot \dfrac{b}{h_\ell} \right) \right]^{0.5}}{(h_\ell) \cdot \ell_{ef}}$$ $\sigma_{m.g.crit} = 74.89 \, \text{N/mm}^2$

Relative slenderness for bending, $\lambda_{rel.m}$ $\lambda_{rel.m} = \sqrt{\dfrac{k_{m.\alpha} \cdot f_{m.g.k}}{\sigma_{m.g.crit}}}$ $\lambda_{rel.m} = 0.6$
(equation (4.10); EC5, *equation (6.30)*)

Lateral stability factor, k_{crit}
(Table 4.3 (EC5, *equation (6.34)*))

$$k_{crit} = 1$$

Design bending strength taking strength reduction factor and lateral torsional buckling effect into account, $f_{mr.y.d}$
(equation (6.10))

$$f_{mr.y.d} = k_{crit} \cdot k_{m.\alpha} \cdot f_{m.g.d}$$

$f_{mr.y.d} = 19.42 \ \text{N/mm}^2$

Bending strength of glulam beam greater than the bending stress and is satisfactory for this loading condition

7. **Shear strength**
   The design load case will be due to a combination of self-weight of the beam plus the permanent and variable loading:

   Design value for end shear force, V_d

   $$V_d = \frac{F_{d.p} \cdot \ell}{2}$$

   $V_d = 28.51 \ \text{kN}$

   Effective shear width of the beam, b_{ef}
   (Equation 4.15(a) (EC5, *equation 6.13a*))

   $$b_{ef} = k_{cr} \cdot b$$

   $b_{ef} = 100.5 \ \text{mm}$

   Design shear stress, $\tau_{v.d}$
   (Equation 4.16 (EC5, *equation 6.60*))

   $$\tau_{v.d} = \frac{3}{2} \cdot \frac{V_d}{b_{ef} \cdot h_s}$$

   $\tau_{v.d} = 1.33 \ \text{N/mm}^2$

   Design shear strength, $f_{v.g.d}$,

   $$f_{v.g.d} = 2.3 \ \text{N/mm}^2$$

   Shear strength is satisfactory

8. **Bearing strength**
   The design load case will be due to a combination of self-weight of the beam plus the permanent and variable loading:

   Design value for end reaction, $Reac_d$

   $$Reac_d = V_d$$

   $Reac_d = 28.51 \ \text{kN}$

   Effective contact area, A_{ef}
   (Equation 4.23 (EC5, *equation 6.4*))

   $$A_{ef} = b \cdot (l_b + 30 \ \text{mm})$$

   $A_{ef} = 2.7 \times 10^4 \ \text{mm}^2$

   Design bearing stress, $\sigma_{c.90.d}$
   (Equation 4.23 (EC5, *equation 6.4*))

   $$\sigma_{c.90.d} = \frac{Reac_d}{A_{ef}}$$

   $\sigma_{c.90.d} = 1.06 \ \text{N/mm}^2$

   Design bearing strength, $f_{c.90.g.d}$

   $$f_{c.90.g.d} = 2.16 \ \text{N/mm}^2$$

   Factored design bearing strength, $k_{c.90} f_{c.90.d}$
   (equation 4.22)

   $$k_{c.90} \cdot f_{c.90.g.d} = 3.78 \ \text{N/mm}^2$$

   Bearing strength is OK

9. **Beam deflection**
   At the serviceability limit states (SLS) the partial safety factor is 1.

   As the member is material having the same creep properties, the mean value of stiffness will be used to derive the instantaneous and the creep deflection of the beam. The greatest value of instantaneous deflection at the SLS will be associated with the largest value of bending moment and will be due to the characteristic combination of the self-weight of the beam plus the permanent and variable loading.

Deflection due to bending and shear effects

Instantaneous deflection due to loading on the beam u_{inst}:

(a) Deflection due to bending under permanent loading $u_{inst,b,dl}$:

Factor a

$$a = \frac{h_\ell}{h_s}$$

Function $k2_{\delta b}$
(Table A6.1.1)

$$k2_{\delta b} = 19.2 \left(\frac{1}{a-1} \right)^4 \cdot \left[3 \cdot (a+1) \cdot \ln\left(\frac{a+1}{2} \right) - (2 \cdot a + 1) \cdot \ln(a) - \frac{(a-1)^2}{2 \cdot (a+1)} \right]$$

$$k2_{\delta b} = 0.24$$

Bending moment at mid-span, $M_{d,SLS}$

$$M_{d,SLS} = \frac{(G_{k,selfwt} + G_{k,p}) \cdot \ell^2}{8}$$

Deflection, $u_{inst,b,dl}$
(Table A6.1.1)

$$u_{inst,b,dl} = \frac{5 \cdot M_{d,SLS} \cdot \ell^2}{48 \cdot E_{0,g,mean} \cdot \left(\frac{b \cdot h_s^3}{12} \right)} \cdot (k2_{\delta b})$$

$$u_{inst,b,dl} = 8.8 \, mm$$

(b) Deflection due to shear under the permanent loading $u_{inst,s,dl}$:

Function $k2_{\delta s}$
(Table A6.1.1)

$$k2_{\delta s} = 2 \cdot \frac{a+1}{a-1} \cdot \ln\left[\frac{(a+1)^2}{4 \cdot a} \right]$$

$$k2_{\delta s} = 0.65$$

Deflection, $u_{inst,s,dl}$
(Table A6.1.1)

$$u_{inst,s,dl} = \frac{1.2 \cdot M_{d,SLS} \cdot k2_{\delta s}}{G_{0,g,mean} \cdot (b \cdot h_s)}$$

$$u_{inst,s,dl} = 0.46 \, mm$$

(c) Deflection due to bending under the variable loading $u_{inst,b,Q}$:

$$u_{inst,b,Q} = \frac{Q_{k,p}}{G_{k,selfwt} + G_{k,p}} \cdot u_{inst,b,dl}$$

$$u_{inst,b,Q} = 9.03 \, mm$$

(d) Deflection due to shear under the variable loading $u_{inst,s,Q}$:

$$u_{inst,s,Q} = \frac{Q_{k,p}}{G_{k,selfwt} + G_{k,p}} \cdot u_{inst,s,dl}$$

$$u_{inst,s,Q} = 0.48 \, mm$$

Instantaneous deflection at mid-span of the beam, u_{inst}

$$u_{inst} = u_{inst,b,dl} + u_{inst,b,Q} + u_{inst,s,dl} + u_{inst,s,Q}$$

$$u_{inst} = 18.77 \, mm$$

Limitation on deflection at the instantaneous state – span/300, w_{inst}
Table 4.6 (EC5, *Table 7.2*)

$$w_{inst} = \frac{\ell}{300}$$

$$w_{inst} = 30 \, mm$$

i.e. OK

Final deflection due to permanent actions, $u_{\text{fin.}G}$
(equation (4.41); EC5, *equation (2.3)*)

$$u_{\text{fin.}G} = (u_{\text{inst,b,dl}} + u_{\text{inst,s,dl}}) \cdot (1 + k_{\text{def}})$$
$$u_{\text{fin.}G} = 14.82 \,\text{mm}$$

Final deflection due to variable and
quasi-permanent actions, $u_{\text{fin.}Q}$
(equation (4.42); EC5, *equation (2.4)*)

$$u_{\text{fin.}Q} = (u_{\text{inst,b,Q}} + u_{\text{inst,s,Q}}) \cdot (1 + \psi_2 \cdot k_{\text{def}})$$
$$u_{\text{fin.}Q} = 9.51 \,\text{mm}$$

Final deflection due to permanent and
quasi-permanent actions, $u_{\text{net.fin}}$

$$u_{\text{net.fin}} = u_{\text{fin.}G} + u_{\text{fin.}Q}$$
$$u_{\text{net.fin}} = 24.33 \,\text{mm}$$

Adopt EC5 limitation on deflection – use
span/250, $w_{\text{net.fin}}$
(Table 4.6 (EC5, *Table 7.2*))

$$w_{\text{net.fin}} = \frac{\ell}{250}$$
$$w_{\text{net.fin}} = 36 \,\text{mm}$$

i.e. OK

Example 6.7.4 A curved glulam beam with a constant cross-section, 175 mm wide, having a profile as shown in Figure E6.7.4, and with an effective span of 18.0 m is to be used in the construction of the roof for a school hall. The beam will be laterally supported along the full length of the compression edge and there will be no load sharing between glulam beams. It is strength class GL 32h in accordance with BS EN 1194:1999, made from 30-mm-thick laminations, and will function in service class 2 conditions. For the design loading condition given below, which includes an allowance for the self-weight of the beam, ignoring SLS requirements, confirm that the beam will comply with the design rules in EC5. The design loading arises from a combination of permanent and short-term variable loading.

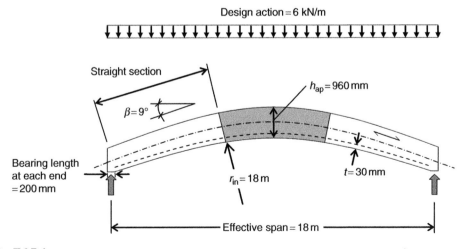

Fig. E6.7.4.

1. **Glulam beam geometric properties**

   Breadth of each beam, b $b = 175 \, \text{mm}$

   Depth of the beam at the apex, h_{ap} $h_{ap} = 960 \, \text{mm}$

   Effective span of beam, ℓ $\ell = 18.0 \, \text{m}$

   Bearing length at each end of the beam, ℓ_b $\ell_b = 200 \, \text{mm}$

   Angle of slope of the straight member lengths of the beam, β $\beta = 9 \cdot \text{deg}$

   Inner radius of the curved beam section, r_{in} $r_{in} = 18 \, \text{m}$

   Thickness of laminations in the beam, t $t = 30 \, \text{mm}$

   Section modulus of the beam about the y–y axis at the apex, W_y $W_y = \dfrac{b \cdot h_{ap}^2}{6}$ $W_y = 2.69 \times 10^7 \, \text{mm}^3$

2. **Glulam properties**

   Table 6.2, homogeneous grade GL 32h

   Characteristic bending strength, $f_{m.g.k}$ $f_{m.g.k} = 32 \, \text{N/mm}^2$

   Characteristic shear strength, $f_{v.g.k}$ $f_{v.g.k} = 3.8 \, \text{N/mm}^2$

   Characteristic bearing strength, $f_{c.90.g.k}$ $f_{c.90.g.k} = 3.3 \, \text{N/mm}^2$

   Characteristic compressive strength, $f_{c.0.g.k}$ $f_{c.0.g.k} = 29 \, \text{N/mm}^2$

   Characteristic tensile strength perpendicular to the grain, $f_{t.90.g.k}$ $f_{t.90.g.k} = 0.5 \, \text{N/mm}^2$

3. **Partial safety factors**

   Table 2.8 (UKNA to BS EN 1990:2002:A1:2005, *Table NA.A1.2(B)*)) for the ULS

   Permanent actions, γ_G $\gamma_G = 1.35$

   Variable actions, γ_Q $\gamma_Q = 1.5$

   Table 2.6 (UKNA to EC5, *Table NA.3*):

   Material factor for glulam at ULS, γ_M $\gamma_M = 1.25$

4. **Actions**

   Design action on the beam, q_d $q_d = 6 \, \text{kN/m}$

5. **Modification factors**

   Factor for short-duration loading and service class 2, $k_{mod.short}$ (Table 2.4 (EC5, *Table 3.1*)) $k_{mod.short} = 0.9$

   Size factor based on maximum depth = 960 mm, k_h (Table 2.11 (EC5, *3.3*)) $k_h = 1$

   Modification factor for the influence of cracks, k_{cr} (Table NA.4 in UKNA to EC5) $k_{cr} = 0.67$

   Compression factor for strain effects, $k_{c.90}$ (for this case, adopt a value of 1) $k_{c.90} = 1.0$

Bearing factor, $k_{c.90}$
(equation (4.22) (EC5, *6.1.5,(1)*))

$k_{c.90} = 1$

Lateral stability factor for beam, k_{crit}
(4.5.1.2 (EC5, *6.3.3*))

$k_{crit} = 1$

Factor applied to obtain bending stress in the apex zone, $k_{curve,b}$
(equation (6.24); EC5, *equation (6.43)*)

$r = r_{in} + 0.5 \cdot h_{ap}$

$$k_{curve,b} = 1 + 0.35 \cdot \left(\frac{h_{ap}}{r}\right) + 0.6 \cdot \left(\frac{h_{ap}}{r}\right)^2$$

$k_{curve,b} = 1.02$

Factor applied to obtain the tensile stress perpendicular to the grain in the apex zone, $k_{curve,t}$
(equation (6.25); EC5, *equation (6.54)*)

$k_{curve,t} = 0.25\left(\frac{h_{ap}}{r}\right)$

$k_{curve,t} = 0.01$

Stressed volume in the apex zone, V
(Table 6.6) (Mathcad adjusted to make it dimensionally correct)

$$V = \frac{\beta \cdot \pi \cdot b}{180. \deg} \cdot (h_{ap}^2 + 2 \cdot r_{in} \cdot h_{ap})$$

$V = 0.98 \, \text{m}^3$

Approximate volume of the beam, V_c

$$V_c = V + 2 \cdot b \cdot h_{ap} \frac{\frac{1}{2} - \left(r_{in} + \frac{h_{ap}}{2}\right) \cdot \sin(\beta)}{\cos(\beta)}$$

$V_c = 3.05 \, \text{m}^3$

Ratio V/V_c
Must be less than 0.67
(equation (6.27) (EC5 *6.4.3(6)*))

$\dfrac{V}{V_c} = 0.32$ less than 0.67 therefore OK

Stress distribution factor for the apex zone, k_{dis}
(equation (6.27); EC5, *equation (6.52)*)

$k_{dis} = 1.4$

Volume factor for the apex zone, k_{vol}
(equation (6.27); EC5, *equation (6.51)*)
(Mathcad adjusted to make it dimensionally correct)

$k_{vol} = \left(\dfrac{0.001 \cdot \text{m}^3}{V}\right)^{0.2}$

$k_{vol} = 0.4$

Reduction factor due to the curvature of the laminations, k_r
(equation (6.28) and (6.26); EC5, *equation (6.49)*)

$\dfrac{r_{in}}{t} = 600$

As the ratio is greater than 240, $k_r = 1$

Load sharing factor, k_{sys}

$k_{sys} = 1.0$

6. Bending and radial strength in apex zone

As the beam is of constant cross-section, the critical condition will occur at the apex, under the action of the design loading:

(a) Bending strength condition

Design bending moment, M_d

$$M_d = \frac{q_d \cdot \ell^2}{8}$$

$M_d = 2.43 \times 10^8 \, \text{N mm}$

Design bending stress, $\sigma_{m,0,d}$
(equation (6.23); EC5, *equation (6.42)*)

$$\sigma_{m,0,d} = \frac{k_{curve.b} \cdot M_d}{W_y}$$

$\sigma_{m,0,d} = 9.22 \, \text{N/mm}^2$

Design bending strength, $f_{m,g,d}$

$$f_{m,g,d} = \frac{k_{mod.short} \cdot k_{sys} \cdot k_h \cdot f_{m,g,k}}{\gamma_M}$$

$f_{m,g,d} = 23.04 \, \text{N/mm}^2$

Design bending strength taking lateral torsional buckling and laminate effect into account, $f_{m,r,y,d}$,

$f_{m,r,y,d} = k_{crit} \cdot k_r \cdot f_{m,g,d}$ $f_{m,r,y,d} = 23.04 \, \text{N/mm}^2$

Bending strength of the glulam beam in apex zone is satisfactory

(b) Radial strength condition
(6.4.2.2.2 (EC5. 6.4.3(6)))

Design radial tensile stress, $\sigma_{t.90.d}$
(equation (6.25); EC5, *equation (6.54)*)

$$\sigma_{t.90.d} = \frac{k_{curve.t} \cdot M_d}{W_y}$$

$\sigma_{t.90.d} = 0.12 \, \text{N/mm}^2$

Design tensile strength perpendicular to the grain, $f_{t.90.d}$

$$f_{t.90.d} = \frac{k_{mod.short} \cdot k_{sys} \cdot f_{t.90.g.k}}{\gamma_M}$$

$f_{t.90.d} = 0.36 \, \text{N/mm}^2$

Design tensile strength perpendicular to the grain taking stress distribution and volume factors into account, $f_{t.r.y.d}$
(equation (6.27); EC5, *equation (6.50)*)

$f_{t.r.y.d} = k_{dis} \cdot k_{vol} \cdot f_{t.90.d}$ $f_{t.r.y.d} = 0.2 \, \text{N/mm}^2$

Tensile strength of the glulam beam in apex zone is greater than the radial tensile stress and is satisfactory

7. Shear strength

The design shear condition due to the design loading:

Design value of the end shear force perpendicular to the grain – ignoring the reduction permitted in EC5 – see 4.5.2.1, V_d

$$V_d = \frac{q_d \cdot \ell \cdot \cos(\beta)}{2}$$

$V_d = 53.34 \, \text{kN}$

Effective shear width of the beam, b_{ef}
(Equation 4.15(a) (EC5, *equation 6.13a*))

$b_{ef} = k_{cr} \cdot b$ $b_{ef} = 117.25 \, \text{mm}$

Design shear stress, $\tau_{v.d}$
(equation 4.16 (EC5, *equation (6.60)*))

$$\tau_{v.d} = \frac{3}{2} \cdot \frac{V_d}{b_{ef} \cdot h_{ap}}$$

$\tau_{v.d} = 0.71 \,\text{N/mm}^2$

Design shear strength, $f_{v.g.d}$

$$f_{v.g.d} = \frac{k_{mod.short} \cdot k_{sys} \cdot f_{v.g.k}}{\gamma_M}$$

$f_{v.g.d} = 2.74 \,\text{N/mm}^2$

Shear strength is satisfactory

8. Bearing strength

The design bearing condition due to the design loading:

Design value of the end reaction, Reac_d $\text{Reac}_d = V_d$ $\text{Reac}_d = 53.34 \,\text{kN}$

Effective contact area, A_{ef} $A_{ef} = b \cdot (lb)$ $A_{ef} = 3.5 \times 10^4 \,\text{mm}^2$
(Equation 4.23 (EC5, *equation 6.4*) – ignore
the 30 mm length on this occassion – safe
approach)

Design bearing stress, $\sigma_{c.90.d}$
(Equation 4.23 (EC5, *equation 6.4*))

$$\sigma_{c.90..d} = \frac{\text{Reac}_d}{A_{ef}}$$

$\sigma_{c.90.d} = 1.52 \,\text{N/mm}^{-2}$

Design bearing strength, $f_{c.90.g.d}$

$$f_{c.90.g.d} = \frac{k_{mod.short} \cdot k_{sys} \cdot f_{c.90.g.k}}{\gamma_M}$$

$f_{c.90.g.d} = 2.38 \,\text{N/mm}^2$

Design compression strength parallel to the grain, $f_{c.0.g.d}$

$$f_{c.0.g.d} = \frac{k_{mod.short} \cdot k_{sys} \cdot f_{c.0.g.k}}{\gamma_M}$$

$f_{c.0.g.d} = 20.88 \,\text{N/mm}^2$

Design compression strength at an angle
β to the grain, $f_{c.\beta.g.d}$
(equation (5.15); EC5, *Equation (6.16)*)

$$f_{c.\beta.g.d} = \frac{f_{c.0.g.d}}{\dfrac{f_{c.0.g.d}}{k_{c.90} \cdot f_{c.90.g.d}} \cdot \sin(90.\deg-\beta)^2 + \cos(90.\deg-\beta)^2}$$

$f_{c.\beta.g.d} = 2.43 \,\text{N/mm}^2$

The bearing stress is less than the bearing
strength; therefore it is acceptable.

9. Combined shear and tension perpendicular to the grain

As the shear stress at the apex will be zero for the design loading condition, the need to check
the combined shear and tension condition given in equation (6.30) can be ignored. At any
other position along the beam the combined stress condition will always be less than 1.

Chapter 7

Design of Composite Timber and Wood-Based Sections

7.1 INTRODUCTION

When designing a structure, the greatest efficiency will be achieved by using structural sections that have a high stiffness and can carry the greatest load for minimum self-weight. The most common types of sections that come into this category are thin webbed beams and thin flanged beams, often referred to as composite I and composite box beams, respectively. These are very efficient sections and the design rules in EC5 [1] cover the two methods that are used for assembly, i.e.:

(a) composite sections formed using glued joints,
(b) composite sections formed using mechanical joints.

The profiles of some of the composite sections that are widely used in timber structures are shown in Figure 7.1. Examples of their use in timber structures are shown in Figure 7.2.

Glued composite sections are, as the name implies, assembled by gluing the elements of the section together to function as a single unit. With these sections, the design rules are formulated on the assumption that no slip will arise between the elements of the section at any of the joint positions and are addressed in EC5, *Section 9*. Mechanically jointed composite sections are assembled by securing the elements together using nails, screws or dowels etc., and with these sections slip will arise between the elements. The design rules in EC5, *Annex B*, take this into account.

It is normal practice for composite sections to be assembled by gluing, and this chapter only covers the design requirements of sections formed this way.

With these sections the webs and flanges are made from different materials; for example, the webs of composite I-beams are usually made from wood-based panel materials such as plywood, oriented strand board (OSB) or particleboard etc., and the flanges from structural timber, LVL or glued-laminated timber. For glued composite sections, a high degree of quality control is required to ensure that sound jointing is achieved and for this reason they are normally factory produced.

In these sections, all of the elements are designed to work very efficiently and because of this only materials with relatively few and minor defects are used.

Structural Timber Design to Eurocode 5, Second Edition. Jack Porteous and Abdy Kermani.
© Jack Porteous and Abdy Kermani 2013. Published 2013 by Blackwell Publishing Ltd.

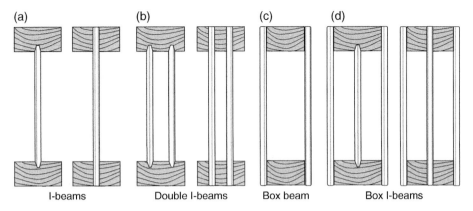

Fig. 7.1. Examples of composite timber and wood-based sections.

The types of glued composite sections referred to in EC5 and covered in this chapter are:

(1) glued thin webbed beams (EC5, *9.1.1*) and
(2) glued thin flanged beams (EC5, *9.1.2*)

with both types of sections being used in situations where they support lateral forces and function as beams. The design rules in EC5 only cover this type of behaviour, however, in the case of glued thin webbed beams, the equations given in this chapter have been adjusted to also include for a design situation where there can be a nominal axial action acting along the longitudinal centroidal axis of the section in addition to the primary bending and shear actions induced by the lateral loading. The theory is only valid for small axial forces (i.e. the section is functioning primarily as a beam) and when instability effects can be ignored.

The general information in 4.3 is relevant to the content of this chapter.

7.2   DESIGN CONSIDERATIONS

Composite sections will function as flexural elements, discussed in Chapter 4, and must comply with the relevant design rules and requirements of EC5 for such elements. The limit states associated with the main design effects are the same as those given in Table 4.1 and where used in residential floors their vibration behaviour must comply with the requirements covered in 4.6.2. Where axial stresses have to be taken into account the relevant requirements of Chapter 5 have to be addressed within the design.

The primary difference between the design process used for composite sections and that used for solid sections is that small deflection bending theory has to be modified to take into account the fact that composite sections are made up of elements having different values of modulus of elasticity. This affects the design process for strength-related properties and is discussed in the chapter.

As well as affecting strength, the variation in modulus of elasticity affects stiffness behaviour and this is also addressed.

The general comments in 4.2 are also applicable to glued composite sections.

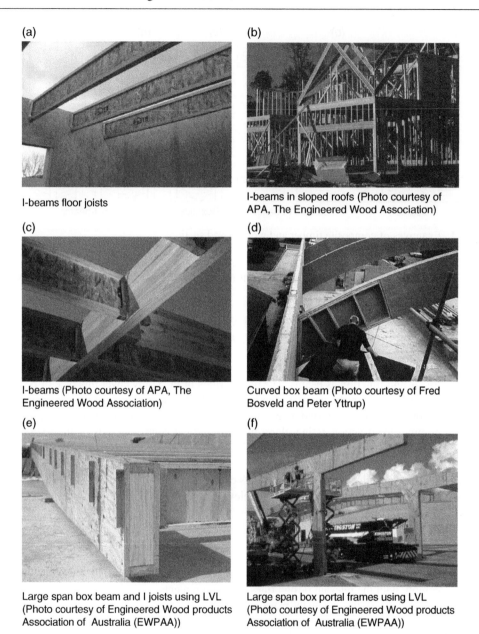

(a)

I-beams floor joists

(b)

I-beams in sloped roofs (Photo courtesy of APA, The Engineered Wood Association)

(c)

I-beams (Photo courtesy of APA, The Engineered Wood Association)

(d)

Curved box beam (Photo courtesy of Fred Bosveld and Peter Yttrup)

(e)

Large span box beam and I joists using LVL (Photo courtesy of Engineered Wood products Association of Australia (EWPAA))

(f)

Large span box portal frames using LVL (Photo courtesy of Engineered Wood products Association of Australia (EWPAA))

Fig. 7.2. Examples of uses of composite timber and wood-based sections.

7.3   DESIGN OF GLUED COMPOSITE SECTIONS

7.3.1   Glued thin webbed beams

Because these sections are designed to function as solid units, the strain is taken to vary linearly over their depth. However, as the value of the modulus of elasticity, E, of each material in the section will normally be different, the theory of bending cannot

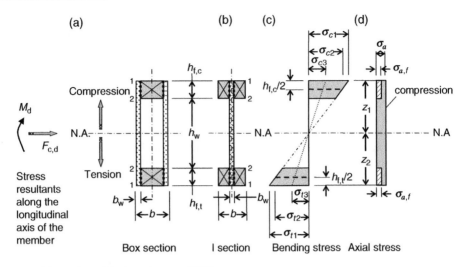

Fig. 7.3. Thin webbed sections and EC5 stresses.

be directly applied. To analyse these sections the *equivalent section method* [2, 3] (commonly referred to as the modular ratio approach or the transformed-section method) can be adopted and has been incorporated into the design procedure used in this chapter.

In this method, one of the materials in the composite section is selected and the others replaced by equivalent areas of the selected material such that when subjected to bending the force in the replaced materials at any distance from the neutral axis will be the same as that taken by the original materials at the same position. By this method, an equivalent homogeneous section is formed and the theory of bending can be applied. The bending stress in the material selected for the section will be found directly from the stress analysis. For the material that has been replaced, the stress is obtained by multiplying the bending stress in the equivalent homogeneous section by the ratio of the modulus of elasticity of the actual material used in the composite section to the modulus of elasticity of the selected material. This ratio is referred to as the *modular ratio*. In the calculation the mean value of modulus of elasticity of each material, E_{mean}, should be used. The approach also applies to axial stress conditions.

The *equivalent section method* is applicable to situations where each material in the section can have a different E value, however, for presentation purposes, the basic theory given in this section has assumed the top and bottom flanges have the same E value (which is typical with this type of section) and the transformed section has been based on the web material being transformed into flange material.

Because the flange width to span ratio in these sections tends to be relatively small, the shear lag effect in the flanges can be ignored and the full flange width is used in the strength calculations.

With these sections, because their lateral stiffness is low, when used in floor construction and subjected to vibrations above the fundamental frequency of the floor, adverse vibration effects will arise. To reduce this effect and also to provide support against lateral and torsional instability, blocking or strutting between the sections, as shown in Chapter 4, is commonly fitted at intervals along the member lengths.

7.3.1.1   *Strength analysis of glued thin webbed beams*

The strength analysis is undertaken at the ultimate limit states (ULS) and the elastic stress distribution in typical thin webbed sections when subjected to bending is shown in Figure 7.3. In addition to this stress condition, which is the stress state covered in EC5, the stress state arising from an axial action along the centroidal axis of the member acting together with bending is also included. The primary function of the floor beam will be to resist the bending stresses and the axial condition being considered is a nominal direct compression stress under which there will still be a significant tensile bending stress in the tension flange in the combined stress state.

Due to the different creep behaviour of the materials in the composite section, the bending stresses within the section will vary with duration of load and moisture content and this effect has to be taken into account in the analysis. The stress distribution is determined at the instantaneous condition and for the condition arising from creep behaviour, as stated in 2.3.4.2(b), the requirement in EC5, 2.2.2(1)P, is that the analysis be undertaken using final mean values of stiffness adjusted to the load component causing the largest stress in relation to strength.

In accordance with the requirements of EC5, 2.2.2(1)P, the value of the design stiffness property at these conditions will be as follows:

(a) At the instantaneous condition,

$$E_{d,ULS} = E_{mean}, \quad G_{d,ULS} = G_{mean} \tag{7.1}$$

(b) At the condition associated with the final mean value of stiffness,

$$E_{d,ULS} = \frac{E_{mean}}{\left(1 + \psi_2 k_{def}\right)}, \quad G_{d,ULS} = \frac{G_{mean}}{\left(1 + \psi_2 k_{def}\right)} \tag{7.2}$$

where the functions are as follows:
- $E_{d,ULS}$ is the design value of the modulus of elasticity at the ULS.
- $G_{d,ULS}$ is the design value of the shear modulus at the ULS.
- E_{mean} is the mean value of the modulus of elasticity of the element.
- G_{mean} is the mean value of the shear modulus of the element.
- k_{def} is the deformation factor for timber and wood-based products and, for connections, it will be as defined in 2.3.2. Values for k_{def} for timber and some wood-related products are given in Table 2.10.
- ψ_2 is the factor for the quasi-permanent value of the action (see Table 2.2) causing the largest stress in relation to the strength. If this is a permanent action, a value of 1 should be used.

Based on the section profiles shown in Figure 7.3, using the same material for both flanges (as stated in 7.3.1) and taking the flange material as the selected material for the equivalent section, from the following assumptions:

(i)   the flange material has a mean value modulus of elasticity of $E_{mean,f}$ with a deformation factor $k_{def,f}$, and

(ii) the web material has a mean value of modulus of elasticity of $E_{mean,w}$ with a deformation factor $k_{def,w}$,

the equivalent cross-sectional area, A_{ef}, and second moment of area, I_{ef}, of the transformed section will be as follows:

(a) At the instantaneous condition:

$$A_{ef,inst} = A_f + \left(\frac{E_{mean,w}}{E_{mean,f}}\right) A_w \qquad (7.3)$$

$$I_{ef,inst} = I_f + \left(\frac{E_{mean,w}}{E_{mean,f}}\right) I_w \qquad (7.4)$$

(b) At the condition associated with the final mean value of stiffness:
(i)   where the permanent action produces the design condition:

$$A_{ef,fin} = A_f + \left(\frac{E_{mean,w}}{E_{mean,f}}\right)\left(\frac{1+k_{def,f}}{1+k_{def,w}}\right) A_w \qquad (7.5)$$

$$I_{ef,fin} = I_f + \left(\frac{E_{mean,w}}{E_{mean,f}}\right)\left(\frac{1+k_{def,f}}{1+k_{def,w}}\right) I_w \qquad (7.6)$$

(ii) where a variable action condition Q_i produces the design condition:

$$A_{ef,fin} = A_f + \left(\frac{E_{mean,w}}{E_{mean,f}}\right)\left(\frac{1+\psi_{2,i}k_{def,f}}{1+\psi_{2,i}k_{def,w}}\right) A_w \qquad (7.7)$$

$$I_{ef,fin} = I_f + \left(\frac{E_{mean,w}}{E_{mean,f}}\right)\left(\frac{1+\psi_{2,i}k_{def,f}}{1+\psi_{2,i}k_{def,w}}\right) I_w \qquad (7.8)$$

Here the symbols are as previously defined and A_f is the total flange area, and for Figure 7.3a $A_f=(b-2b_w)(h_{f,c}+h_{f,t})$, and for Figure 7.3b $A_f=(b-b_w)(h_{f,c}+h_{f,t})$; A_w is the area of the web, and for Figure 7.3a, $A_w=2b_w(h_w+h_{f,c}+h_{f,t})$, and for Figure 7.3b, $A_w=b_w(h_w+h_{f,c}+h_{f,t})$;
$\psi_{2,i}$ is the ψ_2 factor for the quasi-variable value of the variable action Q_i;
I_f is the second moment of area of both flanges about the neutral axis for the condition being considered;
I_w is the second moment of area of the untransformed web about the neutral axis being considered.

When subjected to the design bending moment, M_d, and the modulus of elasticity of the web is less than that of the flange, the bending stresses in the flange will increase and those in the web will decrease with time, which will be the case whether or not the axial stress is applied. For this condition, it will only be necessary to check the stresses in the flanges at the final mean value condition and those in the web at the instantaneous condition.

If, however, the modulus of elasticity of the web is greater than that of the flange, these stresses should be checked at the instantaneous condition in the flanges and at the final mean value condition in the web. When dealing with symmetrical sections, there will not be a significant difference between the respective values of the geometric properties at both conditions, and consequently the differences in stress will also be relatively small.

In the following sections, the stress equations for the flanges and the web at both the instantaneous and the final mean value condition are given for sections subjected to combined moment, axial compression and shear conditions. If an axial compression force is not applied, in the relevant equations, σ_a and $\sigma_{a,f}$ should be set = 0, i.e. $F_{c,d} = 0$. The sign convention used is the same as the generalised convention adopted in EC5, namely, all stresses are shown as positive but where stresses are of opposite sign (e.g. a compression stress is to be added to a tension stress) in the solution to the relevant equation one value will be positive and the other negative. In the example given in the book for such a condition the compression stress has been taken as positive and the tension stress as negative.

It should be noted from EC5, 2.2.2(1)P, only the final condition has to be checked for these types of section however as there can be situations where the instantaneous condition may become critical, the theory given in this chapter covers both states.

7.3.1.1.1 Stresses in the flanges
(a) Bending stresses.
The maximum stress due to bending will arise at the extreme fibre locations at a distance of z_1 or z_2 from the neutral axis, as shown in Figure 7.3c. When the section is symmetrical about the y–y axis, $z_1 = z_2$ and when it is not, z_1 and z_2 will have different values. Also, when the section is not symmetrical about the y–y axis, the position of the neutral axis will be different at the instantaneous and final mean value conditions and the respective values of z_1 and z_2 at these states will also differ. These are referred to in the text as $z_{1,inst}$ and $z_{2,inst}$ for the instantaneous state and $z_{1,fin}$ and $z_{2,fin}$ for the final mean value condition.

The maximum design compressive stress due to the combined bending and axial compression condition will be at z_1, and:
at the instantaneous condition it will be:

$$\sigma_{f,inst,c,max,d} = \sigma_{a,f} + \sigma_{c1} = \frac{F_{c,d}}{A_{ef,inst}} + \left(\frac{M_d}{I_{ef,inst}} z_{1,inst}\right) \tag{7.9a}$$

and at the final mean value condition it will be:

$$\sigma_{f,fin,c,max,d} = \sigma_{a,f} + \sigma_{c1} = \frac{F_{c,d}}{A_{ef,fin}} + \left(\frac{M_d}{I_{ef,fin}} z_{1,fin}\right) \tag{7.9b}$$

The maximum design tensile stress due to the combined bending and axial compression condition will be at z_2, and:
at the instantaneous condition it will be:

$$\sigma_{f,inst,t,max,d} = \sigma_{a,f} + \sigma_{t1} = \frac{F_{c,d}}{A_{ef,inst}} + \left(\frac{M_d}{I_{ef,inst}} z_{2,inst}\right) \tag{7.10a}$$

and at the final mean value condition it will be:

$$\sigma_{f,fin,t,max,d} = \sigma_{a,f} + \sigma_{t1} = \frac{F_{c,d}}{A_{ef,fin}} + \left(\frac{M_d}{I_{ef,fin}} z_{2,fin} \right)$$

(7.10b)

The design requirement in *Section 9* of EC5 is that the design stress in equations (7.9) and (7.10) be less than or equal to the design bending strength, $f_{m,d}$, i.e.:

$$\sigma_{f,inst,c,max,d}, \ \sigma_{f,fin,c,max,d} \quad \text{and} \quad \sigma_{f,inst,t,max,d}, \ \sigma_{f,fin,t,max,d} \leq f_{m,d}$$

(7.11)

where the functions are as described above and $f_{m,d} = k_{mod} \cdot k_h \cdot k_{sys} \cdot f_{m,k}/\gamma_M$, where k_{mod} is the modification factor for load duration and service classes as given in Table 2.4, k_{sys} is the system strength factor discussed in Chapter 2, k_h is the modification factor for member size effect, referred to in Table 2.5 and discussed in 2.3.6, (the effect applies to solid timber (as well as glulam and LVL, when bent flatwise)), $f_{m,k}$ is the characteristic bending strength of the flange material. Strength information for timber and LVL is given in Chapter 1 and for glulam in Chapter 6. γ_M is the partial coefficient for material properties, given in Table 2.6, noting that the value will be dependent on the material being used.

(b) Compression stresses.
Although a full analysis into the lateral torsional instability behaviour of a glued thin webbed section can be undertaken, in EC5 it is conservatively assumed that lateral stability of the section is provided solely by the buckling strength of the compression flange. The design requirement is that the compression stress in the flange must be shown to be less than or equal to the compression strength. The compression stress is taken to be the average value of the compressive stress in the flange due to bending plus the axial compression stress and for this requirement:
at the instantaneous condition it will be:

$$\sigma_{f,inst,c,d} = \sigma_{a,f} + \sigma_{c2} = \frac{F_{c,d}}{A_{ef,inst}} + \left(\frac{M_d}{I_{ef,inst}} \left(z_{1,inst} - \frac{h_{f,c}}{2} \right) \right)$$

(7.12a)

and at the final mean value condition it will be

$$\sigma_{f,fin,c,d} = \sigma_{a,f} + \sigma_{c2} = \frac{F_{c,d}}{A_{ef,fin}} + \left(\frac{M_d}{I_{ef,fin}} \left(z_{1,fin} - \frac{h_{f,c}}{2} \right) \right)$$

(7.12b)

The EC5 design requirement is that the design stress be less than or equal to the modified design compressive strength, i.e. $k_c f_{c,0,d}$, as follows:

$$\sigma_{f,inst,c,d} \quad \text{and} \quad \sigma_{f,fin,c,d} \leq k_c f_{c,0,d}$$

(7.13)

and

$$f_{c,0,d} = \frac{k_{mod} \cdot k_{sys} \cdot f_{c,0,k}}{\gamma_M}$$

(7.14)

where $f_{c,0,d}$ is the design compressive strength of the flange material and $f_{c,0,k}$ is the characteristic compressive strength of the flange material parallel to the grain. Strength information for timber and LVL is given in Chapter 1 and for glulam in Chapter 6. k_c is a factor that takes into account lateral instability, and is derived assuming that the compression flange behaves as a column between adjacent positions of lateral restraint. The section is conservatively equated to a solid rectangular section of depth, b, resulting in a radius of gyration about the z–z axis of the composite beam of $b/\sqrt{12}$. On this basis, the slenderness ratio of the section will be $\sqrt{12}\,(\ell_c\,/b)$, where ℓ_c is the length of the section between the adjacent positions of lateral support. Factor k_c is then derived using the expressions in 5.4.1. Where full lateral restraint is provided by the floor structure, i.e. $k_c = 1$, it is essential that sufficient fixings be used and located to prevent any lateral movement of the beams.

If a special investigation is made with respect to the lateral torsional instability of the beam as a whole, EC5, *9.1.1(3)*, allows k_c to be assumed to be unity.

(c) Tensile stresses.
The flange must also be checked to ensure that the mean design tensile stress in the tension flange at z_2 due to the combined bending and axial compression condition will not exceed the design tension strength and:
at the instantaneous condition it will be:

$$\sigma_{f,inst,t,d} = \sigma_{a,f} + \sigma_{t2} = \frac{F_{c,d}}{A_{ef,inst}} + \left(\frac{M_d}{I_{ef,inst}}\left(z_{2,inst} - \frac{h_{f,t}}{2}\right)\right) \tag{7.15a}$$

and at the final mean value condition it will be:

$$\sigma_{f,fin,t,d} = \sigma_{a,f} + \sigma_{t2} = \frac{F_{c,d}}{A_{ef,fin}} + \left(\frac{M_d}{I_{ef,fin}}\left(z_{2,fin} - \frac{h_{f,t}}{2}\right)\right) \tag{7.15b}$$

The EC5 design requirement is that the design stress be less than or equal to the design tensile strength, $f_{t,0,d}$, i.e.:

$$\sigma_{f,inst,t,d} \quad \text{and} \quad \sigma_{f,fin,t,d} \le f_{t,0,d} \tag{7.16}$$

and

$$f_{t,0,d} = \frac{k_{mod} \cdot k_{sys} \cdot k_n \cdot f_{t,0,k}}{\gamma_M} \tag{7.17}$$

where the functions are as described previously and the rest of the factors are as follows:

- $f_{t,0,d}$ is the design compressive strength of the flange material parallel to the grain.
- $f_{t,0,k}$ is the characteristic tensile strength of the flange material parallel to the grain. Strength information for timber and LVL is given in Chapter 1 and for glulam in Chapter 6.
- k_h is the size effect modification factor for members under tension. It is discussed in Chapter 2 and given in Table 2.11. The largest cross-sectional dimension of

the member should be used to evaluate the factor. When dealing with LVL, it is defined as k_ℓ, and is associated with the length of the member.

7.3.1.1.2 Bending, shear and buckling stresses in the web

Although the primary function of the web is to support the shear stresses in the section, because it is subjected to compressive and tensile stresses it must also be able to withstand these stresses. With proprietary beams, the web material is normally bonded to form a continuous section, however where this is not possible web splice plates will be required to transfer the stress resultants at the junction positions.

Further, the web must be checked to confirm that it will not buckle due to shear stresses and that the glued joints between the web and the flanges will be able to transfer the horizontal shear stresses in the section. If concentrated vertical loads have to be supported by the beam, web stiffeners may be required to prevent axial web buckling, but no design guidance is given in EC5 for this condition.

(a) Bending stresses.
The maximum design stresses due to the combined bending and axial compression condition will arise at the extreme fibre locations at z_1 or z_2 from the neutral axis as shown in Figure 7.3c.

The maximum design stress in the web on the compression side of the section will be at z_1:

at the instantaneous condition it will be:

$$\sigma_{w,inst,c,d} = \sigma_a + \sigma_{c,3} = \left(\frac{F_{c,d}}{A_{ef,inst}} + \frac{M_d}{I_{ef,inst}} z_{1,inst} \right) \left(\frac{E_{mean,w}}{E_{mean,f}} \right) \tag{7.18a}$$

and at the final mean value condition it will be:

$$\sigma_{w,fin,c,d} = \sigma_a + \sigma_{c,3} = \left(\frac{F_{c,d}}{A_{ef,fin}} + \frac{M_d}{I_{ef,fin}} z_{1,fin} \right) \left(\frac{E_{mean,w}}{E_{mean,f}} \right) \left(\frac{1 + \psi_2 k_{def,f}}{1 + \psi_2 k_{def,w}} \right) \tag{7.18b}$$

The maximum design stress in the web on the tension side of the section will be at z_2:
at the instantaneous condition it will be:

$$\sigma_{w,inst,t,d} = \sigma_a + \sigma_{t,3} = \left(\frac{F_{c,d}}{A_{ef,inst}} + \frac{M_d}{I_{ef,inst}} z_{2,inst} \right) \left(\frac{E_{mean,w}}{E_{mean,f}} \right) \tag{7.19a}$$

and at the final mean value condition it will be:

$$\sigma_{w,inst,t,d} = \sigma_a + \sigma_{t,3} = \left(\frac{F_{c,d}}{A_{ef,fin}} + \frac{M_d}{I_{ef,fin}} z_{2,fin} \right) \left(\frac{E_{mean,w}}{E_{mean,f}} \right) \left(\frac{1 + \psi_2 k_{def,f}}{1 + \psi_2 k_{def,w}} \right) \tag{7.19b}$$

As the web has been transformed to flange material in the equivalent section, to obtain the stress in that element the calculated stress must be multiplied by the appropriate modular ratio for the deformation state as shown in equations (7.18) and (7.19).

The design requirement in *Section 9* of EC5 is that the design compressive bending stress from equation (7.18) must be less than or equal to the design compressive bending strength of the web material, $f_{c,w,d}$, and the design tensile bending stress from equation (7.19) must be less than or equal to the design tensile bending strength of the web material, $f_{t,w,d}$, i.e.

$$\sigma_{w,inst,c,d} \quad \text{and} \quad \sigma_{w,fin,c,d} \le f_{c,w,d} \qquad\qquad (7.20)$$

$$\sigma_{w,inst,t,d} \quad \text{and} \quad \sigma_{w,fin,t,d} \le f_{t,w,d} \qquad\qquad (7.21)$$

where

$$f_{c,w,d} = \frac{k_{mod} \cdot k_{sys} \cdot f_{c,w,k}}{\gamma_M} \qquad\qquad (7.22)$$

and

$$f_{t,w,d} = \frac{k_{mod} \cdot k_{sys} \cdot f_{t,w,k}}{\gamma_M} \qquad\qquad (7.23)$$

Here the functions are as previously described and $f_{c,w,k}$ and $f_{t,w,k}$ are the characteristic compressive bending and tensile bending strengths of the web material respectively. If such values are not given, EC5, *9.1.1(5)*, allows the characteristic compression strength and the characteristic tensile strength of the material ($f_{c,0,k}$ or $f_{c,90,k}$, and $f_{t,0,k}$ or $f_{t,90,k}$ as appropriate) to be used. Values for wood-based products are given in Chapter 1.

(b) Web buckling (due to shear) and shear stress check.
A full buckling analysis due to shear can be undertaken to check the buckling resistance of the web(s); however, as an alternative the criteria for buckling resistance given in *9.1.1(7)* of EC5 can be used. The EC5 approach is conservative and simple to apply and the criterion to be met is:

$$h_w \le 70b_w \qquad\qquad \text{(EC5, equation (9.8))} \quad (7.24)$$

It is important to note that the above will only apply where the axial stress condition induces a nominal compression stress. When considering shear stresses in the web the requirements of *6.1.7(2)* can be ignored as $k_{cr} = 1$ for panel material. Although the shear stress in the web will vary, when $h_w \le 35b_w$, assuming a uniform shear stress distribution, the depth of the web taken to be effective in resisting shear is $(h_w + 0.5(h_{f,t} + h_{f,c}))$, as shown in Figure 7.4a. However, within the range $35b_w < h_w \le 70b_w$ the shear resistance is reduced due to buckling instability effects and, for an I (or box) section in this range, again assuming a uniform shear stress distribution, the depth of the web will be dependent on the value of h_w as shown in Figure 7.4b.

The design requirements of EC5 are:

$$F_{v,w,Ed} \le \begin{cases} b_w h_w \left(1 + \dfrac{05(h_{f,t} + h_{f,c})}{h_w} \right) f_{v,0,d} & \text{for } h_w \le 35b_w \\[3mm] 35b_w^2 \left(1 + \dfrac{05(h_{f,t} + h_{f,c})}{h_w} \right) f_{v,0,d} & \text{for } 35b_w \le h_w \le 70b_w \end{cases}$$

$$\text{(EC5, equation (9.9))} \quad (7.25)$$

$$f_{v,0,d} = \frac{k_{mod} \cdot k_{sys} \cdot f_{v,k}}{\gamma_M} \qquad\qquad (7.26)$$

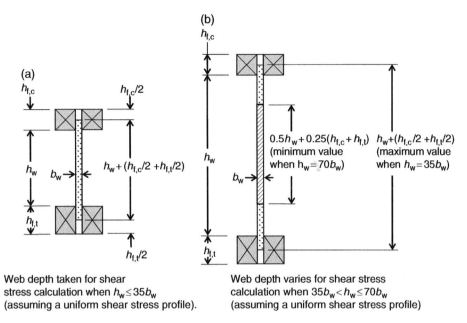

Fig. 7.4. EC5 web strength criteria.

where the symbols are as previously described and/or shown in Figure 7.3 and $F_{v,w,Ed}$ is the design shear force *acting on each web*, $f_{v,0,d}$ is the design panel shear strength of the web, $f_{v,k}$ is the characteristic panel shear strength of the web, and values for different materials are given in Chapter 1.

(c) Horizontal shear stresses in the glued joints between the web and the flanges.
The horizontal shear stresses generated in the flange area are transferred between the web and the flanges through the glued interface connection. The glue in the connection will be able to take these stresses and the limiting design condition will be the rolling shear strength of the web. This type of shear stress is referred to in 4.5.2.2.
   The design shear stress is as follows:

At the instantaneous condition:

$$\tau_{inst,mean,d} = \frac{V_d S_{f,inst}}{I_{ef,inst} \left(nh_f \right)}$$
(7.27a)

At the final mean value condition:

$$\tau_{fin,mean,d} = \frac{V_d S_{f,fin}}{I_{ef,fin} \left(nh_f \right)}$$
(7.27b)

where:

- $\tau_{inst,mean,d}$, $\tau_{fin,mean,d}$ is the design horizontal shear stress at the instantaneous (and final) deformation condition at the glued interface between the flange and the web;

- V_d is the design shear force at the position of maximum shear;
- $S_{f,\,inst}$, $S_{f,\,fin}$ is the first moment of area of the flange (excluding the web area) about the neutral axis at the instantaneous (and final) deformation conditions;
- h_f is the height of the flange, i.e. $h_{f,c}$ or $h_{f,t}$, as appropriate;
- n is the number of glue line interfaces between the flange and the web.

EC5 takes into account the effect of stress concentrations at the web/flange interface in the vicinity of position 2 (see Figure 7.3(a) and (b)) when the height of the flange is greater than $4b_{ef}$ and the design strength requirements are:

$$\tau_{inst,mean,d} \quad and \quad \tau_{fin,mean,d} \le \begin{pmatrix} f_{v,90,d} & for \quad h_f \le 4b_{ef} \\ f_{v,90,d}\left(\dfrac{4b_{ef}}{h_f}\right)^{0.8} & for \quad h_f > 4b_{ef} \end{pmatrix}$$

(EC5, equation (9.10))   (7.28)

with

$$f_{v,90,d} = \frac{k_{mod} \cdot k_{sys} \cdot f_{r,k}}{\gamma_M} \tag{7.29}$$

where the symbols are as previously described and/or shown in Figure 7.3; $f_{v,90,d}$ is the design planar (rolling) shear strength of the web, $f_{r,k}$ is the characteristic planar (rolling) shear strength of the web, and

$$b_{ef} = \begin{pmatrix} b_w & for & box & beams \\ \dfrac{b_w}{2} & for & I\text{-beams} \end{pmatrix}$$

7.3.1.2   *Displacement at the serviceability limit states (SLS)*
Because of the different time-dependent properties of the components in the section, the requirements of 2.3.2 and 2.3.4.1 must be taken into account.

(a) Instantaneous deformation.
At the instantaneous condition the deformation analysis is undertaken using the design value of the combination of actions for the SLS, i.e. either equation (2.24) or (2.25), depending on whether the characteristic or the frequent combination of actions will apply. As the creep behaviour of the member is not relevant at this condition, in accordance with the requirements of EC5, *2.2.3(2)*, the mean value of the appropriate modulus of elasticity, E_{mean}, and shear modulus, G_{mean}, must be used to derive the stiffness properties.

(b) Final deformation.
For the final deformation analysis the requirements are discussed in 2.3.2(b). As stated in 2.3.2(b), for this type of structure it has been accepted by CEN that *2.2.3(3), 2.2.3(4)* and *2.3.2.2(1)* are in conflict and open to misinterpretation and are to be revised as defined in Appendix C. As explained in 2.3.2(b), based on the Appendix C proposal

the creep deformation is to be undertaken using the quasi-permanent combination of actions and the stiffness properties used will be final mean values. The final deformation will be obtained by combining the instantaneous and the creep deformation minus the instantaneous deformation arising from the quasi-permanent combination of actions.

Calculation of the final deformation using this procedure will involve different loading combinations and stiffness properties and for practical analyses, where a conservative value of the deformation will be acceptable (i.e. where a value greater than that derived from the above procedure can be accepted), an alternative and simpler approach is to use the procedure given in the previous edition of this book and also referred to in 2.3.2(b). In that procedure the creep deformation is derived by analysing the structure under the combination of actions used to calculate the instantaneous deformation, with final mean stiffness properties used for the structure elements.

For either option, when calculating the final deformation the final mean values used for stiffness properties are given in equations (2.34) and (2.35) and are:

$$E_{mean,fin} = \frac{E_{mean}}{(1+k_{def})} \qquad (EC5,\ equation\ (2.7)) \quad (2.34)$$

$$G_{mean,fin} = \frac{G_{mean}}{(1+k_{def})} \qquad (EC5,\ equation\ (2.8)) \quad (2.35)$$

where the functions are as follows:

- $E_{mean,fin}$ is the final mean value of the modulus of elasticity;
- E_{mean} is the mean value of the modulus of elasticity;
- $G_{mean,fin}$ is the final mean value of the shear modulus;
- G_{mean} is the mean value of the shear modulus;
- k_{def} is the deformation factor for timber and wood-based products and, for connections, it will be as defined in 2.3.2. Values for k_{def} for timber and some wood-related products are given in Table 2.10.

If the composite section is to be installed at or near its fibre saturation point, but functioning in an environment where it is likely to dry out under load, as required by EC5, 3.2(4), the value of k_{def} obtained from Table 2.10 must be increased by 1.0.

When deriving the deflection due to bending, the second moment of area used in the deflection equation will be the value of the transformed section for the condition being considered, i.e. the value at the instantaneous or the final deformation. When deriving the shear deflection, the shear area will be the *actual* cross-sectional area of the webs and not the transformed area.

Taking the flange material as the material selected for use in the equivalent section approach, on the above basis, the second moment of area to be used in a deformation calculation arising from flexure will be as follows:

(a) The instantaneous condition:

$$I_{ef,inst} = I_f + \left(\frac{E_{mean,w}}{E_{mean,f}}\right) I_w \qquad (7.4)$$

(b) The final deformation condition:

$$I_{ef,fin} = I_f + \left(\frac{E_{mean,w}}{E_{mean,f}}\right)\left(\frac{1+k_{def,f}}{1+k_{def,w}}\right)I_w \tag{7.6}$$

For these structures, the deflection is calculated in the same way as explained in Chapter 4 for rectangular beams, ensuring that shear deformation is taken into account. There are several methods available for deriving the shear deformation of thin web beams and using the approximate method given in *Roark's Formulas for Stress and Strain* [4] in which the form factor for the beam is taken to be unity and all of the shear is considered to be carried solely by the web(s), the shear deflection at the instantaneous condition can be written as:

$$u = \frac{M_d}{G_{w,mean}A_w} \tag{7.30}$$

Where, for simply supported beams, u is the instantaneous shear deformation of the beam at M_d, M_d is the design moment at mid-span, $G_{w,mean}$ is the mean value of the shear modulus of the web material, and A_w is the cross-sectional area of the web(s) in the section (before transformation).

In Table 7.1 equations are given for the deflection of simply supported and cantilever composite beams at the instantaneous and the final deformation conditions due to the effects of bending and shear and taking into account the proposals in Appendix C.

Based on the above, if at the SLS a simply supported thin webbed box beam of span ℓ is subjected to a design loading of q_d (kN m); it will have a creep loading of $q_{d,creep}$; the mean value of the modulus of elasticity of each flange is E_f; the transformed second moment of area is $I_{ef,inst}$ at the instantaneous condition and $I_{ef,fin}$ at the final deformation condition; it has two webs, having a total cross-sectional area A_w, and a mean shear modulus of G_w, where $k_{def,f}$ and $k_{def,w}$ are as defined in equation (7.2), then the deflection at mid-span due to the SLS design load will be as follows:

at the instantaneous condition:

$$u_{inst} = (q_d)\left(\frac{5l^4}{384E_f I_{ef,inst}} + \left(\frac{l^2}{8}\right)\left(\frac{1}{G_w A_w}\right)\right) \tag{7.31}$$

the instantaneous deformation due to creep loading:

$$u_{inst\backslash(creep)} = (q_{d,creep})\left(\frac{5l^4}{384E_f I_{ef,inst}} + \left(\frac{l^2}{8}\right)\left(\frac{1}{G_w A_w}\right)\right) \tag{7.32a}$$

the creep deformation:

$$u_{creep} = (q_{d,creep})\left(\frac{5l^4\left(1+k_{def,f}\right)}{384E_f I_{ef,fin}} + \left(\frac{l^2}{8}\right)\left(\frac{\left(1+k_{def,w}\right)}{G_w A_w}\right)\right) \tag{7.32b}$$

the final deformation:

$$u_{fin} = u_{inst} + u_{creep} - u_{inst(creep)} \tag{7.32c}$$

Table 7.1 Bending and shear deformation of a simply supported or a cantilever composite beam at the instantaneous condition and the creep loading value

Load case	Bending deflection	Shear deflection
udl of q_d kN/m run along the span; and creep load value of $q_{d,creep}$:	At mid-span:	At mid-span:
Instantaneous condition (under q_d):	$\dfrac{5q_d l^4}{384E_{mean,f}I_{ef,inst}}$	$\dfrac{q_d l^2}{8A_w G_{w,mean}}$
Creep deformation:	$\dfrac{5q_{d,creep}l^4(1+k_{def})}{384E_{mean,f}I_{ef,fin}}$	$\dfrac{q_{d,creep}l^2(1+k_{def})}{8A_w G_{w,mean}}$
point load V_d at mid-span; and creep load value of $V_{d,creep}$:	At mid-span:	At mid-span:
Instantaneous condition (under V_d):	$\dfrac{V_d l^3}{48E_{mean,f}I_{ef,inst}}$	$\dfrac{V_d l}{4A_w G_{w,mean}}$
Creep deformation:	$\dfrac{V_{d,creep}l^3(1+k_{def})}{48E_{mean,f}I_{ef,fin}}$	$\dfrac{V_{d,creep}l(1+k_{def})}{4A_w G_{w,mean}}$
point load V_d at the end of a cantilever; and creep load value of $V_{d,creep}$:	At the end of the cantilever:	At the end of the cantilever:
Instantaneous condition (under V_d):	$\dfrac{V_d l^3}{3E_{mean,f}I_{ef,inst}}$	$\dfrac{V_d l}{A_w G_{mean,w}}$
Creep deformation:	$\dfrac{V_{d,creep}l^3(1+k_{def})}{3E_{mean,f}I_{ef,fin}}$	$\dfrac{V_{d,creep}l(1+k_{def})}{A_w G_{mean,w}}$

Note: (1) In the above expressions A_w is the cross-sectional area of the web(s) in the section (before transformation); (2) where a more accurate assessment of the shear deflection is required and the I or box section have flanges of the same thickness and the webs of the box section are both of the same thickness, the deflection expression given in Table 7.1 for the shear deformation should use the area of the section (i.e. the web and the flanges) rather than just the web area and should also be multiplied by the following form factor, F, given in *Roark's Formulas for Stress and Strain* [4],

$$F = \left(1 + \frac{3(D_2^2 - D_1^2)D_1}{2D_2^3}\left(\frac{b_{wt}}{b}-1\right)\right)\frac{4D_2^2}{10r^2}$$

where D_1 is the distance from the neutral axis to the nearest surface of the flange, D_2 is the distance from the neutral axis to the extreme fibre, b is the transformed thickness of the web (or combined web thicknesses in box beams), b_{wt} is the width of the flange, and r is the radius of gyration of the section with respect to the neutral axis.

where $k_{def,f}$ is the deformation factor for the flange material at the relevant service class, and $k_{def,w}$ is the deformation factor for the web material at the relevant service class.

See Example 7.5.1.

7.3.2 Glued thin flanged beams (stressed skin panels)

When a stressed skin panel is subjected to out of plane bending it will function as a thin flanged beam. Stressed skin panels are structural elements in which the web is normally formed using timber sections aligned with the direction of span and the facing panels are formed from wood-based materials such as plywood, OSB or particleboard. The design procedure in EC5, *9.1.2* only covers the design requirements of the composite section in the direction of span of the webs. The panel flanges must also be designed to span across the webs, functioning as beams, as described in Chapters 4 and 5 and this requirement is not addressed in *9.1.2*. It will be this loading condition that will normally determine the thickness of the flange material.

The panels may be on one or both sides of the beam. End blocking between the webs is commonly used to provide lateral torsional restraint at these positions, and where flange splices are required these can be achieved by finger or scarf jointing or by the use of splice plates supported by timber noggings fitted within the box structure.

The connection between the flange and the webs can be glued or formed by mechanical fasteners. If mechanical fasteners are used (e.g. nails, screws, etc.) there will be slip between the flange and the web, and the effect on the design must be taken into account. The more common practice is to use glue and in the following analysis the theory only applies to rigid joints formed by using glued connections, as covered in EC5, *9.1.2*.

As with glued thin webbed beams, glued thin flanged beams also use different materials to form the composite section and the *equivalent section approach* referred to in 7.3.1 is again used in the analyses. Although the method is applicable to situations where each material in the section can have a different E value (which is common with this type of section) for presentation purposes, the basic theory given in this section has assumed the top and bottom flanges have the same E value and the transformed section has been based on the web material being transformed into flange material.

7.3.2.1 *Effective flange width*

When the beam is subjected to bending, stresses are transferred between the web and the flange by shear stresses. Because of shear deformation, the stresses in each flange will reduce as the distance from the web increases and the departure from the uniform stress profile assumed to be generated when using simple bending theory is termed the 'shear lag' effect. To take this into account the concept of the 'effective width' of the flange is used. The effective width is that width of flange over which the stress is taken to be uniform and at the maximum stress value derived from bending theory such that the total force in the flange will equal that carried by the full width of the actual flange.

The effective flange width concept applies to flanges in compression and in tension, and unless a more detailed calculation is carried out, in accordance with the requirements of EC5, *9.1.2(3)*, the effective flange width, b_{ef}, as shown in Figure 7.5, will be as follows:

- For internal I-shaped sections:

$$b_{ef} = b_{c,ef} + b_w \quad \text{or} \quad b_{ef} = b_{t,ef} + b_w \qquad \text{(EC5, equation (9.12))} \quad (7.33)$$

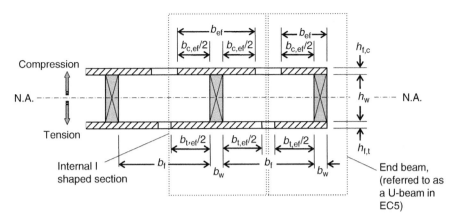

Fig. 7.5. Glued thin flanged beam.

- For U-beams (i.e. the end beams in Figure 7.5):

$$b_{ef} = 0.5b_{c,ef} + b_w \quad \text{or} \quad b_{ef} = 0.5b_{t,ef} + b_w \quad \text{(EC5, equation (9.13))} \quad (7.34)$$

Taking shear lag effects into account, the values of $b_{c,ef}$ and $b_{t,ef}$ must not exceed those given in *Table 9.1* of EC5, reproduced in Table 7.2.

The shear lag effect in structures has been investigated by several researchers and a comparison of the EC5 criteria with the theoretical solution developed by Mohler, Abdel-Sayed and Ehlbeck (referred to in STEP 1 [5]), applied to glued thin flange beams in which plywood is used for the flange material, is shown in Figure 7.6. For the majority of this type of beam, b_f/ℓ will be within the range 0.2–0.3 and as can be seen from Figure 7.6, the effective flange approach will generally give a conservative result.

From this analysis, the thin flanged beam will be divided into I shaped and U shaped beam sections and the question to be asked is whether or not these sections will function as a single lateral element (i.e. the neutral axis will be based on the combined geometry of all of the sections) or as individual elements? As no guidance is given in

Table 7.2 Maximum values to be taken for effective flange widths due to the effects of shear lag and plate buckling[*]

Material used for the flange	Plate buckling[†]	Shear lag[‡]
Oriented strand board (OSB)	$25h_f$	$0.15\,\ell$
Plywood:		
Outer plies parallel to the webs	$20h_f$	$0.1\,\ell$
Outer plies perpendicular to the webs	$25h_f$	$0.1\,\ell$
Particleboard or fibreboard[§]	$30h_f$	$0.2\,\ell$

[*]Based on *Table 9.1* in EC5.
[†]h_f is the thickness of the flange being subjected to compression.
[‡]ℓ is the span of the composite section.
[‡§]With random fibre orientation.

EC5 it is to be assumed the sections are to be designed as individual items however, where the system factor, k_{sys}, associated with the web behaviour can be applied it should be possible to argue that the neutral axis can be calculated assuming single element behaviour.

7.3.2.2 *Plate buckling*

For the compression flange, in addition to complying with the shear lag criteria in Table 7.2, in order to prevent plate buckling from occurring, $b_{c,ef}$ must also comply with the maximum effective width criteria for plate buckling, also given in Table 7.2.

Also, unless a full buckling analysis is undertaken, the clear flange width, b_f, should not be taken to be greater than twice the effective flange width derived from the plate buckling criteria in Table 7.2.

7.3.2.3 *Section properties*

The properties of these sections are determined in the same way as in 7.3.1 for thin webbed beams. Equations (7.3)–(7.8) will apply, but in this case the area of the flanges and the webs will be as follows:

- A_f: the total flange area, $A_f = h_{f,c} (b_{c,ef} + b_w) + h_{f,t} (b_{t,ef} + b_w)$;
- A_w: the area of the web, $A_w = b_w h_w$.

Here the symbols are as shown in Figure 7.5 for an internal I-beam and for an end beam.

When subjected to design conditions, the stresses in the sections are derived assuming elastic theory applies and, because of the creep effect, as with thin webbed beams the stresses will generally be different at the instantaneous and the final mean value conditions.

Under the requirements of *2.2.2(1)P* in EC5 for these types of structure it is only necessary to check the final condition, however, as there can be situations where the instantaneous condition may become critical, in this chapter both states have been covered.

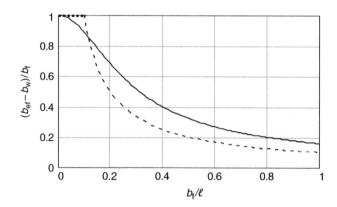

Fig. 7.6. Effective flange width when using plywood flanges (comparing the theoretical solution (solid line) with the EC5 criteria in Table 7.2 (dashed line)).

7.3.2.4   *Stresses in flanges*

A typical glued thin flanged internal I-beam with flanges on the top and bottom faces and subjected to a moment is shown in Figure 7.7(a) and the flange stresses required to be checked in EC5 are shown in Figure 7.7(b).

As the flanges are thin, the stress in each flange due to bending is effectively an axial stress and the design value is taken to be the average value across the flange thickness. With proprietary beams, the flange material is normally bonded to form a continuous section using scarf or finger joints. Where this is not possible, flange splice plates will need to be designed to transfer the stress resultants at the junction position.

The maximum compressive and tensile stress under a design moment M_d will arise at z_1 and z_2, respectively, from the neutral axis as shown in Figure 7.7b. When the areas of the compression and tension flanges are not the same, z_1 and z_2 will have different values. Also, the position of the neutral axis will not be the same at the instantaneous and the final mean value conditions, consequently the respective values of z_1 and z_2 will be different for each condition. Following the approach used for glued thin webbed beams, these are referred to in the text as $z_{1,\,inst}$ and $z_{2,\,inst}$ for the instantaneous state and $z_{1,\,fin}$ and $z_{2,\,fin}$ for the final mean value condition.

(a) Stress in the compression flange.
The mean design compressive stress in the compression flange will be as follows:

- At the instantaneous condition:

$$\sigma_{f,inst,c,max,d} = (\sigma_{c1}) = \left(\frac{M_d}{I_{ef,inst}} z_{1,inst} \right) \qquad (7.35)$$

- At the final mean value condition:

$$\sigma_{f,fin,c,max,d} = (\sigma_{c1}) = \left(\frac{M_d}{I_{ef,fin}} z_{1,fin} \right) \qquad (7.36)$$

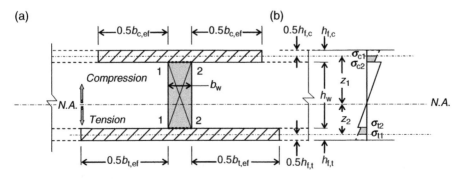

Fig. 7.7. Thin flanged section with a flange on the top and bottom face subjected to bending (the relative values of the stresses will be dependent on the geometry and material E values).

The design requirement in EC5, *9.1.2(7)*, is that the mean flange design compressive stress must be less than or equal to the design compressive strength, i.e.:

$$\sigma_{f,inst,c,max,d} \quad \text{and} \quad \sigma_{f,fin,max,d} \leq f_{c,d} \tag{7.37}$$

and

$$f_{c,d} = \frac{k_{mod} \cdot k_{sys} \cdot f_{c,k}}{\gamma_M} \tag{7.38}$$

where the functions are as previously described and $f_{c,k}$ is the characteristic compressive strength ($f_{c,0,k}$ or $f_{c,90,k}$, as appropriate) of the flange material. Strength information for timber and the commonly used wood-based structural products is given in Chapter 1. k_{sys} is the system factor and for this situation, as the load distribution system with adjacent webs is a single flange panel arrangement, no strength enhancement can arise and the factor will equal 1.0.

(b) Stress in the tension flange.
The mean design tensile stress in the tension flange will be as follows:

• At the instantaneous condition:

$$\sigma_{f,inst,t,mean,d} = (\sigma_{t1}) = \left(\frac{M_d}{I_{ef,inst}} z_{2,inst} \right) \tag{7.39}$$

• At the final mean value condition:

$$\sigma_{f,inst,t,mean,d} = (\sigma_{t1}) = \left(\frac{M_d}{I_{ef,fin}} z_{2,fin} \right) \tag{7.40}$$

The EC5 design requirement is that the design stress be less than or equal to the design tensile strength, i.e.:

$$\sigma_{f,inst,t,mean,d} \quad \text{and} \quad \sigma_{f,fin,t,mean,d} \leq f_{t,d} \tag{7.41}$$

and

$$f_{t,d} = \frac{k_{mod} \cdot k_{sys} \cdot f_{t,k}}{\gamma_M} \tag{7.42}$$

where the functions are as previously described and $f_{t,k}$ is the characteristic tensile strength ($f_{t,0,k}$ or $f_{t,90,k}$, as appropriate) of the flange material. Strength information for timber and the commonly used wood-based structural products is given in Chapter 1.

7.3.2.5   *Stresses in the web*

The design requirements for the web are that it must be able to support the flexural stresses that arise, that the shear stress in the web must be acceptable, and that the glued joints between the web and the flanges must be able to transfer the horizontal shear stresses at the interface.

(a) Bending stresses.

The maximum stresses due to bending will arise at the extreme fibre locations at a distance of $(z_1 - 0.5\,h_{f,c})$ or $(z_2 - 0.5\,h_{f,t})$ from the neutral axis as shown in Figure 7.7b.

The maximum design stress on the compression side of the web will be:

- At the instantaneous condition,

$$\sigma_{w,inst,c,d} = \left(\sigma_{c,2}\right) = \left(\frac{M_d}{I_{ef,inst}}\left(z_{1,inst} - 0.5h_{f,c}\right)\right)\left(\frac{E_{mean,w}}{E_{mean,f}}\right) \tag{7.43}$$

- At the final mean value condition,

$$\sigma_{w,fin,c,d} = \left(\sigma_{c,2}\right) = \left(\frac{M_d}{I_{ef,fin}}\left(z_{1,fin} - 0.5h_{f,c}\right)\right)\left(\frac{E_{mean,w}}{E_{mean,f}}\right)\left(\frac{1+\psi_2 k_{def,f}}{1+\psi_2 k_{def,w}}\right) \tag{7.44}$$

The maximum design stress on the tension side of the web will be:

- At the instantaneous condition,

$$\sigma_{w,inst,t,d} = \left(\sigma_{t,2}\right) = \left(\frac{M_d}{I_{ef,inst}}\left(z_{2,inst} - 0.5h_{f,t}\right)\right)\left(\frac{E_{mean,w}}{E_{mean,f}}\right) \tag{7.45}$$

- At the final mean value condition,

$$\sigma_{w,fin,t,d} = \left(\sigma_{t,2}\right) = \left(\frac{M_d}{I_{ef,fin}}\left(z_{2,fin} - 0.5h_{f,t}\right)\right)\left(\frac{E_{mean,w}}{E_{mean,f}}\right)\left(\frac{1+\psi_2 k_{def,f}}{1+\psi_2 k_{def,w}}\right) \tag{7.46}$$

As the web has been transformed in the equivalent section, to obtain the stress in that element the calculated bending stress must be multiplied by the modular ratio for the appropriate condition.

The design requirement in EC5, *9.1.2(9)*, is that the design stresses obtained from equations (7.43) and (7.44) must be less than or equal to the design compressive bending strength of the web material, $f_{c,w,d}$, and those from equations (7.45) and (7.46) must be less than or equal to the design tensile bending strength, $f_{t,w,d}$. Where timber is used for the web,

$$\sigma_{w,c,d} \le f_{c,w,d} \tag{7.47a}$$

$$\sigma_{w,t,d} \le f_{t,w,d} \tag{7.47b}$$

and

$$f_{c,w,d} = \frac{k_{mod} \cdot k_{sys} \cdot k_h \cdot f_{m,k}}{\gamma_M} \qquad (7.48)$$

and

$$f_{t,w,d} = f_{c,w,d} \qquad (7.49)$$

where the functions are as previously described and $f_{m,k}$ is the characteristic bending strength obtained from Table 1.3; k_{sys} is in accordance with the requirements of 2.3.7 and where the panels are prefabricated, which is the general situation, providing there are four or more equally spaced web members not more than 610 mm apart in a panel and the flange members are continuous within the panel, a factor of 1.1 can be used. If there are open areas in the flanges, a factor of 1.0 should be used.

Where timber is not used for the web, unless compressive and tensile bending strengths are given for the material being used, the in-plane compressive and tensile strengths of the material should be used.

(b) Web shear stress check.
Although there is no stated requirement in EC5 to carry out a shear stress check across the section, the greatest shear stress will arise in the web and the shear strength of this member should be checked. As the requirements of EC5, *6.1.7(2)* will be relevant in the analysis of the web section, rather than apply the k_{cr} factor to the beam width, in the equality given below it has been applied to the web shear strength. The maximum shear stress will arise at the NA position and the value at this position, $\tau_{v,d}$, is determined as follows:

- At the instantaneous condition,

$$\tau_{v,inst,d} = \frac{V_d S_{f,inst,NA}}{I_{ef,inst} b_w} \qquad (7.50)$$

- At the final mean value condition,

$$\tau_{v,fin,d} = \frac{V_d S_{f,fin,NA}}{I_{ef,fin} b_w} \qquad (7.51)$$

where the functions are as previously defined, and V_d is the design shear force at the position of maximum shear, $S_{f,inst,NA}$ and $S_{f,fin,NA}$ are the first moments of the area of the section above the NA about the neutral axis at the instantaneous and at the final mean value conditions, respectively, and b_w is the thickness of the web.

An alternative conservative approach to that given above is to take all shear on the web and to check as a conventional rectangular beam as follows:

$$\tau_v = \frac{3}{2} \frac{V_d}{b_w h_w} \qquad (7.51a)$$

With either option the design requirement for the shear strength will be:

$$\tau_{v,inst,d} \text{ and } \tau_{v,fin,d} \text{ or } \tau_v \leq k_{cr} f_{v,d} \tag{7.52}$$

where $f_{v,d}$ is the design shear strength of the web material, derived as defined in equation (4.18) and in which k_{sys} will be evaluated as for equation (7.48); k_{cr} will be obtained from the UKNA to EC5 [6], and is referred to in 4.5.2.

(c) Horizontal shear stresses in the glued joints between the web and the flanges.
The shear stress in the web is transferred to the flanges in the composite section through the glued interface connection at positions 1–2 shown in Figure 7.7. The glue in the connection will be able to take the stress and the limiting design condition for a flange comprising wood-based panels will be the rolling shear strength of the flange. The shear stress along the interface is assumed to be uniform and will be derived as follows:

- At the instantaneous condition:

$$\tau_{v,inst,d} = \frac{V_d S_{f,inst}}{I_{ef,inst} b_w} \tag{7.53}$$

- At the final mean value condition:

$$\tau_{v,fin,d} = \frac{V_d S_{f,fin}}{I_{ef,fin} b_w} \tag{7.54}$$

where the functions are as previously defined and:
V_d is the design shear force at the position of maximum shear, and $S_{f,inst}$ and $S_{f,fin}$ are the first moments of the area of a flange about the neutral axis at the instantaneous and the final mean value conditions, respectively. For each condition there will be a value for the compression and for the tension flange, and the larger of the values should be used.

For an internal I shaped section the design requirements of EC5 are given in EC5, *9.1.2(6)*, which are:

$$\tau_{mean,d} \leq \begin{pmatrix} f_{v,90,d} & for \quad b_w \leq 8h_f \\ f_{v,90,d} \left(\dfrac{8h_f}{b_w} \right)^{0.8} & for \quad b_w > 8h_f \end{pmatrix} \qquad (EC5, \; equation \; (9.14)) \tag{7.55}$$

In equation (7.55) h_f is the flange thickness associated with the first moment of area used to derive the horizontal shear stress and $f_{v,90,d}$ is the design (planar) rolling shear strength of the flange material obtained as follows:

$$f_{v,90,d} = \frac{k_{mod} \cdot k_{sys} \cdot f_{r,k}}{\gamma_M} \tag{7.56}$$

The functions in equation (7.56) are as previously described and it is to be noted that the value of k_{sys} will $= 1$.

For sections where $b_w \leq 8\,h_f$ the effect of stress concentrations at the glued junction between the flange and the web can be ignored. When $b_w > 8\,h_w$, however, stress concentration effects have to be included for and this is achieved by the power function $(8h_f/b_w)^{0.8}$.

For a U shaped end beam section, the same expressions will apply but with $8h_f$ substituted by $4h_f$.

(d) Panel shear stress check in the flange material at the junctions with the web.
The horizontal shear stresses generated in the flange area must be able to be transferred to the web through the panel shear strength of the flange material. This is unlikely to be a critical design condition and indeed is not referred to as a strength requirement in EC5, however the design procedure for this stress check will be as follows:

at the instantaneous condition:

$$\tau_{inst,mean,d} = \frac{V_d S1_{f,inst}}{I_{ef,inst}\,(nh_f)}$$ (7.57)

at the final mean value condition:

$$\tau_{fin,mean,d} = \frac{V_d S1_{f,fin}}{I_{ef,fin}\,(nh_f)}$$ (7.58)

where functions are as previously defined and;

$\tau_{inst,mean,d}$; $\tau_{fin,mean,d}$;	are the design horizontal shear stress at the instantaneous (and final mean value) condition at the junction with the web;
$S1_{f,inst}$; $S1_{f,fin}$	is the first moment of area of the flange (excluding the area of flange over the web) about the neutral axis at the instantaneous (and final mean value) conditions;
h_f	is the thickness of the flange i.e. $h_{f,c}$ or $h_{f,t}$ as appropriate.
n	is the number of junctions (i.e. 2 for an internal I beam and 1 for an external U-beam).

As the value of the crack modification factor, k_{cr}, referred to in the UKNA, is 1 for wood-based products, it has not been taken into account in this check calculation.

The design requirements of EC5 are:

$\tau_{inst,mean,d}$ and $\tau_{fin,mean,d} \leq f_{v,d}$

where $f_{v,d} = \dfrac{k_{mod} \cdot k_{sys} \cdot f_{v,k}}{\lambda_M}$ and

$f_{v,k}$ is the panel shear strength of the flange material and k_{sys} will be 1.

7.3.2.6   *Deflection at the SLS*

For these structures the deflection is calculated as described for glued thin web beams in Section 7.3.1.2, and the web depth for shear deflection calculations should be based on the depth of the web in the thin flanged beam, h_w.

See Examples 7.5.2 and 7.5.3.

7.4   REFERENCES

1 BS EN 1995-1-1:2004+A1:2008. *Eurocode 5: Design of Timber Structures. Part 1-1: General – Common Rules and Rules for Buildings*, British Standards Institution.
2 Hearn, E. J. *Mechanics of Materials, Volume 1*, 2nd edn. International Series on Materials Science and Technology, Volume 19. BPCC Wheatons Ltd, Exeter, 1985.
3 Gere, J.M., Timoshenko, S.P. *Mechanics of Materials*, 3rd SI edn. Chapman and Hall, London, 1991.
4 Young, W.C. *Roark's Formula's for Stress and Strain*, 6th edn. McGraw-Hill International Editions, New York, 1989.
5 Raadschelders, J.G.M., Blass, H.J. Stressed skin panels. In: Blass, H.J., Aune, P., Choo, B.S., et al. (eds), *Timber Engineering STEP 1*, 1st edn. Centrum Hout, Almere, 1995.
6 NA to BS EN 1995-1-1:2004+A1:2008, Incorporating National Amendment No. 2. *UK National Annex to Eurocode 5: Design of Timber Structures. Part 1-1: General – Common Rules and Rules for Buildings*, British Standards Institution.
7 BS EN 338:2009. *Structural Timber – Strength Classes*, British Standards Institution.
8 BS EN 12369-1:2001. *Wood-Based Panels – Characteristic Values for Structural Design. Part 1: OSB, Particleboards and Fibreboards*, British Standards Institution.
9 NA to BS EN 1990:2002 + A1:2005 *Incorporating National Amendment No1. UK National Annex for Eurocode 0 – Basis of Structural Design*, British Standards Institution.

7.5   EXAMPLES

As stated in 4.3, in order to verify the ultimate and serviceability limit states, each design effect has to be checked and for each effect the largest value caused by the relevant combination of actions must be used.

However, to ensure attention is primarily focused on the EC5 design rules for the timber or wood product being used, only the design load case producing the largest design effect has generally been given or evaluated in the following examples.

Example 7.5.1 The floor in a domestic building comprises glued thin-webbed beams spaced at 450 mm centres with an effective span of 4.0 m. The beams are supported laterally along their full length, and solid timber blocking is also fitted at the ends to ensure lateral torsional rigidity. Allowing for the self-weight of the structure the beams support a characteristic permanent loading of 0.9 kN/m^2 and a characteristic variable medium-term loading of 2.0 kN/m^2. The timber used for the flanges is service class C18 in accordance with BS EN 338:2009 and the web is

12.5 mm thick 5 ply Canadian softwood plywood, with the face ply aligned perpendicular to the direction of span. The web is continuously bonded and no splice plates are required. The cross-section of the beam is shown in Figure E7.5.1. The structure functions in service class 2 conditions.

Excluding the requirement to check the vibration behaviour of the floor, show that the beams will comply with the rules in EC5.

Also show that when each floor beam is subjected to a characteristic variable medium term axial compression load of 0.5 kN along the longitudinal centroidal axis in addition to the floor loading, the behaviour at the ultimate limit state will be acceptable.

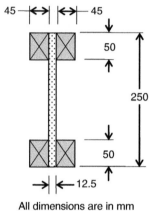

All dimensions are in mm

Fig. E7.5.1.

1. Geometric properties

Effective span of beam, L_e	$L_e = 4.0$ m
Beam depth, H	$H = 250$ mm
Beam spacing, B_s	$B_s = 0.45$ m
Flanges:	
Width of timber flange element, b_f	$b_f = 45$ mm
Height of timber flange element, h_f	$h_f = 50$ mm
Web:	
Thickness of the plywood web, b_w	$b_w = 12.5$ mm
Clear height between the flanges, h_w	$h_w = H - 2 \cdot h_f$ $h_w = 150$ mm
Area of the web, A_w	$A_w = b_w \cdot H$ $A_w = 3.13 \times 10^{-3}$ m²
For the I-beam, b_{ef} (equation (7.29) (EC5 9.1.1(8)))	$b_{ef} = \dfrac{b_w}{2}$ $b_{ef} = 6.25$ mm

2. Material properties

Table 1.3, timber – strength class C18 (BS EN 338:2009, *Table 1*)

Characteristic bending strength, $f_{m.k}$	$f_{m.k} = 18$ N/mm^2
Characteristic compression strength parallel to the grain, $f_{c.0.k}$	$f_{c.0.k} = 18$ N/mm^2
Characteristic tensile strength parallel to the grain, $f_{t.0.k}$	$f_{t.0.k} = 11$ N/mm^2
Mean modulus of elasticity parallel to the grain, E_0, mean	$E_{0.mean} = 9.0$ kN/mm^2

Table 1.13, 12.5 mm thick 5 ply Canadian softwood plywood

Characteristic compression strength, $f_{p.c.90.k}$	$f_{p.c.90.k} = 9.7$ N/mm^2
Characteristic tensile strength, $f_{p.t.90.k}$	$f_{p.t.90.k} = 7.4$ N/mm^2
Characteristic panel shear strength, $f_{p.v.k}$	$f_{p.v.k} = 3.5$ N/mm^2
Characteristic rolling shear strength, $f_{p.r.k}$	$f_{p.r.k} = 0.64$ N/mm^2
Mean modulus of elasticity, $E_{p.c.90.mean}$	$E_{p.c.90.mean} = 3.906$ kN/mm^2
Mean modulus of rigidity, $G_{w.mean}$	$G_{w.mean} = 0.430$ kN/mm^2

In the following analysis, the transformed section is based on the use of the flange material throughout the section.

Note:
1 To comply with the design rules in EC5, it is only necessary for final condition stresses to be considered.
2 In the solution given, initial and final conditions have been considered and as the mean E value of the timber is greater than the mean E value of the plywood, there is only a need to check the bending stresses in the flange at the final condition and in the web at the instantaneous condition. To compare with the EC5 requirement, values derived from the final condition analysis are given in brackets.

3. Partial safety factors

Table 2.8 (UKNA to BS EN 1990:2002+A1:2005, *Table NA.A1.2(B)*)) for the ULS

Permanent actions, γ_G	$\gamma_G = 1.35$
Variable actions, γ_Q	$\gamma_Q = 1.5$

Table 2.2 (UKNA to BS EN 1990:2002+A1:2005, *Table NA.A1.1* – Category A

Factor for the combination value of a variable action for a domestic building, ψ_0	$\psi_0 = 0.7$
Factor for quasi-permanent value of variable action, ψ_2	$\psi_2 = 0.3$

Table 2.6 (UKNA to EC5, *Table NA.3*)

Material factor for solid timber at the ultimate limit states (ULS), γ_M	$\gamma_M = 1.3$
Material factor for plywood web at the ULS, γ_{pM}	$\gamma_{pM} = 1.2$

4. Actions

Characteristic permanent action on the beam, G_k	$G_k = 0.9\,\text{kN/m}^2$
Characteristic variable (imposed) medium-term action on the beam, Q_k	$Q_k = 2.00\,\text{kN/m}^2$
Characteristic variable (imposed) medium term axial action on the beam, $Q1_k$	$Q1_k = 0.5\,\text{kN}$

(i) ULS

(a) Design condition without the axial loading:

Design load due to the critical load combination, F_d

(Table 2.8, equation (c) using the unfavourable condition variable action)

$$F_d = (\gamma_G \cdot G_k + \gamma \cdot Q_k) \cdot B_s \quad F_d = 1.9\,\text{kN/m}$$

Design moment due to the critical load combination, M_d

$$M_d = \frac{F_d \cdot L_e^2}{8}$$

$M_d = 3.79\,\text{kN/m}$

Design shear force due to the critical load combination, V_d

(Ignoring the shear force reduction referred to in 4.5.2.1)

$$V_d = F_d \cdot \frac{L_e}{2}$$

$V_d = 3.79\,\text{kN}$

(b) Design condition with the axial loading - which is an independent action:

Design load due to the axial load, $F_{c.d}$
(This will be the design condition in conjunction with lateral loading)

$$F_{c.d} = \psi_0 \cdot \gamma_Q \cdot Q1_k$$

$F_{c.d} = 0.53\,\text{kN}$

(ii) Serviceability limit states (SLS)

Design load due to permanent action at the SLS, $F_{SLS.G}$

$$F_{SLS.G} = G_k \cdot B_s$$

$F_{SLS.G} = 0.41\,\text{kN/m}$

Design load due to variable action at the SLS, $F_{SLS.Q}$

$$F_{SLS.Q} = Q_k \cdot B_s$$

$F_{SLS.Q} = 0.9\,\text{kN/m}$

5. Modification factors

Factor for permanent duration action and service class 2, $k_{mod.perm}$

(Table 2.4 (EC5, *Table 3.1*))

$k_{mod.perm} = 0.6$

Factor for medium-duration action and service class 2, $k_{mod.med}$

(Table 2.4 (EC5, *Table 3.1*))

$k_{mod.med} = 0.8$

Load sharing factor, k_{sys}
(2.3.7 (EC5, 6.6))
(Note: 1.1 can be used if required)

$$k_{sys} = 1.0$$

Depth factor for solid timber – bending
and axial tension – take the size as h_f, k_h
(Table 2.11 (EC5, *equation (3.1)*))
(the equation incorporates a dimensional
correction factor for Mathcad)

$$k_h = \min\left[\left(\frac{150 \text{ mm}}{h_f}\right)^{0.2}, 1.3\right]$$

$$k_h = 1.25$$

Deformation factor for timber and
service class 2, $k_{def.f}$
(Table 2.10 (EC5, *Table 3.2*))

$$k_{def.f} = 0.8$$

Deformation factor for plywood and
service class 2, $k_{def.w}$
(Table 2.10 (EC5, *Table 3.2*))

$$k_{def.w} = 1.0$$

Buckling resistance condition with beam
laterally supported along its compression
flange, $k_c = 1$
(equation (7.13) (EC5, 6.3.3))

$$k_c = 1$$

6. **Geometric properties – transformed sections**
(Note: Because the section is symmetrical about the y–y axis, the neutral axis will be at mid-depth.)

Instantaneous – transformed section properties:

Transformed web thickness
(into wood), $b_{w.tfd}$

$$b_{w.tfd} = \left(b_w \cdot \frac{E_{p.c.90.mean}}{E_{0.mean}}\right)$$

$$b_{w.tfd} = 5.5 \text{ mm}$$

Second moment of area of the flanges, $I_{ef.f}$

$$I_{ef.f} = \frac{2 \cdot b_f \cdot H^3}{12} - \frac{2 \cdot b_f \cdot (H - 2 \cdot h_f)^3}{12}$$

Second moment of area of the web, $I_{ef.w}$

$$I_{ef.w} = \frac{b_{w.tfd} \cdot H^3}{12}$$

Cross sectional area

$$A_{ef.inst} = 2 \cdot b_f \cdot h_f + b_{w.tfd} \cdot H \qquad A_{ef.inst} = 5.88 \times 10^3 \text{ mm}^2$$

Instantaneous second moment of area
of the transformed section, I_{ef}
(equation (7.4))

$$I_{ef} = I_{ef.f} + I_{ef.w} \qquad I_{ef} = 9.9 \times 10^7 \text{ mm}^4$$

Final – transformed section properties:

Transformed web thickness:

The largest stress to strength ratio will be the larger of the permanent action/$k_{mod.perm}$ and the combined permanent and variable action/$k_{mod.med}$.

(the following analysis is for the condition excluding the axial load.). Let the ratio of one to the other be r:

$$r = \frac{\gamma_G \cdot G_k}{k_{mod.perm}} \cdot \frac{k_{mod.med}}{\gamma_Q \cdot Q_k + \gamma_G \cdot G_k} \qquad\qquad r = 0.38$$

i.e. because the factor is less than 1, the variable loading produces the higher stress/strength ratio, so the factor ψ_2 will be associated with variable loading. This will also be the result for the case which includes for the axial load condition.

Value of the quasi-permanent factor for $\psi_2 = 0.3$
the variable action, ψ_2 (Table 2.2)

$$b_{c.w.tfd} = \left(b_w \cdot \frac{E_{p.c.90.mean}}{E_{0.mean}} \right) \cdot \frac{1 + \psi_2 \cdot k_{def.f}}{1 + \psi_2 \cdot k_{def.w}} \qquad\qquad b_{c.w.tfd} = 5.25\,\text{mm}$$

Cross sectional area at final condition

$$A_{ef.fin} = 2 \cdot b_f \cdot h_f + b_{cw.tfd} \cdot H \qquad\qquad A_{ef.fin} = 5.81 \times 10^3\,\text{mm}^2$$

Second moment of area of the web, $I_{c.ef.w}$ $\quad I_{c.ef.w} = \dfrac{b_{c.w.tfd} \cdot H^3}{12}$

Second moment of area of the $\quad I_{c.ef} = I_{ef.f} + I_{c.ef.w}$
transformed section for the final
condition, $I_{c.ef}$
(equation (7.8)) $\qquad\qquad I_{c.ef} = 9.87 \times 10^7\,\text{mm}^4$

7. Bending stress check in the flanges and the web

Because the mean modulus of elasticity of the flange material is greater than that of the web, only check the stresses in the flanges at the final deformation condition and those in the web at the instantaneous condition.

(a) Design condition without the axial loading:

The critical design load case at the ULS will be due to the combination of permanent and unfavourable medium-term duration variable action:

Stress in flange due to bending – final condition - (the EC5 requirement)

(This is also the critical condition based on the design procedure given in Chapter 7):

Bending stress in the top and bottom
flange, $\sigma_{f.c.max.d.c}$

$$\sigma_{f.c.max.d.c} = \frac{M_d}{I_{c.ef}} \cdot \frac{H}{2} \qquad\qquad \sigma_{f.c.max.d.c} = 4.8\,\text{N/mm}^2$$

Bending strength of the flange, $f_{m.d}$
(equation (7.11))

$$f_{m.d} = \frac{k_{mod.med} \cdot k_h \cdot k_{sys} \cdot f_{m.k}}{\gamma_M} \qquad\qquad f_{m.d} = 13.8\,\text{N/mm}^2$$

bending stress in flanges is OK

Stress in the web due to bending –final condition - (the EC5 requirement)
(Based on the theory given in Chapter 7 the instantaneous condition will be the critical
condition and is given below for comparison with the EC5 value:

EC5 Bending stress in the web, $\sigma_{w,\,fin,c,d}$
(Final condition – Equation 7.18b)

$$\sigma_{w.fin.c.d} = \frac{M_d}{I_{cef}} \cdot \frac{H}{2} \cdot \left[\left(\frac{E_{p.m.90.mean}}{E_{0.mean}} \right) \left(\frac{1+\psi_2 \cdot k_{def.f}}{1+\psi_2 \cdot k_{def.w}} \right) \right]$$

$\sigma_{w.fin.c.d} = 2.02\ \text{N/mm}^2$

Bending stress in the web, $\sigma_{l_{w,\,fin,c,d}}$
(Instantaneous (critical)
condition – Equation 7.18a)

$$\sigma l_{w.fin.c.d} = \frac{M_d}{I_{ef}} \cdot \frac{H}{2} \cdot \left(\left(\frac{E_{p.m.90.mean}}{E_{0.mean}} \right) \right)$$

$\sigma l_{w.fin.c.d} = 2.11\ \text{N/mm}^2$

(i.e. EC5 value is exceeded by the critical
Instantaneous condition by approximately 5%)

Bending strength of the web in
compression, $f_{c.w.d}$ (equation (7.22))

$$f_{c.w.d} = \frac{k_{mod.med} \cdot k_{sys} \cdot f_{p.c.90.k}}{\gamma_{pM}}$$

$f_{c.w.d} = 6.47\ \text{N/mm}^2$

Bending strength of the web in tension, $f_{t.w.d}$
(equation (7.23))

$$f_{t.w.d} = \frac{k_{mod.med} \cdot k_{sys} \cdot f_{p.t.90.k}}{\gamma_{pM}}$$

$f_{t.w.d} = 4.93\ \text{kN/mm}^2$

bending stress in web is OK

Stress in the flange due to axial stress – final condition:
(which is also the EC5 design condition)

Axial stress in the top and bottom flange, $\sigma_{f.fin.c.d}$
(equation (7.12b))

$$\sigma_{f.fin.c.d} = \frac{M_d}{I_{cef}} \cdot \left(\frac{H}{2} - \frac{h_f}{2} \right)$$

$\sigma_{f.fin.c.d} = 3.84\ \text{kN/mm}^2$

Axial strength in compression, $f_{c.0.d}$
(equation (7.14))

$$f_{c.0.d} = \frac{k_{mod.med} \cdot k_{sys} \cdot f_{c.0.k}}{\gamma_M}$$

$f_{c.0.d} = 11.08\ \text{kN/mm}^2$

Axial strength taking lateral
instability into account, $k_c \cdot f_{c.0.d}$
(equation (7.13))

$k_c \cdot f_{c.0.d} = 1.08\ \text{kN/mm}^2$

Axial strength in tension, $f_{t.0.d}$
(equation (7.17))

$$f_{t.0.d} = \frac{k_{mod.med} \cdot k_h \cdot k_{sys} \cdot f_{t.0.k}}{\gamma_M}$$

$f_{t.0.d} = 8.43 \text{ N/mm}^2$

axial stress in flanges is OK

(b) Design condition with the axial loading – which is an independent action:

The critical design load case at the ULS will be due to the combination of permanent and unfavourable medium term duration variable action:

Stress in flange due to bending – final condition – (the EC5 requirement)

(This is also the critical condition based on the design procedure given in Chapter 7):

Bending stress in the top flange, $\sigma\alpha_{f.c.max.d}$
(equation 7.9b)

$$\sigma\alpha_{f.fin.c.max.d} = \left(\frac{F_{c.d}}{A_{ef.fin}} + \frac{M_d}{I_{cef}} \cdot \frac{H}{2} \right)$$

$\sigma\alpha_{f.fin.c.max.d} = 4.89 \text{ N/mm}^2$

bending stress is less than bending strength, OK

Stress in the web due to bending –final condition – (the EC5 requirement)
(Based on the theory given in Chapter 7 the Instantaneous condition will be the critical condition and is given below for comparison with the EC5 value:

EC5 Bending (compression) stress in
the web, $\sigma a_{w.fin.c.d}$
(Final condition – Equation 7.18b)

$$\sigma a_{w.fin.c.d} = \left(\frac{F_{c.d}}{A_{ef.inst}} + \frac{M_d}{I_{ef}} \cdot \frac{H}{2} \right) \cdot \left[\left(\frac{E_{p.m.90.mean}}{E_{0.mean}} \right) \left(\frac{1 + \psi_2 \cdot k_{def.f}}{1 + \psi_2 \cdot k_{def.w}} \right) \right]$$

$\sigma a_{w.fin.c.d} = 2.05 \text{ N/mm}^2$

bending stress is less than bending compression strength, OK

Bending (compression) stress in the web, $\sigma a1_{w.inst.c.d}$
(Instantaneous (critical)
condition – Equation 7.18a)

$$\sigma a1_{w.inst.c.d} = \left(\frac{F_{c.d}}{A_{ef.inst}} + \frac{M_d}{I_{ef}} \cdot \frac{H}{2} \right) \cdot \left(\left(\frac{E_{p.m.90.mean}}{E_{0.mean}} \right) \right)$$

$\sigma a1_{w.inst.c.d} = 2.15 \text{ N/mm}^2$

(i.e. EC5 value exceeded by the critical
Instantaneous condition by approximately 5%)

EC5 Bending (tension) stress in the web, $\sigma a_{w.fin.t.d}$
(Final condition - equation (7.19b)

$$\sigma a_{w.fin.t.d} = \left(\frac{-F_{c.d}}{A_{ef.inst}} + \frac{M_d}{I_{ef}} \cdot \frac{H}{2} \right) \cdot \left[\left(\frac{E_{p.m.90.mean}}{E_{0.mean}} \right) \left(\frac{1 + \psi_2 \cdot k_{def.f}}{1 + \psi_2 \cdot k_{def.w}} \right) \right]$$

$\sigma a_{w.fin.t.d} = 1.97 \text{ N/mm}^2$

bending stress is less than bending tension strength, OK

Bending (tension) stress in the web, $\sigma a_{w.inst.t.d}$
(Instantaneous (critical)
condition – Equation 7.19a)

$$\sigma a_{w.inst.t.d} = \left(\frac{-F_{c.d}}{A_{ef.inst}} + \frac{M_d}{I_{ef}} \cdot \frac{H}{2} \right) \cdot \left(\left(\frac{E_{p.m.90.mean}}{E_{0.mean}} \right) \right)$$

$\sigma a_{w.inst.t.d} = 2.07 \, \text{kN/mm}^2$

(i.e. EC5 value exceeded by the critical
Instantaneous condition by approximately 5%)

Stress in the flange due to axial stress – Final condition:
(which is also the EC5 design condition)

Axial compression stress in the top flange, $\sigma a_{f.fin.c.d}$
(equation (7.12b))

$$\sigma a_{f.fin.c.d} = \frac{F_{c.d}}{A_{ef.fin}} + \frac{M_d}{I_{cef}} \cdot \left(\frac{H}{2} - \frac{hf}{2} \right)$$

$\sigma a_{f.fin.c.d} = 3.93 \, \text{kN/mm}^2$

Axial tension stress in the bottom flange, $f_{c.0.d}$
(equation (7.15b))

$$\sigma a_{f.fin.t.d} = \frac{-F_{c.d}}{A_{ef.fin}} + \frac{M_d}{I_{cef}} \cdot \left(\frac{H}{2} - \frac{hf}{2} \right)$$

$\sigma a_{f.fin.t.d} = 3.75 \, \text{kN/mm}^2$

axial stresses in web are OK

8. Buckling and shear stress check in the web

Buckling condition for the web in EC5:

(equation (7.24); EC5, $\text{ratio} = \dfrac{h_w}{b_w}$ $\text{ratio} = 12$
equation (9.8))

ratio is < 70; therefore buckling OK

Shear strength of the web (As the k_{cr} factor = 1 it is not relevant for the web calculation):

Panel shear strength of the web, $f_{v.0.d}$

$$f_{v.0.d} = \frac{k_{mod.med} \cdot k_{sys} \cdot f_{p.v.k}}{\gamma_{pM}}$$

$f_{v.0.d} = 2.33 \, \text{N/mm}^2$

Design shear force able to be taken by the web
(equation (7.25); EC5, equation (9.9))

$$F_{v.w.Ed} = \begin{vmatrix} b_w \cdot h_w \cdot \left[1 + \dfrac{0.5 \cdot (h_f + h_f)}{h_w} \right] \cdot f_{v.0.d} & \text{if} \quad h_w \leq 35 \cdot b_w \\ \\ 35 \cdot b_w^2 \cdot \left[1 + \dfrac{0.5 \cdot (h_f + h_f)}{h_w} \right] \cdot f_{v.0.d} & \text{if} \quad 35 \cdot b_w \leq h_w \leq 70 \cdot b_w \end{vmatrix}$$

$F_{v.w.Ed} = 5.83 \times 10^3 \, \text{N}$

Design shear force in the web, V_d

$V_d = 3.79 \times 10^3 \, \text{N}$

The design shear strength of the web is greater than
the design shear force in the web; therefore OK

Shear strength of the glued joint between the web and the flanges

At the instantaneous condition:

Design shear force, V_d

$$V_d = 3.79 \times 10^3 \text{ N}$$

First moment of area of a flange about the NA, S_f

$$S_f = 2 \cdot b_f \cdot h_f \cdot \left(\frac{H}{2} - \frac{h_f}{2} \right) \quad S_f = 4.5 \times 10^{-4} \text{ m}^3$$

Total length of the glue line in the flange

$$2 \cdot h_f = 0.1 \text{ m}$$

Shear stress in the glue line, $\tau_{\text{mean.d}}$ (equation (7.27a))

$$\tau_{\text{mean.d}} = \frac{V_d \cdot S_f}{I_{\text{ef}} \cdot 2 \cdot h_f} \quad \tau_{\text{mean.d}} = 0.17 \text{ N/mm}^2$$

At the final condition:

Shear stress in the glue line, $\tau_{\text{c.mean.d}}$ (equation (7.27b))

$$\tau_{\text{c.mean.d}} = \frac{V_d \cdot S_f}{I_{\text{c.ef}} \cdot 2 \cdot h_f} \quad \tau_{\text{c.mean.d}} = 0.17 \text{ N/mm}^2$$

Rolling shear strength of the web material, $f_{v.1.90.d}$ (equation (7.29))

$$f_{v.1.90.d} = \frac{k_{\text{mod.med}} \cdot k_{\text{sys}} \cdot f_{\text{p.r.k}}}{\gamma_{\text{pM}}} \quad f_{v.1.90.d} = 0.43 \text{ N/mm}^2$$

Equation (7.28) (EC5, *equation (9.10)*):

$$f_{v.90.d} = \begin{vmatrix} f_{v.1.90.d} & \text{if} \quad h_f \le 4 \cdot b_{\text{ef}} \\ f_{v.1.90.d} \cdot \left(\frac{4 \cdot b_{\text{ef}}}{h_f} \right)^{0.8} & \text{if} \quad h_{\text{ef}} > 4 \cdot b_{\text{ef}} \end{vmatrix} \quad f_{v.90.d} = 0.25 \text{ N/mm}^2$$

Design rolling shear strength is greater than the mean shear stress in the web; therefore OK

9. Deflection of the beam at the SLS

(a) At the Instantaneous condition:

The instantaneous deflection at mid-span at the SLS, $u_{\text{inst.}}$, will be based on the deflection formulae in Table 7.1:

$$u_{\text{inst.G}} = \frac{5}{384} \cdot \frac{\left(F_{\text{SLS.G}} + F_{\text{SLS.Q}} \right) \cdot L_e^4}{E_{0.\text{mean}} \cdot I_{\text{ef}}} + \left(\frac{F_{\text{SLS.G}} + F_{\text{SLS.Q}}}{8} \right) \cdot L_e^2 \cdot \frac{1}{G_{\text{w.mean}} \cdot A_w} \quad u_{\text{inst.G}} = 6.82 \text{ mm}$$

Allowable instantaneous deflection at mid-span at the SLS, $u_{\text{inst.SLS.all}}$ (Adopt $L_e/300$)

$$u_{\text{inst.SLS.all}} = \frac{L_e}{300} \quad u_{\text{inst.sls.all}} = 13.33 \text{ mm}$$

Instantaneous deflection is O.K.

(b) At the Final condition:

i) Using the proposals in Appendix C (referred to in 2.3.2(b)) to calculate the final deformation:

The stiffness property must be based on the requirements of EC5, 2.3.2.2(1):

$$b1_{c.w.tfd} = \left(b_w \cdot \frac{E_{p.c.90.mean}}{E_{0.mean}} \right) \cdot \frac{1+k_{def.f}}{1+k_{def.w}} \qquad\qquad b1_{c.w.tfd} = 4.95\,\text{mm}$$

Second moment of area of the web, $II_{c.ef.w}$

$$II_{c.ef.w} = \frac{b1_{c.w.tfd} \cdot H^3}{12}$$

Transformed second moment of area of the section for the final deformation condition, $II_{c.ef}$

$$II_{c.ef} = I_{ef.f} + II_{c.ef.w} \qquad\qquad II_{c.ef} = 9.83 \times 10^7\,\text{mm}^4$$

The creep deformation due to the quasi-permanent combination of actions, $u_{c.qp}$:

$$u_{c.qp} = \frac{5}{384} \cdot \frac{\left(F_{sls.G} + \Psi_2 \cdot F_{SLS.Q} \right) \cdot L_e^4}{E_{0.mean} \cdot II_{cef}} \cdot (1+k_{def.f}) + \left(\frac{F_{sls.G} + \Psi_2 \cdot F_{sls.Q}}{8} \right) \cdot L_e^2 \cdot \frac{1+k_{def.w}}{G_{w.mean} \cdot A_w}$$

$$u_{c.qp} = 6.59\,\text{mm}$$

The instantaneous deformation due to the quasi-permanent combination of actions, $u_{inst.qp}$:

$$u_{inst.qp} = \frac{5}{384} \cdot \frac{\left(F_{sls.G} + \Psi_2 \cdot F_{sls.Q} \right) \cdot L_e^4}{E_{0.mean} \cdot I_{ef}} + \left(\frac{F_{sls.G} + \Psi_2 \cdot F_{sls.Q}}{8} \right) \cdot L_e^2 \cdot \frac{1}{G_{w.mean} \cdot A_w}$$

$$u_{inst.qp} = 3.53\,\text{mm}$$

The final deformation, u_{fin}:

$$u_{fin} = u_{c.qp} + u_{inst} - u_{inst.qp} \qquad\qquad u_{fin} = 9.88\,\text{mm}^4$$

ii) Using the alternative procedure given in 2.3.2(b):

The final deformation, u_{fin}:

$$u_{fin} = \frac{5}{384} \cdot \frac{\left(F_{SLS.G} + F_{SLS.Q} \right) \cdot L_e^4}{E_{0.mean} \cdot II_{c.ef}} \cdot (1+k_{def.f}) + \left(\frac{F_{SLS.G} + F_{SLS.Q}}{8} \right) \cdot L_e^2 \cdot \frac{1+k_{def.w}}{G_{w.mean} \cdot A_w}$$

$$u_{fin} = 12.73\,\text{mm}$$

(Note, although the alternative procedure gives a final deformation greater than the value obtained from the proposed revised method referred to in Appendix C, it is still within the allowable limit and is acceptable).

Allowable net final deflection at mid-span at the SLS, $u_{fin.SLS.all}$ (adopting $L_e/250$ from EC5 *Table* 7.2)

$$u_{fin.SLS.all} = \frac{L_e}{250} \qquad\qquad u_{fin.SLS.all} = 16\,\text{mm}$$

The allowable deflection exceeds the actual; therefore OK

Example 7.5.2 A stressed skin panel is used as a flat roof member spanning between two supports 4.5 m apart. The panel is 187 mm deep and its cross-section at an I-shaped beam position is shown in Figure E7.5.2. The clear distance between webs is 585 mm and the panel is glued between the flanges and the web. Including for the self-weight of the structure, the characteristic permanent loading per web is 0.35 kN/m and the characteristic variable short-duration loading is 1.10 kN/m. The timber used for the web is class C22 to BS EN 338:2009 and the flanges are both OSB/3 to BS EN 12369-1:2001, 16 mm and 11 mm thick on the top and bottom faces, respectively, with the faces aligned parallel to the direction of span. The structure functions in service class 2 conditions. Show that the section will comply with the rules in EC5 at the ULS.

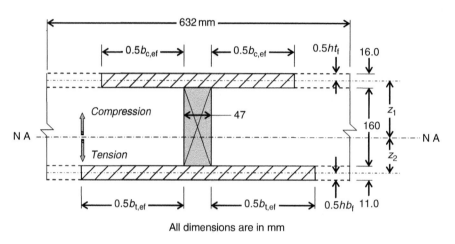

All dimensions are in mm

Fig. E7.5.2.

1. **Panel geometric properties**

    Effective span of the panel, L_e $L_e = 4.5\,\text{m}$

    Clear distance between webs, b_f $b_f = 585\,\text{mm}$

    Web:

        Width of the web, b_w $b_w = 47\,\text{mm}$

        Clear height between the $h_w = 160\,\text{mm}$
        flanges, h_w

        Area of the web, A_w $A_w = b_w \cdot h_w$ $A_w = 7.52 \times 10^{-3}\,\text{m}^2$

    Flanges:

        Top flange thickness, ht_f $ht_f = 16.0\,\text{mm}$

        Bottom flange thickness, hb_f $hb_f = 11.0\,\text{mm}$

        Beam depth, h $h = h_w + ht_f + hb_f$ $h = 187\,\text{mm}$

    Effective flange width of an I-beam section of the panel
    (7.3.2.1 (EC5, *9.1.2*))

In compression, $b1_{c.ef}$

(Table 7.2 (EC5, *Table 9.1*))

$b1_{c.ef} = \min(0.15 \cdot L_e, 25 \cdot ht_f)$ $\qquad\qquad b1_{c.ef} = 400\,\text{mm}$

i.e. the value is based on plate buckling criteria

Check on the buckling length of the compression
flange – based on plate buckling criteria
(7.3.2.2 (EC5, *Clause 9.1.2(5)*)) $\qquad \dfrac{b_f}{2 \cdot b1_{c.ef}} = 0.73$ $\qquad$ the value is less
than 1; therefore OK

In tension, $b1_{t.ef}$

(Table 7.2 (EC5, *Table 9.1*)) $\qquad b1_{t.ef} = 0.15 \cdot L_e$ $\qquad b1_{t.ef} = 675\,\text{mm}$

Check on the clear flange width due
to the geometric constraint of the
section – ratio must not exceed 1: $\qquad \dfrac{b1_{t.ef}}{b_f} = 1.15 \quad \text{Fails}$ $\qquad \dfrac{b1_{c.ef}}{b_f} = 0.68 \quad \text{OK}$

The design sizes of the flanges will be

In compression, $b_{c.ef}$ $\qquad b_{c.ef} = b1_{c.ef}$ $\qquad b1_{c.ef} = 400\,\text{mm}$
(Table 7.2 (EC5, *Table 9.1*))

In tension, $b_{t.ef}$ $\qquad b_{t.ef} = b_f$ $\qquad b_{t.ef} = 585\,\text{mm}$

Effective flange width in compression,
$b_{ef.c}$ $\qquad b_{ef.c} = b_{c.ef} + b_w$ $\qquad b_{ef.c} = 447\,\text{mm}$

Effective flange width in tension, $b_{ef.t}$ $\qquad b_{ef.t} = b_{t.ef} + b_w$ $\qquad b_{ef.t} = 632\,\text{mm}$
(equation (7.33); EC5, *equation (9.12)*)

2. **Material strength properties**

Table 1.3, strength class C22 (BS EN 338:2009, *Table 1*)

Characteristic bending strength, $f_{m.k}$ $\qquad f_{m.k} = 22\ \text{N/mm}^2$

Characteristic shear strength, $f_{v.k}$ $\qquad f_{v.k} = 3.8\ \text{N/mm}^2$

Mean modulus of elasticity parallel
to the grain, $E_{0.mean}$ $\qquad E_{0.mean} = 10.0\ \text{kN/mm}^2$

Table 1.17 – OSB/3 to BS EN 12369-1:2001

Characteristic compression strength,
$f_{OSB.c.0.k}$ $\qquad f_{OSB.c.0.k} = 15.4\ \text{N/mm}^2$

Characteristic tensile strength, $f_{OSB.t.0.k}$ $\qquad f_{OSB.t.0.k} = 9.4\ \text{N/mm}^2$

Characteristic rolling shear strength,
$f_{OSB.r.k}$ $\qquad f_{OSB.r.k} = 1.0\ \text{N/mm}^2$

Characteristic panel shear strength,
$f_{OSB.v.0.k}$ $\qquad f_{OSB.v.0.k} = 6.8\ \text{N/mm}^2$

Mean modulus of elasticity, $E_{OSB.c.0.mean}$ $\qquad E_{OSB.c.0.mean} = 3.8\ \text{kN/mm}^2$

In the following analysis, the transformed section is based on the use of the flange material throughout the section.

Note: As the mean E value of the timber is greater than the mean E value of the OSB, there will only be a need to check the stresses in the flange at the instantaneous condition and the web at the final condition.

3. **Partial safety factors**

Table 2.8 (UKNA to BS EN 1990:2002+A1:2005, *Table NA.A1.2(B)*)) for the ULS

Permanent actions, γ_G	$\gamma_G = 1.35$
Variable actions, γ_Q	$\gamma_Q = 1.5$

Table 2.2 (UKNA to BS EN 1990:2002+A1:2005, *Table NA.A1.1*)

Factor for quasi-permanent value of $\psi_2 = 0.0$
variable action, ψ_2
(roof condition)

Table 2.6 (UKNA to EC5, *Table NA.3*)

Material factor for solid timber, γ_M	$\gamma_M = 1.3$
Material factor for OSB, $\gamma_{OSB.M}$	$\gamma_{OSB.M} = 1.2$

4. **Actions**

Characteristic permanent action on the $G_k = 0.35 \, \text{kN/m}$
structure, G_k

Characteristic variable (imposed) $Q_k = 1.10 \, \text{kN/m}$
short-term action on the structure, Q_k

ULS:

Design load due to the critical load $F_d = (\gamma_G \cdot G_k + \gamma_Q \cdot Q_k)$ $F_d = 2.12 \, \text{kN/m}$
combination, F_d

(Table 2.8, equation (c) using the
unfavourable condition variable action)

Design moment due to the critical load $M_d = \dfrac{F_d \cdot L_e^2}{8}$ $M_d = 5.37 \, \text{kN m}$
combination, M_d

Design shear force due to the critical $V_d = F_d \cdot \dfrac{L_e}{2}$ $V_d = 4.78 \, \text{kN}$
load combination, V_d

5. **Modification factors**

Factor for permanent duration action $k_{\text{mod.perm}} = (0.6 \cdot 0.3)^{0.5}$ $k_{\text{mod.perm}} = 0.42$
and service class 2, $k_{\text{mod.perm}}$
(Table 2.4 (EC5, *Table 3.1*))

Factor for short-duration action and $k_{\text{mod.short}} = (0.9 \cdot 0.7)^{0.5}$ $k_{\text{mod.short}} = 0.79$
service class 2, $k_{\text{mod.short}}$
(Table 2.4 (EC5, *Table 3.1*))

Load sharing factor – k_{sys} $k_{\text{sys}} = 1.0$

(2.3.7 (EC5, *6.6*)) (note: 1.1 can be used if required)

Depth factor for solid timber – take as 1, $k_h = 1$
as the depth is greater than 150 mm, k_h
(Table 2.11 (EC5, *equation (3.1)*))

The crack modification factor for timber, $k_{cr} = 0.67$
k_c (Table 4.5 (4.5.2))

Deformation factor for timber at service class 2 conditions, $k_{def.w}$ $k_{def.w} = 0.8$

(Table 2.10 (EC5, *Table 3.2*))

Deformation factor for OSB at service class 2 conditions, $k_{def.f}$ $k_{def.f} = 2.25$

(Table 2.10 (EC5, *Table 3.2*))

6. Geometric properties

Instantaneous – transformed section properties:

Transformed web thickness (into OSB), $b_{w.tfd}$

$$b_{w.tfd} = \left(b_w \cdot \frac{E_{0.mean}}{E_{O.S.B.c.0mean}} \right) \qquad b_{w.tfd} = 123.68\,\text{mm}$$

Area of flange in compression, $A_{ef.f.c}$ $A_{ef.f.c} = b_{ef.c} \cdot ht_f$ $A_{ef.f.c} = 7.15 \times 10^{-3}\,\text{m}^2$

Area of flange in tension, $A_{ef.f.t}$ $A_{ef.f.t} = b_{ef.t} \cdot hb_f$ $A_{ef.f.t} = 6.95 \times 10^{-3}\,\text{m}^2$

Area of web, $A_{ef.w}$ $A_{ef.w} = b_{w.tfd} \cdot h_w$

Transformed area, A_{ef} $A_{ef} = A_{ef.f.c} + A_{ef.f.t} + A_{ef.w}$ $A_{ef} = 0.03\,\text{m}^2$

First moment of area of the section about the top face:

$$A_{1st} = b_{ef.t} \cdot hb_f \cdot \left(h - \frac{hb_f}{2} \right) + b_{w.tfd} \cdot h_w \cdot \left(\frac{h_w}{2} + ht_f \right) + b_{ef.c} \cdot ht_f \cdot \frac{ht_f}{2} \qquad A_{1st} = 3.22 \times 10^{-3}\,\text{m}^3$$

Neutral axis depth from the top face, z_t $z_t = \dfrac{A_{1st}}{A_{ef}}$ $z_t = 94.97\,\text{mm}$

Second moment of area of the web about the NA, $I_{ef.w}$:

$$I_{ef.w} = \frac{b_{w.tfd} \cdot h_w^3}{12} + A_{ef.w} \cdot \left[z_t - \left(ht_f + \frac{h_w}{2} \right) \right]^2 \qquad I_{ef.w} = 4.22 \times 10^{-5}\,\text{m}^4$$

Second moment of area of the top flange about the NA, $I_{ef.ftf}$

$$I_{ef.f.t.f} = \frac{b_{ef.c} \cdot ht_f^3}{12} + A_{ef.f.c} \cdot \left[z_t - \left(\frac{ht_f}{2} \right) \right]^2 \qquad I_{ef.f.t.f} = 5.42 \times 10^{-5}\,\text{m}^4$$

Second moment of area of the bottom flange about the NA, $I_{ef.fbf}$

$$I_{ef.f.b.f} = \frac{b_{ef.t} \cdot hb_f^3}{12} + A_{ef.f.t} \cdot \left(h - z_t - \frac{hb_f}{2} \right)^2 \qquad I_{ef.f.b.f} = 5.21 \times 10^{-5}\,\text{m}^4$$

Instantaneous second moment of area the transformed section, I_{ef} $I_{ef} = I_{ef.w} + I_{ef.f.t.f} + I_{ef.f.b.f}$ $I_{ef} = 1.49 \times 10^8\,\text{mm}^4$

Final – transformed section properties:

The largest stress to strength ratio will be the larger of the permanent action/$k_{mod.perm}$ and the combined permanent and variable action/$k_{mod.med}$. Let the ratio of one to the other be r:

$$r = \frac{\gamma_G \cdot G_k}{k_{mod.perm}} \cdot \frac{k_{mod.short}}{\gamma_G \cdot G_k + \gamma_Q \cdot Q_k} \qquad\qquad r = 0.42$$

i.e. the variable loading will produce the higher stress/strength ratio, so the factor ψ_2 will be associated with variable loading.

The ψ_2 factor for this loading condition is $\psi_2 = 0$.

Consequently, for this condition the instantaneous and final properties will be the same. We only need to analyse for the instantaneous condition.

7. **Bending stress check in the flanges**

The critical design load case at the ULS will be due to the combination of permanent and unfavourable medium-term duration variable action:

Stress in the flanges due to bending:

Bending stress (compression) in the top flange $\sigma_{f.c.max.d}$

$$\sigma_{f.c.max.d} = \frac{M_d}{I_{ef}} \cdot \left(z_t - \frac{ht_f}{2} \right) \qquad\qquad \sigma_{f.c.max.d} = 3.14 \ \text{N/mm}^2$$

Bending stress (tension) in the bottom flange, $\sigma_{f.t.max.d}$

$$\sigma_{f.t.max.d} = \frac{M_d}{I_{ef}} \cdot \left(h - z_t - \frac{hb_f}{2} \right) \qquad\qquad \sigma_{f.t.max.d} = 3.13 \ \text{N/mm}^2$$

Axial strength of the top flange, $f_{OSB.c.0.d}$

$$f_{OSB.c.0.d} = \frac{k_{mod.short} \cdot k_{sys} \cdot f_{OSB.c.0.k}}{\gamma_{OSB.M}} \qquad\qquad f_{OSB.c.0.d} = 10.19 \ \text{N/mm}^2$$

Axial strength of the bottom flange, $f_{OSB.t.0.d}$

$$f_{OSB.t.0.d} = \frac{k_{mod.short} \cdot k_{sys} \cdot f_{OSB.t.0.k}}{\gamma_{OSB.M}} \qquad\qquad f_{OSB.t.0.d} = 6.22 \ \text{N/mm}^2$$

Stresses in the top and bottom flanges are OK

8. **Bending and shear stress check in the web**

The critical design load case at the ULS will be due to the combination of permanent and unfavourable medium-term duration variable action:

Maximum distance from the NA to the extreme fibre, z_1

$$z_1 = \max[(z_t - ht_f), (h - hb_f - z_t)] \qquad\qquad z_1 = 81.03 \ \text{mm}$$

Bending stress in the web, $\sigma_{w.c.d.c}$

$$\sigma_{w.c.d.c} = \frac{M_d}{I_{ef}} \cdot z_1 \cdot \left(\frac{E_{0.mean}}{E_{OSB.c.0.mean}} \right) \qquad\qquad \sigma_{w.c.d.c} = 7.71 \ \text{kN/mm}^2$$

Bending strength of the web, $f_{w.d}$

$$f_{w.d} = \frac{k_{mod.short} \cdot k_{sys} \cdot k_h \cdot f_{m.k}}{\gamma_M}$$

$f_{w.d} = 13.43 \, \text{kN/mm}^{-2}$

Bending stress in the web is OK

Shear strength of the web:

Shear force in the beam, V_d

$V_d = 4.78 \times 10^3 \, \text{N}$

First moment of area of the section above the NA about the NA, $S_{t.f.NA}$

$$S_{t.f.NA} = b_{ef.c} \cdot ht_f \cdot \left(z_t - \frac{ht_f}{2} \right) + b_{w.tfd} \cdot \frac{(z_t - ht_f)^2}{2}$$

$S_{t.f.NA} = 1.01 \times 10^{-3} \, \text{m}^3$

Shear stress at the NA position, $\tau_{v.d}$ $\tau_{v.d} = \dfrac{V_d \cdot S_{t.f.NA}}{I_{ef} \cdot b_w}$

$\tau_{v.d} = 0.69 \, \text{N/mm}^2$

Shear strength of the web material, $f_{v.d}$

$$f_{v.d} = \frac{k_{cr} \cdot k_{mod.short} \cdot k_{sys} \cdot f_{v.k}}{\gamma_M}$$

$f_{v.d} = 1.55 \, \text{N/mm}^2$

Design shear strength is greater than the shear stress, therefore OK

Shear strength of the glued joint between the web and the flanges:

Shear force in beam, V_d

$V_d = 4.78 \times 10^3 \, \text{N}$

First moment of area of top flange about the NA, S_{tf}

$$S_{t.f} = b_{ef.c} \cdot ht_f \cdot \left(z_t - \frac{ht_f}{2} \right)$$

$S_{t.f} = 6.22 \times 10^{-4} \, \text{m}^3$

First moment of area of the bottom flange about NA, S_{bf}

$$S_{b.f} = b_{ef.t} \cdot hb_f \cdot \left(h - z_t - \frac{hb_f}{2} \right)$$

$S_{b.f} = 6.02 \times 10^{-4} \, \text{m}^3$

Maximum value of first moment of area about NA, S_f

$S_f = \max(S_{t.f}, S_{b.f})$

$S_f = 6.22 \times 10^{-4} \, \text{m}^3$

Mean shear stress in the flange across the glue line, $\tau_{mean.d}$

$$\tau_{mean.d} = \frac{V_d \cdot S_f}{I_{ef} \cdot b_w}$$

$\tau_{mean.d} = 0.43 \, \text{N/mm}^2$

Rolling shear strength of the flange material, $f_{v.1.90.d}$

$$f_{v.1.90.d} = \frac{k_{mod.short} \cdot k_{sys} \cdot f_{OSB.r.k}}{\gamma_{OSB.M}}$$

$f_{v.1.90.d} = 0.66 \, \text{N/mm}^2$

Rolling shear criteria, $f_{v.90.d}$ (equation (7.56) (EC5, 9.1.2(6)))

$$f_{v.90.d} = \begin{cases} f_{v.1.90.d} & \text{if } b_w \le 8 \cdot hb_f \\ f_{v.1.90.d} \cdot \left(\dfrac{8 \cdot hb_f}{b_w} \right)^{0.8} & \text{if } b_w > 8 \cdot hb_f \end{cases}$$

$f_{v.90.d} = 0.66 \, \text{N/mm}^2$

Design rolling shear strength is greater than the mean shear stress in the web across glue line; therefore OK. Also, there is clearly no requirement to check the rolling shear strength of the bottom flange/web connection.

Shear strength of the flange material at the face of the web:

Shear force in beam, V_d

$$V_d = 4.78 \times 10^3 \text{ N}$$

First moment of area of top flange about the NA, SI_{tf}

$$SI_{tf} = (b_{efc} - b_{w.tfd}) \cdot ht_f \cdot \left(z_t - \frac{ht_f}{2} \right)$$

$$SI_{tf} = 4.5 \times 10^{-4} \text{ m}^3$$

First moment of area of the bottom flange about NA, SI_{bf}

$$SI_{bf} = (b_{eft} - b_{w.tfd}) \cdot hb_f \cdot \left(h - z_t - \frac{hb_f}{2} \right)$$

$$SI_{b.f} = 4.48 \times 10^{-4} \text{ m}^3$$

Mean shear stress in the top flange at the web face, $\tau t1_{mean.d}$

$$\tau t1b1_{mean.d} = \frac{V_d \cdot SI_{tf}}{I_{ef} \cdot 2ht_f}$$

$$\tau t1_{mean.d} = 0.45 \text{ N/mm}^2$$

Mean shear stress in the bottom flange at the web face, $\tau b1_{mean.d}$

$$\tau b1_{mean.d} = \frac{V_d \cdot SI_{bf}}{I_{ef} \cdot 2hb_f}$$

$$\tau b1_{mean.d} = 0.71 \text{ N/mm}^2$$

Panel shear strength of the top and the bottom flange material, $f_{osb.v.r.d}$

$$f_{osb.v.0.d} = \frac{k_{mod.short} \cdot k_{sys} \cdot f_{osb.v.0.k}}{\gamma_{osbM}}$$

$$f_{osb.v.0.d} = 4.5 \text{ N/mm}^2$$

Panel shear strength check:

$$\frac{\tau t1_{mean.d}}{f_{osb.v.0.d}} = 0.1 \qquad\qquad \frac{\tau b1_{mean.d}}{f_{osb.v.0.d}} = 0.16 \qquad \text{i.e. less than 1, OK}$$

Design panel shear strength is greater than the mean panel shear stress in the flanges; therefore OK.

Example 7.5.3 The roof structure of an office building is formed by steel beams supporting thin flanged box beam panels as shown in Figure E7.5.3. Each panel comprises 4 No 47 mm thick by 120 mm deep timber webs at 400 mm centre to centre with plywood panels glued to the top and bottom faces and has an effective span of 4000 mm. The plywood is Canadian Douglas fir and is 15.5 mm thick on the top face and 12.5 mm thick on the bottom face and is aligned with the face grain parallel to the direction of span of the panel. To allow roof light structures to be fitted, areas of the plywood flanges between the central ribs in each panel are cut out and this occurs at three positions along the length of each panel as shown in Figure E7.5.3. Each box beam panel is 1272 mm wide and is detailed to fit against adjacent panels, being connected on site

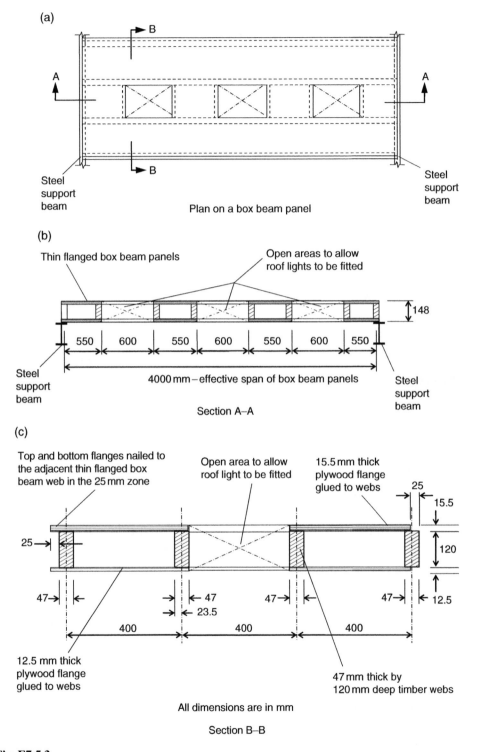

(a)

B

A   A

B

Steel
support
beam

Steel
support
beam

Plan on a box beam panel

(b)

Thin flanged box beam panels

Open areas to allow
roof lights to be fitted

148

550   600   550   600   550   600   550

Steel
support
beam

4000 mm – effective span of box beam panels

Steel
support
beam

Section A–A

(c)

Top and bottom flanges nailed to
the adjacent thin flanged box
beam web in the 25 mm zone

Open area to allow
roof light to be fitted

15.5 mm thick
plywood flange
glued to webs

25

15.5

25

120

47   47   47   12.5

23.5

400   400   400

12.5 mm thick
plywood flange
glued to webs

47 mm thick by
120 mm deep timber webs

All dimensions are in mm

Section B–B

Fig. E7.5.3.

by nailing as shown in Figure E7.5.3. Each panel supports a characteristic permanent load, including self-weight, of $0.81\,kN/m^2$ and a characteristic variable medium-duration load of $0.75\,kN/m^2$. The timber used for the panel ribs is strength class C24 in accordance with BS EN 338:2009, and the properties of the plywood are given below.

The structure functions in service class 2 conditions and the cross-section of a box beam panel is shown in section A–A in Figure E7.5.3.

Check that the panel will comply with the strength rules in EC5 at the ULS and calculate the final deformation of a panel at the SLS when subjected to the combined permanent and variable loading.

1. Panel geometric properties

Effective span of the beam, L_e,	$L_e = 4.0\,m$	
Distance between the beam centre lines, b_{cc}	$b_{cc} = 400\,mm$	

Web:

Width of the plywood web, b_w	$b_w = 47\,mm$	
Clear height between the flanges, h_w	$h_w = 120\,mm$	
Area of the web, A_w	$A_w = b_w \cdot h_w$	$A_w = 5.64 \times 10^{-3}\,m^2$
Clear distance between adjacent webs, b_f	$b_f = b_{cc} - b_w$	$b_f = 353\,mm$

Flanges:

Top flange thickness, ht_f	$ht_f = 15.5\,mm$	
Bottom flange thickness, hb_f	$hb_f = 12.5\,mm$	
Beam depth, h	$h = h_w + (ht_f + hb_f)$	$h = 148\,mm$
Minimum bearing length on the edge beam, b_r	$b_r = b_w - 25\,mm$	$b_r = 22\,mm$

Effective flange width of an I-beam section of the panel (7.3.2.1 (EC5 *9.1.2*))

In compression, $b1_{c.ef}$ (Table 7.2 (EC5, *Table 9.1*))	$b1_{c.ef} = \min(0.1 \cdot L_e,\ 20 \cdot ht_f)$	$b1_{c.ef} = 310\,mm$

i.e. the value is based on plate buckling criteria

Check on buckling length of compression flange – based on plate buckling length 7.3.2.2 (EC5, *9.1.2(5)*)	$\dfrac{b_f}{2 \cdot b1_{c.ef}} = 0.57$	the value is less than unity; therefore OK
Check maximum width allowance in compression, $b_{c.ef}$ (Table 7.2 (EC5, *Table 9.1*))	$b_{c.ef} = \min(b1_{c.ef},\ b_f)$	$b_{c.ef} = 310\,mm$
In tension, $b1_{t.ef}$ (Table 7.2 (EC5, *Table 9.1*))	$b1_{t.ef} = 0.1 \cdot L_e$	$b1_{t.ef} = 400\,mm$

Maximum allowable width, $b_{a.l}$ $b_{a.l} = b_f$ $b_{a.l} = 353\,\text{mm}$

Value to be used in tension, $b_{t.ef}$ $b_{t.ef} = \min{(b1_{t.ef}, b_{a.l})}$ $b_{t.ef} = 353\,\text{mm}$

Effective flange width of the section - with the extent of cross bracing and panel material, treat the section as functioning as a single element: (7.3.2.1)

Effective flange width of a panel in compression, $b_{ef.c}$ $b_{ef.c} = 3 \cdot b_{c.ef} + 4 \cdot b_w - b_{c.ef}$ $b_{ef.c} = 808\,\text{mm}$

(equation (7.33); EC5, *equation (9.12)*)

Effective flange width of a panel in tension, $b_{ef.t}$ $b_{ef.t} = 3 \cdot b_{t.ef} + 4 \cdot b_w - b_j$ $b_{ef.t} = 894\,\text{mm}$

(equation (7.33); EC5, *equation (9.12)*)

Overall width of a panel, b $b = 3 \cdot b_{cc} + b_w$ $b = 1.247 \times 10^3\,\text{mm}$

2. Material strength properties

Table 1.3, strength class C24 (BS EN 338:2009, *Table 1*)

Characteristic bending strength, $f_{m.k}$ $f_{m.k} = 24\,\text{N/mm}^2$

Characteristic shear strength, $f_{v.k}$ $f_{v.k} = 4.0\,\text{N/mm}^2$

Mean modulus of elasticity parallel to the grain, $E_{0.mean}$ $E_{0.mean} = 11.0\,\text{kN/mm}^2$

Mean shear modulus, $G_{0.mean}$ $G_{0.mean} = 0.69\,\text{kN/mm}^2$

Table 1.13, plywood – 12.5 mm Canadian Douglas fir (5 ply)

Characteristic compression strength, $f1_{p.c.0.k}$ $f1_{p.c.0.k} = 20.4\,\text{N/mm}^2$

Characteristic tensile strength, $f1_{p.t.0.k}$ $f1_{p.t.0.k} = 13.5\,\text{N/mm}^2$

Characteristic panel shear strength, $f1_{p.v.0.k}$ $f1_{p.v.0.k} = 3.5\,\text{N/mm}^2$

Characteristic rolling shear strength, $f1_{p.r.k}$ $f1_{p.r.k} = 1.25\,\text{N/mm}^2$

Mean modulus of elasticity, $E1_{p.c.0.mean}$ $E1_{p.c.0.mean} = 7.81\,\text{kN/mm}^2$

Table 1.13, plywood – 15.5 mm Canadian Douglas fir (4 ply)

Characteristic compression strength, $f2_{p.c.0.k}$ $f2_{p.c.0.k} = 19.7\,\text{N/mm}^2$

Characteristic tensile strength, $f2_{p.t.0.k}$ $f2_{p.t.0.k} = 13.1\,\text{N/mm}^2$

Characteristic panel shear strength, $f2_{p.v.0.k}$ $f2_{p.v.0.k} = 3.5\,\text{N/mm}^2$

Characteristic rolling shear strength, $f2_{p.r.k}$ $f2_{p.r.k} = 0.91\,\text{N/mm}^2$

Mean modulus of elasticity, $E2_{p.c.0.mean}$ $E2_{p.c.0.mean} = 7.55\,\text{kN/mm}^2$

In the following analysis, the transformed section is based on the use of the web material throughout the section.

3. **Partial safety factors**

Table 2.8 (UKNA to BS EN 1990:2002+A1:2005, *Table NA.A1.2(B)*)) for the ULS

Permanent actions, y_G	$y_G = 1.35$
Variable actions, y_Q	$y_Q = 1.5$

Table 2.2 (UKNA to BS EN 1990:2002+A1:2005, Table NA.A1.1)

Factor for quasi-permanent value $\psi_2 = 0.0$
of variable action, ψ_2
(Roof condition)

Table 2.6 (UKNA to EC5, *Table NA.3*)

Material factor for solid timber, y_M	$y_M = 1.3$
Material factor for plywood, y_{pM}	$y_{pM} = 1.2$

4. **Actions**

Characteristic permanent action on $G_k = 0.81 \, \text{kN/m}^2$
the panel, G_k

Characteristic variable (imposed) $Q_k = 0.75 \, \text{kN/m}^2$
medium-term action on the panel, Q_k

Design load due to the critical load $F_d = (y_G \cdot G_k + y_Q \cdot Q_k) \cdot b$
combination, F_d

(Table 2.8, equation (c) using the $F_d = 2.77 \, \text{kN/m}$
unfavourable condition)

Design load due to permanent $FP_d = (y_G \cdot G_k) \cdot b$ $FP_d = 1.36 \, \text{kN/m}$
loading, FP_d

Design moment on the panel due to $M_d = \dfrac{F_d \cdot L_e^2}{8}$ $M_d = 5.53 \, \text{kN/m}$
the critical load combination, M_d

Design shear force on the panel due $V_d = F_d \cdot \dfrac{L_e}{2}$ $V_d = 5.53 \, \text{kN}$
to the critical load combination, V_d

The critical design load case at the ULS will be due to the combination of permanent and unfavourable medium-duration variable action.

5. **Modification factors**

Factor for permanent duration action $k_{mod.perm} = 0.6$
and service class 2, $k_{mod.perm}$
(Table 2.4 (EC5, *Table 3.1*))

Factor for medium-duration action $k_{mod.med} = 0.8$
and service class 2, $k_{mod.med}$
(Table 2.4 (EC5, *Table 3.1*))

Load sharing factor – k_{sys} $k_{sys} = 1.0$
(2.3.7 (EC5, 6.6)) (1.1 can be used
for web if required)

Depth factor for solid timber in bending, k_h (Table 2.11 (EC5, *equation (3.1)*))	$k_h = \min\left[\left(\dfrac{150\text{ mm}}{h_w}\right)^{0.2}, 1.3\right]$	$k_h = 1.05$

The crack modification factor for timber, k_{cr} (Table 4.5 (4.5.2))
$$k_{cr} = 0.67$$

Deformation factor for timber at service class 2 conditions, $k_{def.w}$
(Table 2.10 (EC5, *Table 3.2*))
$$k_{def.w} = 0.8$$

Deformation factor for plywood at service class 2 conditions, $k_{def.f}$
(Table 2.10 (EC5, *Table 3.2*))
$$k_{def.f} = 1.0$$

6. Geometric properties

Instantaneous – transformed section properties:

Effective web thickness, $b_{w.eff}$
$$b_{w.eff} = 4 \cdot b_w \qquad b_{w.eff} = 188\text{ mm}$$

Transformed top flange width (into timber), $b_{efc.tfd}$
$$b_{efc.tfd} = \left(b_{efc} \cdot \dfrac{E2_{p.c.0.mean}}{E_{0.mean}}\right) \qquad b_{efc.tfd} = 554.58\text{ mm}$$

Area of top flange in compression, $A_{ef.f.c}$
$$A_{ef.f.c} = b_{ef.c.tfd} \cdot ht_f \qquad A_{ef.f.c} = 8.6 \times 10^{-3}\text{ m}^2$$

Transformed bottom flange width (into timber), $b_{ef.t.tfd}$
$$b_{ef.t.tfd} = \left(b_{ef.t} \cdot \dfrac{E1_{p.c.0.mean}}{E_{0.mean}}\right) \qquad b_{ef.t.tfd} = 634.74\text{ mm}$$

Area of bottom flange in tension, $A_{ef.f.t}$
$$A_{ef.f.t} = b_{ef.t.tfd} \cdot hb_f \qquad A_{ef.f.t} = 7.93 \times 10^{-3}\text{ m}^2$$

Area of the web, $A_{ef.w}$
$$A_{ef.w} = b_{w.eff} \cdot h_w \qquad A_{ef.w} = 2.26 \times 10^4\text{ mm}^2$$

Area of the transformed section, A_{ef}
$$A_{ef} = A_{ef.f.c} + A_{ef.f.t} + A_{ef.w} \qquad A_{ef} = 3.91 \times 10^4\text{ mm}^2$$

First moment of area of the section about the top face A_{1st}
$$A_{1st} = b_{ef.t.tfd} \cdot hb_f \cdot \left(h - \dfrac{hb_f}{2}\right) + b_{w.eff} \cdot h_w \cdot \left(\dfrac{h_w}{2} + ht_f\right) + b_{ef.c.t.fd} \cdot ht_f \cdot \dfrac{ht_f}{2} \qquad A_{1st} = 2.89 \times 10^{-3}\text{ m}^3$$

Neutral axis depth from the top face, z_t
$$z_t = \dfrac{A_{1st}}{A_{ef}} \qquad z_t = 74.05\text{ mm}$$

Second moment of area of the web, $I_{ef.w}$
$$I_{ef.w} = \dfrac{b_{w.eff} \cdot h_w^3}{12} + A_{ef.w} \cdot \left[z_t - \left(ht_f + \dfrac{h_w}{2}\right)\right]^2 \qquad I_{ef.w} = 2.71 \times 10^{-5}\text{ m}^4$$

Second moment of area of the top flange, $I_{ef.ftf}$
$$I_{ef.ftf} = \dfrac{b_{efc.tfd} \cdot ht_f^3}{12} + A_{ef.fc} \cdot \left(z_t - \dfrac{ht_f}{2}\right)^2 \qquad I_{ef.ftf} = 3.8 \times 10^{-5}\text{ m}^4$$

Second moment of area of the bottom flange, $I_{ef.f.b.f}$

$$I_{ef.fbf} = \frac{b_{eft.tfd} \cdot hb_f^3}{12} + A_{ef.ft} \cdot \left(h - z_t - \frac{hb_f}{2}\right)^2 \qquad\qquad I_{ef.f.b.f} = 3.65 \times 10^{-5} \, m^4$$

Instantaneous second moment of area of the transformed panel section, I_{ef}

$$I_{ef} = I_{ef.w} + I_{ef.f.t.f} + I_{ef.f.b.f} \quad I_{ef} = 1.02 \times 10^8 \, mm^4$$

Final – transformed section properties:

The largest stress to strength ratio will be the larger of the permanent action/$k_{mod.perm}$ and the combined permanent and variable action/$k_{mod.short}$. Let the ratio of one to the other be r:

$$r = \frac{\gamma_G \cdot G_k}{k_{mod.perm}} \cdot \frac{k_{mod.med}}{\gamma_G \cdot G_k + \gamma_Q \cdot Q_k} \qquad\qquad r = 0.66$$

i.e. the variable loading will produce the higher stress/strength ratio, so the factor ψ_2 will be associated with variable loading.

The psi factor for this loading condition is $\psi_{2a} = 0$.

Consequently, for this condition the instantaneous and final properties will be the same. We only need to analyse for the instantaneous condition.

7. **Bending stress check in the flange**
   Stress in flange due to bending:
   Bending stress in the top flange, $\sigma_{f.c.max.d.c}$

   $$\sigma_{f.c.max.d.c} = \frac{M_d}{I_{ef}}\left(z_t - \frac{ht_f}{2}\right)\frac{E2_{p.c.0.mean}}{E_{0.mean}} \qquad \sigma_{f.c.max.d.c} = 2.48 \, N/mm^2$$

   Bending stress in the bottom flange, $\sigma_{f.t.max.d.c}$

   $$\sigma_{f.t.max.d.c} = \frac{M_d}{I_{ef}}\left(h - z_t - \frac{hb_f}{2}\right)\frac{E1_{p.c.0.mean}}{E_{0.mean}} \qquad \sigma_{f.t.max.d.c} = 2.62 \, N/mm^2$$

   Axial strength of the top flange, $f_{p.c.0.d}$

   $$f_{p.c.0.d} = \frac{k_{mod.med} \cdot k_{sys} \cdot f2_{p.c.0.k}}{\gamma_{pM}} \qquad\qquad f_{p.c.0.d} = 13.13 \, N/mm^2$$

   Axial strength of the bottom flange, $f_{p.t.0.d}$

   $$f_{p.t.0.d} = \frac{k_{mod.med} \cdot k_{sys} \cdot f1_{p.t.0.k}}{\gamma_{pM}} \qquad\qquad f_{p.t.0.d} = 9 \, N/mm^2$$

   Stresses in the top and bottom flanges are OK

8. **Bending and shear stress check in the web**
   Bending stress in the web:
   Maximum distance from the NA to the extreme fibre, z_1
   $$z_1 = max[(z_t - ht_f), (h - hb_f - z_t)] \qquad\qquad z_1 = 61.45 \, mm$$

$$\sigma_{\text{w.c.d}} = \frac{M_d}{I_{\text{ef}}} \cdot z_1$$

$\sigma_{\text{w.c.d}} = 3.35 \text{ N/mm}^2$

Bending strength of the web, $f_{\text{w.d}}$

$$f_{\text{w.d}} = \frac{k_{\text{mod.med}} \cdot k_{\text{sys}} \cdot k_{\text{h}} \cdot f_{\text{m.k}}}{\gamma_M}$$

$f_{\text{w.d}} = 15.44 \text{ N/mm}^2$

Bending stress in the web is OK

Shear strength of the web:

Shear force in the beam, V_d

$V_d = 5.53 \times 10^3 \text{ N}$

First moment of area of the section above the NA about the NA, $S_{\text{t.f.NA}}$

$$S_{\text{t.f.NA}} = b_{\text{efc.tfd}} \cdot ht_f \cdot \left(z_t - \frac{ht_f}{2} \right) + b_{\text{w.eff}} \cdot \frac{\left(z_t - ht_f \right)^2}{2}$$

$S_{\text{t.f.NA}} = 8.92 \times 10^5 \text{ mm}^3$

Shear stress at the NA position, $\tau_{\text{v.d}}$

$$\tau_{\text{v.d}} = \frac{V_d \cdot S_{\text{t.f.NA}}}{I_{\text{ef}} \cdot b_{\text{w.eff}}}$$

$\tau_{\text{v.d}} = 0.26 \text{ N/mm}^2$

Shear strength of the web material, $f_{\text{v.d}}$

$$f_{\text{v.d}} = \frac{k_{\text{mod.med}} \cdot k_{\text{sys}} \cdot f_{\text{v.k}}}{\gamma_M}$$

$f_{\text{v.d}} = 2.46 \text{ N/mm}^2$

Design shear strength is greater than shear stress; therefore OK

Shear strength of the glued joint between the web and the flanges:

Shear force in the panel, V_d

$V_d = 5.53 \times 10^3 \text{ N}$

First moment of area of the top flange about the NA, S_{tf}

$$S_{\text{tf}} = b_{\text{efc.tfd}} \cdot ht_f \cdot \left(z_t - \frac{ht_f}{2} \right)$$

$S_{\text{tf}} = 5.70 \times 10^5 \text{ mm}^3$

First moment of area of the bottom flange about the NA, S_{bf}

$$S_{\text{bf}} = b_{\text{eft.tfd}} \cdot hb_f \cdot \left(h - z_t - \frac{hb_f}{2} \right)$$

$S_{\text{bf}} = 5.37 \times 10^5 \text{ mm}^3$

Shear width at beam/flange interface, $beff_w$

$beff_w = 3 \cdot b_w + b_r$

Mean shear stress in the top flange above the glue line, $\tau t_{\text{mean.d}}$

$$\tau t_{\text{mean.d}} = \frac{V_d \cdot S_{\text{tf}}}{I_{\text{ef}} \cdot beff_w}$$

$\tau t_{\text{mean.d}} = 0.19 \text{ N/mm}^2$

Mean shear stress in the bottom flange below the glue line, $\tau b_{\text{mean.d}}$

$$\tau b_{\text{mean.d}} = \frac{V_d \cdot S_{\text{bf}}}{I_{\text{ef}} \cdot beff_w}$$

$\tau b_{\text{mean.d}} = 0.18 \text{ N/mm}^2$

Rolling shear strength of the top flange, $f_{\text{v.2.90.d}}$

$$f_{\text{v.2.90.d}} = \frac{k_{\text{mod.med}} \cdot k_{\text{sys}} \cdot f2_{\text{p.r.k}}}{\gamma_{\text{pM}}}$$

$f_{\text{v.2.90.d}} = 0.61 \text{ N/mm}^2$

Rolling shear strength of the bottom flange, $f_{v.1.90.d}$

$$f_{v.1.90.d} = \frac{k_{mod.med} \cdot k_{sys} \cdot f1_{p.r.k}}{\gamma_{pM}}$$

$f_{v.1.90.d} = 0.83 \text{ N/mm}^2$

Rolling shear criteria $f_{v.90.d}$ (equation (7.56)) (EC5, 9.1.2(6))

$$f1_{v.90.d} = \begin{vmatrix} f_{v.1.90.d} & if & b_w \leq 4 \cdot hb_f \\ f_{v.1.90.d} \cdot \left(\dfrac{4 \cdot hb_f}{b_w}\right)^{0.8} & if & b_w > 4 \cdot hb_f \end{vmatrix}$$

$f1_{v.90.d} = 0.83 \text{ N/mm}^2$

$$f2_{v.90.d} = \begin{vmatrix} f_{v.2.90.d} & if & b_w \leq 4 \cdot ht_f \\ f_{v.2.90.d} \cdot \left(\dfrac{4 \cdot ht_f}{b_w}\right)^{0.8} & if & b_w > 4 \cdot ht_f \end{vmatrix}$$

$f2_{v.90.d} = 0.61 \text{ N/mm}^2$

Shear stress/strength ratios

$$\frac{\tau t_{mean.d}}{f2_{v.90.d}} = 0.31$$

$$\frac{\tau b_{mean.d}}{f1_{v.90.d}} = 0.22$$

Design rolling shear strength is greater than the mean shear stress in each flange; therefore OK.

9. Deflection due to combined permanent and variable loading at the SLS – at the final condition

The psi factor for this loading condition is
(2.3.4.1, (EC5, 2.3.2.2))

$$\psi_{2a} = 1$$

Transformed section:

Effective web thickness, $b_{c.w.eff}$ $b_{c.w.eff} = 4 \cdot b_w$ $b_{c.w.eff} = 188 \text{ mm}$

Transformed top flange width, $b_{c.ef.c.tfd}$

$$b_{c.ef.c.tfd} = \left(b_{ef.c} \cdot \frac{E1_{p.c.0.mean}}{E_{0.mean}} \cdot \frac{1 + \psi_{2a} \cdot k_{def.w}}{1 + \psi_{2a} \cdot k_{def.f}}\right)$$

$b_{c.ef.c.tfd} = 516.31 \text{ mm}$

Area of flange in compression, $A_{c.ef.f.c}$ $A_{c.ef.f.c} = b_{c.ef.c.tfd} \cdot ht_f$ $A_{c.ef.f.c} = 8.0 \times 10^{-3} \text{ m}^2$

Transformed bottom flange width, $b_{c.ef.t.tfd}$

$$b_{c.ef.t.tfd} = \left(b_{ef.t} \cdot \frac{E2_{p.c.0.mean}}{E_{0.mean}} \cdot \frac{1 + \psi_{2a} \cdot k_{def.w}}{1 + \psi_{2a} \cdot k_{def.f}}\right)$$

$b_{c.ef.t.tfd} = 525.25 \text{ mm}$

Area of flange in tension, $A_{c.ef.f.t}$ $A_{c.ef.f.t} = b_{c.ef.t.tfd} \cdot hb_f$ $A_{c.ef.f.t} = 6.9 \times 10^{-3} \text{ m}^2$

Area of web, $A_{c.ef.w}$ $A_{c.ef.w} = b_{c.w.eff} \cdot h_w$

Transformed area, $A_{c.ef}$ $A_{c.ef} = A_{c.ef.f.c} + A_{c.ef.f.t} + A_{c.ef.w}$

$A_{c.ef} = 3.75 \times 10^4 \text{ mm}^2$

First moment of area of the transformed section about the top face, $A_{c.1st}$

$$A_{c.1st} = b_{c.ef.t.tfd} \cdot hb_f \cdot \left(h - \frac{hb_f}{2} \right) + b_{c.w.eff} \cdot h_w \cdot \left(\frac{h_w}{2} + ht_f \right) + b_{c.ef.c.tfd} \cdot ht_f \cdot \frac{ht_f}{2}$$

$$A_{c.1st} = 2.74 \times 10^{-3}\ \text{m}^3$$

Neutral axis depth from the top face, $z_{c.t}$

$$z_{c.t} = \frac{A_{c.1st}}{A_{c.ef}}$$

$$z_{c.t} = 73.23\ \text{mm}$$

Second moment of area of the web, $I_{c.ef.w}$

$$I_{c.ef.w} = \frac{b_{c.w.eff} \cdot h_w^3}{12} + A_{c.ef.w} \cdot \left[z_{c.t} - \left(ht_f + \frac{h_w}{2} \right) \right]^2$$

$$I_{c.ef.w} = 2.72 \times 10^{-5}\ \text{m}^4$$

Second moment of area of the top flange, $I_{c.ef.ftf}$

$$I_{c.ef.ftf} = \frac{b_{c.efc.tfd} \cdot ht_f^3}{12} + A_{c.ef.fc} \cdot \left[z_{c.t} - \left(\frac{ht_f}{2} \right) \right]^2$$

$$I_{c.ef.ft.f} = 3.45 \times 10^{-5}\ \text{m}^4$$

Second moment of area of the bottom flange, $I_{c.ef.fbf}$

$$I_{c.ef.fbf} = \frac{b_{c.eft.tfd} \cdot hb_f^3}{12} + A_{c.ef.ft} \cdot \left(h - z_{c.t} - \frac{hb_f}{2} \right)^2$$

$$I_{c.ef.fbf} = 3.25 \times 10^{-5}\ \text{m}^4$$

Instantaneous second moment of area of the transformed section, $I_{c.ef}$

$$I_{c.ef} = I_{c.ef.w} + I_{c.ef.ftf} + I_{c.ef.fbf}$$

$$I_{c.ef} = 9.4162 \times 10^7\ \text{mm}^4$$

Final deflection of the section (Table 7.1), δ_{fin}

The creep deformation due to the quasi-permanent combination of actions, $u_{c.qp}$

$$u_{c.qp} = \frac{5}{384} \cdot \frac{[(G_k + \psi_2 Q_k)b] \cdot L_e^4}{E_{0.mean} \cdot I_{c.ef}} \cdot (1 + k_{def.f}) + \left(\frac{(G_k + \psi_2 Q_k)b}{8} \right) \cdot L_e^2 \cdot \frac{1 + k_{def.w}}{G_{0.mean} \cdot A_{c.ef.w}}$$

$$u_{c.qp} = 6.73\ \text{mm}$$

The instantaneous deformation of the section, u_{inst}

$$u_{inst} = \frac{5}{384} \cdot \frac{[(G_k + Q_k)b] \cdot L_e^4}{E_{0.mean} \cdot I_{ef}} + \left(\frac{(G_k + Q_k)b}{8} \right) \cdot L_e^2 \cdot \frac{1}{G_{0.mean} \cdot A_{c.ef.w}}$$

$$u_{iinst} = 6.06\ \text{mm}$$

The instantaneous deformation due to the quasi-permanent combination of actions, $u_{inst.qp}$

$$u_{inst.qp} = \frac{5}{384} \cdot \frac{[(G_k + \psi_2 \cdot Q_k)b] \cdot L_e^4}{E_{0.mean} \cdot I_{ef}} + \left(\frac{(G_k + \psi_2 \cdot Q_k)b}{8} \right) \cdot L_e^2 \cdot \frac{1}{G_{0.mean} \cdot A_{c.ef.w}}$$

$$u_{inst.qp} = 3.14\ \text{mm}$$

The final deformation, u_{fin}

$$u_{fin} = u_{c.qp} + u_{inst} - u_{inst.qp} \qquad u_{fin} = 9.65\,mm$$

ii) Using the alternative procedure given in 2.3.2(b)

The final deformation, ul_{fin}

$$ul_{fin} = \frac{5}{384} \cdot \frac{\left[(G_k + Q_k) \cdot b\right] \cdot L_e^4}{E_{0.mean} \cdot I_{c.ef}} \cdot (1 + k_{def.f}) + \left[\frac{(G_k + Q_k)b}{8}\right] \cdot L_e^2 \cdot \frac{1 + k_{def.w}}{G_{0.mean} \cdot A_{c.ef.w}}$$

$$ul_{fin} = 12.97\,mm$$

Limiting value for deflection at the net final condition – adopt span/250 (UKNA to EC5, *Table NA.4*), $w_{net,fin}$

$$w_{net,fin} = \frac{L_e}{250} \qquad\qquad w_{net,fin} = 16\,mm \qquad\qquad \text{Deflection is OK}$$

Note: If the timber stiffeners shown in Fig. E7.5.4(a) are not used, load sharing will not occur and the panel should be analysed assuming it will function as a series of U-beams.

Chapter 8
Design of Built-Up Columns

8.1 INTRODUCTION

Columns can be formed using single sawn members or glued-laminated sections or may be built up to form profiles that are structurally more efficient for the type of loading to be supported and/or are required to fit particular situations. The EC5 [1] design requirements for single member columns are discussed in Chapters 5 and 6 for solid and glued-laminated sections respectively and the procedures used for built-up columns are addressed in this chapter.

Built-up columns are columns composed of two or more timber or wood product sections connected by adhesives or by mechanical fasteners (e.g. nails, screws, dowels, punched metal plates, etc.), enabling the combined section to function as a composite element. Such columns can provide higher strength than the sum of the strength of the sections acting alone. When adhesives are used, there will be no slip at the joints and the built-up column will be assumed to behave as a fully composite section, however, with mechanical fasteners there will be some joint slip resulting in a reduction in the load-carrying capacity. Built-up columns can be constructed in a great variety of cross-sections, examples of which are shown in Figures 8.1 and 8.2, and are formed to meet special needs, to provide larger cross-sections than are ordinarily available, or purely for architectural applications.

The design procedure for built-up columns is given in *Annex C* of EC5 and the application of the design rules to composite sections as well as to spaced and lattice type columns are addressed in this chapter. The procedure covers built-up columns formed by gluing or by mechanical fixings.

The general information in 4.3 is relevant to the content of this chapter.

8.2 DESIGN CONSIDERATIONS

Axially loaded built-up columns have to comply with the relevant design rules and requirements of EC5 and the limit states associated with the main design effects for these sections when loaded axially are the same as those given in Table 5.1 for axially loaded columns.

Structural Timber Design to Eurocode 5, Second Edition. Jack Porteous and Abdy Kermani.
© Jack Porteous and Abdy Kermani 2013. Published 2013 by Blackwell Publishing Ltd.

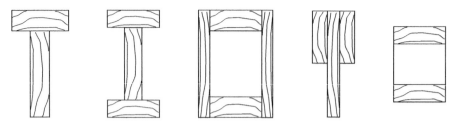

Fig. 8.1. Examples of built-up column sections.

(a)   (b)

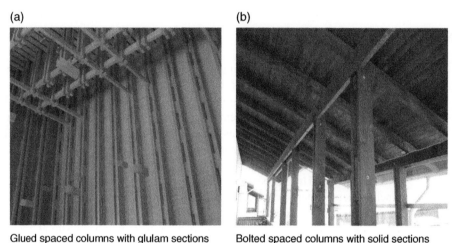

Glued spaced columns with glulam sections   Bolted spaced columns with solid sections

Fig. 8.2. Typical built-up (spaced) columns.

The equilibrium states and strength conditions relate to failure situations and must meet ultimate limit state requirements. The displacement condition relates to normal usage situations, however, as for solid section columns, no guidance is given in EC5 regarding limiting criteria for this state. Where lateral instability of a member can occur, a limitation is set for the maximum deviation from straightness allowed and this is given in EC5, *Section 10*.

Whether a built-up column is formed by gluing or by mechanical fixings, in EC5 the design procedure is only applicable to members that are subjected to axial loading acting through the centre of gravity of the built-up section and to cases where there are only small moments (e.g. due to self-weight) in addition to the axial loading. Only built-up columns complying with these conditions are covered in this chapter.

The general comments in 5.2 are also applicable to built-up columns.

8.3 GENERAL

In addition to the constraint on the type of loading that can be taken by these columns, the design rules in EC5 only apply to members in which the elements of the column are made from full length parts (i.e. there are no joints along the length of the members) and the built-up column is pin jointed and effectively held laterally in position at each end.

8.4 BENDING STIFFNESS OF BUILT-UP COLUMNS

In Chapter 5 it is shown that when dealing with solid section columns, for normal design situations the design strength will generally be determined by the buckling strength of the member. This is also the case for columns made from built-up sections. With built-up sections, however, the determination of the buckling strength is a more involved exercise, for the following reasons:

- Materials having different properties can be combined.
- There will be slip within the section when mechanical fasteners are used.
- There is the added risk of local instability effects with certain sections, e.g. T- or certain I-sections.

If the interfaces of the members in a built-up column are connected by mechanical fasteners, when subjected to flexure there will be slip between adjacent elements leading to a discontinuity in strain at these positions and the curvature of the individual elements will differ. In these situations, conventional bending theory cannot be used to determine the bending stiffness of the section and to analyse this condition the effect of the slip has to be taken into account.

This can be achieved by fully modelling the behaviour of each fastener in the connection and analysing the built-up section using a finite element analysis approach. An alternative, slightly less accurate, method, but suitable for normal design purposes, is to apply conventional bending theory to each element in the built-up column, assume compatibility in the curvature and displacement of adjacent column elements at each interface, and simulate the slip effect at these interfaces by assuming that the fastener resistance in these zones can be represented by linear spring elements.

From this type of analysis a reasonable estimate of the bending stiffness of the composite section will be obtained. When the stiffness of the springs is set equal to infinity, the bending stiffness of a built-up section having glued interfaces will be obtained, and if it is zero, the bending stiffness will equate to a section in which the members are not connected.

An example of the modelling applied to a built-up T-section connected by fasteners is shown in Figure 8.3. For a displacement of the section in the z–z direction, as shown in Figure 8.3a, the resistance offered by the fasteners will equate to the fastener stiffness multiplied by the relative slip between the elements at the interface as shown in Figure 8.3c. In the analysis, a spring arrangement having the same stiffness per unit length as the fastener stiffness is fitted at the interface between the beam and the flange as shown in Figure 8.3d. By applying simple bending theory to the model and using the principles of linear elastic theory, the structure can be analysed and a reasonable estimate of the bending stiffness of the composite column section can be obtained. The method is used in EC5 where the stiffness is referred to as the effective bending stiffness and is defined as $(EI)_{\text{ef}}$.

When designing columns, the design state will be the ultimate limit state (ULS), and where members are connected using mechanical fasteners, the fastener stiffness per unit length is taken to be the ULS slip modulus per shear plane, K_{u}, divided by the fastener spacing. The ULS slip modulus per shear plane is referred to in 10.10 and is obtained from:

$$K_{\text{u}} = \frac{2}{3} K_{\text{ser}} \tag{8.1}$$

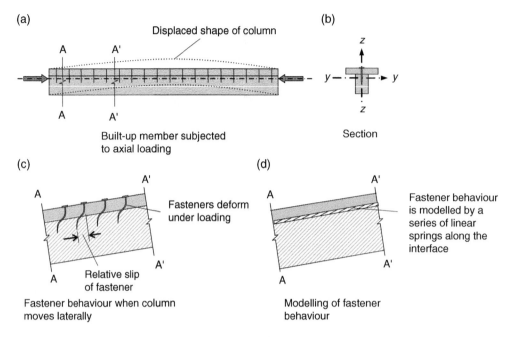

(a)

Displaced shape of column

A   A'

A   A'

Built-up member subjected
to axial loading

(b)

z

y —·— y

z

Section

(c)

A'

A

Fasteners deform
under loading

A'

A

Relative slip
of fastener

Fastener behaviour when column
moves laterally

(d)

A'

A

Fastener behaviour
is modelled by a
series of linear
springs along the
interface

A'

A

Modelling of fastener
behaviour

Fig. 8.3. Modelling slip at interfaces.

where K_{ser} is the slip modulus per shear plane per fastener at the serviceability limit
state (SLS) and the value of this property for different fastener types is obtained from
Table 10.13.

The analysis used in EC5 incorporates the following assumptions:

(a) The member is pin jointed at each end.
(b) The spacing between the fasteners is constant or varies uniformly in accordance
with the shear force distribution between s_{min} and s_{max}, with $s_{max} \leq 4s_{min}$, where s
is the fastener spacing.

Some of the section profiles commonly used for built-up columns are shown in
Figure 8.4.

8.4.1 The effective bending stiffness of built-up sections about the strong (y–y) axis

The effective bending stiffness of columns with the profiles shown in Figures 8.4a–8.4f
when bent about the y–y axis will be obtained from:

$$(EI)_{ef,y} = \sum_{i=1}^{3}(E_i I_i + \gamma_i E_i A_i a_i^2) \qquad \text{(EC5, Annex B, equation (B.1))} \quad (8.2)$$

where the symbols are as defined in Figure 8.4 and:

- E_i is the mean value of the modulus of elasticity of element, i (see also 8.6);
- A_i is the cross-sectional area of element i, i.e. $= b_i \times h_i$;

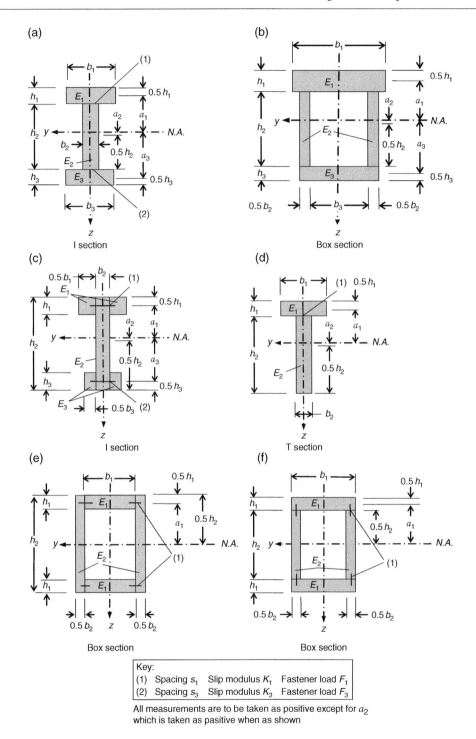

Fig. 8.4. The profiles to which the design rules in EC5 for built-up columns will apply.

- I_i is the second moment of area of element i about its axis of bending, i.e.

  $$I_i = b_i h_i^3 / 12;$$

- γ_i is the connection efficiency factor, where $\gamma_2 = 1$ and for $i = 1$ or 3, then;

  $$\gamma_i = \left[1 + \pi^2 E_i A_i s_i / \left(K_i \ell^2 \right) \right]^{-1} \qquad \text{(EC5, equation (B.5))} \quad (8.3)$$

  where:
  - s_i is the spacing of the fasteners in connection i. Where a variable spacing is used to suit the shear force distribution and $s_{max} \leq 4 s_{min}$, it is proposed by Ceccotti [2] that an effective spacing, $s_{ef} = 0.75 s_{min} + 0.25 s_{max}$ can be used.
  - Where a flange consists of two elements connected to a single web (e.g. Figure 8.4c) or a web consists of two elements connected to a single flange (e.g. Figure 8.4e), s_i will be ½ the fastener spacing per unit length used in each of the joining planes. In other words, the stiffness used for the connection will be twice the fastener stiffness in each of the joining planes.
  - K_i is the stiffness of fastener i per joining plane and equates to $K_{u,i}$ for calculations relating to the ULS (see also 8.6).
  - ℓ is the length of the column.
  - It is to be noted that for a glued interface, i.e. $K_i = \infty$, γ_i will equal 1 and the effective bending stiffness will equate to that derived for a fully composite section. When K_i is 0, γ_i will also be 0, equating to a section in which the elements are unconnected.
- a_i is the distance from the centre of area of element i to the neutral axis of the composite section.

In Figure 8.4, for profiles (e) and (f), $a_2 = 0$ and for profiles (a)–(d), a_2 it will be obtained from:

$$a_2 = \frac{\gamma_1 E_1 A_1 \left(h_1 + h_2 \right) - \gamma_3 E_3 A_3 \left(h_2 + h_3 \right)}{2 \sum_{i=1}^{3} \gamma_i E_i A_i} \qquad \text{(EC5, equation (B.6))} \quad (8.4)$$

The other distances can be calculated from the geometry of the section and for the T-section in profile (d), h_3 (as well as E_3 and A_3) will equal 0 in equation (8.4). All measurements are to be taken as positive except for a_2, which will be positive when as shown in Figure 8.4.

Applying equations (8.2)–(8.4), the effective bending stiffness about the y–y axis for profiles (a)–(f) inclusive will be as given in Table 8.1. It is to be noted that the equations in Table 8.1 (and 8.2) incorporate the effect of the reduction in fastener spacing where a flange consists of two elements connected to a single web.

8.4.2   The effective bending stiffness of built-up sections about the z–z axis

When due to the geometry of the built-up section or the combination of material properties being used the section is asymmetrical about its y–y axis, as in the case of profiles (a)–(d) in Figure 8.4, irrespective of whether the interfaces are glued or joined by mechanical fasteners, it is proposed that the elements of the section are treated as individual members.

Table 8.1 Effective bending stiffness about the y–y axis

Profile in Figure 8.4	Effective bending stiffness about y–y axis, $(EI)_{\text{ef,y}}$
Built-up sections (a) and (b)	$$(EI)_{\text{ef,y}} = E_1\frac{b_1h_1^3}{12} + E_2\frac{b_2h_2^3}{12} + E_3\frac{b_3h_3^3}{12} + \gamma_1 E_1\left(b_1h_1\right)a_1^2$$ $$+ E_2(b_2h_2)a_2^2 + \gamma_3 E_3(b_3h_3)a_3^2$$ where for section (a) $$\gamma_1 = \left[1 + \pi^2 E_1\left(b_1h_1\right)\left(\frac{s_1}{K_1}\right)\frac{1}{\ell^2}\right]^{-1} \quad \text{and} \quad \gamma_3 = \left[1 + \pi^2 E_3\left(b_3h_3\right)\left(\frac{s_3}{K_3}\right)\frac{1}{\ell^2}\right]^{-1}$$ and for section (b) $$\gamma_1 = \left[1 + \pi^2 E_1\left(b_1h_1\right)\left(\frac{s_1}{2K_1}\right)\frac{1}{\ell^2}\right]^{-1} \quad \text{and} \quad \gamma_3 = \left[1 + \pi^2 E_3\left(b_3h_3\right)\left(\frac{s_3}{2K_3}\right)\frac{1}{\ell^2}\right]^{-1}$$
Built-up section (c)	$$(EI)_{\text{ef,y}} = E_1\frac{b_1h_1^3}{12} + E_2\frac{b_2h_2^3}{12} + E_3\frac{b_3h_3^3}{12}$$ $$+ \gamma_1 E_1(b_1h_1)a_1^2 + E_2(b_2h_2)a_2^2 + \gamma_3 E_3(b_3h_3)a_3^2$$ where $$\gamma_1 = \left[1 + \pi^2 E_1\left(b_1h_1\right)\left(\frac{s_1}{2K_1}\right)\frac{1}{\ell^2}\right]^{-1} \quad \text{and} \quad \gamma_3 = \left[1 + \pi^2 E_1\left(b_3h_3\right)\left(\frac{s_3}{2K_3}\right)\frac{1}{\ell^2}\right]^{-1}$$
Built-up section (d)	$$(EI)_{\text{ef,y}} = E_1\frac{b_1h_1^3}{12} + E_2\frac{b_2h_2^3}{12} + \gamma_1 E_1\left(b_1h_1\right)a_1^2 + E_2\left(b_2h_2\right)a_2^2$$ where $$\gamma_1 = \left[1 + \pi^2 E_1\left(b_1h_1\right)\left(\frac{s_1}{K_1}\right)\frac{1}{\ell^2}\right]^{-1}$$
Built-up section (e)	$$(EI)_{\text{ef,y}} = 2E_1\frac{b_1h_1^3}{12} + E_2\frac{b_2h_2^3}{12} + 2\left(\gamma_1\right)E_1\left(b_1h_1\right)\left(\frac{h_2}{2} - \frac{h_1}{2}\right)^2$$ where $$\gamma_1 = \left[1 + \pi^2 E_1\left(b_1h_1\right)\left(\frac{s_1}{2K_1}\right)\frac{1}{\ell^2}\right]^{-1}$$
Built-up section (f)	$$(EI)_{\text{ef,y}} = 2E_1\frac{b_1h_1^3}{12} + E_2\frac{b_2h_2^3}{12} + 2\left(\gamma_1\right)E_1\left(b_1h_1\right)\left(\frac{h_2}{2} + \frac{h_1}{2}\right)^2$$ where $$\gamma_1 = \left[1 + \pi^2 E_1\left(b_1h_1\right)\left(\frac{s_1}{2K_1}\right)\frac{1}{\ell^2}\right]^{-1}$$

The stiffening effect of the built-up section is ignored and the capacity of the column is conservatively taken to be the summation of the strengths of the individual elements in the section.

Table 8.2 Effective bending stiffness about the z–z axis

Profile in Figure 8.4	Effective bending stiffness about z–z axis, $(EI)_{\text{ef},z}$
Built-up sections (a) where $E_1 = E_3$, $b_1 = b_3$ and $h_1 = h_3$	$(EI)_{\text{ef},z} = 2E_1\dfrac{h_1 b_1^3}{12} + E_2\left(\dfrac{h_2 b_2^3}{12}\right)$
Built-up section (c) where $E_1 = E_3$, $b_1 = b_3$ and $h_1 = h_3$	$(EI)_{\text{ef},z} = 4E_1\dfrac{h_1\left(0.5b_1\right)^3}{12} + E_2\dfrac{h_2 b_2^3}{12}$ $+ 4\gamma_1 E_1(h_1 0.5b_1)\left(\dfrac{b_2}{2} + \dfrac{b_1}{4}\right)^2$ where $\gamma_1 = \left[1 + \pi^2 E_1\left(h_1 b_1\right)\left(\dfrac{s_1}{2K_1}\right)\dfrac{1}{\ell^2}\right]^{-1}$
Built-up section (e)	$(EI)_{\text{ef},z} = 2E_1\dfrac{h_1 b_1^3}{12} + 2E_2\dfrac{h_2\left(0.5b_2\right)^3}{12}$ $+ 2\gamma_1 E_2\left(\dfrac{b_2 h_2}{2}\right)\left(\dfrac{b_1}{2} + \dfrac{b_2}{4}\right)^2$ where $\gamma_1 = \left[1 + \pi^2 E_2\left(0.5 b_2 h_2\right)\left(\dfrac{s_1}{2K_1}\right)\dfrac{1}{\ell^2}\right]^{-1}$
Built-up section (f) and (b) where $E_1 = E_3$, $b_1 = (b_2 + b_3)$ and $h_1 = h_3$	$(EI)_{\text{ef},z} = 2E_1\dfrac{h_1 b_1^3}{12} + 2E_2\dfrac{h_2\left(b_2/2\right)^3}{12}$ $+ 2\gamma_1 E_2\left(h_2\dfrac{b_2}{2}\right)\left(\dfrac{b_1}{2} - \dfrac{b_2}{4}\right)^2$ where $\gamma_1 = \left[1 + \pi^2 E_2\left(h_2\dfrac{b_2}{2}\right)\left(\dfrac{s_1}{2K_1}\right)\left(\dfrac{1}{\ell^2}\right)\right]^{-1}$

If, however, the geometry and material properties are such that the section is symmetrical about the y–y axis, the effective bending stiffness of the built-up section can be obtained as in 8.4.1. Profiles (e) and (f) in Figure 8.4 clearly fall into this category and if the flanges in profiles (a)–(c) each have the same cross-section and are made from the same material, the method will also apply. The value of the effective bending stiffness of these profiles about the z–z axis is given in Table 8.2.

8.4.3 Design procedure

When a built-up section is subjected to an axial compression design force, $F_{\text{c,d}}$, the stress in the section will depend on the effective axial stiffness of the composite section, $(EA)_{\text{ef}}$, which is:

$$(EA)_{\text{ef}} = \sum_{i=1}^{n} E_i A_i \tag{8.5}$$

where n represents the number of members in the cross-section, and E_i is the mean modulus of elasticity of member i parallel to the grain and A_i is the cross-sectional area of member i.

In EC5 it is assumed that under the action of the axial design force the section will be subjected to a uniform strain and there will be no shear stress along the interfaces between the members. This will be achieved when the axial force is applied as a rigid body force over the section and the resultant acts through the centroid of the EA diagram of the section. Under this condition, the axial stress in each member i in the section will be:

$$\sigma_{c,0,d,i} = \frac{E_i F_{c,d}}{\sum_{i=1}^{n} E_i A_i} \tag{8.6}$$

where the symbols are as described above and $\sigma_{c,0,d,i}$ is the axial stress in member i of the built-up section.

If the modulus of elasticity is the same for all members in the section, equation (8.6) reduces to:

$$\sigma_{c,0,d,i} = \frac{F_{c,d}}{A_{tot}} \qquad (EC5, \text{ equation (C.2))} \quad (8.7)$$

i.e. the axial stress will be the same in all members and A_{tot} is the total cross-sectional area of the built-up section.

The buckling strength of a built-up section about each axis of bending is dependent on the effective slenderness ratio of the section, λ_{ef}, and, for a built-up column section having a pin jointed length ℓ, it will be:

$$\lambda_{ef} = \frac{\ell}{\sqrt{\dfrac{(EI)_{ef}}{\sum_{i=1}^{n} E_i A_i}}} \tag{8.8}$$

There will be a value for λ_{ef} about the y–y axis and the z–z axis, and the value of the effective bending stiffness, $(EI)_{ef}$, for each axis will be determined in accordance with the requirements of 8.4.1 and 8.4.2.

In EC5, to simplify the design procedure the effective axial stiffness of the built-up section as given in equation (8.5) is replaced by an axial stiffness based on the average E value of the members forming the section, E_{mean}, as follows:

$$\sum_{i=1}^{n} E_i A_i \rightarrow E_{mean} A_{tot} \tag{8.9}$$

where $E_{mean} = \dfrac{1}{n}\sum_{i=1}^{n} E_i$ and n represents the number of members in the section.

This is an approximation that is deemed to be acceptable, and, after substituting in equation (8.8), the effective slenderness ratio of each element in the built-up column becomes:

$$\lambda_{ef} = \frac{\ell}{\sqrt{(EI)_{ef} / (E_{mean} A_{tot})}} \qquad \begin{array}{l} \text{(the combination of } equations \\ (C.3) \text{ and } (C.4) \text{ in EC5)} \end{array} \quad (8.10)$$

In accordance with the approach given in Chapter 5 for the design of a single member axially loaded column, the relative slenderness ratio of member i in the built-up section, $\lambda_{\text{rel},i}$, will be:

$$\lambda_{\text{rel},i} = \frac{\lambda_{\text{ef}}}{\pi} \sqrt{\frac{f_{c,0,k,i}}{E_i}} \tag{8.11}$$

where $f_{c,0,k,i}$ is the characteristic compressive strength of member i loaded parallel to the grain. In EC5 the buckling curves from which the instability factor for each axis of bending is derived are strength related, based on the fifth percentile value of the modulus of elasticity, $E_{0.05,i}$, and for built-up sections the requirement is that the fifth percentile value is used, such that equation (8.11) becomes:

$$\lambda_{\text{rel},i} = \frac{\lambda_{\text{ef}}}{\pi} \sqrt{\frac{f_{c,0,k,i}}{E_{0.05,i}}} \tag{8.12}$$

The k_c stability factors ($k_{c,y}$ and $k_{c,z}$), referred to in Chapter 5 for the design of single member columns, can now be derived for each member in the section in accordance with the procedures in 5.3.1.

Because of the rotation of the member under axial loading, a shear force component, V_d, will be generated along the interface between the members of the built-up column. Where the built-up section is connected using fasteners, the fasteners must be able to withstand this force and if glue is used the shear stress induced at the member interfaces must be checked.

Where fasteners are used, the shear force, F_i, to be taken by each fastener along the interface will be a function of V_d, and is obtained from *equations (B.10)* and *(C.5)* in EC5 as follows:

$$F_i = \frac{\gamma_i E_i A_i a_i s_i}{(EI)_{\text{ef}}} V_d \qquad \text{(EC5, equation (B.10))} \quad (8.13)$$

where V_d is obtained from:

$$V_d = \begin{cases} \dfrac{F_{c,d}}{120 k_c} & \text{for } \lambda_{\text{ef}} < 30 \\[2mm] \dfrac{F_{c,d}\lambda_{\text{ef}}}{3600 k_c} & \text{for } 30 \le \lambda_{\text{ef}} < 60 \\[2mm] \dfrac{F_{c,d}}{60 k_c} & \text{for } \lambda_{\text{ef}} \ge 60 \end{cases} \qquad \text{(EC5, equation (C.5))} \quad (8.14)$$

In the above, $F_{c,d}$ is the design axial load on the column and the other functions are as described in the text.

Where the built-up column is assembled using glued joints, the horizontal shear stress in the members at the glue lines shall be checked using conventional elastic theory and the shear force on the section shall be taken to be V_d. As described in 4.5.2, the shear strength of the members will depend on the direction of the shear force relative to the grain in each member.

8.4.3.1   Design criteria for built-up sections

For built-up sections, the design criteria to be satisfied to comply with the requirements of EC5 are as follows:

(a) *Where* $\lambda_{rel,y}$ *and* $\lambda_{rel,z} \le 0.3$.
Under this condition the built-up section will not buckle, but will fail at the compression strength of the built-up members, i.e.

$$\sigma_{c,0,d,i} \le f_{c,0,d,i} \tag{8.15}$$

where:

- $\sigma_{c,0,d,i}$ is the design compressive stress parallel to the grain in member i and:

$$\sigma_{c,0,d,i} = \frac{E_i F_{c,d}}{\sum_{i=1}^{n} E_i A_i} \tag{8.16}$$

  where $F_{c,d}$ is the design axial load on the column, E_i is the mean value of the modulus of elasticity parallel to the grain of member i, and A_i is the cross-sectional area of member i.
- $f_{c,0,d,i}$ is the design compressive strength parallel to the grain of member i and:

$$f_{c,0,d,i} = \frac{k_{mod,i} \cdot k_{sys} \cdot f_{c,0,k,i}}{\gamma_{M,i}} \tag{8.17}$$

  where:
  - $k_{mod,i}$ is the modification factor for member i for load duration and service classes as given in Table 2.4.
  - k_{sys} is the system strength factor discussed in 2.3.7. When dealing with single column members, this factor is not relevant and will be taken to be unity.
  - $\gamma_{M,i}$ is the partial coefficient for material properties for member i, given in Table 2.6.
  - $f_{c,0,k,i}$ is the characteristic compressive strength of the member i parallel to the grain. Strength information for timber and commonly used wood-based structural products is given in Chapter 1.

(b) *Where either* $\lambda_{rel,z}$ *or* $\lambda_{rel,y}$ *is* > 0.3.
If either $\lambda_{rel,z}$ or $\lambda_{rel,y}$ is > 0.3, the member can buckle and the design requirement will be as follows:

   (i) $\lambda_{rel,z} > 0.3$ *and the built-up section is not symmetrical about the y–y axis.*
       For sections that are not symmetrical about the y–y axis (due to geometry or a combination of materials), the axial load on the column, $F_{c,d}$, must be less than or equal to the summation of the load-carrying capacities of the individual members of the built-up section bent about the respective member neutral axis parallel to the z–z axis of the built-up section.

(ii) $\lambda_{rel,z} > 0.3$ *and the built-up section is symmetrical about the y–y axis.* Under this condition, the built-up section will buckle as a composite section and:

$$\sigma_{c,0,d,i} \leq k_{cz,i} f_{c,0,d,i} \tag{8.18}$$

where:

- $\sigma_{c,0,d,i}$ is the axial stress in member i as defined in equation (8.16).
- $k_{cz,i}$ is the instability factor for member i and is determined from the application of equations (5.4b), (5.5b) and (5.6), with the relative slenderness ratio, $\lambda_{rel,z,i}$, being:

$$\lambda_{rel,z,i} = \frac{\lambda_{ef,z}}{\pi} \sqrt{\frac{f_{c,0,k,i}}{E_{0.05,i}}} \qquad \text{(as equation (8.12))}$$

where $f_{c,0,k,i}$ and $E_{0.05,i}$ are as previously defined and $\lambda_{ef,z}$ is the slenderness ratio of the member about the z–z axis, obtained from:

$$\lambda_{ef,z} = \frac{\ell}{\sqrt{(EI)_{ef,z} / (E_{mean} A_{tot})}}$$

where the symbols are defined in equation (8.10) and $(EI)_{ef,z}$ is the effective bending stiffness about the z–z axis.

Where fasteners are used the fastener design must comply with the requirements of equations (8.13) and (8.14) (*equations (B.10) and (C.5) in EC5*).

(iii) $\lambda_{rel,y} > 0.3$. For this condition, the section must be symmetrical about the z–z axis, and the design condition will be:

$$\sigma_{c,0,d,i} \leq k_{c,y,i} f_{c,0,d,i} \tag{8.19}$$

where:

- $\sigma_{c,0,d,i}$ is the axial stress in member i as defined in equation (8.16).
- $k_{c,y,i}$ is the instability factor for member i and is determined from the application of equations (5.4a), (5.5a) and (5.6) with the relative slenderness ratio, $\lambda_{rel,y,i}$, being:

$$\lambda_{rel,y,i} = \frac{\lambda_{ef,y}}{\pi} \sqrt{\frac{f_{c,0,k,i}}{E_{0.05,i}}} \qquad \text{(as equation 8.12))}$$

where $f_{c,0,k,i}$ and $E_{0.05,i}$ are as previously defined and $\lambda_{ef,y}$ is the slenderness ratio of the member about the y–y axis, obtained from:

$$\lambda_{ef,y} = \frac{\ell}{\sqrt{(EI)_{ef,y} / (E_{mean} A_{tot})}}$$

where the symbols are defined in equation (8.10) and $(EI)_{ef,y}$ is the effective bending stiffness about the y–y axis.

Where fasteners are used the fastener design must comply with the requirements of equations (8.13) and (8.14) (*equations (B.10)* and *(C.5)* in EC5).

(c) *Where, in addition to axial load, small moments due to self-weight also arise.*
The relationship given in *Clause 6.3.2(3)* of EC5 must be satisfied for each element.

See Example 8.8.1.

8.4.4   Built-up sections – spaced columns

A spaced column is a built-up column where there are two or more identical members (referred to as shafts) separated and connected by spacer packs or gusset plates that are either glued or fixed by mechanical fasteners. They are often used in architectural applications, in trusses as compression chords and in frame construction.

Due to the geometry of its cross-section, a spaced column will have a higher load-carrying capacity than a single solid timber member of equivalent volume, and, apart from being economical, they provide suitable construction through which other members such as bracing, beams or trusses can conveniently be inserted and connected, as shown in Figure 8.5.

Where glue is used there will be no slip or relative rotation at the connections between the spacer packs (or gussets) and the shafts, and full composite action will exist. When mechanical fasteners are used, slip will occur and the effect of this on the composite action behaviour must be taken into account. Examples of spaced columns are shown in Figure 8.6.

Fig. 8.5. Beam and post construction using beams inserted between spaced column shafts (photo courtesy of Constructional Timber Limited, a member of the Glued Laminated Timber Association).

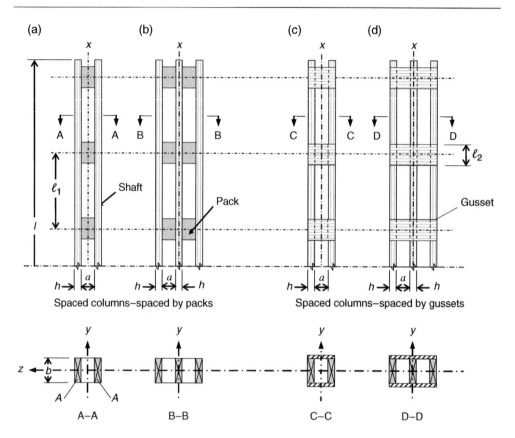

Fig. 8.6. Spaced columns.

The conditions imposed by EC5 for the design of spaced columns with packs or gussets fixed by glue or mechanical fasteners are given in *Annex C, Clause C.3.1*, and summarised as follows:

- The columns are subjected to axial loading acting through the centre of gravity of the spaced column.
- The shafts must be parallel to each other.
- The cross-section of the spaced column must be composed of two, three or four identical shafts.
- The cross-sections of the spaced column must be symmetrical about the y–y and the z–z axes.
- In the height of the column there must be at least three unrestrained bays. To satisfy this the shafts must, at a minimum, be connected at the ends and at the third points.
- $a \leq 3h$ if spacer packs are used and $a \leq 6h$ if gusset plates are used, where a is the free distance between the shafts and h is the thickness of the shaft.
- The joints, packs and gusset plates are designed in accordance with the requirements of *Clause C.3.3* in EC5 (i.e. equations (8.13) and (8.14)).
- The pack length, ℓ_2, must equal or be greater than $1.5a$. At a minimum, there must be four nails or two bolts with connectors in each shear plane. Where nails are used, there must be at least four nails in a row at each end along the direction of the x–x axis of the column.
- The gusset length, ℓ_2 must equal or be greater than $2a$.

8.4.4.1   Design procedure for spaced columns

When considering the buckling behaviour of the spaced column shown in Figure 8.6 about the z–z axis (i.e. deflection is in the y-axis direction), the shafts behave as individual elements and the strength of the column will be the summation of the strength of each shaft about the z–z axis.

For buckling about the y–y axis (i.e. deflection is in the z-axis direction), the shear force in each shaft as well as the shear deformation in the pack/gusset plates causes additional flexural displacement in the shafts and the strength reduction arising from this effect must be taken into account in the design. This is achieved in EC5 by using an effective slenderness ratio, λ_{ef}, which is based on an interaction of the slenderness of the spaced column section (assuming it functions in a fully composite manner) and the slenderness of the individual shafts adjusted to take into account the stiffness of the shaft/pack (or gusset plate) connection as follows:

$$\lambda_{ef} = \sqrt{\lambda^2 + \eta \frac{n}{2}\lambda_1^2} \qquad\qquad (EC5, \; equation \; (C.10)) \quad (8.20)$$

where:

- λ is the slenderness ratio of a solid column having the same length ℓ as the spaced column, the same cross-sectional area ($A_{tot} = nA$, where A is the cross-sectional area of a shaft, bh, and n is the number of shafts), and the same second moment of area about the y–y axis, I_{tot}, i.e:

$$\lambda = \ell\sqrt{A_{tot}/I_{tot}} \qquad\qquad (EC5, \; equation \; (C.11)) \quad (8.21)$$

  and, for the sections shown in Figure 8.6,

$$I_{tot} = \frac{bh^3}{6} + 2bh\left(\frac{a+h}{2}\right)^2 \qquad\qquad \text{for sections A–A, and C–C} \quad (8.22)$$

$$I_{tot} = \frac{bh^3}{4} + 2bh\left(a+h\right)^2 \qquad\qquad \text{for sections B–B and D–D} \quad (8.23)$$

- λ_1 is the slenderness ratio of each shaft, based on the shaft length having the greatest distance between adjacent pack/gusset plate positions:

$$\lambda_1 = \sqrt{12}\,\frac{\ell_1}{h} \qquad\qquad (EC5, \; equation \; (C.12)) \quad (8.24)$$

  If the value obtained is less than 30, a value of at least 30 must be used in equation (8.20).
- n is the number of shafts in the spaced column.
- η is a connection factor for which values are given in Table 8.3. The factor is dependent on whether packs or gusset plates are being used, whether they are glued or mechanically connected, and on the duration of loading. The stiffer the connection, the lower the value of the factor will be.

Table 8.3 Connection factor η^*

		Packs			Gussets	
	Glued	Nailed	Bolted (with connectors)		Glued	Nailed
Permanent/long-term loading	1	4	3.5		3	6
Medium/short-term loading	1	3	2.5		2	4.5

*Based on *Table C.1* in EC5.

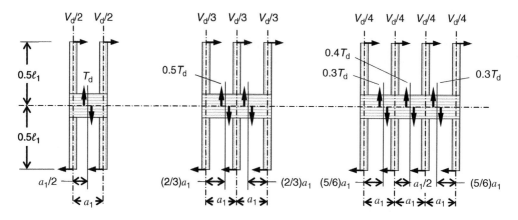

Fig. 8.7. Shear force distribution and loads on packs and gussets (based on *Figure C.2* in EC5).

The load-carrying capacity of the spaced column will be derived using the higher of the slenderness ratios of the column about the y–y and z–z axes and for this condition it has to be verified that:

$$\sigma_{c,0,d} \leq k_c f_{c,0,d} \tag{8.25}$$

where:

- $f_{c,0,d}$ is as previously defined, and:
- $\sigma_{c,0,d}$ is the direct stress on the column, i.e:

$$\sigma_{c,0,d} = \frac{F_{c,d}}{A_{tot}}$$

- where $F_{c,d}$ is the design value of the axial load acting on the spaced column and A_{tot} is as defined against equation (8.20).
  k_c is the instability factor associated with the critical axis (i.e. the axis having the maximum slenderness ratio) and is determined in accordance with 5.3.1.

In addition to checking the strength of the shafts, the strength of the packs, gusset plates and mechanical fixings (when used) must also be checked. The shear forces in the packs or gussets are derived assuming rigid body behaviour and are as shown in Figure 8.7.

From basic statics, the value of the shear force T_d shown in Figure 8.7 is:

$$T_d = \frac{V_d \ell_1}{a_1} \qquad \qquad \text{(EC5, equation (C.13))} \quad (8.26)$$

where a_1 is the distance between the centre lines of adjacent shafts and V_d is obtained from equation (8.27),

$$V_d = \begin{cases} \dfrac{F_{c,d}}{120 k_c} & \text{for} \quad \lambda_{ef} < 30 \\[2mm] \dfrac{F_{c,d}\lambda_{ef}}{3600 k_c} & \text{for} \quad 30 \le \lambda_{ef} < 60 \\[2mm] \dfrac{F_{c,d}}{60 k_c} & \text{for} \quad \lambda_{ef} \ge 60 \end{cases} \qquad \text{(EC5, equation (C.5))} \quad (8.27)$$

where the symbols are as previously defined and $F_{c,d}$ is the design axial load on the spaced column, k_c is the instability factor associated with buckling about the y–y axes, determined in accordance with 5.3.1, and λ_{ef} is the effective slenderness ratio defined in equation (8.20).

The pack or gusset plates and their connections to the shafts must be designed for the effects of the shear force shown in Figure 8.7. The moment on the connection at the shaft is obtained by multiplying the shear force in the pack or gusset plate by the distance to the face of the shaft when packs are used or the distance to the centre line of the shaft when gusset plates are used.

8.4.5 Built-up sections – latticed columns

A latticed column is a built-up column where there are two identical members separated and connected by N or V lattice members fixed to the members by glued or nailed joints. Examples of glued latticed columns with N and V lattice configurations are shown in Figure 8.8.

The conditions imposed by EC5 for the design of lattice columns are given in EC5, *Clause C.4.1, Annex C*, and are as follows:

- The structure must be symmetrical about the y–y and z–z axes.
- The lattice on each side of the lattice column may be staggered relative to each other by a length of $\ell_1/2$, where ℓ_1 is the distance between adjacent nodes.
- There must be at least three bays of latticed column in the column, i.e. $\ell = 3\ell_1$.
- Where the lattice members are nailed to the flanges, there must be at least four nails per shear plane in each diagonal at each nodal point connection.
- Each end of the lattice column structure must be braced, i.e. secured laterally in position.
- For an individual flange between adjacent node connections (i.e. length ℓ_1 in Figure 8.8), the slenderness ratio must not be greater than 60.
- Buckling of the flanges corresponding to the column length ℓ_1 will not occur.

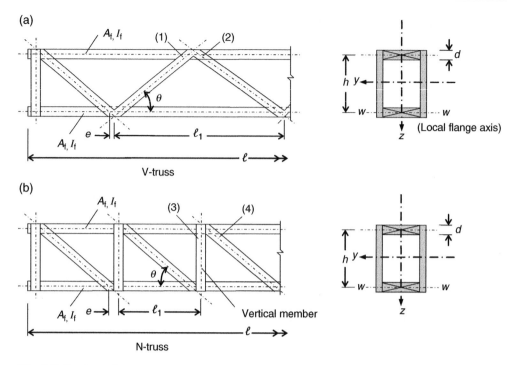

Fig. 8.8. Lattice columns.

- The number of nails in the connection between a vertical (of an N-truss – as shown in Figure 8.8b, position (3)), and the flange must be greater that $n \sin \theta$, where n is the number of nails in the adjacent diagonal connection (Figure 8.8b, position (4)) and θ is the angle of inclination of the diagonal.

8.4.5.1   Design procedure for lattice columns

When the lattice column structures shown in Figure 8.8 buckle about the z–z axis, (i.e. deflection is in the y-axis direction), the flanges behave as individual elements and the strength of the column will be the summation of the strength of the flanges about the z–z axis.

When buckling occurs about the y–y axis (i.e. deflection is in the z-axis direction), the forces in the lattice members lead to additional lateral displacement in the flanges and the strength reduction caused by this is taken into account in the design procedure. As with spaced columns, this is achieved in EC5 by using an effective slenderness ratio, λ_{ef}, derived as follows:

$$\lambda_{ef} = \max \begin{Bmatrix} \lambda_{tot} \sqrt{1+\mu} \\ 1.05\lambda_{tot} \end{Bmatrix} \qquad \text{(EC5, equation (C.14))} \quad (8.28)$$

where:

- λ_{tot} is the slenderness ratio for a solid column having the same length as the latticed column, the same cross-sectional area ($A_{tot} = \sum_{1}^{n} A_{f}$, where A_{f} is the cross-sectional area of each flange, bd), and the same second moment of area I_{tot}, $I_{tot} = (A_{f}/2)((d^2/3) + h^2)$. On this basis:

$$\lambda_{tot} = \frac{2\ell}{\sqrt{h^2 + \dfrac{d^2}{3}}}$$

(8.29)

For most practical column sections, $d^2/3$ is much smaller than h^2 and in EC5 it is ignored, giving

$$\lambda_{tot} = \frac{2\ell}{h}$$

(EC5, *equation (C.15)*)   (8.30)

- μ is a factor that takes into account the stiffness of the connection:
  (a) For glued joints μ will be as follows:

    (i) N-truss:

$$\mu = 4\frac{e^2 A_f}{I_f}\left(\frac{h}{\ell}\right)^2$$

(EC5, *equation (C.16)*)   (8.31)

    (ii) N-truss:

$$\mu = \frac{e^2 A_f}{I_f}\left(\frac{h}{\ell}\right)^2$$

(EC5, *equation (C.17)*)   (8.32)

  where e is the eccentricity of the bracing member at the nodes – as shown in Figure 8.8, A_f is the area of each flange, I_f is the second moment of area of a flange about its own axis (w–w), ℓ is the height of the latticed column, and h is the distance between the centre lines of the flanges.

  (b) For nailed joints μ will be as follows:

    (i) V-truss:

$$\mu = 25\frac{hE_{mean}A_f}{\ell^2 n K_u \sin 2\theta}$$

(EC5, *equation (C.18)*)   (8.33)

    (ii) N-truss:

$$\mu = 50\frac{hE_{mean}A_f}{\ell^2 n K_u \sin 2\theta}$$

(EC5, *equation (C.19)*)   (8.34)

  where the symbols are as described above, and:

  - n is the number of nails in a diagonal – if a diagonal consists of two or more pieces, n is to be taken as the number of nails, not the number of nails per shear plane;
  - E_{mean} is the mean value of modulus of elasticity of the timber;
  - K_u is the ULS slip modulus of one nail.

The load-carrying capacity of the lattice column will be derived using the higher of the slenderness ratios of the column about the y–y and z–z axes, and for this condition it must be verified that equation (8.25) is satisfied. Also, to comply with the EC5 requirement that buckling of the flanges corresponding to a column length ℓ_1 will not occur, using the procedures described in 5.3.1 for an axially loaded column it must be shown that the strength of each flange between adjacent node connections will exceed 50% of the strength of the latticed column.

The bracing members and their connections to the shafts must be designed for the effect of the shear force V_d derived from the equations given in EC5, (C.2.2), i.e.

$$V_d = \begin{cases} \dfrac{F_{c,d}}{120k_c} & \text{for} \quad \lambda_{ef} < 30 \\[2ex] \dfrac{F_{c,d}\lambda_{ef}}{3600k_c} & \text{for} \quad 30 \le \lambda_{ef} < 60 \\[2ex] \dfrac{F_{c,d}}{60k_c} & \text{for} \quad \lambda_{ef} \ge 60 \end{cases} \qquad \text{(EC5, equation (C.5))} \quad (8.35)$$

where the symbols are as previously defined, $F_{c,d}$ is the design axial load acting through the centre of gravity of the lattice column, and λ_{ef} is the effective slenderness ratio defined in equation (8.28).

The above shear force will be taken by the bracing members and their connections as shown in Figure 8.9. The forces in the bracing can be either tension or

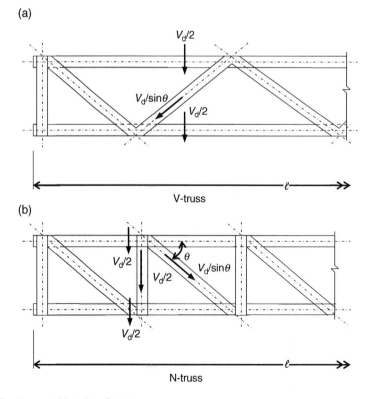

Fig. 8.9. Shear and bracing forces.

compression, and the members and connections should be designed for the more critical design condition.

The design force in the bracing members and at their connections will be:

$$\text{Horizontal bracing } V_d/2 \qquad\qquad\qquad\qquad (8.36)$$

$$\text{Diagonal bracing } V_d/\sin\theta \qquad\qquad\qquad\qquad (8.37)$$

See Example 8.8.3.

8.5   COMBINED AXIAL LOADING AND MOMENT

Whether a built-up column is formed by gluing or by mechanical fixings, the design procedure is only applicable where the column is subjected to axial loading acting through the centre of gravity of the built-up section and to the cases where there are only small moments (e.g. due to self-weight) in addition to the axial loading.

Under such conditions, the design requirements for built-up columns are the same as those for solid sections detailed in 5.4.1(b). Because of the small moment condition permitted, lateral torsional buckling effects can be ignored and on the premise that both λ_{ef} and λ_z will result in relative slenderness ratios exceeding 0.3, the strength equations to be complied with are:

$$\frac{\sigma_{c,0,d}}{k_{c,y}f_{c,0,d}} + \frac{\sigma_{m,y,d}}{f_{m,y,d}} + k_m\frac{\sigma_{m,z,d}}{f_{m,z,d}} \leq 1 \qquad (EC5,\ equation\ (6.23)) \quad (8.38)$$

$$\frac{\sigma_{c,0,d}}{k_{c,z}f_{c,0,d}} + k_m\frac{\sigma_{m,y,d}}{f_{m,y,d}} + \frac{\sigma_{m,z,d}}{f_{m,z,d}} \leq 1 \qquad (EC5,\ equation\ (6.24)) \quad (8.39)$$

$$k_{c,y} = \frac{1}{k_y + \sqrt{k_y^2 - \lambda_{rel,y}^2}} \qquad\qquad (EC5,\ equation\ (6.25)) \quad (8.40)$$

$$k_{c,z} = \frac{1}{k_z + \sqrt{k_z^2 - \lambda_{rel,z}^2}} \qquad\qquad (EC5,\ equation\ (6.26)) \quad (8.41)$$

$$\lambda_{rel,y} = \frac{\lambda_{ef}}{\pi}\sqrt{\frac{f_{c,0,k}}{E_{0.05}}} \quad \text{and} \quad \lambda_{rel,z} = \frac{\lambda_z}{\pi}\sqrt{\frac{f_{c,0,k}}{E_{0.05}}} \qquad\qquad (8.42)$$

where k_z is obtained from equation (5.5b); $\sigma_{c,0,d}$ and $f_{c,0,d}$ are as previously defined and the bending functions are as defined in Chapter 4.

See Example 8.8.3.

8.6 EFFECT OF CREEP AT THE ULS

Where the compression members of a built-up column have different values of modulus of elasticity, and/or the joints in the column are made using fasteners rather than glue, due to creep behaviour the compression stresses within the elements of the built-up column will change with duration of load and change in moisture content. Under such conditions, as required by *Clause 2.2.2* in EC5, when undertaking a first-order linear elastic analysis the effect of creep behaviour must be taken into account.

The analysis procedure in 8.4 relates to the derivation of stresses in the built-up column based on the stiffness properties derived using the mean value of the modulus of elasticity of the materials and the instantaneous value of the slip modulus.

When taking into account the effect of the redistribution of stress due to creep behaviour, EC5 requires that final mean values of stiffness properties as defined in 2.3.4.2(b) be used. The approach gives an approximation to the true behaviour of the column and for this condition the final mean values of the relevant moduli are as follows:

(a) Modulus of elasticity:

$$E_{\text{mean,fin}} = \frac{E_{\text{mean}}}{\left(1 + \psi_2 k_{\text{def}}\right)} \qquad \text{(EC5, equation (2.10))} \quad (8.43)$$

where:

- E_{mean} is the mean value of the modulus of elasticity of the material;
- k_{def} is the creep deformation factor given in Table 2.10 (*Table 3.2* of EC5) at the relevant service class the material will function in;
- ψ_2 is the factor for the quasi-permanent value of the action causing the largest stress in relation to the strength (see Table 2.2). If this is a permanent action a value of 1 should be used.

(b) Mechanical fasteners,

$$K_{\text{fin}} = \frac{K_{\text{ser}}}{\left(1 + \psi_2 k_{\text{def}}\right)} \qquad \text{(EC5, equation (2.12))} \quad (8.44)$$

where the symbols are as described above and K_{ser} is the slip modulus of the fastener per shear plane at the SLS.

Where a fastener is used for the connection, as stated in 2.3.2, if the connection is constituted of elements with the same creep behaviour, the value used for k_{def} is to be twice the value given in Table 2.10 and if the connection comprises two wood-based elements with different creep behaviour, $k_{\text{def,1}}$ and $k_{\text{def,2}}$, the value will be:

$$k_{\text{def}} = 2\sqrt{k_{\text{def,1}} k_{\text{def,2}}} \qquad (8.45)$$

Based on the above the connection efficiency factor, γ_i, referred to in equation (8.3) will become:

$$\gamma_i = \left(1 + \frac{\pi^2 E_i A_i a_i \left(1 + 2\psi_2 k_{def,connection}\right) s_i}{\left(1 + \psi_2 k_{def,i}\right) K_{ser} \ell^2}\right)^{-1}$$

(8.46)

where the symbols are as previously defined and:

- $k_{def,connection,i}$ is the deformation factor for the fastener at connection i, and if the connection comprises two wood-based elements with different creep behaviours, $k_{def,1}$ and $k_{def,2}$, then $k_{def,connection} = \sqrt{k_{def,1} k_{def,2}}$;
- $k_{def,i}$ is the deformation factor for member i;
- K_{ser} is the slip modulus for the fastener type being used.

See Examples 8.8.4 and 8.8.5.

8.7 REFERENCES

1 BS EN 1995-1-1:2004+A1:2008. *Eurocode 5: Design of Timber Structures. Part 1-1: General – Common Rules and Rules for Buildings*, British Standard Institution.
2 Ceccotti, A. Composite structures. In: Thelandersson, S., Larsen, H.J. (eds), *Timber Engineering*. Wiley, London, 2003.
3 BS EN 338:2009. *Structural Timber – Strength Classes*, British Standards Institution.
4 NA to BS EN 1990:2002+A1:2005. *Incorporating National Amendment No 1. UK National Annex for Eurocode 0 – Basis of Structural Design*, British Standards Institution.
5 NA to BS EN 1995-1-1:2004+A1.:2008, *Incorporating National Amendment No.2. UK National Annex to Eurocode 5: Design of Timber Structures. Part 1-1: General – Common Rules and Rules for Buildings*, British Standards Institution.
6 BS EN 1991-1-1:2002. *Eurocode 1: Actions on Structures. Part 1-1: General Actions – Densities, Self-Weight and Imposed Loads for Buildings*, British Standards Institution.

8.8 EXAMPLES

As stated in 4.3, to be able to verify the ultimate and serviceability limit states, each design effect has to be checked and for each effect the largest value caused by the relevant combination of actions must be used.

However, to ensure attention is primarily focused on the EC5 design rules for the timber or wood product being used, only the design load case producing the largest design effect has generally been given or evaluated in the following examples.

Example 8.8.1 The glued built-up box section, shown in Figure E8.8.1, is made from 200 mm by 44 mm solid sections of strength class C16 timber to BS EN 338:2009, functions under service class 2 conditions, and supports the axial compression loading given below. The column is 4.50 m high, and is pinned and held laterally in position at each end.

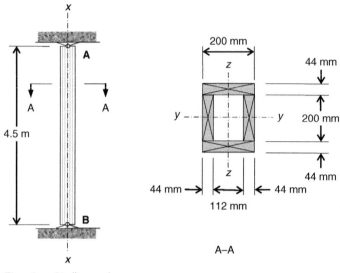

Elevation of built-up column

Fig. E8.8.1.

Check that the section complies with the strength requirements of EC5 at the ultimate limit states (ULS).

The axial loading is along the centroidal axis and:

The characteristic permanent action is 40 kN;

The characteristic variable action (medium term) is 80 kN.

1. **Column geometric properties**

Column length, L $\qquad$ $L = 4.5\,\text{m}$

Effective length about the y–y axis, $L_{e.y}$ $\qquad$ $L_{e.y} = 1.0 \cdot L$ $\qquad$ i.e. $L_{e.y} = 4.5\,\text{m}$

Effective length about the z–z axis, $L_{e.z}$ $\qquad$ $L_{e.z} = 1.0 \cdot L$ $\qquad$ i.e. $L_{e.z} = 4.5\,\text{m}$

Adopting the symbols used for profile (f) in Figure 8.4

Width of solid section member 1, b_1 mm $\qquad$ $b_1 = 200\,\text{mm}$

Thickness of solid section member 1, h_1 mm $\qquad$ $h_1 = 44\,\text{mm}$

Width of solid section member 2, h_2 mm $\qquad$ $h_2 = 200\,\text{mm}$

Combined thickness of members 2, b_2 mm $\qquad$ $b_2 = 88\,\text{mm}$

Cross-sectional area of section, A_{tot} $\qquad$ $A_{tot} = 2 \cdot \left(b_1 \cdot h_1 + h_2 \cdot \dfrac{b_2}{2} \right)$

$$A_{tot} = 3.52 \times 10^4\,\text{mm}^2$$

Second moment of area about the z–z axes, I_z (Table 8.2)

Taking the connection efficiency factor $= 1$

$$I_z = \frac{2h_1 \cdot b_1^3}{12} + \frac{2 \cdot h_2 \cdot \left(\frac{b_2}{2}\right)^3}{12} + 2h_2 \cdot \frac{b_2}{2} \cdot \left(\frac{b_1}{2} - \frac{b_2}{4}\right)^2 \qquad\qquad I_z = 1.69 \times 10^8 \, \text{mm}^4$$

Radius of gyration about the z–z axis, i_z $\qquad i_z = \sqrt{\dfrac{I_z}{A_{tot}}} \qquad\qquad i_z = 69.21 \, \text{mm}$

Slenderness ratio about the z–z axis, λ_z $\qquad \lambda_z = \dfrac{L_{e.z}}{i_z} \qquad\qquad \lambda_z = 65.02$

Second moment of area about the y–y axis, I_y (Table 8.1)

$$I_y = \frac{2 \cdot b_1 \cdot h_1^3}{12} + \frac{b_2 \cdot h_2^3}{12} + 2b_1 \cdot h_1 \cdot \left(\frac{h_2}{2} + \frac{h_1}{2}\right)^2 \qquad\qquad I_y = 3.23 \times 10^8 \, \text{mm}^4$$

Radius of gyration about the y–y axis, i_y $\qquad i_y = \sqrt{\dfrac{I_y}{A_{tot}}} \qquad\qquad i_y = 95.86 \, \text{mm}$

Slenderness ratio about the y–y axis, λ_y $\qquad \lambda_y = \dfrac{L_{e.y}}{i_y} \qquad\qquad \lambda_y = 46.94$

Critical design condition $\qquad\qquad \lambda = \max(\lambda_y, \lambda_z) \qquad \lambda = 65.02$
i.e. the critical
condition is λ_z

2. Timber properties
Table 1.3, strength class C16 (BS EN 338:2009(E), *Table 1*)

Characteristic compression strength parallel to the grain, $f_{c.0.k}$ $\qquad\qquad f_{c.0.k} = 17 \, \text{N/mm}^2$

Fifth-percentile modulus of elasticity parallel to the grain, $E_{0.05}$ $\qquad\qquad E_{0.05} = 5.4 \, \text{kN/mm}^2$

3. Partial safety factors
Table 2.8 (UKNA to BS EN 1990:2002+A1:2005, *Table NA.A1.2(B)*) for the ULS

Permanent actions, γ_G $\qquad\qquad\qquad\qquad \gamma = 1.35$

Variable actions, γ_Q $\qquad\qquad\qquad\qquad\quad \gamma_Q = 1.5$

Table 2.6 (UKNA to EC5, *Table NA.3*)

Material factor for solid timber, γ_M $\qquad\qquad \gamma_M = 1.3$

4. Actions
Characteristic permanent action, G_k $\qquad\qquad G_k = 40 \, \text{kN}$

Characteristic variable (imposed) action, Q_k $\qquad\qquad Q_k = 80 \, \text{kN}$

Design compressive action for the critical load combination, N_d (Table 2.8, equation (c) using the unfavourable condition) $\qquad N_d = \gamma_G \cdot G_k + \gamma_Q \cdot Q_k$

$\qquad\qquad\qquad\qquad\qquad\qquad\qquad\qquad N_d = 1.74 \times 10^5 \, \text{N}$

5. Modification factors

Factor for medium-duration loading and service class 2, $k_{mod.med}$
(Table 2.4 (EC5, *Table 3.1*))

$$k_{mod.med}=0.8$$

System strength factor, k_{sys} – not relevant

$$k_{sys}=1.0$$

6. Compression strength of column

The critical design load case at the ULS will be due to the combination of permanent and unfavourable medium-duration variable action:

Design compression stress, $\sigma_{c.0.d}$
(equation (8.7))

$$\sigma_{c.0.d}=\frac{N_d}{A_{tot}}$$

$\sigma_{c.0.d}=4.94$ N/mm²

Design compression strength, $f_{c.0.d}$

$$f_{c.0.d}=\frac{k_{mod.med}\cdot k_{sys}\cdot f_{c.0.k}}{\gamma_M}$$

$$f_{c.0.d}=10.46 \text{ N/mm}^2$$

Buckling resistance condition
(5.3.1 (EC5, *6.3.2*))

Relative slenderness about the z–z axis, $\lambda_{rel.z}$
(equation (8.12); EC5, *equation (6.22)*)

$$\lambda_{rel.z}=\frac{\lambda_z}{\pi}\cdot\sqrt{\frac{f_{c.0.k}}{E_{0.05}}}$$

$\lambda_{rel.z}=1.16$

As the relative slenderness ratio is greater than 0.3, conditions in 5.3.1(b) apply (EC5, *6.3.2(3)*):

Factor β_c for solid timber
(equation (5.6); EC5, *equation (6.29)*)

$$\beta_c=0.2$$

Instability factor, k_z
(equation (5.5b); EC5, *equation (6.28)*)

$$k_z=0.5\cdot\left[1+\beta_c\cdot\left(\lambda_{rel.z}-0.3\right)+\lambda_{rel.z}^2\right]$$

$$k_z=1.26$$

Instability factor, $k_{c.z}$
(equation (5.4b); EC5, *equation (6.26)*)

$$k_{c.z}=\frac{1}{k_z+\sqrt{k_z^2-\lambda_{rel.z}^2}}$$

$$k_{c.z}=0.57$$

Instability factor condition (equation (8.18); EC5, *equation (6.23)*) – with bending stresses equal to zero)

$$\frac{\sigma_{c.0.d}}{k_{c.z}\cdot f_{c.0.d}}=0.83$$

The relationship is less than 1; therefore the built-up section in strength class C16 is satisfactory. (A check at the glued connections will also demonstrate that the shear stress is also acceptable.)

Example 8.8.2 A glued spaced column is fabricated from two shafts of equal cross-section, 60 mm thick by 194 mm deep as shown in Figure E8.8.2. The spaced column is pin jointed at each end and held laterally in position at these locations. It is fabricated from C22 timber to BS EN 338:2009 and functions under service class 1 conditions. The packs are glued to the shafts,

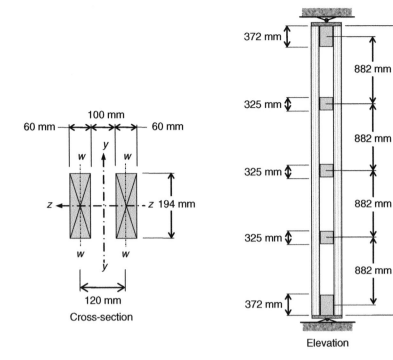

Fig. E8.8.2.

are the same depth as each shaft, and are aligned with their grain running parallel to the grain of the shafts.

Check the adequacy of the spaced column to support a combined characteristic permanent compression action of 25 kN and a characteristic medium-duration variable compression action of 35 kN applied axially through the centroid of the section.

1. Column geometric properties

Column length, L_c	$L_c = 3.9\,\text{m}$	
Number of shafts, n	$n = 2$	
Depth of each shaft, b	$b = 194\,\text{mm}$	
Thickness of each shaft, h	$h = 60\,\text{mm}$	
Space between shafts, a	$a = 100\,\text{mm}$	
Maximum free distance permitted by EC5, a_{EC5} (8.4.4 (EC5, *C3.1*))	$a_{EC5} = 3 \cdot h$	$a_{EC5} = 180\,\text{mm}$
Distance is acceptable if a/a_{EC5} < 1 or = 1	$\dfrac{a}{a_{EC5}} = 0.56$	i.e. OK
Distance between centres of the shafts, a_1	$a_1 = a + 2 \cdot \left(\dfrac{h}{2}\right)$	$a_1 = 160\,\text{mm}$

Length of end pack, $\ell_{2.e}$	$\ell_{2.e} = 372\,\text{mm}$	
Length of intermediate pack $\ell_{2.i}$	$\ell_{2.i} = 325\,\text{mm}$	
Length of pack for design, ℓ_2	$\ell_2 = \min\left(\ell_{2.e},\,\ell_{2.i}\right)$	$\ell_2 = 325\,\text{mm}$
Minimum pack length permitted, $\ell_{2.\min}$ (8.4.4 (EC5, C.3.1))	$\ell_{2.\min} = 1.5 \cdot a$	$\ell_{2.\min} = 150\,\text{mm}$
Pack length acceptable if $\ell_2 / \ell_{2.\min} > 1$ or $= 1$	$\dfrac{\ell_2}{\ell_{2.\min}} = 2.17$	i.e. OK
Depth of intermediate pack, w_2	$w_2 = b$	
Cross-sectional area of each shaft, A_{shaft}	$A_{\text{shaft}} = b \cdot h$	$A_{\text{shaft}} = 1.16 \times 10^4\,\text{mm}^2$
Section modulus of each pack (based on design length ℓ_2), W_{pack}	$W_{\text{pack}} = \dfrac{w_2 \cdot \ell_2^2}{6}$	$W_{\text{pack}} = 3.42 \times 10^6\,\text{mm}^3$
Total cross-sectional area of the spaced column, A_{tot}	$A_{\text{tot}} = n \cdot A_{\text{shaft}}$	$A_{\text{tot}} = 2.33 \times 10^4\,\text{mm}^2$
Cross-sectional area of a pack for design, A_{pack}	$A_{\text{pack}} = w_2 \cdot \ell_2$	$A_{\text{pack}} = 6.31 \times 10^4\,\text{mm}^2$
Second moment of area of the spaced column about the z–z axes, I_z	$I_z\,\dfrac{n \cdot h \cdot b^3}{12}$	$I_z = 7.3 \times 10^7\,\text{mm}^4$
Second moment of area of the spaced column about the y–y axis, I_{tot} (equation (8.22))	$I_{\text{tot}} = \left[\dfrac{b \cdot h^3}{6} + 2 \cdot b \cdot h \cdot \left(\dfrac{a+h}{2}\right)^2\right]$	$I_{\text{tot}} = 1.56 \times 10^8\,\text{mm}^4$
Second moment of area of a shaft about the local w–w axis, I_w	$I_w = \dfrac{b \cdot h^3}{12}$	$I_w = 3.49 \times 10^6\,\text{mm}^4$
Radius of gyration of the spaced column about the z–z axis, i_z	$i_z = \sqrt{\dfrac{I_z}{A_{\text{tot}}}}$	$i_z = 56\,\text{mm}$
Slenderness ratio of the spaced column about the z–z axis, λ_z	$\lambda_z\,\dfrac{L_c}{i_z}$	$\lambda_z = 69.64$
Radius of gyration of a shaft about the w–w axis, i_w	$i_w = \sqrt{\dfrac{I_w}{A_{\text{shaft}}}}$	$i_w = 17.32\,\text{mm}$
Effective length of a shaft about its own axis (w–w) – based on longest length of shaft, L_1	$L_{\text{int}} = 882\,\text{mm}$ $L_1 = \max(L_{\text{int}},\,L_{\text{end}})$	$L_{\text{end}} = 882\,\text{mm}$ $L_1 = 882\,\text{mm}$
Slenderness ratio of a shaft about its w–w axis, λ_1	$\lambda_1 = \dfrac{L_1}{i_w}$	$\lambda_1 = 50.92$

Minimum value to be used in equation (8.20) must be greater than 30 (equation (8.24); EC5, *equation (C.12)*)

As the value of λ_1 is greater than 30, we will use this value in, equation (8.20)

Slenderness ratio of the solid column of area A_{tot} and second moment of area I_{tot}, λ
(equation (8.21); EC5, *equation (C.11)*)

$$\lambda = L_c \cdot \left(\frac{A_{tot}}{I_{tot}} \right)^{0.5}$$

$\lambda = 47.65$

The value of the connection factor for a pack with a glued joint, η
(Table 8.3 (EC5, *Table C.1*))

$\eta = 1$

Slenderness ratio of the column about the y–y axis, λ_{ef}
(equation (8.20); EC5, *equation (C10)*)

$$\lambda_{ef} = \left(\lambda^2 + \eta \cdot \frac{n}{2} \cdot \lambda_1^2 \right)^{0.5}$$

$\lambda_{ef} = 69.74$

Critical design condition

$\lambda_{max} = \max(\lambda_{ef}, \lambda_z)$

$\lambda_{max} = 69.74$

2. Timber strength properties

Table 1.3, strength class C22 (BS EN 338:2009(E), *Table 1*)

Characteristic compression strength parallel to the grain, $f_{c.0.k}$

$f_{c.0.k} = 20 \text{ N/mm}^2$

Fifth-percentile modulus of elasticity parallel to the grain, $E_{0.05}$

$E_{0.05} = 6.7 \text{ kN/mm}^2$

Characteristic tensile strength perpendicular to the grain, $f_{t.90.k}$

$f_{t.90.k} = 0.4 \text{ N/mm}^2$

Characteristic shear strength across the grain, $f_{v.k}$

$f_{v.k} = 3.8 \text{ N/mm}^2$

3. Partial safety factors

Table 2.8 (UKNA to BS EN 1990:2002+A1:2005, *Table NA.A1.2(B)*) for the ULS

Permanent actions, γ_G

$\gamma_G = 1.35$

Variable actions, γ_Q

$\gamma_Q = 1.5$

Table 2.6 (UKNA to EC5, *Table NA.3*)

Material factor for solid timber, γ_M

$\gamma_M = 1.3$

4. Actions

Characteristic permanent action, G_k

$G_k = 25 \text{ kN}$

Characteristic medium-duration variable (imposed) action, Q_k

$Q_k = 35 \text{ kN}$

Design compressive action for the critical load combination, N_d
(Table 2.8, equation (c) using the unfavourable condition)

$N_d = \gamma_G \cdot G_k + \gamma_Q = Q_k$

$N_d = 8.63 \times 10^4 \text{ N}$

5. **Modification factors**

Factor for medium-duration loading and service class 1, $k_{\text{mod.med}}$
(Table 2.4 (EC5, *Table 3.1*)) $k_{\text{mod.med}} = 0.8$

System strength factor, k_{sys} – not relevant $k_{\text{sys}} = 1.0$

Modification factor for the influence of cracks, k_{cr}
(Table NA.4 in UKNA to EC5) $k_{\text{cr}} = 0.67$

6. **Strength of column and packs**

The critical design load case at the ULS will be due to the combination of permanent and unfavourable medium-duration variable action:

Direct stress on the spaced column:

Design compression stress, $\sigma_{c.0.d}$
(equation (8.25)) $$\sigma_{c.0.d} = \frac{N_d}{A_{tot}}$$ $\sigma_{c.0.d} = 3.7 \text{ N/mm}^2$

Design compression strength, $f_{c.0.d}$ $$f_{c.0.d} = \frac{k_{\text{mod.med}} \cdot k_{\text{sys}} \cdot f_{c.0.k}}{\gamma_M}$$

$$f_{c.0.d} = 12.31 \text{ N/mm}^2$$

Buckling resistance condition for a spaced column (5.3.1 (EC5, *6.3.2*))

Relative slenderness about the y–y axis:
(equation (5.3); EC5, *equation (6.21)*) $$\lambda_{rel.y} = \frac{\lambda_{ef}}{\pi}\sqrt{\frac{f_{c.0.k}}{E_{0.05}}}$$ $\lambda_{rel.y} = 1.21$

As relative slenderness ratio is greater than 0.3, conditions in 5.3.1 (EC5, *6.3.2(3)*) will apply:

Factor β_c for solid timber
(equation (5.6); EC5, *equation (6.29)*) $\beta_c = 0.2$

Instability factor, k_y
(equation (5.5a); EC5, *equation (6.27)*) $k_y = 0.5 \cdot \lfloor 1 + \beta_c \cdot (\lambda_{rel.y} - 0.3) + \lambda_{rel.y}^2 \rfloor$

$$k_y = 1.33$$

Instability factor, $k_{c.y}$
(equation (5.4a); EC5, *equation (6.25)*) $$k_{c.y} = \frac{1}{k_y + \sqrt{k_y^2 - \lambda_{rel.y}^2}}$$

$$k_{c.y} = 0.54$$

Instability factor condition
(equation (5.11a); EC5, *equation (6.23)* – with bending stresses equal to zero) $$\frac{\sigma_{c.0.d}}{k_{c.y} \cdot f_{c.0.d}} = 0.56$$ i.e. OK

Strength of the packs:

Shear force across the shafts, V_d
(equation (8.14); EC5, *equation*
(C.5))

$$V_d = \begin{vmatrix} \dfrac{N_d}{120 \cdot k_{c.y}} & \text{if } \lambda_{ef} < 30 \\[3mm] \dfrac{N_d \cdot \lambda_{ef}}{3600 \cdot k_{c.y}} & \text{if } 30 \le \lambda_{ef} 60 \\[3mm] \dfrac{N_d}{60 \cdot k_{c.y}} & \text{if } 60 \le \lambda_{ef} \end{vmatrix}$$

$$V_d = 2.68 \times 10^3 \text{ N}$$

Shear condition:

Design shear force on the pack, T_d
(equation (8.26); EC5, equation (C.13))

$$T_d = \frac{V_d \cdot L_1}{a_1}$$

$$T_d = 1.48 \times 10^4 \text{ N}$$

Design shear stress across the pack, τ_d

$$\tau_d = \frac{3}{2} \cdot \frac{T_d}{k_{cr} A_{pack}}$$

$$\tau_d = 0.52 \text{ N/mm}^2$$

Design shear strength of a pack, $f_{v.d}$

$$f_{v.d} = \frac{k_{mod.med} \cdot k_{sys} \cdot f_{v.k}}{\gamma_M}$$

$$f_{v.d} = 2.34 \text{ N/mm}^2$$

i.e. OK in shear

Bending condition (critical condition will be the pack and the shaft in tension perpendicular to the grain due to the bending stresses at the glued interface with the shaft):

Bending moment at the junction with
the shaft, M_d

$$M_d = T_d \cdot \frac{a}{2}$$

$$M_d = 7.39 \times 10^5 \text{ N mm}$$

Design stress on the pack due to the
bending moment, $\sigma_{m.d}$

$$\sigma_{m.d} = \frac{M_d}{W_{pack}}$$

$$\sigma_{m.d} = 0.22 \text{ N/mm}^2$$

Design strength of the pack based on
the tension strength perpendicular to
the grain, $f_{t.90.d}$

$$f_{t.90.d} = \frac{k_{mod.med} \cdot k_{sys} \cdot f_{t.90.k}}{\gamma_M}$$

$$f_{t.90.d} = 0.31 \text{ N/mm}^2$$

i.e. OK

Section is satisfactory as a spaced column in strength class C22 timber.

Example 8.8.3 A glued lattice column with N lattice bracing is assembled from two shafts of equal cross-section, 63 mm thick by 225 mm deep, as shown in Figure E8.8.3. The column is pin jointed at each end and held laterally in position at these locations. It is fabricated from C22 timber to BS EN 338:2009 and functions under service class 1 conditions. The eccentricity of the diagonal at each connection is 150 mm.

Check the adequacy of the latticed column to support a combined characteristic permanent axial compression action of 45 kN, a characteristic short-duration variable axial compression action of 60 kN and a characteristic bending moment due to self-weight of 1.0 kN/m about the z–z axis. Also calculate the design loads to be taken by the bracing and their connections.

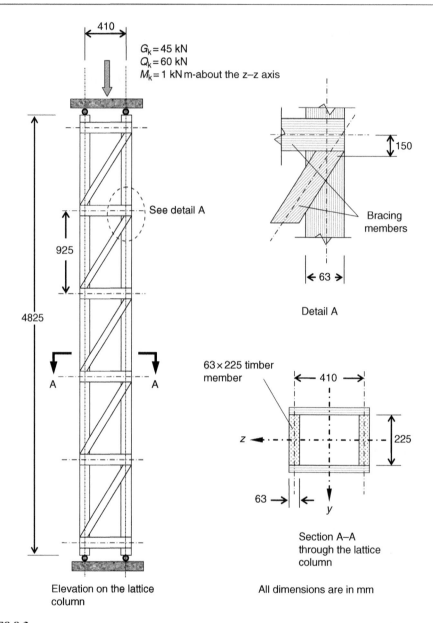

$G_k = 45$ kN
$Q_k = 60$ kN
$M_k = 1$ kN m-about the z–z axis

See detail A

925

4825

A   A

Detail A

150

Bracing
members

← 63 →

63×225 timber
member

← 410 →

z

225

63 → ←

y

Section A–A
through the lattice
column

All dimensions are in mm

Elevation on the lattice
column

Fig. E8.8.3.

1. Column geometric properties

Column length, L_c	$L_c = 4.825$ m
Depth of each flange, b	$b = 225$ mm
Thickness of each flange, t	$t = 63$ mm
Distance between the centre lines of the flanges, h	$h = 410$ mm

Eccentricity of the connection between a diagonal bracing member and a shaft, e	$e = 150$ mm	
Section modulus of the latticed column about the z–z axis, W_z	$W_z = \dfrac{2 \cdot t \cdot b^2}{6}$	$W_z = 1.06 \times 10^6 \, \text{mm}^3$
Cross-sectional area of each flange, A_f	$A_f = b \cdot t$	$A_f = 1.42 \times 10^4 \, \text{mm}^2$
Cross-sectional area of the latticed column, A_{tot}	$A_{tot} = 2 \cdot A_f$	$A_{tot} = 2.84 \times 10^4 \, \text{mm}^2$
Second moment of area of the latticed column about the z–z axes, I_z	$I_z = \dfrac{2 \cdot t \cdot b^3}{12}$	$I_z = 1.2 \times 10^8 \, \text{mm}^4$
Second moment of area of a flange about its local w–w axis, I_f	$I_f = \dfrac{b \cdot t^3}{12}$	$I_f = 4.69 \times 10^6 \, \text{mm}^4$
Radius of gyration of the latticed column about the z–z axis, i_z	$i_z = \sqrt{\dfrac{I_z}{A_{tot}}}$	$i_z = 64.95$ mm
Slenderness ratio of the latticed column about the z–z axis, λ_z	$\lambda_z = \dfrac{L_c}{i_z}$	$\lambda_z = 74.29$
Radius of gyration of a flange about its w–w axis, i_w	$i_w = \sqrt{\dfrac{I_f}{A_f}}$	$i_w = 18.19$ mm
Effective length of a flange about its own axis (w–w) – based on longest length of flange, L_1	$L_1 = 925$ mm	
Slenderness ratio of a flange about the w–w axis, λ_1 (maximum value to be < 60 (8.4.5 (EC5, C.4.1(2)))	$\lambda_1 = \dfrac{L_i}{i_w}$	
	$\lambda_1 = 50.86$	i.e. O.K.
Effective length of the latticed column about the y–y axis, $L_{e.y}$	$L_{e.y} = L_c$	$L_{e.y} = 4.825$ m
Slenderness ratio of the latticed column of the same area and same I, λ_{tot} (equation (8.30); EC5, *equation (C.15)*)	$\lambda_{tot} = \dfrac{2 \cdot L_c}{h}$	$\lambda_{tot} = 23.54$
The connection factor for a glued joint in an N-truss, μ (equation (8.32); EC5, *equation (C.17)*)	$\mu = \dfrac{e^2 A_f}{I_f} \cdot \left(\dfrac{h}{L_c}\right)^2$	$\mu = 0.49$
Slenderness ratio of the latticed column about the y–y axis, λ_{ef} (equation (8.28); EC5, *equation (C14)*)	$\lambda_{ef} = \max \lfloor \lambda_{tot} \cdot (1+\mu)^{0.5}, 1.05 \cdot \lambda_{tot} \rfloor$	
		$\lambda_{ef} = 28.74$
Critical design condition, λ_{max}	$\lambda_{max} = \max (\lambda_{ef}, \lambda_1, \lambda_z)$	$\lambda_{max} = 74.29$ i.e. λ_z
Check also on the flange strength condition, i.e. $\dfrac{\lambda_1}{\lambda_{max}}$ must be less than 1	$\dfrac{\lambda_1}{\lambda_{max}} = 0.68$	i.e. the ratio is less than 1; therefore OK

(if the ratio is 1 the lattice column profile must be changed)

2. Timber strength properties

Table 1.3, strength class C22 (BS EN 338:2009(E), *Table 1*)

Characteristic compression strength parallel to the grain, $f_{c.0.k}$ $f_{c.0.k} = 20$ N/mm^2

Characteristic bending strength, $f_{m.0.k}$ $f_{m.0.k} = 22$ N/mm^2

Fifth-percentile modulus of elasticity parallel to the grain, $E_{0.05}$ $E_{0.05} = 6.7$ kN/mm^2

3. Partial safety factors

Table 2.8 (UKNA to BS EN 1990:2002 + A1:2005, *Table NA.A1.2(B)*) for the ULS

Permanent actions, γ_G $\gamma_G = 1.35$

Variable actions, γ_Q $\gamma_Q = 1.5$

Table 2.6 (UKNA to EC5, *Table NA.3*)

Material factor for solid timber, γ_M $\gamma_M = 1.3$

4. Actions

Characteristic bending moment about the z–z axis, M_k $M_k = 1.0$ kN m

Characteristic permanent axial action, G_k $G_k = 45$ kN

Characteristic medium-duration variable axial action, Q_k $Q_k = 60$ kN

Design compressive axial action for the critical load combination, N_d $N_d = \gamma_G \cdot G_k + \gamma_Q \cdot Q_k$

(Table 2.8, equation (c) using the unfavourable condition) $N_d = 1.51 \times 10^5$ N

Design moment on the column about the z–z axis, M_d $M_d = \gamma_G \cdot M_k$ $M_d = 1.35$ kN m

5. Modification factors

Factor for short-duration loading and service class 1, $k_{mod.short}$ (Table 2.4 (EC5, *Table 3.1*)) $k_{mod.short} = 0.9$

System strength factor, k_{sys} – not relevant $k_{sys} = 1.0$

6. Strength of the column and the packs

The critical design load case at the ULS will be due to the combination of permanent and unfavourable short-duration variable action:

Direct stress on the latticed column:

Design compression stress, $\sigma_{c.0.d}$ (equation (8.25)) $\sigma_{c.0.d} = \dfrac{N_d}{A_{tot}}$ $\sigma_{c.0.d} = 5.32$ N/mm^2

Design compression strength, $f_{c.0.d}$ (equation (8.25)) $f_{c.0.d} = \dfrac{k_{mod.short} \cdot k_{sys} \cdot f_{c.0.k}}{\gamma_M}$ $f_{c.0.d} = 13.85$ N/mm^2

Buckling resistance condition for a latticed column (5.3.1 (EC5, *6.3.2*))

Relative slenderness about the z–z axis (equation (5.3); EC5, *equation (6.22)*)
$$\lambda_{rel.z} = \frac{\lambda_z}{\pi} \cdot \sqrt{\frac{f_{c.0.k}}{E_{0.05}}} \qquad \lambda_{rel.z} = 1.29$$

As relative slenderness ratio is greater than 0.3, conditions in 5.3.1 (EC5, *6.3.2(3)*) apply:

Factor β_c for solid timber (equation (5.6); EC5, *equation (6.29)*)
$$\beta_c = 0.2$$

Instability factor, k_z (equation (5.5b); EC5, *equation (6.28)*)
$$k_z = 0.5 \cdot \left\lfloor 1 + \beta_c \cdot \left(\lambda_{rel.z} - 0.3 \right) + \lambda_{rel.z}^2 \right\rfloor$$
$$k_z = 1.43$$

Instability factor, $k_{c.z}$ (equation (5.4b); EC5, *equation (6.26)*)
$$k_{c.z} = \frac{1}{k_z + \sqrt{k_z^2 - \lambda_{rel.z}^2}}$$
$$k_{c.z} = 0.49$$

Instability factor condition (equation (5.11b); EC5, *equation (6.24)* – with bending stresses equal to zero)
$$\frac{\sigma_{c.0.d}}{k_{c.z} \cdot f_{c.0.d}} = 0.79 \qquad \text{i.e. OK}$$

Combined bending and direct stress condition on latticed column

Design bending stress in the latticed column flanges – about the z–z axis, $\sigma_{m.z}$
$$\sigma_{m.z} = \frac{M_d}{W_z} \qquad \sigma_{m.z} = 1.27 \text{ N/mm}^2$$

Bending strength of column, $f_{m.z}$ (size factor = 1)
$$f_{m.z} = \frac{k_{mod.short} \cdot k_{sys} \cdot f_{m.0.k}}{\gamma_M}$$
$$f_{m.z} = 15.23 \text{ N/mm}^2$$

Strength condition when bending about the z–z axis is: (equation (5.22); EC5, *equation (6.24)*)
$$\frac{\sigma_{c.0.d}}{k_{c.z} \cdot f_{c.0.d}} + \frac{\sigma_{m.z}}{f_{m.z}} = 0.87$$

As the result is less than 1 the section is satisfactory as a latticed column in strength class C22 timber

Forces in the bracing members and their connections

Shear force across the flanges, V_d (equation (8.35); EC5, *equation (C.5)*)

$$V_d = \left| \begin{array}{l} \dfrac{N_d}{120 \cdot k_{c.z}} \quad \text{if} \quad \lambda_{ef} < 30 \\[2ex] \dfrac{N_d \cdot \lambda_{ef}}{3600 \cdot k_{c.z}} \quad \text{if} \quad 30 \le \lambda_{ef} < 60 \\[2ex] \dfrac{N_d}{60 \cdot k_{c.z}} \quad \text{if} \quad 60 \le \lambda_{ef} \end{array} \right.$$

$$V_d = 2.58 \times 10^3 \text{ N}$$

Member and connection forces – noting bracing is on both faces:

Design force in each horizontal bracing member, T_d
$$T_d = \frac{1}{4} \cdot V_d \qquad T_d = 645.52 \text{ N}$$

Design force in each diagonal bracing member, T_b
$$T_b = \frac{\dfrac{2 \cdot T_d}{h}}{\left[h^2 + \left(L_1 - 2 \cdot e \right)^2 \right]^{0.5}} \qquad T_b = 2.35 \times 10^3 \text{ N}$$

The design forces in the end connections will be T_d and T_b

Example 8.8.4 The built-up box section, shown in Figure E8.8.1, is to be assembled using nailed joints rather than glue and will be as shown in Figure E8.8.4. It is made from 200 mm by 44 mm solid sections of strength class C16 timber to BS EN 338:2009, functioning under service class 2 conditions. The column is 4.50 m high, is pinned and held laterally in position at each end, and is in a Category C area in accordance with BS EN 1991-1-1:2002. The nails are smooth round, 3.00 mm diameter at 75 mm c/c, the stiffness per nail at the SLS is 928 N/mm, and the lateral design strength of each nail is 421 N.

Check if this section complies with the strength requirements of EC5 at the ULS, when subjected to the loading applied to the column in Example 8.8.1, taking creep effects into account.

The axial compression loading is along the centroidal axis and:

The characteristic permanent action is 40 kN;

The characteristic variable action (medium term) is 80 kN.

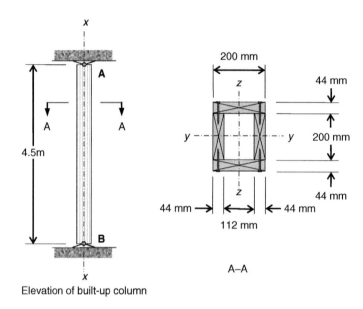

Elevation of built-up column

Fig. E8.8.4.

1. **Material and nail properties**

   Table 1.3, strength class C16 (BS EN 338:2009(E), *Table 1*)

   Characteristic compression strength parallel to the grain, $f_{c.0.k}$ $f_{c.0.k} = 17$ N/mm^2

   Mean modulus of elasticity parallel to the grain, $E_{0.mean}$ $E_{0.mean} = 8.0$ kN/mm^2

   Fifth-percentile modulus of elasticity parallel to the grain, $E_{0.05}$ $E_{0.05} = 5.4$ kN/mm^2

   Stiffness of each nail fastener at the SLS, K_{ser} $K_{ser} = 928$ N/mm

   Spacing of each fastener, s_p $s_p = 75$ mm

2. **Partial safety factors**
   Table 2.8 (UKNA to BS EN 1990:2002 + A1(2005), *Table NA.A1.2(B)*) for the ULS

Permanent actions, γ_G	$\gamma_G = 1.35$

   Variable actions, γ_Q $\gamma_Q = 1.5$

   Table 2.2 (UKNA to BS EN 1990:2002 + A1:2005 (E), *Table NA.A1.1*)

   Combination factor for the quasi-permanent value of a variable action, ψ_2 $\psi_2 = 0.6$

   Table 2.6 (UKNA to EC5, *Table NA.3*)

   Material factor for solid timber, γ_M $\gamma_M = 1.3$

   Material factor for connection, $\gamma_{M.connection}$ $\gamma_{M.connection} = 1.3$

3. **Actions**
   Characteristic permanent action, G_k $G_k = 40$ kN

   Characteristic variable (imposed) action, Q_k $Q_k = 80$ kN

   Design compressive action for the permanent action, G_d $G_d = \gamma_G \cdot G_k$ $G_d = 5.4 \times 10^4$ N

   Design compressive action for the variable (imposed) action, Q_d $Q_d = \gamma_Q \cdot Q_k$ $Q_d = 1.2 \times 10^5$ N

   Design compressive action for the critical load combination, N_d (Table 2.8, equation (c) using the unfavourable condition) $N_d = G_d + Q_d$ $N_d = 1.74 \times 10^5$ N

4. **Modification factors**
   Factor for permanent duration loading and service class 2, $k_{mod.med}$ (Table 2.4 (EC5, *Table 3.1*)) $k_{mod.perm} = 0.6$

   Factor for medium-duration loading and service class 2, $k_{mod.med}$ (Table 2.4 (EC5, *Table 3.1*)) $k_{mod.med} = 0.8$

   System strength factor, k_{sys} – not relevant $k_{sys} = 1.0$

   Deformation factor for timber and service class 2, k_{def} (Table 2.10 (EC5, *Table 3.2*)) $k_{def} = 0.8$

5. **Column geometric properties**
   Column length, L $L = 4.5$ m

   Effective length about the y–y axis, $L_{e.y}$ $L_{e.y} = 1.0 \cdot L$ i.e. $L_{e.y} = 4.5$ m

   Effective length about the z–z axis, $L_{e.z}$ $L_{e.z} = 1.0 \cdot L$ i.e. $L_{e.z} = 4.5$ m

   Adopting the symbols used for profile (f) in Figure 8.2

   Width of solid section member 1, b_1 $b_1 = 200$ mm

   Thickness of solid section member 1, h_1 $h_1 = 44$ mm

Width of solid section member 2, h_2 $h_2 = 200$ mm

Combined thickness of members 2, b_2 $b_2 = 88$ mm

Cross-sectional area of section, A_{tot} $A_{tot} = 2 \cdot \left(b_1 \cdot h_1 + h_2 \cdot \dfrac{b_2}{2} \right)$

$$A_{tot} = 3.52 \times 10^4 \text{ mm}^2$$

Second moment of area about z–z axes, I_z (Table 8.2)

Evaluation of the connection efficiency factor, $\gamma1_1$
(equation (8.3); EC5, *equation (B.5)*), adopting the final mean value for stiffness property)
The largest stress to strength ratio will be the larger of the permanent action/$k_{mod.perm}$ and
the combined permanent and variable action/$k_{mod.med}$. Let the ratio of one to the other be r:

$$r = \frac{\gamma_G \cdot G_k}{k_{mod.perm}} \cdot \frac{k_{mod.med}}{\gamma_G \cdot G_k + \gamma_Q \cdot Q_k} \qquad r = 0.41$$

i.e. because the factor is less than 1, the variable loading produces the higher stress/
strength ratio, so the factor ψ_2 will be associated with variable loading.

To simplify the connection efficiency equations:

$$E1 = \frac{E_{0.mean}}{1 - \psi_2 \cdot k_{def}} \qquad\qquad K1 = \frac{K_{ser}}{1 - \psi_2 s_2 \cdot 2 \cdot k_{def}}$$

Connection efficiency factor, $\gamma1_1$

$$\gamma1_1 = \left[1 + \pi^2 \cdot E1 \cdot \frac{(h_2 \cdot b_2) \cdot s_p}{4 \cdot K1 \cdot L_{e.z}^2} \right]^{-1} \qquad\qquad \gamma1_1 = 0.35$$

Effective bending stiffness about the z–z axis, $EI_{c.ef.z}$ Table 8.2

$$EI_{c.ef.z} = E1 \cdot \left[\left(\frac{2 \cdot h_1 \cdot b_1^3}{12} \right) + \frac{2 \cdot h_2 \cdot \left(\dfrac{b_2}{2} \right)^3}{12} + 2 \cdot \gamma1_1 \cdot h_2 \cdot \frac{b_2}{2} \left(\frac{b_1}{2} - \frac{b_2}{4} \right)^2 \right]$$

$$EI_{c.ef.z} = 5.37 \times 10^{11} \text{ N mm}$$

Second moment of area about the z–z axis, $I_{ef.z}$

$$I_{ef.z} = \frac{EI_{c.ef.z}}{E1} \qquad\qquad I_{ef.z} = 9.93 \times 10^7 \text{ mm}^4$$

Effective slenderness ratio about the z–z axis, $\lambda_{ef.z}$ (equation (8.10))

$$\lambda_{ef.z} = L_{e.z} \cdot \sqrt{\frac{A_{tot}}{I_{ef.z}}} \qquad\qquad \lambda_{ef.z} = 84.74$$

Second moment of area about the y–y axis, I_y (Table 8.1)

Evaluation of the connection efficiency factor, $\gamma2_1$
(equation (8.3); EC5 *equation (B.5)*), adopting the final mean value for the stiffness property)

$$\gamma2_1 = \left[1 + \pi^2 \cdot E1 \cdot \frac{(b_1 \cdot h_1) \cdot s_p}{2 \cdot K1 \cdot L_{e.y}^2} \right]^{-1} \qquad\qquad \gamma_1^2 = 0.35$$

Effective bending stiffness about the y–y axis, $EI_{ef.y}$ (Table 8.2)

$$EI_{ef.y} = E1 \cdot \left[\left(\frac{2 \cdot b_1 \cdot h_1^3}{12} \right) + \left(\frac{b_2 \cdot h_2^3}{12} \right) + 2 \cdot \gamma 2_1 \cdot b_1 \cdot h_1 \left(\frac{h_1}{2} + \frac{h_2}{2} \right)^2 \right]$$

$EI_{ef.y} = 8.32 \times 10^{11}$ N mm^2

Second moment of area about the y–y axis, $I_{ef.y}$

$$I_{ef.y} = \frac{EI_{ef.y}}{\dfrac{E_{0.mean}}{1 + \psi_2 \cdot k_{def}}}$$

$I_{ef.y} = 1.54 \times 10^8$ mm^4

Effective slenderness ratio about the y–y axis, $\lambda_{ef.y}$ (equation (8.10))

$$\lambda_{ef.y} = L_{e.y} \cdot \sqrt{\frac{A_{tot}}{I_{ef.z}}}$$

$\lambda_{ef.y} = 68.06$

Critical effective slenderness ratio, $\lambda_{ef.z}$ (large of $\lambda_{ef.y}$ and $\lambda_{ef.z}$)

$\lambda_{ef.z} = 87.74$

6. Compression strength of column

The critical design load case at the ULS will be due to the combination of permanent and unfavourable medium-duration variable action:

Design compression stress, $\sigma_{c.0.d}$ (equation (8.7))

$$\sigma_{c.0.d} = \frac{N_d}{A_{tot}}$$

$\sigma_{c.0.d} = 4.94$ N/mm^2

Design compression strength, $f_{c.0.d}$

$$f_{c.0.d} = \frac{k_{mod.med} \cdot k_{sys} \cdot f_{c.0.k}}{\gamma_M}$$

$f_{c.0.d} = 10.46$ N/mm^2

Buckling resistance condition (5.3.1 (EC5, 6.3.2))

Relative slenderness about the z–z axis, $\lambda_{rel.z}$ (equation (8.11); EC5, *equation (6.22)*)

$$\lambda_{rel.z} = \frac{\lambda_{ef.z}}{\pi} \cdot \sqrt{\frac{f_{c.0.k}}{E_{0.05}}}$$

$\lambda_{rel.z} = 1.51$

As the relative slenderness ratio is greater than 0.3, conditions in 5.3.1 apply (EC5, *6.3.2(3)*):

Factor β_c for solid timber (equation (5.6); EC5, *equation (6.29)*)

$\beta_c = 0.2$

Instability factor, k_z (equation (5.5b); EC5, *equation (6.28)*)

$$k_z = 0.5 \cdot \left[1 + \beta_c \cdot \left(\lambda_{rel.z} - 0.3 \right) + \lambda_{rel.z}^2 \right]$$

$k_z = 1.77$

Instability factor, $k_{c.z}$ (equation (5.4b); EC5, *equation (6.26)*)

$$k_{c.z} = \frac{1}{k_z + \sqrt{k_z^2 - \lambda_{rel.z}^2}}$$

$k_{c.z} = 0.37$

Instability factor condition (equation (8.18); EC5, *equation (6.23)* – with bending stresses equal to zero)

$$\frac{\sigma_{c.0.d}}{k_{c.z} \cdot f_{c.0.d}} = 1.27$$

i.e. the section is not acceptable

Check the capacity of the fasteners

Force on fastener, $F1$:

Critical condition will be due to the shear force when bending about z–z axis, $V2_d$ (equation (8.14); EC5, *equation (C.5)*)

$$V2_d = \frac{N_d}{60k_{c.z}}$$

$V2_d = 7.77 \times 10^3$ N

The load to be taken by each fastener bending about the z–z axis, $F1$ (equation (8.13); EC5, *equation (B.10)*)

$$F1 = \frac{\gamma_1 \cdot E1 \cdot \left(\frac{b_2}{2} \cdot h_2\right) \cdot \left(\frac{b_1}{2} - \frac{b_2}{4}\right) \cdot \frac{s_p}{2} \cdot V2_d}{EI_{c.ef.z}}$$

$F1 = 710.05$ N

Design strength of a 3.00 mm diameter round nail, $F_{v,Rd}$ (as nail spacing exceeds $14d$, no reduction in strength is required)

$F_{v,Rd} = 421$ N

i.e. not acceptable

Load reduction to achieve an acceptable design:

The design axial load on the column, N_d $N_d = 1.74 \times 10^5$ N

(a) Percentage of axial load to comply with the axial strength condition

$$red_{ax} = \frac{1.100}{\frac{\sigma_{c.0.d}}{k_{c.z} \cdot f_{c.0.d}}}$$

$red_{ax} = 79.03\%$

(b) Percentage of axial load to comply with the nail strength condition

$$red_n = \frac{f_{v,Rd} \cdot 100}{F1}$$

$red_n = 59.29\%$ i.e. (b) is the critical condition

Comparison with the strength of the same column, but glued rather than nailed, as given in Example 8.8.1:

The maximum axial load able to be supported by the glued column, N_g

$$N_g = N_d \cdot \frac{1}{0.83}$$

$N_g = 2.1 \times 10^5$ N

The maximum axial load able to be supported by the nailed column, N_n

$$N_n = N_d \cdot \frac{421}{710.05}$$

$N_n = 1.03 \times 10^5$ N

i.e. under combined permanent and medium-term axial loading, the nailed column using 3 mm diameter smooth wire nails at 75 mm c/c is only 49.2% as strong as the equivalent glued column.

$$\frac{N_n}{N_g} \cdot 100 = 49.2$$

Example 8.8.5   The box section column, shown in Figure E8.8.5, is made from solid sections of timber, strength class C18 to BS EN 338:2009, faced with 12-mm-thick Finnish 'combi' plywood, fixed by nails at 50 mm c/c. The column is 3.0 m high and functions under service class 2 conditions. It is pinned and held laterally in position at each end. The stiffness of each fastener is 900 N/mm and they are fixed at 50 mm c/c along each flange. The building is a Category C area in accordance with BS EN 1991-1-1:2002 and the timber and plywood face grain are in the same direction.

Check that the column is able to support a combined characteristic permanent loading of 15 kN and a characteristic medium duration variable load of 25 kN at the final condition. There is no requirement to check the strength of the fasteners.

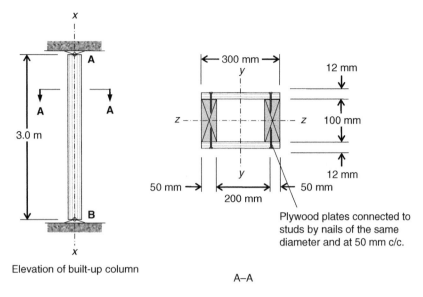

Fig. E8.8.5

1. **Material and nail properties**

   Table 1.3, strength class C18 (BS EN 338:2009(E), *Table 1*)

   Characteristic compression strength parallel to the grain, $f1_{c.0.k}$ $f1_{c.0.k} = 18 \text{ N/mm}^2$

   Fifth-percentile modulus of elasticity parallel to the grain, $E1_{0.05}$ $E1_{0.05} = 6.0 \text{ kN/mm}^2$

   Mean value of modulus of elasticity parallel to the grain, $E1_{0.mean}$ $E1_{0.mean} = 9.0 \text{ kN/mm}^2$

   Table 1.10, 12-mm-thick Finnish 'combi' plywood

   Characteristic compression strength, $f2_{p.c.0.k}$ $f2_{p.c.0.k} = 21.5 \text{ N/mm}^2$

   Mean value of modulus of elasticity parallel to the grain, $E2_{p.0.mean}$ $E2_{p.0.mean} = 7.758 \text{ kN/mm}^2$

   Fifth-percentile modulus of elasticity parallel to the grain, taken to be $0.75 E2_{p.0.mean}$, $E2_{p.0.05}$ $E2_{p.0.05} = 0.75 \cdot E2_{p.0.mean}$

   $E2_{p.0.05} = 5.82 \text{ kN/mm}^2$

   Shear modulus of the panel, $G2_{p.mean}$ $G2_{p.mean} = 589 \text{ N/mm}^2$

   Mean value of modulus of elasticity of the section $E_{0.mean}$

   $$E_{0.mean} = 2 \cdot \frac{E2_{p.0.mean} + E1_{0.mean}}{4}$$ $E_{0.mean} = 8.38 \text{ kN/mm}^2$

   Stiffness of each fastener at the SLS, K_{ser} $K_{ser} = 900 \text{ N/mm}$

2. **Partial safety factors**
   Table 2.8 (UKNA to BS EN 1990:2002+A1:2005, *Table NA.A1.2(B)*) for the ULS

Permanent actions, γ_G	$\gamma_G = 1.35$
Variable actions, γ_Q	$\gamma_Q = 1.5$

   Table 2.2 (UKNA to BS EN 1990:2002+A1:2005, *Table NA.A1.1*) – Category C

   Factor for quasi-permanent value of $\psi_2 = 0.6$
   variable action, ψ_2

   Table 2.6 (UKNA to EC5, *Table NA.3*)

Material factor for plywood, γ_{M1}	$\gamma_{M1} = 1.2$
Material factor for solid timber, γ_{M2}	$\gamma_{M2} = 1.3$
Connection factor, $\gamma_{M.connection}$	$\gamma_{M.connection} = 1.3$

3. **Actions**

Characteristic permanent action, G_k	$G_k = 15$ kN	
Characteristic variable (imposed) action, Q_k	$Q_k = 25$ kN	
Design compressive action for the critical load combination, N_d (Table 2.8, equation (c) using the unfavourable condition)	$N_d = \gamma_G \cdot G_k + \gamma_Q \cdot Q_k$	$N_d = 5.78 \times 10^4$ N

4. **Modification factors**

Factor for permanent duration loading and service class 2, $k_{mod.perm}$ (Table 2.4 (EC5, *Table 3.1*))	$k_{mod.perm} = 0.6$
Factor for medium-duration loading and service class 2, $k_{mod.med}$ (Table 2.4 (EC5, *Table 3.1*))	$k_{mod.med} = 0.8$
System strength factor, k_{sys} – not relevant	$k_{sys} = 1.0$
Deformation factor for timber and service class 2, $k_{def.t}$ (Table 2.10 (EC5, *Table 3.2*))	$k_{def.t} = 0.8$
Deformation factor for plywood and service class 2, $k_{def.p}$ (Table 2.10 (EC5, *Table 3.2*))	$k_{def.p} = 1.0$

5. **Geometric properties**

Column length, L	$L = 3.0$ m	
Effective length about the y–y axis, $L_{e.y}$	$L_{e.y} = 1.0 \cdot L$	i.e. $L_{e.y} = 3$ m
Effective length about the z–z axis, $L_{e.z}$	$L_{e.z} = 1.0 \cdot L$	i.e. $L_{e.z} = 3$ m
Spacing of each fastener, s_p	$s_p = 50$ mm	

For the creep condition:

The largest stress to strength ratio will be the larger of the permanent action/$k_{mod.perm}$ and the combined permanent and variable action/$k_{mod.med}$. Let the ratio of one to the other be r:

$$r = \frac{\gamma_G \cdot G_k}{k_{mod.perm}} \cdot \frac{k_{mod.med}}{\gamma_Q \cdot Q_k + \gamma_Q \cdot Q_k} \qquad r = 0.36$$

i.e. because the factor is less than 1, the variable loading produces the higher stress/strength ratio, so the factor ψ_2 will be associated with variable loading.

Final mean value of the modulus of elasticity parallel to the grain for timber, $E1_{mean.fin}$

$$E1_{mean.fin} = \frac{E1_{0.mean}}{1 + \psi_2 \cdot k_{def.t}}$$

$$E1_{mean.fin} = 6.08 \times 10^3 \text{ N/mm}^2$$

Final mean value of the modulus of elasticity parallel to the grain for plywood, $E2_{p.mean.fin}$

$$E2_{p.mean.fin} = \frac{E2_{p.0.mean}}{1 + \psi_2 \cdot k_{def.p}}$$

$$E2_{p.mean.fin} = 4.85 \times 10^3 \text{ N/mm}^2$$

Mean of the final mean value of the modulus of elasticity of section, $E_{mean.fin}$

$$E_{mean.fin} = 2 \cdot \frac{E2_{p.mean.fin} + E1_{mean.fin}}{4}$$

$$E_{mean.fin} = 5.46 \text{ kN/mm}^2$$

Final mean value of the stiffness of each fastener, K_{fin} (equation 2.40; EC5, *equations (2.12) and (2.13)*)

$$K_{fin} = \frac{K_{ser}}{1 + \psi_2 \cdot 2 \cdot \left(k_{def.p} \cdot k_{def.t}\right)^{0.5}}$$

$$K_{fin} = 434.09 \text{ N/mm}$$

Adopting the symbols used for profile (e) in Figure 8.2

Width of plywood section, member 2, h_2	$h_2 = 300$ mm	
Thickness of each plywood flange, member 2, h	$h = 12$ mm	
Width of the timber section, member 1, h_1	$h_1 = 50$ mm	
Clear distance between the inside faces of members 2, b	$b = h_2 - 2 \cdot h_1$	$b = 200$ mm

Check the effective flange width of the panel (7.3.2.1 (EC5, *9.1.2*))

In compression, $b1_{c.ef}$ (Table 7.2 (EC5 *Table 9.1*))

$$b1_{c.ef} = \min(0.1 \cdot L_{e.z}, 20 \cdot h)$$

$$b1_{c.ef} = 240 \text{ mm}$$

i.e. the value is based on plate buckling criteria

As the width exceeds the clear distance between members 2, the full width of the section flange can be used in the design.

Also, check on the buckling length of the compression flange – based on plate buckling criteria Section 7.3.2.2 (EC5, *Clause 9.1.2(5)*)

$$\frac{b}{2 \cdot b} = 0.5$$

the value is less than 1; therefore OK

Combined thickness of plywood members 2, h_1

$b_2 = 24$ mm

Depth of timber member, 1, b_1

$b_1 = 100$ mm

ΣEA of the section, $EA_{c.tot}$

$EA_{c.tot} = (2 \cdot E1_{mean.fin} \cdot b_1 \cdot h_1 + E2_{p.mean.fin} \cdot b_2 \cdot h_2)$

$EA_{c.tot} = 9.57 \times 10^7$ N

Properties of section about the y–y axis:

Connection efficiency factor, γ_{y1}
(equation 8.3; EC5, *equation (B.5)*)

$$\gamma_{y1} = \left[1 + \pi^2 E1_{mean.fin} \cdot \frac{(b_1 \cdot h_1) \cdot \dfrac{s_p}{2}}{K_{fin} \cdot L_{e.y}^2} \right]^{-1}$$

$\gamma_{y1} = 0.34$

Effective bending stiffness about the y–y axis, $EI_{c.ef.y}$ (Table 8.1)

$$EI_{c.ef.y} = 2 \cdot E1_{mean.fin} \cdot \frac{b_1 \cdot h_1^3}{12} + E2_{p.mean.fin} \cdot \frac{b_2 \cdot h_2^3}{12} + 2 \cdot \gamma_{y1} \cdot E1_{mean.fin} \cdot b_1 \cdot h_1 \cdot \left(\frac{h_2}{2} - \frac{h_1}{2} \right)^2$$

$EI_{c.ef.y} = 6 \times 10^{11}$ N mm^2

Effective slenderness ratio about the y–y axis, $\lambda_{c.ef.y}$ (equation (8.10))

$\lambda_{c.ef.y} = L_{e.y} \cdot \sqrt{\dfrac{EA_{c.tot}}{EI_{c.ef.y}}}$

$\lambda_{c.ef.y} = 37.9$

Second moment of area of the section about the z–z axis, I_z (Table 8.2)

Connection efficiency factor per nail line, γ_{z1}

$$\gamma_{z1} = \left[1 - \pi^2 E2_{p.mean.fin} \cdot \frac{\left(\dfrac{b_2}{2} \cdot (h_2) \right) \cdot (s_p)}{2 \cdot K_{fin} \cdot L_{e.z}^2} \right]^{-1}$$

$\gamma_{z1} = 0.48$

Effective bending stiffness about the z–z axis, $EI_{c.ef.z}$ (Table 8.2)

$$EI_{c.ef.z} = 2 \cdot E1_{mean.fin} \cdot \frac{h_1 \cdot b_1^3}{12} + 2 \cdot E2_{p.mean.fin} \cdot \frac{\left[h_2 \cdot \left(\dfrac{b_2}{2} \right)^3 \right]}{12} + 2 \cdot \gamma_{z1} \cdot E2_{p.mean.fin} \cdot \left(h_2 \cdot \frac{b_2}{2} \right) \cdot \left(\frac{b_1}{2} + \frac{b_2}{4} \right)^2$$

$EI_{c.ef.z} = 1.03 \times 10^{11}$ N mm

Effective slenderness ratio about the z–z axis, $\lambda_{c.ef.z}$ (equation (8.10))

$\lambda_{c.ef.z} = L_{e.z} \cdot \sqrt{\dfrac{EA_{c.tot}}{EI_{c.ef.z}}}$

$\gamma_{c.ef.z} = 91.38$

Critical slenderness ratio for the instantaneous condition, $\lambda_{c,ef}$

$$\lambda_{c,ef} = \max(\lambda_{c.ef.y}, \lambda_{c.ef.z})$$

$$\lambda_{c,ef} = 91.38$$
$$\text{i.e. } \lambda_{c.ef.z}$$

The design condition will relate to properties of the section about the z–z axes

6. **Strength of the column**

The critical design load case at the ULS will be due to the combination of permanent and unfavourable medium-duration variable action:

Design compression stress:

Design compression stress in the plywood, $\sigma 1_{p.c.0.d}$ (equation 8.6)

$$\sigma 1_{p.c.0.d} = \frac{E2_{p.mean.fin} \cdot N_d}{EA_{c.tot}}$$

$$\sigma 1_{p.c.0.d} = 2.93 \text{ N/mm}^2$$

Design compression stress in the timber, $\sigma 2_{c.0.d}$ (equation (8.6))

$$\sigma 2_{c.0.d} = \frac{E1_{mean.fin,} N_d}{EA_{c.tot}}$$

$$\sigma 2_{c.0.d} = 3.67 \text{ N/mm}^2$$

Design compression strength of the plywood, $f_{p.c.0.d}$

$$f_{p.c.0.d} = \frac{k_{mod.med} \cdot k_{sys} \cdot f2_{c.0.k}}{\gamma_{M1}}$$

$$f1_{p.c.0.d} = 14.33 \text{ N/mm}^2$$

Design compression strength of the timber, $f2_{c.0.d}$

$$f2_{c.0.d} = \frac{k_{mod.med} \cdot k_{sys} \cdot f1_{c.0.k}}{\gamma_{M2}}$$

$$f2_{c.0.d} = 11.08 \text{ N/mm}^2$$

Design compression strength about the critical z–z axis:

Buckling resistance condition for each element (5.3.1 (EC5, *6.3.2*))

Member 1: Plywood

Relative slenderness about the z–z axis, $\lambda_{rel.z,1}$ (equation 8.11; EC5, *equation (6.21)*)

$$\gamma_{rel.z,1} = \frac{\gamma_{c.ef.z}}{\pi} \cdot \sqrt{\frac{f2_{p.c.0.k}}{E2_{p,0.05}}}$$

$$\lambda_{rel.z,1} = 1.77$$

The relative slenderness ratio is greater than 0.3, hence conditions in EC5, *6.3.2(3)* apply:

Factor β_c – using the solid timber value (equation (5.6); EC5, *equation (6.29)*)

$$\beta_c = 0.2$$

Instability factor, $k_{z,1}$ (equation (5.5a); EC5, *equation (6.27)*)

$$k_{z,1} = 0.5 \cdot \lfloor 1 + \beta_c \cdot (\lambda_{rel.z,1} - 0.3) + \gamma^2_{rel.z,1} \rfloor$$

$$k_{z,1} = 2.21$$

Instability factor, $k_{c.z,1}$ (equation (5.4a); EC5, *equation (6.25)*)

$$k_{c.z,1} = \frac{1}{k_{z,1} + \sqrt{k_{z,1}^2 - \lambda^2_{rel.z,1}}}$$

$$k_{c.z,1} = 0.28$$

Strength of element 1 $\qquad$ $k_{c.z.1} \cdot f1_{p.c.0.d} = 4.05 \text{ N/mm}^2$

Member 2: Timber

Relative slenderness about the z–z axis, $\lambda_{rel.z.2}$ (EC5, *equation (6.21)*)
$$\gamma_{rel.z.2} = \frac{\gamma_{c.ef.z}}{\pi} \cdot \sqrt{\frac{f1_{c.0.k}}{E1_{0.05}}} \qquad \gamma_{rel.z.2} = 1.59$$

The relative slenderness ratio is greater than 0.3, hence conditions in 5.3.1 (EC5, *6.3.2(3)*) apply:

Instability factor, k_{z2} (equation (5.5a) (EC5, *equation (6.27)*)
$$k_{z.2} = 0.5 \cdot \left[1 + \beta_c \cdot \left(\lambda_{rel.z.2} - 0.3\right) + \gamma_{rel.z.2}^2\right]$$
$$k_{z.2} = 1.9$$

Instability factor, $k_{c.z.2}$ (equation (5.4a); EC5, *equation (6.25)*)
$$k_{c.z.2} = \frac{1}{k_{z.2} + \sqrt{k_{z.2^2} - \lambda_{rel.z.2}^2}}$$
$$k_{c.z.2} = 0.34$$

Strength of element 2 $\qquad$ $k_{c.z.2} \cdot f2_{c.0.d} = 3.78 \text{ N/mm}^2$

Stress/strength ratio of element 1, i.e. plywood, at the final condition
$$\frac{\sigma1_{p.c.0.d}}{k_{c.z.1} \cdot f1_{p.c.0.d}} = 0.72 \qquad \text{i.e. less than 1, OK}$$

Stress/strength ratio of element 2, i.e. timber, at the final condition
$$\frac{\sigma2_{c.0.d}}{k_{c.z.2} \cdot f2_{c.0.d}} = 0.97 \qquad \text{i.e. less than 1, OK}$$

Chapter 9

Design of Stability Bracing, Floor and Wall Diaphragms

9.1 INTRODUCTION

In this chapter, the design of structural elements that provide lateral stability or act as diaphragms to transfer lateral actions through the structure is addressed.

There are several ways in which stability or lateral bracing can be provided to a structure and those most commonly used in timber structures are as follows:

(a) provision of lateral bracing members;
(b) the use of roof or floor diaphragms;
(c) the use of wall diaphragms.

When an element in a structure is subjected to compression due to a direct force or by a bending moment and is insufficiently stiff to prevent lateral instability or excessive lateral deflection, lateral bracing of the member is likely to be required. This is particularly relevant to the design of columns and beams acting as individual members or as members in a braced system, and the methodology used in such cases is given in *9.2.5*, EC5 [1].

Examples of the types of situations where lateral stability bracing is commonly required in timber structures are shown in Figure 9.1.

When lateral forces at floor or roof level in timber-framed buildings have to be resisted, this is commonly achieved by the use of the floor and/or roof structure functioning as diaphragms and the end reaction forces are typically provided by wall diaphragms as illustrated in Figure 9.2. The wall forces are then transferred to the foundation structure through the racking resistance of the diaphragms. For clarity, diaphragms or equivalent bracing structures to resist lateral forces when they act along the length of the building, i.e. at right angles to those shown in Figure 9.2, have not been shown. The design procedures for these elements are covered in *9.2.3* and *9.2.4* of EC5.

In this chapter, the design of structural elements that provide stability or act as diaphragms to transfer actions through the structure is addressed.

The general information in 4.3 is relevant to the content of this chapter.

Structural Timber Design to Eurocode 5, Second Edition. Jack Porteous and Abdy Kermani.
© Jack Porteous and Abdy Kermani 2013. Published 2013 by Blackwell Publishing Ltd.

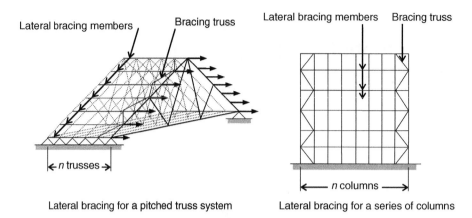

Fig. 9.1. Examples where lateral bracing of compression members are required.

9.2   DESIGN CONSIDERATIONS

Bracing members and diaphragm structures have to satisfy the relevant design rules and requirements of EC5 and the limit states associated with the main design effects are given in Table 9.1. The equilibrium states and strength conditions relate to failure situations and are therefore ultimate limit states (ULS). There is no displacement criterion for diaphragm structures, and for bracing structures the criterion is associated with the limitation of movement of the bracing system under the effect of the combined bracing force with any other external loading (e.g. wind loading) at the ULS. There is no serviceability limit states (SLS) displacement criterion.

9.3   LATERAL BRACING

9.3.1   General

The design procedures in EC5 cover the situation where compression members function as single elements in a structure as well as the condition where they form part of a bracing system as indicated in Figure 9.1.

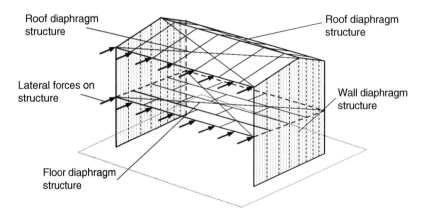

Fig. 9.2. Floor, roof and wall diaphragms (for lateral forces acting as shown).

Table 9.1 Main design requirements for bracing and diaphragm structures and the associated EC5 limit states

Element	Design or displacement effect	EC5 limit states
Bracing members	Axial stress, including the effect of lateral instability	ULS
	Deflection	ULS
Diaphragm floors and roofs	Retention of static equilibrium (sliding, uplift (for roofs))	ULS
	Bending stress	ULS
	Shear stress	ULS
Diaphragm walls	Retention of static equilibrium (sliding, uplift)	ULS
	Racking resistance	ULS

In general, the design procedure incorporates the maximum out of straightness allowances permitted in EC5, *Section 10*, i.e. $(1/300)\ell$ for solid timber and $(1/500)\ell$ for glued-laminated timber or laminated veneer lumber, where ℓ is the distance between the supports of the member.

The function of lateral bracing is to provide lateral support to a member in order to prevent it from moving laterally at the bracing position and by so doing increase the buckling strength of the member. The approach used for the design of bracing providing lateral support to single members in compression is to design each bracing member for the force and stiffness required to effectively hold the compression member in position at each bracing member location. Where several members or assemblies (e.g. roof trusses) have to be braced laterally and a bracing system is to be used, the system must be designed to withstand the lateral loading arising from the bracing members. The forces are determined on the assumption that each compression member will deform in a single wave sinusoidal mode, and to comply with the assumptions used in the analysis, the maximum deformation of the bracing system under this loading together with any other external loading (e.g. wind loading) the system supports must not exceed span/500.

The design procedure for members subjected to a direct compression force is different to that used for members subjected to a compression force caused by a bending moment. For members subjected to direct compression, the design force in the bracing is derived from the full value of the design compression force in the member, ignoring any compressive resistance provided by it in the unbraced condition. In the case of members subjected to bending moment, the unrestrained lateral torsional buckling strength is taken into account and the design force to be taken by the bracing is the design compressive force in the member less this resistance.

For a rectangular member of depth h subjected to a design moment, M_d, with a lateral buckling instability factor k_{crit} derived in accordance with the procedures described in 4.5.1.2, the approximate solution used in EC5 to determine the design value of the compression force in the member, N_d, is:

$$N_d = \left(1 - k_{crit}\right)\frac{M_d}{h} \qquad\qquad (EC5,\ equation\ (9.36)) \quad (9.1)$$

When dealing with trusses used in domestic construction, formed from members having the same thickness and all joined in the same plane by punched metal plate

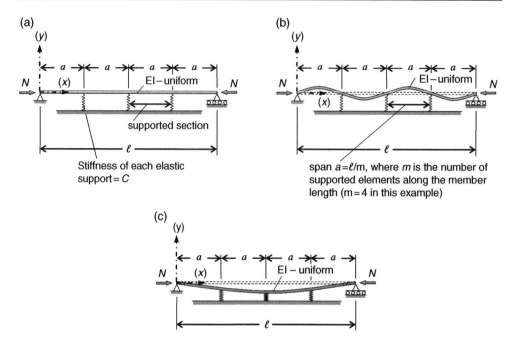

Fig. 9.3. Plan view on the lateral buckling modes of an elastically supported member.

fasteners, although the general principles for the design of bracing members will apply, the specific requirements to be adopted in the UK are given in PD6693-1 [2].

9.3.2 Bracing of single members (subjected to direct compression) by local support

The method detailed in *9.2.5.2* of EC5 is applicable to the design of bracing members laterally supporting a single member subjected to direct compression, e.g. the compression boom member of a truss, or a single beam under bending moment.

The function of the bracing is to prevent the member from buckling laterally at the bracing positions and by so doing increase its lateral buckling strength. Consequently, when determining the buckling strength of the single member its buckling length should be taken to be the distance between adjacent bracing members.

Consider a compression member of length ℓ braced laterally by elastic supports spaced at equal intervals along its length, each having the same axial stiffness. A plan view on such a member is shown in Figure 9.3a.

Where failure can occur due to the effects of lateral instability, the compression member will buckle about its weak z–z axis and for a perfectly straight member, where it is supported rigidly at bracing positions equally spaced along the length of the member, the failure mode will be as shown in Figure 9.3b. For this condition, the stiffness to be provided at each bracing position will be theoretically infinite and the force in the bracing members and its connections will be zero. Where the stiffness of the bracing and the connections is relatively small, the member will displace laterally but in so doing will reduce the buckling load that can be supported by the member.

To optimise the strength of the member, the design condition to be achieved by the bracing is equivalent to that shown in Figure 9.3b, and this can be simulated by

Table 9.2 Value of k_s against the number of bays, m, in the compression member

$m = \ell/a^*$	2	3	4	5	6	8	10	∞
$k_s = 1/\gamma$	2	3	3.41	3.62	3.73	3.80	3.90	4

$*\ell$ is the length of the compression member and a is the length between lateral supports.

increasing the stiffness of the bracing members for the mode shown in Figure 9.3c until they have reached a value that will effectively hold the member in position at each support.

Based on classical elastic stability theory, assuming a perfectly straight member, Timoshenko and Gere [3] have shown that for such a condition the minimum spring stiffness, C, to be provided by each bracing member will be:

$$C = \frac{mN_e}{\ell\gamma} \tag{9.2}$$

where:
- C is the minimum axial stiffness of each spring;
- m is the number of supported elements of the compression member along its length (see Figure 9.3b where, for the example used, $m = 4$);
- N_e is the critical buckling load of the member, i.e. the Euler buckling load for a member of length ℓ/m, and for a timber or wood-based product $N_e = \pi^2 EI/(\ell/m)^2$, where EI is the flexural rigidity about the z–z axis of the member and E is the 5th-percentile value of the modulus of elasticity parallel to the grain;
- ℓ is the overall length of the member;
- γ is a numerical factor that depends on the number of supported sections of the laterally supported member and $= 1/(2(1 + \cos(\pi/m)))$.

Substituting for γ, and letting $\ell/m = a$, as shown in Figure 9.3b, equation (9.2) can be written as:

$$C = \frac{2(1+\cos(\pi/m))N_e}{a} \tag{9.3}$$

For loads less than N_e the assumption is made that the stiffness can be allowed to reduce linearly and under the action of the mean design load in the member at the ULS, N_d, the minimum spring stiffness required is written as:

$$C = k_s \frac{N_d}{a} \qquad\qquad \text{(EC5, equation (9.34))} \tag{9.4}$$

where $k_s = 1/\gamma$, is a modification factor and the other functions are as described in EC5, N_d, is the mean design compressive force. Where the force varies along the member length, N_d will equate to the uniform design force that produces the same member strain as will occur under the actual design load.

The value of k_s for increasing values of m is given in Table 9.2, achieving a maximum value of 4 when m is essentially greater than 10. A range of values is given for k_s in EC5, however as it is a nationally determined parameter, the requirement in the UKNA to EC5 [4] is that $k_s = 4$, i.e. the largest theoretical value.

In determining the stiffness of the bracing, the effect of deviation from straightness of the compression member has been ignored, however, in determining the force to be

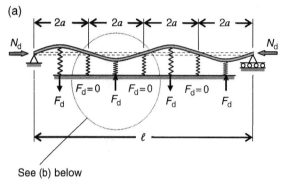

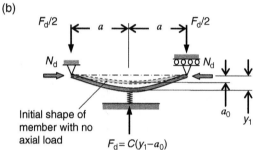

Fig. 9.4. Buckling mode adopted in EC5 for the strength analysis of the bracing member.

taken by each bracing member, EC5 takes this into account. For the buckling mode shown in Figure 9.3d, EC5 assumes that the member will have an initial deviation from straightness, a_0, as shown in Figure 9.4b.

The axial force in the bracing, F_d, arising from the design compression force in the member, N_d, will be increased due to the initial deviation and from a second-order linear elastic analysis, as stated in STEP 1 [5], a conservative value of F_d can be shown to be:

$$F_d \leq 2.6 N_d \frac{a_0}{a}$$ (9.5)

Adopting the maximum value of initial deviation from straightness between supports permitted in EC5, *10.2*, i.e. $(2a)/300$ and $(2a)/500$ for solid timber and for glulam or LVL, respectively, and applying these to equation (9.5), the design force in each bracing member will be as follows:

- For solid timber:

$$F_d \leq 5.2 N_d \frac{2a}{300} \times \frac{1}{2a} = \left(\frac{1}{57.69} \right) N_d$$

- For glulam or LVL:

$$F_d \leq 5.2 N_d \frac{2a}{500} \times \frac{1}{2a} = \left(\frac{1}{96.15} \right) N_d$$

In EC5 the factors applied to the design force N_d are defined as $1/k_{f,1}$ for solid timber and $1/k_{f,2}$ for glued-laminated timber or LVL, and the design force is obtained as follows:

- For solid timber:

$$F_d = \frac{N_d}{k_{f,1}}$$ (EC5, *equation (9.35)*)   (9.6a)

- For glulam or LVL:

$$F_d = \frac{N_d}{k_{f,2}}$$ (EC5, *equation (9.35)*)   (9.6b)

As with k_s, referred to above, $k_{f,1}$ and $k_{f,2}$ are nationally determined parameters and the requirement in *NA.2.10* of the UKNA to EC5 is that they shall be taken to be 60 and 100 respectively.

Applying the principle of static equilibrium to the beam section shown in Figure 9.4b, when the stiffness of the bracing, C, is as given in equation (9.4) the following relationship will exist between the design compression force in the member, N_d, and the lateral force, F_d, in each bracing member:

$$F_d = \frac{a_0}{\left(a/2N_d - 1/C\right)} = \frac{4N_d a_0}{a}$$ (9.7)

The lateral force in a bracing member when using equation (9.7) will be greater than that obtained from *equation (9.35)* in EC5, the difference arising because deviation from straightness has not been taken into account when determining the bracing stiffness equation (9.4). It is to be presumed that taking into account other inherent conservatisms in the analysis methodology, EC5 does not consider this omission to be significant for practical design situations.

As the force in the bracing member can be a tensile or a compressive action, depending on the bracing system being used the bracing member and its end connections must be designed to resist both conditions.

See Example 9.7.1.

9.3.3   Bracing of single members (subjected to bending) by local support

The method is applicable to the design of bracing members when fitted to a single member subjected to compression forces arising from bending, e.g. the compression force in the compression zone of a beam. The function of the bracing is to prevent the single member from buckling laterally at the bracing positions and by so doing increase the lateral torsional buckling strength of the member. Consequently, when determining the lateral torsional buckling strength of the braced member, its effective length should be taken to be the distance between adjacent bracing members.

With beams, the relationships given in equations (9.4) and (9.6) for laterally braced single members subjected to direct compression will still apply. However, the value to be used for N_d, i.e. the mean design compressive force in the member, must be determined as described in 9.3.1.

When dealing with a rectangular member of depth h, N_d is derived in accordance with equation (9.1), i.e.:

$$N_d = \left(1 - k_{crit}\right)\frac{M_d}{h} \qquad \text{(EC5, equation (9.36))} \quad (9.1)$$

where k_{crit} is derived from EC5, 6.3.3(4), for the unbraced member length, as described in Chapter 4, M_d is the maximum design moment acting on the beam, and h is the depth of the rectangular beam.

If k_{crit} is unity, function $(1 - k_{crit})$ will be zero, meaning that the beam will not buckle and there will be no requirement for lateral bracing along its length. When k_{crit} is less than unity, lateral torsional buckling can arise and N_d will be derived from equation (9.1). It is to be noted that for beams the design compressive force is derived using the maximum value of the design moment in the section, whereas for members primarily subjected to direct compression (referred to in 9.3.2) the mean value of the design compressive force is used. Also, to satisfy the theory, the bracing must be positioned such that the lateral support is provided at the compression edge of the member.

Having determined the design force, N_d, the bracing stiffness and the design axial force in each bracing member is derived as described in 9.3.2.

See Example 9.7.2.

9.3.4   Bracing for beam, truss or column systems

Where a bracing system is required to provide lateral stability to a series of compression or bending members (e.g. columns, trusses or beams), this is effectively achieved by providing lateral stiffness using truss or plate action within the plane of the bracing structure.

For the general case of a series of similar compression members that require to be braced at positions along their length, the approach used in EC5 is to assume that the deflected shape of each compression member under load will be a sinusoidal form between its supports and will include for the maximum initial out of straightness permitted in EC5, 10.2(1). Although the lateral stiffness of the structure will be a combination of the lateral stiffness of the members and of the bracing system, in EC5 the member stiffness is ignored and also the effect of shear deformations is not taken into account.

Each member in the structure, including the bracing system, is assumed to have an initial sine-shaped deformed profile, $y = a_0 \sin\left(\pi x / \ell\right)$ as shown in Figure 9.5, where ℓ is the member length in m and a_0 is the maximum deviation from straightness at mid-length of the members as well as the bracing system, also in m. Where n members are to be braced and each member is subjected to a compression force N_d, assuming that all members contribute to the loading to be taken by the bracing system, the total compression load will be nN_d. Taking the deflection of the bracing system under this load to be y_1 at mid-length, the bending moment along the system will also be a sinusoidal function and the maximum value will be $nN_d(a_0 + y_1)$.

As has been shown by Timoshenko and Gere, the above problem can alternatively be analysed by replacing the effect of the initial deviation from straightness on the deflection behaviour of the bracing system by the effect of an equivalent lateral load acting on the bracing system when in an initially straight condition, such that the

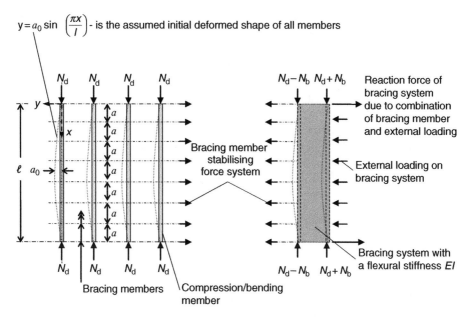

$y = a_0 \sin\left(\dfrac{\pi x}{l}\right)$ - is the assumed initial deformed shape of all members

Fig. 9.5. Bracing system for a series of compression or bending members.

bending moment diagram in each case will be the same. As the equivalent lateral load must also be a sine function, it can be expressed as:

$$q = q_\mathrm{d} \sin\left(\frac{\pi x}{\ell}\right) \tag{9.8}$$

where q_d is the maximum value of the equivalent lateral load per unit length on the system.

Under the effect of q the additional lateral deflection of the bracing system, y_q, will be:

$$y_\mathrm{q} = q_\mathrm{d} \frac{\ell^4}{\pi^4 EI} \sin\left(\frac{\pi x}{\ell}\right) \tag{9.9}$$

and the deflection at mid-span will be:

$$y_1 = q_\mathrm{d} \frac{\ell^4}{\pi^4 EI}$$

The maximum bending moment due to the equivalent lateral load will be $q_\mathrm{d}\ell^2/\pi^2$ and equating this to the maximum bending moment in the axially loaded bracing system:

$$\frac{q_\mathrm{d}\ell^2}{\pi^2} = nN_\mathrm{d}\left(a_0 + y_1\right) \tag{9.10}$$

Although the additional deflection y_1 is a function of q_d, if it is restricted to a small enough value that its effect can be considered to be negligible, equation (9.10) reduces to:

$$q_\mathrm{d} = \frac{\pi^2}{\ell^2} nN_\mathrm{d} a_0 \tag{9.11}$$

Equation (9.11) gives the value to be used in equation (9.8) for the maximum lateral load per unit length acting on the bracing system at mid-length. Because of the approximation used in the derivation of equation (9.11) the lateral force on the bracing system will be underestimated. However, as it can also be argued that all of the compression members in the system are unlikely to have the maximum value of initial out of straightness permitted in EC5, the lateral loading will be reduced, offsetting the effect of the underestimate.

In EC5, q_d is referred to as the internal stability load per unit length and is applied as a uniform load along the full length of the bracing system. It is obtained from EC5, *equation (9.37)*, incorporating a modification factor $k_{f,3}$, which can be considered to relate to function $(\pi^2/\ell)a_0$ in equation (9.11), as well as an additional factor k_ℓ:

$$q_d = k_\ell \frac{nN_d}{k_{f,3}\ell} \qquad\qquad \text{(EC5, equation (9.37))} \quad (9.12)$$

where:

- q_d is the uniformly distributed internal stability load per unit length to be imposed on the bracing system.
- n is the number of compression or bending members to be supported by the system.
- N_d is the mean design compressive force in each compression or bending member. Where the members are rectangular beams subjected to bending moment, equation (9.1) will apply.
- k_ℓ is a factor that adjusts the out of tolerance allowance for members greater than 15 m long and is equal to:

$$\min \left\{ \begin{array}{c} 1 \\ \sqrt{\frac{15}{\ell}} \end{array} \right\}$$

- $k_{f,3}$ is a modification factor and, as required by *NA.2.10* of the UKNA to EC5, will be:
    - 50 – when compression or bending members are spaced at ≤600 mm c/c,
    - 40 – when compression of bending members are spaced at >600 mm c/c.
- ℓ is the span of the bracing system in m.

The modification factor, k_ℓ, has been introduced to cover for cases where the span of the compression or bending members is greater than 15 m as, in such cases, it has been concluded that the visual impact caused by compliance with the deviation from straightness limits in *10.2(1)* would be unacceptable. The introduction of this factor reduces the maximum deformation to a value judged to be acceptable and, as this is a design imposed criterion, the responsibility will rest with the designer to ensure the contract will incorporate a deflection limit compatible with the use of the modification factor. For example, when dealing with a bracing system involving glued-laminated timber members with a span greater than 15 m, based on the maximum deviation from straightness of $\ell/500$ the deflection condition required to be achieved on site will be $\delta = (\ell/500)(15/\ell)^{0.5} = (\ell/16667)^{0.5}$ m, rather than $\ell/500$ m.

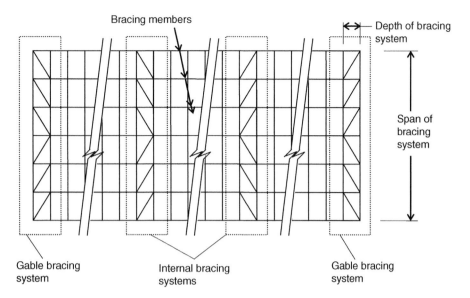

Fig. 9.6. Multiple bracing systems for a series of compression or bending members.

Where compression or bending members form part of the bracing system, for example as shown in Figure 9.5, these members will be subjected to additional axial forces, N_b, arising from the effect of the lateral loading on the bracing system. For the loading configuration shown, the mean design compression force in the bracing system members will be adjusted as indicated.

Depending on the number and location of bracing systems being used to provide the lateral stability, the bracing members may be in tension or compression and the bracing system must also be designed for the combination of internal and external loading that will result in the maximum deflection/stress condition. For the example shown in Figure 9.5, where only one bracing system is used and the external loading is acting in the direction indicated, the loading combination leading to the greatest lateral deflection condition will be when the bracing members are functioning in tension.

It is normally more economic to design bracing members to function in tension rather than compression, and this can be achieved by providing a bracing system at each end of the system of braced members together with internal bracing system(s), if required, as shown in Figure 9.6.

With this type of configuration, the gable bracing systems are designed to support external loading in addition to an element of lateral stability loading, and the internal bracing system(s) provide support structure designed to enable the attached bracing members to function in tension.

To conform with the assumption used in the theoretical approach that the lateral deformation is kept small such that its effect can be ignored, EC5 requires that the maximum deflection in a bracing system due to q_d combined with any other external loading the system has to withstand (e.g. wind loading) does not exceed $\ell/500$.

See Example 9.7.3.

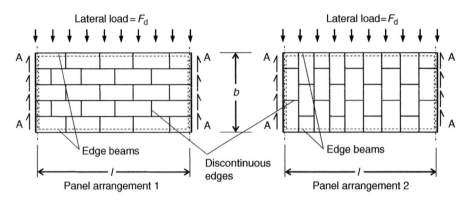

Fig. 9.7. Loading and supports to horizontal diaphragms.

9.4 FLOOR AND ROOF DIAPHRAGMS

The design procedure given in EC5 for floor and roof diaphragms is a simplified approach to the design of such structures and applies only to diaphragms assembled from wood-based panels fixed to a timber-framed structure by mechanical fasteners, e.g. nails or screws.

In deriving the lateral strength of the mechanical fixings at the edges of the panels, EC5, *9.2.3.1(2)*, allows the capacity to be increased by 20% over the value calculated in accordance with the procedures described in Chapter 10.

9.4.1 Limitations on the applicability of the method

Typical horizontal floor or roof diaphragms are shown in Figure 9.7 using alternative arrangements of staggered panels. The structure has a span ℓ (mm) and a width b (mm) and acts as a deep beam between end supports along lines A–A when subjected to lateral loading as shown.

To be able to apply the simplified method given in EC5 the following conditions have to apply:

(a) The span ℓ must lie between $2b$ and $6b$.
(b) The design condition at the ULS must be by failure of the fasteners and not failure of the beams or panels.
(c) The panels are fixed to their supports. All unsupported edges shall be fixed to adjacent panels as defined in EC5, *10.8.1*. The fixings shall be screws or nails (other than smooth nails) compliant with EN 14592 [6] and the maximum spacing along the panel edges shall be 150 mm and 300 mm along the supports.

9.4.2 Simplified design procedure

The diaphragm must be checked for bending, shear and for the lateral design strength of the fixings of the structure to the supporting timber frame along lines A–A. The design procedures are given for the condition where there is a design lateral load, F_d, along length ℓ of the diaphragm as shown in Figure 9.7 and the diaphragm width is b.

(a) *Bending strength (assuming that the edge beams are supported vertically and they only have to resist the effects of the lateral loading).*

All of the bending is assumed to be taken by the edge beams and consequently they must be continuous or detailed to be able to transfer the tensile/compression loading at adjacent sections. For the condition where the edge beam is continuous, has a width w (mm) and depth h (mm), the weaker strength will be its tensile strength and the condition to be satisfied will be

$$\frac{F_d \ell}{8(wh)b} \leq f_{t.0.d} \tag{9.13}$$

where the symbols are as defined above and:

- $f_{t.0.d}$ is the design tension strength of the timber or wood-based material in the edge beam and will be obtained from:

$$f_{t.0.d} = \frac{k_{mod} k_h f_{t,0,k}}{\gamma_M}$$

- k_{mod} is the modification factor for load duration and service classes as given in Table 2.4.

where:

- k_h is the size effect modification factor for members under tension, as discussed in Chapter 2. It is referred to in Table 2.11 and the largest cross-sectional dimension of the member should be used to evaluate the factor. When dealing with LVL, the factor is defined as k_ℓ, and is associated with member length.
- γ_M is the partial coefficient for material properties, given in Table 2.6.
- $f_{t,0,k}$ is the characteristic tensile strength of the timber or wood-based product parallel to the grain. Strength information for timber and the commonly used wood-based structural products is given in Chapter 1.

(b) *Shear strength of the diaphragm along edges A–A.*

All of the shear must be taken by the panel material. The shear stress is assumed to be uniform across the width of the diaphragm and where the panel material is t (mm) thick, the condition to be satisfied will be:

$$\frac{F_d}{2(bt)} \leq f_{v,d} \tag{9.14}$$

where the symbols are as defined above and:

- where $f_{v,d}$ is the design panel shear strength of the panel material and will be obtained from:

$$f_{v,d} = \frac{k_{mod} f_{v,k}}{\gamma_M}$$

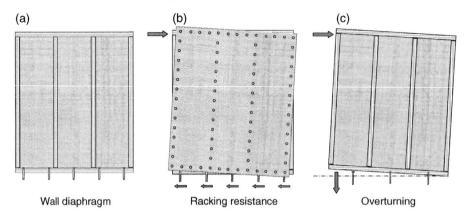

(a)   (b)   (c)

Wall diaphragm   Racking resistance   Overturning

Fig. 9.8. Wall diaphragm resisting racking loads.

• $f_{v,k}$ is the characteristic panel shear strength of the timber or wood-based product. Strength information for timber and the commonly used wood-based structural products is given in Chapter 1.

(c) *Lateral strength of the panel to support structure fixings along edges A–A.*
  The end reactions in the diaphragm must be transferred to the supporting structure by the fasteners which secure the panels to the structure. Where the detailing at each end is the same and the fasteners are at a spacing of s (mm) along the length of the support, the condition to be satisfied will be:

$$\frac{F_d}{2} \le \frac{R_d b}{s} \tag{9.15}$$

where the functions are as described above and:

• R_d is the design lateral strength of a fastener.

9.5   THE IN-PLANE RACKING RESISTANCE OF TIMBER WALLS UNDER HORIZONTAL AND VERTICAL LOADING

The stud walls associated with timber-framed buildings are usually sheathed on one or both faces with the sheathing securely fixed to the studs, enabling the wall to act as a rigid diaphragm. The in-plane resistance of the diaphragm to an external applied lateral load, as shown in Figure 9.8(b), referred to as its racking resistance, is provided by the shear and buckling strength of the wall panelling; the lateral strength of the panel to stud fixings; the axial strength of the studs and the bearing strength of the header and sole plates in the diaphragms. Horizontal sliding of the structure is resisted by fixings between the diaphragm sole plate and its support structure and overturning about its leeward side has to be prevented by base fixings or vertical loading on the wall or a combination of both, as shown in Figure 9.8c.
  In most timber-framed buildings, beams and floors are generally designed as simply supported elements on pin jointed walls and the lateral in-plane strength and stability of the structure is provided by use of wall diaphragms. Examples of some diaphragm

(a)   (b)

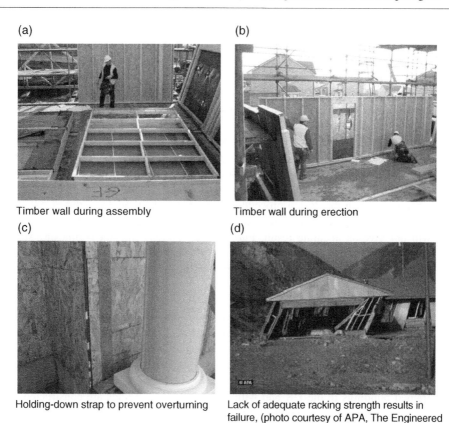

Timber wall during assembly   Timber wall during erection

(c)   (d)

Holding-down strap to prevent overturning   Lack of adequate racking strength results in
failure, (photo courtesy of APA, The Engineered
Wood Association)

Fig. 9.9. Examples of timber shear walls.

walls are shown in Figure 9.9 and in Figure 9.9c a typical failure resulting from the
lack of adequate racking strength is illustrated.

From the initial conceptual analysis of the building, the load paths of the forces
through the structure to the foundations will be determined and the diaphragms to be
designed to provide racking, sliding and overturning resistance can be identified.

The racking resistance of a diaphragm can be obtained either by tests or by calculation
and in EC5, *9.2.4*, two calculation methods, Method A (*9.2.4.2*) and Method B (*9.2.4.3*),
are given. Method A has been developed in Europe and Method B is a soft conversion of
the procedure developed in the United Kingdom for racking strength, given in BS 5268
[7]. Method A can only be applied to wall diaphragms in which the leading stud is
anchored (either by tie-down structure or by the effect of axial load on the stud) to its
supporting underlying structure and progressively down the structure at this position to
the foundation structure and is the most common method used in domestic construction
in European countries. The diaphragm must be shown not to overturn and/or slide and
the racking strength must not be less than the applied racking force on the diaphragm.

In UK domestic construction it is not common practice to anchor the leading stud
in walls and also the structural layouts commonly used mitigate against axial loads
being large enough at this position to provided adequate anchorage. Because of this,
Method B, which has been developed for use with these types of structure, has been

more appropriate for use with UK designed structures. With this method the wall is not taken to be anchored at the leading stud, but anchorage can be introduced if required. The approach is generally to try and design the racking wall such that there will be no uplift along its base and under this condition the racking force is resisted by the lateral design capacity of the fasteners connecting the sheathing to the header plate (or sole plate, as the same fixings must be used in both locations) adjusted by modification factors.

As stated above, Method B has been derived from the method used to determine the racking resistance of wall diaphragms given in BS 5268. The BS 5268 method is based on a permissible stress methodology and was developed using a semi-empirical approach, taking into account the results of racking tests on walls. Unfortunately, when converting the design procedure to limit state methodology, the EC5 codifiers incorrectly interpreted some important factors in the UK procedure [8] and the method will not give an accurate result. Further, the codifiers did not permit any strength benefit to be gained from the use of use of plasterboard in racking walls and the methodology can only be applied to walls sheathed with wood-based products as defined in EC5, 3.5.

Recognising the deficiencies in the methodology and that neither Method A nor Method B fully cover all design issues associated with racking walls, the UK has developed a third method, which is defined in PD6693-1 and referred to as the 'Simplified analysis of wall diaphragms'. It is able to be used for racking walls in timber framed buildings constructed using the platform frame method of construction. The platform frame method is the most common construction method used for timber framed buildings in the UK and with this approach buildings are constructed using storey height timber framed wall elements (these are generally prefabricated) which support wood based floor structures as well as the building roof structure.

The method allows for the effect of openings on racking strength as well as the strength contribution from the use of plasterboard as a sheathing material. The method is included in PD 6693-1, and, as stated in the UKNA to EC5, for wall diaphragms connected to underlying timber construction or foundations by bottom rail connections or a combination of bottom rail connections and tiedowns, this method should be used in preference to Method B. For wall diaphragms held down at their ends to underlying supporting construction or foundations by tiedowns, (or equivalent, e.g. permanent actions), the UKNA requirement is that Method A should be used.

An explanation of some of the background to the development of the PD Method as well as discussion on the requirements of Method A, is given in Chapter 13.

9.6 REFERENCES

1 BS EN 1995-1-1:2004+A1:2008. *Eurocode 5: Design of Timber Structures. Part 1-1: General – Common Rules and Rules for Buildings*, British Standard Institution.

2 PD6693-1: 2012, Incorporating Corrigendum No1: PUBLISHED DOCUMENT – *Recommendations for the design of timber structures: Design of Timber Structures – Part 1-1: General – Common Rules and Rules for Buildings*.

3 Timoshenko, S.P., Gere, J.M. *Theory of Elastic Stability. International Student Edition*, 2nd edn. McGraw-Hill Book Company, New York, 1961.

4 NA to BS EN 1995-1-1:2004+A1:2008, Incorporating National Amendment No. 2. *UK National Annex to Eurocode 5: Design of Timber Structures. Part 1-1: General – Common Rules and Rules for Buildings*, British Standards Institution.

5 Bruninghoff, H., Bracing – design. In: Blass, H.J., Aune, P., Choo, B.S., et al. (eds), *Timber Engineering STEP 1*, 1st edn. Centrum Hout, Almere, 1995.

6 EN 14592:2008. *Timber Structures – Dowel-type Fasteners – Requirements*.

7 BS 5268-6.1:1996. *The Structural Use of Timber. Part 6: Code of Practice for Timber Framed Walls. Section 6.1: Dwellings not Exceeding 4 Storeys*, British Standards Institution. BS 5268-6.2:1996. *The Structural Use of Timber. Part 6: Code of Practice for Timber Framed Walls. Section 6.2: Buildings Other than Dwellings not Exceeding 4 Storeys*, British Standards Institution.

8 Griffiths, B., Enjily, V., Blass, H., Kallsner, B. A unified design method for the racking resistance of timber framed walls for inclusion in Eurocode 5. *CIB-W18/38-15-11*, Karlsruhe, Germany, 2005.

9 BS EN 338:2009. *Structural Timber – Strength Classes*, British Standards Institution.

10 NA to BS EN 1990:2002 + A1:2005. *Incorporating National Amendment No1. UK National Annex for Eurocode 0 – Basis of Structural Design*, British Standards Institution.

9.7   EXAMPLES

As stated in 4.3, to be able to verify the ultimate and serviceability limit states, each design effect has to be checked and for each effect the largest value caused by the relevant combination of actions must be used.

However, to ensure attention is primarily focused on the EC5 design rules for the timber or wood product being used, only the design load case producing the largest design effect has generally been given or evaluated in the following examples.

Example 9.7.1 To enable a solid timber column 6 m long to withstand a design axial compression load of 100 kN, it is laterally braced about its weaker axis at mid-height as shown in Figure E9.7.1. What is the minimum stiffness to be provided by the bracing member and also the stabilising force in the member?

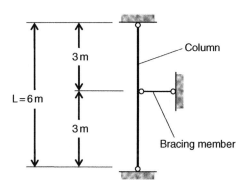

Fig. E9.7.1.

1. Geometric properties

Column:

Length of the column, L	$L = 6\,\text{m}$	
Bay length for bracing analysis, a (Figure 9.3b (EC5, *Figure 9.9*))	$a = \dfrac{L}{2}$	$a = 3\,\text{m}$

2. Actions

Column – ultimate limit state (ULS)

Design axial load on the column, N_d $N_d = 100\,\text{kN}$

3. Modification factors

Modification factor for bracing stiffness, k_s
(equation (9.4) (UKNA to EC5, *Table NA.7*)) $k_s = 4$

Modification factor, $k_{f,1}$
(equation (9.6a) (UKNA to EC5, *Table NA.7*)) $k_{f,1} = 60$

4. Stiffness of the bracing member

Minimum stiffness to be provided by the bracing
member, C

$$C = k_s \cdot \frac{N_d}{a} \qquad C = 133.33\,\text{N/mm}$$

(equation (9.4); EC5, *equation (9.34)*)

5. Stabilising force in the bracing member

Stabilising axial force to be provided by the
bracing member, F_d

$$F_d = \frac{N_d}{k_{f,1}} \qquad F_d = 1.67\,\text{kN}$$

(equation (9.6a); EC5, *equation (9.35)*)

Example 9.7.2 A 50 mm wide by 300 mm deep sawn timber beam is laterally braced against lateral torsional buckling at the positions shown in Figure E9.7.2. The beam supports a characteristic permanent action of 2.25 kN and a characteristic variable medium-duration action of 3.35 kN at mid-span and has an effective span of 3.9 m. It is of strength class C22 to BS EN 338:2009 and functions in service class 2 conditions. The bracing members are 38 mm by 100 mm deep, 1.8 m long, of strength class C16 to BS EN 338:2009 and pin jointed at each end.

Carry out a design check to confirm that the bracing members will meet the requirements of EC5, assuming that the axial stiffness of each end connection on each bracing member is equivalent to 50% of the axial stiffness of the member.

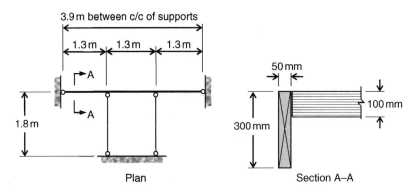

Plan   Section A–A

Fig. E9.7.2.

1. Geometric properties

Beam:

Breadth of the beam, b	$b = 50$ mm
Depth of the beam, h	$h = 300$ mm
Beam span, L_1	$L_1 = 3.9$ m
Bay length for the bracing analysis, a (Figure 9.3b (EC5, *Figure 9.9*))	$a = \dfrac{L_1}{3}$

Bracing member:

Breadth of each bracing member, b_b	$b_b = 38$ mm	
Depth of each bracing member, h_b	$h_b = 100$ mm	
Effective length of each bracing member, ℓ_b	$\ell_b = 1800$ mm	
Section modulus of each bracing member about the y–y axis, Wb_y	$Wb_y = \dfrac{b_b \cdot h_b^2}{6}$	$Wb_y = 6.33 \times 10^4$ mm^3
Second moment of area of each bracing member about the z–z axis, Ib_z	$Ib_z = \dfrac{h_b \cdot b_b^3}{12}$	$Ib_z = 4.57 \times 10^5$ mm^4
Radius of gyration of each bracing member about the z–z axis, ib_z	$ib_z = \sqrt{\dfrac{Ib_z}{b_b \cdot h_b}}$	$ib_z = 10.97$ mm
Slenderness ratio of each bracing member about the z–z axis, λb_z	$\lambda b_z = \dfrac{\ell_b}{ib_z}$	$\lambda b_z = 164.09$
Second moment of area of each bracing member about the y–y axis, Ib_y	$Ib_y = \dfrac{b_b \cdot h_b^3}{12}$	$Ib_y = 3.17 \times 10^6$ mm^4
Radius of gyration of each bracing member about the y–y axis, ib_y	$ib_y = \sqrt{\dfrac{Ib_y}{b_b \cdot h_b}}$	$ib_y = 28.87$ mm
Slenderness ratio of each bracing member about the y–y axis, λb_y	$\lambda b_y = \dfrac{\ell_b}{ib_y}$	$\lambda b_y = 62.35$

2. Timber properties

Table 1.3 (BS EN 338:2009, *Table 1*)

For the main beam – C22:

Characteristic bending strength, $f_{m,k}$	$f_{m.k} = 22$ N/mm^2
Fifth-percentile modulus of elasticity parallel to grain, $E_{0.05}$	$E_{0.05} = 6.7$ kN/mm^2
Mean modulus of elasticity parallel to grain, $E_{0,mean}$	$E_{0,mean} = 10.0$ kN/mm^2
Mean density of the beam timber, ρ_m	$\rho_m = 410$ kg/m^3

For each bracing member – C16:

Characteristic bending strength, $fb_{m,k}$	$fb_{m.k} = 16$ N/mm^2
Characteristic compressive strength, $fb_{c,0,k}$	$fb_{c.0.k} = 17$ N/mm^2
Fifth-percentile modulus of elasticity parallel to the grain, $Eb_{0.05}$	$Eb_{0.05} = 5.4$ kN/mm^2
Mean modulus of elasticity parallel to the grain, $Eb_{0.mean}$	$Eb_{0.mean} = 8$ kN/mm^2
Mean density of each bracing member, ρb_m	$\rho b_m = 370$ kg/m^3

3. Partial safety factors

Table 2.8 (UKNA to BS EN 1990:2002+A1:2005, *Table NA.A1.2(B)*) for the ULS

Permanent actions, γ_G	$\gamma_G = 1.35$
Variable actions, γ_Q	$\gamma_Q = 1.5$

Table 2.6 (UKNA to EC5, *Table NA.3*)

Material factor for solid timber, γ_M	$\gamma_M = 1.3$

4. Actions

Beam:

Self-weight of the beam, w	$w = b \cdot h \cdot g \cdot \rho_m$	$w = 0.06$ kN/m
Design action from the self-weight of the beam, W_d	$W_d = \gamma_G \cdot w$	$W_d = 0.08$ kN/m
Characteristic permanent action from the point load, G_k	$G_k = 2.25$ kN	
Characteristic variable (imposed) action from the point load, Q_k	$Q_k = 3.35$ kN	
Design action from the point load, F_d (Table 2.8, equation (c) using the unfavourable condition variable action)	$F_d = \gamma_G \cdot G_k + \gamma_Q \cdot Q_k$	$F_d = 8.06 \times 10^3$ N

Bracing members:

Self-weight of each bracing member, wb	$wb = b_b \cdot h_b \cdot g \cdot \rho b_m$	$wb = 0.01$ kN/m
Design action from the self-weight of each bracing member, Wb_d	$Wb_d = \gamma_G \cdot wb$	$Wb_d = 0.02$ kN/m

5. Modification factors

Factor for medium-duration loading and service class 2, $k_{mod,med}$ (Table 2.4 (EC5, *Table 3.1*))	$k_{mod.med} = 0.8$

Factors for the beam:

Size factor for the beam, k_h (Table 2.11 (EC5, *3.2*))	$k_h = 1.0$
Lateral stability of the beam, k_{crit} (4.5.1.2 (EC5, *6.3.3*))	

Effective length of the unbraced beam, L_{ef}

(Table 4.2 (EC5, *Table 6.1*))

$$L_{ef} = 0.8 \cdot L_1 + 2 \cdot h \qquad L_{ef} = 3.72 \text{ m}$$

Critical bending stress, $\sigma_{m,crit}$
(equation (4.9); EC5, *equation (6.32)*)

$$\sigma_{m.crit} = \frac{0.78 \cdot b^2}{h \cdot L_{ef}} \cdot E_{0.05} \qquad \sigma_{m.crit} = 11.71 \text{ N/mm}^2$$

Relative slenderness for bending, $\lambda_{rel,m}$ (equation (4.10); EC5, *equation (6.30)*)

$$\lambda_{rel.m} = \sqrt{\frac{f_{m.k}}{\sigma_{m.crit}}} \qquad \lambda_{rel.m} = 1.37$$

Lateral stability factor, k_{crit}
(Table 4.3 (EC5, *equation (6.34)*))

$$k_{crit} = \begin{vmatrix} 1 & if & \lambda_{rel.m} \leq 0.75 \\ 1.56 - 0.75 \cdot \lambda_{rel.m} & if & 0.75 < \lambda_{rel.m} \leq 1.4 \\ \dfrac{1}{\lambda_{rel.m}^2} & otherwise \end{vmatrix} \qquad k_{crit} = 0.53$$

Factors for each bracing member:
Size factor for a bracing member, kb_h
(Table 2.11 (EC5, *3.2*))

$$kb_h = \min\left[\left(\frac{150 \text{mm}}{h_b}\right)^{0.2}, 1.3\right] \qquad kb_h = 1.08$$

Lateral stability of each bracing member, kb_{crit}

(4.5.1.2 (EC5, *6.3.3*))
Effective length of bracing member, ℓb_b
(Table 4.2 (EC5, *Table 6.1*))

$$\ell b_b = 0.9 \cdot \ell_b \qquad \ell b_b = 1.62 \text{ m}$$

Critical bending stress, $\sigma b_{m.crit}$
(equation (4.9a); EC5, *equation (6.32)*)

$$\sigma b_{m.crit} = \frac{0.78 \cdot b_b^2}{h_b \cdot \ell b_b} \cdot Eb_{0.05}$$
$$\sigma b_{m.crit} = 37.54 \text{ N/mm}^2$$

Relative slenderness for bending, $\lambda b_{rel.m}$ (equation (4.10); EC5, *equation (6.30)*)

$$\lambda b_{rel.m} = \sqrt{\frac{fb_{m.k}}{\sigma b_{m.crit}}} \qquad \lambda b_{rel.m} = 0.65$$

Lateral stability factor for each bracing member, kb_{crit}
(Table 4.2 (EC5, *equation (6.34)*))

$$kb_{crit} = \begin{vmatrix} 1 & if & \lambda b_{rel.m} \leq 0.75 \\ 1.56 - 0.75 \cdot \lambda b_{rel.m} & if & 0.75 < \lambda b_{rel.m} \leq 1.4 \\ \dfrac{1}{\lambda b_{rel.m}^2} & otherwise \end{vmatrix} \qquad kb_{crit} = 1$$

As $kb_{crit} = 1$, lateral torsional buckling is not relevant

Modification factor for bracing stiffness, k_s
(equation (9.4) (*UKNA to EC5, Table NA.6*))

$$k_s = 4$$

Modification factor, $k_{f.1}$
(equation (9.6a)) (*UKNA to EC5,*
Table NA.6))
$$k_{f.1} = 60$$

Load sharing factor, k_{sys} is not relevant $k_{sys} = 1$

Redistribution factor for a solid
timber rectangular section, k_m
(4.5.1.1 (EC5, *6.1.6(2)*))
$$k_m = 0.7$$

6. Stiffness of each bracing member
The critical design load case at the ULS will be due to the combination of permanent and unfavourable medium-duration variable action:

Design bending moment in the main beam, M_d

$$M_d = \frac{F_d \cdot L_1}{4} + \frac{W_d \cdot L_1^2}{8} + \frac{Wb_d \cdot \ell_b}{2} \cdot \frac{L_1}{3} \qquad M_d = 8.04 \text{ kN m}$$

The mean design stabilising force in
each bracing member, N_d
(equation (9.1); EC5, *equation (9.36)*)
$$N_d = (1 - k_{crit}) \cdot \frac{M_d}{h} \qquad N_d = 1.25 \times 10^4$$

Minimum stiffness of each bracing
member, C
(equation (9.4); EC5, *equation (9.34)*)
$$C = k_s \cdot \frac{N_d}{a} \qquad C = 38.59 \text{ kN/m}$$

Axial stiffness of each bracing
member, $C_a = AE/L$
$$C_a = \frac{b_b \cdot h_b \cdot Eb_{0,mean}}{\ell_b} \qquad C_a = 1.69 \times 10^4 \text{ kN/m}$$

Axial stiffness of each bracing member
plus connections, C_{af}
$$C_{af} = \frac{1}{\dfrac{1}{C_a} + \dfrac{2}{0.5 \cdot C_a}} \qquad C_{af} = 3.38 \times 10^2 \text{ kN/m}$$

Actual stiffness exceeds minimum stiffness; therefore bracing stiffness is OK

7. Strength of each bracing member
Design each bracing member as a strut subjected to bending due to its self-weight:

Design bending moment in each
bracing member about the
$y-y$ axis, Mb_d
$$Mb_d = \frac{Wb_d \cdot \ell_b^2}{8} \qquad Mb_d = 7.54 \times 10^{-3} \text{ kNm}$$

Design bending stress about the
$y-y$ axis, $\sigma b_{m.d}$
$$\sigma b_{m.d} = \frac{Mb_d}{Wb_y} \qquad \sigma b_{m.d} = 0.12 \text{ N/mm}^2$$

Design bending strength about the
$y-y$ axis, $fb_{m.d}$
$$fb_{m.d} = \frac{k_{mod.med} \cdot k_{sys} \cdot kb_{crit} \cdot kb_h \cdot fb_{m.k}}{\gamma_M} \qquad fb_{m.d} = 10.68 \text{ N/mm}^2$$

Compression strength of bracing
member
Design stabilising force in each
bracing member, Nb_d
(equation (9.6a); EC5,
equation (9.35))
$$Nb_d = \frac{N_d}{k_{f.1}} \qquad Nb_d = 209.04 \text{ N}$$

Design compression stress in each bracing member, $\sigma b_{c.0.d}$

$$\sigma b_{c.0.d} = \frac{N b_d}{b_b \cdot h_b}$$

$\sigma b_{c.0.d} = 0.06 \text{ N m}$

Design compressive strength of each bracing member, $fb_{c.0.d}$

$$fb_{c.0.d} = \frac{k_{mod.med} \cdot k_{sys} \cdot fb_{c.0.k}}{\gamma_M}$$

$fb_{c.0.d} = 10.46 \text{ N mm}$

Buckling resistance condition (5.3.1 (EC5, 6.3.2))

Relative slenderness of each bracing member about the z–z axis, $\lambda b_{rel.z}$ (equation (5.3); EC5, *equation (6.22)*)

$$\lambda b_{rel.z} = \frac{\lambda b_z}{\pi} \cdot \sqrt{\frac{fb_{c.0.k}}{Eb_{0.05}}}$$

$\lambda b_{rel.z} = 2.93$

Factor β_c for solid timber (equation (5.6); EC5, *equation (6.29)*)

$\beta_c = 0.2$

Instability factor, kb_z (equation (5.5b); EC5, *equation (6.28)*)

$$kb_z = 0.5 \cdot [1 + \beta_c \cdot (\lambda b_{rel.z} - 0.3) + \lambda b_{rel.z}^2]$$

$kb_z = 5.06$

Instability factor about the z–z axis of each bracing member, $kb_{c.z}$ (equation (5.4b); EC5, *equation (6.26)*)

$$kb_{c.z} = \frac{1}{kb_z + \sqrt{kb_z^2 - \lambda b_{rel.z}^2}}$$

$kb_{c.z} = 0.11$

Relative slenderness of each bracing member about the y–y axis, $\lambda b_{rel.y}$ (equation (5.3); EC5, *equation (6.21)*)

$$\lambda b_{rel.y} = \frac{\lambda b_y}{\pi} \cdot \sqrt{\frac{fb_{c.0.k}}{Eb_{0.05}}}$$

$\lambda b_{rel.y} = 1.11$

Instability factor, kb_y (equation (5.5a); EC5, *equation (6.27)*)

$$kb_y = 0.5 \cdot \left[1 + \beta_c \cdot \left(\lambda b_{rel.y} - 0.3\right) + \lambda b_{rel.y}^2\right]$$

$kb_y = 1.2$

Instability factor about the y–y axis of each bracing member, $kb_{c.y}$ (equation (5.4a); EC5, *equation (6.25)*)

$$kb_{c.y} = \frac{1}{kb_y + \sqrt{kb_y^2 - \lambda b_{rel.y}^2}}$$

$kb_{c.y} = 0.61$

As lateral torsional buckling does not occur, the combined stress conditions to be met will be 5.21, 5.22 and 5.23 (EC5, *equations (6.23), (6.24) and (6.35)*):

$$\frac{\sigma b_{c.0.d}}{kb_{c.y} \cdot fb_{c.0.d}} + \frac{\sigma b_{m.d}}{fb_{m.d}} = 0.02$$

$$\frac{\sigma b_{c.0.d}}{kb_{c.z} \cdot fb_{c.0.d}} + \frac{k_m \cdot \sigma b_{m.d}}{fb_{m.d}} = 0.06$$

$$\left(\frac{\sigma b_{m.d}}{kb_{crit} \cdot fb_{m.d}}\right)^2 + \frac{\sigma b_{c.0.d}}{kb_{c.z} \cdot fb_{c.0.d}} = 0.05$$

All less than 1; therefore OK

The 38 mm by 100 mm sawn section timber in strength class C16 is satisfactory for the bracing members.

Example 9.7.3 The roof structure of a timber educational building 52 m long, 19.2 m wide and 10 m high is shown in Figure E9.7.3. It comprises homogeneous glued-laminated beams, class GL 28 h in accordance with EN 1194:1999. The beams are 1200 mm deep by 155 mm wide, spaced at 4 m centre to centre and function under service class 2 conditions. The bracing structure is shown in Figure E9.7.3 and is to be designed on the basis that horizontal bracing members will function as ties.

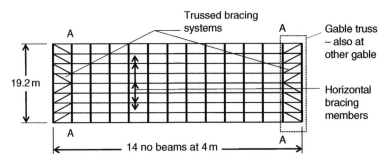

Fig. E9.7.3.

Determine the lateral force to be taken by the bracing truss at each gable end from the roof bracing members.

The vertical loading on the structure is:

Characteristic permanent loading	4.2 kN/m along the direction of each beam
Characteristic variable loading	5.0 kN/m along the direction of each beam – (medium-term action)

1. Roof beam geometric properties

Breadth of each beam, b $b = 155$ mm

Depth of each beam, h $h = 1200$ mm

Beam spacing, J_s $J_s = 4000$ mm

Span of each beam, ℓ $\ell = 19200$ mm

Length of the building, L_b $L_b = 52$ m

Number of beams contributing to the loading, n (each end beam only contributing 50% of internal beam loading) $n = \dfrac{L_b}{Js} + 1$ $n = 14$

Number of bays in the beam span, n_1 $n_1 = 8$

Laterally restrained length of the beam, ℓ_c $\ell_c = \dfrac{\ell}{n_1}$ $\ell_c = 2.4$ m

Section modulus of the beam about the y–y axis, W_y $W_y = \dfrac{b \cdot h^2}{6}$ $W_y = 3.72 \times 10^7 \, \text{mm}^3$

2. Glulam strength properties

Table 6.2 (BS EN 1194:1999, *Table 1 – GL 28h*)

Characteristic bending strength, $f_{m.g.k}$ $f_{m.g.k} = 28$ N/mm²

Fifth-percentile modulus of elasticity parallel to the grain, $E_{0.g.05}$ $E_{0.g.05} = 10.2$ kN/mm²

Fifth-percentile shear modulus
$(G_{0.g.05} = E_{0.g.mean}/16), G_{0.g.05}$

$$G_{0.g.05} = \frac{E_{0.g.05}}{16} \qquad G_{0.g.05} = 0.64 \text{ kN/mm}^2$$

3. Partial safety factors

Table 2.8 (UKNA to BS EN 1990:2002+A1:2005, *Table NA.A1.2(B)*)) for the ULS

Permanent actions, γ_G	$\gamma_G = 1.35$
Variable actions, γ_Q	$\gamma_Q = 1.5$

4. Actions

Total characteristic permanent action, G_k $\qquad$ $G_k = 4.2 \text{ kN/m}$

Total characteristic variable (imposed) action, Q_k $\qquad$ $Q_k = 5.0 \text{ kN/m}$

Design action, F_d (Table 2.8, equation (c) using the unfavourable condition variable action)

$$F_d = \gamma_G \cdot G_k + \gamma_Q \cdot Q_k$$

$$F_d = 13.17 \text{ kN/m}$$

5. Modification factors

Factor for medium-duration loading and service class 2, $k_{mod.med}$ (Table 2.4 (EC5, *Table 3.1*))

$$k_{mod.med} = 0.8$$

Size factor – beam depth greater than 600 mm, k_h (Table 2.11 (EC5, *3.3*))

$$k_h = 1.0$$

Lateral stability of the beam, k_{crit} (4.5.1.2 (EC5, *6.3.3*))

Effective length of each beam, ℓ_{ef} (Table 4.2 (EC5, *Table 6.1*))

$$\ell_{ef} = 0.9 \cdot \ell + 2 \cdot h \qquad \ell_{ef} = 19.68 \text{ m}$$

Critical bending moment, M_{crit} (equation (4.7b))

$$M_{crit} = \pi b^3 \cdot h \cdot \frac{\left[E_{0.g.05} \cdot G_{0.g.05} \cdot \left(1 - 0.63 \cdot \frac{b}{h} \right) \right]^{0.5}}{6 \ell_{ef}}$$

$$M_{crit} = 290.58 \text{ kNm}$$

Critical bending stress, $\sigma_{m.crit}$ (equation (4.7c); EC5, *equation (6.31)*)

$$\sigma_{m.crit} = \frac{M_{crit}}{W_y} \qquad \sigma_{m.crit} = 7.81 \text{ N/mm}^2$$

Relative slenderness for bending, $\lambda_{rel.m}$ (equation (4.10); EC5, *equation (6.30)*)

$$\lambda_{rel.m} = \sqrt{\frac{f_{m.g.k}}{\sigma_{m.crit}}} \qquad \lambda_{rel.m} = 1.89$$

Lateral stability factor, k_{crit} (Table 4.3 (EC5, *equation (6.34)*))

$$k_{crit} = \begin{vmatrix} 1 & \text{if} & \lambda_{rel.m} \leq 0.75 \\ 1.56 - 0.75 \cdot \lambda_{rel.m} & \text{if} & 0.75 < \lambda_{rel.m} \leq 1.4 \\ \dfrac{1}{\lambda_{rel.m}^2} & \text{otherwise} & \end{vmatrix}$$

$$k_{crit} = 0.28$$

Modification factor for bracing stiffness, k_s (equation (9.4) (UKNA to EC5, *Table NA.6*))

$$k_s = 4$$

Modification factor, $k_{f.2}$
(equation (9.6b)) (UKNA to EC5, *Table NA.6*))

$k_{f.2} = 100$

Modification factor, $k_{f.3}$
(equation (9.12)) (UKNA to EC5, *Table NA.6*))

$k_{f.3} = 40$

Modification factor, k_{ℓ}
(equation (9.12); *EC5, equation (9.38)*)

$k_{\ell} = \min\left[1, \left(\frac{15 \cdot m}{\ell}\right)^{0.5}\right]$ $k_{\ell} = 0.88$

6. Bracing actions

The critical design load case at the ULS will be due to the combination of permanent and unfavourable medium-duration variable action:

Design bending moment in each glulam beam, M_d

$M_d = \dfrac{F_d \cdot \ell^2}{8}$ $M_d = 606.87$ kNm

Compression force in the glulam beam, N_d
(equation (9.1); *EC5, equation (9.36)*)

$N_d = (1 - k_{crit}) \cdot \dfrac{M_d}{h}$ $N_d = 3.65 \times 10^5$ N

Design requirements for the bracing trusses:
The gable trusses are to be designed to allow the bracing to function as tension members and the bracing force applied at the gable member in each truss will be taken by the associated bracing truss.

Internal stability load per unit length on member A–A, q_d
(equation (9.12); *EC5, equation (9.37)*)

$q_d = k_{\ell} \cdot \dfrac{(n - 1.5) \cdot N_d}{k_{f.3} \cdot \ell}$ $q_d = 5.25$ kN/m

Design load on member A–A from each bracing member, $F_{d.q}$

$F_{d.q} = q_d \cdot \ell_c$ $F_{d.q} = 12.59$ kN

Internal stability load per unit length on each gable member, $q_{d.1}$ (equation (9.12); *EC5, equation (9.37)*)
(Note: If the bracing trusses have also to resist external loading, this will also have to be taken into account on this member.)

$q_{d.1} = k_{\ell} \cdot \dfrac{(0.5) \cdot N_d}{k_{f.3} \cdot \ell}$ $q_{d.1} = 0.21$ kN/m

Additional force in member A-A and the adjacent gable member at mid-span arising from the bracing system loading, N_b

$N_b = \dfrac{(q_d + q_{d.1}) \cdot \ell^2}{8 \cdot J_s}$ $N_b = 6.28 \times 10^4$ N

Chapter 10

Design of Metal Dowel-type Connections

10.1 INTRODUCTION

It is commonly stated that 'a structure is a constructed assembly of joints separated by members' [1] and in timber engineering the joint will generally be the critical factor in the design of the structure. The strength of the structure will normally be determined by the strength of the connections; its stiffness will greatly influence the displacement behaviour, and member sizes are often determined by the number and physical characteristics of the type of connector being used rather than by the strength requirements of the member material.

The most common form of connector used in timber connections is the mechanical type, of which there are two main groups:

- Metal dowel-type fasteners – where the load is transferred by dowel action, e.g. nails, screws, dowels and bolts, staples, etc.
- Bearing-type connectors – where the load is primarily transferred by bearing onto the timber near the surface of the member, e.g. punched metal plate, split-ring, etc.

In this chapter, the design of metal dowel-type connections subjected to lateral loading and/or axial loading is considered and bearing-type connectors are covered in Chapter 11. Connections subjected to the effect of a moment are addressed in Chapter 12.

The general information in 4.3 is relevant to the content of this chapter.

10.1.1 Metal dowel-type fasteners

10.1.1.1 Nails

Nailing is the most commonly used method for attaching members in timber frame construction. Nails are straight slender fasteners, usually pointed and headed, and are available in a variety of lengths, cross-sectional shapes and areas. The primary standards covering the types of nail that will comply with the design rules in EC5 [2] are EN 14592 [3] and EN 10230-1 [4]. Nails are classified in EN 14592 as plain shank (smooth nails) or threaded nails and the different forms available are shown in EN 10230-1.

Structural Timber Design to Eurocode 5, Second Edition. Jack Porteous and Abdy Kermani.
© Jack Porteous and Abdy Kermani 2013. Published 2013 by Blackwell Publishing Ltd.

The most frequently used type is the plain shank nail (commonly referred to as a smooth nail), which is formed from drawn non-alloy steel rods, has a minimum tensile strength of 600 N/mm², is normally circular in cross-section and is available in standard sizes, which for structural use generally range from 2.7 mm to 8 mm in diameter. Smooth nails having a square or grooved profile are also available but are less commonly used.

The performance of a nail both under lateral and withdrawal loading may be enhanced by mechanically deforming the nail shank to form *annular ringed shank* or *indented shank* nails. The processes used to form such nails not only modify the nail surface but also work-harden the steel wire, increasing the withdrawal, tensile and bending resistance over plain shank nails of the same nominal diameter. These types of nail are referred to in EN 14592 as threaded nails and in EC5 as 'nails other than smooth nails'. To be classed as a threaded nail, *clause 6.1.4.3* in EN 14592 requires the shank to be profiled over a minimum length of 4.5 times the nominal diameter of the nail and $f_{ax,k} \geq 4.5$ N/mm², where $f_{ax,k}$ is the characteristic withdrawal parameter ($f_{ax,k}$ is referred to in 10.8.1). The value of $f_{ax,k}$ must be derived using timber having a characteristic density of 350 kg/m² and conditioned to constant mass at 20 °C and 65% relative humidity. Although *8.3.2* in EC5 gives different criteria, the requirements of BS EN 14592 should be followed if categorising is required. Other forms of threaded nail are obtained by spiral rolling the shank of round nails or twisting the shank of square cross-sectioned nails and the alternative shank profiles available are shown in BS EN 10230-1.

In EC5 the nominal diameter of the nail, *d*, is used to derive design properties and in BS EN 14592 this is defined as the minimum outer cross-sectional diameter of the smooth round nail wire or the side dimension of the minimum cross-section of a square nail. For spiral rolled or annular ring shank threaded nails it is the same as for a smooth round nail and for the remaining threaded types it is the minimum cross-section of the original wire rod from which the profiled nail has been produced. Also, to be able to be used as a fastener in timber or wood-based structures, for any type the area of the nail head must not be less than $2.5d^2$ (when $d < 3.8$ mm). Some alternative types of nail together with the associated nominal diameter *d* are shown in Figure 10.1.

Nails can be plain, electroplated or galvanised to suit the finish required and the environment within which they are to be used. Corrosion protection can also be

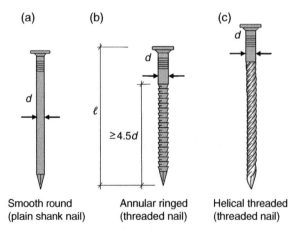

(a)   (b)   (c)

Smooth round   Annular ringed   Helical threaded
(plain shank nail)   (threaded nail)   (threaded nail)

Fig. 10.1. Types of nails.

obtained by using nails formed from austenitic stainless steel rods. The requirements for resin coated nails are still to be included in BS EN 14592.

Nailed connections are simple to form and are suitable for lightly loaded structures and where connections are formed from relatively thin members. They are commonly used in framing, walls, decks, floors and roofs and in nearly every construction that involves light loads and simple elements.

10.1.1.2 Screws

Screws are used in place of nails in applications requiring higher capacities, in particular in situations where a greater withdrawal capacity is required. They can be used for timber-to-timber joints but are especially suitable for steel-to-timber and panel-to-timber joints.

The requirements for screws are given in BS EN 14592 and the standard defines the two methods of production that can be used to produce screws that will comply with the design rules in EC5. They can be formed by having their threaded part turned down from the original rod diameter over a length of at least 4 times the nominal diameter of the screw (producing traditional smooth shank wood screws and coach screws), or they can be formed by rolling or forging the wire rod, generally over the full length of the screw but can also be over selected lengths (producing self tapping types of screw). With smooth shank screws the length of the threaded portion of the shank is generally taken to be about 60% of the shank length.

Both types of screw are produced from mild steel or carbon steel wire drawn from rods or from austenitic stainless steel wire drawn from rods which comply with the standards referred to in BS EN 14592.

As with nails, in EC5 the nominal diameter of the screw, d, is used to derive design properties and in BS EN 14592 this is defined as the maximum outer cross-section diameter of the threaded part of the screw. For screws formed by threading down the original wire, it will equate to the diameter of the unthreaded wire, and for screws formed by rolling or forging it will be the diameter across the threaded section of the screw. The inner threaded diameter of the screw, d_1, is also relevant as some strength properties in EC5 are dependent on the ratio d_1/d and indeed to be able to be used as a fastener in timber or wood-based structures, the ratio of d_1/d must not be less than 60% or more than 90%. When dealing with smooth shank wood screws this ratio is commonly taken as 0.7. BS EN 14592 also states that the nominal diameter of screws must not be less than 2.4 mm nor exceed 24 mm.

Screws should always be fixed by being threaded into the timber, not by being hammered into position, and the characteristic strengths given for screws in EC5 are based on this assumption. Where screws are used in softwood connections and the smooth shank diameter of the screw is 6 mm or less, providing the requirements of *8.3.1.2(6)* are satisfied, pre-drilling is not required. Where the diameter is greater than 6 mm in softwood connections and for screws of any diameter in hardwood connections, pre-drilling must be used and the following requirements stated in *10.4.5* of EC5 will apply:

- The pre-drilled hole for the shank should have a diameter equal to the shank diameter and be of the same depth as the shank length.
- The pre-drilled hole for the threaded portion of the screw should have a diameter of approximately 70% of the shank diameter.

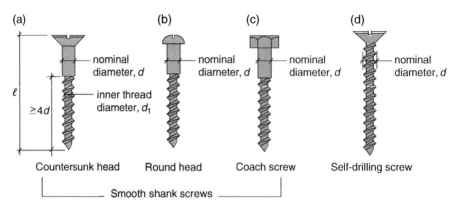

Fig. 10.2. Common types of wood screws.

- Where pre-drilling is to be used for self-drilling screws, the lead hole diameter shall not be greater than the inner thread diameter, d_1.
- Where the timber density exceeds $500\,\text{kg/m}^3$ larger pre-drilled holes may be required and this will have to be determined by testing. For particularly dense timbers a diameter equal to the nominal diameter may be required.

Examples of some screw types are illustrated in Figure 10.2.

10.1.1.3 Dowels and bolts

Dowels are generally cylindrical rods made of steel. They have a smooth or striated surface, are available in diameters from 6 mm to a maximum of 30 mm and, where the surface is irregular, the inside diameter of the dowel must be less than 95% of its outer diameter.

Bolts are circular and uniform in cross-section, made from carbon or alloy steels, and the property classes permitted by BS EN 14592 for use in timber structures are 4.6, 4.8, 5.8 or 8.8. They will have hexagonal heads and nuts, or, if not hexagonal, as stated in BS EN 14592 they must have the equivalent performance to a bolt having the same size with a hexagonal head and nut. As with dowels, bolts sizes can range from 6 mm up to 30 mm in diameter. The nominal diameter of the bolt will be the diameter of the screw thread, which will equal the diameter of the unthreaded bolt shank.

Dowels and bolts are commonly used in connections that require a higher lateral load carrying capacity than will be possible from the use of nails or screws. Bolts can also be used in axially loaded tension connections but dowels cannot.

Dowels and bolts can be used in two-member connections but the most common connection type involves three or more members in a multiple shear arrangement. The side members can be either timber or steel. When bolts are used, washers are required under the bolt head and under the nut to distribute the loads and, when tightened, a minimum of one complete thread on the bolt should protrude from the nut. Typical examples of dowels and bolts are illustrated in Figure 10.3.

Joints made with dowels are easy to fabricate. They are inserted into pre-drilled holes having a diameter not greater than the dowel diameter. With a bolted connection, the diameter of the pre-drilled hole in the timber must not be more than 1 mm greater

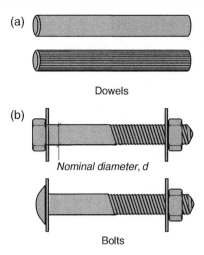

Dowels

(b)

Nominal diameter, d

Bolts

Fig. 10.3. Typical dowels and bolts.

Table 10.1 Main design requirements for metal dowel-type connections under lateral or axial loading and the associated EC5 limit states

Design or displacement effect	EC5 limit states
Lateral and/or axial loading strength	ULS
Displacement	SLS

than the bolt diameter. Where steel plates are used in the connection the tolerance on the hole diameter in the steel plate (i.e. the difference between the pre-drilled hole size in the plate and the bolt diameter) is given in *10.4.3(1)* and will influence the characteristic strength of the connection. This is discussed in 10.3.

Some examples of connections formed with metal dowel-type fasteners are shown in Figure 10.4.

10.2   DESIGN CONSIDERATIONS

Metal dowel-type connections have to satisfy the relevant design rules and requirements of EC5 and the limit states associated with the main design effects are given in Table 10.1. The strength conditions relate to failure situations and are therefore ultimate limit states (ULS) and the displacement condition relates to serviceability limit states (SLS). No displacement limit is given in EC5 for connections, however it is a requirement that the movement at connections in a structure must be taken into account when determining the instantaneous and final displacements of the structure. Also, where connection movement will affect the stiffness distribution in a structure, to be able to determine the stress resultant distribution at the ULS the effect of this movement on stiffness properties has to be included for.

(a)

Nailed holding down strap

(b)

Nailed OSB gusset joints

(c)

Bolted timber to timber connection

(d)

Bolted timber to timber and bolted
steel hanger support

(e)

Fin-plate bolted connection (steel
plates in slots cut in timber)

(f)

Fin-plate connection with nearly
hidden dowels

(g)

Dowelled connection (circular profile), (photo
courtesy of Axis Timber Ltd., a member
of Glued Laminated Timber Association)

(h)

Bolted steel to timber connection
(photo courtesy of Donaldson &
McConnell Midlands Ltd., a member of
Glued Laminated Timber Association)

Fig. 10.4. Examples of connections formed using metal dowel-type fasteners.

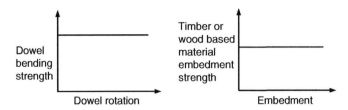

Fig. 10.5. Strength/strain relationships used for dowel connections.

Reduction in the cross-sectional area of a member due to the fitting of fasteners has also to be taken into account in the member strength verification. From EC5, *5.2(3)*, the effect of loss of cross-sectional area may be ignored where nails and screws with a diameter of 6 mm or less driven without pre-drilling are used. Also holes in the compression area of members can be ignored if the holes are filled with a material of higher stiffness than wood. However, as required by *5.2(4)*, all holes (that are not excluded by the above rules) within a distance of half the minimum fasteners spacing measured parallel to the grain from a given cross-section should be taken as occurring at that cross-section.

10.3 FAILURE THEORY AND STRENGTH EQUATIONS FOR LATERALLY LOADED CONNECTIONS FORMED USING METAL DOWEL FASTENERS

When subjected to lateral loading, a connection formed using metal dowel fasteners may fail in a brittle or a ductile mode and the design rules in EC5 have been developed to try and ensure that failure will be in a ductile rather than a brittle manner. There will be connections where both type of failure mode can arise and it is recommended that in such situations the design is adjusted to ensure failure will be in the ductile rather than the brittle mode.

The minimum spacings, edge and end distances given in EC5 when using these fasteners have been derived to prevent splitting failure when the connection is being formed and when it is subjected to lateral load. Also, for any softwood member in a connection subjected to a design force at an angle to the grain, a procedure is given in EC5 to ensure that the splitting capacity of the member will exceed the component of the design tension force in the member perpendicular to the grain.

'Dowel' is the generic term used for a fastener that transfers load between connected members by a combination of flexure and shear in the dowel and shear and bearing (referred to as embedment) in the timber. The ductile failure theory used for connections formed with dowels is that the fastener and the timber or wood-based material being connected will behave as essentially rigid plastic materials in accordance with the strength–displacement relationships shown in Figure 10.5.

This assumption considerably simplifies the analysis and using such relationships, based on the possible alternative ductile failure modes that can occur in a connection, Johansen [5] derived the strength equations for connections formed using metal dowel-type fasteners in timber. When using such fasteners, the possible failure modes that can arise in timber-to-timber and wood panel to timber connections are shown in Table 10.2 and, for timber-to-steel connections, in Table 10.3.

Table 10.2 Characteristic load carrying capacity per fastener per shear plane for timber-timber and timber-wood based connections*

Connections in single shear

	(a)	(b)	(c)	(d)	(e)	(f)	
Failure modes (Fastener axes to be aligned as shown across)	t_1 t_2						Member axes 90° 90° Fastener axis
(EYM Mode type)	1	1	1	2	2	3)	

Characteristic load-carrying capacity per fastener per shear plane, $F_{v,Rk}$, is the minimum value from the mode failure equations: (EC5 equations (8.6))

$$F_{v,Rk} = f_{h,1,k} \cdot t_1 \cdot d \qquad \text{mode (a)} \quad (10.1)$$

$$F_{v,Rk} = f_{h,2,k} \cdot t_2 \cdot d \qquad \text{mode (b)} \quad (10.2)$$

$$F_{v,Rk} = \frac{f_{h,1,k} \cdot t_1 \cdot d}{1+\beta} \left[\sqrt{\beta + 2\beta^2 \left[1 + \frac{t_2}{t_1} + \left(\frac{t_2}{t_1}\right)^2 \right] + \beta^3 \left(\frac{t_2}{t_1}\right)^2} - \beta \left(1 + \frac{t_2}{t_1}\right) \right] + \frac{F_{ax,Rk}}{4} \qquad \text{mode (c)} \quad (10.3)$$

$$F_{v,Rk} = 1.05 \frac{f_{h,1,k} \cdot t_1 \cdot d}{2+\beta} \left[\sqrt{2\beta(1+\beta) + \frac{4\beta(2+\beta)M_{y,Rk}}{f_{h,1,k} \cdot t_1^2 \cdot d}} - \beta \right] + \frac{F_{ax,Rk}}{4} \qquad \text{mode (d)} \quad (10.4)$$

$$F_{v,Rk} = 1.05 \frac{f_{h,1,k} \cdot t_2 \cdot d}{1+2\beta} \left[\sqrt{2\beta^2(1+\beta) + \frac{4\beta(1+2\beta)M_{y,Rk}}{f_{h,1,k} \cdot t_2^2 \cdot d}} - \beta \right] + \frac{F_{ax,Rk}}{4} \qquad \text{mode (e)} \quad (10.5)$$

$$F_{v,Rk} = 1.15 \sqrt{\frac{2\beta}{1+\beta}} \sqrt{2M_{y,Rk} \cdot f_{h,1,k} \cdot d} + \frac{F_{ax,Rk}}{4} \qquad \text{mode (f)} \quad (10.6)$$

(*continued*)

Table 10.2 (*continued*)

Connections in double shear

Failure modes (Fastener axes to be as shown for single shear condition)	(g) $t_1 \;\; t_2 \;\; t_1$	(h)	(j)	(k)
(EYM Mode type	1	1	2	3)

Characteristic load-carrying capacity per fastener **per shear plane**, $F_{v,Rk}$ is minimum value from the mode failure equations: (EC5 *equations* (8.7))	$F_{v,Rk} = f_{h,1,k} \cdot t_1 \cdot d$	mode (g) (10.7)
	$F_{v,Rk} = 0.5 f_{h,2,k} \cdot t_2 \cdot d$	mode (h) (10.8)
	$F_{v,Rk} = 1.05 \dfrac{f_{h,1,k} \cdot t_1 \cdot d}{2+\beta}\left[\sqrt{2\beta(1+\beta)+\dfrac{4\beta(2+\beta)M_{y,Rk}}{f_{h,1,k}\cdot t_1^2 \cdot d}} - \beta\right] + \dfrac{F_{ax,Rk}}{4}$	mode (j) (10.9)
	$F_{v,Rk} = 1.15 \sqrt{\dfrac{2\beta}{1+\beta}}\sqrt{2M_{y,Rk}\cdot f_{h,1,k}\cdot d} + \dfrac{F_{ax,Rk}}{4}$	mode (k) (10.10)

* Based on 8.2.2 in EC5

Table 10.3 Characteristic load carrying capacity per fastener per shear plane for steel–timber connections*

Connections in single shear – modes (a) and (b) have thin plates and modes (c), (d), (e) have thick plates

Failure modes (Fastener axes to be as shown on Table 2)	(a)	(b)	(c)	(d)	(e)
	$t_1 \longrightarrow \longleftarrow t_{steel}$				
(EYM Mode type	1	2	1	2	3)

Characteristic load-carrying capacity per fastener per shear plane, $F_{v,Rk}$, is the minimum value from the mode failure equations: (EC5 equations (8.9), (8.10))	$F_{v,Rk} = 0.4 f_{h,k} \cdot t_1 \cdot d$	mode (a) (10.11)
	$F_{v,Rk} = 1.15\sqrt{2M_{y,Rk} \cdot f_{h,k} \cdot d} + \dfrac{F_{ax,Rk}}{4}$	mode (b) (10.12)
	$F_{v,Rk} = f_{h,k} \cdot t_1 \cdot d$	mode (c) (10.13)
	$F_{v,Rk} = f_{h,k} \cdot t_1 \cdot d \left[\sqrt{2 + \dfrac{4M_{y,Rk}}{f_{h,k} \cdot t_1^2 \cdot d} - 1} \right] + \dfrac{F_{ax,Rk}}{4}$	mode (d) (10.14)
	$F_{v,Rk} = 2.3\sqrt{M_{y,Rk} \cdot f_{h,k} \cdot d} + \dfrac{F_{ax,Rk}}{4}$	mode (e) (10.15)

(*continued*)

Table 10.3 (continued)

Joints in double shear – steel plates are of any thickness unless stated otherwise below

	(f)	(g)	(h)	(j/l)	(k)	(m)
Failure modes (Fastener axes as shown on Table 10.2)	$t_1 \rightarrow$ t_{steel} t_1			t_{steel} t_2 t_{steel}	Thin plates	Thick plates
(EYM Mode type)	1	2	3	1 2 3	2	3

Characteristic load-carrying capacity per fastener *per shear plane*, $F_{v,Rk}$, is the minimum value from the mode failure equations: (EC5 *equations (8.11)*, *(8.12)*, *(8.13)*)

$$F_{v,Rk} = f_{h,1,k} \cdot t_1 \cdot d \qquad \text{mode (f)} \quad (10.16)$$

$$F_{v,Rk} = f_{h,1,k} \cdot t_1 \cdot d \left[\sqrt{2 + \frac{4M_{y,Rk}}{f_{h,1,k} \cdot t_1^2 \cdot d}} - 1 \right] + \frac{F_{ax,Rk}}{4} \qquad \text{mode (g)} \quad (10.17)$$

$$F_{v,Rk} = 2.3\sqrt{M_{y,Rk} \cdot f_{h,1,k} \cdot d} + \frac{F_{ax,Rk}}{4} \qquad \text{mode (h)} \quad (10.18)$$

$$F_{v,Rk} = 0.5 f_{h,2,k} \cdot t_2 \cdot d \qquad \text{mode (j/l)} \quad (10.19)$$

$$F_{v,Rk} = 1.15\sqrt{2M_{y,Rk} \cdot f_{h,2,k} \cdot d} + \frac{F_{ax,Rk}}{4} \qquad \text{mode (k)} \quad (10.20)$$

$$F_{v,Rk} = 2.3\sqrt{M_{y,Rk} \cdot f_{h,2,k} \cdot d} + \frac{F_{ax,Rk}}{4} \qquad \text{mode (m)} \quad (10.21)$$

* Based on EC5, *Clause 8.2.3*.

The Johansen approach is also commonly referred to as the European Yield Model and the ductile failure modes he developed are referred to in research papers as modes type 1, 2 and 3. Mode type 1 is where failure is solely by embedment of the connection material and there is no yielding of the fastener; mode type 2 is where failure is by a combination of embedment failure in the materials and a single yield failure in the fastener and mode type 3 is where there is a combination of embedment failure and double yield failure in the fastener. The mode failures are defined in EC5 by letters (*Figure 8.2* and *Figure 8.3*) and are included on Table 10.2 and 10.3 as well as the categorisation using the European Yield Model (EYM).

The associated connection strength equations are dependent on the geometry of the connection, the embedment strength of the timber or wood-based material, the bending strength of the fastener and on the basis that the fastener will not withdraw from the connection. They are also only valid for connections in which the longitudinal axis of the fastener is effectively at right angles to member axes. When dealing with screws as the fastener, it is common for connections to be formed with the axis of the screws at an angle to the grain direction, and for such connections the strength should be derived using the EC5 rules for the axial strength of the connection, referred to in 10.8.4.

In the case of timber-to-steel connections, the strength of the steel plates must also be shown to exceed the connection strength and this should be carried out in accordance with the requirements of BS EN 1993-1-1 [6] and BS EN 1993-1-8 [7].

Further, in regard to timber-to-steel connections, where the steel plate thickness is less than or equal to $0.5 \times$ the dowel diameter (d), in EC5 the plate is classified as a thin plate and when it is equal to or greater than d and the tolerance allowance for the dowel hole is less than $0.1d$, it is classified as a thick plate. Strength equations have been derived for connections using each type of plate, and for those formed using steel plates with a thickness between these limits, the strength is obtained by linear interpolation between the limiting values based on thin and thick plate arrangements.

Since Johansen's equations were derived they have been slightly modified and added to by other researchers to enhance the connection strength, and the formulae now used in EC5 for metal dowel-type fasteners are given in Tables 10.2 and 10.3 for the relevant failure modes that can occur. The equations given for double shear connections only apply to symmetrical assemblies and if non-symmetrical arrangements are used, new equations have to be developed or approximate solutions can be used.

Connections can be formed with fasteners in single or double shear and examples of each type are shown in Figure 10.6. In the single shear connection, there is one shear plane per fastener and in the double shear connection there are two shear planes per fastener. It is important to note that the equations given in Tables 10.2 and 10.3 refer to the characteristic load-carrying capacity of a fastener per shear plane.

For connections in single shear, the characteristic load-carrying capacity per shear plane per fastener, $F_{v,Rk}$, will be the minimum value equation for the relevant single shear cases given in Tables 10.2 and 10.3. Because there is only one shear plane this value will also equate to the load-carrying capacity per fastener in the connection and the failure mode will be the mode associated with the minimum value equation, as shown in the relevant table.

For symmetrical connections in double shear, the characteristic load-carrying capacity per shear plane per fastener, $F_{v,Rk}$, will be the minimum value equation for the relevant double shear cases given in Tables 10.2 and 10.3, and the failure mode will be

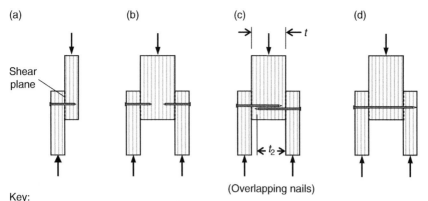

Key:
Connections (a),(b),(c) are examples of single shear with one shear plane per fastener.
Where nails are used they can overlap as shown in (c) provided: $(t-t_2) > 4 \times$ nail diameter (in mm).
Connection (d) is an example of double shear with two shear planes per fastener.
Where screws are being used, it is normal that they will only be used for single shear connections

Fig. 10.6. Metal dowel-type fasteners loaded laterally in single and double shear.

the mode associated with that equation. However, because there are two shear planes per fastener, the characteristic load-carrying capacity per fastener will be:

$$\text{Characteristic load} - \text{carrying capacity of a fastener in double shear} = 2 \cdot F_{v,Rk}$$

$$(10.22)$$

In this book the assumption has been made that the reader has access to computer software, Mathcad being used in the examples given in the book, or can prepare Excel or equivalent spreadsheets. On this basis the equations given in the Tables 10.2 and 10.3 can be readily programmed for general use to obtain the minimum strength value. Where no such facilities are available, to speed up the design process the equations can be simplified by allowing for some conservatism in the analyses, and examples of such approaches are given in [8] and [9].

The main functions used in the strength equations are the diameter of the dowel, d, the characteristic fastener yield moment, $M_{y,Rk}$ and the characteristic embedment strength, $f_{h,i,k}$, of the connected member i, and the significance of these functions is discussed in the following sections.

10.3.1   Dowel diameter

The value to be used for the diameter of a dowel-type fastener in the strength equations in Tables 10.2 and 10.3 is dependent on the type of fastener being used in the connection and is given for different types in Table 10.4.

The value used in the equations will be the nominal diameter except when dealing with screws. Differing views have been used by researchers (e.g. Blass, H.J., [10], Larsen, H.J., [11]) for calculating the embedment strength and the yield moment when using screws as the fastener, and EC5 has also not been clear on how these functions should be calculated. It has recently been accepted by the code authorities that the information relating to screws in EC5, *8.7.1* requires clarification and the requirements

Table 10.4 EC5 requirement for the diameter of a fastener

Fastener	d (mm)
Nails:	Nominal diameter:
Smooth round nails (EC5, *8.3.1.1(4)*)	The minimum outer cross-sectional diameter of the round wire nail (EN 14592) – see Figure 10.1.
Square and grooved nails (EC5, *8.3.1.1(3)*)	The side dimension of the minimum cross-section.
Threaded nails	Spiral rolled or annular ringed types - as smooth round nails; For other types, the minimum cross-sectional diameter of the original wire rod from which the profiled nail is produced (EN 14592) – see Figure 10.1.
Staples:	
With round legs (EC5, *8.4(1)*)	The leg diameter.
With a rectangular cross section (EC5, *8.4(2)*)	The square root of the product of both dimensions.
Bolts: (EC5, *8.5.1.1(2)*)	The nominal diameter is the bolt diameter.
Dowels: (EC5, *8.6(1)*)	The nominal diameter is the dowel diameter.
Screws: (EC5, *8.7.1 and Appendix C*)	
(a) General – for use as defined in EC5	(a) The nominal diameter is the maximum outer cross-section diameter of the threaded part of the screw. (see Figure 10.2 and 10.1.1.2.)
(b) When determining the embedment strength and the yield moment of the screw the **effective diameter** shall be used and its value determined as follows:	
(i) when the requirements of EC5, *8.7.1(2)* are satisfied.	(b) (i) The effective diameter, d_{ef} is the nominal diameter (i.e. the solid shank diameter).
(ii) where the conditions in (i) are not met.	(b) (ii) The effective diameter, d_{ef} is 1.1 times the thread root diameter.
(c) When determining spacings, end and edge distances and the effective number of screws.	(c) The nominal diameter as defined in (a) above.

are to be amended as defined in Appendix C. The revision will change the content of *8.7.1(1), (4), (5) and (6)* and will effectively state:

(1) When determining the load carrying capacity of a screw (e.g. when using the Johansen equations in EC5), the effect of the threaded part of the screw shall be taken into account by using an effective diameter, d_{ef}, when determining the yield capacity and the embedment strength. The value used for d_{ef} will be as defined in the existing clauses *8.7.1(2)* and *8.7.1(3)*. Also, it will be stated that

when determining spacing, edge and end distances and the effective number of screws, the outer thread diameter, d, shall be used.

(2) When the value of d_{ef} is greater than 6 mm, the rules in *8.5.1* will apply.

(3) When the value of d_{ef} is less than or equal to 6 mm, the rules in *8.3.1* will apply.

(4) The current (*6*) will be revised as stated in the Appendix.

In the Johansen equations, the value used for d will be d_{ef}.

The UK view is that the proposed change will not adequately clarify points of detail and a more detailed statement of requirement has been incorporated into PD6693-1 [12].

10.3.2   Characteristic fastener yield moment ($M_{y,Rk}$)

In Johansen's original equations, the yield moment was taken to be the moment at the elastic limit of the fastener and was derived from the product of the yield strength and the elastic modulus of the fastener. This gave a lower bound strength value and in the subsequent development of his theory by other researchers the elasto-plastic strength has been used. This takes account of the amount of rotation at the failure state for different types of fastener; the tensile strength of the fastener, including the effect of strain hardening where relevant and of variation in material strength.

From these investigations the characteristic yield moment, $M_{y,Rk}$, for the different types of metal fasteners referred to in EC5 is summarised in Table 10.5. Because of strain hardening and the varying effects of profiling associated with threaded nails, $M_{y,Rk}$ for these types cannot be derived by calculation and testing is required.

When using screws the yield moment of the solid shank of the screw will be different to the yield moment of the threaded shank and to take this into account when calculating the yield moment of the screw you are told in EC5 to use the effective diameter, d_{ef}, of the screw. Where yielding occurs in the solid shank length, d_{ef} = the smooth shank diameter (d) and when it occurs in the threaded portion, d_{ef} = 1.1 times the thread root diameter, as stated in Table 10.4. For mode 3 type failure, there is the possibility of one hinge forming in the solid shaft zone and the other in the threaded length and for these types of failure the yield moment in the threaded part of the screw will be smaller than the solid shank section. The rules in EC5 require that d_{ef}, based on 1.1 times the threaded root diameter, is used in the relevant strength equations, resulting in a slight conservatism in the fastener strength. This condition will generally only arise when using smooth shank screws and to assist in determining which value of d_{ef} should be used in such cases EC5 states that providing the smooth shank penetrates into the member containing the point of the screw by not less than $4d$, then d_{ef} = the smooth shank diameter, i.e. yielding of the screw for modes 2 or 3 failures will occur in the solid shank. If it penetrates less than $4d$, the yield moment is to be calculated using d_{ef} = 1.1 times the thread root diameter i.e. yielding will occur in the threaded shank length. From an analysis of the Johansen equations for mode 2 and 3 failure equations where the smooth shank penetrates into the member containing the point of the screw by $4d$, it can be shown there are cases where yield occurs in the threaded

Table 10.5 EC5 requirements for determining the characteristic yield moment, $M_{y,Rk}$

Fastener	$M_{y,Rk}$ (N mm)	
Nails:		
Smooth round nails (EC5 *equation (8.14)*)	$0.3f_u d^{2.6}$	(10.23)
Smooth square and grooved nails (EC5 *equation (8.14)*)	$0.45f_u d^{2.6}$	(10.24)
Threaded nails	As stated in BS EN 14592, by testing in accordance with BS EN 409 [13].	
Staples (EC5 *equation (8.29)*)	$240d^{2.6}$ (the proposed change in Appendix C is $150d^3$)	(10.25)
Bolts (EC5 *equation (8.30)*)	$0.3f_{u,k} d^{2.6}$	(10.26)
Dowels (EC5, *8.6.1*)	As for bolts	
Screws:		
effective diameter $d_{ef} \le 6$ mm	as for nails	
effective diameter $d_{ef} > 6$ mm	as for bolts	

Where:
d is the nominal diameter of the nail, bolt, staple or dowel in mm. For staples with rectangular sections, d is the square root of the product of the leg dimensions and for screws d_{ef} is the effective diameter referred to in Table 10.4;
f_u is the tensile strength of the nail wire (or screw material), in N/mm²;
$f_{u,k}$ is the characteristic tensile strength of the bolt (or screw material), in N/mm².

shank length beyond this position, particularly when thick steel gusset plates are being used, however, the $4d$ rule in EC5 is to be followed.

With staples it has been concluded that EC5 *equation (8.29)* is wrong and, as stated in Appendix C and noted on Table 10.5, the equation is to be revised.

10.3.3 Characteristic embedment strength ($f_{h,k}$)

The embedment strength of timber or a wood-based product, f_h, is the average compressive strength of the timber or wood-based product under the action of a stiff straight dowel loaded as shown in Figure 10.7. For a piece of timber t (mm) thick,

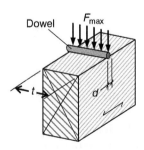

Fig. 10.7. Embedment strength of timber or wood-based material.

loaded with a nail d (mm) in diameter, under the maximum load able to be taken, $F_{\max}$ (N), the embedment strength is defined as:

$$f_{h} = \frac{F_{\max}}{d \cdot t} \quad (\text{N/mm}^2)$$
(10.27)

The embedment is the distance the dowel depresses the timber or wood-based product. Because of the complex cellular nature of timber and wood-based products, the embedment strength is not a material property. It is a system property dependent on several factors, including the type of fastener being used. From the results of investigations by various researchers, the value derived for the characteristic embedment strength of timber and wood products, $f_{h,k}$, given in EC5 when using different types of metal dowel fasteners is summarised in the following sub-sections.

10.3.3.1   *Characteristic embedment strength when using nails (diameter ≤8mm)*
The embedment strength varies depending on the diameter of nail being used, the types of material used in the connection, and whether or not pre-drilling is adopted. Embedment strengths for the possible variations arising in design are given in Table 10.6 and when using threaded nails, as for smooth nails, the nominal diameter shall be used.

10.3.3.2   *Characteristic embedment strength when using staples*
The embedment strength is derived using the expressions given in Table 10.6 for nails and where the staple has a rectangular cross-section the diameter d should be taken as the square root of the product of its leg dimensions.

Table 10.6 Characteristic embedment strength when using nails

Condition	$f_{h,k}$ (N/mm²)	
For timber and for LVL connections using nails up to 8 mm diameter:		
Without predrilled holes (EC5 *equation (8.15)*)	$f_{h,k} = 0.082 \rho_k d^{-0.3}$	(10.28)
With predrilled holes (EC5 *equation (8.16)*)	$f_{h,k} = 0.082(1 - 0.01d) \rho_k$	(10.29)
For timber and for LVL connections using nails greater than 8 mm diameter:	The expressions in Table 10.7 for bolts will apply.	
For panel to timber connections with nails having a head diameter of at least 2d and where the panel material is:		
Plywood (EC5 *equation (8.20)*):	$f_{h,k} = 0.11 \rho_k d^{-0.3}$	(10.30)
Hardboard (in accordance with EN 622-2 [14]) (EC5 *equation (8.21)*);	$f_{h,k} = 30 d^{-0.3} t^{0.6}$	(10.31)
Particle board or O.S.B. (EC5 *equation (8.22)*).	$f_{h,k} = 65 d^{-0.7} t^{0.1} \text{ N/mm}^2$	(10.32)

Where:
d is the nominal diameter of the nail, in mm,
ρ_k is the characteristic density of the timber, LVL or panel material, in kg/m³,
t is the panel thickness (or the length of the nail in the member if less than the panel thickness), in mm.

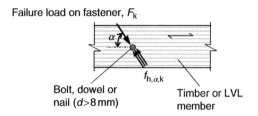

Fig. 10.8. Embedment strength of a bolt, dowel or nail ($d > 8$ mm) fastener loaded at an angle α to the grain.

10.3.3.3 Characteristic embedment strength when using bolts, nails (diameter > 8 mm) and dowels

When using timber to timber or LVL, the embedment strength of bolts, nails (with a diameter greater than 8 mm) and dowels is dependent on the direction of the applied load relative to the grain and for such fasteners the embedment strength is determined by using Hankinson's equation [15], referred to in 5.3.2.

Applying the Hankinson equation to determine the characteristic embedment strength, $f_{h,\alpha,k}$, when the fastener is loaded at an angle α to the grain, as shown in Figure 10.8, the equation can be written as:

$$f_{h,\alpha,k} = \frac{f_{h,0,k} \cdot f_{h,90,k}}{f_{h,0,k} \sin^2\alpha + f_{h,90,k} \cos^2\alpha} \tag{10.33}$$

where $f_{h,\alpha,k}$ is the characteristic embedment strength at an angle α to the grain, $f_{h,0,k}$ is the characteristic embedment strength parallel to the grain, and $f_{h,90,k}$ is the characteristic embedment strength perpendicular to the grain.

In EC5, the above equation is reduced to:

$$f_{h,\alpha,k} = \frac{f_{h,0,k}}{k_{90} \sin^2\alpha + \cos^2\alpha} \qquad\text{(EC5, equation (8.31))} \tag{10.34}$$

where $k_{90} = f_{h,0,k}/f_{h,90,k}$ and values for softwood, LVL and hardwoods are given in Table 10.7.

For bolted panel-to-timber connections, the embedment strength in the panel material is the same for all angles of load resulting in factor k_{90} equalling unity and equation (10.34) simplifying to the embedment strength of the panel material.

The characteristic embedment strength for bolts and dowels from 6 mm up to 30 mm diameter, nails greater than 8 mm diameter, in timber, LVL and panel product connections, is given in Table 10.7.

10.3.3.4 Characteristic embedment strength when using screws

As stated in 10.3.1 the characteristic embedment strength is derived using the effective diameter, d_{ef}, of the screw (also refer to 10.3.2 and Table 10.4) rather than the nominal diameter and is calculated as follows:

- when $d_{ef} \leq 6$ mm – as for nails (see Table 10.6) – (where $d = d_{ef}$ in all relevant embedment equations);
- when $d_{ef} > 6$ mm – as for bolts (see Table 10.7) – (where $d = d_{ef}$ in *equations (8.32) and (8.33)*.

Table 10.7 Characteristic embedment strength using bolts, dowels, nails (greater than 8 mm in diameter) or dowels

Condition	$f_{h,k}$ (N/mm²)	
For timber and LVL:		
Loaded parallel to grain (EC5 *equation (8.32)*)	$f_{h,0,k} = 0.082(1 - 0.01d)\rho_k$	(10.35)
Loaded at an angle α to grain (EC5 *equation (8.31)*)	$f_{h,\alpha,k} = \dfrac{f_{h,0,k}}{k_{90}\sin^2\alpha + \cos^2\alpha}$	(10.36)
For panel to timber loaded at any angle to the face grain where the panel material is plywood, the characteristic embedment strength in the panel material will be: (EC5 *equation (8.36)*)	$f_{h,\alpha,k} = f_{h,k} = 0.11(1 - 0.01d)\rho_k$	(10.37)
For panel to timber loaded at any angle to the face grain where the panel material is either particle board or O.S.B, the characteristic embedment strength in the panel material will be: (EC5 *equation (8.37)*)	$f_{h,\alpha,k} = f_{h,k} = 50d^{-0.6}t^{0.2}$	(10.38)

Where:
d is the nominal diameter of the fastener;
ρ_k is the characteristic density of the material being considered;
t is the thickness of the panel material (or the length of the fastener in the member if less than the panel thickness, e.g. when using dowels);
α is the angle of the load in the fastener relative to the grain.
k_{90} $=(1.35 + 0.015d)$ for softwoods; $(1.3 + 0.015d)$ for LVL; $(0.9 + 0.015d)$ for hardwoods.

This means that for screws having an effective diameter of 6 mm or less the embedment strength in a timber connection will be the same irrespective of the angle of load relative to the grain direction and will have the same value as a connection of equivalent configuration formed using nails with the same nominal diameter. However, when $d_{ef} > 6$ mm, the embedment strength in timber will be a function of the angle of load relative to the grain and unless the load is being applied along the grain, the embedment strength will be smaller than that obtained from a similarly configured nailed connection formed from nails of equivalent diameter.

When using smooth shank screws, as explained in 10.3.2, d_{ef} will equal the diameter of the solid shank of the screw if the smooth shank penetrates into the member containing the point of the screw by not less than $4d$. Where this criterion is not met, d_{ef} is to be taken as 1.1 times the thread root diameter, which will result in a smaller value and a lower embedment strength. The rules in EC5 require that this value is used to derive the embedment strengths of the connection materials, and will result in conservatism when evaluating some mode strengths. This will be clear when considering mode 1 type failures in timber to timber and certain panel to timber connections. If such modes in a connection are shown to be critical and an increased strength is required, providing the solid shank of the smooth shank screw extends beyond member 1 into member 2, the embedment strength for member 1 derived from d_{ef} based on the solid shank diameter can theoretically be used.

10.3.4   Member thickness, t_1 and t_2

In a connection the members are classified as member 1 and member 2 as shown in Tables 10.2 and 10.3 and defined in the following sub-sections.

10.3.4.1 t_1 and t_2 for a nail connection
t_1 is:

- the nail headside material thickness where the connection is in single shear;
- the minimum of the nail headside material thickness and the nail pointside penetration in a double shear connection.

t_2 is:

- the nail pointside penetration where the connection is in single shear;
- the central member thickness for a connection in double shear.

'Nail headside material thickness' is the thickness of the member containing the nail head and 'nail pointside thickness' is the distance the pointed end of the nail penetrates into a member.

10.3.4.2 t_1 and t_2 for a staple connection
t_1 and t_2 are as shown in Figure 10.9.

10.3.4.3 t_1 and t_2 for a bolt connection
t_1 is the bolt headside member thickness where the connection is in single shear or double shear (assuming the connection is symmetrical).

t_2 is the bolt threaded end member thickness when the connection is in single shear and the central member thickness when in double shear.

Where the bolt (head and/or nut) is recessed into timber members the relevant value of t_1 and/or t_2 will equal $(t - t_r)$, where t is the member thickness and t_r is the depth of the recess. This equates to the length of bolt bearing against the connection material.

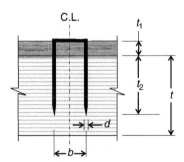

Fig. 10.9. Staple dimension.

10.3.4.4 t_1 and t_2 for a dowel connection

t_1 is:

- the dowel length in member 1 when the connection is in single shear;
- the minimum dowel length in the outer members when the connection is in double shear.

t_2 is:

- the dowel length in member 2 when the connection is in single shear;
- the central member thickness for a connection in double shear.

10.3.4.5 t_1 and t_2 for a screw connection

The definitions given for nails will apply.

10.3.5   Friction effects and axial withdrawal of the fastener

The basic Johansen yield equation for each failure mode for connections in single or double shear can be derived by the use of a static analysis (as shown in STEP 1 [16]) or by the virtual work approach commonly used in the plastic analysis of steel structures [17]. To simplify the equations, the ratio of the characteristic embedment strength of member 2 in the connection, $(f_{h,2,k})$, to the characteristic embedment strength of member 1, $(f_{h,1,k})$, is used and written as:

$$\beta = \frac{f_{h,2,k}}{f_{h,1,k}} \qquad\qquad (EC5,\ equation\ (8.8))\quad (10.39)$$

In deriving these equations, friction forces between the members of the connection are ignored as well as the withdrawal resistance of the fasteners. In EC5 the Johansen yield equations form the basis of the strength equations, however for those failure modes that involve yielding of the fastener, the equations have, where relevant, been modified to include for friction and withdrawal effects.

There are two types of friction effects that can arise in a connection. One will develop if the members are in contact on assembly and the other will arise when the fasteners yield and pull the members together when the fasteners deform under lateral load. The former type of friction will be eliminated if there is shrinkage of the timber or wood products when in service and because of this it is not included for in the EC5 strength equations. The latter type of friction will, however, always arise in failure modes that involve yielding of the fasteners and this has been included for in the EC5 equations relating to such modes.

Consider, for example, a single shear connection formed with a plywood gusset plate and a timber member connected by a single dowel-type fastener as shown in Figure 10.10. Assume that under the lateral shear force on the joint the fastener yields in the gusset plate and in the timber member allowing it to rotate by an angle θ, as shown, and that the coefficient of friction between the gusset plate and the timber is μ. In addition to being subjected to bending, the fastener will be subjected to a tension force N_d due to the withdrawal effect during loading. Force N_d will have a vertical component, $N_d \sin \theta$, and a horizontal component, $N_d \cos \theta$, the latter compressing the

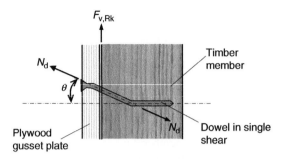

Fig. 10.10. Connection with dowel-type fastener in single shear at the failure condition.

gusset plate onto the timber and inducing an additional vertical resistive force, $\mu\,N_d$ cos θ, due to friction. The force in the fastener, $F_{v,Rk}$, will equate to the sum all of the vertical forces in the connection as follows:

$$F_{v,Rk} = N_d\,(\sin\theta + \mu\cos\theta) + \text{Johansen's yield load for the joint } (F_{y,Rk})$$

$$(10.40)$$

At the failure condition, N_d will be the withdrawal capacity of the fastener and in EC5 the component $N_d \sin\theta$ is taken to be $F_{ax,Rk}/4$, where $F_{ax,Rk}$ is the fastener's characteristic withdrawal capacity. Component $N_d\mu\cos\theta$ is equated to a percentage of $F_{y,k}$, the Johansen yield load. Taking these effects into account, the characteristic lateral load carrying capacity of a fastener, $F_{v,Rk}$, in EC5 is written in the following format:

$$F_{v,Rk} = \text{friction factor} \times \text{Johansen yield load} + (\text{withdrawal capacity}/4)$$

In EC5 the values used for the friction factor are 5% where the fastener partially yields (e.g. modes (d) and (e) in Table 10.2) and 15% where the fastener fully yields (e.g. mode (f) in Table 10.2). The use of these factors can be seen in equations (10.4)–(10.6), (10.9), (10.10) in Table 10.2 and in equations (10.12) and (10.15), (10.18), (10.20), (10.21) in Table 10.3. In equations (10.15), (10.18) and (10.21) in Table 10.3 the numerical coefficient incorporates a factor of 1.15 for this effect. There are however some type 2 modes associated with steel–timber connections where the prior risk of pointside pull out effects offset the possibility of friction pick-up and a friction factor is not permitted in such cases (e.g. mode (d) and (g) in Table 10.3).

To discriminate between the Johansen yield load and the combined withdrawal and friction forces in a connection, the latter are commonly referred to as the *rope effect* forces, however in EC5 reference is only made to the term $F_{ax,Rk}/4$ as the contribution from this effect.

As previously stated, $F_{ax,Rk}$ is the characteristic axial withdrawal capacity of the fastener and is defined in EC5, *8.3.2*, for nails and is also applicable to staples. It is defined in EC5, *8.5.2*, for bolts and *8.7.2* for screws, and is discussed in 10.8. For those fasteners that are potentially susceptible to withdrawal, the minimum penetration permitted in timber is specified in the following sub-sections.

10.3.5.1 *Minimum penetration when using nails (EC5, 8.3.1.2 and 8.3.2)*

1. Smooth nails – the minimum pointside penetration (i.e. the penetration of the pointed end of the nail into the timber) is 8d, however at this value the pointside withdrawal capacity of the nail is taken to be zero. Where the pointside penetration is at least 12d, the full characteristic value of the withdrawal strength given in EC5, *equation (8.25)*, can be used and between 8d and 12d the withdrawal strength should be multiplied by $(t_{pen}/4d-2)$, where t_{pen} is the pointside penetration length, as discussed in 10.8.1.

2. Other nails (threaded nails as defined in BS EN 14592) – the minimum pointside penetration of the threaded part is 6d and at this value the pointside withdrawal capacity of the nail is taken to be zero. With these nails, when the pointside penetration is at least 8d the full characteristic value of the withdrawal strength can be used and between 6d and 8d the withdrawal strength should be multiplied by $(t_{pen}/2d-3)$, as discussed in 10.8.1.

3. For nails in end grain special rules apply, as given in EC5, *8.3.1.2(4) and 8.3.2(3)*.

10.3.5.2 *Minimum penetration when using staples (EC5, 8.4 (3))*

1. The minimum pointside penetration (see dimension t_2 in Figure 10.5) is 14×the staple diameter.

10.3.5.3 *Minimum penetration of screws (EC5, 8.7.2(3))*

1. The minimum pointside penetration length of the *threaded part* of the screw must be 6 times the outer diameter of the screw.

In EC5, *8.2.2(2)*, an upper limit is also set for the value of $F_{ax,Rk}/4$. It is taken to be a percentage of the first term of the relevant strength equations given in Tables 10.2 and 10.3 (i.e. a percentage of the Johansen yield load ($F_{y,k}$) enhanced by the friction factor associated with the rope effect) as follows:

$$perentage \leq \begin{cases} 15\% & \text{Round} & \text{nails} \\ 25\% & \text{Square} & \text{nails} \\ 50\% & \text{Other} & \text{nails} \\ 100\% & \text{Screws} \\ 25\% & \text{Bolts} \\ 0\% & \text{Dowels} \end{cases}$$

The maximum percentage increase is dependent on the type of fastener being used and screws will achieve the greatest enhancement. In single shear connections when using nails, $F_{ax,Rk}$ will be the lower of the fastener head pull-through strength (including the withdrawal strength associated with the headside penetration of the fastener) and the pointside withdrawal strength, which are discussed in 10.8.1. When dealing with bolts, the resistance provided by the washers, which is defined in EC5, *10.4.3*, should be taken into account.

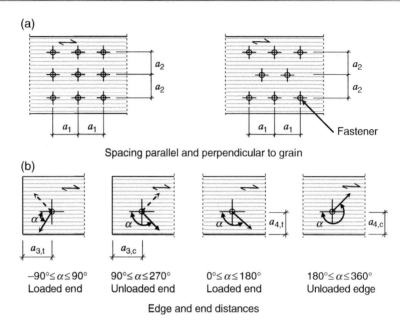

Fig. 10.11. Fastener spacings and distances.

10.3.6   Brittle failure

The EC5 strength equations in Tables 10.2 and 10.3 are only valid if there is no premature splitting or shearing of the timber resulting in a brittle-type failure. To try and eliminate the risk of such failures minimum edge, end and spacing criteria for use with dowel-type fasteners have been developed from testing programmes and the requirements for nails, staples, bolts, dowels and screws are given in Tables 10.8 and 10.9. The spacings and distances referred to in these tables are as shown in Figure 10.11. Also, to prevent splitting in timber when using nails or screws greater than 6 mm nominal diameter, pre-drilling must be used and the particular require-ments for these fasteners are given in *Section 10* of EC5. It is to be noted that pre-drilling may also be necessary to allow fixings to be formed in dense timber and this is a requirement for nails driven in timber with a characteristic density greater than 500 kg/m³ (EC5, *8.3.1.1*) and for all screws in hardwood (EC5, *10.4.5*). The above and other related requirements in EC5 are summarised in the notes accompany-ing Tables 10.8 and 10.9.

EC5 rules for spacing, edge and end distances for joints formed with nails and/or staples are summarised in Table 10.8 and the following requirements also apply.

Additional requirements in EC5
(a) Laterally loaded nailed connections

General
8.3.1.1(2) – the timber should be pre-drilled when

- its characteristic density > 500 kg/m³;
- the nail diameter exceeds $6d$.

Table 10.8 Minimum spacings and edge and end distances using nails and/or staples of diameter d for timber-to-timber connections[*]

	Minimum spacing or distance			
	Nails (see Figure 10.11)			Staples (see Figures 10.11 and 10.17)
	Without pre-drilled holes		With pre-drilled holes	
Spacing or distance and α[†]	$\rho_k \leq 420\,\text{kg/m}^3$	$420\,\text{kg/m}^3 < \rho_k \leq 500\,\text{kg/m}^3$		
Spacing parallel to the grain – a_1 $0° \leq \alpha \leq 360°$	$d < 5$ mm: $(5+5|\cos\alpha|)d$ $d \geq 5$ mm: $(5+7|\cos\alpha|)d$	$(7+8|\cos\alpha|)d$	$(4+|\cos\alpha|)d$	for $\theta \geq 30°$: $(10+5|\cos\alpha|)d$ for $\theta < 30°$: $(15+5|\cos\alpha|)d$
Spacing perpendicular to the grain – a_2 $0° \leq \alpha \leq 360°$	$5d$	$7d$	$(3+|\sin\alpha|)d$	$15d$
Loaded end distance $a_{3,t}$ $-90° \leq \alpha \leq 90°$	$(10+5\cos\alpha)d$	$(15+5\cos\alpha)d$	$(7+5\cos\alpha)d$	$(15+5|\cos\alpha|)d$
Unloaded end $a_{3,c}$ $90° \leq \alpha \leq 270°$	$10d$	$15d$	$7d$	$15d$
Loaded edge distance $a_{4,t}$ $0° \leq \alpha \leq 180°$	$d < 5$ mm: $(5+2\sin\alpha)d$ $d \geq 5$ mm: $(5+5\sin\alpha)d$	$d < 5$ mm: $(7+2\sin\alpha)d$ $d \geq 5$ mm: $(7+5\sin\alpha)d$	$d < 5$ mm: $(3+2\sin\alpha)d$ $d \geq 5$ mm: $(3+4\sin\alpha)d$	$(15+5|\sin\alpha|)d$
Unloaded edge distance $a_{4,c}$ $180° \leq \alpha \leq 360°$	$5d$	$7d$	$3d$	$10d$

[*] Based on *Tables 8.2* and *8.3* in EC5.
[†] α is the angle between the direction of the nail force and the grain.

Table 10.9 Minimum spacings and edge and end distances for bolts and/or dowels in timber-to-timber, panel-to-timber and steel-to-timber connections[*]

Spacing or distance (see Figure 10.11) and α[†]	Minimum spacing or distance	
	Bolts	Dowels
Spacing parallel to the grain: a_1 $0° \leq \alpha \leq 360°$	$(4+\mid\cos\alpha\mid)d$	$(3+2\mid\cos\alpha\mid)d$
Spacing perpendicular to the grain: a_2 $0° \leq \alpha \leq 360°$	$4d$	$3d$
Loaded end distance: $a_{3,t}$ $-90° \leq \alpha \leq 90°$	max $(7d; 80\,\text{mm})$	max $(7d; 80\,\text{mm})$
Unloaded end distance: $a_{3,c}$ $90° \leq \alpha < 150°$ $150° \leq \alpha < 210°$ $210° \leq \alpha \leq 270°$	max $(1+6\sin\alpha)d$ $4d$ max $(1+6\sin\alpha)d$	$(a_{3,t}\mid\sin a\mid)$ max$(3.5d; 40\,\text{mm})$ $(a_{3,t}\mid\sin a\mid)$
Loaded edge distance: $a_{4,t}$ $0° \leq \alpha \leq 180°$	max $[(2+2\sin\alpha)d; 3d]$	max $[(2+2\sin\alpha)d; 3d]$
Unloaded edge distance: $a_{4,c}$ $180° \leq \alpha \leq 360°$	$3d$	$3d$

[*] Based on *Tables 8.4* and *8.5* in EC5 and Appendix C (for dowels)
[†] α is the angle between the direction of the nail force and the grain.

Nailed timber-to-timber connections
8.3.1.2(6) – when the thickness of the timber members is smaller than t derived from EC5, *equation (8.18)*, given below, the members should be pre-drilled:

$$t = \max \left\{ \begin{array}{l} 7d \\ (13d - 30)\rho_k/400 \end{array} \right\} \qquad \text{(EC5, } equation\ (8.18)\text{)}$$

where t is the thickness of timber member to avoid pre-drilling (in mm), ρ_k is the characteristic timber density (in kg/m³) and d is the nail diameter (in mm).
8.3.1.2(7) – the requirement of *NA.2.9* in the UKNA to EC5 [18] is that this rule should not be applied to nailed joints.

(b) Nailed panel-to-timber connections
8.3.1.3(1) – minimum nail spacings for all nailed panel-to-timber connections are those given in Table 10.8 multiplied by a factor of 0.85. The end/edge distances remain unchanged unless specifically defined in EC5.
8.3.1.3(2) – minimum edge distances and end distances in plywood members should be taken as $3d$ for an unloaded edge (or end) and $(3+4\sin\alpha)d$ for a loaded edge (or end), where α is the angle between the direction of the load and the loaded edge (or end).

(c) Nailed steel-to-timber connections
8.4.1.4(1) – minimum nail spacings are those given in Table 10.8 multiplied by a factor of 0.7. The minimum edge and end distances remain unchanged.

EC5 rules for spacing, edge and end distances for joints formed with bolts and/or dowels are summarised in Table 10.9.

EC5 rules for spacing, edge and end distances for joints formed with screws (EC5, *8.7.1*) are as follows:

- For smooth shank screws with a nominal diameter greater than 6 mm, the rules for bolts will apply.
- For smooth shank screws with a nominal diameter d of 6 mm or less, the rules for nails given in Table 10.8 will apply.

Where multiple dowel-type connections near the end of the timber member in a steel-to-timber connection are loaded with a force component acting parallel to the grain in a loaded end configuration, there is a risk of a brittle-type failure due to block shear and plug shear. This type of failure is referred to in EC5, *8.2.3(5)*, and in *Annex A (informative)*, and the guidance in *NA.3.1* of the UKNA to EC5 is that the annex should only be used in connections where:

- there are ten or more metal dowel-type fasteners with a diameter ≤6 mm in line parallel to the grain;
- there are five or more metal dowel-type fasteners with a diameter >6 mm in line parallel to the grain.

In the approach in *Annex A* the characteristic strength of the connection will be the greater of the shear strength and the tensile strength of the timber around the perimeter of the fastener area. These failure modes are independent of each other and the greater strength dictates the connection strength and mode of failure. A requirement of EC5, *10.4.1* is that defects are limited in the connection region so that the load-carrying capacity of the connection is not reduced and on this basis the tensile strength of the timber can be taken to be relatively defect free and higher than that used in BS EN 338. In *equation (A1)* a factor of 1.5 is used when determining the tensile strength. For the shear strength condition there is no stated requirement to include for the effect of cracks in the shear area, which could be argued to be relevant around the fastener perimeter zone. When applying the brittle strength equations given in *Annex A* it should also be noted that when dealing with a plug type failure based on the type 2 failure modes (*d*) or (*g*), referred to in the annex under *equation (A7)*, the equation has been incorrectly derived from *equations (8.10)(d)* and *8.10(g)* and, as stated in Appendix C in the book, should read:

$$t_{ef} = t_1 \left[\sqrt{2 + \frac{4M_{y,Rk}}{f_{h,k}dt_1^2}} - 1 \right]$$

Where it is necessary to check a connection for block shear or plug shear-type failure it is good design practice to try and ensure the ductile strength of the connection (i.e. the strength derived from applying the relevant Johansen equations in EC5, *8.2.3(3)*) will be less than the fracture strength obtained from the equations in EC5, *Annex A*.

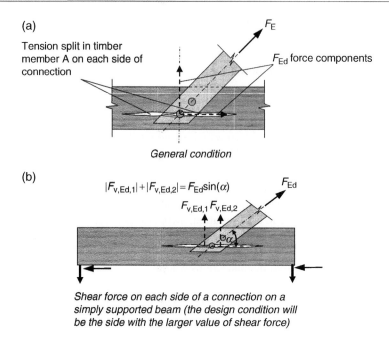

(a)

Tension split in timber member A on each side of connection

General condition

F_E

F_{Ed} force components

(b)

$$|F_{v,Ed,1}| + |F_{v,Ed,2}| = F_{Ed}\sin(\alpha)$$

$F_{v,Ed,1}$ $F_{v,Ed,2}$

F_{Ed}

Shear force on each side of a connection on a simply supported beam (the design condition will be the side with the larger value of shear force)

Fig. 10.12. Member loaded in tension at an angle to the grain.

10.3.6.1 *Brittle failure due to connection forces at an angle to the grain*

This form of brittle failure can arise when fasteners apply a force at an angle to the timber grain resulting in the possibility of splitting caused by the force component perpendicular to the grain, as shown in Figure 10.12. This failure can occur at a load less than the design capacity of the fasteners.

Preliminary issues of Eurocode 5 were written on the premise that by using the minimum spacings, end and edge distances given in the code, splitting would be prevented and joint failure would always be by ductile failure of the fasteners. The strength of the timber was based solely on a check of the shear strength of the member. This was later recognised as an incorrect representation of the failure behaviour of the timber and was replaced by a design check on the splitting resistance of the connection members where the forces in the fasteners were able to induce a tension component perpendicular to the grain in the connection members.

The strength equation in EC5 has been developed from the application of linear elastic fracture mechanics to a splitting failure mechanism formed in a timber member in a connection loaded at an angle to the grain of the member, and the design requirements are covered in the strength expressions given in EC5, *8.1.4*. An outline of the method together with references to the fracture mechanics model from which the EC5 strength equation has been derived is given in reference [19].

When a member is subjected to a lateral force perpendicular to its grain through a dowel-type connection the largest tension force in the member will occur at the dowel(s) positioned furthest from the loaded edge (e.g. position h_e on Figure 10.13a). This is the position where splitting of the member can arise and the strength relationships in *8.1.4* have been developed on the basis that cracks will spread along the grain rather than perpendicular to the grain. If a connection occurs within a member span,

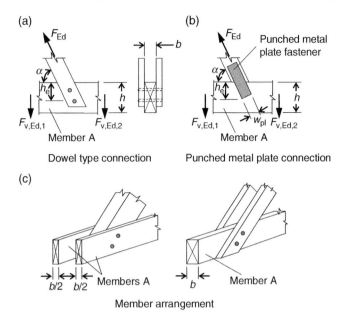

Fig. 10.13. Inclined force transmitted by a connection.

as indicated on Figure 10.12b, the design lateral force on each side of the connection will equate to the design shear force at the respective positions, i.e. $F_{v,Ed,1}$ and $F_{v,Ed,2}$ arising from the component of the splitting force perpendicular to the grain (i.e. $F_{ed}\sin(\alpha)$) and splitting failure will occur when the larger of these shear forces exceeds the design value of characteristic splitting capacity of the member. The characteristic splitting capacity equation (*equation (8.4)*) is the force required to cause splitting failure on *one side* of the member in the connection.

Based on the above, the design procedure requirement in EC5 is:
At the ULS,

$$F_{v,Ed} \leq F_{90,Rd} \tag{10.41a}$$

and

$$F_{v,Ed} = \max\left(F_{v,Ed,1}, F_{v,Ed,2}\right) \tag{10.41b}$$

where:

- $F_{90,Rd}$ is the design splitting capacity, calculated from:

$$F_{90,Rd} = k_{mod}\frac{F_{90,Rk}}{\gamma_M} \tag{10.41c}$$

  and $F_{90,Rk}$ for softwood is defined below.
- $F_{v,Ed,1}$ and $F_{v,Ed,2}$ are the design shear forces as shown in Figure 10.12b and, in more detail, on Figure 10.13.

Timber trusses are normally fabricated using softwood species and *equation (8.4)* has been calibrated against test data from connections in which the splitting member is softwood. On this basis, as indicated in EC5 *8.1.3(3)* the characteristic splitting capacity of softwood member(s) A, shown in Figure 10.13, is:

$$F_{90,\mathrm{Rk}} = 14bw\sqrt{\frac{h_e}{(1-(h_e/h))}} \qquad \text{(EC5, equation (8.4))} \quad (10.41\mathrm{d})$$

where $w=\max ((w_{pl}/100)^{0.35},\ 1)$ for punched metal plate fasteners and $w=1$ for all other fasteners, $F_{90,\mathrm{Rk}}$ is the characteristic splitting capacity (in N), w is a modification factor, h_e is the distance from the most distant connector or the edge of the punched metal plate fastener to the loaded edge (in mm), h is the member height (in mm), b is the member thickness (in mm), and w_{pl} is the width of the punched metal plate fastener parallel to the grain (in mm). It is important to stress that equation (10.41d) can only be used in connections formed using softwood members.

10.4   MULTIPLE DOWEL FASTENERS LOADED LATERALLY

In EC5, a number of fasteners lying along a line running parallel to the grain direction, as shown in Figure 10.14, are referred to as a *row of fasteners* parallel to the grain.

Where there is only a single fastener in the row, based on ductile failure the design strength of the row per shear plane will be the design lateral load carrying capacity of the fastener per shear plane and where there are r such rows, the design strength of the connection parallel to the grain per shear plane will be:

Design lateral capacity of connection per shear plane

= $r \times$ Design lateral capacity of the fastener per shear plane

Where there is more than one fastener per row parallel to the grain the strength of the row depends on the stiffness of the fastener and the strength of the bedding material and in general the stiffer the fastener the greater the design strength of the row. Several researchers have investigated this effect, and the effective characteristic load-carrying capacity of a row of fasteners parallel to the grain, $F_{v,\mathrm{ef,Rk}}$, in EC5 is:

$$F_{v,\mathrm{ef,Rk}} = n_{\mathrm{ef}} F_{v,\mathrm{Rk}} \qquad \text{(EC5, equation (8.1))} \quad (10.42)$$

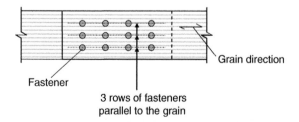

Fig. 10.14. Rows of fasteners.

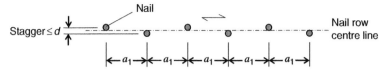

A row formed using single nails with a stagger $\leq d$
(or $2d$ if overlapping nails are used)

Fig. 10.15. Nails in a row (nail spacing parallel to the grain – see Table 10.8).

where $F_{v,ef,Rk}$ is the effective characteristic lateral load carrying capacity per shear plane of one row of fasteners parallel to the grain, n_{ef} is the effective number of fasteners per shear plane in the row parallel to the grain, and $F_{v,Rk}$ is the characteristic lateral load carrying capacity per shear plane of the fastener type being used.

10.4.1   The effective number of fasteners

The effective number of fasteners in a connection is dependent on the type of fastener and the direction of loading relative to the grain, and is covered in the following sub-sections.

10.4.1.1   *Nails*

(a) Loaded parallel to the grain.
Where non-overlapping nails are staggered (i.e. are out of alignment) in a row, providing the stagger distance, as shown in Figure 10.15, does not exceed the nail diameter, all of the nails can be considered to be in the same row with a nail spacing of a_1. When the stagger exceeds the nail diameter, two separate rows will be formed (each with a nail spacing of $2a_1$), but, this will not be a configuration that will comply with the rules in EC5. To function as two rows, EC5 requires that the 'stagger' distance between the rows will be compliant with the minimum spacing distance a_2 as defined in EC5, *Table 8.2*. It is better to think of the 'd' requirement shown in Figure 10.15 as the maximum tolerance the fabricator/assembler will be permitted to form a row of nails of spacing a_1 and if 'd' is exceeded the detail will be non-compliant. This information should be passed to the fabricator/assembler as part of the joint specification requirement.

For nails in a row loaded parallel to the grain in a connection:

- when using single nails in single or double shear:

$$n_{ef} = n^{k_{ef}}$$ (EC5, *equation (8.17)*)   (10.43)

- when using overlapping nails:

$$n_{ef} = n_p^{k_{ef}}$$ (10.44)

Here

- n_{ef} is the effective number of nails in a row parallel to the grain;
- n is the number of nails in a row parallel to the grain (for single nails);

Table 10.10 Values for exponent k_{ef} in equations (10.43) and (10.44)*

Spacing†	k_{ef}	
	Pre-drilled	Not pre-drilled
$a_1 = 14d$	1.0	1.0
$a_1 = 12d$	0.925	0.925
$a_1 = 10d$	0.85	0.85
$a_1 = 9d$	0.8	0.8
$a_1 = 8d$	0.75	0.75
$a_1 = 7d$	0.7	0.7
$a_1 = 4d$	0.5	–

* Based on *Table 8.1* in EC5.
† Linear interpolation of k_{ef} is permitted for spacings between the stated values; spacing a_1 is as shown in Figure 10.15.

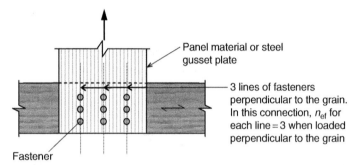

Fig. 10.16. Lines of fasteners perpendicular to the grain.

- n_p is the number of overlapping nails in the row parallel to the grain (noting that an overlapping nail is formed from two nails);
- k_{ef} is an exponent that is dependent on the nail spacing and whether or not pre-drilling is used, and is given in Table 10.10.

(b) Loaded perpendicular to the grain.
When loading nails perpendicular to the grain in a single or double shear connection, as shown in Figure 10.16, the effective number of nails, n_{ef}, in each line of nails shall be taken to equal the actual number of nails, n, when using single nails or the number of overlapping nails (as defined in equation (10.44)) when using overlapping nails. As explained in 10.3.6.1, when loaded this way there is the risk of splitting of the connection material, and the connection capacity will be the lesser of the splitting capacity of the member(s) subjected to the tension force and the connection strength derived from the summation of the strength of the fasteners.

(c) Loaded at an angle to the grain.
When the nails in a single or double shear connection are laterally loaded at an angle to the grain, the force components parallel and perpendicular to the grain have to be derived, and:

(i) the component of the design force acting parallel to the grain must not exceed the load-carrying capacity based on the use of the effective number of nails per row in the connection as defined in 10.4.1.1(a));

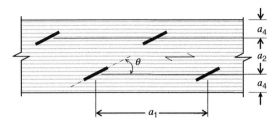

Fig. 10.17. Spacing for staples.

(ii)   the component of the design force acting perpendicular to the grain must not exceed the load-carrying capacity as defined in 10.4.1.1(b)).

10.4.1.2   Staples

With staples, n is equal to 2 times the number of staples in the row parallel to the grain, and where the angle θ between the crown of the staple and the direction of the grain of the timber, as shown in Figure 10.17, is greater than $30°$, equation (10.43) will apply.

If the angle θ is less than $30°$, the above will again apply but the lateral load carrying capacity should be multiplied by a factor of 0.7.

For multiple staples in a row, $n_{ef} = n$.

10.4.1.3   Bolts and dowels

(a) Loaded parallel to the grain.

As bolts/dowels are stiffer than nails or staples, for connections in single or double shear, the reduction in row capacity parallel to the grain is less than with these fasteners and:

$$n_{ef} = \min\left\{ n, n^{0.9} \sqrt[4]{\frac{a_1}{13d}} \right\}$$ (EC5, *equation (8.34)*)   (10.45)

where n_{ef} is the effective number of bolts or dowels in a row parallel to the grain, a_1 is the bolt/dowel spacing in the grain direction, d is the diameter of the bolt/dowel and n is the number of bolts/dowels in the row.

(b) Loaded perpendicular to the grain.

For loading perpendicular to the grain, in a single or double shear connection there is no reduction and:

$$n_{ef} = n$$ (EC5, *equation (8.35)*)   (10.46)

(c) Loaded at an angle to the grain.

The load-carrying capacity parallel or perpendicular to the grain shall be determined in the same way as for nails, as described in 10.4.1(c). As an alternative, when loaded an angle $(0° < \theta < 90°)$, n_{ef} may be determined by linear interpolation between equations (10.45) and (10.46).

10.4.1.4   Screws

Where the effective diameter of the smooth shank of the screw is 6 mm or less, the rules for nails will apply and where it is greater than 6 mm, the rules for bolts will apply.

10.4.2   Alternating forces in connections

Where a connection is subjected to short-term or instantaneous alternating forces, the characteristic load-carrying capacity of the connection will not be affected.

If, however, the connection is subjected to alternating forces due to long- or medium-term actions, the characteristic load-carrying capacity of the connection will be reduced. In such circumstances, if the design force on the connection alternates between a tensile value, $F_{t,Ed}$, and a compressive value, $F_{c,Ed}$, the connection must be designed for:

$$\text{Design tensile force} = (F_{t,Ed} + 0.5F_{c,Ed}) \tag{10.47}$$

$$\text{Design compressive force} = (F_{c,Ed} + 0.5F_{t,Ed}) \tag{10.48}$$

10.5   DESIGN STRENGTH OF A LATERALLY LOADED METAL DOWEL CONNECTION

The strength equations given in the following sub-sections assume that in the connection the design shear strength of the fasteners will always exceed the design capacity derived from the timber/panel strength equations for the relevant fastener type. Where the shear strength of the fastener is required, it must be determined in accordance with the requirements of EN 1993-1-1, Eurocode 3, *Design of Steel Structures, General Rules and Rules for Buildings*.

10.5.1   Loaded parallel to the grain

The design strength of a laterally loaded single fastener, $F_{v,Rd}$, is obtained from the characteristic load-carrying capacity of the laterally loaded fastener as follows:

$$F_{v,Rd} = \frac{k_{mod} \cdot F_{v,Rk}}{\gamma_M} \tag{10.49}$$

- k_{mod} is the modification factor referred to in 2.2.20, and where the connection comprises two timber elements $k_{mod,1}$ and $k_{mod,2}$ the value used in the equation will be $k_{mod} = \sqrt{k_{mod,1} k_{mod,2}}$, as required by EC5, *equation (2.6)*.
- γ_M is the partial factor for connections given in Table 2.6. Except when determining the plate strength of punched metal plate fasteners, the factor value will be 1.3.
- $F_{v,Rk}$ is the characteristic load-carrying capacity of the fastener per shear plane when loaded laterally, i.e. the lowest value determined from the relevant equations given in Tables 10.2 and 10.3.

For a connection containing r_{pl} rows of fasteners laterally loaded parallel to the grain, with each row containing n equally spaced fasteners of the same size, each with a design strength per shear plane, $F_{v,Rd}$, the effective lateral load design capacity of the connection parallel to the grain, $F_{v,ef,Rd}$, will be:

$$F_{v,ef,Rd} = n_{sp} \cdot r_{pl} \cdot n_{ef} \cdot F_{v,Rd} \qquad (10.50)$$

where n_{ef} is the effective number of fasteners in the connection in each row parallel to the grain and n_{sp} is the number of shear planes in the connection.

When using metal dowel-type fasteners in steel to timber connections loaded with a force component acting parallel to the grain in a loaded end configuration, there is a risk of block shear and plug shear failure and this condition has to be considered for the cases referred to in 10.3.6. The design load carrying capacity of fracture for this condition, $F_{bs,Rd}$, will be obtained from equation (10.49) using $F_{bs,Rk}$ (derived from EC5, *equations (A1)*) in place of $F_{v,Rk}$. For such connections, the design strength will be the lesser of $F_{v,ef,Rd}$ and $F_{bs,Rd}$ and, as recommended in this book, the designer should try and ensure $F_{v,ef,Rd} < F_{bs,Rd}$.

10.5.2 Loaded perpendicular to the grain

Where loads are imposed on the timber by fasteners loaded perpendicular to the grain, there are two possible forms of failure:

(a) By the timber splitting in tension and this condition is covered in 10.3.6.1.
(b) By ductile yielding of the fastener and for this condition, where there are r_{pr} lines of fasteners with each line containing n fasteners, all of the same size:

$$F_{v,ef,Rd} = n_{sp} \cdot r_{pr} \cdot n \cdot F_{v,Rd} \qquad (10.51)$$

where:

- $F_{v,ef,Rd}$ is the effective design strength of the fastener per shear plane when loaded laterally and perpendicular to the grain.
- n_{sp} is the number of shear planes in the connection.
- n is the number of fasteners in each line of fasteners perpendicular to the grain. If overlapping nails are being used, n will be the number of overlapping nails as defined in equation (10.44).
- $F_{v,Rd}$ is the design load-carrying capacity of a laterally loaded single fastener per shear plane when loaded perpendicular to the grain. For nails and staples ≤ 8 mm in diameter, as well as screws with an effective diameter ≤ 6 mm, the capacity will be the same as for the fastener loaded parallel to the grain. For bolts and dowels as well as screws with an effective diameter > 6 mm and nails > 8 mm in diameter in connections using timber or LVL, the capacity derived from the strength equations in Tables 10.2 and 10.3 will have to take account of the requirements of equation (10.34) where the characteristic embedment strength of the timber or LVL in the connection will become:

$$f_{h,90,k} = \frac{f_{h,0,k}}{k_{90}} \qquad \text{(equation (10.34) with } \theta = 90°) \quad (10.52)$$

From the above, the design load-carrying capacity of a connection loaded perpendicular to the grain is the condition which has the largest ratio of design load to design strength as follows:

$$\text{Design connection} = \max\left(F_{v,\text{ed}}/F_{90,\text{Rd}}, F_{d}/F_{v,\text{ef,Rd}}\right) \tag{10.53}$$

where $F_{v,\text{ed}}$ is the largest design shear force at the connection; $F_{90,\text{Rd}}$ is the design splitting capacity of the timber member, and for softwood is obtained from equation (10.41c); F_{d} is the design value of the lateral force on the connection.

If the larger design ratio is derived from the tension splitting condition it is recommended in this book that where possible the joint detail is adjusted until the ductile failure condition becomes critical.

10.6   EXAMPLES OF THE DESIGN OF CONNECTIONS USING METAL DOWEL-TYPE FASTENERS

See Examples 10.13.1 and 10.13.2.

10.7   MULTIPLE SHEAR PLANE CONNECTIONS

Where a connection involves multiple shear planes and the connected members are at varying angles to each other, it is not possible to directly apply the joint strength equations given in Tables 10.2 and 10.3.

For such situations, the guidance in EC5, *8.1.3(1)*, is that the resistance of each shear plane is derived on the assumption that the shear plane in the connection forms part of a series of three-member connections and the connection strength is derived by combining the strength values of those failure modes that are compatible with each other. EC5 requires that failure modes (a), (b), (g) and (h) in Table 10.2 or modes (c), (f) and (*j/ℓ*) in Table 10.3 must not be combined with the other failure modes.

Using this approach, to evaluate the strength of each shear plane in a multiple-plane connection, three-member connections are formed by working from one side of the joint (say the left side) and setting up a series of three-member *symmetrical* connections such that the central member of each connection is the actual joint member and the outer members are its adjacent members. Where the adjacent members have different properties (i.e. material, cross-sectional or directional) the member on the right side of the connection is replaced by the left-side member to form the symmetrical connection. The value and direction of the design force to be transmitted at the shear plane between the members are then derived by statics, assuming the fastener to be rigid. Knowing the direction of the design force, for each shear plane set-up, the mode failure strength can be determined from the application of the strength equations in Table 10.2 (or 10.3). Commencing with the left-side shear plane, the shear plane strength will be the mode strength having the lowest value and after deleting the incompatible and unacceptable combinations of failure mode in the remaining shear planes, the joint strength can be determined. It must be shown that the strength of each shear plane is equal to or greater than its associated design force. The method produces an approximate and safe result and simplifies what is otherwise a relatively complex problem.

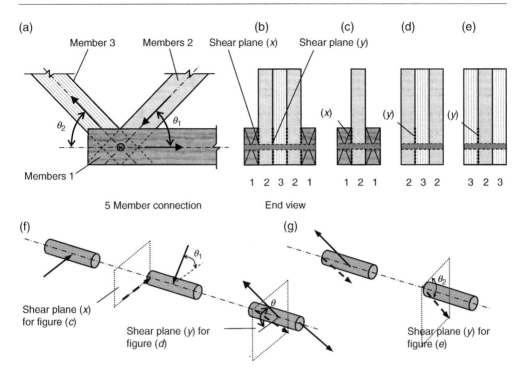

Fig. 10.18. Multiple shear plane connection procedure.

As an example of the application of the method, consider the five-member single bolt joint subjected to design axial forces acting in the directions shown in Figure 10.18a. Members 1, 2 and 3 have different properties. From the application of statics, the value of the design force and its direction in each shear plane can be derived as shown in Figures 10.18f–10.18g.

The three-member connections are now formed, working from the left side of the joint. Consider the shear plane between members 1 and 2 (shear plane x) and form a three-member double shear connection by adding member 1 to the other side as shown in Figure 10.18c. The other members are ignored. The strength of shear plane x is now calculated using the strength equations, ensuring that the embedment strength of each connected member takes account of the angle of the shear plane load relative to its grain direction. For example, in Figure 10.18c member 1 will be loaded parallel to the grain but member 2 will be loaded at an angle θ_1 to the grain.

Shear plane y between members 2 and 3 is now considered. A three-member symmetrical connection is formed around member 3 using member 2 as the outer member, as shown in Figure 10.18d, and the other members are ignored. Note that on this occasion the actual joint members can be used as the joint is already symmetrical. The strength of the shear plane is then derived as above. It will be found that the value of the resultant force in shear plane y will be half the value of the force in the central member and it will act along the direction of the centroidal axis of that member. To consider all possible failure modes for shear plane y, a double shear connection involving member 2 as the central member with member 3 on either side, as shown in Figure 10.18e, must also be analysed as for the previous simulations.

Table 10.11 Compatibility of failure modes across shear planes in the connection

Double shear failure mode	Lateral load capacity of shear plane (x) (kN)	Lateral load capacity of shear plane (y) (kN)	Lateral load capacity of alternative shear plane (y) (kN)
Mode (g)	14.46	7.44	19.99
Mode (h)	4.5	10	3.72
Mode (j)	26.68	(38.07)	(47.09)
Mode (k)	37.36	(7.44)	(3.72)

From these analyses, the minimum strength of shear plane x and of shear plane y will be determined, ensuring that compatibility of the failure modes across the joint is taken into account. For the joint being considered an example of the requirement is shown in Table 10.11. The mode with the minimum strength in shear plane x is mode (h), consequently the shear strength of shear plane x will be 4.5 kN. Because modes (j) and (k) are not compatible with mode (h), they need not be evaluated (see values in brackets in the table), and mode (h) in alternative shear plane y becomes the y plane shear failure mode as it has the lowest value.

In addition to the above check, if the design force in a shear plane results in any member in a timber connection being subjected to a force component at an angle to the grain, the timber must also be checked for compliance with the requirements of 10.3.6.1.

See Examples 10.13.3 and 10.13.4.

10.8   AXIAL LOADING OF METAL DOWEL CONNECTION SYSTEMS

The strength equations given in the following sub-sections assume that the tensile strength of fasteners will always exceed their withdrawal capacity from the connection. If, however, there is a need to evaluate the tensile strength of the fastener, it should be carried out in accordance with the requirements of EN 1993-1-1.

10.8.1   Axially loaded nails

The withdrawal capacity of nails loaded axially is dependent on the type of nail being used. Smooth round wire nails give the poorest result and with threaded nails the capacity is greatly increased. However, no matter the type, nails are not considered capable of sustaining axial load in end grain.

Also, EC5 does not permit axially loaded smooth nails to be used in situations involving permanent or long-term loading, and where threaded nails are used, only the threaded part of the nail is to be taken as relevant for determining the nail strength. Only threaded nails are allowed to be used to resist permanent or long-term axial loading.

Ignoring tension failure of the nail, there are two possible failure modes when subjected to axial loading:

- pointside withdrawal of the nail;
- pull-through of the nail head.

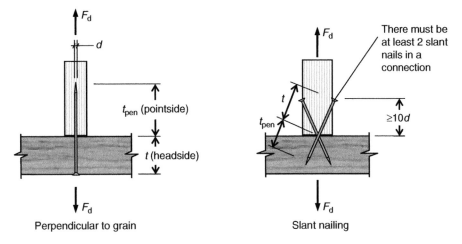

Fig. 10.19. Nailing in tension.

Nails can be driven perpendicular to the grain and/or at a slant, as shown in Figure 10.19, and will fail in the mode with the lesser capacity. For threaded nails compliant with BS EN 14592, only the threaded part of the nail in the pointside member is taken to resist axial load, and the headside capacity can only utilise the head pull-through resistance. With smooth nails, there will be withdrawal resistance for the pointside penetration and the headside resistance takes both the headside pull-through strength and the shank friction resistance on the headside of the nail into account.

When a nail is subjected to an axial force, F_d, the following condition has to be satisfied:

$$F_d \leq F_{ax,Rd} \tag{10.54}$$

where $F_{ax,Rd}$ is the design withdrawal capacity of the nail.

The design withdrawal capacity of the nail is obtained from the characteristic withdrawal capacity, $F_{ax,Rk}$, as follows:

$$F_{ax,Rd} = \frac{k_{mod} F_{ax,Rk}}{\gamma_M} \tag{10.55}$$

where the functions are as previously defined and $F_{ax,Rk}$ is the characteristic withdrawal capacity of a nail derived as follows:

For nails, other than smooth wire nails, as defined in BS EN 14592:

$$F_{ax,Rk} = \min \begin{pmatrix} f_{ax,k} d t_{pen} \\ f_{head,k} d_h^2 \end{pmatrix} \qquad \text{(EC5, } equation \text{ (8.23))} \quad (10.56)$$

For smooth wire nails:

$$F_{ax,Rk} = \min \begin{pmatrix} f_{ax,k} d t_{pen} \\ f_{ax,k} d t + f_{head,k} d_h^2 \end{pmatrix} \qquad \text{(EC5, } equation \text{ (8.24))} \quad (10.57)$$

where:

- $f_{ax,k}$ is the characteristic pointside withdrawal strength;
- $f_{head,k}$ is the characteristic headside pull-through strength;
- d is the nail diameter, and for square or grooved nails d is the side dimension;
- t_{pen} is the pointside penetration or the length of the threaded part in the pointside member, excluding the point length (as stated in Appendix C);
- t is the thickness of the headside member;
- d_n is the diameter of the nail head. For smooth wire nails, in accordance with BS EN 10230, this is $2.25d$ for nails of nominal diameter 2.7 mm up to 3.4 mm and $2d$ for all nails with nominal diameters greater than 3.4 mm.

Values for $f_{ax,k}$ and $f_{head,k}$ can be determined by testing, and EC5 gives the following values for smooth nails with a pointside penetration of at least $12d$:

$$f_{ax,k} = 20 \times 10^{-6} \rho_k^2 \quad N/mm^2 \qquad \text{(EC5, equation (8.25))} \quad (10.58)$$

$$f_{head,k} = 70 \times 10^{-6} \rho_k^2 \quad N/mm^2 \qquad \text{(EC5, equation (8.26))} \quad (10.59)$$

where ρ_k is the characteristic timber density in kg/m^3.

If the pointside nail penetration is less than $12d$ the withdrawal capacity of the nail has to be linearly reduced by multiplying by the factor $((t_{pen}/4d) - 2)$. When the minimum nail penetration of $8d$ for smooth nails is used, the factor will be zero and there will be no axial withdrawal strength. This procedure must also be applied to the headside penetration of the nail to determine the strength contribution from shank friction towards the headside strength.

When threaded nails are used, the rules in EC5 require that the threaded or deformed length of the nail shank shall penetrate the pointside member by at least $6d$. For a threaded pointside nail penetration of $8d$ the full value of the characteristic withdrawal strength can be used and for values less than this the strength is reduced linearly by multiplying by $((t_{pen}/2d) - 3)$.

Because of the variation in the nature of the surface finish over the threaded or deformed lengths of the different types of threaded nail that are manufactured, no expressions are given in EC5 for the characteristic pointside withdrawal strength, $f_{ax,k}$, and the characteristic headside pull-through strength, $f_{head,k}$ for these nails. The respective strengths have to be derived for the particular nail type and connection material being used by tests carried out in accordance with the requirements of BS EN 1382 [20] for $f_{ax,k}$ and BS EN 1383 [21] for $f_{head,k}$. It is to be expected that nail manufacturers will in due course publish verified data for these strengths.

Where structural timber has been designed to function under service class 1 or 2 conditions but will possibly be installed at or near the fibre saturation point, the values of $f_{ax,k}$ and $f_{head,k}$ must be multiplied by 2/3 to take account of the reduction in the respective strengths when drying out.

The spacings, end and edge distances for axially loaded nails are the same as those given in EC5 for laterally loaded nails.

See Example 10.13.5.

10.8.2   Axially loaded bolts

With axially loaded bolts, the strength of the connection is dependent on the tensile strength of the bolt and the bearing strength of the material onto which the bolt washer beds. An approximate value for the characteristic tensile strength of the bolt is obtained from $f_{u,k}A_{net}$ where $f_{u,k}$ and A_{net} are the tensile strength and the cross-sectional area at the threaded end of the bolt respectively. The more accurate value is derived using the strength equations given in EN 1993-1-8 and is $0.9f_{u,k}A_{net}$ for normal bolts or $0.65f_{u,k}A_{net}$ for countersunk headed bolts, also taking into account that when deriving the design value the partial material factor for this type of steel connection will be 1.25, not 1.3 as used for timber.

When bearing onto timber or wood products, the bearing capacity below the washer should be calculated assuming a 300% increase in the characteristic strength of the timber perpendicular to the grain over the contact area, i.e. $f_{c,k}=3.0\times f_{c,90,k}$. The washer size will be obtained from EC5, *10.4.3(2)*, having a diameter (or side length) of at least $3d$ with a thickness of at least $0.3d$.

When using a steel plate, the bearing capacity per bolt should not exceed that of a circular washer with a diameter that is the lesser of $12t$ (where t is the plate thickness) or $4d$ (where d is the bolt diameter).

10.8.3   Axially loaded dowels

Dowel fasteners cannot be used to take tensile loading.

10.8.4   Axially loaded screws

With screws it is stated in EC5 that there are six possible failure modes:

- withdrawal of the threaded part of the screw;
- when used with steel plates, there is the risk of tearing off the screw head and the requirement is that the tear-off resistance of the screw head is greater than the tensile strength of the screw;
- pull-through failure of the screw head;
- the screw failing in tension;
- buckling failure of the screw when loaded in compression;
- failure along the circumference of a group of screws used in conjunction with steel plates (a block shear or plug shear failure).

Failure modes in the steel or in the timber around the screw are brittle-type modes and the significance of this should be taken into account in the design of a connection subjected to axial loading. Also, buckling failure can only arise where screws have a threaded length in each compression member in the connection and there is an unconstrained length of shank that will permit unconstrained lateral instability movement of the shank. This will not be a common connection detail.

To control block failures and ensure that the minimum withdrawal resistance is achieved, minimum spacing and penetration requirements are specified for axially

Table 10.12 Minimum spacings and edge distances as illustrated in Figure 10.20, when using axially loaded screws[*]

Minimum screw spacing in a plane parallel to the grain a_1	Minimum screw spacing in a perpendicular to a plane parallel to the grain a_2	Minimum end distance of the centre of gravity of the threaded part of the screw in the member $a_{1,CG}$	Minimum edge distance of the centre of gravity of the threaded part of the screw in the member $a_{2,CG}$
$7d$	$5d$	$10d$	$4d$

[*] Based on *Table 8.6* in EC5.
d is the nominal diameter of the screw diameter and the minimum pointside penetration of the threaded part of the screw must be $6d$.
The above values are only valid provided the timber thickness, t, is $\geq 12d$

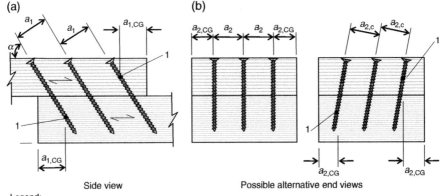

Legend:
1 is the centre of gravity of the threaded part of the screw in the member

Fig. 10.20. Spacings, edge and end distances for screws (based on *Figure 8.11a* in EC5).

loaded screws. The minimum pointside penetration of the threaded length of the screw must be $6d$, where d is the nominal diameter and the spacing criteria are as given in Table 10.12 and shown in Figure 10.20 for screws driven at an angle α to the grain. It is to be noted the values given can only be used provided the timber thickness of the connection members is $\geq 12d$. No guidance is given for member thicknesses less than $12d$.

When a screwed connection is subjected to an axial design force F_d, the following condition has to be satisfied:

$$F_d \leq \min(F_{ax,\alpha,Rd}, F_{t,Rd}) \tag{10.60}$$

Where:

- $F_{ax,\alpha,Rd}$ is the lesser of the design withdrawal capacity of the connection when loaded axially at an angle α to the grain or the design pull-through capacity of the connection when loaded at an angle α to the grain;
- $F_{t,Rd}$ is the design tensile resistance of the connection (i.e. based on the tensile capacity of the shank).

The design tensile strength of the connection is obtained from the relevant characteristic value of the tensile as follows:

$$F_{ax,\alpha,Rd} = \frac{k_{mod} F_{ax,\alpha,Rk}}{\gamma_M} \qquad F_{t,Rd} = \frac{k_{mod} F_{t,Rk}}{\gamma_{M,steel}} \qquad (10.61)$$

where the functions are as previously defined and:

- $F_{ax,\alpha,Rk}$ is the lesser of the characteristic withdrawal capacity or characteristic pull-through resistance of the connection;
- $F_{t,Rk}$ is the characteristic tensile resistance of the connection;
- γ_M is the material factor for the connection – 1.3, in the UKNA to EC5 [22];
- $\gamma_{M,steel}$ is the material factor for the screw – obtained from BS EN 1993-1-8.

The derivation of the strength equations in EC5, *8.7.2(4)* and *(5)* have been based on test results using connections formed from softwood timbers and it is not clear if they can be applied where hardwoods are to be used.

(a) Calculation of the characteristic withdrawal capacity of a connection loaded at an angle α to the grain.
A screw cannot be used in a structural connection unless it complies with the requirements of BS EN 14592 and the ratio of its inner threaded diameter (see 10.1.1.2) to its nominal diameter, d_1/d, is not less than 60% or more than 90%. Design rules are given in EC5 for the characteristic withdrawal capacity of screws within the range $0.6 \le d_1/d \le 0.75$ (the range which will cover smooth shank screws) and for screws where $0.75 < d_1/d \le 0.9$, but the rules only apply to screws within the range $6\,\text{mm} \le d \le 12\,\text{mm}$.

For connections in softwood timber using screws within the range $6\,\text{mm} \le d \le 12\,\text{mm}$ and $0.6 \le d_1/d \le 0.75$, $F_{ax,\alpha,Rk}$ (in N) is obtained using EC5, *equation (8.38)* as follows:

$$F_{ax,\alpha,Rk} = \frac{n_{ef} f_{ax,k} d \ell_{ef} k_d}{1.2 \cos^2 \alpha + \sin^2 \alpha} \qquad \text{(EC5, equation (8.38))} \qquad (10.62)$$

$$f_{ax,k} = 0.52 d^{-0.5} \ell_{ef}^{-0.1} \rho_k \qquad \text{(EC5, equation (8.39))} \qquad (10.63)$$

and

$$k_d = \min \begin{cases} \dfrac{d}{8} \\ 1 \end{cases} \qquad \text{(EC5, equation (8.40))} \qquad (10.64)$$

where:

- d is the nominal diameter of the screw, in mm;
- ℓ_{ef} is the penetration of the *threaded part*, in mm;
- $f_{ax,k}$ is the characteristic withdrawal strength perpendicular to the grain, in N/mm^2;
- ρ_k is the characteristic density, in kg/m^3;

- α is the angle between the screw axis and the grain direction, with $\alpha \geq 30°$;
- n_{ef} is the effective number of screws and for a connection with n screws acting in the connection all loaded by a force component parallel to the shank, it will be:

$$n_{ef} = n^{0.9} \qquad \qquad (EC5, \text{ equation } (8.41)) \qquad (10.65)$$

For the condition where $\alpha = 90°$ to the grain, equation (10.62) reduces to:

$$F_{ax,\alpha,Rk} = n_{ef} f_{ax,k} d\ell_{ef} k_d \qquad \qquad (10.62a)$$

For connections using screws having a nominal diameter less than 6 mm and $0.6 \leq d_1/d \leq 0.75$, which covers the majority of sizes used in domestic construction, a value for $F_{ax,\alpha,Rk}$, for a single screw, is given in PD6693-1.

For connections using screws within the range $6\,mm \leq d \leq 12\,mm$ and $0.75 < d_1/d \leq 0.9$, $F_{ax,\alpha,Rk}$ (in N) should be taken as:

$$F_{ax,\alpha,Rk} = \frac{n_{ef} f_{ax,k} d\ell_{ef}}{1.2\cos^2 \alpha + \sin^2 \alpha} \left(\frac{\rho_k}{\rho_a}\right)^{0.8} \qquad (EC5, \text{ equation } (8.40a)) \qquad (10.66)$$

and, where $\alpha = 90°$ to the grain:

$$F_{ax,\alpha,Rk} = n_{ef} f_{ax,k} d\ell_{ef} \left(\frac{\rho_k}{\rho_a}\right)^{0.8} \qquad \qquad (10.66a)$$

where

- $f_{ax,k}$ is the characteristic withdrawal strength perpendicular to the grain for the screw in material of density ρ_a, in N/mm², derived in accordance with the requirements of BS EN 1382;
- ρ_a is the characteristic density of the material used for the withdrawal strength tests; and the remaining symbols are as noted above.

(b) Calculation of the characteristic pull-through resistance a connection loaded at an angle α to the grain.

The characteristic pull-through resistance of a connection with axially loaded screws at an angle of α to the grain, $F_{ax,\alpha,Rk}$ (in N), is obtained from:

$$F_{ax,\alpha,Rk} = n_{ef} f_{head,k} d_h^2 \left(\frac{\rho_k}{\rho_a}\right)^{0.8} \qquad (EC5, \text{ equation } (8.40b)) \qquad (10.67)$$

where the functions are as stated above and:

- $f_{head,k}$ is the characteristic pull-through capacity of the screw in material of density ρ_a, in N, derived in accordance with the requirements of BS EN 1383;
- d_h is the head diameter of the screw, in mm.

$F_{ax,\alpha,Rk}$ can only be derived if a value for $f_{head,k}$ is provided from test results or from information from manufacturers (also derived from tests in accordance with BS EN 1383) and when the density of the material used in the tests equates the material being used in the connection:

$$F_{ax,a,Rk} = n_{ef} f_{head,k} d_h^2$$

For connections using screws having a nominal diameter less than 12 mm and $d_h/d \leq 2.5$, a value for $F_{ax,\alpha,Rk}$ for a single screw is given in PD6693-1.

(c) Calculation of the characteristic tensile resistance of the connection.
Screws can fail due to tension failure in the shank, normally in the inner thread zone, or by tearing off of the screw head. The characteristic tensile resistance of the connection, $F_{t,Rk}$, in N, is to Be taken as:

$$F_{t,Rk} = n_{ef} f_{tens,k} \qquad \text{(EC5, equation (8.40c))} \qquad \text{(10.68)}$$

where n_{ef} is as defined in equation (10.65) and:

- $f_{tens,k}$ is the characteristic tensile capacity of the screw, in N, determined in accordance with the requirements of BS EN 14592 and BS EN 1383.

Where screws are fixed at an angle to the grain (e.g. as shown in Figure 10.20(a)) and loaded laterally along the grain direction, a conservative value for the lateral strength of the connection can be taken to be:

$$F_{v,Rd} = \min (F_{ax,\alpha,Rd}, F_{t,Rd}) \cos(\alpha)$$

For the condition where the screws are fixed at right angles to the grain direction and the connection is subjected to lateral loading, the design rules for laterally loaded screwed connections will apply and to derive $F_{ax,Rk}$, used to calculate the rope effect (referred to in EC5, *8.2.3*), the characteristic withdrawal capacity of a screw will be derived as follows:

$$F_{v,Rk} = \min (F_{ax,\alpha,Rk,1}, F_{t,Rk,1})$$

Where:

- $F_{ax,\alpha,Rk,1}$ is the characteristic withdrawal capacity of a screw, calculated using the relevant equations in 10.8.4(a) and (b) but with $n_{ef}=1$ and $\alpha=90°$.
- $F_{t,Rk,1}$ is the characteristic tensile resistance of a screw, calculated using equation (10.68) but with $n_{ef}=1$.

10.9   COMBINED LATERALLY AND AXIALLY LOADED METAL DOWEL CONNECTIONS

When nailed or screwed connections are subjected to the combination of a lateral design load, $F_{v,Ed}$, and an axial design load, $F_{ax,Ed}$, they must comply with the interaction relationships shown in Figure 10.21. No guidance is given for bolted connections that are loaded in this manner.

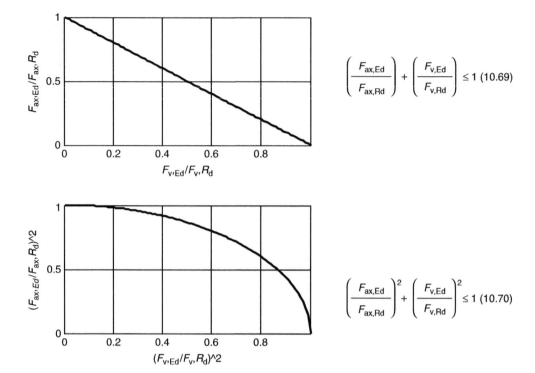

$$\left(\frac{F_{ax,Ed}}{F_{ax,Rd}}\right) + \left(\frac{F_{v,Ed}}{F_{v,Rd}}\right) \leq 1 \; (10.69)$$

$$\left(\frac{F_{ax,Ed}}{F_{ax,Rd}}\right)^2 + \left(\frac{F_{v,Ed}}{F_{v,Rd}}\right)^2 \leq 1 \; (10.70)$$

Fig. 10.21. Interaction diagrams for combined axially and laterally loaded connections. The linear relationship is for smooth nails and the curved graph is for nails (other than smooth nails) and screws.

For smooth nails the combined design force to strength ratios must stay within the elastic range, resulting in a linear relationship in accordance with equation (10.69). The combined ratios for other nail types and for screws can extend beyond the elastic limit but must comply with the power function given in equation (10.70) (*equations (8.27)* and *(8.28)* respectively in EC5), where $F_{ax,Rd}$ is the design strength of the connection loaded axially, and $F_{v,Rd}$ is the design strength of the connection loaded laterally.

See Example 10.13.6.

10.10   LATERAL STIFFNESS OF METAL DOWEL CONNECTIONS AT THE SLS AND ULS

Because of tolerance allowances in the assembly process of a connection or yielding of the fasteners and/or the timber or wood product in the connection, or through a combination of both of these factors, joints formed with mechanical fasteners will slip when subjected to lateral load. The amount of slip will vary depending on the fastener type being used, and typical load–slip curves for a nailed or screwed connection and for a bolted connection are shown in Figure 10.22.

With bolts, because of the tolerance required to enable the bolt to be fitted and the bedding in process of the bolt onto the surface of the pre-drilled hole when subjected to the lateral load, there is an immediate slip when loaded and this is shown in Figure 10.22b.

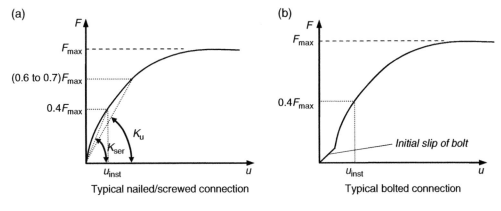

Legend:
F_{max} is the maximum load taken by the connection;
u_{inst} is the instantaneous slip at the SLS.

Fig. 10.22. Typical instantaneous load–slip behaviour of a metal dowel fastener connections.

Timber has a relatively low stiffness to strength ratio resulting in comparatively flexible structural systems, and although the significance of ensuring that joint strength criteria are fully met is well understood, for most structures failure by non-compliance with stiffness criteria is likely to be the more common reason for problems arising during the design life of a timber structure.

The stiffness of a fastener is defined as the ratio of its lateral load per shear plane divided by its slip, and knowing this relationship the slip under any load can be obtained. In EC5 this stiffness property is referred to as the slip modulus. When timber design codes were based on a permissible stress design approach, stiffness criteria were given at the working load condition and in most codes the slip limit for joints was set at 0.15 inches, nominally 0.4 mm. In EC5, different values of stiffness are given for the SLS, K_{ser} (and the ULS, K_u), but no limit is set for joint slip. It is left to the designer to decide on the value that will be acceptable for the structure being designed.

In EC5 the instantaneous slip modulus for design at the SLS, K_{ser}, is taken to be the secant modulus of the load–displacement curve at a load level of approximately 40% of the maximum load able to be taken by the fastener [23], and is shown for a nailed connection in Figure 10.22a. It is seen that the use of a straight-line relationship up to this limit will give a reasonably good approximation to the actual load–slip behaviour of the fastener.

From the results of many tests on joints, the instantaneous slip at approximately 40% of the maximum load has been determined by various researchers. Also, by adopting failure modes that entail full yielding of the fasteners and the timber/wood product (referred to as type 3 failure modes, e.g. modes (f) and (k) in Table 10.2), which are the most common failure modes in an efficiently designed connection, the joint strength can be evaluated. Multiplying the result by 0.4, the SLS strength is obtained. From this data, the slip modulus per shear plane per fastener under service load for different metal dowel-type fasteners has been derived and the relationships are given in Table 10.13.

Table 10.13 Values for K_{ser} for fasteners (in N/mm) in timber-to-timber and wood-based panel-to-timber connections*

Type of fastener used	Serviceability limit state slip modulus K_{ser}
Nails	
Without pre-drilling	$\rho_m^{1.5} d^{0.8}/30$
With pre-drilling	$\rho_m^{1.5} d/23$
Staples	$\rho_m^{1.5} d^{0.8}/80$
Screws	$\rho_m^{1.5} d/23$
Bolts with or without clearance†	$\rho_m^{1.5} d/23$
Dowels	$\rho_m^{1.5} d/23$

* Based on *Table 7.1* in EC5.
† Where there is a clearance allowance for the bolt, this should be added to the connection slip.

K_{ser} is based on the diameter of the fastener, d (in mm), and where the same timber or wood-based product is used for all of the joint members, it is based on the mean density, ρ_m (in kg/m³) of the material. Where the connection involves members of different densities, ρ_{m1} and ρ_{m2}, the ρ_m to be used in the expressions in Table 10.13 will be:

$$\rho_m = \sqrt{\rho_{m1} \cdot \rho_{m2}} \qquad \text{(EC5, equation (7.1))} \quad (10.71)$$

The instantaneous slip in a connection, u_{inst}, is a summation of the slip in the respective members forming the connection, and for the single shear timber-to-timber (or wood-based product) connection in Figure 10.23a, the instantaneous slip will be as shown in Figure 10.23b. There will be slip in member 1 (u_{inst1}) and in member 2 (u_{inst2}) and u_{inst} will be:

$$u_{inst} = u_{inst1} + u_{inst2} \qquad (10.72)$$

Where the members have the same properties and, say, member 2 is used,

$$u_{inst} = u_{inst2} + u_{inst2} = 2u_{inst2} \qquad (10.73)$$

If one of the members is steel, for the same applied load the slip in the steel member will be effectively zero while the slip in member 2 will remain as before and for this situation:

$$u_{inst} = 0 + u_{inst2} = +u_{inst2}$$

For the steel-to-timber connection, the instantaneous slip will be half the value of the timber-to-timber connection and consequently its stiffness will theoretically be twice the slip modulus of the timber-to-timber connection, i.e. $2 \times K_{ser}$. This is an approximation to the real behaviour as it ignores the effect of clearance between the

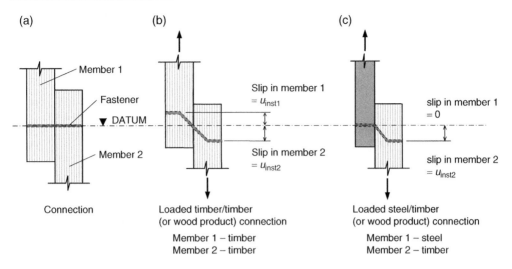

(a)

Member 1

Fastener

▼ DATUM

Member 2

Connection

(b)

Slip in member 1
= u_{inst1}

Slip in member 2
= u_{inst2}

Loaded timber/timber
(or wood product) connection

Member 1 – timber
Member 2 – timber

(c)

slip in member 1
= 0

slip in member 2
= u_{inst2}

Loaded steel/timber
(or wood product) connection

Member 1 – steel
Member 2 – timber

Fig. 10.23. Instantaneous slip in a connection.

fastener and the steel, rotation of the fastener in the steel member and yielding of the steel member where it is in contact with the fastener, and will result in an overestimate of the stiffness. Although EC5, *7.1(3)*, states that the slip modulus for steel-to-timber and concrete-to-timber connections may be taken to be $2 \times K_{ser}$, if the structure being designed is sensitive to deflection behaviour it is recommended that a smaller value be used and for timber-to-concrete connections from the research undertaken by Dias et al. [24] a factor of $1.7 \times K_{ser}$ to $1.8 \times K_{ser}$ is proposed.

When deriving the strength of a multiple metal dowel-type fastener connection loaded perpendicular to the grain, the actual number of fasteners, n, in the connection is used in the joint strength equation, and when deriving the strength parallel to the grain, the effective number of fasteners, n_{ef}, must be used. No clear guidance is given in EC5 on the value to be used to determine the stiffness of a connection and, irrespective of the angle of load relative to the grain, for single and double shear connections the actual number of fasteners, n, should be used.

The stiffness values given in Table 10.13 for K_{ser} are the stiffness per shear plane per fastener, and for connections with n fasteners per shear plane, the connection stiffness, $K_{ser,sc}$ for single and $K_{ser,dc}$ for double shear configurations are given in Figure 10.24.

Where a clearance is required to permit the fastener to be fitted, e.g. with bolts, there will be additional slip in the connection caused by the take-up of this allowance. If the clearance (tolerance) provided is c (mm), usually taken to be 1 mm for bolts, the instantaneous slip for a connection with a single fastener and a single shear plane will be:

$$u_{inst} = \frac{F}{K_{ser}} + c \qquad (10.74)$$

where F is the SLS load on the connection, and K_{ser} is the slip modulus of the fastener per shear plane at the service condition.

When a structure is subjected to a characteristic permanent action G_k, a dominant characteristic variable action $Q_{k,1}$ and accompanying unrelated characteristic variable actions $Q_{k,i}$, the design load on a connection at the SLS, F_d, will be obtained by

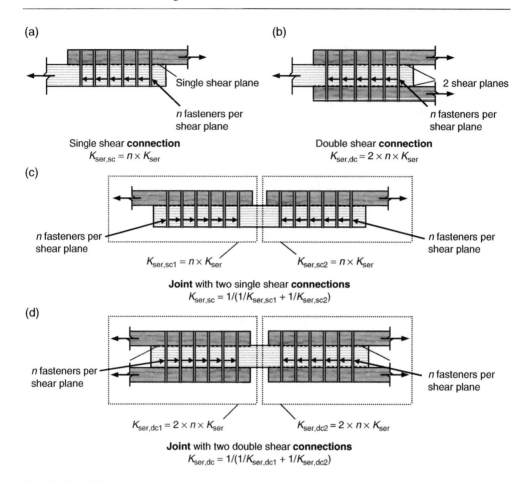

Fig. 10.24. Stiffness of single and double shear connections and joints.

analysing the structure under the characteristic loading combination referred to in 2.2.25.2, i.e.:

$$\sum_{j\geq1}G_{kj} + Q_{k,1} + \sum_{i>1}\psi_{0,i}Q_{k,i} \Rightarrow F_{d} \tag{10.75}$$

where $\psi_{0,i}$ is the combination value factor for the variable action $Q_{k,i}$, obtained from the UKNA to EN 1990:2002 [25].

The creep loading will be derived from the quasi-permanent combination, referred to in 2.2.25.2(c), equation (2.26), i.e.

$$\sum_{j\geq1}G_{k} + \sum_{i\geq1}^{n}\psi_{2,i}Q_{k,i} \Rightarrow F_{d,\text{creep}} \tag{10.76}$$

To obtain the final deformation of a connection, the method of analysis will depend on whether or not the connections have the same time-dependent properties as the members or components in the structure, and the appropriate methodology for each condition is given in 2.3.2.

For structures that comprise members, components and connections with the same creep behaviour, the final deformation is obtained by applying the combined characteristic and quasi-permanent combination of actions to the structure. In this situation, the design load on the connection due to the characteristic combination, F_{cd}, will be obtained by analysing the structure under the combined characteristic and quasi-permanent loading combinations referred to in 2.2.25.2, i.e.:

$$\sum_{j\geq1}G_{k,j}(1+k_{def})+Q_{k,1}(1+\psi_{2,1}k_{def})+\sum_{i>1}Q_{k,i}(\psi_{o,i}+\psi_{2,i}k_{def}) \Rightarrow F_{cd} \qquad (10.77)$$

where the symbols are as previously defined, $\psi_{2,i}$ is the quasi-permanent value factor for the quasi-permanent combination of actions obtained from the UKNA to EN 1990:2002 and k_{def} is the deformation factor for timber and wood-based materials for quasi-permanent actions. Values for k_{def} for timber and some wood-related products are given in Table 2.10.

The final deformation, u_{fin}, at the SLS of a connection having a stiffness $K_{ser,scl}$ per shear plane, with a single shear plane and in a structure comprising members, components and connections having the same creep behaviour will be:

$$u_{fin}=\frac{F_d}{K_{ser,scl}}+c \qquad (10.78)$$

For structures that comprise members, components and connections having different creep behaviour and where the creep effect will influence the stiffness distribution, the above approach cannot be applied. In these situations, as stated in 2.3.2(b), the procedure for determining the final deformation is open to misinterpretation and for this reason is to be revised as stated in Appendix C. For this type of condition the final deformation is to be obtained by calculating the creep deformation under the action of the quasi-permanent loading combination and then to add the instantaneous deformation less the instantaneous deformation arising from the quasi-permanent combination of actions (the reader is referred to 2.3.2(b) for a fuller explanation of the development of the requirement). For the creep deformation analysis the stiffness properties of the members, components and connections are reduced in value and final mean values, referred to in EC5, 2.3.2.2(1) and given in equations (2.34), (2.35) and (2.36), are to be used as follows:

$$E_{mean,fin}=\frac{E_{mean}}{(1+k_{def})} \qquad (EC5,\ equation\ (2.7)) \qquad (2.34)$$

$$G_{mean,fin}=\frac{G_{mean}}{(1+k_{def})} \qquad (EC5,\ equation\ (2.8)) \qquad (2.35)$$

$$K_{ser,fin}=\frac{K_{ser}}{(1+k_{def})} \qquad (EC5,\ equation\ (2.9)) \qquad (2.36)$$

The functions are as described in 2.3.4.1(c).

Considering the creep deformation, the creep deformation of a connection in a joint, $u_{creep,fin}$, having a final mean stiffness $K_{ser,fin,scl}$ per shear plane, with a single shear plane and in a structure comprising members, components and connections having different creep behaviour will be obtained from:

$$u_{\text{creep,fin}} = \frac{F_{d,\text{creep}}}{K_{\text{ser,fin,sc1}}}$$

(10.79)

where the final mean stiffness of the connection in the joint will be obtained as defined in equation (2.36).

The final deformation of the connection, u_{fin}, will then be:

$$u_{\text{fin}} = \frac{F_{cd}}{K_{\text{ser,sc1}}} + u_{\text{creep,fin}} - \frac{F_{d,\text{creep}}}{K_{\text{ser,sc1}}} + c$$

(10.80)

where the functions are as defined above.

An alternative but more conservative approach for such a structure, which is also referred to in 2.3.2(b), is to analyse the structure under the action of the design load arising from the characteristic combination of actions and use final mean values of stiffness in accordance with equations (2.34) to (2.36). From such an analysis the final deformation of the same connection will be:

$$u_{\text{fin}} = \frac{F_{cd}}{K_{\text{ser,fin,sc1}}} + c$$

(10.81)

where the functions are as previously defined.

In the above analyses the values of k_{def}, for timber and some wood-related products in service class 1, 2 and 3 conditions, are given in Table 2.10, and when dealing with connections the factor must be modified as described in 2.3.2. For completeness, the modifications to k_{def} are as follows:

(i)   Where the connection comprises timber elements with the same time-dependent behaviour, k_{def} should be taken to be 2 times the value given in Table 2.10.

(ii)   Where the connection comprises two wood-based elements having different time-dependent behaviour, k_{def} should be taken to be:

$$k_{\text{def}} = 2\sqrt{k_{\text{def,1}} \cdot k_{\text{def,2}}}$$

(EC5, *equation (2.13)*)   (10.82)

where $k_{\text{def,1}}$ and $k_{\text{def,2}}$ are the deformation factors for the respective wood-based elements.

(iii) When timber is being used and installed at or near its fibre saturation point, but is to function in an environment where it was likely to dry out under load, the value of k_{def} in Table 2.10 must be increased by 1.0.

The use of $2k_{\text{def}}$ as the deformation factor when dealing with joints has arisen because it has been concluded that when k_{def} is used there is a tendency to underestimate the deflection in a joint. It is to be stated, however, that this is not fully supported by the findings from all researchers.

For structures at the ULS, the analysis procedure will depend on whether or not the distribution of the stress resultants in the members of the structure is affected by the stiffness distribution, and the methodology to be adopted for each alternative is given in 2.3.4.2. Unless the joints in the structure are designed to be pin jointed, the distribution of the stress resultants in the members will be affected by the stiffness distribution, and in a stress analysis at the ULS the final mean value of the stiffness properties in equations (2.38), (2.39) and (2.40) will be used as follows:

$$E_{mean,fin} = \frac{E_{mean}}{(1 + \psi_2 k_{def})} \qquad \text{(EC5, equation (2.10))} \qquad (2.38)$$

$$G_{mean,fin} = \frac{G_{mean}}{(1 + \psi_2 k_{def})} \qquad \text{(EC5, equation (2.11))} \qquad (2.39)$$

$$K_{ser,fin} = \frac{K_{ser}}{(1 + \psi_2 k_{def})} \qquad \text{(EC5, equation (2.12))} \qquad (2.40)$$

where the functions are as previously defined and ψ_2 is the factor for the quasi-permanent value of the action causing the largest stress in relation to strength.

At this state, the instantaneous slip modulus, K_u, is taken to be the secant modulus of the load–slip curve at a load between 60 and 70% of the maximum load [23] and is shown for a nailed or screwed connection in Figure 10.22(a). From the analysis of test results EC5 states that for design purposes, K_u is to be taken as:

$$K_u = \frac{2}{3} K_{ser} \qquad \text{(EC5, equation (2.1))} \qquad (10.83)$$

where K_{ser} is the instantaneous slip modulus per shear plane per fastener given in Table 10.13.

At the ULS, it should only be necessary to determine the deformation of structural elements where there is a risk that the displacement at this state could lead to unacceptable consequences, e.g. confirming the deformation of bracing systems or checking deformations that could lead to an unacceptable failure resulting in a hazard to the users of the facility.

See Example 10.13.7.

10.11   FRAME ANALYSIS INCORPORATING THE EFFECT OF LATERAL MOVEMENT IN METAL DOWEL FASTENER CONNECTIONS

In 10.10 the stiffness of connections at the SLS and the ULS are defined and a method for determining the displacement at the connection when it is subjected to a lateral load is also given.

The displacement effect caused by slip in connections in a structure must be taken into account in any displacement analysis of the structure. Unless the structure is statically determinate, the slip at the connections will also affect the stress distribution around the structure and this must also be taken into account. Further, where the connections have different creep behaviour, from the content of 10.10 it will be understood that as well as having an effect on the displacement behaviour of the structure at the final deformation condition, the stress resultants at the ULS will also be affected and need to be addressed in the analysis.

Where a computer-based frame analysis is to be used, slip behaviour at connections can be included for by introducing a linear elastic spring element at each affected connection with an axial stiffness equal to the lateral stiffness of the connection. Spring elements with rotational stiffness simulating semi-rigid rotational behaviour can also be modelled, and this is discussed in Chapter 12.

In modelling these additional elements care has to be taken to ensure that the stability of the structure is retained, that the shear and flexural properties of the elements properly represent the connection behaviour, and that the size of the elements used will not result in ill-conditioned equations.

Guidance on the loading and stiffness requirements to be used in this type of analysis is given in 2.3.4.1 and 2.3.4.2, respectively.

10.12   REFERENCES

1 McLain, T.E. Connectors and fasteners: research needs and goals. In: Fridley, K.J. (ed), *Wood Engineering in the 21st Century; Research Needs and Goals*. ASCE, Reston, VA, 1998, pp. 56–69.

2 BS EN 1995-1-1:2004+A1:2008. Eurocode 5: Design of Timber Structures. Part 1-1: General – Common Rules and Rules for Buildings, British Standards Institution.

3 BS EN 14592:2008. *Timber Structures – Dowel-type Fasteners – Requirements*, British Standards Institution.

4 EN 10230-1:2000. *Steel Wire Nails – Part 1: Loose Nails for General Applications.*

5 Johansen, K.W. *Theory of Timber Connections, IABSE*, Publication No. 9, Bern, Switzerland, 1949, pp. 249–262.

6 BS EN 1993-1-1:2005. *Eurocode 3: Design of Steel Structures. Part 1-1: General Rules and Rules for Buildings*, British Standards Institution.

7 BS EN 1993-1-8:2005. *Eurocode 3: Design of Steel Structures. Part 1-8: Design of Joints*, British Standards Institution.

8 Blass H. J., Ehlbeck J., Rouger F. 'Simplified design of joints with dowel-type fasteners'. *Proceedings of PTEC'99, Pacific Timber Engineering Conference*, Rotorua, New Zealand, Vol. 3, 1999, pp. 275–279.

9 Ballerini, M. 'A EYM Based simplified design formula for the load-carrying capacity of dowel-type connections'. *Proceedings of 40th CIB-W18 Timber Structures Meeting*, paper 40-7-4, Bled, Slovenia, 2007.

10 Blass, H, J., Joints with dowel-type fasteners. In: Thelandersson, S. and Larsen, H. J. (eds), *Timber Engineering*, Wiley, London, 2003.

11 Larsen, H, J., Enjily, V. *Practical Design of Timber Structures to Eurocode 5*, Thomas Telford, London.

12 PD6693-1-1:2012, Incorporating Corrigendum No1: *PUBLISHED DOCUMENT – Recommendations for the design of timber structures: Design of Timber Structures – Part 1-1: General – Common Rules and Rules for Buildings*, British Standards Institution.

13 BS EN 409: 2009. *Timber Structures – Test Methods – Determination of the Yield Moment of Dowel-type Fasteners*. British Standards Institution.

14 BS EN 622-2, -3:2004, -4:2005, -5:2004. *Fibreboards – Specifications*. British Standards Institution.

15 Hankinson, R.L. Investigation of crushing strength of spruce at varying angles to the grain. *Air Service Information Circular*, Vol. 3, No. 259 (Material Section Paper No. 130), 1921.

16 Hilson, B.O. Joints with dowel-type fasteners – theory. In: Blass, H.J., Aune, P., Choo, B.S., et al. (eds), *Timber Engineering STEP 1*, 1st edn. Centrum Hout, Almere, 1995.

17 Heyman, J. *Basic Structural Theory*, Cambridge University Press, Mass.

18 NA to BS EN 1995-1-1:2004. *UK National Annex to Eurocode 5: Design of Timber Structures. Part 1-1: General – Common Rules and Rules for Buildings*, British Standards Institution.

19 Leijten, A., Van der Put, T. 'Splitting strength of beams loaded perpendicular to grain by connections, a fracture mechanical approach', Proceedings of the 8th World Conference on Timber Engineering, Lahti, Finland, June 2004, Volume 1, pp. 269–274.

20 BS EN 1382: 1999. *Timber structures – Test Methods – Withdrawal Capacity of Timber Fasteners*. British Standards Institution.

21 BS EN 1383: 1999. *Timber Structures – Test Methods – Pull-through Resistance of Timber Fasteners*. British Standards Institution.

22 NA to BS EN 1995-1-1:2004+A1:2008, *Incorporating National Amendment No 2: Design of Timber Structures. Part 1-1: General – Common Rules and Rules for Buildings*, British Standards Institution.

23 Ehlbeck, J., Larsen, H.J. EC5 – design of timber structures: joints. In: *Proceedings of International Workshop on Wood Connectors*, No. 7361, Forest Products Society, Madison, WI, 1993, pp. 9–23.

24 Dias, A.M.P.G., Cruz, H.M.P., Lopes, S.M.R., van de Kuilen, J.W. 'Stiffness of dowel-type fasteners in timber-concrete joints', *Structures and Buildings* 163, Issue 584, ICE Publishing.

25 NA to BS EN 1990:2002+A1:2005. *Incorporating National Amendment No1. UK National Annex for Eurocode 0 – Basis of Structural Design*, British Standards Institution.

26 BS EN 338:2009. *Structural Timber – Strength Classes*, British Standards Institution.

10.13 EXAMPLES

As stated in 4.3, in order to verify the ultimate and serviceability limit states, each design effect has to be checked and for each effect the largest value caused by the relevant combination of actions must be used.

However, to ensure attention is primarily focused on the EC5 design rules for the timber or wood product being used, only the design load case producing the largest design effect has generally been given or evaluated in the following examples.

Example 10.13.1 A timber-to-timber tension splice connection functioning in service class 2 conditions is required to connect two 60 mm by 145 mm timber members as shown in Figure E10.13.1(a). Two 35 mm by 145 mm timber side members will be used, connected by 3.4 mm nominal diameter smooth round nails 80 mm long without pre-drilling and driven from both sides to overlap in the central member. The nails have a tensile strength of 600 N/mm². The joint is subjected to a characteristic permanent action of 2.5 kN and a characteristic medium-term variable action of 3.5 kN and all timber is strength class C22 to BS EN 338:2009.

Determine the number of nails and the nailing pattern required to comply with the rules in EC5.

(a)

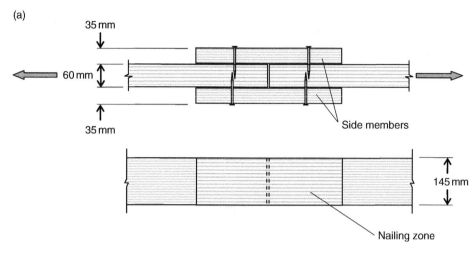

Fig. E10.13.1(a).

1. Geometric properties

Thickness of the side members, t_1	$t_1 = 35\,\text{mm}$	
Thickness of the inner members, t_2	$t_2 = 60\,\text{mm}$	
Width of the members, h	$h = 145\,\text{mm}$	
Cross sectional area of each of the side members, A_s	$A_s = h \cdot t_1$	$A_s = 5.08 \times 10^{-3}\,\text{m}^2$
Cross sectional area of the inner member, A_{in}	$A_{in} = h \cdot t_2$	$A_{in} = 8.7 \times 10^3\,\text{mm}^2$
Nail diameter, d	$d = 3.4\,\text{mm}$	
Nail head diameter, d_h	$d_h = 2.25 \cdot d$	$d_h = 7.65\,\text{mm}$
Nail length, ℓ_{nail}	$\ell_{nail} = 80\,\text{mm}$	
Nail pointside penetration (allowing 2.5d for nail point), t_{point}	$t_{point} = (\ell_{nail} - 2.5\,d) - t_1$	$t_{point} = 36.5\,\text{mm}$
Maximum nail pointside penetration, $t_2 - t_{point} > 4d$	$t_2 - t_{point} = 23.5\,\text{mm}$	
(Figure 10.6(c) (EC5, 8.3.1.1.(7)))	$4 \cdot d = 13.6\,\text{mm}$	i.e. OK
Minimum penetration of the nail in the inner timber member, $t_{point.pen}$	$t_{point.pen} = 8 \cdot d$	$t_{point.pen} = 27.2\,\text{mm}$
(10.3.5.1(1) (EC5, 8.3.1.2(1)))	$\dfrac{t_{point.pen}}{t_{point}} = 0.75$	ratio less than 1, i.e. OK

Both maximum and minimum nail penetration criteria met

2. Timber and nail properties

Table 1.3, strength class C22 (BS EN 338:2009(E), *Table 1*)

Characteristic tension strength parallel to the grain, $f_{t.0.k}$ $f_{t.0.k} = 13\,\text{N/mm}^2$

Characteristic density of the timber, ρ_k	$\rho_k = 340\,kg/m^3$
Tensile strength of each nail, f_u	$f_u = 600\,N/mm^2$

Check to validate that no predrilling is acceptable:

Density must be less than $500\,kg/m^3$ $\quad rp = \dfrac{\rho k}{500\,\dfrac{kg}{m^3}}$ $\quad rp = 0.68$ $\quad$ i.e. less than 1 therefore O.K.

Diameter of nail must be less than $6\,mm$ $\quad rd = \dfrac{d}{6\,mm}$ $\quad rd = 0.57$ $\quad$ i.e. less than 1 therefore O.K.

The timber thickness must not be less than t $\quad t = \max\left[7d, (13d - 30\,mm)\dfrac{\rho k}{400\,\dfrac{kg}{m^3}}\right]$

$t = 23.8\,mm$ $\qquad\qquad$ i.e. less than member thickness, therefore O.K.

3. Partial safety factors
Table 2.8 (UKNA to BS EN 1990:2002+A1:2005, *Table NA.A1.2(B)*) for the ULS

Permanent actions, γ_G	$\gamma_G = 1.35$
Variable actions γ_Q	$\gamma_Q = 1.5$

Table 2.6 (UKNA to EC5, *Table NA.3*)

Material factor for timber, γ_M,	$\gamma_M = 1.3$
Material factor for connections, $\gamma_{M.connection}$	$\gamma_{M.connection} = 1.3$

4. Actions
Characteristic permanent action, G_k	$G_k = 2.5\,kN$	
Characteristic variable action, Q_k	$Q_k = 3.5\,kN$	
Design tensile action for the critical load combination, F_d	$F_d = \gamma_G \cdot G_k + \gamma_Q \cdot Q_k$	$F_d = 8.63\,kN$

(Table 2.8, equation (c) using the unfavourable condition variable action)

5. Modification factors
Factor for medium-duration loading and service class 2, $k_{mod.med}$ (Table 2.4 (EC5, *Table 3.1*))	$k_{mod.med} = 0.80$
Tension member width factor for solid timber, k_h (Table 2.11 (EC5, *3.2*))	$k_h = 1$

6. Tension strength of the timber

The critical design load case at the ULS will be due to the combination of permanent and unfavourable medium-term duration variable action:

Nails are driven without pre-drilling and are less than 6 mm in diameter. In accordance with 5.3.3 (EC5, *5.2(3)*), the loss of area effect can be ignored.

Design tensile stress parallel to the grain in the side members, $\sigma1_{t.0.d}$

$$\sigma1_{t.0.d} = \frac{F_d}{2 \cdot A_s}$$

$\sigma1_{t.0.d} = 0.85\,\text{N/mm}^2$

Design tensile stress parallel to the grain in the inner member, $\sigma2_{t.0.d}$

$$\sigma2_{t.0.d} = \frac{F_d}{A_{in}}$$

$\sigma2_{t.0.d} = 0.99\,\text{N/mm}^2$

Design tensile strength parallel to grain, $f_{t.0.d}$

$$f_{t.0.d} = \frac{k_{mod.med} \cdot k_h \cdot f_{t.0.k}}{\gamma_M}$$

$f_{t.0.d} = 8\,\text{N/mm}^2$

Design strength exceeds design stress in members; therefore OK

7. Embedment strength of timber

Characteristic embedment strength of timber parallel to the grain, $f_{h \cdot k}$ (equation (10.28); EC5, *equation (8.15)*)

The equation incorporates dimension factors for Mathcad:

$$f_{h.k} = 0.082 \cdot \left(\rho_k \cdot \frac{m^3}{kg} \right) \cdot \left(d \cdot mm^{-1} \right)^{-0.3} \text{N/mm}^2$$

$f_{h.k} = 19.31\,\text{N/mm}^2$

Characteristic embedment strength of the headside timber, $f_{h.1.k}$

$f_{h.1.k} = f_{h.k}$

$f_{h.1.k} = 19.31\,\text{N/mm}^2$

Characteristic embedment strength of the pointside timber, $f_{h.2.k}$

$f_{h.2.k} = f_{h.k}$

$f_{h.2.k} = 19.31\,\text{N/mm}^2$

8. Yield moment of a nail

Characteristic yield moment of a nail, $M_{y.Rk}$ (Table 10.5 (EC5, *equation (8.14)*).

$$M_{y.Rk} = 0.3 \cdot \left(f_u \cdot \frac{mm^2}{N} \right) \cdot \left(d \cdot mm^{-1} \right)^{2.6} \cdot \text{Nmm}$$

$M_{y.Rk} = 4.34 \times 10^3 \text{Nmm}$

(The equation incorporates dimension factors for Mathcad.)

9. Withdrawal resistance

Nail pointside penetration/nail diameter: (10.8.1 (EC5, *8.3.2(7)*))

$$\frac{t_{point}}{d} = 10.74$$

i.e. less than 12 d

Nail headside penetration/nail diameter (10.8.1 (EC5, *8.3.2(7)*))

$$\frac{t_1}{d} = 10.29$$

i.e. less than 12 d

Characteristic pointside withdrawal strength, $fp_{ax.k}$ (10.8.1 (EC5, *8.3.2(6)&(7)*), (The equation incorporates dimensional factors for Mathcad.)

Pointside penetration factor, Dp

$$Dp = \left(\frac{t_{point}}{4 \cdot d} - 2 \right)$$

$$fp_{ax.k} = 20 \times 10^{-6} \cdot \left(\rho_k \cdot \frac{m^3}{kg} \right)^2 \cdot (Dp) \cdot N \cdot mm^{-2} \qquad\qquad fp_{ax.k} = 1.58\,N/mm^{-2}$$

Characteristic withdrawal strength in headside member, $fh_{ax.k}$ (10.8.1 (EC5, *8.3.2(6)&(7)*), (The equation incorporates dimensional factors for Mathcad.)

Pointside penetration factor, Dh $\qquad\qquad Dh = \left(\dfrac{t_1}{4 \cdot d} - 2 \right)$

$$fh_{ax.k} = 20 \times 10^{-6} \cdot \left(\rho_k \cdot \frac{m^3}{kg} \right)^2 \cdot Dh \cdot N \cdot mm^{-2} \qquad\qquad fh_{ax.k} = 1.33\,N/mm^2$$

Characteristic headside pull through strength, $f_{head.k}$ (equation (*10.59*) (EC5, *(8.26)*)) (The equation incorporates dimensional factors for Mathcad.)

$$f_{head.k} = 70 \times 10^{-6} \cdot \left(\rho_k \cdot \frac{m^3}{kg} \right)^2 \cdot N \cdot mm^{-2} \qquad\qquad f_{head.k} = 8.09\,N/mm^{-2}$$

Characteristic withdrawal capacity of nail,
$F_{ax.Rk}$, is lesser of equations (10.57)
(EC5, *equations (8.24)*)

$$F_{ax.Rk1} = fp_{ax.k} \cdot d \cdot t_{point} \qquad\qquad F_{ax.Rk1} = 196.2\,N$$

and $\quad F_{ax.Rk2} = fh_{ax.k} \cdot d \cdot t_1 + f_{head.k} \cdot d_h^2 \qquad\qquad F_{ax.Rk2} = 631.36\,N$

therefore $\quad F_{ax.Rk} = min(F_{ax.Rk1}, F_{ax.Rk2}) \qquad\qquad F_{ax.Rk} = 196.2\,N$

10. Load-carrying capacity of connection

For a timber-to-timber joint with nails in single shear, the characteristic lateral resistance per shear plane is the lesser of equations (10.1) to (10.6) in Table 10.2 (EC5, *equations (8.6)*) where

$$\beta = \frac{f_{h.2.k}}{f_{h.1.k}} \qquad \beta = 1 \quad and \quad t_2 = t_{point}$$

Failure mode (a): $\quad F_{v.Rk.a} = f_{h.1.k} \cdot t_1 \cdot d \qquad\qquad F_{v.Rk.a} = 2.3 \times 10^3\,N$

Failure mode (b): $\quad F_{v.Rk.b} = f_{h.2.k} \cdot t_2 \cdot d \qquad\qquad F_{v.Rk.b} = 2.4 \times 10^3\,N$

Failure mode (c):

$$F_{v.Rk.c} = \frac{f_{h.1.k} \cdot t_1 \cdot d}{1+\beta} \cdot \left[\sqrt{\beta + 2 \cdot \beta^2 \cdot \left[1 + \frac{t_2}{t_1} + \left(\frac{t_2}{t_1} \right)^2 \right] + \beta^3 \cdot \left(\frac{t_2}{t_1} \right)^2} - \beta \cdot \left[1 + \left(\frac{t_2}{t_1} \right) \right] \right] + \frac{F_{ax.Rk}}{4}$$

$$F_{v.Rk.c} = 1.02 \times 10^3\,N$$

Failure mode (d):

$$F_{v.Rk.d} = 1.05 \cdot \frac{f_{h.1.k} \cdot t_1 \cdot d}{2+\beta} \cdot \left[\sqrt{2 \cdot \beta \cdot (1+\beta) + \frac{4 \cdot \beta \cdot (2+\beta) \cdot M_{y.Rk}}{f_{h.1.k} \cdot t_1^2 \cdot d}} - \beta \right] + \frac{F_{ax.Rk}}{4}$$

$$F_{v.Rk.d} = 978.65\,N$$

Failure mode (e):

$$F_{v.Rk.c} = 1.05 \cdot \frac{f_{h.1.k} \cdot t_2 \cdot d}{1+2 \cdot \beta} \cdot \left[\sqrt{2 \cdot \beta^2 \cdot (1+\beta) + \frac{4 \cdot \beta \cdot (1+2 \cdot \beta) \cdot M_{y.Rk}}{f_{h.1.k} \cdot t_2^2 \cdot d}} - \beta \right] + \frac{F_{ax.Rk}}{4}$$

$$F_{v.Rk.e} = 1.01 \times 10^3 \, N$$

Failure mode (f):

$$F_{v.Rk.f} = 1.15 \cdot \sqrt{\frac{2 \cdot \beta}{1+\beta}} \cdot \left(2 \cdot M_{y.Rk} \cdot f_{h.1.k} \cdot d\right) + \frac{F_{ax.Rk}}{4}$$

$$F_{v.Rk.f} = 916.88 \, N$$

Limiting $F_{ax.Rk}/4$ to 15% of the Johansen part of the relevant equations – consider only modes (e) and (f) as possible limiting values (10.3.5.3, (EC5, 8.2.2(2)))

Failure mode (e):

$$F_{v.Rk.ee} = 1.15 \cdot 1.05 \cdot \frac{f_{h.1.k} \cdot t_2 \cdot d}{1+2 \cdot \beta} \cdot \left[\sqrt{2 \cdot \beta^2 \cdot (1+\beta) + \frac{4 \cdot \beta \cdot (1+2 \cdot \beta) \cdot M_{y.Rk}}{f_{h.1.k} \cdot t_2^2 \cdot d}} - \beta \right]$$

$$F_{v.Rk.ee} = 1.1 \times 10^3 \, N$$

Failure mode (f):

$$F_{v.Rk.ff} = 1.15 \cdot 1.15 \cdot \sqrt{\frac{2 \cdot \beta}{1+\beta}} \cdot \left(2 \cdot M_{y.Rk} \cdot f_{h.1.k} \cdot d\right)$$

$$F_{v.Rk.ff} = 998.01 \, N$$

The characteristic lateral resistance per shear plane per nail, $F_{v.Rk}$, will be

$$F_{v.Rk} = \min \left(F_{v.Rk.a}, F_{v.Rk.b}, F_{v.Rk.c}, F_{v.Rk.d}, F_{v.Rk.e}, F_{v.Rk.f}, F_{v.Rk.ee}, F_{v.Rk.ff} \right)$$

$$F_{v.Rk} = 916.88 \, N$$

i.e. failure mode (f)

The design resistance per nail per shear plane, $F_{v.Rd}$

$$F_{v.Rd} = \frac{k_{mod.med} \cdot F_{v.Rk}}{\gamma_{M.connection}}$$

$$F_{v.Rd} = 564.24 \, N$$

Number of nails required per side, N_{nails}

$$N_{nails} = \frac{F_d}{F_{v.Rd} \cdot 2}$$

$$N_{nails} = 7.64$$

Adopt nine nails per side, $N1_{nails}$

$$N1_{nails} = \text{ceil}(N_{nails}) + 1 \qquad N1_{nails} = 9$$

11. **Nail spacing (see Figure E10.13.1(b))**
    Table 10.8 (EC5, *Table 8.2*)

Angle of load relative to the grain, $\qquad \alpha = 0$

Minimum spacing parallel to the grain for $d < 5$ mm and $\rho_k < 420$ kg/m^3, $a1_1$ $\qquad a1_1 = (5+5 | \cos(\alpha)|) \cdot d \quad a1_1 = 34$ mm

To eliminate the effect of nail spacing in a row (i.e. to make $k_{ef} = 1$ in equation (10.43) (EC5, *equation (8.17)*)), from Table 10.12 (EC5, *Table 8.1*), a_1: $\qquad a_1 = 14 \cdot d \qquad a_1 = 47.6$ mm

Minimum spacing perpendicular to the grain, a_2 $a_2 = 5 \cdot d$ $a_2 = 17\,\mathrm{mm}$

Minimum loaded end distance, $a_{3.t}$ $a_{3.t} = (10 + 5\cos(\alpha)) \cdot d$ $a_{3.t} = 51\,\mathrm{mm}$

Minimum unloaded edge distance, $a_{4.c}$ $a_{4.c} = 5 \cdot d$ $a_{4.c} = 17\,\mathrm{mm}$

Adopt the layout as shown in Figure E10.13.1(b).

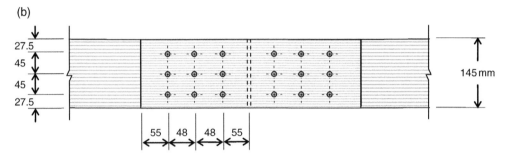

Fig. E10.13.1(b).

Example 10.13.2 A timber-plywood gusset plate apex joint for the connection shown in Figure E10.13.2 is to be designed using 12-mm-thick Finnish birch plywood with a characteristic density of $630\,\mathrm{kg/m^3}$ and fixed with the face grain horizontal. The joint fasteners are 3.00 mm nominal diameter by 50 mm long smooth round wire nails, fixed without pre-drilling, with a tensile strength of $650\,\mathrm{N/mm^2}$ and act in single shear. The timber members are strength class C18 to BS EN 338:2009 and the sizes are as shown in Figure E10.13.2. The joint will function under service class 2 conditions. The joint is subjected to design loading as shown, arising from a combination of permanent and medium-term variable actions.

Design the connection to comply with the requirements of EC5. (There is no requirement to check the strength of the timber.)

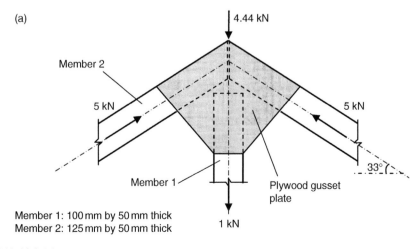

Fig. E10.13.2 (a).

1. Geometric properties

Plywood:

Thickness of each plywood gusset plate, t_p	$t_p = 12.0\,mm$	
Timber:		
Thickness of the timber, t_t	$t_t = 50\,mm$	
Width of timber member 1, h_1	$h_1 = 100\,mm$	
Width of timber member 2, h_2	$h_2 = 125\,mm$	
Cross-sectional area of member 1, A_1	$A_1 = h_1 \cdot t_t$	$A_1 = 5 \times 10^{-3}\,m^2$
Cross-sectional area of member 2, A_2	$A_2 = h_2 \cdot t_t$	$A_2 = 6.25 \times 10^{-3}\,m^2$
Nail diameter, d	$d = 3.0\,mm$	
Nail head diameter, d_h	$d_h = 2.25 \cdot d$	$d_h = 6.75\,mm$
Nail length, ℓ_{nail}	$\ell_{nail} = 50\,mm$	
Nail pointside penetration (allowing 2.5d for nail point), t_{point}	$t_{point} = (\ell_{nail} - 2.5\,d) - t_p$	$t_{point} = 30.5\,mm$
Minimum penetration of nail in inner member, $t_{point.pen}$ (10.3.5.1(1) (EC5, 8.3.1.2(1)))	$t_{point.pen} = 8 \cdot d$	$t_{point.pen} = 24\,mm$
	$\dfrac{t_{point.pen}}{t_{point}} = 0.79$	Ratio less than 1, i.e. OK

Minimum nail penetration criteria met

2. Timber and nail properties

Table 1.3, strength class C18 (BS EN 338:2009(E), *Table 1*)

Characteristic density of the timber, ρ_k $\rho_k = 320\,kg/m^3$

Table 1.10, 12-mm-thick Finnish birch plywood

Characteristic tensile strength at 90° to the face grain direction, $f_{p.t.90.k}$	$f_{p.t.90.k} = 35.0\,N/mm^2$
Characteristic compressive strength parallel to the face grain, $f_{p.c.0.k}$	$f_{p.c.0.k} = 27.7\,N/mm^2$
Characteristic compressive strength perpendicular to the face grain, $f_{p.c.90.k}$	$f_{p.c.90.k} = 24.3\,N/mm^2$
Characteristic density of the plywood, ρ_{pk}	$\rho_{pk} = 630\,kg/m^3$
Nails:	
Tensile strength of a nail, f_u	$f_u = 650\,N/mm^2$

Check to validate that no predrilling is acceptable:

Density must be less than 500 kg/m³	$rp = \dfrac{\rho_k}{500 \cdot \dfrac{kg}{m^3}}$	$rp = 0.64$	i.e. less than 1 therefore O.K.
Diameter of nail must be less than 6 mm	$rp = \dfrac{d}{6 \cdot mm}$	$rd = 0.5$	i.e. less than 1 therefore O.K.

The timber thickness must not be less than t

$$t = max\left[7 \cdot d, (13 \cdot d - 30 \cdot mm) \cdot \cfrac{\rho_k}{400 \cdot \cfrac{kg}{m^3}}\right]$$

$t = 21\,mm$ i.e. less than timber member thickness, therefore O.K.

3. Partial safety factors
Table 2.6 (UKNA to EC5, *Table NA.3*)

Material factor for plywood $\gamma_{M.plywood} = 1.2$

Material factor for connections $\gamma_{M.connection} = 1.3$

4. Actions
Design action in member 1, $F_{d.1}$ $F_{d.1} = 1\,kN$

Design action in member 2, $F_{d.2}$ $F_{d.2} = 5\,kN$

5. Modification factors
Factor for medium-duration loading and service class 2, $k_{mod.med}$
(Table 2.4 (EC5, *Table 3.1*))

$k_{mod.med} = 0.80$

Bearing factor, $k_{c.90}$
(equation (4.22) (EC5, *6.1.5 (1)*))

$k_{c.90} = 1$

6. Embedment strength of timber
Characteristic embedment strength of timber parallel to the grain
(equation (10.28); EC5, *equation (8.15)*)
(The equation incorporates dimension factors for Mathcad.)

$$f_{h.k} = 0.082 \cdot \left(\rho_k \cdot \frac{m^3}{kg}\right) \cdot \left(d \cdot mm^{-1}\right)^{-0.3} \cdot N \cdot mm^{-2}$$ $f_{h.k} = 18.87\,N/mm^2$

Characteristic embedment strength of the plywood, $f_{h.p.k}$
(Table 10.6 (EC5, *equation (8.20)*)

$$f_{h.p.k} = 0.11 \cdot \left(\rho_{pk} \cdot \frac{m^3}{kg}\right) \cdot \left(d \cdot mm^{-1}\right)^{-0.3} \cdot N \cdot mm^{-2}$$ $f_{h.p.k} = 49.84\,N/mm^2$

7. Yield moment of a nail
Characteristic yield moment of a nail, $M_{y.Rk}$
(Table 10.5 (EC5, *equation (8.14)*))
(The equation incorporates dimension factors for Mathcad.)

$$M_{y.Rk} = 0.3 \cdot \left(f_u \cdot \frac{mm^2}{N}\right) \cdot \left(d \cdot mm^{-1}\right)^{2.6} \cdot N \cdot mm$$

$M_{y.Rk} = 3.39 \times 10^3\,N\,mm$

8. Withdrawal resistance
Nail pointside penetration/nail diameter (10.8.1 (EC5, *8.3.2(7)*))

$\dfrac{t_{point}}{d} = 10.17$ i.e. more than $8d$ and less than $12d$

Nail headside penetration/nail diameter (10.8.1 (EC5, *8.3.2(7)*))

$\dfrac{t_p}{d} = 4$ i.e. less than $8d$

Characteristic pointside withdrawal strength, $fp_{ax.k}$
(10.8.1 (EC5, *8.3.2(6), (7)*))
(The equation incorporates dimensional factors for Mathcad.)

Pointside penetration factor, Dp $\qquad Dp = \left(\dfrac{t_{point}}{4 \cdot d} - 2 \right)$

$$fp_{ax.k} = 20 \times 10^{-6} \cdot \left(\rho_k \cdot \dfrac{m^3}{kg} \right)^2 \cdot Dp \cdot N \cdot mm^{-2} \qquad\qquad fp_{ax.k} = 1.11\,N/mm^{-2}$$

Characteristic withdrawal strength in headside member, $fh_{ax.k}$
(10.8.1 (EC5, *8.3.2(6), (7)*)) (The equation incorporates dimensional factors for Mathcad.)

Headside penetration factor, Dh $\qquad Dh = \left(\dfrac{t_p}{4 \cdot d} - 2 \right)$

$$fh_{ax.k} = 20 \times 10^{-6} \cdot \left(\rho_{\rho k} \cdot \dfrac{m^3}{kg} \right)^2 \cdot Dh \cdot N \cdot mm^{-2} \qquad\qquad fh_{ax.k} = 1.11\,N/mm^2$$

Characteristic headside pull-through strength, $f_{head.k}$
(equation (10.59) (EC5, *(8.26)*) (The equation incorporates dimensional factors for Mathcad.

$$f_{headx.k} = 70 \times 10^{-6} \cdot \left(\rho_{\rho k} \cdot \dfrac{m^3}{kg} \right)^2 \cdot Dh \cdot N \cdot mm^{-2} \qquad\qquad f_{head.k} = 27.78\,N/mm^2$$

Characteristic withdrawal capacity of nail, $F_{ax.Rk}$,
is lesser of equations (10.57) (EC5, *equations (8.24)*)

$$F_{ax.Rk.1} = fp_{ax.k} \cdot d \cdot t_{point} \qquad\qquad F_{ax.Rk.1} = 101.5\,N$$
and $\qquad\qquad F_{ax.Rk.2} = fh_{ax.k} \cdot d \cdot t_p + f_{head.k} \cdot d_h^2 \qquad\qquad F_{ax.Rk.2} = 1.27 \times 10^3\,N$
therefore $\qquad\qquad F_{ax.Rk} = \min\,(F_{ax.Rk.1},\, F_{ax.Rk.2}) \qquad\qquad F_{ax.Rk} = 101.5\,N$

9. Load-carrying capacity of connection

For a plywood-to-timber joint with nails in single shear, the characteristic lateral resistance per shear plane is the smallest value of equations (10.1)–(10.6) in Table 10.2 (EC5, *equations (8.6)*) where:

$\beta = \dfrac{f_{h.k}}{f_{h.p.k}} \qquad \beta = 0.38 \quad \text{and} \quad t_2 = t_{point} \quad \text{and} \quad t_1 = t_p$

Failure mode (a): $\qquad\qquad F_{v.Rk.a} = f_{h.p.k} \cdot t_1 \cdot d \qquad\qquad F_{v.Rk.a} = 1.79 \times 10^3\,N$

Failure mode (b): $\qquad\qquad F_{v.Rk.b} = f_{h.k} \cdot t_2 \cdot d \qquad\qquad F_{v.Rk.b} = 1.73 \times 10^3\,N$

Failure mode (c):

$$F_{v.Rk.c} = \dfrac{f_{h.p.k} \cdot t_1 \cdot d}{1 + \beta} \cdot \left[\sqrt{ \beta + 2 \cdot \beta^2 \cdot \left[1 + \dfrac{t_2}{t_1} + \left(\dfrac{t_2}{t_1} \right)^2 \right] + \beta^3 \cdot \left(\dfrac{t_2}{t_1} \right)^2 } [-\beta \cdot \left[1 + \left(\dfrac{t_2}{t_1} \right) \right] \right] + \dfrac{F_{ax.Rk}}{4}$$

$$F_{v.Rk.c} = 748.51\,N$$

Failure mode (d):

$$F_{v.Rk.d} = 1.05 \cdot \frac{f_{h.p.k} \cdot t_1 \cdot d}{2+\beta} \cdot \left[\sqrt{2 \cdot \beta \cdot (1+\beta) + \frac{4 \cdot \beta \cdot (2+\beta) \cdot M_{y.Rk}}{f_{h.p.k} \cdot t_1^2 \cdot d}} - \beta\right] + \frac{F_{ax.Rk}}{4}$$

$$F_{v.Rk.d} = 731.01\,N$$

Failure mode (e):

$$F_{v.Rk.e} = 1.05 \cdot \frac{f_{h.p.k} \cdot t_2 \cdot d}{1+2 \cdot \beta} \cdot \left[\sqrt{2 \cdot \beta^2 \cdot (1+\beta) + \frac{4 \cdot \beta \cdot (1+2 \cdot \beta) \cdot M_{y.Rk}}{f_{h.p.k} \cdot t_2^2 \cdot d}} - \beta\right] + \frac{F_{ax.Rk}}{4}$$

$$F_{v.Rk.e} = 842.22\,N$$

Failure mode (f):

$$F_{v.Rk.f} = 1.15 \cdot \sqrt{\frac{2 \cdot \beta}{1+\beta}} \cdot \left(2 \cdot M_{y.Rk} \cdot f_{h.p.k} \cdot d\right) + \frac{F_{ax.Rk}}{4} \qquad F_{v.Rk.f} = 883.9\,N$$

Limiting $F_{ax.Rk}/4$ to 15% of the Johansen part of the relevant equations – consider only mode (d) as the limiting value (10.3.5.3, (EC5, 8.2.2(2)))

Failure mode (d):

$$F_{v.Rk.dd} = 1.15 \cdot 1.05 \cdot \frac{f_{h.p.k} \cdot t_2 \cdot d}{2+\beta} \cdot \left[\sqrt{2 \cdot \beta \cdot (1+\beta) + \frac{4 \cdot \beta \cdot (2+\beta) \cdot M_{y.Rk}}{f_{h.p.k} \cdot t_1^2 \cdot d}} - \beta\right]$$

$$F_{v.Rk.dd} = 811.48\,N$$

The characteristic lateral resistance per shear plane per nail will be

$$F_{v.Rk} = \min(F_{v.Rk.a}, F_{v.Rk.b}, F_{v.Rk.c}, F_{v.Rk.d}, F_{v.Rk.e}, F_{v.Rk.f}, F_{v.Rk.dd}) \qquad F_{v.Rk} = 731.01\,N$$

i.e. failure mode (d)

The design resistance per nail per shear plane, $F_{v.Rd}$

$$F_{v.Rd} = \frac{k_{mod.med} \cdot F_{v.Rk}}{\gamma_{M.connection}} \qquad F_{v.Rd} = 449.85\,N$$

Number of nails for member 1 connection per shear plane:

Number of nails required per side, $N1_{nails}$

$$N1_{nails} = \frac{F_{d.1}}{F_{v.Rd} \cdot 2} \qquad N1_{nails} = 1.11$$

For a symmetrical nailing pattern adopt 2 nails per side

$$N1_{nails} = \text{ceil}(N1_{nails}) \qquad N1_{nails} = 2$$

Number of nails for member 2 connection per shear plane:

Number of nails required per side, $N2_{nails}$

$$N2_{nails} = \frac{F_{d.2}}{F_{v.Rd} \cdot 2} \qquad N2_{nails} = 5.56$$

For a symmetrical nailing pattern adopt six nails per side

$$N2_{nails} = \text{ceil}(N2_{nails}) \qquad N2_{nails} = 6$$

10. Check strength of plywood gusset plates

Strength of gusset plates at connection with member 1:
(can ignore the loss of area due to nail holes)

Tension force taken by gusset plates, $F_{d.1}$

$$F_{d.1} = 1 \times 10^3 \, \text{N}$$

Design tensile stress in gusset plates – assuming a conservative tensile stress area $= 2(t_p \times h_1)$, $\sigma p1_{t.90.d}$

$$\sigma p1_{t.90.d} = \frac{F_{d.1}}{2 \cdot (t_p \cdot h_1)}$$

$$\sigma p1_{t.90.d} = 0.42 \, \text{N/mm}^2$$

Design tensile strength of gusset plates, $fp1_{t.90.d}$

$$fp1_{t.90.d} = \frac{k_{mod.med} \cdot f_{p.t.90.k}}{\gamma_{M.plywood}}$$

$$fp1_{t.90.d} = 23.33 \, \text{N/mm}^2$$

i.e. stress is less than strength; OK

Strength of gusset plates at connection with member 2:
(can ignore the loss of area due to nail holes)

Compression force taken by gusset plates, $F_{d.2} \cos(33°)$

$$F2_d = F_{d.2} \cdot \cos(33°)$$

$$F2_d = 4.19 \times 10^3 \, \text{N}$$

Design compressive stress in gusset plates – assuming a conservative stress area $= 2(t_p \times h_2)$, $\sigma2_{t.90.d}$

$$\sigma p2_{t.90.d} = \frac{F2_d}{2 \cdot (t_p \cdot h_2)}$$

$$\sigma p2_{t.90.d} = 1.4 \, \text{N/mm}^2$$

Design compressive strength of gusset plates, $fp_{c.\alpha.d}$ (equation (5.14); EC5, *equation (6.16)*)

$$fp2_{c.90.d} = \frac{k_{mod.med} \cdot f_{p.c.90.k}}{\gamma_{M.plywood}}$$

$$fp2_{c.90.d} = 16.2 \, \text{N/mm}^2$$

$$fp2_{c.0.d} = \frac{k_{mod.med} \cdot f_{p.c.0.k}}{\gamma_{M.plywood}}$$

$$fp2_{c.0.d} = 18.47 \, \text{N/mm}^2$$

$$fp_{c.\alpha.d} = \frac{fp2_{c.0.d}}{\dfrac{fp2_{c.0.d}}{k_{c.90} \cdot fp2_{c.90.d}} \cdot \sin(33 \cdot \text{deg})^2 + \cos(33 \cdot \text{deg})^2}$$

$$fp_{c.\alpha.d} = 17.73 \, \text{N/mm}^2$$

i.e. stress is less than strength, OK

11. Nail spacing (see Figure 10.13.2(b))

Table 10.8 (EC5, *Table 8.2*) incorporating the requirements of EC5, *8.3.1.3(1) and (2)* for nailed panel connections:

Angle to timber grain

$$\alpha = 0°$$

Angle to plywood face grain relative to timber grain

$$\alpha_p = 33°$$

Minimum spacing parallel to grain for $d < 5$ mm and $\rho_k < 420 \, \text{kg/m}^3$, $a1_1$

$$a1_1 = 0.85 \cdot (5 + 5 \mid \cos(\alpha) \mid) \cdot d$$

$$a1_1 = 25.5 \, \text{mm}$$

To eliminate the effect of nail spacing in a row (i.e. to make $k_{ef}=1$ in equation (10.43) (EC5, *equation (8.17)*)), from Table 10.12 (EC5, *Table 8.1*), a_1	$a_1 = 14 \cdot d$	$a_1 = 42\,\text{mm}$
Minimum spacing perpendicular to the timber grain and the plywood grain, a_2	$a_2 = 0.85.5 \cdot d$	$a_2 = 12.75\,\text{mm}$
Minimum loaded end distance for timber, $a_{3.t}$	$a_{3.t} = (10+5\cos(\alpha)) \cdot d$	$a_{3.t} = 45\,\text{mm}$
Minimum unloaded end distance for timber, $a_{3.c}$	$a_{3.c} = 10 \cdot d$	$a_{3.c} = 30\,\text{mm}$
Minimum unloaded edge distance for timber, $a_{4.c}$	$a_{4.c} = 5 \cdot d$	$a_{4.c} = 15\,\text{mm}$
Minimum loaded end and edge distance for plywood, $ap_{3.c}$	$ap_{3.c} = (3+4\sin(\alpha_p)) \cdot d$	$ap_{3.c} = 9\,\text{mm}$
Minimum loaded edge distance for plywood when loaded at 90° to the grain, $ap1_{3.t}$	$ap1_{3.t} = (3+4\sin(90°)) \cdot d$	$ap1_{3.t} = 21\,\text{mm}$
Minimum unloaded end and edge distance for plywood, $ap_{4.t}$	$ap_{4.t} = 3 \cdot d$	$ap_{4.t} = 9\,\text{mm}$

(b)

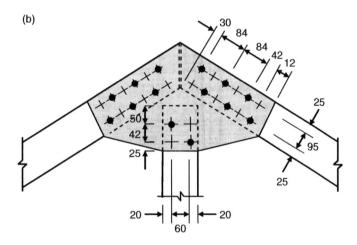

Fig. E10.13.2(b).

Example 10.13.3 The five-member single-bolt (12 mm diameter) joint of a truss functioning in service class 2 conditions and supporting medium-term variable loading is shown in Figure E10.13.3. All members are 150 mm wide C18 timber to BS EN 338:2009 and the member thicknesses are as detailed below. The design forces in the members are a combination of permanent and medium-term variable actions.

Check that the joint complies with the design requirements of EC5. (There is no requirement to check loaded end and edge distances.)

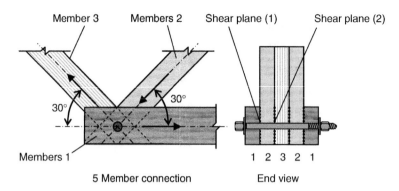

Fig. E10.13.3.

Data:

Member	Thickness	Force	Design value
1	$t_1 = 50$ mm	In each member	2.5 kN in tension
2	$t_2 = 50$ mm	In each member	1.444 kN in compression
3	$t_3 = 70$ mm	In member	2.887 kN in tension

Tensile strength of the bolt $f_{u.k} = 400$ N/mm²
Characteristic density of the timber $\rho_k = 320$ kg m³

1. Geometric properties

Thickness of members 1, t_1	$t_1 = 50$ mm
Thickness of members 2, t_2	$t_2 = 50$ mm
Thickness of member 3, t_3	$t_3 = 70$ mm
Member width (all members), h	$h = 150$ mm
Loaded edge distance, h_e	$h_e = \dfrac{h}{2}$ $h_e = 75$ mm
Bolt diameter, d	$d = 12$ mm
Tensile stress area of the bolt, A_{bt}	$A_{bt} = 84.3$ mm²

2. Timber and bolt properties
Table 1.3, strength class C18 (BS EN 338:2009(E), *Table 1*)

Characteristic strength perpendicular to the grain, $f_{c.90.k}$	$f_{c.90.k} = 2.2$ N/mm²
Characteristic density of the timber, ρ_k	$\rho_k = 320$ kg/m³
Characteristic strength of the bolt, $f_{u.k}$	$f_{u.k} = 400$ N/mm²

3. **Partial safety factors and modification factor**
   Table 1.6 (UKNA to BS EN 1995-1-1, Table NA.3)

   Material factor for connection, $\gamma_{M.connection}$ $\gamma_{M.connection} = 1.3$

   Material factor, γ_M $\gamma_M = 1.3$

   Factor for medium-duration loading $k_{mod.med} = 0.80$
   and service class 2, $k_{mod.med}$
   (Table 2.4 (EC5, *Table 3.1*))

4. **Actions**
   Design action in member 1, $F_{v.d.1}$ $F_{v.d.1} = 2.5\,\text{kN}$

   Design action in member 2, $F_{v.d.2}$ $F_{v.d.2} = 1.444\,\text{kN}$

   Design action in member 3, $F_{v.d.3}$ $F_{v.d.3} = 2.887\,\text{kN}$

5. **Strength of shear planes in joint**
   Embedment factor for softwood, k_{90} $k_{90} = 1.35 + 0.015 \cdot d \cdot \text{mm}^{-1}$
   (Table 10.7, (EC5, *equation (8.32)*)) $k_{90} = 1.53$

   Characteristic embedment strength of timber parallel to the grain, $f_{h.0.k}$
   (equation (10.35); EC5, *equation (8.32)*))
   (The equation incorporates dimension factors for Mathcad.)

   $$f_{h.0.k} = 0.082 \cdot (1 - 0.01 \cdot d \cdot \text{mm}^{-1}) \cdot \left(\rho_k \cdot \frac{\text{m}^3}{\text{kg}} \right) \cdot \text{N} \cdot \text{mm}^{-2} \qquad\qquad f_{h.0.k} = 23.09\,\text{N/mm}^2$$

   Shear plane force and its angle relative to the grain:

   Design force in shear plane 1, $F_{v.sp.1} = F_{v.d.1}$ $F_{v.sp.1} = 2.5 \times 10^3\,\text{N}$

   Angle of shear plane 1 force relative to $\theta_{1.2} = 0 \cdot \text{deg}$
   the grain direction in member 1 in the
   joint comprising members 1,2,1, $\theta_{1.2}$

   Angle of shear plane 1 force relative to $\theta_{2.1} = 30 \cdot \text{deg}$
   the grain direction in member 2 in the
   joint comprising members 1,2,1, $\theta_{2.1}$

   Design force in shear plane 2, $F_{v.sp.2}$

   $$F_{v.sp.2} = [(F_{v.sp.1} - F_{v.d.2} \cdot \cos(30 \cdot \text{deg}))^2 + (F_{v.d.2} \cdot \sin(30 \cdot \text{deg}))^2]^{0.5} \qquad\qquad F_{v.sp.2} = 1.44\,\text{kN}$$

   To obtain angle of shear plane 2 force
   relative to the horizontal plane:

   Horizontal component

   $$F_{v.sp.2.h} = F_{v.d.1} - F_{v.d.2} \cdot \cos(30 \cdot \text{deg}) \qquad\qquad F_{v.sp.2.h} = 1.25\,\text{kN}$$

   Vertical component

   $$F_{v.sp.2.v} = -F_{v.d.2} \cdot \sin(30 \cdot \text{deg}) \qquad\qquad F_{v.sp.2.v} = -0.72\,\text{kN}$$

Angle of shear plane 2 force relative to the horizontal plane

$$\text{angle}\,(F_{\text{v.sp.2.h}},F_{\text{v.sp.2.v}})\cdot\frac{180}{\pi}=330$$

Angle of the shear plane 2 force relative to the grain of member 2 in the joint comprising members 2,3,2, $\theta_{2.3.a}$

$\theta_{2.3.a}=60\cdot\deg$

Angle of the shear plane 2 force relative to the grain of member 3 in the joint comprising members 2,3,2, $\theta_{3.2.a}$

$\theta_{3.2.a}=0\cdot\deg$

Angle of the shear plane 2 force relative to the grain of member 3 in the joint comprising members 3,2,3: $\theta_{3.2.b}$

$\theta_{3.2.b}=0\cdot\deg$

Angle of the shear plane 2 force relative to the grain of member 2 in the joint comprising members 3,2,3: $\theta_{2.3.b}$

$\theta_{2.3.b}=60\cdot\deg$

Embedment strength of the timber:

Of member 1 in joint 1,2,1, $f_{h.\theta.1.2.k}$

$$f_{h.\theta.1.2.k}=\frac{f_{h.0.k}}{k_{90}\cdot\sin(\theta_{1.2})^2+\cos(\theta_{1.2})^2}$$

$f_{h.\theta.1.2.k}=23.09\,\text{N/mm}^2$

Of member 2 in joint 1,2,1, $f_{h.\theta.2.1.k}$

$$f_{h.\theta.2.1.k}=\frac{f_{h.0.k}}{k_{90}\cdot\sin(\theta_{2.1})^2+\cos(\theta_{2.1})^2}$$

$f_{h.\theta.2.1.k}=20.39\,\text{N/mm}^2$

Of member 2 in joint 2,3,2, $f_{h.\theta.2.3.a.k}$

$$f_{h.\theta.2.3.a.k}=\frac{f_{h.0.k}}{k_{90}\cdot\sin(\theta_{2.3.a})^2+\cos(\theta_{2.3.a})^2}$$

$f_{h.\theta.2.3.a.k}=16.52\,\text{N/mm}^2$

Of member 3 in joint 2,3,2, $f_{h.\theta.3.2.a.k}$

$$f_{h.\theta.3.2.a.k}=\frac{f_{h.0.k}}{k_{90}\cdot\sin(\theta_{3.2.a})^2+\cos(\theta_{3.2.a})^2}$$

$f_{h.\theta.3.2.a.k}=23.09\,\text{N/mm}^2$

Of member 3 in joint 3,2,3, $f_{h.\theta.3.2.b.k}$

$$f_{h.\theta.3.2.b.k}=\frac{f_{h.0.k}}{k_{90}\cdot\sin(\theta_{3.2.b})^2+\cos(\theta_{3.2.b})^2}$$

$f_{h.\theta.3.2.b.k}=23.09\,\text{N/mm}^2$

Of member 2 in joint 3,2,3, $f_{h.\theta.2.3.b.k}$

$$f_{h.\theta.2.3.b.k}=\frac{f_{h.0.k}}{k_{90}\cdot\sin(\theta_{2.3.b})^2+\cos(\theta_{2.3.b})^2}$$

$f_{h.\theta.2.3.b.k}=16.52\,\text{N/mm}^2$

Characteristic yield moment of the bolt, $M_{y.Rk}$
(Table 10.5 (EC5, *equation (8.14)*))
(The equation incorporates dimension factors for Mathcad.)

$$M_{y.Rk}=0.3\cdot\left(f_{u.k}\cdot\frac{\text{mm}^2}{\text{N}}\right)\cdot(d\cdot\text{mm}^{-1})^{2.6}\cdot\text{N}\cdot\text{mm}$$

$M_{y.Rk}=7.67\times10^4\cdot\text{N}\,\text{mm}$

Design strength of the bolt, $F1_{ax.Rd}$:
(BS EN 1993-1-8.)

$$F1_{ax.Rd} = 0.9 \cdot f_{u.k} \cdot \frac{1}{1.25} \cdot A_{bt}$$

$$F1_{ax.Rd} = 2.43 \times 10^4 \, \text{N mm}$$

Washer diameter used, d_w

$$d_w = 3 \cdot d \qquad\qquad d_w = 36 \, \text{mm}$$

Design capacity of the washer $F2_{ax.Rd}$:

$$F2_{ax.Rd} = 3 \cdot f_{c.90.k} \cdot \frac{\pi}{4} [d_w^2 - (d+1 \cdot \text{mm})^2] \cdot \frac{k_{mod.med}}{\gamma_{M.connection}}$$

(10.8.2 (EC5, 8.5.2(2)))

$$F2_{ax.Rd} = 3.6 \times 10^3 \, \text{N}$$

As failure is by bearing, characteristic axial withdrawal capacity of the bolt, $F_{ax.Rk}$:

$$F_{ax.Rk} = F2_{ax.Rd} \cdot \frac{\gamma_{M.connection}}{k_{mod.med}}$$

$$F_{ax.Rk} = 5.84 \times 10^3 \, \text{N}$$

Shear plane 1 – strength assessment:

Double shear joint formed from members 1,2,1. For a timber-to-timber joint with a bolt in double shear, the characteristic lateral resistance per shear plane is the smallest value of equations (10.7)–(10.10) in Table 10.2 (EC5, *equations (8.7)*), where:

Ratio of the embedment strength of member 2 to member 1, $\beta_{2.1}$

$$\beta_{2.1} = \frac{f_{h.\theta.2.1.k}}{f_{h.\theta.1.2.k}} \qquad\qquad \beta_{2.1} = 0.88$$

Failure mode (g):

$$F_{v.Rk.1} = f_{h.\theta.1.2.k} \cdot t_1 \cdot d \qquad\qquad F_{v.Rk.1} = 13.85 \, \text{kN}$$

Failure mode (h):

$$F_{v.Rk.2} = 0.5 \cdot f_{h.\theta.2.1.k} \cdot t_2 \cdot d \qquad\qquad F_{v.Rk.2} = 6.12 \, \text{kN}$$

Failure mode (j):

$$F_{v.Rk.3} = 1.05 \cdot \frac{f_{h.\theta.1.2.k} \cdot t_1 \cdot d}{2+\beta_{2.1}} \cdot \left[\sqrt{2 \cdot \beta_{2.1} \cdot (1+\beta_{2.1}) + \frac{4 \cdot \beta_{2.1} \cdot (2+\beta_{2.1}) \cdot M_{y.Rk}}{f_{h.\theta.1.2.k} \cdot t_1^2 \cdot d}} - \beta_{2.1} \right] + \frac{F_{ax.Rk}}{4}$$

$$F_{v.Rk.3} = 7.65 \, \text{kN}$$

Failure mode (k):

$$F_{v.Rk.4} = 1.15 \cdot \sqrt{\frac{2 \cdot \beta_{2.1}}{1+\beta_{2.1}}} \cdot (2 \cdot M_{y.Rk} \cdot f_{h.\theta.1.2.k} \cdot d) + F_{ax.Rk} \qquad\qquad F_{v.Rk.4} = 13.11 \, \text{kN}$$

Check mode (j) based on 25% of Johansen load

$$F_{v.Rk.3.3} = 1.25 \cdot 1.05 \cdot \frac{f_{h.\theta,1.2.k} \cdot t_1 \cdot d}{2+\beta_{2.1}} \cdot \left[\sqrt{2 \cdot \beta_{2.1} \cdot (1+\beta_{2.1}) + \frac{4 \cdot \beta_{2.1} \cdot (2+\beta_{2.1}) \cdot M_{y.Rk}}{f_{h.\theta.1.2.k} \cdot t_1^2 \cdot d}} - \beta_{2.1} \right]$$

$$F_{v.Rk.3.3} = 7.74 \, \text{kN}$$

All modes are possible and the minimum characteristic strength of shear plane 1 is:

$$F_{v.Rk.1.2} = \min (F_{v.Rk.1}, F_{v.Rk.2}, F_{v.Rk.3}, F_{v.Rk.4}, F_{v.Rk.3.3}) \qquad F_{v.Rk.1.2} = 6.12 \, \text{kN} \quad \text{(mode (h))}$$

Design force in shear plane 1	$F_{v.sp.1} = 2.5 \times 10^3 \, \text{N}$	

Design capacity of shear plane 1

$$F_{v.R.d.sp.1} = \frac{k_{mod.med}}{\gamma_{M.connection}} \cdot F_{v.Rk.1.2}$$

$$F_{v.R.d.sp.1} = 3.76 \, \text{kN OK}$$

Design capacity of shear plane 1 exceeds the design force in shear plane 1; therefore OK.

Shear plane 2 strength assessment (based on 2,3,2 configuration):

Double shear joint formed from members 2,3,2. For a timber-to-timber joint with a bolt in double shear, the characteristic lateral resistance per shear plane is the smallest value of equations (10.7)–(10.10) in Table 10.2 (EC5, *equations (8.7)*), where:

Ratio of the embedment strength of member 3 to member 2, $\beta_{3.2}$

$$\beta_{3.2} = \frac{f_{h.\theta.3.2.a.k}}{f_{h.\theta.2.3.a.k}}$$

$$\beta_{3.2} = 1.4$$

Failure mode (g):

$$F_{v.Rk.1.a} = f_{h.\theta.2.3.a.k} \cdot t_2 \cdot d$$

$$F_{v.Rk.1.a} = 9.91 \, \text{kN}$$

Failure mode (h):

$$F_{v.Rk.2.a} = 0.5 \cdot f_{h.\theta.3.2.a.k} \cdot t_3 \cdot d$$

$$F_{v.Rk.2.a} = 9.7 \, \text{kN}$$

Failure modes (j) and (k) are not compatible with $F_{v.Rk.1.2}$ and are not considered (10.7 (EC5, 8.1.3(2)))

Minimum value for shear plane 2

$$F_{v.Rk.2.3} = \min(F_{v.Rk.1.a}, F_{v.Rk.2.a})$$

$$F_{v.Rk.2.3} = 9.7 \, \text{kN}$$
(mode (h))

Shear plane 2 strength assessment (based on 3,2,3 configuration):

Double shear joint formed from members 3,2,3. For a timber-to-timber joint with a bolt in double shear, the characteristic lateral resistance per shear plane is the smallest value of equations (10.7)–(10.10) in Table 10.2 (EC5, *equations (8.7)*), where:

Ratio of the embedment strength of member 2 to member 3, $\beta_{2.3}$

$$\beta_{2.3} = \frac{f_{h.\theta.2.3.b.k}}{f_{h.\theta.3.2.b.k}}$$

$$\beta_{2.3} = 0.72$$

Failure mode (g):

$$F_{v.Rk.1.b} = f_{h.\theta.3.2.b.k} \cdot t_3 \cdot d$$

$$F_{v.Rk.1.b} = 19.4 \, \text{kN}$$

Failure mode (h):

$$F_{v.Rk.2.b} = 0.5 \cdot f_{h.\theta.2.3.b.k} \cdot t_2 \cdot d$$

$$F_{v.Rk.2.b} = 4.96 \, \text{kN}$$

Failure modes (j) and (k) are not compatible with $F_{v.Rk.1.2}$ and not considered (10.7 (EC5, 8.1.3(2)))

Minimum value for the shear plane

$$F_{v.Rk.3.2} = \min(F_{v.Rk.1.b}, F_{v.Rk.2.b})$$

$$F_{v.Rk.3.2} = 4.96 \, \text{kN}$$
(mode (h))

Strength of shear plane 2 (based on minimum value for 2,3,2 and 3,2,3 configurations):

Minimum characteristic strength of shear plane 2, $F_{v.Rk.2.3.f}$

$$F_{v.Rk.2.3.f} = \min(F_{v.Rk.2.3}, F_{v.Rk.3.2})$$

$$F_{v.Rk.2.3.f} = 4.96 \, \text{kN}$$
(mode (h))

Design force in shear plane 2

$$F_{v.sp.2} = 1.44 \times 10^3 \, \text{N}$$

Design capacity of shear plane 2

$$F_{v.R.d.sp.2} = \frac{k_{mod.med}}{\gamma_{M.connection}} \cdot F_{v.Rk.2.3.f}$$

$$F_{v.R.d.sp.2} = 3.05 \, \text{kN}$$

Design capacity of shear plane 2 exceeds the design force in shear plane 2; therefore OK.

6. Splitting capacity of timber

As the direction of shear plane 1 force relative to member 1 and of shear plane 2 force relative on member 3 is parallel to the grain of these members, timber splitting will not be relevant to these members.

Splitting strength will, however, need to be investigated for member 2 in joint 1,2,1 (and member 2 in joint 3,2,3, which is subjected to the same loading condition):

Splitting force in member 2 for each joint set-up, $F_{v.Ed}$ $F_{v.Ed}=2\cdot F_{v.sp.2}\cdot \sin(\theta_{2.3.b})$ $F_{v.Ed}=2.5\times10^3\,N$

Design splitting capacity of member 2, $F_{90.Rd}$ (Equation 10.41(c) (EC5, *8.1.4*))

$$F_{90.Rd}=\left(14\cdot t_2\cdot mm^{-1}\cdot\sqrt{\frac{h_e\cdot mm^{-1}}{1-\frac{h_e}{h}}}\cdot\frac{k_{mod.med}}{\gamma_M}\cdot N\right)\qquad F_{90.Rd}=5.28\times10^3\,N$$

Splitting capacity of member exceeds the design force in the member; therefore OK.

Example 10.13.4 The timber truss shown in Figure E10.13.4 functions in service class 2 conditions and supports a combination of permanent and medium-term variable loading. The loading from the roof structure is applied uniformly along the top boom members of the truss and the effect of the self-weight of the truss can be ignored. The truss joint at node B comprises five members connected by two 16 mm diameter mild steel bolts fitted with washers compliant with the requirements of EC5. Members 1 are 225 mm deep by 47 mm thick, members 2 are 175 mm deep by 47 mm thick, and member 3 is 200 mm deep by 37 mm thick. Members 1 are of strength class C22, and members 2 and 3 are of strength class C18, in accordance with BS EN 338:2009. The contribution to the load-carrying capacity of the joint by each bolt due to the rope effect can be taken to be 15% of the Johansen part of the relevant strength equations given in EC5.

Confirm that the joint will comply with the requirements of EC5 at the ultimate limit states (ULS). (There is no requirement to check spacings, edge and end distances.)

Data:

Member	Thickness	Force	Design value
1	$t_1=47\,mm$	In each member AB	13.92 kN in compression
		In each member BC	8.7 kN in compression
2	$t_2=47\,mm$	In each member	7.382 kN in compression
3	$t_3=37\,mm$	In member	3.48 kN in tension

Design load at node B from the loading on the roof members = 6.96 kN
Characteristic density of member 1 timber $\rho1_k=340\,kg/m^3$
Characteristic density of member 2 timber $\rho2_k=320\,kg/m^3$
Characteristic tensile strength of the bolt $f_{u.k}=400\,N/mm^2$

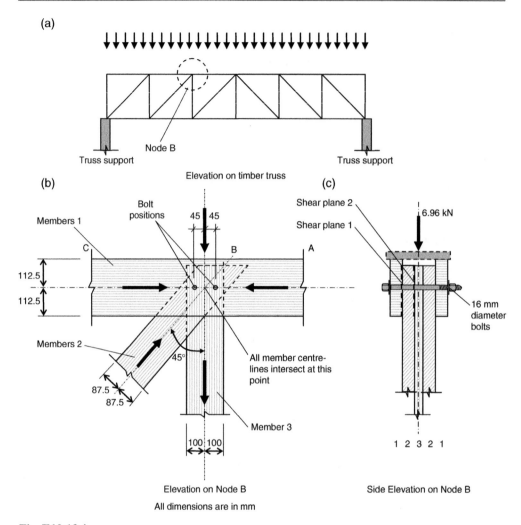

Fig. E10.13.4.

1. Geometric properties

Thickness of members 1, t_1	$t_1 = 47$ mm
Thickness of members 2, t_2	$t_2 = 47$ mm
Depth of member 1, h_1	$h_1 = 225$ mm
Depth of member 2, h_2	$h_2 = 175$ mm
Thickness of member 3, t_3	$t_3 = 37$ mm
Bolt diameter d	$d = 16$ mm
Number of bolts, n	$n = 2$

2. Timber and bolt strength properties

Table 1.3, strength class C22 and C18 timber (BS EN 338:2009(E), *Table 1*)

Characteristic density of member 1 timber, ρ_k	$\rho 1_k = 340$ kg/m³

Characteristic density of members 2 and 3 timber, $\rho 2_k$

$\rho 2_k = 320\,\text{kg/m}^3$

Stength of bolt:

Characteristic tensile strength of bolt, $f_{u.k}$

$f_{u.k} = 400\,\text{N/mm}^2$

3. **Partial safety factors and modification factor**

Table 1.6 (UKNA to BS EN 1995-1-1, *Table NA.3*)

Material factor for connection, γ_M
EC5: Part 1.1, *Table 3.1*

$\gamma_M = 1.3$

Factor for medium-duration loading and service class 2, $k_{mod.med}$
(Table 2.4 (EC5, *Table 3.1*))

$k_{mod.med} = 0.80$

4. **Actions**

Design action in member AB, $FAB_{v.d.1}$

$FAB_{v.d.1} = 13.92\,\text{kN}$

Design action in member BC, $FBC_{v.d.1}$

$FBC_{v.d.1} = 8.7\,\text{kN}$

Resultant horizontal design action in each member 1 at B, $F_{v.d.1}$

$F_{v.d.1} = FAB_{v.d.1} - FBC_{v.d.1}$

$F_{v.d.1} = 5.22\,\text{kN}$

Design vertical action in each member 1 at B, $FV_{v.d.1}$

$FV_{v.d.1} = 3.48\,\text{kN}$

Design action in each member 2, $F_{v.d.2}$

$F_{v.d.2} = 7.382\,\text{kN}$

Design action in member 3, $F_{v.d.3}$

$F_{v.d.3} = 3.48\,\text{kN}$

5. **Strength of shear planes in the joint**

Embedment factor for softwood timber, k_{90}
(Table 10.7, (EC5, *equation (8.32)*))

$k_{90} = 1.35 + 0.015 \cdot d\,\text{mm}^{-1}$
$k_{90} = 1.59$

Characteristic embedment strength of member 1 parallel to the grain, $f1_{h.0.k}$
(equation (10.35); EC5, *equation (8.32)*)
(The equation incorporates dimension factors for Mathcad.

$$f1_{h.0.k} = 0.082 \cdot (1 - 0.01 \cdot d \cdot \text{mm}^{-1}) \cdot \left(\rho1_k \cdot \frac{\text{m}^3}{\text{kg}} \right) \cdot \text{N} \cdot \text{mm}^{-2} \qquad f1_{h.0.k} = 23.42\,\text{N/mm}^2$$

Characteristic embedment strength of members 2 and 3 parallel to the grain, $f2_{h.0.k}$
(equation (10.35); EC5, *equation (8.32)*)

$$f2_{h.0.k} = 0.082 \cdot (1 - 0.01 \cdot d \cdot \text{mm}^{-1}) \cdot \left(\rho2_k \cdot \frac{\text{m}^3}{\text{kg}} \right) \cdot \text{N} \cdot \text{mm}^{-2} \qquad f2_{h.0.k} = 22.04\,\text{N/mm}^2$$

Shear plane force and its angle relative to the grain (in radians):

Design force in shear plane 1 (between member 1 and member 2), $F_{v.sp.1}$

$$F_{v.sp.1} = (F_{v.d.1}^2 + FV_{v.d.1}^2)^{0.5}$$

$F_{v.sp.1} = 6.27\,\text{kN}$

Angle of shear plane 1 force relative to the grain direction in member 1 in the joint comprising members 1,2,1, $\theta_{1.2}$ (in degrees)

$$\theta_{1.2} = \text{angle}(F_{v.d.1}, FV_{v.d.1}) \cdot \frac{180}{\pi}$$

$$\theta_{1.2} = 33.69$$

Angle of shear plane 1 force relative to the grain direction in member 2 in the joint comprising members 1,2,1, $\theta_{2.1}$ (in degrees)

$$\theta_{2.1} = 45 - \theta_{1.2}$$

$$\theta_{2.1} = 11.31$$

Design force in shear plane 2, $F_{v.sp.2}$

$$F_{v.sp.2} = \left[(F_{v.sp.1} \cdot \cos(\theta_{1.2} \cdot \deg) - F_{v.sp.2} \cdot \cos(45 \cdot \deg))^2 + (F_{v.sp.1} \cdot \sin(\theta_{1.2} \cdot \deg) - F_{v.sp.2} \cdot \sin(45 \cdot \deg))^2 \right]^{0.5}$$

$$F_{v.sp.2} = 1.74\,\text{kN}$$

Angle of shear plane 2 force relative to the horizontal plane in the joint comprising members 2,3,2

Horizontal component, $F_{v.sp.2.h}$

$$F_{v.sp.2.h} = F_{v.sp.1} \cdot \cos(\theta_{1.2} \cdot \deg) - F_{v.d.2} \cdot \cos(45 \cdot \deg)$$

$$F_{v.sp.2.h} = 1.38 \times 10^{-4}\,\text{kN}$$

Vertical component, $F_{v.sp.2.v}$

$$F_{v.sp.2.v} = F_{v.sp.1} \cdot \sin(\theta_{1.2} \cdot \deg) - F_{v.d.2} \cdot \sin(45 \cdot \deg)$$

$$F_{v.sp.2.v} = -1.74\,\text{kN}$$

Angle of shear plane 2 force relative to the horizontal plane (in degrees)

$$\text{angle}\left(F_{v.sp.2.h}, F_{v.sp.2.v} \right) \cdot \frac{180}{\pi} = 270$$

Angle of the shear plane 2 force relative to the grain of member 2 in the joint comprising members 2,3,2, $\theta_{2.3.a}$

$$\theta_{2.3.a} = 45$$

Angle of the shear plane 2 force relative to the grain of member 3 in the joint comprising members 2,3,2, $\theta_{3.2.a}$

$$\theta_{3.2.a} = 0$$

Angle of the shear plane 2 force relative to the grain of member 3 in the joint comprising members 3,2,3, $\theta_{3.2.b}$

$$\theta_{3.2.b} = 0$$

Angle of the shear plane 2 force relative to the grain of member 2 in the joint comprising members 3,2,3, $\theta_{2.3.b}$

$$\theta_{2.3.b} = 45$$

Embedment strength of the timber:

Of member 1 in joint 1,2,1, $f_{h.\theta.1.2.k}$

$$f_{h.\theta.1.2.k} = \frac{f1_{h.0.k}}{k_{90} \cdot \sin(\theta_{1.2} \cdot \deg)^2 + \cos(\theta_{1.2} \cdot \deg)^2}$$

$$f_{h.\theta.1.2.k} = 19.82\,\text{N/mm}^2$$

Of member 2 in joint 1,2,1, $f_{h.\theta.2.1.k}$

$$f_{h.\theta.2.1.k} = \frac{f2_{h.0.k}}{k_{90} \cdot \sin(\theta_{2.1} \cdot \deg)^2 + \cos(\theta_{2.1} \cdot \deg)^2}$$

$$f_{h.\theta.2.1.k} = 21.55\,\text{N/mm}^2$$

Of member 2 in joint 2,3,2, $f_{h.\theta2.3.a.k}$

$$f_{h.\theta2.3.a.k} = \frac{f2_{h.0.k}}{k_{90} \cdot \sin(\theta_{2.3.a} \cdot \deg)^2 + \cos(\theta_{2.3.a} \cdot \deg)^2} \qquad f_{h.\theta2.3.a.k} = 17.02\,\text{N/mm}^2$$

Of member 3 in joint 2,3,2, $f_{h.\theta3.2.a.k}$

$$f_{h.\theta3.2.a.k} = \frac{f2_{h.0.k}}{k_{90} \cdot \sin(\theta_{3.2.a} \cdot \deg)^2 + \cos(\theta_{3.2.a} \cdot \deg)^2} \qquad f_{h.\theta3.2.a.k} = 22.04\,\text{N/mm}^2$$

Of member 3 in joint 3,2,3, $f_{h.\theta3.2.b.k}$

$$f_{h.\theta3.2.b.k} = \frac{f2_{h.0.k}}{k_{90} \cdot \sin(\theta_{3.2.b} \cdot \deg)^2 + \cos(\theta_{3.2.b} \cdot \deg)^2} \qquad f_{h.\theta3.2.b.k} = 22.04\,\text{N/mm}^2$$

Of member 2 in joint 3,2,3, $f_{h.\theta2.3.b.k}$

$$f_{h.\theta2.3.b.k} = \frac{f2_{h.0.k}}{k_{90} \cdot \sin(\theta_{2.3.b} \cdot \deg)^2 + \cos(\theta_{2.3.b} \cdot \deg)^2} \qquad f_{h.\theta2.3.b.k} = 17.02\,\text{N/mm}^2$$

Characteristic yield moment of
a bolt, $M_{y.Rk}$
(Table 10.5 (EC5, *equation (8.14)*)).
(The equation incorporates dimension
factors for Mathcad.)

$$M_{y.Rk} = 0{,}3 \cdot \left(\frac{f_{u.k}}{\text{N}\cdot\text{mm}^{-2}}\right) \cdot \left(\frac{d}{\text{mm}}\right)^{2.6} \cdot \text{N}\cdot\text{mm}$$

$$M_{y.Rk} = 1.62\times10^5\,\text{Nmm}$$

Shear plane 1 – strength assessment:

Double shear joint formed from members 1,2,1. For a timber-to-timber joint with a bolt in double shear, the characteristic lateral resistance per shear plane is the smallest value of equations (10.7)–(10.10) in Table 10.2 (EC5, *equations (8.7)*), where:

Ratio of the embedment strength of
member 2 to member 1, $\beta_{2.1}$

$$\beta_{2.1} = \frac{f_{h.\theta2.1.k}}{f_{h.\theta1.2.k}} \qquad \beta_{2.1} = 1.09$$

Failure mode (g): $\qquad F_{v.Rk.1} = f_{h.\theta1.2.k} \cdot t_1 \cdot d \qquad F_{v.Rk.1} = 14.91\,\text{kN}$

Failure mode (h): $\qquad F_{v.Rk.2} = 0.5 \cdot f_{h.\theta2.1.k} \cdot t_2 \cdot d \qquad F_{v.Rk.2} = 8.1\,\text{kN}$

Failure mode (j), incorporating 15% for the axial strength of the bolt

$$F_{v.Rk.3} = (1.15)\cdot(1.05)\cdot\frac{f_{h.\theta1.2.k}\cdot t_1 \cdot d}{2 + \beta_{2.1}}\cdot\left[\sqrt{2\cdot\beta_{2.1}\cdot(1+\beta_{2.1}) + \frac{4\cdot\beta_{2.1}\cdot(2+\beta_{2.1})\cdot M_{y.Rk}}{f_{h.\theta1.2.k}\cdot t_1^2 \cdot d}} - \beta_{2.1}\right]$$

$$F_{v.Rk.3} = 9.78\,\text{kN}$$

Failure mode (k), incorporating 15% for the axial strength of the bolt

$$F_{v.Rk.4} = (1.15)\cdot(1.15)\cdot\sqrt{\frac{2\cdot\beta_{2.1}}{1+\beta_{2.1}}}\cdot(2\cdot M_{y.Rk}\cdot f_{h.\theta1.2.k}\cdot d)$$

$$F_{v.Rk.4} = 13.69\,\text{kN}$$

All modes are possible and the
minimum characteristic strength of
shear plane 1 per bolt

$$F_{v.Rk.1.2} = \min(F_{v.Rk.1}, F_{v.Rk.2}, F_{v.Rk.3}, F_{v.Rk.4})$$

$$F_{v.Rk.1.2} = 8.1\times10^3\,\text{N}$$
(mode (h))

Effective number of bolts when loaded parallel to the grain direction in member 1, n_{ef}:

Spacing of bolts parallel to the grain, a_1 $a_1 = 90\,\text{mm}$

Effective number of bolts based on equation (10.45), (EC5, *equation (8.34)*), n_{ef}

$$n_{ef} = \min\left[n, n^{0.9}\left(\frac{a_1}{13 \cdot d}\right)^{0.25}\right] \quad n_{ef} = 1.51$$

Design capacity of shear plane 1, $F_{v.R.d.sp.1}$ (10.4.1.3c (EC5, 8.5.1.1(4)))

$$F_{v.R.d.sp.1} = \frac{k_{mod.med}}{\gamma_M} \cdot F_{v.Rk.1.2} \cdot \left(n_{ef} + (n - n_{ef}) \cdot \frac{\theta_{1.2}}{90}\right) \qquad F_{v.R.d.sp.1} = 8.46\,\text{kN}$$

Design force on shear plane 1, $F_{v.sp.1}$ $F_{v.sp.1} = 6.27\,\text{kN}$

Design capacity of shear plane 1 exceeds the design force in shear plane 1; therefore OK

Check horizontal component in shear plane 1, using the effective number of bolts:

Horizontal component of shear plane 1 force in the connection, $F_{v.d.1}$: $F_{v.d.1} = 5.22\,\text{kN}$

Strength of connection parallel to the grain, adopting failure mod h, $F_{v.Rk}$ $F_{v.Rk} = 0.5 \cdot f_{h.\theta.2.3.a.k} \cdot t_2 \cdot d \cdot n_{ef}$

Design capacity $F_{v.Rd} = \dfrac{k_{mod.med}}{\gamma_M} \cdot F_{v.Rk}$

$$F_{v.Rd} = 5.96 \times 10^3\,\text{N}$$

Design capacity when loaded parallel to the grain exceeds the horizontal component of the design force in shear plane 1; therefore OK.

Shear plane 2 strength assessment (based on 2,3,2 configuration):

Double shear joint formed from members 2,3,2. For a timber-to-timber joint with a bolt in double shear, the characteristic lateral resistance per shear plane is the smallest value of equations (10.7)–(10.10) in Table 10.2 (EC5, *equations (8.7)*), where

Ratio of the embedment strength of member 3 to member 2, $\beta_{3.2}$ $\beta_{3.2} = \dfrac{f_{h.\theta.3.2.a.k}}{f_{h.\theta.2.3.a.k}}$ $\beta_{3.2} = 1.3$

Failure mode (g): $F_{v.Rk.1.a} = f_{h.\theta.2.3.a.k} \cdot t_2 \cdot d$

$$F_{v.Rk.1.a} = 12.8\,\text{kN}$$

Failure mode (h): $F_{v.Rk.2.a} = 0.5 \cdot f_{h.\theta.3.2.a.k} \cdot t_3 \cdot d$

$$F_{v.Rk.2.a} = 6.52\,\text{kN}$$

Failure modes (j) and (k) are not compatible with $F_{v.Rk.1.2}$ and are not considered (10.7 (EC5, *8.1.3(2)*))

Minimum value for shear plane 2 per bolt, $F_{v.Rk.2.3}$

$$F_{v.Rk.2.3} = \min\left(F_{v.Rk.1.a}, F_{v.Rk.2.a}\right) \qquad F_{v.Rk.2.3} = 6.52\,\text{kN (mode (h))}$$

Shear plane 2 strength assessment (based on the 3,2,3 joint configuration):

Double shear joint formed from members 3,2,3. For a timber-to-timber joint with a bolt in double shear, the characteristic lateral resistance per shear plane is the smallest value of equations (10.7)–(10.10) in Table 10.2 (EC5, *equations (8.7)*), where:

Ratio of the embedment strength of member 2 to member 3, $\beta_{2.3}$

$$\beta_{2.3} = \frac{f_{h.\theta.2.3.b.k}}{f_{h.\theta.3.2.b.k}}$$

$\beta_{2.3} = 0.77$

Failure mode (g):

$$F_{v.Rk.1.b} = f_{h.\theta.3.2.b.k} \cdot t_3 \cdot d$$

$F_{v.Rk.1.b} = 13.05 \, \text{kN}$

Failure mode (h):

$$F_{v.Rk.2.b} = 0.5 \cdot f_{h.\theta.2.3.b.k} \cdot t_2 \cdot d$$

$F_{v.Rk.2.b} = 6.4 \, \text{kN}$

Minimum value for shear plane per bolt

$$F_{v.Rk.3.2} = \min(F_{v.Rk.1.b}, \, F_{v.Rk.2.b})$$

$F_{v.Rk.3.2} = 6.4 \, \text{kN (mode (h))}$

Strength of shear plane 2 (based on the minimum value for 2,3,2 and 3,2,3 configurations):

Minimum characteristic strength of shear plane 2 per bolt, $F_{v.Rk.2.3.f}$

$$F_{v.Rk.2.3.f} = \min(F_{v.Rk.2.3}, \, F_{v.Rk.3.2})$$

$$F_{v.Rk.2.3.f} = 6.4 \, \text{kN (mode (h) in configuration 3,2,3)}$$

Effective number of bolts acting in shear plane 2, n:

Design capacity of shear plane 2 based on n bolts, $F_{v.Rd.sp.2}$

$$F_{v.Rd.sp.2} = \frac{k_{mod.med}}{\gamma_M} \cdot F_{v.Rk.2.3.f} \cdot n$$

$F_{v.Rd.sp.2} = 7.88 \, \text{kN}$

Design force in shear plane 2

$F_{v.sp.2} = 1.74 \, \text{kN}$

Design capacity of shear plane 2 exceeds the design force in shear plane 2; therefore OK.

6. **Splitting capacity of the connection**
   As the direction of shear plane 2 force relative to member 3 in configuration 3,2,3 is parallel to the grain of the member, timber splitting will not be relevant to member 3. The splitting strength will, however, need to be investigated for members 1 and 2 in joint 1,2,1. There is no need to check member 2 in joint 3,2,3 as the loading and geometrical configuration of the connection will be the same as for member 2 in 1,2,1.

   Shear plane (1) – failure mode (h):

   1. Resultant force in shear plane (1) is at $\theta_{1.2}$ to the axis of member AB

Design shear force on each member 1 – from the roof loading arrangement, the shear force in each member on each side of the connection will be

$$F1x_d = \frac{FV_{v.d.1}}{2}$$

$F1x_d = 1.74 \, \text{kN}$

Design splitting capacity of member 1, $F1_{90.Rd}$ (equation 10.41(c) (EC5, *8.1.4*))

Distance from extreme bolt position to the loaded face of member 1, $h1_e$

$$h1_e = \frac{h_1}{2}$$

$h1_e = 112.5 \, \text{mm}$

$$F1_{90.Rd} = \frac{k_{\text{mod.med}}}{\gamma_M} \cdot 14 \cdot t_1 \cdot \sqrt{\frac{h1_e}{1 - \frac{h1_e}{h_1}}} \, (\text{mm}^{-1.5} \cdot \text{N})$$
$$F1_{90.Rd} = 6.07 \, \text{kN}$$

i.e. $F1_{90.Rd}$ exceeds $F1x_d$; therefore OK

2. Resultant force in shear plane (1) is at $(45 - \theta_{12}) \cdot \deg$ to the axis of member 2

Design shear force on member 2 from the resultant force in shear plane (1), $F2x_d$

$$F2x_d = 2 \cdot F_{\text{v.sp.1}} \cdot \sin(45 \cdot \deg - \theta_{12} \cdot \deg)$$
$$F2x_d = 2.46 \times 10^3 \, \text{N}$$

Design splitting capacity of member 2, $F2_{90.Rd}$
(equation (10.41(c)) (EC5, 8.1.4))

Distance from extreme bolt position to the loaded face of member 1, $h2_e$
$$h2_e = \frac{h_2}{2} + 0.5 \cdot a_1 \cdot \cos(45 \cdot \deg)$$
$$h2_e = 119.32 \, \text{mm}$$

$$F2_{90.Rd} = \frac{k_{\text{mod.med}}}{\gamma_M} \cdot 14 \cdot t_2 \cdot \sqrt{\frac{h2_e}{1 - \frac{h2_e}{h_2}}} \, (\text{mm}^{-1.5} \cdot \text{N})$$
$$F2_{90.Rd} = 7.84 \times 10^3 \, \text{N}$$

Splitting capacity of member 2 exceeds the design shear force on the member; therefore OK.

Example 10.13.5 A 15.0-mm-thick plywood fascia panel with a characteristic density of 550 kg/m³ is supported vertically from a rigid base and is fixed to vertical 97 mm by 47 mm timber members by nails as shown in Figure E10.13.5. It is subjected to wind loading and the design suction force at the ULS on the panel at the timber support is 600 N/m run. The nails are smooth round wire, 2.7 mm nominal diameter and 40 mm long, and the timber is strength class C18 to BS EN 338:2009. Determine the spacing of the nails assuming the connection will function under service class 3 conditions.

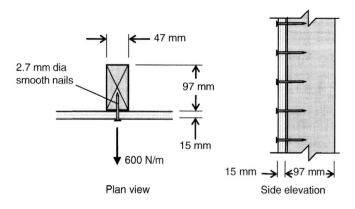

Fig. E10.13.5.

1. Geometric properties

Thickness of plywood fascia, t_1 $t_1 = 15\,mm$

Width of timber support member, t_2 $t_2 = 47\,mm$

Depth of timber support member, h $h = 97\,mm$

Nail diameter, d $d = 2.7\,mm$

Nail head diameter, d_h $d_h = 2.25 \cdot d$ $d_h = 6.08\,mm$

Nail length, ℓ_{nail} $\ell_{nail} = 50\,mm$

Nail pointside penetration, t_{pen} $t_{pen} = (\ell_{nail} - 2.5d) - t_1$ $t_{pen} = 28.25\,mm$

Minimum penetration of nail in inner member, $t_{point.pen}$
(10.3.5.1(1) (EC5, 8.3.1.2(1)))

$t_{point.pen} = 8 \cdot d$ $t_{point.pen} = 21.6\,mm$

$\dfrac{t_{point.pen}}{t_{pen}} = 0.76$ ratio less than 1, i.e. OK

Minimum nail penetration criteria met

Pentration of nail into timber support member in terms of nail diameter, coeff

$coeff = \dfrac{t_{pen}}{d}$ $coeff = 10.46$

i.e. $> 8d$ but is less than $12d$

2. Timber and plywood properties
Table 1.3, strength class C18 (BS EN 338:2009(E), *Table 1*)

Characteristic density of the timber, ρ_k $\rho_k = 320\,kg/m^3$

Plywood:
Characteristic density of the plywood $\rho_{pk} = 550\,kg/m^3$

3. Partial safety factors
Table 2.6 (UKNA to EC5, *Table NA.3*)

Material factor for connections $\gamma_{M.connection} = 1.3$

4. Actions
Design action per metre length of the plywood at the support, $F_{ax,d}$ $F_{ax,d} = 600\,N$

5. Modification factors
Factor for instantaneous duration loading and service class 3, $k_{mod,inst}$
(Table 2.4 (EC5, *Table 3.1*)) $k_{mod,inst} = 0.9$

6. Tensile strength of the connection
Because there is no permanent or long-term axial loading on the connection, smooth nails can be used (EC5, *8.3.2(1)*)

Characteristic pointside withdrawal strength, $f_{ax.k}$
(10.8.1 (EC5, *8.3.2(6), (7)*))

Assuming the connection may be in a condition where it is installed near its fibre saturation point. (The equation incorporates dimension factors for Mathcad.)

Pointside penetration factor, Dp $Dp = \left(\dfrac{t_{pen}}{4 \cdot d} - 2 \right)$ $D_p = 0.62$

$$f_{ax.k} = \dfrac{2}{3} \cdot 20 \times 10^{-6} \cdot \left(\rho_k \cdot \dfrac{m^3}{kg} \right)^2 Dp \cdot N \cdot mm^{-2}$$ $f_{ax.k} = 0.84\, N/mm^2$

Characteristic headside withdrawal strength, $fh_{ax.k}$
(10.8.1 (EC5, *8.3.2(6),(7)*))
(The equation incorporates dimensional factors for Mathcad.)

Headside penetration factor, Dh $Dh = \left(\dfrac{t_1}{4 \cdot d} - 2 \right)$

$$fh_{ax.k} = \dfrac{2}{3} \cdot 20 \times 10^{-6} \cdot \left(\rho_{\rho k} \cdot \dfrac{m^3}{kg} \right)^2 Dh \cdot N \cdot mm^{-2}$$ $fh_{ax.k} = 0\, N/mm^2$

Characteristic headside pull-through strength, $f_{head.k}$
(equation (10.59); EC5, *equation (8.26)*)
Assume the connection may be in a condition where it is installed near its fibre saturation point. (The equation incorporates dimension factors for Mathcad.)

$$f_{head.k} = \dfrac{2}{3} \cdot 70 \times 10^{-6} \cdot \left[\left(\rho_{\rho k} \cdot \dfrac{m^3}{kg} \right)^2 \cdot N/mm^2 \right]$$ $f_{head.k} = 14.12\, N/mm^2$

Characteristic withdrawal capacity of nail, $F_{ax.Rk}$, is lesser of equations (10.57)
(EC5, *equations (8.24)*)

$$F_{ax.Rk.1} = f_{ax.k} \cdot d \cdot t_{pen}$$ $F_{ax.Rk.1} = 64.12\, N$

and $$F_{ax.Rk.2} = fh_{ax.k} \cdot d \cdot t_2 + f_{head.k} \cdot d_h^2$$ $F_{ax.Rk.2} = 520.98\, N$

therefore $$F_{ax.Rk} = \min(F_{ax.Rk.1}, F_{ax.Rk.2})$$ $F_{ax.Rk} = 64.12\, N$

Design strength of a nail in tension, $$F_{ax.Rd} = \dfrac{k_{mod,inst}}{\gamma_{M,correction}} \cdot F_{ax.Rk}$$ $F_{ax.Rd} = 44.39\, N$

Number of nails required per metre, n $n = \text{ceil}\left(\dfrac{F_{ax,d}}{F_{ax.Rd}} \right) + 1$ $n = 15$

Nails spacing to be used, $nail_{sp}$ $nail_{sp} = \text{floor}\left(\dfrac{1000}{n} \right)$ $nail_{sp} = 66\, mm$, c/c

Table 10.8 (EC5, *Table 8.2*) incorporating the requirements of EC5, *8.3.1.3(1)* and *(2)*, for nailed panel connections:

Minimum nail spacing parallel $nail_{sp,min} = 0.85 \cdot 10\, d$
to the grain, $nail_{sp,min}$
                              $nail_{sp,min} = 22.52\, mm$

Ratio of actual to minimum nail $\dfrac{nail_{sp} \cdot mm}{nail_{sp,min}} = 2.88$ i.e. OK
spacing must exceed unity

Minimum unloaded edge distance for timber, $nail_{spe,min}$	$nail_{spe,min} = 5\,d$	$nail_{spe,min} = 13.5\,mm$
Minimum width of timber for connection, $width_{min}$	$width_{min} = 2 \cdot nail_{spe,min}$	$width_{min} = 27\,mm$
Ratio of actual to minimum width of timber member must exceed unity	$\dfrac{t_2}{width_{min}} = 1.74$	i.e. OK

The nail connection using 2.7 mm diameter smooth round nails, 50 mm long, at 65 mm c/c, will comply with the requirements of EC5.

Example 10.13.6 A metal bracket made from 8-mm-thick mild steel is secured to a timber post by four coach screws as shown in Figure E10.13.6. The bracket provides support to a bracing member and is subjected to a design load at the ULS of 15 kN from a short-term tensile variable action at an angle of 45° to the vertical. The coach screws are 8 mm nominal diameter by 130 mm long with a tensile strength, f_{ub}, of 400 N/mm² and pre-drilling is used. The timber is strength class C22 to BS EN 338:2009 and the joint functions under service class 2 conditions. The tolerance on the hole diameter in the bracket for each coach screw is less than 0.5 mm.

Check that screw fixings comply with the requirements of EC5. There is no requirement to check the strength of the steel bracket.

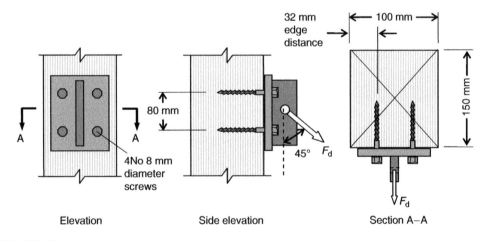

Fig.E10.13.6.

1. **Geometric properties**

Thickness of steel bracket, t	$t_1 = 8\,mm$
Width of timber member, t_2	$t_2 = 100\,mm$
Depth of timber member, h	$h = 150\,mm$
Coach screw diameter, d	$d = 8\,mm$

Coach screw diameter for tensile strength, d_t	$d_t = 0.7\,d$	$d_t = 5.6 \times 10^{-3}\,\text{m}$
Coach screw length, ℓ_{screw}	$\ell_{screw} = 130\,\text{mm}$	
Penetration of the solid shaft into the timber, t_{pen}	$t_{pen} = 0.4 \cdot \ell_{screw} - t_1$	$t_{pen} = 44\,\text{mm}$
Ratio of penetration of solid shaft to coach screw diameter:	$\dfrac{t_{pen}}{d} = 5.5$	

i.e. the ratio is greater than 4; therefore the full diameter can be used for flexure and because $d > 6\,\text{mm}$, the coach screws will be classed as bolts for lateral strength assessment.
(Table 10.4 (EC5, 8.7.1(2)))

Coach screw pointside threaded penetration, ℓ_{ef}	$\ell_{ef} = 0.6 \cdot \ell_{screw}$	$\ell_{ef} = 78\,\text{mm}$
Coach screw penetration coefficient must exceed 6
(10.3.5.3 (EC5, 8.7.2(3)))	$t_{coeff} = \dfrac{\ell_{ef} - t_1}{d}$	$t_{coeff} = 15.25$
i.e. minimum screw penetration criteria met
Number of coach screws, n	$n = 4$	
Lateral spacing of each coach screw parallel to the grain in the direction of the lateral loading, $sp\ell$	$sp\ell = 80\,\text{mm}$	(greater than the minimum value of $5d$ i.e. O.K.)
(Table 10.9)
Lateral spacing of coach screw for tension loading direction, sph	$sph = 36\,\text{mm}$	

2. Timber and coach screw properties
Table 1.3, strength class C22 (BS EN 338:2009(E), *Table 1*)

Characteristic density of timber, ρ_k	$\rho_k = 340\,\text{kg/m}^3$

Coach screws:

Tensile strength of each coach screw, f_{uk}	$f_{uk} = 400\,\text{N/mm}^2$

3. Partial safety factors
Table 2.6 (UKNA to EC5, *Table NA.3*)

Material factor for connections	$\gamma_{M.connection} = 1.3$

4. Actions
Design load on the connection, F_d	$F_d = 15\,\text{kN}$	
Design axial load on the connection, $F_{ax.d}$	$F_{ax.d} = F_d \cdot \sin(45 \cdot \deg)$	$F_{ax.d} = 10.61\,\text{kN}$
Design lateral load on the connection, $F_{v.d}$	$F_{v.d} = F_d \cdot \cos(45 \cdot \deg)$	$F_{v.d} = 10.61\,\text{kN}$

5. **Modification factors**

Factor for short-duration loading
and service class 2, $k_{mod.short}$
(Table 2.4 (EC5, *Table 3.1*))

$k_{mod.short}=0.90$

6. **Embedment strength of timber**

Characteristic embedment strength of the timber parallel to the grain, $f_{h.k}$:
(equation (10.35); EC5, *equation (8.32)*)
(The equation incorporates dimension factors for Mathcad.)

$$f_{h.k} = 0.082 \cdot \left(1-0.01 \cdot \frac{d}{mm}\right) \cdot \left(\rho_k \cdot \frac{m^3}{kg}\right) \cdot N \cdot mm^{-2}$$

$f_{h.k}=25.65\,N/mm^2$

7. **Yield moment of a coach screw**

Characteristic yield moment of a coach screw (equates to a bolt):
(equation (10.26); EC5, *equation (8.30)*)
(The equation incorporates dimension factors for Mathcad.)

$$M_{y.Rk} = 0.3\left(f_{uk} \cdot \frac{mm^2}{N}\right)\left(\frac{d}{mm}\right)^{2.6} \cdot N \cdot mm$$

$M_{y.Rk}=2.67\times10^4 \cdot Nmm$

8. **Tension strength of the screwed connection**

Minimum spacing of the coach
screws, *spha*
(Table 10.12 (EC5, *Table 8.6*))

$spha=4\,d$

$spha=32\,mm$

Check the spacing ratio

$\dfrac{spha}{sph} = 0.89$

Ratio of minimum to actual <1; therefore
spacing for tension stresses is OK

Characteristic pointside withdrawal strength, $f_{ax.k}$ (equation (10.63); EC5, *equation (8.39)*)
(The equation incorporates dimension factors for Mathcad.)

$$f_{ax.k} = 0.52\left(\frac{d}{mm}\right)^{-0.5}\left(\frac{\ell_{ef}}{mm}\right)^{-0.1}\left(\rho_k \cdot \frac{m^3}{kg}\right) \cdot N \cdot mm^{-2}$$

$f_{ax.k}=12.6\,N/mm^{-2}$

Factor, k_d
(equation 10.64, (EC5, *equation (8.40)*) $k_d = min\left(\dfrac{\dfrac{d}{mm}}{8}, 1\right)$

$k_d=1$

Design strength of connection in tension in softwood, based on the withdrawal
strength,, $F1_{ax.Rd}$ (equation 10.62; EC5, *equation (8.38)*)
The equation incorporates dimension factors for Mathcad:

$$F1_{ax.Rd} = \frac{n^{0.9} \cdot \left(\dfrac{f_{ax.k}}{N \cdot mm^{-2}}\right)\left(\dfrac{d}{mm}\right)\left(\dfrac{l_{ef}}{mm}\right) \cdot k_d}{1.2 \cdot cos(45\,deg)^2 + sin(45\,deg)^2} \cdot (k_{mod.short}/\gamma_{M.connection}) \cdot (N)$$

$F1_{ax.Rd}=$
$1.72\times10^4\,N$

Design tensile strength of the coach screws, $Fs_{\text{ax.Rd}}$ (Based on the strength of a bolt in tension in EN1993-1-8 -(head tear off will not occur)

$$Fs_{\text{ax.Rd}} = n \cdot 0.9 \cdot f_{\text{uk}} \cdot \left(\frac{1}{1.25}\right) \cdot \left[\frac{\pi \cdot (0.7 \cdot d)^2}{4}\right] \qquad Fs_{\text{ax.Rd}} = 2.84 \times 10^4\,\text{N}$$

Because of the thickness of the steel bracket, its failure strength will exceed the tensile strength of the coach screws, so head pull-through failure will not be considered.

As withdrawal strength is critical, characteristic strength of the connection in tension, $F_{\text{ax.Rk}}$

$$F_{\text{ax.Rk}} = \frac{\gamma_{\text{M.connection}}}{k_{\text{mod .short}}} \cdot F_{\text{ax.Rd}}$$

$$F_{\text{ax.Rk}} = 2.49 \times 10^4\,\text{N}$$

9. Lateral load-carrying capacity of the screws

Steel plate thickness factor (10.3 (EC5, *8.2.3(1)*))
$$\text{coeff} = \frac{t_1}{d} \qquad \text{coeff} = 1$$

coefficient $= 1$ and tolerance for screw hole is less than $0.1d$; therefore it is a thick steel plate

For a steel-to-timber connection with coach screws in single shear, the characteristic lateral resistance per shear plane is the smallest value of equations (10.13)–(10.15) in Table 10.3 (EC5, *equations (8.10)*), where:

Let
$$t_1 = \ell_{\text{screw}} - t_1 \qquad t_1 = 122\,\text{mm}$$

Failure mode (c):

$$F_{\text{v.Rk.c}} = f_{\text{h.k}} \cdot t_1 \cdot d \cdot \left(\sqrt{2 + \frac{4 \cdot M_{\text{y.Rk}}}{f_{\text{h.k}} \cdot d \cdot t_1^2}} - 1\right) + \frac{F_{\text{ax.Rk}}}{4 \cdot n} \qquad F_{\text{v.Rk.c}} = 1.22 \times 10^4\,\text{N}$$

Failure mode (d):

$$F_{\text{v.Rk.d}} = 2.3 \cdot \sqrt{M_{\text{y.Rk}} \cdot f_{\text{h.k}} \cdot d} + \frac{F_{\text{ax.Rk}}}{4 \cdot n} \qquad F_{\text{v.Rk.d}} = 6.94 \times 10^3\,\text{N}$$

Failure mode (e):
$$F_{\text{v.Rk.e}} = f_{\text{h.k}} \cdot t_1 \cdot d \qquad F_{\text{v.Rk.e}} = 2.5 \times 10^4\,\text{N}$$

10.3.5 requires $F_{\text{ax}}/4$ to be less than 100% of the Johansen yield equation (EC5, *8.2.2(2)*):

$$F_{\text{v.Rk.cc}} = 2 \cdot f_{\text{h.k}} \cdot t_1 \cdot d \cdot \left(\sqrt{2 + \frac{4 \cdot M_{\text{y.Rk}}}{f_{\text{h.k}} \cdot d \cdot t_1^2}} - 1\right) \qquad F_{\text{v.Rk.cc}} = 2.14 \times 10^4\,\text{N}$$

$$F_{\text{v.Rk.dd}} = 2 \cdot 2.3 \cdot \sqrt{M_{\text{y.Rk}} \cdot f_{\text{h.k}} \cdot d} \qquad F_{\text{v.Rk.dd}} = 1.08 \times 10^4\,\text{N}$$

On this basis, the characteristic lateral resistance per shear plane per coach screw will be:

$$F_{v.Rk} = \min (F_{v.Rk.c,} \ F_{v.Rk.d,} \ F_{v.Rk.e,} \ F_{v.Rk.cc,} \ F_{v.Rk.dd}) \qquad\qquad R_{v.Rk} = 6.94 \times 10^3 \, \text{N}$$

Effective number of coach screws in a line parallel to the grain, n_{ef} (equation (10.45)) (EC5, *8.5.1.1(4)*))

$$n_{ef} = \min\left[\frac{n}{2}, \left(\frac{n}{2}\right)^{0.9} \cdot \left(\frac{sp\ell}{13d}\right)^{0.25} \right] \qquad\qquad n_{ef} = 1.75$$

The design lateral strength of the connection (based on two lines of coach screws), $F_{v.Rd}$

$$F_{v.Rd} = \frac{2 \cdot n_{ef} \cdot k_{mod.short} \cdot F_{v.Rk}}{\gamma_{M.connection}}$$

$$F_{v.Rd} = 1.68 \times 10^4 \, \text{N}$$

Design lateral force on the connection, $F_{v.d}$

$$F_{v.d} = 1.06 \times 10^4 \, \text{N} \qquad\qquad \text{i.e. OK}$$

10. Combined lateral and axial capacity of connection

The combined design condition for lateral and axial loading to be satisfied (equation (10.70); EC5, *equation (8.28)*))

$$\left(\frac{F_{ax,d}}{F_{ax.Rd}}\right)^2 + \left(\frac{F_{v.d}}{F_{v.Rd}}\right)^2 = 0.78$$

The combined design condition < 1; therefore coach screw design is OK

Example 10.13.7 A timber-plywood gusset plate joint as shown in Figure E10.13.7 is subjected to a lateral design load of 3.48 kN at the serviceability limit states (SLS) and comprises 1.39 kN from permanent actions and 2.09 kN from a medium term variable action. The plywood gusset plates are cut from 12.5-mm-thick Canadian Douglas fir plywood, and have a mean density of 460 kg/m³. The timbers are strength class C18 to BS EN 338:2009 and the joint will function under service class 2 conditions. There are 6 No 3.00 mm diameter by 50 mm long smooth wire nails acting in single

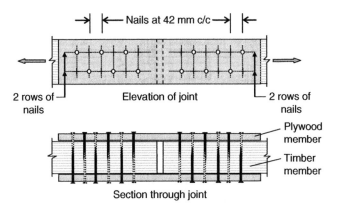

Figure E10.13.7.

shear on each side in each connection forming the joint. The nails are driven with pre-drilling, do not overlap, and are spaced as shown in the figure. The joint forms part of a floor structure within an office building and functions within a service class 2 environment.

Determine the lateral stiffness and instantaneous and final slip of the joint at the SLS based on the requirements of EC5.

1. Geometric properties

Canadian Douglas fir plywood:

Thickness of the plywood gusset plates, t_p	$t_p = 12.5\,\text{mm}$	

Timber:

Thickness of the timber, t_t	$t_t = 50\,\text{mm}$	
Nail diameter, d	$d = 3.0\,\text{mm}$	
Nail length, ℓ_{nail}	$\ell_{nail} = 50\,\text{mm}$	
Nail pointside penetration (allowing $2.5d$ for nail point), t_{point}	$t_{point} = (\ell_{nail} - 2.5 \cdot d) - t_p$	$t_{point} = 30\,\text{mm}$
Number of rows of nails per connection per shear plane, r	$r = 2$	
Number of nails in single shear per row per connection per shear plane, n	$n = 3$	
Number of shear planes in each connection, shp	$shp = 2$	
Minimum penetration of the nail in the timber, $t_{point.pen}$ (10.3.5.1 (EC5, 8.3.1.2(1)))	$t_{point.pen} = 8\,d$	$t_{point.pen} = 24\,\text{mm}$
Ratio of minimum allowable penetration to the actual penetration	$\dfrac{t_{point.pen}}{t_{point}} = 0.8$	Ratio less than 1; therefore OK

2. Timber and nail strength properties

Table 1.3, strength class C18 (BS EN 338:2009(E), *Table 1*)

Mean density of the timber, ρ_m	$\rho_m = 380\,\text{kg/m}^3$	

Table 1.13, Canadian Douglas fir plywood

Mean density of plywood, ρ_{pm}	$\rho_{pm} = 460\,\text{kg/m}^3$	
Deformation factor for the timber in the joint, $k_{def.t}$ (EC5, *Table 3.2*)	$k_{def.t} = 0.8$	
Deformation factor for the plywood in the joint, $k_{def.pl}$ (EC5, *Table 3.2*)	$k_{def.pl} = 1.0$	
Deformation factor for each connection per shear plane, $k_{def.pl}$ (equation (10.82), (EC5, *equation (2.13)*))	$k_{def.c} = 2 \cdot (k_{def.t} \cdot k_{def.pl})^{0.5}$	$k_{def.c} = 1.79$

3. Actions

Design action at the SLS, $F_{v.d}$

$F_{v.d} = 3.48 \, \text{kN}$

Design permanent action $FG_{v.d}$

$F_{v.d} = 1.39 \cdot \text{kN}$

Design variable action, $FQ_{v.d}$

$FQ_{v.d} = 2.09 \cdot \text{kN}$

Combination factor for variable load in an office area, ψ_2

$\psi_2 = 0.3$

(NA to BS EN 1990:2002+A1:2005, *Table NA.A1.1*)

Design load arising from quasi-permanent combination of action, $FQP_{v.d}$

$FQP_{v.d} = FG_{v.d} + \psi_2 \cdot FQ_{v.d}$ $FQP_{v.d} = 2.02 \, \text{kN}$

4. Stiffness of the joint at the SLS

Stiffness of each connection:

Mean density of each connection, ρ_{mean} (equation (10.71); EC5, *equation (7.1)*)

$\rho_{mean} = \sqrt{\rho_m \cdot \rho_{pm}}$ $\rho_{mean} = 418.09 \, \text{kg/m}^3$

Slip modulus/shear plane/fastener at the SLS, for nails with pre-drilling, K_{ser} (Table 10.15 (EC5, *Table 7.1*)) (The equation incorporates dimension factors for Mathcad.)

$$K_{ser} = \left(\rho_{mean} \cdot \frac{\text{m}^3}{\text{kg}} \right)^{1.5} \cdot \frac{d \cdot \text{mm}^{-1}}{23} \cdot \text{N} \cdot \text{mm}^{-1}$$ $K_{ser} = 1.115 \times 10^3 \, \text{N/mm}$

Lateral stiffness per connection with pre-drilling at the instantaneous condition, $K_{ser.c}$

$K_{ser.c} = r \cdot n \cdot K_{ser} \cdot shp$

(Figure 10.24)

$K_{ser.c} = 1.34 \times 10^4 \, \text{N/mm}$

Lateral stiffness of the joint at the instantaneous condition, $K_{ser.j}$ (Figure 10.24)

$K_{ser.j} = \dfrac{K_{ser.c}}{2}$ $K_{ser.j} = 6.69 \times 10^3 \, \text{N/mm}$

Final mean value of slip modulus/shear plane/fastener at the final deformation condition, for nails with pre drilling, $K_{ser.fin}$ (equation (2.36), (EC5, *equation (2.9)*))

The equation incorporates dimension factors for mathcad:

$K_{ser.fin} = \dfrac{K_{ser}}{1 + k_{def.c}}$ $K_{ser.fin} = 399.828 \, \text{N/mm}$

Lateral stiffness per connection with pre-drilling at the final condition, $K_{ser.c.fin}$ (Figure 10.24)

$K_{ser.fin.c} = r \cdot n \cdot K_{ser\,fin} \cdot shp$ $K_{ser.fin.c} = 4.8 \times 10^3 \, \text{N/mm}$

Lateral stiffness of the joint at the final condition, $K_{ser.fin.j}$ (Figure 10.24)

$K_{ser.fin.j} = \dfrac{K_{ser.fin.c}}{2}$ $K_{ser.fin.c} = 2.4 \times 10^3 \, \text{N/mm}$

5. Joint slip at the SLS

Instantaneous slip of the joint, u_{inst} $\qquad u_{inst} = \dfrac{F_{v.d}}{K_{ser.j}}$ $\qquad u_{inst} = 0.52\,\text{mm}$

Final deformation of the joint, u_{fin} $\qquad u_{fin} = u_{inst} + \dfrac{FQP_{v.d}}{K_{ser.fin.j}} - \dfrac{FQP_{v.d}}{K_{ser.j}}$

$$u_{fin} = 1.06\,\text{mm}$$

[Final deformation of the joint – based on the alternative conservative approach (equation 10.81), $u1_{fin}$] $\qquad u1_{fin} = \dfrac{F_{v.d}}{K_{ser.fin.j}}$ $\qquad u1_{fin} = 1.45\,\text{mm}$

Chapter 11

Design of Joints with Connectors

11.1 INTRODUCTION

Connectors generally comprise toothed plates, discs or rings that are partly embedded into the adjacent timber members of a connection and are normally held in place by a connecting bolt. They are used for connections subjected to lateral loading and the load is primarily transferred between the joint members by bearing near the surface of the member. By increasing the bearing area they can take substantially increased loads over the metal dowel-type fasteners referred to in Chapter 10, which function primarily by dowel action. Generally, the connector bolts hold the members together and do not contribute to the load-carrying capacity of the connection, but with some types of connector they do.

Several types of timber connector are available and the ones most commonly used in timber joint design are the toothed-plate, the split-ring or the shear-plate form. The configuration and material requirements for these connectors are defined in BS EN 912 (2011) [1] and BS EN 14545 2008 [2], with bolt sizes complying with BS EN 13271:2002 [3]. The design rules given in EC5 [4] for each type are discussed in the following sections.

The general information in 4.3 is relevant to the content of this chapter.

11.2 DESIGN CONSIDERATIONS

The design of joints formed using connectors subjected to lateral loading is covered in this chapter and the design of connectors subjected to a moment is addressed in Chapter 12.

Connections fitted with connectors have to satisfy the relevant design rules and requirements of EC5 and the limit states associated with the main design effects are given in Table 11.1. The strength and displacement requirements for connectors are the same as those referred to in 10.1 for metal dowel-type connections.

Structural Timber Design to Eurocode 5, Second Edition. Jack Porteous and Abdy Kermani.
© Jack Porteous and Abdy Kermani 2013. Published 2013 by Blackwell Publishing Ltd.

Table 11.1 Main design requirements for connectors subjected to lateral loading and the associated EC5 limit states

Design or displacement effect	EC5 limit states
Lateral loading strength	ULS
Displacement	SLS

11.3 TOOTHED-PLATE CONNECTORS

11.3.1 Strength behaviour

Toothed-plate connectors are available in circular, square, octagonal and other shapes with sizes ranging from 38 to 165 mm. There are 11 types of connectors defined in BS EN 912, referred to as type C1, C2, etc., to C11, each type categorised by its shape, by the material used for the toothed plate and whether it is single or double sided. Guidance on minimum and maximum diameter bolts to be used with each type of connector is given in BS EN 13271. Depending on the connector type, the toothed plate will be made from cold rolled uncoated low-carbon steel, continuous hot dipped galvanised mild steel or malleable cast iron. With steel connectors, the edges of the connector plate are bent over to form triangular projecting teeth, and with cast iron connectors, conical spikes having a blunted point project from one or both faces. An example of a circular toothed-plate connector formed from steel plate having projecting teeth on one or both sides is shown in Figure 11.1.

Single-sided toothed-plate connectors can be used to connect timber to steel or, if connections are required to be demountable in timber-to-timber joints, they can be used in pairs back to back. Double-sided toothed plates are suitable where non-demountable timber-to-timber connections are required.

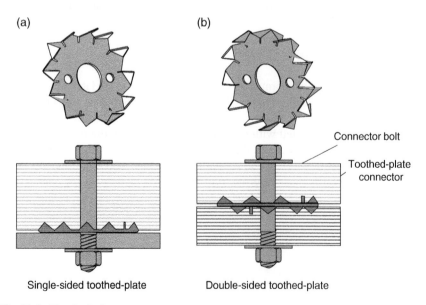

(a)   (b)

Connector bolt

Toothed-plate connector

Single-sided toothed-plate   Double-sided toothed-plate

Fig. 11.1. Toothed-plate connectors.

To assemble, the bolt hole is formed in the members, the connector is located in position, and the connection is pressed together mechanically or, if relatively low-density timber is being used, by the use of a high-strength bolt fitted with large washers. After pressing into position, the connector bolt and its washers are fitted. The connector must be pressed fully into the timber to develop its design capacity. Because the resistance offered by the timber to the penetration of the connector teeth will increase as the density of the timber increases, toothed-plate connectors are not suitable for use with timber having a characteristic density greater than about 500 kg/m³.

The design requirements for toothed-plate connectors are covered in EC5, *8.10*. In a connection formed with a double-sided toothed plate, the load is passed by embedment stresses from one of the members to the teeth of the connector and from there by shear stresses within the connector plate to the teeth on the opposite side and into the other member through embedment stresses. Embedment slip at the connector teeth also results in the timber members bearing onto the bolt and the joint strength is taken to be a combination of the force carried by the toothed-plate connector and the bolt. The bolt also holds the members together to ensure that the connector(s) remain fully embedded in the timber. Because the shear strength of the bolt considerably exceeds the embedment strength or the strength of any brittle failure mechanism in the timber, shear failure of the bolt is not considered in the strength assessment equations given in EC5.

Where a single-sided toothed-plate connector is used, the load transfer mechanism from the toothed side of the connector will be the same as for the double-sided connector but thereafter the load is passed directly to the bolt by bearing stresses between the connector and the bolt and also by bearing of the member onto the bolt due to embedment slip at the connector teeth. From there the bolt functions as a normal dowelled connection. Again, with this connection, the joint strength is a combination of the strength of the toothed-plate connector and the bolt.

Ductile failure mechanisms as described above will only arise if brittle failure is prevented and, as with metal dowel-type fasteners referred to in Chapter 10, this is achieved in EC5 by specifying minimum spacings, end and edge distances as well as minimum member thicknesses. The minimum spacings etc., for the toothed-plate connector, are given in Table 11.2 and the minimum spacing requirements for the connector bolts are as given in Table 10.9. With regard to member thicknesses, the minimum requirements are:

$$t_1 \geq 2.25 h_e \qquad\qquad\qquad \text{(EC5, } Clause\ 8.10(3)) \quad (11.1)$$

$$t_2 \geq 3.75 h_e \qquad\qquad\qquad \text{(EC5, } Clause\ 8.10(3)) \quad (11.2)$$

The thickness is as shown in Figure 11.2 and t_1 is the thickness of the outer timber member(s) in the connection. t_2 is the thickness of the inner timber member in the connection, and h_e is the embedment depth of the toothed-plate connector teeth (in mm) in the member being considered, i.e. $= (h_c - t)/2$ for a double-sided connector and $(h_c - t)$ for a single-sided connector, where h_c is the connector height and t is the thickness of the metal used for the connector.

The characteristic strength of a connection formed using a toothed-plate connector is:

$$F_{v,\mathrm{Rk,connection}} = F_{v,\mathrm{Rk}} + F_{v,\mathrm{Rk,bolt}} \qquad\qquad\qquad (11.3)$$

Table 11.2 Minimum spacing, edge and end distances for toothed-plate connector types C1 to C11 and split-ring and shear-plate connectors as referenced in BS EN 912[*]

Spacings or distances and $\alpha^{\dagger}$ (see Figures 10.11 and 11.3)	Minimum spacings and distances		
	Toothed-plate connectors		Split-ring and shear-plate connectors
	Types C1 to C9	Types C10 and C11	
Spacing parallel to the grain: a_1 $0° \leq \alpha \leq 360°$	$(1.2+0.3|\cos \alpha|)d_c$	$(1.2+0.8|\cos \alpha|)d_c$	$1.2+0.8|\cos \alpha|)d_c$
Spacing perpendicular to the grain: a_2 $0° \leq \alpha \leq 360°$	$1.2\,d_c$	$1.2\,d_c$	$1.2\,d_c$
Loaded end distance $a_{3,t}$ $-90° \leq \alpha \leq 90°$	$1.5\,d_c^{**}$	$2.0\,d_c$	$2.0 d_c^{**}$
Unloaded end distance $a_{3,c}$ $90° \leq \alpha \leq 150°$ $150° \leq \alpha \leq 210°$ $210° \leq \alpha \leq 270°$	$(0.9+0.6|\sin \alpha|)d_c$ $1.2d_c$ $(0.9+0.6|\sin \alpha|)d_c$	$(0.4+1.6|\sin \alpha|)d_c$ $1.2d_c$ $(0.4+1.6|\sin \alpha|)d_c$	$(0.4+1.6|\sin \alpha|)d_c$ $1.2d_c$ $(0.4+1.6|\sin \alpha|)d_c$
Loaded edge distance $a_{4,t}$ $0° \leq \alpha \leq 180°$	$(0.6+0.2|\sin \alpha|)d_c$	$(0.6+0.2|\sin \alpha|)d_c$	$(0.6+0.2|\sin \alpha|)d_c$
Unloaded edge distance $a_{4,c}$ $180° \leq \alpha \leq 360°$	$0.6\,d_c$	$0.6\,d_c$	$0.6\,d_c$

[*]Based on *Tables 8.7, 8.8* and *8.9* in EC5 but noting for $a_{3,t}$ for Type C1 and C9 connectors and for split-ring and shear-plate connectors the values are wrong and in this table have been changed (see values marked [**]) in accordance with the requirements of Appendix C.

[†]α is the angle between the direction of the connector force and the grain.

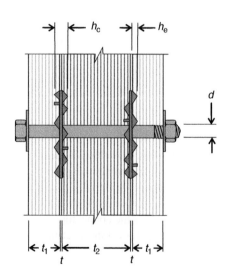

Fig. 11.2. Dimensions for connections with toothed-plate connectors.

where $F_{v,Rk,connection}$ is the characteristic strength of the toothed-plate connection (in N), $F_{v,Rk}$ is the characteristic load-carrying capacity of the toothed-plate connector (in N), and $F_{v,Rk,bolt}$ is the characteristic strength of the connector bolt (in N).

Type C2, C4, C7, C9 and C11 toothed-plate connectors are single sided, type C1, C3, C5, C6, C8 and C10 are double sided. It is to be noted that only a limited number of tests were carried out with toothed-plate connectors loaded at an angle to the grain and from these it was concluded that the strength of the toothed-plate connector would be taken as independent of the load–grain angle. However, as the lateral strength of the connector bolt is a function of the angle of load relative to the grain, the strength of the toothed-plate connection will be dependent on the load–grain angle.

The characteristic load-carrying capacity of the connector, $F_{v,Rk}$, has been determined from tests on joints [5], and to ensure that a brittle failure does not occur the minimum member thicknesses permitted in EC5 are:

- outer timber members: $2.25h_e$
- inner timber member: $3.75h_e$

where h_e is the embedment depth and is shown for a double-sided toothed-plate connector in Figure 11.2.

Density and loaded end distance criteria were also included in the test programme and from a review of the results EC5 has adopted the relationships for the characteristic load-carrying capacity, $F_{v,Rk}$, per toothed-plate connector given in equation (11.4):

$$F_{v,Rk} = \begin{cases} 18k_1k_2k_3d_c^{1.5} & \text{for types C1 to C9} \qquad (EC5, \ equation \ (8.72)) \\ 25k_1k_2k_3d_c^{1.5} & \text{for types C10 and C11} \qquad\qquad\qquad\qquad (11.4) \end{cases}$$

where:

- $F_{v,Rk}$ is· the characteristic load-carrying capacity of the toothed-plate connector (in N).
- d_c is:
  - the toothed-plate connector diameter for types C1, C2, C6, C7, C10 and C11 (in mm),
  - the toothed-plate connector side length for types C5, C8 and C9 (in mm),
  - the square root of the product of both side lengths for types C3 and C4 (in mm).
- k_1 is a modification factor for the effect of member thickness:

$$k_1 = \min \begin{cases} 1 \\ t_1/3h_e \\ t_2/5h_e \end{cases} \qquad (EC5, \ equation \ (8.73)) \quad (11.5)$$

where t_1 and t_2 are as defined in Figure 11.2 and h_e is the tooth penetration depth (in mm). The strength of the connection will reduce to 75% of its full value if the minimum permitted values of t_1 and t_2 are used. Also, when t_1 and t_2 exceed $3h_e$ and $5h_e$ respectively, there will be no increase in joint strength due to this factor.

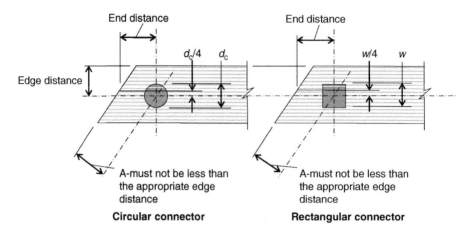

Fig. 11.3. End and edge distance for connectors (Based on BS 5268-2).

- k_2 is the modification factor for the loaded end distance, $a_{3,t}$, and, for a member with a sloping end, based on the requirements of BS5268-2 [6], the distance can be taken to be that shown in Figure 11.3. The factor only applies to members having a loaded end and is dependent on the type of toothed-plate connector being used as follows.

  For types C1–C9, the minimum loaded end distance is stated in EC5 *Table 8.8* to be $2.0d_c$ however this is an error and as stated on Appendix C the distance should be $1.5d$, and has been included as this value in Table 11.2. A distance less than this value is permitted in the strength equation provided it is not less than $1.1d_c$ or the minimum loaded end distance criteria allowed for bolts (see Table 10.9), whichever is the greater. If a loaded end distance less than $1.5d_c$ is used, the value of k_2 will be:

$$k_2 = \min \begin{cases} 1 \\ a_{3,t}/1.5d_c \end{cases}$$ (EC5, *equation (8.74)*) (11.6)

where d_c is as previously defined and:

$$a_{3,t} = \max \begin{cases} 1.1d_c \\ 7d \\ 80\,\text{mm} \end{cases}$$ (EC5, *equation (8.75)*) (11.7)

Here d is the bolt diameter and both d and d_c are in mm.

Provided the minimum loaded end distance required for the connector bolt is greater than the minimum loaded end distance required for the connector, k_2 will remain equal to 1 for a connector loaded end distance greater than $1.5d_c$. It will only be less than 1 when the distance is smaller than $1.5d_c$, reducing to a value of 0.73 at the minimum permitted end distance of $1.1d_c$.

With types C10 and C11, a similar situation applies but for these connectors the loaded end distance, $a_{3,t}$, may only be reduced to $1.5\ d_c$, i.e.:

$$k_2 = \min \begin{cases} 1 \\ a_{3,t}/2.0d_c \end{cases}$$ (EC5, *equation (8.76))*   (11.8)

with

$$a_{3,t} = \max \begin{cases} 1.5d_c \\ 7d \\ 80\,\text{mm} \end{cases}$$ (EC5, *equation (8.77))*   (11.9)

where d_c and d are as previously defined.

With these connectors, provided the minimum loaded end distance required for the connector bolt is less than the loaded end distance required for the toothed-plate connector, k_2 will equal 1 when the connector loaded end distance is $2.0d_c$ (or greater) and will be smaller than 1 when the distance is less than $2.0d_c$. It will reduce to a value of 0.75 at the minimum permitted loaded end distance of $1.5d_c$.

- k_3 is a modification factor for timber density, and:

$$k_3 = \min \begin{cases} 1.5 \\ \rho_k/350 \end{cases}$$ (EC5, *equation (8.78))*   (11.10)

where ρ_k is the characteristic density of the timber in the connection (in kg/m³).

The reference strength equations (11.4) are based on a connection where the timber has a characteristic density of 350 kg/m³, i.e. strength class C24 in BS EN 338:2009 [7]. Factor k_3 will increase the strength of the connector when higher strength class timber is used, achieving a maximum increase of 50% when the characteristic density is 525 kg/m³. This will cover all of the strength classes for softwood in BS EN 338 as well as hardwood strength class D30. Although it is likely that strength classes D35 and D40 will be able to be used with this type of connector, D50 to D70 will not because of the resistance offered by the timber to the penetration of the connector teeth.

As stated previously, the minimum spacings, edge and end distances are given in Table 11.2 for the different types of connector and the symbols are the same as those shown in Figures 10.11 and 11.3.

Where types C1, C2, C6 and C7 connectors (which are circular in shape) are being used and they are staggered, as shown in Figure 11.4, the minimum spacings a_1 and a_2 in Table 11.2 may be reduced by the use of reduction factors k_{a1} and k_{a2}, respectively, so that:

- the minimum spacing parallel to the grain $= k_{a1}a_1$
- the minimum spacing perpendicular to the grain $= k_{a2}a_2$

provided the reduction factors k_{a1} and k_{a2} comply with the following criteria:

$$(k_{a1})^2 + (k_{a2})^2 \geq 1 \quad \text{with} \quad \begin{cases} 0 \leq k_{a1} \leq 1 \\ 0 \leq k_{a2} \leq 1 \end{cases}$$ (EC5, *equation (8.69))*   (11.11)

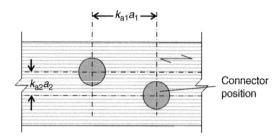

Fig. 11.4. Reduced distances for staggered toothed-plate connectors (based on EC5 *Figure 8.13*).

Equation (11.11) will only apply when using the above referenced connectors. Where the connection contains a row of connectors parallel to the grain, EC5 allows the full strength of each toothed-plate connector in the row to be used when loaded in this direction. For connections in single or double shear, irrespective of the type of connector being used, when determining the strength of a connection the number of connectors per shear plane shall be taken to equal the number of connector bolts. This will apply when the connectors are loaded parallel or perpendicular to the grain. When determining the strength of the connector bolts, however, the effective number of bolts as described in 10.4.1.3 shall be used when loaded parallel to the grain and the actual number of connector bolts when loaded perpendicular to the grain.

The requirements of bolts and washers shall fully comply with criteria given in *10.4.3* in EC5.

See Example 11.12.1.

11.4   RING AND SHEAR-PLATE CONNECTORS

11.4.1   Strength behaviour

Ring and shear-plate connectors are circular in shape and manufactured from aluminium alloy, hot rolled or temper rolled steel strip, hot rolled steel alloy strip, grey cast iron or cast metal, in accordance with the requirements of BS EN 912. They fit into preformed grooves in the timber members that accurately profile the connector and are capable of taking much greater loads than are achievable with toothed-plate connectors.

The strength equations for these types of connectors are given in EC5, *8.9*, and are only applicable to connectors with a diameter no larger than 200 mm. With the exception of three type A5 split-ring connectors in BS EN 912 where d_c is 216 mm, 236 mm and 260 mm, all of the others in the standard comply with this limit.

Ring connectors are of solid cross-section and where they are formed with a cut across the section they are referred to as split-ring connectors. The split allows the connector to be easier to fit and also to be relatively flexible to accommodate distortion that may arise in the joint after assembly due to changes in moisture content. The design rules in EC5 are the same for the ring and the split-ring forms.

These connectors are only suitable for timber-to-timber connections and are shown in Figure 11.5. They are referenced as type A1 to A6 in BS EN912 and those to which

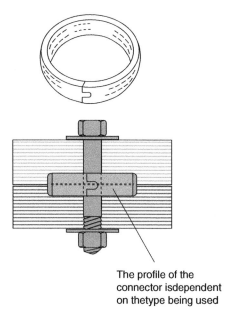

The profile of the
connector isdependent
on thetype being used

Fig. 11.5. Split-ring connector.

the EC5 strength equations will apply range from 60 mm to a maximum of 200 mm in diameter. The connectors are held in place by bolts and washers, but unlike toothed-plate connectors, the bolts and washers will not contribute to the lateral strength of the connection. The bolt diameter requirements for these types of connector are given in BS EN 13271.

In a connection formed using a ring connector, the load is passed from one member onto the ring by embedment stresses, and after shear transfer to the part of the ring in the second member, passes into that member by embedment stresses. The function of the connector bolt is to ensure that the ring remains fully embedded in the members. An example of this type of connector is shown in Figure 11.5.

Shear-plate connectors are used where there is a requirement for a timber-to-steel (or concrete) connection, and where the joint is to be demountable or where connections are to be formed on site.

In BS EN 912 shear-plate connectors are referenced as type B1 to B4, range in diameter from 65 to 190 mm, and are held in place by bolts and washers. With these types of connectors the shear strength of the bolt is a key element in the transfer of lateral load across the connection. However, because the shear strength of the bolt specified for use with these connectors will always exceed the lateral strength of the connector, EC5 does not require the bolt shear strength to be checked. Also, as with a ring connector, the connector bolt is not considered to contribute to the lateral strength of the connector.

In a shear-plate connection, the load passes from one member into the shear plate by embedment stresses and the bolt is then loaded through bearing stresses between the shear plate and the bolt. From there it is transferred through the bolt by shear stresses to the second shear plate in the case of a timber-to-timber joint, or, in the case of a timber-to-steel (or concrete) joint, directly into the steel (or concrete) member. An example of the use of this type of connector is shown in Figure 11.6.

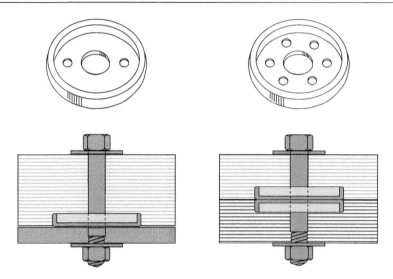

Fig. 11.6. Shear-plate connectors.

The design requirements for split-ring and shear-plate connectors are covered in EC5, *8.9.* For a joint using these connectors the strength model adopted assumes that when the connector is loaded parallel to the grain the joint strength will be the lesser of the strength due to embedment failure of the timber at the connector or shear block failure of the timber at the loaded end of the connection.

The timber within the area bounded by the connector is assumed to have sheared off before the failure load of the connection is achieved and does not contribute to the strength. Assuming that a shear-plate connection is being used, the design conditions to be satisfied for member B in a two-member connection are shown in Figure 11.7.

Two modes of failure are possible. These are embedment failure and shear block failure as shown on the loaded face of the connector at its loaded end. If the embedment resistance of the timber exceeds the resistance offered by the shear block, shear block failure will dictate the strength, and if it is smaller, embedment failure will occur. As shear block failure is a brittle failure mechanism, failure by the ductile embedment failure mode is the preferred design condition. Where shear block failure will not occur, as in the case of member A in Figure 11.7, only embedment failure will be relevant when determining the strength of that member.

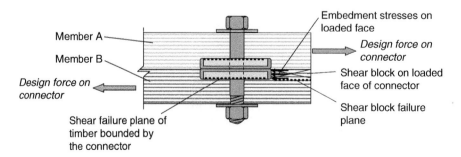

Fig. 11.7. Resistance of connection at the loaded end.

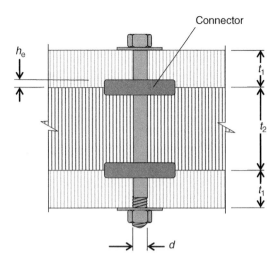

Fig. 11.8. Relevant dimensions used for connections fitted with split-ring and shear plate connectors (Based on EC5 *Figure 8.12*).

To prevent other forms of brittle failure in this type of connection, EC5 specifies minimum spacings, end and edge distances as well as minimum member thicknesses, and the respective criteria are the same for shear-plate and ring connectors. The requirements for spacings and distances are as given in Table 11.2.

Minimum member thicknesses are given in EC5, *8.9(2)*, and the criteria to be met are the same as for toothed-plate connectors. The requirements are summarised as follows:

$$t_1 \geq 2.25 h_e \tag{11.12}$$

$$t_2 \geq 3.75 h_e \tag{11.13}$$

where the thicknesses are as shown in Figure 11.8 and t_1 is the thickness of the outer timber member(s) in the connection (in mm), t_2 is the thickness of the inner timber member in the connection (in mm), and h_e is the embedment depth of the ring or shear-plate (in mm).

The same equations are used to determine the characteristic strength of a split-ring and a shear-plate connector, based on a connection formed with timber having a characteristic density of $350 \, \text{kg/m}^3$, a loaded end distance of $2d_c$, a side member thickness of $3h_e$, and a central member thickness of $5h_e$:

- From *equation 8.61(a)* in EC5, the characteristic strength of the shear block (in N) at the loaded end when loaded in tension parallel to the grain is:

$$F1_{v,0,Rk} = 35 d_c^{1.5} \tag{11.14}$$

  where d_c is the connector diameter (in mm).
- From *equation (8.61b)* in EC5, the characteristic embedment strength of the timber (in N) at the loaded face of the connector is:

$$F2_{v,0,Rk} = 31.5 d_c h_e \tag{11.15}$$

  where h_e is the embedment depth (in mm) of the connector type being used.

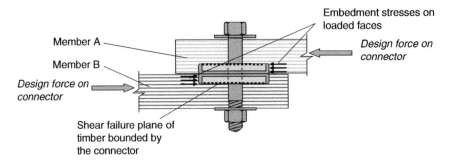

Fig. 11.9. Connection using split ring connectors loaded in an 'unloaded' end condition.

The connector strength will be the minimum value obtained from equations (11.14) and (11.15). For a connection in which the connector is loaded in an 'unloaded' end condition as shown in Figure 11.9 (i.e. $150° \leq \alpha \leq 210°$), shear block failure will not arise and the strength will only be based on equation (11.15).

To take account of the effect of variation in member thickness, loaded end distance and characteristic density as well as the increase in shear block strength when there is a steel member in the joint, modification factors must be applied to equations (11.14) and (11.15). The characteristic load-carrying capacity parallel to the grain (in N), $F_{v,0,Rk}$, per connector, per shear plane, incorporating these factors is:

$$F_{v,0.Rk} = \min \begin{cases} k_1 k_2 k_3 k_4 (35 d_c^{1.5}) & (a) \quad \text{(block shear)} & \text{(EC5, equation (8.61))} \\ k_1 k_3 (315 d_c h_e) & (b) \quad \text{(embedment)} \end{cases} \quad (11.16)$$

where:

- d_c is the connector diameter (in mm).
- k_1 is a modification factor taking into account the effect of member thickness

$$k_1 = \min \begin{cases} 1 \\ t_1 / 3h_e \\ t_2 / 5h_e \end{cases} \quad \text{(EC5, equation (8.62))} \quad (11.17)$$

where t_1 and t_2 are as defined in Figure 11.8 and h_e is the embedment depth (in mm). The strength of the connection will be reduced by 25% if the minimum permitted values of t_1 and t_2 are used. Also, when t_1 and t_2 exceed $3h_e$ and $5h_e$, respectively, there will be no increase in joint strength.

- k_2 only applies to the block shear strength at a loaded end and is 1.0 unless the angle of load is between $-30° \leq \alpha \leq 30°$, in which case:

$$k_2 = \min \begin{cases} k_a \\ a_{3,t} / 2d_c \end{cases} \quad \text{(EC5, equation (8.63))} \quad (11.18)$$

where $a_{3,t}$, the loaded end distance, is obtained from Table 11.2, or, for members with a sloping end, measured as shown in Figure 11.3; d_c is as previously defined and $k_a = 1.25$ for connections with one connector per shear plane. For connections with more than one connector per shear plane, $k_a = 1.0$.

As stated in Appendix C, the value given for the minimum loaded end distance in EC ($1.5d_c$) is to be changed to $2.0d_c$ and this value has been inserted in Table 11.2

- k_3 is a modification factor for timber density and:

$$k_3 = \min \begin{cases} 1.75 \\ \rho_k/350 \end{cases} \qquad \text{(EC5, equation (8.65))} \quad (11.19)$$

where ρ_k is the characteristic density of the timber in the joint (in kg/m^3).

As previously stated, the reference strength equations (11.16) are based on a connection in which the timber has a characteristic density of 350 kg/m^3 (i.e. strength class C24, as specified in EN 338). As in the case of connections using toothed-plate connectors, for connections using ring or shear-plate connectors, factor k_3 will increase the strength of the connector when higher strength class timber is used. With these types of connectors a maximum increase of 75% will be achieved when the characteristic density is at least 613 kg/m^3. This will cover the requirements for all strength classes of softwood in BS EN 338 as well as hardwood strength classes D30–D40, but there will be no increase beyond this limit.

- k_4 is a modification factor that depends on the materials connected and only applies to the shear block failure equation (11.16, equation a). It is obtained from

$$k_4 = \begin{cases} 1.0 & \text{for timber-to-timber connections} \\ 1.1 & \text{for steel-to-timber connections} \end{cases}$$

$$\text{(EC5, equation (8.66))} \quad (11.20)$$

Where the connector is loaded at an angle α to the grain, the characteristic load-carrying capacity, $F_{v,\alpha,Rk}$, per connector per shear plane is determined using Hankinson's equation [8], referred to in 5.3.2, and is:

$$F_{v,\alpha,Rk} = \frac{F_{v,0,Rk}}{k_{90} \sin^2 \alpha + \cos^2 \alpha} \qquad \text{(EC5, equation (8.67))} \quad (11.21)$$

with

$$k_{90} = 1.3 + 0.001d_c \qquad \text{(EC5, equation (8.68))} \quad (11.22)$$

where $F_{v,0,Rk}$ is defined in equation (11.16), and d_c is as previously defined.

The minimum spacings, edge and end distances for ring and shear-plate connectors will be as given in Table 11.2 with the symbols being the same as those in Figures 10.11 and 11.3.

Where ring or shear-plate connectors are staggered, as for toothed-plate connectors, the minimum spacings parallel and perpendicular to the grain can be reduced by complying with the following rules:

$$(k_{a1})^2 + (k_{a2})^2 \geq 1 \quad \text{with} \quad \begin{cases} 0 \leq k_{a1} \leq 1 \\ 0 \leq k_{a2} \leq 1 \end{cases} \qquad \text{(EC5, equation (8.69))} \quad (11.23)$$

where k_{a1} and k_{a2} are reduction factors applied to the minimum distance a_1 parallel to the grain and a_2 perpendicular to the grain, respectively, as shown in Figure 11.4.

If required, the spacing parallel to the grain, $k_{a1}a_1$ can be further reduced by multiplying by a factor $k_{s,red}$, provided $0.5 \le k_{s,red} \le 1.0$ and the load-carrying capacity of each connector, $F_{v,0,Rk,connector}$, is also reduced as follows:

$$F_{v,0,Rk,connector} = k_{r,red} \times F_{v,0,Rk} \tag{11.24}$$

where

$$k_{r,red} = 0.2 + 0.8 k_{s,red} \qquad (EC5, equation (8.70))$$

Using the largest permitted reduction, i.e. $k_{s,red} = 0.5$, the strength of each connector will be reduced to 60% of its full strength.

Where there is a row of split-ring or shear-plate connectors parallel to the grain and more than two connectors are in the row, the strength of the connectors per shear plane when loaded parallel to the grain will be reduced. For this condition, the effective number of connectors, n_{ef}, should be taken to be as follows:

For connections in single or double shear, irrespective of the type of connector being used, the number of connectors shall be taken to equal the number of connector bolts in the row and:

$$n_{ef} = 2 + \left(1 - \frac{n}{20}\right)(n-2) \qquad (EC5, equation (8.71)) \quad (11.25)$$

where n is the number of connector bolts in the row parallel to the grain. It should also be noted that when $k_{a2}a_2 < 0.5\ k_{a1}a_1$, connectors should be considered to be in the same row.

If the connection has more than one row of connectors parallel to the grain, the effective number of connectors shall be taken to be the sum of the effective number of connectors in each row. When the connectors are loaded perpendicular to the grain the effective number of connectors will equal the number of connector bolts in the connection.

Although the bolts and washers used for ring or shear-plate connectors do not contribute to the lateral strength of the connection, they are required to hold the connectors in position, and must comply with the requirements of EC5, *10.4.3*. The minimum and the maximum diameter of bolt permitted for use with these types of connector are given in Table 11.3, where d_c is the diameter of the connector (in mm), d is the diameter of the bolt (in mm), and d_1 is the diameter of the centre hole of the connector (in mm).

See Example 11.12.2.

Table 11.3 Requirements for the size of bolts used with ring and shear-plate connectors (as referenced in BS EN912)[*]

	Connector type		
	A1 to A6	A1, A4, A6	B
Connector diameter, d_c	≤ 130	>130	
Minimum bolt diameter, d_{min}	12	$0.1\ d_c$	$d_1 - 1$
Maximum bolt diameter, d_{max}	24	24	d_1

[*]Based on *Table 10.1* in EC5 and values are in mm.

11.5   MULTIPLE SHEAR PLANE CONNECTIONS

Where a connection involves multiple shear planes and the connected members are at varying angles to each other, it is not possible to directly apply the joint strength equations given in 11.3 for toothed-plate connectors and in 11.4 for ring or split-ring connectors.

For this situation, the guidance given in EC5, *8.1.3(1)*, is that the resistance of each shear plane in the connection is derived on the assumption that the shear plane forms part of a series of three-member connections and the connection strength is derived by combining the strength values of those failure modes that are compatible with each other. When dealing with ring and shear-plate connectors, all failure modes will be compatible. However, when considering the failure modes associated with the toothed-plate connector bolt in toothed-plate connections, EC5 requires that failure modes (a), (b), (g) and (h) in Table 10.2 or modes (c), (f) and (j/l) in Table 10.3 are not combined with the other failure modes.

The procedure to be followed to determine the joint strength in multiple member joints formed with connectors will, in principle, be the same as that used for multiple member joints formed using metal dowel-type fasteners, as described in 10.7.

11.6   BRITTLE FAILURE DUE TO CONNECTION FORCES AT AN ANGLE TO THE GRAIN

This form of brittle failure can arise when connectors apply a force at an angle to the timber grain resulting in the possibility of splitting caused by the force component perpendicular to the grain, as shown in Figure 11.10. The failure can occur at a load less than the design capacity of the fasteners.

This form of failure is discussed in 10.3.6.1 and the splitting strength of a connector in which a tension component can occur at right angles to the grain is obtained from equation (10.41d).

11.7   ALTERNATING FORCES IN CONNECTIONS

Where a connection is subjected to short-term or instantaneous alternating forces, the characteristic load-carrying capacity of the connection will not be affected. If, however, the connection is subjected to alternating forces due to long- or medium-term actions, the characteristic load-carrying capacity of the connection will be reduced.

The design requirements for such situations will be as described in 10.4.2.

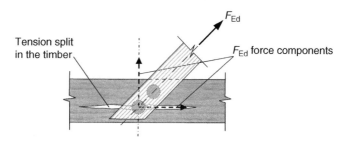

Fig. 11.10. Member loaded in tension at an angle to the grain.

11.8   DESIGN STRENGTH OF A LATERALLY LOADED CONNECTION

11.8.1   Loaded parallel to the grain

The design strength of a laterally loaded single connector, $F_{v,Rd}$, is obtained from the characteristic load-carrying capacity of the laterally loaded connector as follows:

$$F_{v,Rd} = \frac{k_{mod} \cdot F_{v,Rk}}{\gamma_M} \qquad (11.26)$$

where:

- k_{mod} is the modification factor referred to in 2.2.20, and where the connection comprises two timber elements, $k_{mod,1}$ and $k_{mod,2}$, the value used in the equation will be $k_{mod} = \sqrt{k_{mod,1} k_{mod,2}}$, as required by EC5, *equation (2.6)*.
- γ_M is the partial factor for connections given in Table 2.6. Except when determining the plate strength of punched metal plate fasteners, the factor value will be 1.3.
- $F_{v,Rk}$ is the characteristic load-carrying capacity of the connector per shear plane when loaded laterally, i.e. the lowest value determined from the relevant strength equations in 11.3 for toothed-plate connectors or in 11.4 for ring or shear-plate connectors.

For a connection in single or double shear containing r_{pl} rows of connectors laterally loaded parallel to the grain, with each row containing n equally spaced connectors of the same type and size, each with a design strength, $F_{v,Rd}$, the effective lateral load design capacity of the connection per shear plane parallel to the grain, $F_{v,ef,Rd}$, will be:

$$F_{v,ef,Rd} = n_{sp} r_{pl} \cdot n_{ef} \cdot F_{v,Rd} \qquad (11.27)$$

where n_{ef} is the effective number of connectors in the connection in each row parallel to the grain, n_{sp} is the number of shear planes in the connection, and:

(a) *Effective number of toothed-plate connectors.*
  The effective number of toothed-plate connectors in a row shall equal the number of connector bolts in the row. For the strength contribution from the bolts in the connection, the effective number of bolts shall be derived as described in 10.4.1.3.

(b) *Effective number of ring or shear-plate connectors.*
  The effective number of ring or shear-plate connectors in a row is obtained from equation (11.25).

11.8.2 Loaded perpendicular to the grain

Where loads are imposed on the timber by connectors loaded perpendicular to the grain, there are two possible forms of failure:

(a) By the timber splitting in tension and this condition is covered in 11.6 and 10.3.6.1.
(b) By ductile yielding of the connector and for this condition, where there are r_{pr} lines of connectors with each line containing n connectors all of the same size,

$$F_{v,ef,Rd} = n_{sp} r_{pr} \cdot n_{cb} \cdot F_{v,Rd} \qquad (11.28)$$

where:

- $F_{v,ef,Rd}$ is the effective design strength of the connector per shear plane when loaded laterally and *perpendicular* to the grain.
- n_{sp} is the number of shear planes in the connection.
- n_{cb} is the number of connector bolts per line *perpendicular* to the grain.
- $F_{v,Rd}$ is the design load-carrying capacity of a laterally loaded single connector per shear plane when loaded perpendicular to the grain, i.e. the lowest value determined from the relevant strength equations in 11.3 for toothed-plate connectors or in 11.4 for ring and shear-plate connectors.

From the above the design load-carrying capacity of the connection *perpendicular* to the grain will be obtained using the same procedure as defined for dowelled connections, referred to in equation (10.53).

11.8.3 Loaded at an angle to the grain

When connectors are laterally loaded at an angle to the grain, the force components parallel and perpendicular to the grain have to be derived, and:

(i) the component of the design force acting parallel to the grain must not exceed the load-carrying capacity of the connection determined as defined in Section 11.8.1, and
(ii) the component of the design force acting perpendicular to the grain must not exceed the lesser of load-carrying capacities of the connection determined as defined of the load-carrying capacity in 11.8.2.

11.9 STIFFNESS BEHAVIOUR OF TOOTHED-PLATE, RING AND SHEAR-PLATE CONNECTORS

The lateral stiffness of toothed-plate, split-ring and shear-plate connectors is determined in the same way as discussed in 10.10 for metal dowel-type fasteners.

The content of 10.10 is applicable to connections formed using these types of connector and although some matters of detail have been repeated in the following paragraphs for clarity, generally the content relates to areas where differences arise.

The slip modulus K_{ser} per shear plane per connector at the serviceability limit state is given in Table 11.4. Where the joint comprises two jointed wood-based members having mean densities $\rho_{m,1}$ and $\rho_{m,2}$ respectively, the ρ_m used in the equations should be:

$$\rho_m = \sqrt{\rho_{m,1}\rho_{m,2}} \qquad\qquad (EC5,\ equation\ (7.1)) \quad (11.29)$$

Also, if the connection is steel to timber or concrete to timber, as stated in EC5, *7.1(3)*, K_{ser} should be based on the mean density, ρ_m, of the timber member. Connectors are stiffer than metal dowel-type fasteners and when they are of types able to be used with steel or concrete, to allow for the lack of slip in the steel or concrete member, the value of K_{ser} may be multiplied by 2.

As for metal dowel-type fasteners, when designing at the ultimate limit state the instantaneous slip modulus, K_u, is taken to be:

$$K_u = \frac{2}{3}K_{ser} \qquad\qquad (EC5,\ equation\ (2.1)) \quad (11.30)$$

Where a connection is formed using single-sided connectors positioned back to back, e.g. when using single-sided toothed-plate connectors or shear-plate connectors, the stiffness of each pair of connectors (per connecting bolt) should be taken to be the stiffness of the single-sided connector as given in Table 11.4.

Due to the need to include for some tolerance in the diameter of the toothed-plate connector bolt hole to permit the bolt to be fitted, joints made with single-sided connectors will exhibit a small initial slip when being loaded and some codes suggest an allowance should be made for this in any deflection calculation. There is no requirement in EC5 for such an allowance to be made with these types of connector.

No guidance is given in EC5 on whether the effective number or the full number of connectors should be used to derive the stiffness of a connection formed using toothed-plate connectors. It is recommended that, as with connections formed using metal dowel-type fasteners, the connection stiffness should be calculated assuming that all connectors will contribute their full stiffness irrespective of the angle of loading.

See Examples 11.12.1 and 11.12.2.

Table 11.4 Values for K_{ser} for connectors in N/mm in timber to timber and wood-based panel-to-timber connections, (based on *Table 7.1* in EC5)

Type of connector used (in accordance with BS EN 912)	Serviceability limit state slip modulus K_{ser}

Toothed-plate connectors:
  Connector-types C1 to C9
  Connector-type C10 and C11
Split-ring connectors type A
	$1.5\rho_m d_c/4$
	$\rho_m d_c/2$
	$\rho_m d_c/2$

Shear-plate connectors type B	$\rho_m d_m/2$

11.10 FRAME ANALYSIS INCORPORATING THE EFFECT OF LATERAL MOVEMENT IN CONNECTIONS FORMED USING TOOTHED-PLATE, SPLIT-RING OR SHEAR-PLATE CONNECTORS

Where the slip effect in connections formed with toothed-plate, split-ring or shear-plate connectors has to be incorporated into the analysis of a structure, the content of 10.11 will apply.

11.11 REFERENCES

1 BS EN 912:2011. *Timber Fasteners – Specifications for Connectors for Timber,* British Standards Institution.

2 BS EN 14545:2008. *Timber Fasteners – Connectors – Requirements,* British Standards Institution.

3 BS EN 13271:2002. *Timber Fasteners – Characteristic Load-Carrying Capacities and Slip-Moduli for Connector Joints,* British Standards Institution.

4 BS EN 1995-1-1:2004. *Eurocode 5: Design of Timber Structures. Part 1-1:2004 + A1:2008 General – Common Rules and Rules for Buildings,* British Standards Institution.

5 Blass, H.J. Toothed-plate connector joints. In: Blass, H.J., Aune, P., Choo, B.S., et al. (eds), *Timber Engineering STEP 1,* 1st edn. Centrum Hout, Almere, 1995.

6 BS 5268-2-2002. *Structural Use of Timber. Part 2: Code of Practice for Permissible Stress Design, Materials and Workmanship,* British Standards Institution.

7 BS EN 338:2009. *Structural Timber – Strength Classes,* British Standards Institution.

8 Hankinson, R.L. Investigation of crushing strength of spruce at varying angles of grain. *Air Service Information Circular,* Vol. 3, No. 259 (Materials Section Paper No. 130), 1921.

9 BS EN 1991-1-1:2002. *Eurocode 1: Actions on Structures. Part 1-1: General Actions – Densities, Self-Weight and Imposed Loads for Buildings,* British Standards Institution.

10 NA to BS EN 1990:2002 + A1:2005 *Incorporating National Amendment No1. UK National Annex for Eurocode 0 – Basis of Structural Design,* British Standards Institution.

11 BS EN 1993-1-8:2005. *Eurocode 3: Design of Steel Structures. Part 1-8: Design of Joints,* British Standards Institution.

11.12 EXAMPLES

As stated in 4.3, in order to verify the ultimate and serviceability limit states, each design effect has to be checked and for each effect the largest value caused by the relevant combination of actions must be used.

However, to ensure attention is primarily focused on the EC5 design rules for the timber or wood product being used, only the design load case producing the largest design effect has generally been given or evaluated in the following examples.

Example 11.12.1 A timber-to-timber tension splice connection in a statically indeterminate floor structure, which functions in service class 2 conditions in a Category A loaded area

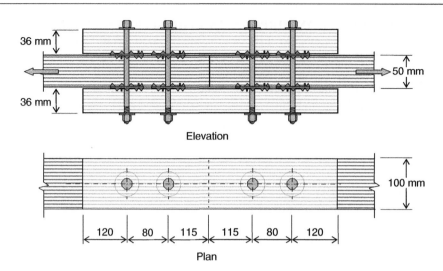

Elevation

Plan

Fig. E11.12.1.

(in accordance with BS EN 1991-1-1), is required to connect two 50 mm by 100 mm members, and is shown in Figure E11.12.1. Two 36 mm by 100 mm side members will be used and the connection will be formed using two type C1 toothed-plate connectors to BS EN 912:2011, diameter $d_c = 50$ mm, spaced 80 mm apart parallel to the grain. The connector bolt is 16 mm in diameter and has a tensile strength of 400 N/mm². The joint is subjected to a characteristic permanent tensile lateral action of 5.5 kN and a characteristic medium-term variable tensile lateral action of 12.5 kN and all timber is of strength class C22 to BS EN 338:2009.

Confirm that the joint complies with the rules in EC5 and determine the instantaneous and final deflection of the joint at the serviceability limit states (SLS), on the basis that the connection forms part of a structure where the creep behaviour of the connections is different from that of the members.

1. **Geometric properties of the joint**

Thickness of each side member, t_1	$t_1 = 36$ mm	
Thickness of central members, t_2	$t_2 = 50$ mm	
Width of all timber members, h	$h = 100$ mm	
Bolt diameter, d	$d = 16$ mm	
Tensile stress area of a bolt, A_{bt}	$A_{bt} = 157$ mm²	
Diameter of toothed-plate connector, d_c (BS EN 912:2011)	$d_c = 50$ mm	
Thickness of metal in connector, t	$t = 1.0$ mm	
Height of each toothed-plate connector, h_c	$h_c = 13$ mm	
Loaded end distance of the connection, $ac_{3.t}$	$ac_{3.t} = 115$ mm	
Minimum loaded end distance for a type C1 connector, $acmin_{3.t}$ (equation (11.7); EC5, *equation (8.75)*)	$acmin_{3.t} = 1.1 \cdot d_c$	$acmin_{3.t} = 55$ mm

| Minimum loaded end distance for the bolt, $abmin_{3.t}$ (Table 10.9 (EC5, *Table 8.4*)) | $abmin_{3.t} = \max{(7 \cdot d, \ 80\,mm)}$ |
| | $abmin_{3.t} = 112\,mm$ |

Minimum loaded end distance permitted for this type of connector, $a_{3.t}$

$$a_{3.t} = \max(acmin_{3.t}, \ abmin_{3.t})$$
$$a_{3.t} = 112\,mm$$

The loaded end distance is greater than the minimum distance; therefore OK

Angle of loading on the connection relative to the grain, α, in degrees

$\alpha = 0°$

Bolt spacing parallel to the grain, a_1

$a_1 = 80\,mm$

Minimum bolt spacing parallel to the grain based on Table 11.2, $a_{1c.min}$

$$a_{1c.min} = (1.2 + 0.3 \cdot l \cos{(\alpha)l}) \cdot d_c$$
$$a_{1c.min} = 75\,mm$$

Minimum bolt spacing parallel to the grain based on Table 10.9, $a_{1b.min}$

$$a_{1b.min} = (4 + l\cos(\alpha)\ l) \cdot d$$
$$a_{1b.min} = 80\,mm$$

Minimum bolt spacing parallel to the grain permitted for this connector, $a_{1b.min}$

$$a_{1.min} = \max{(a_{1b.min}, a_{1c.min})}$$
$$a_{1.min} = 80\,mm$$

i.e. actual bolt spacing is equal to the minimum permitted spacing, therefore OK

Embedment depth per member, h_e

$$h_e = \frac{h_c - t}{2}$$
$$h_e = 6\,mm$$

Minimum thickness of outer member, t_o (equation (11.1) (EC5, *8.10.3*))

$t_o = 2.25\,h_e$ $t_o = 13.5\,mm$

Minimum thickness of inner member, t_i (equation (11.2) (EC5, *8.10.3*))

$t_i = 3.75\,h_e$ $t_i = 36\,mm$

Each member thickness exceeds the minimum value, therefore OK

Cross-sectional area of the side members – taking account of loss of area due to the bolt hole plus a tolerance of 1 mm, A_s

$$A_s = h \cdot t_1 - (d + 1\,mm) \cdot t_1$$
$$A_s = 2.99 \times 10^{-3}\,m^2$$

Cross-sectional area of the inner members, with same tolerance allowance, A_{in}

$$A_{in} = h \cdot t_2 - (d - 1\,mm) \cdot t_1$$
$$A_{in} = 4.15 \times 10^3\,mm^2$$

Number of toothed-plate connectors, per shear plane per row, n_{tp}

$n_{tp} = 2$

Number of bolts per shear plane per row, n

$n = 2$

Number of effective bolts per shear plane per row, $n_{ef.b}$ (equation (10.45); EC5, *equation (8.34)*)

$$n_{ef.b} = \min\left(n, n^{0.9} \sqrt[4]{\frac{a_1}{13 \cdot d}}\right)$$
$$n_{ef.b} = 1.47$$

2. Timber and bolt properties

Table 1.3, strength class C22 (BS EN 338:2009(E), *Table 1*)

Characteristic tension strength parallel to the grain, $f_{t.0.k}$

$f_{t.0.k} = 13\,N/mm^2$

Characteristic bearing strength perpendicular to the grain, $f_{c.90.k}$ $f_{c.90.k} = 2.4\,\text{N/mm}^2$

Characteristic density of the timber, ρ_k $\rho_k = 340\,\text{kg/m}^3$

Mean modulus of elasticity parallel to the grain, $E_{0.\,\text{mean}}$ $E_{0.\text{mean}} = 10\,\text{kN/mm}^2$

Mean density of the timber, ρ_m $\rho_m = 410\,\text{kg/m}^3$

Tensile strength of each bolt, $f_{u.k}$ $f_{u.k} = 400\,\text{N/mm}^2$

3. **Partial safety factors**

Table 2.8 (UKNA to BS EN 1990:2002+A1(2005), *Table NA.A1.2(B)*) for the ULS

Permanent actions, $\gamma_{G.\text{ULS}}$ $\gamma_{G.\text{ULS}} = 1.35$

Variable actions, $\gamma_{Q.\text{ULS}}$ $\gamma_{Q.\text{ULS}} = 1.5$

Table 2.6 (UKNA to EC5, *Table NA.3*)

Material factor for timber, $\gamma_{M.\text{timber}}$ $\gamma_{M.\text{timber}} = 1.3$

Material factor for connections, $\gamma_{M.\text{connection}}$ $\gamma_{M.\text{connection}} = 1.3$

4. **Actions**

(i) Ultimate limit states (ULS)

Characteristic permanent tensile action, G_k $G_k = 5.5\,\text{kN}$

Characteristic variable tensile action, Q_k $Q_k = 12.5\,\text{kN}$

Design value of the tensile force in the central member at the ULS, F_d $F_d = \gamma_{G.\text{ULS}} \cdot G_k + \gamma_{Q.\text{ULS}} \cdot Q_k$

(Table 2.8, equation (c) using the unfavourable condition variable action) $F_d = 26.18\,\text{kN}$

(ii) Serviceability limit states (SLS)

Combination factor for variable load in a domestic area, ψ_2 (NA to BS EN 1990:2002+A1:2005, *Table NA.A1.1*) $\psi_2 = 0.3$

Design value of the tensile force in the central member at the SLS, $F_{d.\text{SLS}}$ $F_{d.\text{SLS}} = G_k + Q_k$

Based on the characteristic combination (equation (2.24); EC0, *equation (6.14b)*) $F_{d.\text{SLS}} = 18\,\text{kN}$

Design value of the quasi-permanent tensile force in the central member at the SLS, $F_{d.\text{creep.sls}}$
Based on the characteristic combination, (Equation 2.24 (EC0, Equation 6.14b)) $F_{d.\text{creep.sls}} = G_k + \psi_2 \cdot Q_k$ $F_{d.\text{creep.sls}} = 9.25 \times 10^3\,\text{N}$

5. **Modification factors**

Factor for medium-duration loading and service class 2, $k_{mod.med}$ (Table 2.4 (EC5, *Table 3.1*))

$k_{mod.med} = 0.80$

Size factor for width less than 150 mm, k_h, Table 2.11 (EC5, *equation (3.1)*)

$$k_h = \begin{vmatrix} 1.0 & if\ h \ge 150\,mm \\ \left(\dfrac{150\,mm}{h}\right)^{0.2} & if\ 1.3 > \left(\dfrac{150\,mm}{h}\right)^{0.2} \\ 1.3 & otherwise \end{vmatrix}$$

$k_h = 1.08$

Deformation factor for timber and service class 2, k_{def} (Table 2.10 (EC5, *Table 3.2*))

$k_{def} = 0.80$

6. **Tension strength of timber – ULS condition**

The design load case at the ULS will be due to a combination of the permanent and variable design loads.

Design tensile stress parallel to grain in the side members, $\sigma_{t.0.d.s}$

$$\sigma_{t.0.d.s} = \frac{F_d}{2 \cdot A_s}$$

$\sigma_{t.0.d.s} = 4.38\,N/mm^2$

Design tensile stress parallel to grain in the central members, $\sigma_{t.0.d.i}$

$$\sigma_{t.0.d.i} = \frac{F_d}{A_{in}}$$

$\sigma_{t.0.d.i} = 6.31\,N/mm^2$

Design tensile strength of the timber in the joint parallel to grain, $f_{t.0.d}$

$$f_{t.0.d} = \frac{k_{mod.med} \cdot k_h \cdot f_{t.0.k}}{\gamma_{M.timber}}$$

$f_{t.0.d} = 8.68\,N/mm^2$

Tensile strength of members is OK

7. **Embedment strength of timber**

Characteristic embedment strength of timber parallel to grain, $f_{h.k}$ (equation (10.35); EC5, *equation (8.32)*)
(The equation incorporates dimension factors for Mathcad.)

$$f_{h.k} = 0.082 \cdot (1 - 0.01 \cdot d \cdot mm^{-1}) \cdot \left(\rho_k \cdot \frac{m^3}{kg}\right) N/mm^2$$

$f_{h.k} = 23.42\,N/mm^2$

Characteristic embedment strength of the side timber members, $f_{h.1.k}$

$f_{h.1.k} = f_{h.k}$

$f_{h.1.k} = 23.42\,N/mm^2$

Characteristic embedment strength of the central timber members: $f_{h.2.k}$

$f_{h.2.k} = f_{h.k}$

$f_{h.2.k} = 23.42\,N/mm^2$

8. **Yield moment of bolts**

Characteristic yield moment of a bolt, $M_{y.Rk}$ (Table 10.5 (EC5, *equation (8.30)*)).
(The equation incorporates dimension factors for Mathcad.)

$$M_{y.Rk} = 0.3 \cdot \left(f_{u.k} \cdot \frac{mm^2}{N}\right) \cdot (d \cdot mm^{-1})^{2.6}\,N\,mm$$

$M_{y.Rk} = 1.62 \times 10^5\,N\,mm$

9. Withdrawal resistance of bolt

Design strength of the bolt, $F1_{ax.Rd}$
(BS EN 1993-1-8)

$$F1_{ax.Rd} = 0.9 \cdot f_{u.k} \cdot \left(\frac{1}{1.25}\right) \cdot A_{bt}$$

$$F1_{ax.Rd} = 4.52 \times 10^4 \text{ N}$$

Washer diameter used, d_w $d_w = 3 \cdot d$ $d_w = 48 \text{ mm}$

Design capacity of the washer, $F2_{ax.Rd}$
(10.8.2 (EC5, 8.5.2(2)))

$$F2_{ax.Rd} = 3 \cdot f_{c.90.k} \cdot \frac{\pi}{4}\left[d_w^2 - (d+1\cdot\text{mm})^2\right] \cdot \frac{k_{mod.med}}{\gamma_{M\cdot\text{Connection}}}$$

$$F2_{ax.Rd} = 7.01 \times 10^3 \text{ N}$$

As failure is by bearing, characteristic axial withdrawal capacity of the bolt, $F_{ax.Rk}$

$$F_{ax.Rk} = F2_{ax.Rd} \cdot \frac{\gamma_{M.\text{Connection}}}{k_{mod.med}}$$

$$F_{ax.Rk} = 1.14 \times 10^4 \text{ N}$$

10. Load-carrying capacity

(i) Bolt strength

For a timber-to-timber joint with toothed-plate connectors, the characteristic lateral resistance per shear plane per connector bolt is the lesser of equations (10.7)–(10.10) in Table 10.2 (EC5, *equations (8.7)*), where:

$$\beta = \frac{f_{h.2.k}}{f_{h.1.k}} \qquad \beta = 1$$

and

Failure mode (g): $F_{v.Rk.g} = f_{h.1.k} \cdot t_1 \cdot d$ $F_{v.Rk.g} = 1.35 \times 10^4 \text{ N}$

Failure mode (h): $F_{v.Rk.h} = 0.5 \cdot f_{h.2.k} \cdot t_2 \cdot d$ $F_{v.Rk.h} = 9.37 \times 10^3 \text{ N}$

Failure mode (j):

$$F_{v.Rk.j} = 1.05 \cdot \frac{f_{h.1.k} \cdot t_1 \cdot d}{2+\beta}\left[\sqrt{2\cdot\beta\cdot(1+\beta)+\frac{4\cdot\beta\cdot(2+\beta)\cdot M_{y.Rk}}{f_{h.1.k}\cdot t_1^2 \cdot d}} - \beta\right] + \frac{F_{ax.Rk}}{4}$$

$$F_{v.Rk.j} = 1.15 \times 10^4 \text{ N}$$

Failure mode (k)

$$F_{v.Rk.k} = 1.15 \cdot \sqrt{\frac{2\cdot\beta}{1+\beta}} \cdot \left(2\cdot M_{y.Rk}\cdot f_{h.1.k} \cdot d\right) + \frac{F_{ax.Rk}}{4}$$ $F_{v.Rk.k} = 1.55 \times 10^4 \text{ N}$

Limiting $F_{ax.Rk}/4$ to 25% of the Johansen part of the relevant equations:

(10.3.5.3, (EC5, 8.2.2(2))

Failure mode (j):

$$F_{v.Rk.jj} = 1.25 \cdot 1.05 \cdot \frac{f_{h1k} \cdot t_1 \cdot d}{2+\beta} \cdot \left[\sqrt{2\cdot\beta\cdot(1+\beta)+\frac{4\cdot\beta\cdot(2+\beta)\cdot M_{y.Rk}}{f_{h1k}\cdot t_1^2 \cdot d}} - \beta\right]$$

$$F_{v.Rk.jj} = 1.08 \times 10^4 \text{ N}$$

Failure mode (k):

$$F_{v.Rk.kk} = 1.25 \cdot 1.15 \cdot \sqrt{\frac{2 \cdot \beta}{1 + \beta} \cdot \left(2 \cdot M_{y.Rk} \cdot f_{hlk} \cdot d \right)} \qquad\qquad F_{v.Rk.kk} = 1.55 \times 10^4 \, \text{N}$$

The characteristic lateral resistance per shear plane per bolt will be:

$$F_{v.Rk.bolt} = \min(F_{v.Rk.g}, \, F_{v.Rk.h}, \, F_{v.Rk.j}, \, F_{v.Rk.k}, \, F_{v.Rk.jj}, \, F_{v.Rk.kk})$$

$$F_{v.Rk.bolt} = 9.37 \times 10^3 \, \text{N}$$

i.e. failure mode (h)

The characteristic lateral resistance per shear plane provided by the bolts, $F_{v.Rk.bolts}$

$$F_{v.Rk.bolts} = F_{v.Rk.bolt} \cdot n_{ef.b}$$

$$F_{v.Rk.bolts} = 1.38 \times 10^4 \, \text{N}$$

(ii) Connector strength

The characteristic lateral resistance per shear plane per toothed-plate connector will be $F_{v.Rk}$

Modification factors:

Factor k_1
(equation (11.5); EC5,
equation (8.73))

$$k_1 = \min\left(1, \frac{t_1}{3 \cdot h_e}, \frac{t_2}{5 \cdot h_e} \right) \qquad k_1 = 1$$

Factor k_2
(equation (11.6); EC5,
equation (8.74))

$$k_2 = \min\left(1, \frac{ac_{3.t}}{1.5 d_c} \right) \qquad k_2 = 1$$

Factor k_3
(equation (11.10); EC5, *equation (8.78)*)

(The equation incorporates dimension factors for Mathcad.)

$$k_3 = \min\left(1.5, \frac{\rho_k}{350 \cdot \frac{\text{kg}}{\text{m}^3}} \right) \qquad k_3 = 0.97$$

Characteristic lateral resistance per shear plane per toothed-plate connector in the row, $F_{v.Rk}$ (Equation (11.4); EC5, *equation (8.72)*)

The equation incorporates dimension factors for Mathcad:

$$F_{v.Rk} = 18 \cdot k_1 \cdot k_2 \cdot k_3 \cdot \left(\frac{d_c}{\text{mm}} \right)^{1.5} \cdot \text{N}$$

$$F_{v.Rk} = 6.18 \times 10^3 \cdot \text{N}$$

Characteristic lateral resistance per shear plane provided by the toothed-plate connectors in the row, $F_{v.Rk.n}$

$$F_{v.Rk.n} = F_{v.Rk} \cdot n_{tp}$$

$$F_{v.Rk.n} = 12.36 \, \text{kN}$$

Characteristic lateral resistance per shear plane at each connection provided by the toothed-plate connectors plus the bolts in the row: (Equation 11.3)

$$F_{v.Rk.connection} = F_{v.Rk.n} + F_{v.Rk.bolts}$$

$$F_{v.Rk.connection} = 26.13 \, \text{kN}$$

The design resistance of each connection per shear plane per row, $F_{\text{v.Rd}}$	$F_{\text{v.Rd}} = \dfrac{k_{\text{mod.med}} \cdot F_{\text{v.Rk.connection}}}{\gamma_{\text{M.connection}}}$
	$F_{\text{v.Rd}} = 16.08 \text{ kN}$

The design strength of each connection, $F_{\text{v.Rd.joint}}$

$$F_{\text{v.Rd.joint}} = 2 \cdot F_{\text{v.Rd}}$$

$$F_{\text{v.Rd.joint}} = 32.16 \text{ kN}$$

The design load on the joint, F_{d}

$$F_{\text{d}} = 26.18 \text{ kN}$$

The design load is less than the design strength of the connection; therefore OK

11. The instantaneous and final deflection of the joint at the SLS

The design condition is taken to be based on the characteristic combination of actions at the SLS

The slip modulus of a connector per shear plane at the SLS, K_{ser} (Table 11.4) (EC5, *Table 7.1*) (The equation incorporates dimension factors for Mathcad.)

$$K_{\text{ser}} = 1.5 \cdot \dfrac{\left(\rho_{\text{m}} \cdot \dfrac{\text{m}^3}{\text{kg}} \right) \cdot d_{\text{c}} \text{ mm}^{-1}}{4} \text{ N/mm}$$

$$K_{\text{ser}} = 7.69 \times 10^3 \text{ N mm}^{-1}$$

The stiffness of each connection at the SLS, $K_{\text{ser.connection}}$ (2 No shear planes; n_{tp} connectors per shear plane)

$$K_{\text{ser.connection}} = 2 \cdot n_{\text{tp}} \cdot K_{\text{ser}}$$

$$K_{\text{ser.connection}} = 3.08 \times 10^4 \text{ N/mm}$$

The stiffness of the joint at the SLS, $K_{\text{ser.joint}}$ (Figure 10.24)

$$K_{\text{ser.joint}} = \dfrac{K_{\text{ser.connection}}}{2}$$

$$K_{\text{ser.joint}} = 1.54 \times 10^4 \text{ N/mm}$$

The instantaneous deflection of the joint at the SLS, δ_{inst} (equation (10.74) also applies to connectors)

$$\delta_{\text{inst}} = \dfrac{F_{\text{d.sls}}}{K_{\text{ser.joint}}}$$

$$\delta_{\text{inst}} = 1.17 \text{ mm}$$

At the final condition, the deformation factor for a connection with the same time-dependent elements is $2k_{\text{def}}$ (EC5, *2.3.2.2(3)*)

$$k_{\text{def.j}} = 2 \cdot k_{\text{def}}$$

The stiffness of the joint at the final condition at the SLS, $K_{\text{ser.fin.joint}}$ (equation (2.36))

$$K_{\text{ser.fin.joint}} = \dfrac{K_{\text{ser.joint}}}{1 + k_{\text{def.j}}}$$

The final deflection of the joint at the SLS (equation (10.80)), δ_{fin}

$$\delta_{\text{fin}} = \dfrac{F_{\text{d.creep.sls}}}{K_{\text{ser.fin.joint}}} + \delta_{\text{inst}} - \dfrac{F_{\text{d.creep.sls}}}{K_{\text{ser.joint}}}$$

$$\delta_{\text{fin}} = 2.13 \text{ mm}$$

[Final deformation of the joint - based on the alternative conservative approach (equation (10.81)), $\delta 1_{\text{fin}}$]

$$\delta 1_{\text{fin}} = \dfrac{F_{\text{d.sls}}}{K_{\text{ser.fin.joint}}}$$

$$\delta 1_{\text{fin}} = 3.04 \text{ mm}$$

Example 11.12.2 A timber-to-timber tension splice connection incorporating spacer timbers and connected by type A2 split-ring connectors, as specified in BS EN912:2011, is shown in Figure E11.12.2. The main members are 50 mm thick by 150 mm wide. The central spacer timber is 70 mm thick. The split-ring connectors are held in place by 12-mm-diameter bolts fitted with washers compliant with the requirements of EC5, *10.4.3*. The split-ring connectors are 150 mm apart. The connection is subjected to a tensile design force of 60 kN at the ULS and a tensile design force of 40 kN at the SLS, both arising from a combination of a permanent and medium-term variable actions. The connection functions in service class 2 conditions and all of the timber is strength class C22 to BS EN 338:2009.

Confirm the joint complies with the rules in EC5 and determine the instantaneous deflection of the connection at the SLS.

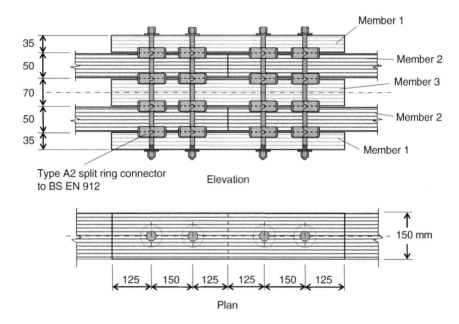

Fig. E11.12.2.

1. Geometric properties of the joint

Thickness of member 1, t_1	$t_1 = 35\,\text{mm}$
Width of member 1, h_1	$h_1 = 150\,\text{mm}$
Thickness of member 2, t_2	$t_2 = 50\,\text{mm}$
Width of member 2, h_2	$h_2 = 150\,\text{mm}$
Thickness of member 3, t_3	$t_3 = 70\,\text{mm}$
Width of member 3, h_3	$h_3 = 150\,\text{mm}$
Split-ring connector, type A2, diameter, d_c (BS EN 912:2011)	$d_c = 72.0\,\text{mm}$

Bolt diameter to be used with the connector, d (Table 11.3 (EC5, Table 10.1))	$d = 12$ mm Satisfies the criteria in EC5, *10.4.3(4)*
Loaded end distance of all members, $a_{3.t}$	$a_{3.t} = 125$ mm
Minimum loaded end distance, $amin_{3.t}$ (Table 11.2 (EC5, *Table 8.7*))	$amin_{3.t} = 1.5 \cdot d_c$ $amin_{3.t} = 108$ mm Loaded end distance exceeds minimum permitted value; therefore OK
Unloaded edge distance of all members, $a_{4.c}$	$a_{4.c} = \dfrac{h_1}{2}$ $a_{4.c} = 75$ mm
Minimum unloaded edge distance, $amin_{4.c}$ (Table 11.2)	$amin_{4.c} = 0.6 \cdot d_c$ $amin_{4.c} = 43.2$ mm Unloaded edge distance exceeds minimum permitted value; therefore OK
Connector spacing parallel to the grain, a_1	$a_1 = 150$ mm
Angle of connector force parallel to the grain, α	$\alpha = 0°$
Minimum connector spacing permitted parallel to the grain, $amin_1$ Table 11.3 (EC5 *Table 8.7*)	$amin_1 = (1.2 + 0.8 \cdot \lvert \cos(\alpha) \rvert) \cdot d_c$ $amin_1 = 144$ mm Distance between connectors exceeds the minimum value; therefore OK
Thickness of metal in connector, t (BS EN912:2011)	$t = 4.1$ mm
Height of split-ring connector, h_c (BS EN912:2011)	$h_c = 19.0$ mm
Embedment depth of split-ring connector, h_e (BS EN912:2011)	$h_e = \dfrac{h_c}{2}$ $h_e = 9.5$ mm
Minimum thickness of side member, $t_{1.min}$, (equation (11.12))	$t_{1.min} = 2.25 \cdot h_e$ $t_{1.min} = 21.37$ mm
Minimum thickness of members 2 and 3, t_i (equation (11.13))	$t_i = 3.75 \cdot h_e$ $t_i = 35.62$ mm i.e. the thickness of members 2 and 3 exceed the minimum values; OK

Cross-sectional area of each member 1 – taking account of loss of area due to the bolt hole and the split ring, A_1

$$A_1 = h_1 \cdot t_1 - (d_c \cdot h_e) - (d + 1\,\text{mm}) \cdot (t_1 - h_e) \qquad\qquad A_1 = 4.23 \times 10^3\,\text{mm}^2$$

Cross-sectional area of each member 2 – taking account of loss of area due to the bolt hole and split ring, A_2

$$A_2 = h_2 \cdot t_2 - 2 \cdot (d_c \cdot h_e) - (d + 1\,\text{mm}) \cdot (t_2 - 2 \cdot h_e) \qquad\qquad A_2 = 5.73 \times 10^3\,\text{mm}^2$$

Cross-sectional area of member 3 – taking account of loss of area due to the bolt hole and split ring, A_3

$A_3 = 2 \cdot A_1$

Number of shear-plate connectors parallel to the grain in each shear plane of each connection, n

$n = 2$

Effective number of shear-plate connectors parallel to the grain in each shear plane of each connection, n_{ef} (equation (11.25); EC5, *equation (8.71)*)

$n_{ef} = 2 + \left(1 - \dfrac{n}{20}\right) \cdot (n-2)$

$n_{ef} = 2$

2. Timber strength properties

Table 1.3, strength class C22 (BS EN 338:2009(E), *Table 1*)

Characteristic tensile strength of the timber, $f_{t.0.k}$

$f_{t.0.k} = 13 \, \text{N/mm}^2$

Characteristic density of timber, ρ_k

$\rho_k = 340 \, \text{kg/m}^3$

Mean density of timber, ρ_m

$\rho_m = 410 \, \text{kg/m}^3$

3. Partial safety factors

Table 2.6 (UKNA to EC5, *Table NA.3*)

Material factor for connection, $\gamma_{M.connection}$

$\gamma_{M.connection} = 1.3$

Material factor for members, $\gamma_{M.timber}$

$\gamma_{M.timber} = 1.3$

4. Actions

Design force on the connection at the ULS, derived from a combination of permanent and medium-term variable loading, $F_{d.ULS}$

$F_{d.ULS} = 60 \, \text{kN}$

Design force on the connection at the SLS, derived from a combination of permanent and medium-term variable loading, $F_{d.SLS}$

$F_{d.SLS} = 40 \, \text{kN}$

5. Modification factors

Factor for medium-duration loading and service class 2, $k_{mod.med}$, (Table 2.4 (EC5, *Table 3.1*))

$k_{mod.med} = 0.80$

As tension members are 150 mm wide, $k_h = 1$ (Table 2.11 (EC5, *equation (3.1)*))

$k_h = 1$

6. Design forces and the strength of the shear planes in the connection

The connection is taken to comprise two double shear connections, one on either side of the longitudinal centre line

Shear plane 1 (shear plane between members 1 and 2 (taking member 3 to be 35 mm thick):

Capacity of the split-ring connector between members 1 and 2:

Factor $k1_1$
(equation (11.17); EC5,
equation (8.62))

$$k1_1 = \min\left(1, \frac{t_1}{3h_e}, \frac{t_2}{5h_e}\right)$$

$$k1_1 = 1$$

Factor $k1_2$
(equation (11.18); EC5,
equation (8.63)) where $k_a = 1$

$$k1_2 = \min\left(1.00, \frac{a_{3.t}}{2 \cdot d_c}\right)$$

$$k1_2 = 0.87$$

Factor $k1_3$
(equation (11.19); EC5,
equation (8.65))

$$k1_3 = \min\left(1.75, \frac{\rho_k}{350 \cdot \frac{kg}{m^3}}\right)$$

$$k1_3 = 0.97$$

Factor $k1_4$:
Timber-to-timber $k_4 = 1$
(equation (11.20); EC5, *equation (8.66))*

$$k1_4 = 1$$

Characteristic load-carrying capacity of a split-ring connector between members 1 and 2, $F1_{v.Rk}$
(equation (11.16); EC5, *equations (8.61))*

Block shear failure

$$F1a_{v.Rk} = k1_1 \cdot k1_2 \cdot k1_3 \cdot k1_4 \cdot 35 \cdot \left(\frac{d_c}{mm}\right)^{1.5} \cdot N$$

$$F1a_{v.Rk} = 18.03 \text{ kN}$$

Embedment failure

$$F1b_{v.Rk} = k1_1 \cdot k1_3 \cdot \frac{h_e}{mm} \cdot 31.5 \cdot \frac{d_c}{mm} \cdot N$$

$$F1b_{v.Rk} = 20.93 \text{ kN}$$

Characteristic load-carrying capacity of the split-ring connector between members 1 and 2, $F1_{v.Rd}$

$F1_{v.Rk} = \min(F1a_{v.Rk}, F1b_{v.Rk})$
$F1_{v.Rk} = 18.03 \text{ kN}$
i.e. failure condition will be due to block shear

Design capacity of the split-ring connection in the shear plane between members 1 and 2, $F1_{v.Rd}$

$$F1_{v.R.d} = \frac{k_{mod.med}}{\gamma_{M.connection}} \cdot F1_{v.Rk} \cdot n_{ef}$$

$F1_{v.R.d} = 22.19 \text{ kN}$

Design capacity of the connection, $F_{v.Rd}$
Based on four shear planes per connection

$F_{v.Rd} = 4 F1_{v.R.d}$ $F_{v.Rd} = 88.77 \text{ kN}$

Design force on the connection, $F_{d.ULS}$

$F_{d.ULS} = 60 \text{ kN}$
i.e. the design force is less than the design capacity of the connection; therefore OK

7. Tensile strength of the timber members – ULS condition

Apportioning the force in each member on the basis of the design strength of the connector(s) connected to it:

Design force in member 1, $F1_d$

$$F1_d = \frac{F_{d.ULS}}{4}$$

Design force in member 2, $F2_d$

$$F2_d = \frac{F_{d.ULS}}{2}$$

Design force in member 3, $F3_d$

$$F3_d = \frac{F_{d.ULS}}{2}$$

Design tensile stress parallel to the grain in member 1, $\sigma 1_{t.0.d}$

$$\sigma 1_{t.0.d} = \frac{F1_d}{A_1} \qquad \sigma 1_{t.0.d} = 3.54\,\text{N/mm}^2$$

Design tensile stress parallel to the grain in member 2, $\sigma 2_{t.0.d}$

$$\sigma 2_{t.0.d} = \frac{F2_d}{A_2} \qquad \sigma 2_{t.0.d} = 5.24\,\text{N/mm}^2$$

Design tensile stress parallel to the grain in member 3, $\sigma 3_{t.0.d}$

$$\sigma 3_{t.0.d} = \frac{F3_d}{A_3} \qquad \sigma 3_{t.0.d} = 3.54\,\text{N/mm}^2$$

Design tensile strength of the timber parallel to grain, $f_{t.0.d}$

$$f_{t.0.d} = \frac{k_{mod.med} \cdot k_h \cdot f_{t.0.k}}{\gamma_{M,timber}}$$

$$f_{t.0.d} = 8\,\text{N/mm}^2$$

i.e. the tensile strength of the timber exceeds the design stress; therefore OK

8. The instantaneous deflection of the joint at the SLS

The slip modulus of a connector per shear plane at the SLS, K_{ser} (Table 11.4 (EC5, *Table 7.1*). The equation incorporates dimension factors for Mathcad.)

$$K_{ser} = \frac{\left(\rho_m \cdot \dfrac{\text{m}^3}{\text{kg}}\right) \cdot \left(\dfrac{d_c}{\text{mm}}\right) \cdot \text{Nmm}^{-1}}{2}$$

$$K_{ser} = 1.48 \times 10^4\,\text{N/mm}$$

The stiffness of each connection at the SLS, i.e 4 No shear planes; n connectors per shear plane; $K_{ser.connection}$

$$K_{ser.connection} = 4 \cdot n \cdot K_{ser}$$

$$K_{ser.connection} = 1.18 \times 10^5\,\text{N/mm}$$

The stiffness of the joint at the SLS, $K_{ser.joint}$ (Figure 10.24)

$$K_{ser.joint} = \frac{K_{ser.connection}}{2}$$

$$K_{ser.joint} = 5.9 \times 10^4\,\text{N/mm}$$

The instantaneous deflection of the joint at the SLS, δ_{inst}

$$\delta_{inst} = \frac{F_{d.SLS}}{K_{ser.joint}}$$

$$\delta_{inst} = 0.68\,\text{mm}$$

Chapter 12

Moment Capacity of Connections Formed with Metal Dowel Fasteners or Connectors

12.1 INTRODUCTION

When a connection in a structure is subjected to rotation, provided there is no relative movement between the members of the connection, all elements will be subjected to the same rotation. Under this condition, the connection is classified as rotationally rigid and will provide maximum rotational stiffness to the structure. At the other extreme where the members in a connection are held in position but are free to rotate relative to each other, no moment can be transferred and this condition is classified as rotationally pinned. When the degree of fixity in the joint is between these states, it is classified as rotationally semi-rigid.

In timber design, as stated in EC5, *5.4.2(7)*, connections may be assumed to be rotationally rigid if their deformation has no significant effect on the distribution of member forces and moments, otherwise they may be assumed to be rotationally pinned. This is the practice followed by most design engineers. However the reality is that, in service, connections will, to varying degrees, exhibit semi-rigid behaviour. When a connection is subjected to a moment, the stress resultants in the connection are transferred between the connection members by the fasteners/connectors and movement in the connection will arise from a combination of movement due to the take-up of tolerance allowances, embedment of the fasteners/connectors in the connection members, and, where the fasteners being used are relatively flexible (e.g. nails, bolts, etc.), by deformation of the fasteners. As the stiffness of the members is generally large relative to the stiffness of the fastener configuration, their flexural, axial and shear deformations are normally small compared to the deformation arising from the combined embedment and fastener/connector deformation, and are normally neglected.

In this chapter, a design procedure for rigid and semi-rigid connections formed with metal dowel-type fasteners or connectors and subjected to moment induced by lateral loading is addressed.

Unless otherwise stated, where reference is made to the use of fasteners, this is to be interpreted to mean either metal dowel-type fasteners or connectors and when referring to the number of fasteners, when connectors are being used this is to be understood to mean the number of connector bolts.

The general information in 4.3 is relevant to the content of this chapter.

Structural Timber Design to Eurocode 5, Second Edition. Jack Porteous and Abdy Kermani.
© Jack Porteous and Abdy Kermani 2013. Published 2013 by Blackwell Publishing Ltd.

Table 12.1 Main design requirements for connections subjected to moment due to lateral loading and the associated EC5 limit states

Design or displacement effect	EC5 limit states
Moment strength due to lateral loading	ULS
Displacement (rotation)	SLS

12.2 DESIGN CONSIDERATIONS

Connections subjected to moment have to satisfy the relevant design rules and requirements of EC5 [1] and the limit states associated with the main design effects are given in Table 12.1.

The strength and displacement requirements for connectors are the same as those required for metal dowel-type connections referred to in 10.1, and reference to displacement effect relates to rotational movement within the connection.

12.3 THE EFFECTIVE NUMBER OF FASTENERS IN A ROW IN A MOMENT CONNECTION

When a connection is subjected to a moment the direction of the force in each fastener in the connection will vary and where the fastener configuration is such that:

(a) no row exists where a force component from two or more fasteners in the row is parallel to the grain in any of the connected members,
(b) or there are rows where the above situation will arise but the spacing between the fasteners parallel to the member grain complies with the criteria given in Table 12.2,

the full number of fasteners in the connection can be used.

Table 12.2 Criteria at which the full number of fasteners in a row can be used in strength calculations

Fastener type	Criteria for fasteners in a row with a force component parallel to the grain of a member
Nails (≤8 mm in diameter) and screws (≤6 mm in diameter)	The spacing between adjacent fasteners having a component parallel to the grain must be ≥ $14d$ (Table 10.10)
Nails (>8 mm in diameter), screws (>6 mm in diameter), bolts or dowels	The function $n^{0.9} \sqrt[4]{\dfrac{a_1}{13d}}$ must be ≥ 1 (equation 10.45)
Toothed-plate connectors	Always use the full number of connectors
Ring and shear plate connectors	Where there are no more than two connectors per shear plane in a row having a component parallel to the grain.

d is the nominal diameter of the fastener (in mm); n is the number of fasteners in the row with a component parallel to the grain; a_1 is the distance parallel to the grain between adjacent fasteners in the row.

Table 12.3 Minimum edge, end and spacing criteria for fasteners used in moment connections using timber to timber*

Fastener	Edge distance** $(a_{4,t})$	Loaded end distance** $(a_{3,t})$	Lateral spacing** $(a_1$ and $a_2)$
Nails, screws[†]: $\rho_k \leq 420\,\text{kg/m}^3$	$d \geq 5\,\text{mm}; 10d$ $d < 5\,\text{mm}; 7d$	$d \geq 5\,\text{mm}; 15d$ $d < 5\,\text{mm}; 15d$	$d \geq 5\,\text{mm}; 12d$ $d < 5\,\text{mm}; 10d$
Nails, screws[†]: $420\,\text{kg/m}^3 < \rho_k \leq 550\,\text{kg/m}^3$	$d \geq 5\,\text{mm}; 12d$ $d < 5\,\text{mm}; 9d$	$d \geq 5\,\text{mm}; 20d$ $d < 5\,\text{mm}; 20d$	$d \geq 5\,\text{mm}; 15d$ $d < 5\,\text{mm}; 15d$
Nails, screws[†]: with pre-drilled holes	$d \geq 5\,\text{mm}; 7d$ $d < 5\,\text{mm}; 5d$	$d \geq 5\,\text{mm}; 12d$ $d < 5\,\text{mm}; 12d$	$d \geq 5\,\text{mm}; 5d$ $d < 5\,\text{mm}; 5d$
Bolts, dowels and screws[‡]	$4d$	The greater of $7d$ or $80\,\text{mm}$	$6d$
Toothed-plate connectors	d_c	$2d_c$	$2d_c$
Ring or shear plate connectors	d_c	$2d_c$	$2d_c$

ρ_k is the characteristic density of the timber (in kg/m³); d is the nominal diameter of the nail, screw, bolt or dowel (in mm); d_c is the dimension of the fastener given in BS EN 912:2000 [2] (in mm).
*When using panel material or steel gusset plates connected by nails, it is proposed that the criteria given in Chapter 10 for nailed timber-to-timber connections should apply.
[†]Only applies to screws $\leq 6d$.
[‡]Only applies to screws $> 6d$.
**Spacing symbols are as shown in Fig 10.11.

Where the above conditions do not apply, for all design checks on the capacity of fasteners loaded parallel to the grain of a member, the effective number of fasteners in a row, as defined in Chapter 10 for metal dowel-type fasteners and Chapter 11 for connectors, must be used.

12.4   BRITTLE FAILURE

The ductile strength of a connection subjected to a moment can only be achieved if there is no brittle failure caused by premature splitting or shearing of the timber.

In order to prevent this type of occurrence, EC5 gives minimum edge, end and spacing criteria for use with metal dowel-type fasteners and connectors loaded laterally. These are to be considered as recommended minimum values and, if anything, larger spacings and distances should be adopted. From the EC5 requirements and adjusting as considered appropriate, Table 12.3 gives suggested criteria for the types of fasteners commonly used in moment connections.

Also, to prevent splitting in timber when using nails with a nominal diameter greater than 6 mm, pre-drilling must be used and the particular requirements for these fasteners are given in *Section 10* of EC5. Pre-drilling is also necessary to allow nails to be fixed in dense timber and is a requirement for nails if the characteristic density is $> 500\,\text{kg/m}^3$. It is also a requirement where the thickness of the timber members will not comply with the requirements of *equation (8.18)* in EC5, *8.3.1.2(6)*. When using screws, pre-drilling is always required in hardwoods and, in softwoods, pre-drilling is also required when the nominal diameter is greater than 6 mm, as stated in EC5, *10.4.5.*

12.5   MOMENT BEHAVIOUR IN TIMBER CONNECTIONS: RIGID MODEL BEHAVIOUR

Validation of the strength of a connection required to transfer a moment in a structure is undertaken in two stages. There is the analysis of the structure to determine the stress resultants the connection will be subjected to and this is followed by the design procedure to demonstrate that the connection is strong enough to resist these forces.

Unless a structure is statically determinate, depending on whether connections behave in a rigid or a semi-rigid manner, the force distribution in a structure in which connections are designed to transfer moment will differ. When the connections are rigid in behaviour the structure will be at its stiffest and when they are semi-rigid the stiffness will be reduced and the stress resultant distribution around the structure will change.

The elastic analysis of indeterminate timber structures fitted with connections that exhibit rigid behaviour can readily be undertaken using traditional methods or by common software applications. The connections are considered to be rigid and the stiffness properties of the members will be determined as defined in 2.3.4.2. Where the connections in the structure are semi-rigid in behaviour, the analysis procedure has to be modified, as discussed in 12.6.

12.5.1   Assumptions in the connection design procedure

When designing a timber connection subjected to moment, because the members of the connection are generally stiff in comparison with the stiffness behaviour of the connection fasteners, to simplify the analysis it is normal practice to assume that the members behave as rigid elements. This is a reasonable assumption provided that there is no significant bending of the members over the length of the connection and should be achieved by ensuring that the bending stresses in the connection members are relatively small.

All movement is taken to be due to displacement at the fastener position, and in the rigid model approach, the following assumptions are made:

(a) The position of the centre of rotation of the fasteners in the connection remains fixed. When a connection is subjected to a moment as well as lateral loading, the forces induced in the fasteners will depend on whether or not the centre of rotation of the connection is fixed or changes as the loading increases. In timber connections, the centre of rotation will normally change as the loading is applied. However, because of the relatively high rigidity of the fastener configurations normally used in structural connections, the change in position will generally be small and in the rigid model approach the assumption is made that the centre of rotation is fixed. On this basis, the centre of rotation is taken to be the centroid of the fastener group. Where the fasteners are all the same size, which is normal practice in a timber connection and has been assumed for the design procedures in this chapter, the centre of rotation will be the geometric centre of the group.

(b) From the assumption in (a), when lateral shear forces act on a connection, each fastener will take an equal share of the force. Hence, if the shear force per shear plane in the connection is H_d and there are n fasteners per shear plane, the lateral force taken by each fastener in the shear plane, $F_{h,d}$, will be:

$$F_{h,d} = \frac{H_d}{n}$$ (12.1)

(c) Adopting the conservative approximation that all fasteners in the connection will have the same linear load-stiffness behaviour, for a rigid connection condition the forces in the fasteners can be derived using either a rigid model approach or by assuming that there will be a small rotation between the adjacent connection members. The same result will be obtained from either approach and the latter method has been used in the analysis.

Although the above assumption means that the strength of the connection will be independent of the fastener stiffness, the ultimate limit states (ULS) slip modulus will be taken to apply in the development of the solution.

Assuming a small rotation of the member in the connection when loaded by the action of a moment, the fasteners will rotate about the centre of rotation and transfer load by embedment stresses to the other members in the connection.

The further the fastener is from the centre of rotation, C, the greater will be the associated displacement and the maximum displacement will occur at the fastener furthest from the centroid, as shown in Figure 12.1. When the connection is subjected to a rotation, ϑ, fastener, i_{max}, at the greatest distance from the centre of rotation, r_{max}, will have the largest displacement, δ_{max}, and the displacement of any intermediate fastener i at a distance r_i from the centroid will be:

$$\delta_i = \frac{r_i}{r_{max}} \delta_{max}$$ (12.2)

(d) The linear load-stiffness behaviour of each fastener complies with the stiffness requirements of EC5 and the direction of the force in each fastener is taken to be at right angles to the line joining the fastener and the centre of rotation in the unstressed condition.

(e) The spacing of fasteners will be such that premature splitting of the timber or wood-related product will not occur. Guidance on this is given in 12.4.

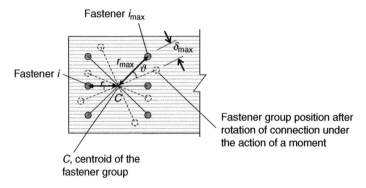

Fig. 12.1. The rotational behaviour of fasteners in a connection.

Incorporating the above assumptions into the design procedure, elastic theory can be used to analyse the moment behaviour and a conservative assessment of the connection strength should be achieved.

12.5.2   Connection design procedure

Consider a single or double shear connection with the generalised fastener configuration shown in Figure 12.2, in which there are n fasteners of the same size in a regular pattern.

At the ULS design condition each shear plane of the connection is subjected to a combination of moment, M_d, and lateral forces, H_d and V_d.

The fastener at the greatest distance from the centroid will have the maximum movement (and hence force) under the action of the moment, and when including for the additional effect of the lateral forces H_d and V_d, fastener A in Figure 12.2a will be subjected to the greatest combined force.

Consider the connection subjected to a moment per shear plane, M_d, causing a rotation ϑ and a lateral displacement of δ_{max} in fastener A, as shown in Figures 12.2b and 12. 2c.

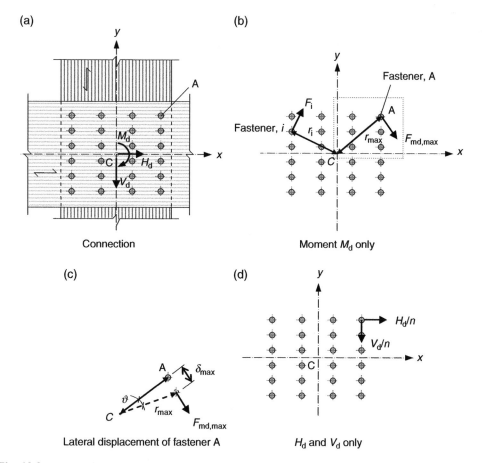

Fig. 12.2. Connection subjected to a combination of moment and lateral forces.

At the ULS, assuming that the slip modulus of each fastener is, say, K, for this loading configuration the maximum force, $F_{m,d,max}$, will be in fastener A and can be written as

$$F_{m,d,max} = K\delta_{max} = Kr_{max}\theta \tag{12.3}$$

where ϑ is the rotation of the connection under the action of M_d and r_{max} is the radial distance from the centroid to fastener A.

Similarly, from the relationship in equation (12.2), the force in any intermediate fastener, i, will be:

$$F_i = K\delta_i = Kr_{max}\vartheta\frac{r_i}{r_{max}} \tag{12.4}$$

where δ_i is the lateral movement of fastener i, r_i is the radial distance of fastener i from the centroid, and the moment taken by fastener i will be:

$$M_i = F_i r_i = Kr_{max}\vartheta\frac{r_i^2}{r_{max}} = F_{m,d,max}\frac{r_i^2}{r_{max}} \tag{12.5}$$

The moment taken by the n fasteners per shear plane in the connection will be:

$$M_d = \frac{F_{m,d,max}}{r_{max}}\sum_{i=1}^{n}r_i^2 \tag{12.6}$$

From the above equation, it is to be noted that, as stated in 12.5.1, the moment in the connection is independent of the stiffness of the fastener and from this relationship the maximum force per shear plane, $F_{m,d,max}$, under the action of the moment will be:

$$F_{m,d,max} = M_d\frac{r_{max}}{\displaystyle\sum_{i=1}^{n}r_i^2} \tag{12.7}$$

If the lateral horizontal and vertical forces per shear plane in each fastener are now considered,

$$F_{h,d} = \frac{H_d}{n} \quad\text{and}\quad F_{v,d} = \frac{V_d}{n} \tag{12.8}$$

When M_d, H_d and V_d are acting on the connection, the largest force will be in fastener A, and will be obtained from the vector sum of the forces calculated from equations (12.7) and (12.8). The vector forces are shown in Figure 12.3 and the maximum force in this fastener per shear plane, F_d, will be:

$$F_d = \sqrt{\left(F_{v,d} + F_{m,d,max}\cos\beta\right)^2 + \left(F_{h,d} + F_{m,d,max}\sin\beta\right)^2} \tag{12.9}$$

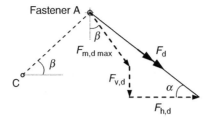

Fig. 12.3. Vector sum of all forces on fastener A.

and the angle, α, of F_d from the horizontal axis will be:

$$\alpha = \arccos\left(\frac{F_{h,d} + F_{m,d,max}\sin\beta}{F_d}\right) \tag{12.10}$$

When a circular pattern is used for the fasteners, which is common practice for moment resisting connections, the above equations can be simplified. If there are fasteners lying on the horizontal axis through the group centroid, as shown in Figure 12.4, the fastener with the greatest load will be fastener A and the force per shear plane in that fastener, F_d, and the angle of inclination of this force to the horizontal, α, will be:

$$F_d = \sqrt{\left(F_{v,d} + F_{m,d}\right)^2 + F_{h,d}^2} \tag{12.11}$$

$$\alpha = \arccos\left(\frac{F_{h,d}}{F_d}\right) \tag{12.12}$$

and

$$F_{m,d} = \frac{M_d}{nr} \tag{12.13}$$

where n is the number of fasteners in the circle and r is the radius of the fastener group, and $F_{v,d}$ and $F_{h,d}$ are as previously defined.

The force in the maximum loaded fastener per shear plane must now be shown to be no greater than the lateral design strength per shear plane of the fastener, obtained

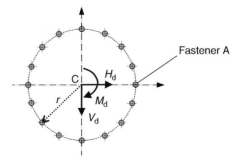

Fig. 12.4. Circular group of fasteners in the connection.

as described in Chapters 10 and 11 for metal dowel-type fasteners and connectors, respectively.

Where the load carrying capacity of the type of fastener or connector being used in the connection is a function of the angle of load relative to the grain direction of any member in the connection, the strength of each fastener will be dependent on the angle of the lateral load relative to grain direction and it has to be shown that the largest ratio of the design lateral force to the associated design lateral strength of the fastener or connector will not exceed 1. A reasonable approximation to determine this condition is, in addition to checking the maximum loaded fastener/connector as stated above, to also check the fastener furthest from the group centroid along each member axis. At such positions the strength of the fastener or connector will be at its minimum value and must be shown to be greater than the associated design force. When dealing with timber to timber connections this requirement will generally be covered by the need to have to check both members in the connection to determine the connection strength, as explained later in the Chapter, but when dealing with timber to steel or to panel material gusset plate connections it will be an additional check as shown in the examples at the end of the Chapter.

12.5.3   Shear strength and force component checks on connections subjected to a moment and lateral forces

12.5.3.1   Shear strength

In the typical connection discussed in 12.5.2 the vertical and horizontal components of the forces in the fasteners will generate shear forces in the connection members and where these act perpendicular to the grain in a member its shear strength must be checked. Where a rectangular rather than a circular grid of fasteners is to be used, because of the forces in the fasteners in the corner zones it is to be noted that there will be a greater risk of timber splitting. For the connection shown in Figure 12.5, there will be force components from the fastener loads perpendicular to the grain in member 1 and in member 2 and strength checks are required for both members.

When considering member 1, under the force configuration shown in Figure 12.5, for most practical joint configurations the maximum shear force perpendicular to the grain within the boundary area of the connection will occur to the left of the centre of rotation and will envelope the fasteners within the shear zone area shown in the figure.

Consider each shear plane in the connection subjected to moment M_d and lateral forces V_d and H_d. Under the action of the design moment only, lateral forces will be induced in the fasteners and the force in fastener i in the shear zone area will be as shown in Figure 12.5a. The radius of the fastener from the centroid of the connection is r_i and is at an angle ϕ_i to the horizontal. The vertical design force per shear plane in the fastener, $F_{M,v,i,d}$, shown in Figure 12.5b, can be written as:

$$F_{M,v,i,d} = F_i \cos \phi_i = F_{m,d,max} \frac{r_i}{r_{max}} \cdot \frac{|x_i|}{r_i} = F_{m,d,max} \frac{|x_i|}{r_{max}} \tag{12.14}$$

where $|x_i|$ is the absolute value of the x coordinate of fastener i. The origin of the coordinate system is the centre of rotation of the fastener configuration in the connection.

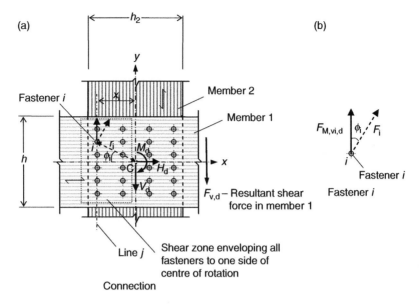

Fig. 12.5. Shear force analysis of member 1 subjected to moment M_d per shear plane.

The summation of the vertical component of the forces in the fasteners in each line j to the left side of the centre of rotation arising from the design moment, M_d, will be:

$$F_{M,v,d,j} = \frac{F_{m,d,max}}{r_{max}} \sum_{}^{\text{line } j} |x_i| = \frac{F_{m,d,max}}{r_{max}} n_j |x_i| \tag{12.15}$$

where the functions are as previously defined, and $F_{M,v,d,j}$ is the vertical shear force per shear plane in each line j in the zone arising from the applied moment M_d and n_j is the number of fasteners in the shear plane in line j.

If the connection is now subjected to the vertical design force, V_d, in accordance with the assumption in 12.5.1(b), the vertical lateral force taken by each fastener in the shear plane, $F_{V,v,i,d}$, shown in Figure 12.6, will be:

$$F_{V,v,i,d} = \frac{V_d}{n} \tag{12.16}$$

where n is the number of fasteners in the shear plane.

When the connection is subjected to the horizontal design force, H_d, the horizontal lateral force taken by each fastener in the shear plane, $F_{H,h,i,d}$, shown in Figure 12.7, will be

$$F_{H,h,i,d} = \frac{H_d}{n} \tag{12.17}$$

When considering only member 1, the forces contributing to the vertical shear force in the member will be those arising from the application of the design moment, M_d, and the vertical design shear force, V_d.

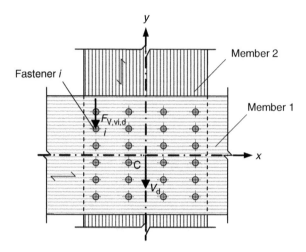

Fig. 12.6. Shear force in each fastener due to vertical force, V_d, per shear plane.

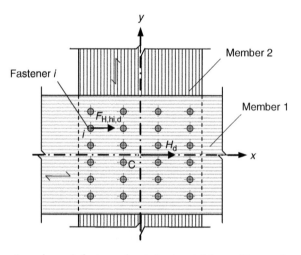

Fig. 12.7. Shear force in each fastener due to horizontal force, H_d, per shear plane.

The vertical design shear force per shear plane on member 1, $F_{V,v,d}$, will be the maximum shear force in the shear zone. This is obtained by starting at the left side of the zone and evaluating the cumulated shear force across the zone. The maximum value will depend on the fastener configuration and the value of the vertical shear design force, V_d. For the case shown in Figure 12.5 the maximum value, $F_{V,v,d}$, will be as follows:

- In the first line of fasteners:

$$F1_{V,v,d} = \frac{F_{m,d,max}}{r_{max}} n_{\ell 1} |x_1| - \frac{n_{\ell 1}}{n} V_d \qquad (12.18)$$

- In the second line of fasteners:

$$F2_{V,v,d} = F1_{V,v,d} + \frac{F_{m,d,max}}{r_{max}} n_{\ell 2} \left| x_2 \right| - \frac{n_{\ell 2}}{n} V_d \qquad (12.19)$$

- The maximum value will be:

$$F_{V,v,d} = \max \left\{ F1_{V,v,d}, F2_{V,v,d} \right\} \qquad (12.20)$$

(Note that for the example used the first line is the line furthest from the centroid and the second line is the line closest to the centroid. For connections with greater numbers of lines of fasteners, the process must be extended to cover all of the lines in the zone.)

In equation (12.18) and (12.19), n is as defined in equation (12.16) and the number of fasteners in the shear plane is $n_{\ell 1}$ in the first line and $n_{\ell 2}$ in the second line.

The shear force in member 1 will generate splitting forces in the member within the boundary area of the connection, however, the splitting equation in EC5, *equation (8.4)*, does not apply to this loading condition. Assuming tension splitting will not be critical, the strength check will relate to a confirmation that the shear strength of the member within the boundary area is not exceeded and for member 1 this will be

$$F_{V,v,d} \leq \frac{2}{3} \frac{\left(k_{cr} bh_{ef} \right) k_{mod} f_{v,k}}{\gamma_{Mconnection}} \qquad (12.21)$$

where, for a member with a rectangular cross-section, b (mm) thick and h (mm) deep; k_{cr} is the crack factor referred to in Table 4.5; k_{mod} is the modification factor for load duration and service classes as given in Table 2.4; γ_M is the partial coefficient for material properties, given in Table 2.6, and $f_{v,k}$ is the characteristic shear strength of the member and values for the shear strength of timber and wood-based structural products are given in Chapter 1; h_{ef} is the effective depth, allowing for fasteners holes (based on depth h).

When considering member 2, for the loading configuration shown, the maximum shear force perpendicular to the grain within the boundary area of the connection will occur in the shear zone below the centre of rotation as shown in Figure 12.8a.

As for member 1, consider the connection to be subjected to the moment and the lateral forces separately. Under the action of the design moment M_d, the lateral force in fastener i in the zone area will be as shown in Figure 12.8a. The radius of the fastener from the centroid of the connection is r_i and is at an angle β_i to the vertical. The horizontal design force per shear plane in the fastener, $F_{M,hi,d}$, shown in Figure 12.8b, will be

$$F_{M,h,i,d} = F_i \cos \beta_i = F_{m,d,max} \frac{r_i}{r_{max}} \cdot \frac{\left| y_i \right|}{r_i} = F_{m,d,max} \frac{\left| y_i \right|}{r_{max}} \qquad (12.22)$$

where $|y_i|$ is the absolute value of the y coordinate of fastener i.

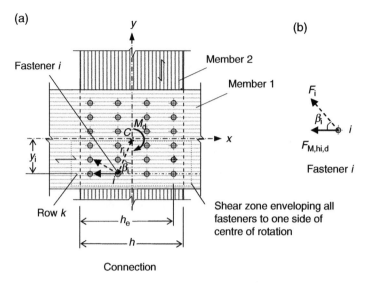

Fig. 12.8. Shear force analysis of member 2 subjected to moment M_d per shear plane.

The summation of the horizontal component of the forces in the fasteners in each row k acting perpendicular to the grain in member 2 arising from the design moment, M_d, will be:

$$F_{M,h,d,k} = \frac{F_{M,d,max}}{r_{max}} \sum^{row\,k} |y_i| = \frac{F_{M,d,max}}{r_{max}} n_k |y_i| \tag{12.23}$$

where the functions are as previously defined, and $F_{M,h,d,k}$ is the horizontal shear force per shear plane in row k in the zone arising from the applied moment M_d and n_k is the number of fasteners per shear plane in row k.

If the connection is now subjected to the horizontal design force, H_d, in accordance with the assumption in 12.5.1(b), the horizontal lateral force taken by each fastener in the shear plane, $F_{H,h,i,d}$, shown in Figure 12.9, will be:

$$F_{H,h,i,d} = \frac{H_d}{n} \tag{12.24}$$

Although there will be vertical forces in the fasteners due to the vertical design force V_d, these will not influence the horizontal shear force and need not be considered. The horizontal design shear force per shear plane on member 2, $F_{H,h,d}$, will be the maximum shear force in the zone. It is obtained by starting at the bottom line in the zone and progressively evaluating the cumulated shear force within the zone. The maximum value will depend on the fastener configuration and the value of the applied horizontal shear force. For the case shown in Figure 12.5 the maximum value, $F_{H,h,d}$, will be as follows:

- In the first row of fasteners,

$$F1_{H,h,d} = \frac{F_{M,d,max}}{r_{max}} n_{r1} |y_1| - \frac{n_{r1}}{n} H_d \tag{12.25}$$

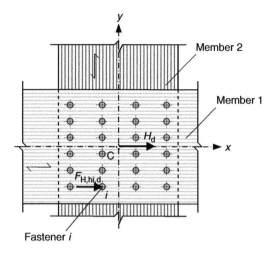

Fig. 12.9. Shear force in each fastener due to horizontal force, H_d, per shear plane.

- In the second row of fasteners,

$$F2_{\text{H,h,d}} = F1_{\text{h,d}} + \frac{F_{\text{M,d,max}}}{r_{\text{max}}} n_{\text{r2}} \left| y_2 \right| - \frac{n_{\text{r2}}}{n} H_{\text{d}}$$ (12.26)

In the third row of fasteners,

$$F3_{\text{H,h,d}} = F2_{\text{H,h,d}} + \frac{F_{\text{M,d,max}}}{r_{\text{max}}} n_{\text{r3}} \left| y_3 \right| - \frac{n_{\text{r3}}}{n} H_{\text{d}}$$ (12.27)

- The maximum value will be

$$F_{\text{H,h,d}} = \max \left\{ F1_{\text{H,h,d}}, F2_{\text{H,h,d}}, F3_{\text{H,h,d}} \right\}$$ (12.28)

(Note that the first row is the row furthest from the centroid, the second row is next closest to the centroid, and the third row is the row closest to the centroid. For connections with greater numbers of rows of fasteners, the process is extended to cover all of the rows in the zone.)

In equations (12.25) to (12.28) n is as defined in equation (12.16) and the number of fasteners in the shear plane in each row is n_{r1} in the first row, n_{r2} in the second row and n_{r3} in the third row.

As for member 1, since no guidance is given in EC5 to enable the splitting capacity of member 2 to be checked, the shear strength of the member within the boundary area of the connection should be checked as follows:

$$F_{\text{V,v,d}} \leq \frac{2}{3} \frac{\left(k_{\text{cr}} b_2 h_{\text{2ef}} \right) k_{\text{mod}} f_{\text{v,k}}}{\gamma_{\text{M,connection}}}$$ (12.29)

where the member thickness is b_2 (mm) and the depth is h_2 (mm) and the other symbols are as previously defined.

It should be noted that if there are n_{sp} shear planes in the connection (e.g. in a three-member connection formed by another member 1 fastened to the other face of member 2 using the same fastener configuration, $n_{sp} = 2$), the design force to be taken by member 2 will be the force per shear plane multiplied by n_{sp}.

12.5.3.2 Force component checks in a row of fasteners parallel to the grain

As stated in 12.3, where the fastener configuration in a connection is such that:

(a) no row exists where a force component from two or more fasteners in the row is parallel to the grain in any of the connected members, or,
(b) there are rows where the above situation will arise but the spacing between the fasteners parallel to the member grain complies with the criteria given in Table 12.2,

the full number of fasteners in the connection can be used.

However, where the above criteria are not met EC5 rules require that where a row of fasteners is acted on at an angle to the grain, it must be verified that the force component parallel to the row will be less than or equal to the load-carrying capacity of the row based on the use of the effective number of fasteners in the row, n_{ef}. The effective number of fasteners is defined in Chapter 10 for metal dowel-type fasteners and in Chapter 11 for connectors.

For this situation, the reduced capacity of each fastener in a row will be obtained by multiplying the design lateral capacity of the fastener when loaded parallel to the grain by n_{ef}/n and the result must be shown equal to or greater than the value of fastener force component in the row acting parallel to the grain.

Consider, for example, a connection subjected to combined moment and lateral forces where the design forces in a row of fasteners in member 1 are as shown in Figure 12.10. The force in fastener i, $F_{i,d}$, will have the following value of component acting parallel to the grain:

$$F_{h,i,d} = F_{i,d} \cos \phi_i \qquad (12.30)$$

The capacity of each fastener in the row, $F_{v,d}$, will be:

$$F_{v,d} = \frac{n_{ef}}{n} \cdot \frac{k_{mod} F_{v,Rk}}{\gamma_{M,connection}} \qquad (12.31)$$

where the symbols are as previously defined and $F_{v,Rk}$ is the characteristic load-carrying capacity of the fastener, derived as described in Chapters 10 and 11 for metal dowel-type fasteners and connectors, respectively.

To comply with the requirements of EC5 the design condition to be satisfied will be

$$\max \left\{ F_{h,i,d} \cos \phi_i \right\} \le F_{v,d}$$

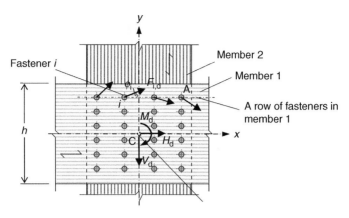

Fig. 12.10. Forces in a row of fasteners in a connection subjected to combined moment and lateral loading.

where $\max\{F_{h,i,d}\cos\phi_i\}$ is the component of force with the largest value acting along the direction of the row. For a regular pattern, the value will be the same for all fasteners.

When the connection comprises timber, LVL or glulam members the above will in effect also cover for the requirement to check the strength of the fastener furthest from the group centroid along each member axis, as referred to in 12.5.2. When steel or panel material gusset plates are used to form the connection, the strength check of the fastener furthest from the group centroid along each member axis, as referred to in 12.5.2 will be required.

See Example 12.8.1.

12.6   THE ANALYSIS OF STRUCTURES WITH SEMI-RIGID CONNECTIONS

If the connections in a statically indeterminate structure exhibit semi-rigid rather than fully fixed rotational behaviour, there will be a reduction in the stiffness of the structure leading to smaller moments at connections, an increase in member span moments and an increase in those displacements that are affected by the moment distribution in the structure. By taking the effect of semi-rigidity into account, in the analysis process a more realistic assessment of the moment distribution and displacement behaviour of the structure will be obtained.

EC5 allows the structural model in an analysis to take into account the effects of deformations of connections having adequate ductility and by using the slip properties of the fastener types given in EC5 a methodology can be developed to accommodate semi-rigid behaviour.

For semi-rigid rotational behaviour, the moment in the shear plane of a connection, M, can be related to the relative rotation ϑ_r between the adjacent connection members by the following relationship,

$$M = k\vartheta_r \tag{12.32}$$

where k is the secant rotational stiffness, and will be referred to as the rotational stiffness of the connection. The rotational stiffness of a rigid, pinned or semi-rigid connection is shown in Figure 12.11.

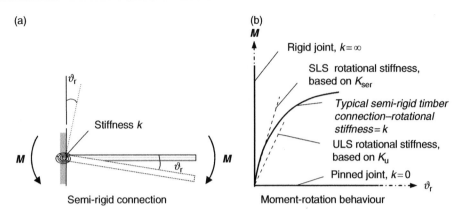

Fig. 12.11. Different types of rotational stiffness, k, in a timber connection.

With a rigid connection, irrespective of the value of the moment, the relative rotation of the connection members will be zero and the $M - \vartheta_r$ relationship will be a vertical line acting through the origin as shown in Figure 12.11b. The rotational stiffness for this condition will be infinity. For a pinned connection, no moment can develop and the $M - \vartheta_r$ relationship will be a horizontal line acting through the origin. The rotational stiffness for this condition will be zero. All conditions between these extremes will constitute semi-rigid rotational behaviour.

12.6.1   The stiffness of semi-rigid moment connections

A possible method for taking into account the effect of semi-rigid moment connections on the behaviour of a structure, which has been widely used in the analysis of steel structures but to a lesser degree in timber structures, is to represent the rotational stiffness of each connection by a spring.

Where the rotational stiffness varies, an iteration procedure has to be incorporated into the analysis to take account of the non-linear behaviour. However, where the rotational stiffness can be considered to be linear this procedure is not required. As stated in 10.10 and 11.8, in EC5 the lateral stiffness at the serviceability limit states (SLS) and the ULS of a metal dowel-type fastener or a connector are linear relationships. Consequently, the rotational stiffness will also be linear and, depending on the limit states being considered, will be based on either the slip modulus at the SLS, K_{ser}, or at the ULS, K_u, and the rotational stiffness at these states is as indicated in Figure 12.11b.

Using the assumptions in 12.5.1, adopting a rotational stiffness based on K_{ser} will realistically cover the behaviour of the fastener up to the SLS. Where the fastener is loaded beyond this state, its rotational stiffness will reduce and at the ULS it will be based on K_u. For connections formed using several fasteners, it is likely that at the ULS many fasteners will only be stressed to levels approximating the SLS condition, and in the following approach the rotational stiffness of all fasteners in the connection is assumed to be based on K_{ser}.

Consider a single or double shear connection within a structure having a regular pattern of fasteners, as shown in Figure 12.12, in which M_i is the moment per shear plane to be taken by fastener i at the ULS design condition. As the distribution of forces in the structure will be affected by the rotational behaviour of the connection, as stated in 2.3.4.2, for the stress resultant analysis the stiffness properties

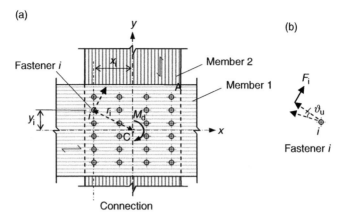

Fig. 12.12. Semi-rigid connection subjected to a design moment, M_d.

must be based on final mean values adjusted to the load component causing the largest stress in relation to strength, and are given in equations (2.38), (2.39) and (2.40). Adopting the relationship in equation (2.40) for the stiffness of each fastener and assuming a rotation ϑ_u in the connection, the moment rotation relationship given in equation (12.5) can be expressed as:

$$M_i = \left(\frac{K_{ser}}{\left(1+\psi_2 k_{def}\right)} \right) \vartheta_u r_i^2 \tag{12.33}$$

where the functions are as previously defined, and k_{def} is the deformation factor discussed in 2.3.2 and 10.10 and ψ_2 is the factor for the quasi-permanent value of the action causing the largest stress in relation to the strength.

Summing the contribution of all fasteners in the connection the moment per shear plane, M_{sp}, taken by the connection can be written as:

$$M_{sp} = \left(\frac{K_{ser}}{\left(1+\psi_2 k_{def}\right)} \right) \sum_{i=1}^{n} r_i^2 \vartheta_u \tag{12.34}$$

or

$$M_{sp} = \left(\frac{K_{ser}}{\left(1+\psi_2 k_{def}\right)} \right) \sum_{i=1}^{n} \left(x_i^2 + y_i^2\right) \vartheta_u \tag{12.35}$$

where x_i and y_i are the coordinates of fastener i relative to the axes shown in Figure 12.12 and n is the number of fasteners per shear plane in the connection. The rotational stiffness of the connection per shear plane will be as follows:

- For the ULS design condition,

$$k_u = \left(\frac{K_{ser}}{\left(1+\psi_2 k_{def}\right)} \right) \sum_{i=1}^{n} \left(x_i^2 + y_i^2\right) \tag{12.36}$$

- For the SLS design condition,

$$k_{ser} = \left(\frac{K_{ser}}{\left(1+k_{def}\right)}\right)\sum_{i=1}^{n}\left(x_i^2 + y_i^2\right) \tag{12.37}$$

Here the symbols are as previously defined, and k_u is the rotational stiffness of the connection per shear plane at the ULS design condition; k_{ser} is the rotational stiffness of the connection per shear plane at the SLS design condition, and K_{ser} is the slip modulus per fastener per shear plane at the SLS, given in Chapter 10 for metal dowel-type fasteners and in Chapter 11 for connectors.

When the connection is also subjected to vertical and lateral design forces, the lateral stiffness of fastener i at the ULS can be written as:

$$V_i = \frac{K_{ser}}{\left(1+\psi_2 k_{def}\right)}\delta_{v,i,u} \quad \text{and} \quad H_i = \frac{K_{ser}}{\left(1+\psi_2 k_{def}\right)}\delta_{h,i,u} \tag{12.38}$$

The lateral stiffness per shear plane of the connection at the ULS for horizontal displacement, $K_{H,u}$, and for vertical displacement, $K_{V,u}$ will be:

$$K_{H,u} = n\frac{K_{ser}}{\left(1+\psi_2 k_{def}\right)} \quad \text{and} \quad K_{V,u} = n\frac{K_{ser}}{\left(1+\psi_2 k_{def}\right)} \tag{12.39}$$

where n is the number of fasteners per shear plane in the connection.

Using the above expressions the semi-rigid behaviour of connections can be taken into account in the structural analysis.

Where a computer-based frame analysis is to be used, lateral slip effects can be represented by the use of linear elastic spring elements at connection positions, as discussed in 10.11. To simulate rotational slip, linear elastic rotational springs can be used, each having the rotational stiffness of the connection being modelled. As stated in 10.11, when modelling these additional elements care has to be taken to ensure that stability of the structure is retained, that the shear and flexural properties of the elements properly represent the connection behaviour, and that the size of the elements used will not result in ill-conditioned equations.

Guidance on the loading and stiffness requirements to be used in this type of analysis is given in 2.3.4.1 and 2.3.4.2 respectively.

12.6.2 The analysis of beams with semi-rigid end connections

In the following analysis, the behaviour of beams with end connections that exhibit semi-rigid rotational behaviour and in which lateral displacement effects can be ignored is considered. This situation commonly arises in timber construction and the method of analysis is such that it can readily be undertaken using Mathcad or equivalent software.

Consider within a structure any prismatic member ab of length L, flexural rigidity EI and single or double shear connections at each end. The rotational spring stiffness at each connection is k_1 and k_2 per shear plane and each is of negligible length, as shown in Figure 12.13. Instability effects due to axial loading are ignored.

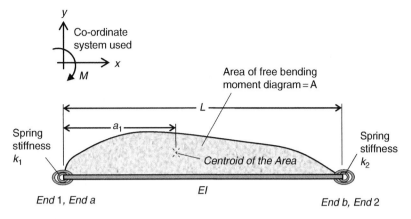

Fig. 12.13. Member with rotational springs at each end.

Ends 1 and 2 are the positions at which the members of the connection are rigidly attached to the structure and 1,a and b,2 are the rotational spring elements simulating the behaviour of the fixings between the connection members and the beam at each end. Under the action of end moments from the structure, the connection members will rotate and there will also be a relative rotation ϑ_r between these members and the beam ends, represented by rotation of the spring elements. The joint rotations at ends 1 and 2 are designated ϑ_1 and ϑ_2, respectively, and the relative rotation of the springs at each end will be ϑ_{r1} and ϑ_{r2} respectively.

The moment required to be applied to the end of a prismatic member to cause a unit rotation at that end when the other end is fixed in position is referred to as the member stiffness. For a member of length L, having a deformation factor $k_{def,m}$, a modulus of elasticity $E = \dfrac{E_{mean}}{\left(1 + \psi_2 k_{def,m}\right)}$ and second moment of area about the axis of bending I, it can be shown from basic principles that the member stiffness will equal $4EI/L$. To align with the member stiffness definition, the spring rotational stiffness at each end is written in a similar format as follows:

$$k_1 = \frac{EI}{L}\beta_a, \ k_2 = \frac{EI}{L}\beta_b \tag{12.40}$$

where $k_1 = M_1/\vartheta_{r1}$ and $k_2 = M_2/\vartheta_{r2}$, and M_1 and M_2 are the end moments on the beam, ϑ_{r1} and ϑ_{r2} are the rotations of the spring at end a and b, respectively, β_a and β_b are the secant rotational stiffness coefficients of the joint connection at a and b, respectively, and include for all of the shear planes in each of the connections.

From equations (12.36) and (12.37), for a connection in timber structures the secant rotational stiffness coefficient can be written as;

$$\beta = \frac{L}{EI}n_{sp}K\sum_{i=1}^{n}\left(x_i^2 + y_i^2\right) \tag{12.41}$$

where the symbols are as previously defined and:

K is the relevant slip modulus of the fastener type being used in the connection for the design condition being addressed, i.e. either $\frac{K_{ser}}{(1+\psi_2 k_{def,c})}$ when analysing at the ULS or $\frac{K_{ser}}{(1+k_{def,c})}$ at the SLS. If the connection at each end is different, the value of K at each end will also differ. The value of the deformation factor(s), $(k_{def,c})$, will be obtained from Table 2.10 and reference should be made to 2.3.2 to determine the adjustments to be applied to suit the configuration being used.

n_{sp} is the number of shear planes in the connection.

Consider the beam to be loaded in a generalised manner where the area of the free bending moment diagram is A with its centre of area at a distance a_1 from end a. If the rotation at end 1 of the member is ϑ_1 and at end 2 is ϑ_2, using unit load theory, a combination of flexibility coefficients and the stiffness method of analysis, also assuming no relative movement in the direction of the y-axis between ends 1 and 2, it will be shown that:

$$M_1 = \left(\frac{1}{1+(3EI/k_1 L)}\right)\left(\frac{3EI\theta_1}{L} + \frac{M_2}{2} - \frac{3A(L-a_1)}{L^2}\right) \tag{12.42}$$

$$M_2 = \left(\frac{1}{1+(3EI/k_2 L)}\right)\left(\frac{3EI\theta_2}{L} + \frac{M_1}{2} + \frac{3Aa_1}{L^2}\right) \tag{12.43}$$

The functions $\left(\frac{1}{1+(3EI/k_1 L)}\right)$ and $\left(\frac{1}{1+(3EI/k_2 L)}\right)$ modify the prismatic member relationship to take account of the effects of the end springs and are referred to by Monforton and Wu [3] as the fixity factors and by Kermani [4] as the rigidity factors. They are dimensionless parameters and are functions of the respective spring rotational stiffness and the stiffness properties of the beam.

Expressing the rigidity factors in terms of the symbol γ, they can be written as follows:

- At end 1 of the beam,

$$\gamma_1 = \left(\frac{1}{1+(3EI/k_1 L)}\right) = \frac{1}{1+\dfrac{3EI}{L\left(n_{sp1} K_1 \sum_{i=1}^{n_1}\left(x_i^2 + y_i^2\right)\right)}} \tag{12.44}$$

- At end 2 of the beam,

$$\gamma_2 = \left(\frac{1}{1+(3EI/k_2 L)}\right) = \frac{1}{1+\dfrac{3EI}{L\left(n_{sp2} K_2 \sum_{i=1}^{n_2}\left(x_i^2 + y_i^2\right)\right)}} \tag{12.45}$$

Here, for a single or double shear connection, $n_{sp,1}$ and n_1 are the number of shear planes and the number of fasteners in a shear plane at end a, and $n_{sp,2}$ and n_2 are the number of shear planes and the number of fasteners in a shear plane at end b. K_1 is the slip modulus of the connection at end a, and K_2 is the slip modulus of the connection at end b, referred to in equation (12.41).

After solving (12.42) and (12.43) for M_1 and M_2 the equations can be written in terms of the rigidity factors:

$$M_1 = \left(\frac{6\gamma_1}{(4-\gamma_1\gamma_2)}\right)\frac{EI}{L}\left(2\theta_1 + \gamma_2\theta_2 - \frac{A}{EIL}\left(2(L-a_1) - \gamma_2 a_1\right)\right) \qquad (12.46)$$

$$M_2 = \left(\frac{6\gamma_2}{(4-\gamma_1\gamma_2)}\right)\frac{EI}{L}\left(2\theta_2 + \gamma_1\theta_1 + \frac{A}{EIL}\left(2a_1 - \gamma_1(L-a_1)\right)\right) \qquad (12.47)$$

The above are referred to as the modified slope equations and are simplifications of the more generalised format developed by Monforton and Wu that allow for relative lateral movement at the ends of each member. Those elements in the equations containing the free bending moment are the modified end fixing moments on the beam and can be written as

$$M_{F1} = \frac{6A}{L^2}\frac{\gamma_1}{(4-\gamma_1\gamma_2)}\left(2(L-a_1) - \gamma_2 a_1\right) \qquad (12.48)$$

$$M_{F2} = \frac{6A}{L^2}\frac{\gamma_2}{(4-\gamma_1\gamma_2)}\left(2a_1 - \gamma_1(L-a_1)\right) \qquad (12.49)$$

where M_{F1} and M_{F2} are the modified end fixing moments at ends 1 and 2 of the beam. Where the end rotations of the beam, θ_1 and θ_2, are zero, equations (12.48) and (12.49) will give the end moments in the beam.

The above are the general relationships applicable to prismatic members with rotational springs at each end resting on unyielding supports. If the fixing detail between the connection members and the beam end is rigid, the associated value of k will be infinity and the rigidity factor γ will be unity. If it functions as a pinned connection, both k and the rigidity factor will be zero.

In Table 12.4 the area of the free bending moment diagram and the position of its centroid from the left hand support are given for commonly occurring load cases.

An example of a beam designed at the ULS condition, with end connections analysed as semi-rigid and also as fully fixed, is given in Example 12.8.2.

With the semi-rigid analysis, it is shown that for a relatively small end rotation the end moment in a semi-rigid connection can be well below the fully fixed value, resulting in a more economic connection design. A more realistic assessment of the deflection of the beam will be obtained and the strength of the member at mid-span can be checked against the increased span moment. It should also be noted that even if the connection is designed to resist the fully fixed moment, there will still be rotation at the connection, and although it will be less than that obtained had the connection been designed as semi-rigid, additional moment will nevertheless be transferred to the beam span and there will also be an increase in its mid-span deflection. These effects will not be taken into account in the rigid analysis.

Table 12.4 The area of the free bending moment and the position of its centroid for typical load cases (see Figure 12.13 for symbols)

Loading condition (beam span L (m))	Area of free bending moment diagram: A (kN m²)	Distance to centroid of the area: a_1 (m)
Uniformly distributed load, where the mid-span moment is M (in kN m)	$\dfrac{2}{3}ML$	$\dfrac{L}{2}$
Point load at mid-span, where the mid-span moment is M (in kN m)	$\dfrac{M}{2L}$	$\dfrac{L}{2}$
Point load at a distance x from the left end of the beam and the moment at the point load is M (in kN m)	$\dfrac{M}{2L}$	$\dfrac{1}{3}(L+x)$

See Examples 12.8.2 and 12.8.3.

12.7   REFERENCES

1 BS EN 1995-1-1:2004. *Eurocode 5: Design of Timber Structures. Part 1-1:2004+A1:2008 General – Common Rules and Rules for Buildings*, British Standards Institution.

2 BS EN 912:2011. *Timber Fasteners – Specifications for Connectors for Timber*, British Standards Institution.

3 Monforton, G.R., Wu, T.S. Matrix analysis of semi-rigidly connected frames. *Journal of the Structural Division, Proceedings of the American Society of Civil Engineers*, Vol. 89, Issue ST6, 1963, pp. 13–42.

4 Kermani, A. A study of semi-rigid and non-linear behaviour of nailed joints in timber portal frames. *Journal of Forest Engineering*, Vol. 7, No. 2, 1996, pp. 17–33.

5 BS EN 338:2009. *Structural Timber – Strength Classes*, British Standards Institution.

6 BS EN 1993-1-8:2005. *Eurocode 3: Design of Steel Structures. Part 1-8: Design of Joints*, British Standards Institution.

7 NA to BS EN 1990:2002+A1:2005 *Incorporating National Amendment No1. UK National Annex for Eurocode 0 – Basis of Structural Design*, British Standards Institution.

12.8   EXAMPLES

As stated in 4.3, in order to verify the ultimate and serviceability limit states, each design effect has to be checked and for each effect the largest value caused by the relevant combination of actions must be used.

However, to ensure attention is primarily focused on the EC5 design rules for the timber or wood product being used, only the design load case producing the largest design effect has generally been given or evaluated in the following examples.

Example 12.8.1 At the ULS a bolted connection is subjected to a design moment of 3.25 kN m and a design vertical action of 10.05 kN as well as a design horizontal action of 1 kN applied through the timber member arising from combined permanent and short-term loading, as shown in Figure E12.8.1. The connection comprises two 12-mm-thick steel side plates and an inner timber member of strength class C24 to BS EN 338:2009, connected by 12-mm-diameter bolts. The bolts have a characteristic strength of 400 N/mm² and the tolerance on the bolt hole diameter in the steel plate is 1 mm. The joint functions under service class 2 conditions.

On the understanding the strength of the steel plates and the bending strength of the timber member are satisfactory, check the adequacy of the connection.

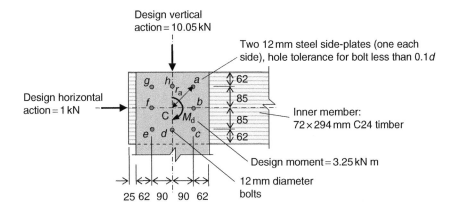

Fig. E12.8.1.

1. Geometric properties

Bolt diameter, d	$d = 12$ mm
Area of bolt in tension, $A_{b.t}$	$A_{b.t} = 84.3$ mm²
Number of bolts per shear plane, n_{bolt}	$n_{bolt} = 8$
Number of shear planes in the connection, n_{sp}	$n_{sp} = 2$

Bolting pattern:

Horizontal distance to the bolt a, x	$x = 90$ mm
Vertical distance to bolt a, y	$y = 85$ mm
Thickness of side members, t_{steel}	$t_{steel} = 12$ mm
Side member thickness coefficient, coef	$\mathrm{coef} = \dfrac{t_{steel}}{d}$ $\qquad$ $\mathrm{coef} = 1$

when coef equals 1 and hole tolerance $< 0.1d$ the plate is a thick plate (10.3 (EC5, *8.2.3(1)*))

Thickness of the timber member, t_2	$t_2 = 72$ mm

Depth of the timber member, h	$h = 294\,mm$
Distance from the timber loaded edge to the extreme bolt, h_e	$h_e = 232\,mm$

2. Timber and bolt properties

Table 1.3, strength class C24 (BS EN 338:2009(E), *Table 1*)

Characteristic shear strength, $f_{v.k}$	$f_{v.k} = 4.0\,N/mm^2$
Characteristic bearing strength, $f_{c.90.k}$	$f_{c.90.k} = 2.5\,N/mm^2$
Characteristic density of the timber, ρ_k	$\rho_k = 350\,kg/m^3$
Characteristic strength of a bolt, $f_{u.k}$	$f_{u.k} = 400\,N/mm^2$

3. Partial safety factors

Table 2.6 (UKNA to EC5, *Table NA.3*)

Material factor	$\gamma_M = 1.3$
Connection factor	$\gamma_{M.connection} = 1.3$

4. Modification factors

Factor for short-duration and service class 2, $k_{mod.short}$	$k_{mod.short} = 0.90$
(Table 2.4 (EC5, *Table 3.1*))	$k_{cr} = 0.67$
Modification factor for the influence of cracks, k_{cr} (Table NA.4 in UKNA to EC5)	

5. Actions

Design vertical load at the connection, V_d	$V_d = 10.05\,kN$	
Design moment at the connection, M_d	$M_d = 3.25\,kN\,m$	
Design horizontal load at the connection, H_d	$H_d = 1.0\,kN$	
Radial distance from dowels to centre of rotation	$r_{max} = \sqrt{x^2 + y^2}$	$r_{max} = 123.79\,mm$

$$r_b = x \qquad r_f = x \qquad r_c = r_{max}$$
$$r_e = r_{max} \qquad r_g = r_{max} \qquad r_d = y$$
$$r_h = y$$

Sum of squares of distances, $D = sum(r^2)$	$D = 4 \cdot r_{max}^2 + 2 \cdot r_b^2 + 2 \cdot r_d^2$	$D = 9.2 \times 10^4\,mm^2$
Force acting on bolt a per shear plane, $F_{m.d.max}$ (equation (12.7))	$F_{m.d.max} = \dfrac{M_d \cdot r_{max}}{n_{sp} \cdot D}$	
	$F_{m.d.max} = 2.19 \times 10^3\,N$	
Vertical force acting on each bolt per shear plane due to vertical load, $F_{v.d}$	$F_{v.d} = \dfrac{V_d}{n_{sp} \cdot n_{bolt}}$	$F_{v.d} = 628.13\,N$
Horizontal force acting on each bolt per shear plane due to horizontal load $F_{h.d}$	$F_{h.d} = \dfrac{H_d}{n_{sp} \cdot n_{bolt}}$	$F_{h.d} = 62.5\,N$

Resultant force acting on bolt *a* per shear plane (*a* is the maximum loaded bolt in the group under the loading configuration being used)
(equation (12.9))

$$F_d = \sqrt{\left(F_{v.d} + F_{m.d.max} \cdot \frac{x}{r_{max}}\right)^2 + \left(F_{h.d} + F_{m.d.max} \cdot \frac{y}{r_{max}}\right)^2} \qquad F_d = 2.71 \times 10^3 \, N$$

Angle of F_d to horizontal, i.e. the angle to the grain in the timber
(equation (12.10))

$$\alpha = \left(\arccos \left(\frac{F_{h.d} + F_{m.d.max} \cdot \frac{y}{r_{max}}}{F_d} \right) \right)$$

$$\alpha_{deg} = \frac{\alpha \cdot 180}{\pi} \qquad \qquad \alpha_{deg} = 54.81 \; degrees$$

Resultant force acting on bolt *b* per shear plane, to check the condition where the bolt is loaded at a greater angle to the grain than bolt *a*:

$$Fb_d = \sqrt{\left(F_{v.d} + F_{m.d.max} \cdot \frac{x}{r_{max}}\right)^2 + F_{h.d}^{\,2}} \qquad Fb_d = 2.22 \times 10^3 \, N$$

Angle of Fb_d to horizontal - i.e. the angle to the grain in the timber:

$$\alpha b = \left(\arccos\left(\frac{F_{h.d}}{Fb_d} \right) \right) \qquad \alpha b_{deg} = \frac{\alpha b \cdot 180}{\pi} \qquad \alpha b_{deg} = 88.39 \; degrees$$

6. Embedment strength of the timber
Characteristic embedment strength of the timber parallel to the grain, $f_{h.0.k}$
(equation (10.35); EC5, *equation (8.32)*).
(The equation incorporates dimension factors for Mathcad.)

$$f_{h.0.k} = 0.082 \cdot \left(1 - 0.01 \cdot \frac{d}{mm}\right) \cdot \left(\rho_k \frac{m^3}{kg}\right) \cdot \left(N \cdot mm^{-2}\right) \qquad f_{h.0.k} = 25.26 \, N/mm^2$$

Embedment factor for softwood, k_{90}
(Table 10.7 (EC5, *equation (8.33)*))

$$k_{90} = 1.35 + 0.015 \cdot d \cdot mm^{-1} \qquad k_{90} = 1.53$$

Characteristic embedment strength of the timber at an angle α to the grain $f_{h.\alpha.k}$:
(Equation (10.36); EC5, *equation (8.31)*)

$$f_{h.\alpha.k} = \frac{f_{h.0.k}}{k_{90} \cdot \sin(\alpha)^2 + \cos(\alpha)^2}$$

$$f_{h.\alpha.k} = 18.65 \, N/mm^2$$

Characteristic embedment strength of
the timber at an angle αb to the grain,
$f_{h.\alpha b.k}$:
(Equation (10.36); (EC5, *equation 8.31*))

$$f_{h.\alpha b.k} = \frac{f_{h.0.k}}{k_{90} \cdot \sin(\alpha b)^2 + \cos(\alpha b)^2}$$

$$f_{h.\alpha b.k} = 16.51 \, \text{N/mm}^2$$

7. Yield moment of a bolt

Characteristic yield moment of a
bolt, $M_{y.Rk}$

$$M_{y.Rk} = 0.3 \cdot \left(f_{u.k} \cdot \frac{\text{mm}^2}{N} \right) \cdot \left(\frac{d}{\text{mm}} \right)^{2.6} \cdot N \cdot \text{mm}$$

(Table 10.5 (EC5, *equation (8.30)*).
(The equation incorporates dimension
factors for Mathcad.)

$$M_{y.Rk} = 7.67 \times 10^4 \cdot \text{N mm}$$

8. Withdrawal resistance of a bolt, EC5 8.5.2

Design strength of the bolt, $F1_{ax.Rd}$:
(BS EN 1993-1-8)

$$F1_{ax.Rd} = 0.9 \cdot f_{u.k} \cdot \left(\frac{1}{1.25} \right) \cdot A_{bt}$$

$$F1_{ax.Rd} = 2.43 \times 10^4 \, N$$

Bearing diameter used for the plate, d_w:
(10.8.2 (EC5, *8.5.2(3)*))

$$d_w = \min(12 \cdot t_{steel}, \ 4 \cdot d, \ 85 \, \text{mm})$$

$$d_w = 48 \, \text{mm}$$

Design capacity of steel plate, $F2_{ax.Rkd}$
(10.8.2 (EC5, *8.5.2(2)*))

$$F2_{ax.Rd} = 3 \cdot f_{c.90.k} \cdot \frac{\pi}{4} \left[d_w^2 - (d + 1 \cdot \text{mm})^2 \right] \cdot \frac{k_{mod.short}}{\gamma_{M.connection}}$$

$$F2_{ax.Rd} = 8.71 \times 10^3 \, N$$

As bearing is critical, characteristic axial withdrawal capacity of the bolt, $F_{ax.Rk}$

$$F_{ax.Rk} = F2_{ax.Rd} \cdot \frac{\gamma_{M.connection}}{k_{mod.short}}$$

$$F_{ax.Rk} = 1.26 \times 10^4 \, N$$

9. Load-carrying capacity of the connection

a) Check bolt *a*

For a thick steel plate to timber joint with bolts in double shear, the characteristic
lateral resistance per shear plane is the lesser of equations (10.19) and (10.21)
(EC5, *equations (8.13(ℓ))* and *(8.13(m))*):

Failure mode (ℓ):

$$F_{v.Rk.a} = 0.5 \cdot f_{h.\alpha.k} \cdot t_2 \cdot d$$

$$F_{v.Rk.a} = 8.06 \times 10^3 \, N$$

Failure mode (*m*):

$$F_{v.Rk.b} = 2.3 \cdot \sqrt{M_{y.Rk} \cdot f_{h.\alpha.k} \cdot d} + \frac{F_{ax.Rk}}{4}$$

$$F_{v.Rk.b} = 1.27 \times 10^4 \, N$$

10.3.5, (EC5, *8.2.2(2)*) requires $F_{ax.Rk}/4$ to be less than 25% of the Johansen yield equation:

$$\frac{\dfrac{F_{ax.Rk}}{4}}{F_{v.Rk.b} - \dfrac{F_{ax.Rk}}{4}} \cdot 100 = 32.98 \qquad \text{i.e. too great; cannot exceed 25\%}$$

Limiting the axial resistance to 25% of the Johansen yield equation:

$$F1_{v.Rk.b} = (1.25) \cdot 2.3 \cdot \sqrt{M_{y.Rk} \cdot f_{h.\alpha.k} \cdot d}$$

$$F1_{v.Rk.b} = 1.19 \times 10^4 \, N$$

The characteristic lateral resistance per shear plane per bolt, $F_{v.Rk}$

$$F_{v.Rk} = min(F_{v.Rk.a}, F1_{v.Rk.b})$$

$$F_{v.Rk} = 8.06 \times 10^3 \, N$$

i.e. failure mode (ℓ) – bearing in timber

The design load carrying capacity per bolt per shear plane, $F_{v.Rd}$:

$$F_{v.Rd} = \frac{k_{mod.sort} \cdot F_{v.Rk}}{\gamma_{M.connection}}$$

$$F_{v.Rd} = 5.58 \times 10^3 \, N$$

The design load on the bolt per shear plane, F_d:

$$F_d = 2.71 \times 10^3 \, N$$

The design load is less than the capacity of the bolt; therefore OK

b) Check bolt *b*

Failure mode (ℓ):

$$Fb_{v.Rka} = 0.5 \cdot f_{h.\alpha b.k} \cdot t_2 \cdot d$$

$$Fb_{v.Rka} = 7.13 \times 10^3 \, N$$

Failure mode (*m*):

$$Fb_{v.Rkb} = 2.3 \cdot \sqrt{M_{y.Rk} \cdot f_{h.\alpha b.k} \cdot d} + \frac{F_{ax.Rk}}{4}$$

$$Fb_{v.Rkb} = 1.21 \times 10^4 \, N$$

Limiting the axial resistance to 25% of the Johansen yield equation:

$$Fb_{v.Rkb} = (1.25) \cdot 2.3 \cdot \sqrt{M_{y.Rk} \cdot f_{h.\alpha b.k} \cdot d}$$

$$Fb_{v.Rkb} = 1.12 \times 10^4 \, N$$

The characteristic lateral resistance per shear plane per bolt, $Fb_{v.Rk}$:

$$Fb_{v.Rk} = min(Fb_{v.Rka}, Fb_{v.Rkb})$$

$$Fb_{v.Rk} = 7.13 \times 10^3 \, N$$

i.e. failure mode (ℓ)-bearing in timber

The design load carrying capacity per bolt per shear plane, $Fb_{v.Rd}$:

$$Fb_{v.Rd} = \frac{k_{mod.short} \cdot Fb_{v.Rk}}{\gamma_{M.connection}}$$

$$Fb_{v.Rd} = 4.94 \times 10^3 \, N$$

The design load on the bolt per shear plane, F_d:

$$Fb_d = 2.22 \times 10^3 \, N$$

The design load, Fb_d is less than the capacity of the bolt; therefore OK

Check the capacity of the row of bolts subjected to the maximum forces parallel to the grain, i.e. the bolts in the row positioned 85 mm above the centroid.
(Note: No need to check rows perpendicular to grain as there is no strength reduction in that direction.)
The characteristic load-carrying capacity per shear plane per bolt based on failure mode ℓ, $F1_{v.Rk.a}$ (from consideration of the strength equations this will still be the failure condition)

$$F1_{v.Rk.a} = 0.5 \cdot f_{h.0.k} \cdot t_2 \cdot d$$

$$F1_{v.Rk.a} = 1.09 \times 10^4 \, N$$

Number of bolts per shear plane in the row, n_b:

$$n_b = 3$$

Effective number of bolts per shear plane in the row, n_{ef} (equation (10.45); EC5, *equation (8.34)*)
(The equation incorporates dimension factors for Mathcad.)

$$n_{ef} = \min\left[n_b, n_b^{0.9} \cdot \left(\frac{90 \cdot mm}{13 \cdot d}\right)^{0.25}\right]$$

$$n_{ef} = 2.34$$

Design force capacity per bolt per shear plane parallel to the grain taking bolt spacing effect into account, $F1_h$:

$$F1_h = n_{ef} \cdot \frac{1}{n_b} \cdot F1_{v.Rk.a} \cdot \frac{k_{mod.short}}{\gamma_{M.connection}}$$

$$F1_h = 5.9 \times 10^3 \, N$$

Design force component per bolt per shear plane parallel to the grain in each of the bolts in the row, $F_{H,a}$

$$F_{H,a} = \frac{M_d \cdot r_{max} \cdot \dfrac{y}{r_{max}}}{n_{sp} \cdot D} + \frac{H_d}{n_{bolt} \cdot n_{sp}}$$

$$F_{H,a} = 1.56 \times 10^3 \, N$$

Actual design force component in each bolt parallel to the grain is less than the design capacity per bolt per shear plane; therefore OK
(NB: An analysis of the fasteners in the connection will show that the critical stress to strength ratio will be bolt *a*)

10. **Shear strength of the timber**
    The design shear force on the timber:

Summation of '*x*' distances of bolts in shear zone, S_x

$$S_x = 3 \cdot x$$

Maximum design shear force – based on the forces in two shear planes in the line of bolts to the extreme left of the connection, $F1_{V,vd}$:
(equation (12.18))

$$F1_{V.vd} = 2 \cdot F_{m.d.max} \cdot \frac{S_x}{r_{max}} - \frac{3 \cdot V_d}{n_{bolt}}$$

$$F1_{V.vd} = 5.77 \, kN$$

Design shear force at the end of the beam, V_d

$$V_d = 10.05 \, kN$$

Design shear stress within the connection area, $\tau_{c.s}$:

$$\tau_{c.s} = \frac{3}{2} \cdot \frac{F1_{V.v.d}}{k_{cr} \cdot t_2 \cdot \left[h - 3 \cdot (d + 1 \cdot mm)\right]}$$

$$\tau_{c.s} = 0.7 \, N/mm^2$$

Shear stress in the beam at the connection, $\tau_{b.s}$:

$$\tau_{b.s} = \frac{3}{2} \cdot \frac{V_d}{k_{cr} \cdot t_2 \cdot h}$$

$$\tau_{b.s} = 1.06 \, N/mm^2$$

Design shear strength, $f_{v.d}$:

$$f_{v.d} = \frac{k_{mod.short} \cdot f_{v.k}}{\gamma_M}$$

$$f_{v.d} = 2.77 \, N/mm^2$$

Design shear stress is less than the design shear strength; therefore OK

11. Minimum bolt spacing
   Using the data in Table 12.3:

Bolts spacing criteria	$6d=72\,\text{mm}$	85 mm/90 mm provided, OK
Loaded edge criteria	$e_{\ell}=4\cdot d\quad e_{\ell}=48\,\text{mm}$	62 mm provided, OK
Loaded end criteria	$\max(7\cdot d, 80\,\text{mm})=84\,\text{mm}$	87 mm provided, OK

Example 12.8.2 Each end of a timber beam is connected to plywood gusset plates, which are glued to a rigid supporting structure in an office area. The beam is strength class C27 to BS EN 338:2009, spans 3.5 m and supports a design load of 10.0 kN/m at the ultimate limit states (ULS) and 6.83 kN/m at the serviceability limit states (SLS) from a combination of permanent and medium-term actions. It is bolted to the plywood gusset plates at each end as shown in Figure E12.8.2. The connection comprises two 24-mm-thick Finnish birch plywood gusset plates connected to the timber beam by 12-mm bolts acting in double shear. The bolts have a characteristic tensile strength of 400 N/mm² and the connection functions under service class 2 conditions. The ψ_2 factor is 0.3 and the strength of the plywood gusset plates will comply with the requirements of EC5.

   Compare the semi-rigid behaviour of the connection with the assumed fully rigid behaviour at the ULS and check the adequacy of the connection on the basis of semi-rigid behaviour. Also check the instantaneous deflection of the beam at mid-span for each condition.

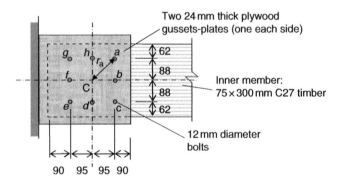

All dimensions are in mm

Fig. E12.8.02.

1. Geometric properties

Bolt diameter, d	$d=12\,\text{mm}$
Tensile stress area of a bolt, A_{bt}	$A_{bt}=84.3\,\text{mm}^2$
Number of bolts per shear plane, n_{bolt}	$n_{bolt}=8$
Bolting pattern:	
Horizontal distance to extreme bolt, x	$x=95\,\text{mm}$
Vertical distance to extreme bolt, y	$y=88\,\text{mm}$
Thickness of each plywood gusset plates, t_{gusset}	$t_{gusset}=24\,\text{mm}$
Thickness of timber member, t_2	$t_2=75\,\text{mm}$

Depth of inner timber member, h	$h = 300\,\text{mm}$	
Moment of inertia of the beam, I_b	$I_b = \dfrac{t_2 \cdot h^3}{12}$	$I_b = 1.69 \times 10^{-4}\,\text{m}^4$
Span of the beam, ℓ_b	$\ell_b = 3500\,\text{mm}$	

2. **Timber and bolt strength properties**

Table 1.3, strength class C27 (BS EN 338:2009(E), *Table 1*)

Characteristic shear strength, $f_{v.k}$	$f_{v.k} = 4.0\,\text{N/mm}^2$
Characteristic bearing strength, $f_{c.90.k}$	$f_{c.90.k} = 2.6\,\text{N/mm}^2$
Characteristic density of the timber, ρ_k	$\rho_k = 370\,\text{kg/m}^3$
Mean density of the timber, ρ_m	$\rho_m = 450\,\text{kg/m}^3$
Mean modulus of elasticity of the timber parallel to the grain, $E_{0.mean}$	$E_{0.mean} = 11.5 \cdot 10^3\,\text{N/mm}^2$
Mean shear modulus of the timber parallel to the grain, $G_{0.mean}$	$G_{0.mean} = 0.72 \times 10^3\,\text{N/mm}^2$

Table 1.10 – 24 mm nominal thickness Finnish birch plywood:

Characteristic compression strength, $f_{p.c.90.k}$	$f_{p.c.90.k} = 25.1\,\text{N/mm}^2$
Characteristic shear strength, $f_{p.v.k}$	$f_{p.v.k} = 9.5\,\text{N/mm}^2$
Characteristic density of the plywood, $\rho_{p.k}$	$\rho_{p.k} = 630\,\text{kg/m}^3$
Mean density of the plywood, $\rho_{p.mean}$	$\rho_{p.mean} = 680\,\text{kg/m}^3$

Slip modulus at SLS, K_{ser}
(Table 10.13 (EC5, *Table 7.1*))

(Mathcad adjusted to make it dimensionally correct.)

$$K_{ser} = \left[\left(\rho_{p.mean} \cdot \frac{\text{m}^3}{\text{kg}} \cdot \rho_m \cdot \frac{\text{m}^3}{\text{kg}}\right)^{0.5}\right]^{1.5} \cdot \frac{d \cdot \text{mm}^{-1}}{23} \cdot \text{Nmm}^{-1} \qquad K_{ser} = 6.79 \times 10^3\,\text{N/mm}$$

Deformation factor for the timber, $k_{def.m}$ Table 2.10 (EC5, *Table 3.2*)	$k_{def.m} = 0.8$
Deformation factor for the plywood, $k_{def.ply}$ (complies with BS EN 636: Part 2)	$k_{def.ply} = 1.0$
Deformation factor for the connection, $k_{def.c}$	$k_{def.c} = 2 \cdot (k_{def.m} \cdot k_{def.ply})^{0.5} \qquad k_{def.c} = 1.79$

(Complies with BS EN 636 Part 2)
(equation 10.78 (EC5, *2.3.2.2 (4)*))

Characteristic strength of each bolt, $f_{u.k}$	$f_{u.k} = 400\,\text{N/mm}^2$

3. **Partial safety factors**

Table 2.6 (UKNA to EC5, *Table NA.3*)

Material factor for timber, γ_M	$\gamma_M = 1.3$
Connection factor, $\gamma_{M.connection}$	$\gamma_{M.connection} = 1.3$

Table NA.A1.1 (UKNA to BS EN 1990:2002+A1:2005) – Category A

Factor for quasi-permanent value of the variable action, ψ_2	$\psi_2 = 0.3$

4. **Modification factors**

Factor for medium-duration and service class 2, $k_{mod.med}$ $\qquad k_{mod.med} = 0.8$

(Table 2.4 (EC5, *Table 3.1*))

Modification factor for the
influence of cracks, k_{cr}
(Table NA.4 in UKNA to EC5) $\qquad\qquad k_{cr} = 0.67$

5. **Actions**

Design loading on the beam, $F_{b.d}$ $\qquad F_{b.d} = 10$ kN/m

Horizontal loading on the beam, H_d $\qquad H_d = 0$ kN

Vertical loading at each connection $\qquad V_d = F_{b.d} \cdot \dfrac{\ell_b}{2}$ $\qquad V_d = 1.75 \times 10^4$ N

Assuming full fixity conditions at each end:

Design moment due to the design loading, M_d $\qquad M_d = \dfrac{F_{b.d} \cdot \ell_b^2}{12}$ $\qquad M_d = 10.21$ kN m

Assuming semi-rigid conditions at each end:

Radial distance from each bolt to the centre of rotation
$$r_{max} = \sqrt{x^2 + y^2} \qquad r_{max} = 129.5 \, mm$$

$$r_b = x \qquad r_f = x \qquad r_c = r_{max}$$
$$r_e = r_{max} \qquad r_g = r_{max} \qquad r_d = y$$
$$r_h = y$$

Sum of the squares of the radial distances, D
$$D = 4 \cdot r_{max}^2 + 2 \cdot r_b^2 + 2 \cdot r_d^2 \quad D = 1.01 \times 10^5 \, mm^2$$

Number of shear planes per end connection, n_{sp} $\qquad n_{sp} = 2$

Rotational stiffness of each end connection, k_u
(equation (12.36))
$$k_u = n_{sp} \cdot \dfrac{K_{ser} \cdot D}{1 + \psi_2 \cdot k_{def.c}} \qquad k_u = 888.91 \text{ kN m}$$

Rigidity factor of each end connection, γ
(equation (12.44))

$$\gamma = \cfrac{1}{1 + \left(\cfrac{3 \cdot \cfrac{E_{0.mean}}{1 + \psi_2 \cdot k_{def.m}} \cdot I_b}{\ell_b \cdot k_u} \right)} \qquad\qquad \gamma = 0.4$$

Area of free bending moment diagram, $A_{b.f}$
(Table 12.4)

$$A_{b.f} = \frac{2}{3} \cdot \ell_b \cdot \left(\frac{F_{b.d} \cdot \ell_b^2}{8} \right)$$

$A_{b.f} = 3.57 \times 10^{10} \, \text{N mm}^2$

Distance from centroid of $A_{b.f}$ to the centre of the end connection, a_1
(Table 12.4)

$$a_1 = \frac{\ell_b}{2}$$

$a_1 = 1.75 \times 10^3 \, \text{mm}$

End moment on each connection, $M_{SR.m.d}$
(equation (12.48)):

$$M_{SR.m.d} = \frac{6 \cdot A_{b.f}}{\ell_b^2} \cdot \left(\frac{\gamma}{4 - \gamma^2} \right) \cdot \left[2 \cdot (\ell_b - a_1) - \gamma \cdot a_1 \right]$$

$M_{SR.m.d} = 5.09 \times 10^6 \, \text{N mm}$

Ratio of semi-rigid to fixed end moments, R_{end}

$$R_{end} = \frac{M_{SR.m.d}}{M_d}$$

$R_{end} = 0.5$

Ratio of semi-rigid to fixed span moments, R_{mid}

$$R_{mid} = \frac{F_{b.d} \cdot \dfrac{\ell_b^2}{8} - M_{SR.m.d}}{F_{b.d} \cdot \dfrac{\ell_b^2}{8} - M_d}$$

$R_{mid} = 2$

Rotation of each semi-rigid joint, θ_{rads}

$$\theta_{rads} = \frac{M_{SR.m.d}}{k_u}$$

$\theta_{rads} = 5.72 \times 10^{-3}$

$$\theta_{degrees} = \theta_{rads} \cdot \frac{180}{\pi}$$

$\theta_{degrees} = 0.33$

The movement of the extreme fastener under the applied load, δ_e

$$\delta_e = \theta_{rads} \cdot r_{max}$$

$\delta_e = 0.74 \, \text{mm}$

Assuming the beam is fully fixed, the design actions at the support will be:

$$M_d = 10.21 \, \text{kN m} \qquad V_d = 17.5 \, \text{kN}$$

Assuming the beam is semi-rigid at the ends, the design actions at the support will be:

$$M_{SR.m.d} = 5.09 \, \text{kN m} \qquad V_d = 17.5 \, \text{kN}$$

Bolt forces – based on semi-rigid joint behaviour:

Force acting on bolt a per shear plane, $F_{m.d.max}$
(equation (12.7))

$$F_{m.d.max} = \frac{M_{SR.m.d} \cdot r_{max}}{n_{sp} \cdot D}$$

$F_{m.d.max} = 3.27 \times 10^3 \, \text{N}$

Vertical force acting on each bolt per shear plane due to vertical load, $F_{v.d}$:

$$F_{v.d} = \frac{V_d}{n_{sp} \cdot n_{bolt}}$$

$F_{v.d} = 1.09 \times 10^3 \, \text{N}$

Horizontal force acting on each bolt per shear plane due to horizontal load, $F_{h.d}$:

$$F_{h.d} = \frac{H_d}{n_{sp} \cdot n_{bolt}} \qquad F_{h.d} = 0$$

Resultant force acting on bolt a per shear plane (a is the maximum loaded bolt in the group – and will have the same load as bolt c as there is no horizontal load on the connection) under the loading configuration being used) :
(equation (12.9))

$$F_d = \sqrt{\left(F_{v.d} + F_{m.d.max} \cdot \frac{x}{r_{max}}\right)^2 + \left(F_{h.d} + F_{m.d.max} \cdot \frac{y}{r_{max}}\right)^2} \qquad F_d = 4.14 \times 10^3 \, N$$

Angle of F_d to the grain in the timber, α
(equation (12.10))

$$\alpha = \arccos\left(\frac{F_{h.d} + F_{m.d.max} \cdot \frac{y}{r_{max}}}{F_d}\right)$$

$$\alpha_{deg} = \frac{\alpha \cdot 180}{\pi} \qquad \alpha_{deg} = 57.52 \, degrees$$

Resultant force acting on bolt b per shear plane, to check the condition where the bolt is loaded at a greater angle to the grain than bolt a:

$$Fb_d = \sqrt{\left(F_{v.d} + F_{m.d.max} \cdot \frac{x}{r_{max}}\right)^2 + F_{h.d}^2} \qquad Fb_d = 3.5 \times 10^3 \, N$$

Angle of Fd_d to horizontal – i.e. the angle to the grain in the timber:

$$\alpha b = \left(\arccos\left(\frac{F_{hd}}{Fb_d}\right)\right)$$

$$\alpha b_{deg} = \frac{\alpha b \cdot 180}{\pi} \qquad \alpha b_{deg} = 90 \, degrees$$

6. Embedment strength of the timber and the plywood

Characteristic embedment strength of the timber parallel to grain, $f_{h.0.k}$;
(equation (10.35); EC5, *equation (8.32)*)
(The equation incorporates dimension factors for Mathcad.)

$$f_{h.0.k} = 0.082 \cdot \left(\rho_k \frac{m^3}{kg}\right) \cdot \left(1 - 0.01 \cdot \frac{d}{mm}\right) \cdot (N \cdot mm^{-2}) \qquad f_{h.0.k} = 26.7 \, N/mm^2$$

Embedment factor for softwood, k_{90}
(Table 10.7 (EC5, *equation (8.33)*)) $k_{90} = 1.35 + 0.015 \cdot d \cdot mm^{-1}$ $k_{90} = 1.53$

Characteristic embedment strength of timber at an angle α to the grain, $f_{h.\alpha.k}$
(equation (10.36); EC5, *equation (8.31)*)

$$f_{h.\alpha.k} = \frac{f_{h.0.k}}{k_{90} \cdot \sin(\alpha)^2 + \cos(\alpha)^2} \qquad f_{h.\alpha.k} = 19.39 \, N/mm^2$$

Characteristic embedment strength of
timber at an angle αb to the grain, $f_{h.\alpha b.k}$:
(equation (10.36); EC5, *equation (8.31)*)

$$f_{h.\alpha b.k} = \frac{f_{h.0.k}}{k_{90} \cdot \sin(\alpha b)^2 + \cos(\alpha b)^2}$$
$\qquad f_{h.\alpha b.k} = 17.45 \, \text{N/mm}^2$

Characteristic embedment strength of plywood, $f_{h.p.k}$
Table 10.6 (EC5, *equation (8.20)*)

(The equation incorporates dimensional factors for Mathcad.)

$$f_{h.p.k} = 0.11 \cdot \left(\rho_{pk} \cdot \frac{\text{m}^3}{\text{kg}} \right) \cdot \left(d \cdot \text{mm}^{-1} \right)^{-0.3} \cdot \text{N} \cdot \text{mm}^{-2}$$
$\qquad f_{h.p.k} = 32.88 \, \text{N/mm}^2$

7. Yield moment of a bolt

Characteristic yield moment of
a bolt, $M_{y.Rk}$
(Table 10.5 (EC5, *equation (8.30)*)).

$$M_{y.Rk} = 0.3 \cdot \left(f_{u.k} \cdot \frac{\text{mm}^2}{\text{N}} \right) \cdot \left(\frac{d}{\text{mm}} \right)^{2.6} \cdot \text{N} \cdot \text{mm}$$

(The equation incorporates dimension
factors for Mathcad.)
$\qquad M_{y.Rk} = 7.67 \times 10^4 \, \text{Nmm}$

8. Withdrawal resistance of a bolt

Design strength of the bolt, $F1_{ax.Rd}$:
(BS EN 1993-1-8)

$$F1_{ax.Rd} = 0.9 \cdot f_{uk} \cdot \left(\frac{1}{1.25} \right) \cdot A_{bt}$$

$\qquad F1_{ax.Rd} = 2.43 \times 10^4 \, \text{N}$

Washer diameter used, d_w
$\qquad d_w = 3 \cdot d$ $\qquad d_w = 36 \, \text{mm}$

Design capacity of the washer (using
the lower bearing strength of the timber
and the plywood and ignoring any load
dispersion effect – safe assessment),
$F2_{ax.Rd}$
(10.8.2 (EC5, *8.5.2(2)*))

$$F2_{ax.Rd} = 3 \cdot f_{c.90.k} \cdot \frac{\pi}{4} \left[d_w^2 - (d + 1 \cdot \text{mm})^2 \right]$$
$$\cdot \frac{k_{mod.med}}{\gamma_{M.connection}}$$

$\qquad F2_{ax.Rd} = 4.25 \times 10^3 \, \text{N}$

As bearing is critical, characteristic axial
withdrawal capacity of the bolt, $F_{ax.Rk}$:

$$F_{ax.Rk} = F2_{ax.Rd} \cdot \frac{\gamma_{M.connection}}{k_{mod.med}}$$
$\qquad F_{ax.Rk} = 6.9 \times 10^3 \, \text{N}$

9. Load-carrying capacity of the connection
a) Check bolt *a*

For a plywood to timber joint with a bolt in double shear, the characteristic lateral
resistance per shear plane is the smallest value of equations (10.7)–(10.10) in Table 10.2
(EC5, *equations (8.7)*), where:

Ratio of the embedment strength of
member 2 to member 1, $\beta_{2,1}$

$$\beta_{2,1} = \frac{f_{h.\alpha.k}}{f_{h.p.k}}$$
$\qquad \beta_{2,1} = 0.59$

$$t_1 = t_{gusset}$$

Failure mode (g): $\qquad F_{v.Rk.g} = f_{h.p.k} \cdot t_1 \cdot d \qquad F_{v.Rk.g} = 9.47 \text{ kN}$

Failure mode (h): $\qquad F_{v.Rk.h} = 0.5 \cdot f_{h.\alpha.k} \cdot t_2 \cdot d \qquad F_{v.Rk.h} = 8.72 \text{ kN}$

Failure mode (j):

$$F_{v.Rk.j} = 1.05 \cdot \frac{f_{h.p.k} \cdot t_1 \cdot d}{2 + \beta_{2,1}} \cdot \left[\sqrt{2 \cdot \beta_{2,1} \cdot \left(1 + \beta_{2,1}\right) + \frac{4 \cdot \beta_{2,1} \cdot \left(2 + \beta_{2,1}\right) \cdot M_{y.Rk}}{f_{h.\alpha.k} \cdot t_1^2 \cdot d}} - \beta_{2,1} \right] + \frac{F_{ax.Rk}}{4}$$

$$F_{v.Rk.j} = 8.36 \text{ kN}$$

Failure mode (k):

$$F_{v.Rk.k} = 1.15 \cdot \sqrt{\frac{2 \cdot \beta_{2,1}}{1 + \beta_{2,1}} \cdot \left(2 \cdot M_{y.Rk} \cdot f_{h.\alpha.k} \cdot d\right)} + \frac{F_{ax.Rk}}{4} \qquad F_{v.Rk.k} = 7.64 \text{ kN}$$

Limiting $F_{ax.Rk}/4$ to 25% of the
Johansen part of the relevant equations:

(10.3.5.3, (EC5, 8.2.2(2)))

$$F_{v.Rk.jj} = 1.25 \cdot 1.05 \cdot \frac{f_{h.p.k} \cdot t_1 \cdot d}{2 + \beta_{2,1}} \cdot \left[\sqrt{2 \cdot \beta_{2,1} \cdot \left(1 + \beta_{2,1}\right) + \frac{4 \cdot \beta_{2,1} \cdot \left(2 + \beta_{2,1}\right) \cdot M_{y.Rk}}{f_{h.\alpha.k} \cdot t_1^2 \cdot d}} - \beta_{2,1} \right]$$

$$F_{v.Rk.jj} = 8.3 \text{ kN}$$

$$F_{v.Rkkk} = 1.25 \cdot 1.15 \cdot \sqrt{\frac{2 \cdot \beta_{21}}{1 + \beta_{21}} \cdot \left(2 \cdot M_{y.Rk} \cdot f_{h.\alpha.k} \cdot d\right)} \qquad F_{v.Rkkk} = 7.4 \text{ kN}$$

The minimum characteristic strength
of the bolt is:

$$F_{v.Rk} = \min \left(F_{v.Rk.g}, \ F_{v.Rk.h}, \ F_{v.Rk.j}, \ F_{v.Rk.k}, \ F_{v.Rk.jj}, \ F_{v.Rk.kk} \right) \qquad F_{v.Rk} = 7.4 \text{ kN (mode (k))}$$

The design load-carrying capacity per
bolt per shear plane, $F_{v.Rd}$ $\qquad F_{v.Rd} = \dfrac{k_{mod.med} \cdot F_{v.Rk}}{\gamma_{M.connection}} \qquad F_{v.Rd} = 4.55 \times 10^3 \text{ N}$

The design load on bolt (a) per shear
plane, F_d $\qquad F_d = 4.14 \times 10^3 \text{ N}$

The design load is less than the capacity of the
bolt; therefore OK

b) Check bolt *b*

Ratio of the embedment strength of
member 2 to member 1, β_{21} $\qquad \beta_{21} = \dfrac{f_{h.\alpha b.k}}{f_{h.p.k}} \qquad \beta_{21} = 0.53$

$$t_1 = t_{gusset}$$

Failure mode (g) $\qquad Fb_{v.Rkg} = f_{h.p.k} \cdot t_1 \cdot d \qquad Fb_{v.Rkg} = 9.47 \text{ kN}$

Failure mode (h) $\qquad Fb_{v.Rkh} = 0.5 \cdot f_{h.\alpha b.k} \cdot t_2 \cdot d \qquad Fb_{v.Rkh} = 7.85 \text{ kN}$

Failure mode (j)

$$Fb_{v.Rkj} = 1.05 \cdot \frac{f_{h.p.k} \cdot t_1 \cdot d}{2 + \beta_{21}} \cdot \left[\sqrt{2 \cdot \beta_{21} \cdot \left(1 + \beta_{21}\right) + \frac{4 \cdot \beta_{21} \cdot \left(2 + \beta_{21}\right) \cdot M_{y.Rk}}{f_{h.\alpha b.k} \cdot t_1^2 \cdot d}} - \beta_{21} \right] + \frac{F_{ax.Rk}}{4}$$

$$Fb_{v.Rkj} = 8.46 \text{ kN}$$

Failure mode (k)

$$Fb_{v.Rkk} = 1.15 \cdot \sqrt{\frac{2 \cdot \beta_{21}}{1 + \beta_{21}}} \cdot \left(2 \cdot M_{y.Rk} \cdot f_{h.\alpha b.k} \cdot d\right) + \frac{F_{ax.Rk}}{4} \qquad F_{v.Rk.k} = 7.16 \text{ kN}$$

Limiting $F_{ax,Rk}/4$ to 25% of the Johansen part of the relevant equations:
(10.3.5.3, (EC5, 8.2.2(2)))

$$Fb_{v.Rkjj} = 1.25 \cdot 1.05 \cdot \frac{f_{h.p.k} \cdot t_1 \cdot d}{2 + \beta_{21}} \left[\sqrt{2 \cdot \beta_{21} \cdot \left(1 + \beta_{21}\right) + \frac{4 \cdot \beta_{21} \cdot \left(2 + \beta_{21}\right) \cdot M_{y.Rk}}{f_{h.\alpha b.k} \cdot t_1^2 \cdot d}} - \beta_{21} \right]$$

$$Fb_{v.Rkjj} = 8.42 \text{ kN}$$

$$Fb_{v.Rkkk} = 1.25 \cdot 1.15 \cdot \sqrt{\frac{2 \cdot \beta_{21}}{1 + \beta_{21}}} \cdot \left(2 \cdot M_{y.Rk} \cdot f_{h.\alpha b.k} \cdot d\right) \qquad Fb_{v.Rkkk} = 6.79 \text{ kN}$$

The minimum characteristic strength of the bolt is:

$$Fb_{v.Rk} = \min \left(Fb_{v.Rkg}, Fb_{v.Rkh}, Fb_{v.Rkj}, Fb_{v.Rkk}, Fb_{v.Rkjj}, Fb_{v.Rkkk}\right) \qquad Fb_{v.Rk} = 6.79 \text{ kN (mode k)}$$

The design load-carrying capacity per bolt per shear plane, $F_{v.Rd}$:

$$Fb_{v.Rd} = \frac{k_{mod.med} \cdot Fb_{v.Rk}}{\gamma_{M.connection}}$$

$$Fb_{v.Rd} = 4.18 \times 10^3 \text{ N}$$

The design load on bolt (a) per shear plane, Fb_d:

$$Fb_d = 3.5 \times 10^3 \text{ N}$$

The design load is less than capacity of the bolt therefore OK

Check the capacity of the row of bolts subjected to the maximum forces parallel to the grain of the timber beam, i.e. the bolts in the row positioned 88 mm above the centroid. (Note: No need to check rows perpendicular to timber grain as there is no strength reduction for rows in that direction.)

The characteristic load-carrying capacity per shear plane per bolt based on failure mode (m), $F1_{v.Rk.b}$ parallel to the grain: (from consideration of the strength equations this will still be the failure condition)

$$\beta 1_{2.1} = \frac{f_{h.0.k}}{f_{h.p.k}} \qquad\qquad F1_{v.Rk.b} = 1.25 \cdot 2.3 \cdot \sqrt{\frac{\beta 1_{2.1}}{1 + \beta 1_{2.1}}} \cdot M_{y.Rk} \cdot f_{h.0.k} \cdot d$$

Number of bolts per shear plane in the row, n_b

$$n_b = 3$$

Effective number of bolts per shear plane in the row
(equation (10.45); EC5, equation (8.34))
(Mathcad adjusted to make it dimensionally correct.)

$$n_{ef} = \min \left[n_b, n_b^{0.9} \cdot \left(\frac{95 \cdot \text{mm}}{13 \cdot d} \right)^{0.25} \right]$$

$$n_{ef} = 2.37$$

Design force capacity per bolt per shear plane parallel to the grain taking bolt spacing effect into account, $F1_h$

$$F1_h = n_{ef} \cdot \frac{1}{n_b} \cdot F1_{v.Rk.b} \cdot \frac{k_{mod.med}}{\gamma_{M.connection}}$$

$$F1_h = 4.65 \times 10^3 \, N$$

Design force component per bolt per shear plane parallel to the grain in each of the bolts in the row, $F_{H.a}$

$$F_{H.a} = \frac{M_{SR.m.d} \cdot r_{max} \cdot \dfrac{y}{r_{max}}}{n_{sp} \cdot D}$$

$$F_{H.a} = 2.23 \times 10^3 \, N$$

Design force component in each bolt parallel to the grain is less than the permitted design force per bolt; therefore OK

(NB: An analysis of the fasteners in the connection shows that the critical stress to strength ratio will be bolt *a* (and will have the same ratio as bolt *c*).)

10. Shear strength of the timber

The design shear force on the timber:
Summation of '*x*' distances of the bolts in the shear zone, S_x

$$S_x = 3 \cdot x$$

Maximum shear force – based on the forces in two shear planes in the line of bolts to the extreme left of the connection, $F1_{v.d}$: (equation (12.18))

$$F1_{V.v.d} = 2 \cdot F_{m.d.max} \cdot \frac{S_x}{r_{max}} - \frac{3 \cdot V_d}{n_{bolt}}$$

$$F1_{V.v.d} = 7.85 \times 10^3 \, N$$

The design shear force at the end of the beam due to vertical loading:

$$V_d = 1.75 \times 10^4 \, N$$

The timber beam only considered (as plywood gussets will have a greater shear capacity):

Design shear stress in timber beam within the connection area, $\tau_{c.s}$:

$$\tau_{c.s} = \frac{3}{2} \cdot \frac{F1_{V.v.d}}{k_{cr} \cdot t_2 \cdot [h - 3 \cdot (d + 1mm)]}$$

$$\tau_{c.s} = 0.9 \, N/mm^2$$

Shear stress in the beam at the connection, $\tau_{b.s}$:

$$\tau_{b.s} = \frac{3}{2} \cdot \frac{V_d}{k_{cr} \cdot t_2 \cdot h}$$

$$\tau_{b.s} = 1.74 \, N/mm^2$$

Design shear strength of the timber beam, $f_{v.d}$:

$$f_{v.d} = \frac{k_{mod.med} \cdot f_{v.k}}{\gamma_M}$$

$$f_{v.d} = 2.46 \, N/mm^2$$

Design shear stress is less than the design shear strength, therefore OK. (Design bending checks will also show the elements to be satisfactory.)

11. Minimum bolt spacing

Using the data in Table 12.3:

Bolts spacing criteria	$6 \cdot d = 72 \, mm$	88 mm/95 mm provided, OK
Loaded edge criteria	$e_t = (4 \cdot d)$	
	$e\ell = 48 \, mm$	62 mm provided, OK
loaded end criteria	max $(7 \cdot d, 80mm) = 84 \, mm$	90 mm, provided, OK

12. Instantaneous deflection at mid-span at the SLS

Design loading on the beam at the SLS, $F_{b.d.SLS}$ $\qquad\qquad F_{b.d.SLS} = 6.83$ kN/m

Vertical loading at each connection $\qquad V_{d.SLS} = F_{b.d.SLS} \cdot \dfrac{\ell_b}{2}$ $\qquad V_{d.SLS} = 1.2 \times 10^4$ N

Rotational stiffness of each end connection at the SLS at the instantaneous condition, $k1_u$ (equation (12.37))

$$k1_u = n_{sp} \cdot K_{ser} \cdot D$$

$$k1_u = 1.37 \times 10^9 \, \text{N mm}$$

Rigidity factor of each end connection, γ_{SLS}

$$\gamma_{SLS} = \dfrac{1}{1 + \dfrac{3 \cdot E_{0.mean} \cdot I_b}{\ell_b \cdot k1_u}}$$

$$\gamma_{SLS} = 0.45$$

Area of the free bending moment diagram, $A_{b.f.SLS}$ (Table 12.4)

$$A_{b.f.SLS} = \dfrac{2}{3} \cdot \ell_b \cdot \left(\dfrac{F_{b.d.SLS} \cdot \ell_b^2}{8} \right)$$

$$A_{b.f.SLS} = 2.44 \times 10^{10} \, \text{N mm}^2$$

End moment on each connection, $M_{SR.m.d.SLS}$ (equation (12.48))

$$M_{SR.m.d.SLS} = \dfrac{6 \cdot A_{b.f.SLS}}{\ell_b^2} \cdot \left(\dfrac{\gamma_{SLS}}{4 - \gamma_{SLS}^2} \right) \cdot \left[2 \cdot (\ell_b - a_1) - \gamma_{SLS} \cdot a_1 \right] \qquad M_{SR.m.d.SLS} = 3.85 \times 10^6 \, \text{mm}$$

Instantaneous deflection of the semi-rigid beam at mid-span, δ_{inst}

$$\delta_{inst} = \dfrac{\ell_b^2}{E_{0.mean} \cdot I_b} \cdot \left(\dfrac{5 \cdot F_{b.d.SLS} \cdot \ell_b^2}{384} - \dfrac{M_{SR.m.d.SLS}}{8} \right) + \dfrac{1.2 \cdot V_{d.SLS} \cdot \ell_b}{4 G_{0.mean} \cdot (h \cdot t_2)} \qquad \delta_{inst} = 4.62 \, \text{mm}$$

Instantaneous deflection of the rigid beam at mid-span, $\delta 1_{inst}$

$$\delta 1_{inst} = \dfrac{\ell_b^2}{E_{0.mean} \cdot I_b} \cdot \left(\dfrac{5 \cdot F_{b.d.SLS} \cdot \ell_b^2}{384} - \dfrac{F_{b.d.SLS} \cdot \ell_b^2}{8.12} \right) + \dfrac{1.2 \cdot V_{d.SLS} \cdot \ell_b}{4 G_{0.mean} \cdot (h \cdot t_2)} \qquad \delta 1_{inst} = 2.15 \, \text{mm}$$

i.e. fully rigid beam deflection is only about 50% of the semi-rigid value

Example 12.8.3 A timber roof structure is formed from 300 mm deep by 75 mm wide timber beams supported by columns as shown in Figure E12.8.3. Each beam is connected at its ends to the adjacent beams by steel brackets, such that the beam can be considered to form a continuous semi-rigid structure. At each end of the roof structure the steel bracket is fixed to a rigid support on adjacent structures.

The connection detail between adjacent beams at the ridge level is shown in Figure E12.8.3. It is an 8-mm-thick steel plate fitted within a notch in the beams and the connection is formed using 8 No 12-mm-diameter mild steel bolts and the bolt spacings comply with the requirements of Table 12.3. A similar bracket is used to connect adjacent beams at the eaves level and

(a)

8.5 kN/m design loading

See detail A

3700 mm effective span
of beam AB

A

30°

7100 mm

B

Tie bars supported at mid
span to prevent sagging

Columns

75 mm by 300 mm C24
timber beams

Roof structure elevation

(b)

End A

8 no12 mm diameter
bolts per connection

100
100

Connection detail is the
same as shown across

75 mm by 300 mm
C24 timber beam

a

h

g

b

55

f

c

95

e

d

95

30°

55

8 mm thick
steel plate

3700 mm to the centre
line of the connection
at End B of the beam

Detail A

(c)

33.5

33.5

8 mm thick
steel plate

12 mm diameter
bolt

75 mm by 300 mm
C24 timber beams

Plan on detail A

All dimensions are in mm

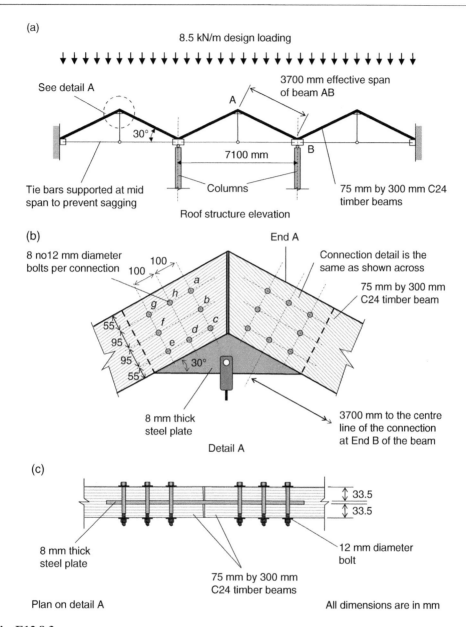

Fig. E12.8.3.

the rigidity factor for that connection is only 40% of the rigidity factor of the beam connection
at the ridge. At the eaves level, the steel brackets are connected across the structure by steel ties
and these can be considered to be axially rigid. The beams are strength class C24 to BS EN
338:2009 and the effective span of each beam along the direction of its longitudinal axis can be
taken to be 3700 mm. Each beam is restrained from lateral buckling and lateral torsional movement at the ends and at the mid-span positions. The design loading acting vertically along the
length of the beam axis, which includes for the effect of the self-weight of the beam, is 8.5 kN/m,
and arises from a combination of permanent and medium-term variable loading. The largest

stress in relation to strength will be due to the variable loading and in accordance with the requirements of EC5, *2.3.2.2(2)*; ψ_2 will be 0. The structure functions under service class 2 conditions.

Taking semi-rigid behaviour into account, check that the connection at the ridge will comply with the strength requirements of EC5 and that the bending, axial and shear strength of the beams will be acceptable.

1. Geometric properties

Bolt diameter, d	$d = 12\,\text{mm}$	
Tensile stress area of a bolt, $A_{b.t}$	$A_{b.t} = 84.3\,\text{mm}^2$	
Number of bolts per shear plane, n_{bolt}	$n_{bolt} = 8$	

Bolt distances from the centroid of the group – relative to the grain:

Horizontal distance along grain to the extreme bolt a, x	$x = 100\,\text{mm}$	
Vertical distance perpendicular to the grain to the extreme bolt a, y	$y = 95\,\text{mm}$	
Thickness of the steel plate in the timber member, t_{steel}	$t_{steel} = 8\,\text{mm}$	
Thickness of the timber member, t_2	$t_2 = 75\,\text{mm}$	
Thickness of the side timbers at the joint, t_s	$t_s = \dfrac{t_2 - t_{steel}}{2}$	$t_s = 33.5\,\text{mm}$
Depth of the timber member, h	$h = 300\,\text{mm}$	
Moment of inertia of the beam, I_b	$I_b = \dfrac{t_2 \cdot h^3}{12}$	$I_b = 1.69 \times 10^{-4}\,\text{m}^4$
Section modulus of the beam, W_b	$W_b = \dfrac{t_2 \cdot h^2}{6}$	$W_b = 1.12 \times 10^6\,\text{mm}^3$
Span of the beam, ℓ_b	$\ell_b = 3700\,\text{mm}$	
Clear span of each bent, ℓ	$\ell = 7100\,\text{mm}$	
Effective length for lateral buckling about the y–y axis, $L_{e,y}$ (based on the content of Table 5.2, adopt a factor of $0.9\ell_b$)	$L_{e,y} = 0.9 \cdot \ell_b$	$L_{e,y} = 3.33\,\text{m}$
Effective length for lateral buckling about the z–z axis, $L_{e,z}$ (based on the content of Table 5.2, adopt a factor of $0.5\ell_b$)	$L_{e,z} = 0.5 \cdot \ell_b$	$L_{e,z} = 1.85\,\text{m}$
Cross-sectional area, A	$A = t_2 \cdot h$	$A = 2.25 \times 10^4\,\text{mm}^2$
Second moment of area about the y–y axes, I_y	$I_y = \dfrac{t_2 \cdot h^3}{12}$	$I_y = 1.69 \times 10^8\,\text{mm}^4$

Section modulus about the y–y axes, W_y	$W_y = \dfrac{2I_y}{h}$	$W_y = 1.12 \times 10^6 \, \text{mm}^3$
Radius of gyration about the y–y axis, i_y	$i_y = \sqrt{\dfrac{I_y}{A}}$	$i_y = 86.6 \, \text{mm}$
Slenderness ratio about the y–y axis, λ_y	$\lambda_y = \dfrac{L_{e,y}}{i_y}$	$\lambda_y = 38.45$
Second moment of area about the z–z axes, I_z	$I_z = \dfrac{h \cdot t_2^3}{12}$	$I_z = 1.05 \times 10^7 \, \text{mm}^4$
Radius of gyration about the z–z axis, i_z	$i_z = \sqrt{\dfrac{I_z}{A}}$	$i_z = 21.65 \, \text{mm}$
Slenderness ratio about the z–z axis, λ_z	$\lambda_z = \dfrac{L_{e,z}}{i_z}$ $\lambda_z = 85.45$	

2. Timber and bolt strength properties

Table 1.3, strength class C24 (BS EN 338:2009(E), *Table 1*)

Characteristic bending strength, $f_{m.k}$	$f_{m.k} = 24.0 \, \text{N/mm}^2$
Characteristic shear strength, $f_{v.k}$	$f_{v.k} = 4.0 \, \text{N/mm}^2$
Characteristic compression strength parallel to the grain, $f_{c.0.k}$	$f_{c.0.k} = 21 \, \text{N/mm}^2$
Characteristic compression strength perpendicular to the grain, $f_{c.90.k}$	$f_{c.90.k} = 2.5 \, \text{N/mm}^2$
Characteristic density of the timber, ρ_k	$\rho_k = 350 \, \text{kg/m}^3$
Mean density of the timber, ρ_m	$\rho_m = 420 \, \text{kg/m}^3$
Mean modulus of elasticity of the timber parallel to the grain, $E_{0.mean}$	$E_{0.mean} = 11.0 \times 10^3 \, \text{N/mm}^2$
Fifth percentile modulus of elasticity of the timber parallel to grain, $E_{0.05}$	$E_{0.05} = 7.4 \times 10^3 \, \text{N/mm}^2$

Slip modulus at the SLS, K_{ser}
(Table 10.11 (EC5, *Table 7.1*))
(Not applying the factor of 2 permitted in EC5)

$$K_{ser} = \left(\rho_m \cdot \frac{\text{m}^3}{\text{kg}} \right)^{1.5} \cdot \frac{d \cdot \text{mm}^{-1}}{23} \cdot \text{N} \cdot \text{mm}^{-1}$$

(Mathcad adjusted to make it dimensionally correct.)

$$K_{ser} = 4.49 \times 10^3 \, \text{N/mm}$$

Deformation factor for the timber, $k_{def.m}$ $k_{def.m} = 0.8$
(EC5, *Table 3.2*)

Deformation factor for the connection, $k_{def.c}$ $k_{def.c} = 0.8$

(EC5, *Table 3.2* with no factor applied – there is only one timber member per shear plane)

Characteristic tensile strength of each bolt, $f_{u.k}$	$f_{u.k}=400\,\text{N/mm}^2$

3. **Partial safety factors**

   Table 2.6 (UKNA to EC5, *Table NA.3*)

Material factor	$\gamma_M=1.3$
Connection factor	$\gamma_{M.connection}=1.3$

   Table NA.A1.1 (UKNA to BS EN 1990:2002+A1:2005) – Category H – roof

   Factor for the quasi-permanent value of the action to be taken to be 0, ψ_2 $\psi_2=0$

   (Note: Because ψ_2 is 0 the stiffness values used in the solution will equate to mean values.)

4. **Modification factors**

   Factor for medium-duration and service class 3, $k_{mod.med}$ $k_{mod.med}=0.8$

   (Table 2.4 (EC5, *Table 3.1*))

   System strength factor, k_{sys} $k_{sys}=1$
   (2.3.7 (EC5, *6.7*))

   Lateral stability of the beam, k_{crit}
   (4.5.1.2 (EC5, *6.3.3*))

   Effective length of the beam, ℓ_{ef}, (safe approximation) (Table 4.2)

   $$\ell_{ef}=0.9\cdot\frac{\ell_b}{2}+2\cdot h \qquad \ell_{ef}=2.27\,\text{m}$$

   Critical bending stress, $\sigma_{m.crit}$
   (equation (4.8); EC5, *equation (6.32)*)

   $$\sigma_{m.crit}=\frac{0.78\cdot t_2^2}{h\cdot\ell_{ef}}\cdot E_{0.05} \qquad \sigma_{m.crit}=47.78\,\text{N/mm}^2$$

   Lateral torsional buckling condition (5.3.1 (EC5, *6.3.3*))

   Relative slenderness for bending, $\lambda_{rel.m}$ (equation (4.10); EC5, *equation (6.30)*)

   $$\lambda_{rel.m}=\sqrt{\frac{f_{m.k}}{\sigma_{m.crit}}} \qquad \lambda_{rel.m}=0.71$$

   Lateral stability factor, k_{crit} (Table 4.3 (EC5, *equation (6.34)*))

   $$k_{crit}=\begin{vmatrix}1 & \text{if} \quad \lambda_{rel.m}\le 0.75 \\ 1.56-0.75\cdot\lambda_{rel.m} & \text{if} \quad 0.75<\lambda_{rel.m}\le 1.4 \\ \dfrac{1}{\lambda_{rel.m}^2} & \text{otherwise}\end{vmatrix}$$

   $$k_{crit}=1$$

   Modification factor for the influence of cracks, k_{cr}
   (Table NA.4 in UKNA to EC5) $k_{cr}=0.67$

5. Actions on beam AB and beam strength

Vertical design loading along the sloping axis of the beam, $FV_{b.d}$

$$FV_{b.d} = 8.5 \cdot kN \cdot mm^{-1}$$

Projected design load along the beam span, $F_{b.d}$

$$F_{b.d} = FV_{b.d} \cdot \cos(30°) \qquad F_{b.d} = 7.36 \text{ kN/m}$$

Loading on the connection at A acting parallel to the grain of the beam, H_d

$$H_d = FV_{b.d} \cdot \frac{1}{4 \cdot \cos(60°)} \qquad H_d = 3.02 \times 10^4 \text{ N}$$

Assuming full fixity conditions at the ends of the beam:

Design moment at ends A and B due to the design loading along the beam, M_d

$$M_d = \frac{F_{b.d} \cdot \ell_b^2}{12} \qquad M_d = 8.4 \text{ kN m}$$

Assuming semi-rigid conditions at the ends of the beam:

At end A, radial distance from the centroid of the connection to each dowel, r

$$r_{max} = \sqrt{x^2 + y^2} \qquad r_{max} = 137.93 \text{ mm}$$

$$r_b = x \qquad r_f = x \qquad r_c = r_{max}$$

$$r_e = r_{max} \qquad r_g = r_{max} \qquad r_d = y \qquad r_h = y$$

Sum of the squares of the distances at end A, D

$$D = 4 \cdot r_{max}^2 + 2 \cdot r_b^2 + 2 \cdot r_d^2 \qquad D = 1.14 \times 10^5 \text{ mm}^2$$

Number of shear planes at end A connection, n_{sp}

$$n_{sp} = 2$$

Rotational stiffness of end A connection, k_u (equation (12.36))

$$k_u = n_{sp} \cdot \frac{K_{ser} \cdot D}{1 + \psi_2 \cdot k_{def.c}} \qquad k_u = 1.03 \times 10^3 \text{ kN m}$$

Rigidity factor of end A connection, γ_A (equation (12.44))

$$\gamma_A = \frac{1}{1 + \left(\dfrac{3 \cdot \dfrac{E_{0.mean}}{1 + \psi_2 \cdot k_{def.m}} \cdot I_b}{\ell_b \cdot k_u} \right)} \qquad \gamma_A = 0.41$$

Rigidity factor of end B connection, γ_B

$$\gamma_B = 0.4 \cdot \gamma_A \qquad \gamma_B = 0.16$$

Area of the free bending moment diagram on beam AB, $A_{b.f}$ (Table 12.4)

$$A_{b.f} = \frac{2}{3} \cdot \ell_b \cdot \left(\frac{F_{b.d} \cdot \ell_b^2}{8} \right) \qquad A_{b.f} = 3.11 \times 10^{10} \text{ N mm}^2$$

Distance from the centroid of the bending moment diagram to the centroid of the end connection, a_1 (Table 12.4)

$$a_1 = \frac{\ell_b}{2} \qquad a_1 = 1.85 \times 10^3 \text{ mm}$$

The end moment on each connection, $MA_{SR.m.d}$ and $MB_{SR.m.d}$ at ends A and B respectively (equations (12.48) and (12.49)):

$$MA_{SR.m.d} = \frac{6 \cdot A_{b.f}}{\ell_b^2} \cdot \left(\frac{\gamma_A}{4 - \gamma_A \cdot \gamma_B} \right) \cdot \left[2 \cdot (\ell_b - a_1) - \gamma_B \cdot a_1 \right] \qquad MA_{SR.m.d} = 4.77 \times 10^3 \, \text{N mm}$$

$$MB_{SR.m.d} = \frac{6 \cdot A_{b.f}}{\ell_b^2} \cdot \left(\frac{\gamma_B}{4 - \gamma_A \cdot \gamma_B} \right) \cdot \left[2 \cdot a_1 - \gamma_A \cdot (\ell_b - a_1) \right] \qquad MB_{SR.m.d} = 1.66 \times 10^3 \, \text{N mm}$$

Value of end A reaction, RA

$$RA = \frac{(MA_{SR.m.d} - MB_{SR.m.d}) + F_{b.d} \cdot \dfrac{\ell_b^2}{2}}{\ell_b}$$

$$RA = 1.45 \times 10^4 \, \text{N}$$

The position of zero shear force from the centroid of the connection at end A, xA

$$xA = \frac{RA}{F_{b.d}}$$

$$xA = 1.96 \, \text{m}$$

Maximum span moment on beam AB, M_{span}

$$M_{span} = RA \cdot xA - MA_{SR.m.d} - F_{b.d} \cdot \frac{xA^2}{2}$$

$$M_{span} = 9.43 \times 10^3 \, \text{N m}$$

Check the combined bending and axial stress condition on the beam:

Bending stress at the position of maximum bending moment on the beam, σ_m

$$\sigma_m = \frac{M_{span}}{W_b}$$

$$\sigma_m = 8.38 \, \text{N/mm}^2$$

Bending strength of beam AB

$$f_{m.d} = \frac{k_{mod.med} \cdot k_{sys} \cdot f_{m.k}}{\gamma_M}$$

$$f_{m.d} = 14.77 \, \text{N/mm}^2$$

Design axial compression stress in beam AB, $\sigma_{c.0.d}$

$$\sigma_{c.0.d} = \frac{H_d}{A}$$

$$\sigma_{c.0.d} = 1.34 \, \text{N/mm}^2$$

Design axial compression strength of beam AB, $f_{c.0.d}$

$$f_{c.0.d} = \frac{k_{mod.med} \cdot k_{sys} \cdot f_{c.0.k}}{\gamma_M}$$

$$f_{c.0.d} = 12.92 \, \text{N/mm}^2$$

Axial buckling resistance condition (5.3.1 (EC5, 6.3.2))

Relative slenderness about the y–y axis, $\lambda_{rel.y}$ (equation (5.3); EC5, *equation* *(6.21)*)

$$\lambda_{rel.y} = \frac{\lambda_y}{\pi} \cdot \sqrt{\frac{f_{c.0.k}}{E_{0.05}}}$$

$$\lambda_{rel.y} = 0.65$$

Relative slenderness about the z–z axis, $\lambda_{rel.z}$ (equation (5.3); EC5, *equation* *(6.22)*)

$$\lambda_{rel.z} = \frac{\lambda_z}{\pi} \cdot \sqrt{\frac{f_{c.0.k}}{E_{0.05}}}$$

$$\lambda_{rel.z} = 1.45$$

As both relative slenderness ratios are greater than 0.3, conditions in 5.3.1(b) apply (EC5, *6.3.2(3)*):

Redistribution factor for a rectangular section, k_m (see equation (4.4) (EC5, *6.1.6*))

$k_m = 0.7$

Factor β_c for solid timber (equation (5.6); EC5, *equation (6.29)*)

$\beta_c = 0.2$

Factor, k_y (equation 5.5a); EC5, *equation (6.27)*)

$$k_y = 0.5 \cdot \left[1 + \beta_c \cdot \left(\lambda_{rel.y} - 0.3 \right) + \lambda_{rel.y}^2 \right]$$

$k_y = 0.75$

Instability factor about the y–y axis, $k_{c.y}$ (equation (5.4a); EC5, *equation (6.25)*)

$$k_{c.y} = \frac{1}{k_y + \sqrt{k_y^2 - \lambda_{rel.y}^2}}$$

$k_{c.y} = 0.9$

Factor, k_z (equation (5.5b); EC5, *equation (6.28)*)

$$k_z = 0.5 \cdot \left[1 + \beta_c \cdot \left(\lambda_{rel.z} - 0.3 \right) + \lambda_{rel.z}^2 \right]$$

$k_z = 1.66$

Instability factor about the z–z axis (equation (5.4b); EC5, *equation (6.26)*)

$$k_{c.z} = \frac{1}{k_z + \sqrt{k_z^2 - \lambda_{rel.z}^2}}$$

$k_{c.z} = 0.4$

Combined bending and axial stress condition:

Combined axial and bending stress condition about the y–y axis – functioning as a column (equation (5.21) (EC5, *6.3.2(3)*)):

$$\frac{\sigma_{c.0.d}}{k_{c.y} \cdot f_{c.0.d}} + \frac{\sigma_m}{f_{m.d}} = 0.68$$

Compression stress condition about the z–z axis (equation (5.22) (EC5 *6.3.2(3)*))

$$\frac{\sigma_{c.0.d}}{k_{c.z} \cdot f_{c.0.d}} + k_m \cdot \frac{\sigma_m}{f_{m.d}} = 0.66$$

Combined bending and axial stress condition functioning as a beam (equation (5.23); EC5, *equation (6.35)*):

$$\left(\frac{\sigma_m}{k_{crit} \cdot f_{m.d}} \right)^2 + \frac{\sigma_{c.0.d}}{k_{c.z} \cdot f_{c.0.d}} = 0.58$$

The stress to strength ratios are all less than 1; therefore OK

Check the end shear condition:

Shear stress at end A of the beam – taken within the zone of the steel bracket (conservative approach using RA as shear force), τ_d

$$\tau_d = \frac{3}{2} \cdot \frac{RA}{2 \cdot k_{cr} \cdot t_s \cdot [h - 3 \cdot (d + 1 \cdot mm)]}$$

$\tau_d = 1.85 \, N/mm^2$

Shear strength at end A of the beam, τ_d

$$f_{v.d} = \frac{k_{mod.med} \cdot f_{v.k}}{\gamma_M}$$

$f_{v.d} = 2.46 \, N/mm^2$ OK

Bolt forces at end A connection – based on semi-rigid joint behaviour:

Force acting on the maximum loaded bolt (bolt g) per shear plane, $F_{m.d.max}$ (equation (12.7))

$$F_{m.d.max} = \frac{MA_{SR.m.d} \cdot r_{max}}{n_{sp} \cdot D}$$

$$F_{m.d.max} = 2.88 \times 10^3 \, N$$

Force on each bolt per shear plane from the projected load acting along the beam span, $F_{v.d}$

$$F_{v.d} = \frac{RA}{n_{sp} \cdot n_{bolt}}$$

$$F_{v.d} = 903.73 \, N$$

Force on each bolt per shear plane parallel to the grain of the beam, $F_{h.d}$

$$F_{h.d} = \frac{H_d}{n_{sp} \cdot n_{bolt}}$$

$$F_{h.d} = 1.89 \times 10^3 \, N$$

Resultant force acting on bolt g per shear plane (g is the maximum loaded bolt in the group under the loading configuration being used) F_d (equation (12.9))

$$F_d = \sqrt{\left(F_{v.d} + F_{m.d.max} \cdot \frac{x}{r_{max}}\right)^2 + \left(F_{H.d} + F_{m.d.max} \cdot \frac{y}{r_{max}}\right)^2}$$

$$F_d = 4.89 \times 10^3 \, N$$

Angle of F_d to the grain in the timber, α (equation (12.10))

$$\alpha = \arccos\left(\frac{F_{h.d} + F_{m.d.max} \cdot \frac{y}{r_{max}}}{F_d}\right)$$

$$\alpha_{deg} = \frac{\alpha \cdot 180}{\pi}$$

$$\alpha_{deg} = 37.71 \, degrees$$

Resultant force acting on bolt f per shear plane, to check the condition where the bolt is loaded at a greater angle to the grain than bolt g:

$$Ff_d = \sqrt{\left(F_{v.d} + F_{m.d.max} \cdot \frac{x}{r_{max}}\right)^2 + F_{h.d}^2}$$

$$Ff_d = 3.54 \times 10^3 N$$

Angle of Ff_d to horizontal - i.e. the angle to the grain in the timber:

$$\alpha f = \left(acos\left(\frac{F_{hd}}{Ff_d}\right)\right)$$

$$\alpha f_{deg} = \frac{\alpha f \cdot 180}{\pi}$$

$$\alpha f_{deg} = 57.78 \, degrees$$

6. Embedment strength of the timber

Characteristic embedment strength of the beam timber parallel to the grain, $f_{h.0.k}$; (equation (10.35); EC5, *equation (8.32)*)

(The equation incorporates dimension factors for Mathcad.)

$$f_{h.0.k} = 0.082 \cdot \left(\rho_k \cdot \frac{m^3}{kg}\right) \cdot \left(1 - 0.01 \cdot \frac{d}{mm}\right) \cdot N \cdot mm^{-2}$$

$$f_{h.0.k} = 25.26 \, N/mm^2$$

Embedment factor for softwood, k_{90}, (Table 10.7 (EC5, *equation (8.33)*))

$$k_{90} = 1.35 + 0.015 \cdot d \cdot mm^{-1} \quad k_{90} = 1.53$$

Characteristic embedment strength of timber at an angle to the grain, $f_{h.\alpha.k}$ (equation (10.36); EC5, equation (8.31))

$$f_{h.\alpha.k} = \frac{f_{h.0.k}}{k_{90} \cdot \sin(\alpha)^2 + \cos(\alpha)^2}$$

$$f_{h.\alpha.k} = 21.08 \, \text{N/mm}^2$$

Characteristic embedment strength of the timber at an angle αf to the grain, $f_{h.\alpha b.k}$ (equation (10.36); EC5, equation (8.31))

$$f_{h.\alpha f.k} = \frac{f_{h.0.k}}{k_{90} \cdot \sin(\alpha f)^2 + \cos(\alpha f)^2}$$

$$f_{h.\alpha f.k} = 18.31 \, \text{N/mm}^2$$

7. Yield moment of a bolt

Characteristic yield moment of a bolt, $M_{y.Rk}$ (Table 10.5 (EC5, equation (8.30))) (The equation incorporates dimension factors for Mathcad.)

$$M_{y.Rk} = 0.3 \cdot \left(f_{u.k} \cdot \frac{\text{mm}^2}{\text{N}} \right) \cdot \left(\frac{d}{\text{mm}} \right)^{2.6} \cdot \text{N} \cdot \text{mm}$$

$$M_{y.Rk} = 7.67 \times 10^4 \, \text{N mm}$$

8. Withdrawal resistance of a bolt

Design strength of the bolt, $F1_{ax.Rd}$ (BS EN 1993-1-8)

$$F1_{ax.Rd} = 0.9 \cdot f_{uk} \left(\frac{1}{1.25} \right) \cdot A_{bt}$$

$$F1_{ax.Rd} = 2.43 \times 10^4 \, \text{N}$$

Washer diameter used, d_w

$$d_w = 3 \cdot d$$

$$d_w = 36 \, \text{mm}$$

Design capacity of the washer, $F2_{ax.Rd}$ (10.8.2 (EC5, 8.5.2(2)))

$$F2_{ax.Rd} = 3 \cdot f_{c.90.k} \cdot \frac{\pi}{4} \cdot \left[d_w^2 - (d+1 \cdot \text{mm})^2 \right]$$

$$\cdot \frac{k_{mod.med}}{\gamma_{M.connection}}$$

$$F2_{ax.Rd} = 4.09 \times 10^3 \, \text{N}$$

As bearing is critical, characteristic axial withdrawal capacity of the bolt, $F_{ax.Rk}$

$$F_{ax.Rk} = F2_{ax.Rd} \cdot \frac{\gamma_{M.connection}}{k_{mod.med}}$$

$$F_{ax.Rk} = 6.64 \times 10^3 \, \text{N}$$

9. Load-carrying capacity of the connection at end A of beam AB

a) Check bolt g

For a central steel plate to timber joint with bolts in double shear, the characteristic lateral resistance per shear plane is the lesser of equations (10.16) (10.17) and (10.18) (EC5, equations (8.11)):

Failure mode (f):

$$F_{v.Rk.f} = f_{h.\alpha.k} \cdot t_s \cdot d$$

$$F_{v.Rk.f} = 8.47 \times 10^3 \, \text{N}$$

Failure mode (g):

$$F_{v.Rk.g} = f_{h.\alpha.k} \cdot t_s \cdot d \cdot \left[\left(2 + 4 \cdot \frac{M_{y.Rk}}{f_{h.\alpha.k} \cdot d \cdot t_s^2} \right)^{0.5} - 1 \right] + \frac{F_{ax.Rk}}{4}$$

$$F_{v.Rk.g} = 8.06 \times 10^3 \, \text{N}$$

Failure mode (h):

$$F_{v.Rk.h} = 2.3 \cdot \sqrt{M_{y.Rk} \cdot f_{h.\alpha.k} \cdot d} + \frac{F_{ax.Rk}}{4}$$

$$F_{v.Rk.h} = 1.18 \times 10^4 \, \text{N}$$

Maximum value of $\dfrac{F_{ax,Rk}}{4}$ is 25% of Johansen part, $Fm_{ax.Rk}$ (10.3.5.3, (EC5, 8.2.2(2)))

$$Fm_{ax.Rk} = f_{h.\alpha.k} \cdot t_s \cdot d \cdot \left[\left(2 + 4 \cdot \frac{M_{y.Rk}}{f_{h.\alpha.k} \cdot d \cdot t_s^2}\right)^{0.5} - 1\right]$$

$Fm_{ax.Rk} = 6.4 \times 10^3\,\text{N}$

i.e. $Fm_{ax.Rk} < F_{ax.Rk}$ therefore the axial strength must be limited to 25% of the Johansen part:

Failure mode (g):

$$F1_{v.Rk.g} = 1.25 \cdot f_{h.\alpha.k} \cdot t_s \cdot d \cdot \left[\left(2 + 4 \cdot \frac{M_{y.Rk}}{f_{h.\alpha.k} \cdot d \cdot t_s^2}\right)^{0.5} - 1\right]$$

$F1_{v.Rk.g} = 8 \times 10^3\,\text{N}$

Failure mode (h):

$$F1_{v.Rk.h} = 1.25 \cdot 2.3 \cdot \sqrt{M_{y.Rk} \cdot f_{h.\alpha.k} \cdot d}$$

$F1_{v.Rk.h} = 1.27 \times 10^4\,\text{N}$

The characteristic lateral resistance per shear plane per bolt, $F_{v.Rk}$

$F_{v.Rk} = \min(F_{v.Rk.f}, F_{v.Rk.g}, F_{v.Rk.h}, F1_{v.Rk.g}, F1_{v.Rk.h})$
$F_{v.Rk} = 8 \times 10^3\,\text{N}$ i.e. failure mode (g)

The design load-carrying capacity per bolt per shear plane, $F_{v.Rd}$

$$F_{v.Rd} = \frac{k_{mod.med} \cdot F_{v.Rk}}{\gamma_{M.connection}}$$

$F_{v.Rd} = 4.92 \times 10^3\,\text{N}$

The design load on bolt g per shear plane, F_d

$F_d = 4.89 \times 10^3\,\text{N}$

The design load is less than the capacity of the bolt; therefore OK

b) Check bolt f

Failure mode (f):

$Ff_{v.Rkf} = f_{h.\alpha f.k} \cdot t_s \cdot d$

$Ff_{v.Rkf} = 7.36 \times 10^3\,\text{N}$

Failure mode (g):

$$Ff_{v.Rkg} = f_{h.\alpha f.k} \cdot t_s \cdot d \cdot \left[\left(2 + 4 \cdot \frac{M_{y.Rk}}{f_{h.\alpha f.k} \cdot d \cdot t_s^2}\right)^{0.5} - 1\right] + \frac{F_{ax.Rk}}{4}$$

$Ff_{v.Rkg} = 7.56 \times 10^3\,\text{N}$

Failure mode (h):

$$Ff_{v.Rk.h} = 2.3 \cdot \sqrt{M_{y.Rk} \cdot f_{h.\alpha f.k} \cdot d} + \frac{F_{ax.Rk}}{4}$$

$Ff_{v.Rkh} = 1.11 \times 10^4\,\text{N}$

Maximum value of $\dfrac{F_{ax.Rk}}{4}$ is 25% of Johansen part, $Fm_{ax.Rk}$ (10.3.5.3, (EC5, 8.2.2(2)))

$$Ffm_{ax.Rk} = f_{h.\alpha f.k} \cdot t_s \cdot d \cdot \left[\left(2 + 4 \cdot \frac{M_{y.Rk}}{f_{h.\alpha f.k} \cdot d \cdot t_s^2}\right)^{0.5} - 1\right]$$

$Ffm_{ax.Rk} = 5.9 \times 10^3\,\text{N}$

i.e. $Fbm_{ax.Rk} < F_{ax.Rk}$ therefore the axial strength must be limited to 25% of the Johansen part:

Failure mode (g):
$$Ff1_{v.Rkg} = 1.25 \cdot f_{h.\alpha f.k} \cdot t_s \cdot d \cdot \left[\left(2 + 4 \cdot \frac{M_{y.Rk}}{f_{h.\alpha f.k} \cdot d \cdot t_s^2} \right)^{0.5} - 1 \right]$$

$$Ff1_{v.Rkg} = 7.37 \times 10^3 \, \text{N}$$

Failure mode (h):
$$Ff1_{v.Rkh} = 1.25 \cdot 2.3 \cdot \sqrt{M_{y.Rk} \cdot f_{h.\alpha f.k} \cdot d}$$

$$Ff1_{v.Rkh} = 1.18 \times 10^4 \, \text{N}$$

The characteristic lateral resistance per shear plane per bolt, $F_{v.Rk}$:
$$Ff_{v.Rk} = \min (Ff_{v.Rkf}, Ff_{v.Rkg}, Ff_{v.Rkh}, Ff1_{v.Rkg}, Ff1_{v.Rkh})$$

$$Ff_{v.Rk} = 7.36 \times 10^3 \, \text{N}$$

i.e. failure mode (f)

The design load-carrying capacity per bolt per shear plane, $Ff_{v.Rd}$:
$$Ff_{v.Rd} = \frac{k_{mod.med} \cdot Ff_{v.Rk}}{\gamma_{M.connection}}$$

$$Ff_{v.Rd} = 4.53 \times 10^3 \, \text{N}$$

The design load on bolt f per shear plane, Ff_d:

$$Ff_d = 3.54 \times 10^3 \, \text{N}$$

The design load is less than the capacity of the bolt; therefore OK

Check the capacity of the row of bolts subjected to the maximum forces parallel to the grain, i.e. the bolts in the row positioned 95 mm above the centroid.

(Note: No need to check rows perpendicular to grain as there is no strength reduction for rows in that direction.)

The characteristic load-carrying capacity per shear plane per bolt based on failure mode (g), $FH_{v.Rk.g}$, parallel to the grain:

(from consideration of the strength equations this will still be the failure condition)

$$FH_{v.Rk.g} = (1.25) \cdot f_{h.0.k} \cdot t_s \cdot d \cdot \left[\left(2 + 4 \cdot \frac{M_{y.Rk}}{f_{h.0.k} \cdot d \cdot t_s^2} \right)^{0.5} - 1 \right]$$

$$FH_{v.Rk.g} = 8.93 \times 10^3 \, \text{N}$$

Number of bolts per shear plane in the row, n_b
$$n_b = 3$$

Effective number of bolts per shear plane in the row, n_{ef}
$$n_{ef} = \min \left[n_b, n_b^{0.9} \cdot \left(\frac{100 \cdot \text{mm}}{13 \cdot d} \right)^{0.25} \right]$$

$$n_{ef} = 2.41$$

(equation (10.45); EC5, *equation (8.34)*)

(Mathcad adjusted to make it dimensionally correct.)

Design force capacity per bolt per shear plane parallel to the grain taking bolt spacing effect into account, FH_h
$$FH_h = n_{ef} \cdot \frac{1}{n_b} \cdot FH_{v.Rk.g} \cdot \frac{k_{mod.med}}{\gamma_{M.connection}}$$

$$FH_h = 4.41 \times 10^3 \, \text{N}$$

Design force component per bolt per shear plane parallel to the grain in each of the bolts in the row, FH_g

$$FH_g = \frac{MA_{SR.m.d} \cdot r_{max} \cdot \frac{y}{r_{max}}}{n_{sp} \cdot D} + \frac{H_d}{n_{sp} \cdot n_{bolt}}$$

$$FH_g = 3.87 \times 10^3 \, N$$

Actual design force component in each bolt parallel to the grain is less than the permitted design force per bolt; therefore OK

10. Shear strength of the timber at end A of beam AB

The design shear force on the timber:

Summation of 'x' distances of the bolts in the shear zone, S_x

$$S_x = 3 \cdot x$$

Maximum shear force – based on the forces in two shear planes in the line of bolts to the extreme left of the connection, $Fl_{V.v.d}$ (equation (12.19a))

$$Fl_{V.v.d} = 2 \cdot F_{m.d.max} \cdot \frac{S_x}{r_{max}} - \frac{3 \cdot RA}{n_{bolt}}$$

$$Fl_{V.v.d} = 7.11 \times 10^3 \, N$$

Design shear stress within the connection area, $\tau_{c.s}$:

$$\tau_{c.s} = \frac{3}{2} \cdot \frac{Fl_{V.v.d}}{2 \cdot k_{cr} \cdot t_s \cdot \left[h - 3 \cdot (d + 1 \cdot mm) \right]}$$

$$\tau_{c.s} = 0.91 \, N/mm^2$$

Design shear stress is less than the design shear strength and a check on the design bending stress will also show it is less than the design bending strength; therefore satisfactory.

Chapter 13

Racking Design of Multi-storey Platform Framed Wall Construction

13.1 INTRODUCTION

In this chapter, the racking strength of platform framed walls in multi-storey timber framed construction is addressed. Two methods for determining racking strength are considered: 'Method A', referred to in EC5, *9.2.4.2* and the 'Simplified analysis of wall diaphragms', referred to in PD6693-1 [1]. As it is a requirement of the UKNA to EC5 [2] that the simplified analysis method, referred to in this chapter as the PD Method, should be used in preference to EC5, *9.2.4.3*, Method B, this design procedure has not been considered.

In the platform frame form of construction, commencing from ground floor level, storey height stud walls (as referred to in Chapter 5) faced with sheathing material are erected either in situ or more commonly using prefabricated panels, and the first floor deck structure is connected to the top of these elements. This structure then forms the building platform supporting the next level of storey height wall panels onto which the second floor deck structure is connected, and by repeating the storey height wall and floor/roof erection procedure the building structure is constructed. Guidance on the detail requirements of this method of construction and on typical floor to wall details in buildings assembled this way is given in Timber Frame Construction [3]. A typical external wall in a timber framed building constructed using the platform frame method is shown in Figure 13.1.

With this form of construction, connections between walls and floors function basically as pin joints and resistance to lateral forces from wind actions has to be derived by using in-plane sheathed walls designed to function as diaphragm structures, as referred to in 9.5.

13.2 CONCEPTUAL DESIGN

With platform construction, vertical actions from the floor/roof structures are supported by the walls and where walls are aligned above each other throughout the height of the building there will be direct load paths for these actions to transfer to foundation level. To resist the effects of horizontal wind loading, as stated in 13.1, the designer must select sheathed walls that can be designed to provide adequate racking resistance at each floor level. The floor structures are designed as diaphragm

Structural Timber Design to Eurocode 5, Second Edition. Jack Porteous and Abdy Kermani.
© Jack Porteous and Abdy Kermani 2013. Published 2013 by Blackwell Publishing Ltd.

(a)

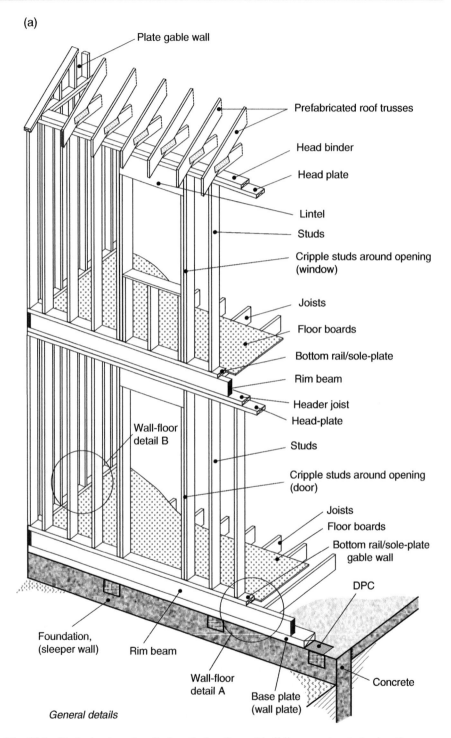

Plate gable wall

Prefabricated roof trusses

Head binder

Head plate

Lintel

Studs

Cripple studs around opening (window)

Joists

Floor boards

Bottom rail/sole-plate

Rim beam

Header joist

Head-plate

Wall-floor detail B

Studs

Cripple studs around opening (door)

Joists

Floor boards

Bottom rail/sole-plate gable wall

DPC

Foundation, (sleeper wall)

Rim beam

Wall-floor detail A

Base plate (wall plate)

Concrete

General details

Fig. 13.1. Typical external walls in a timber framed building constructed using the platform frame method.

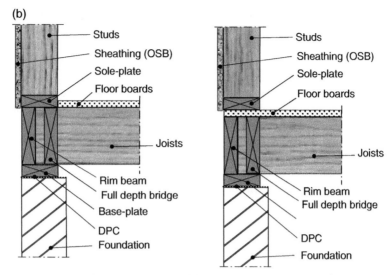

Wall-floor junction – Detail A (two typical options illustrated)

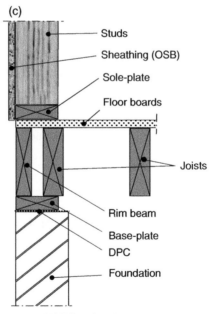

Wall-floor junction – Detail B

Fig. 13.1. (*continued*)

structures (referred to in Chapter 9) capable of transferring the wind induced lateral forces to the supporting racking walls and should there be a situation where the designer is forced to accept that a racking wall between floors is not vertically aligned with the racking wall providing vertical support from the floor below, the floor structure between these walls must be designed to resist the racking forces from the upper wall and transfer these to the supporting racking wall. As well as introducing complications in the design process, this will generally result in reduced lateral stiffness of the building structure and is something that is not taken into account in the racking procedures referred to in 13.1. In this chapter the understanding is that the plane of a racking wall will not vary throughout the height of the building.

The structure has to be designed to resist wind loading applied in two orthogonal directions and racking walls must be selected and designed to resist the loading in each of these directions. It has also to be noted that in each orthogonal direction wind loading can act from left to right and right to left configurations. For situations where the value of the wind action will be the same in either configuration and there is symmetrical vertical loading as well as a symmetrical racking wall arrangement in the orthogonal direction being considered, because of symmetry, the racking analysis need only be undertaken for wind applied in one of the configurations to obtain the design requirements applicable to both directions. If there is no symmetry in loading and/or racking wall arrangements, racking strength checks for each orthogonal direction with wind applied in both configurations will be required.

13.3 DESIGN REQUIREMENTS OF RACKING WALLS

Although different terminology is used to define a racking wall in EC5 for Method A and in PD 6693-1 for the PD Method, they essentially have the same meaning. In Method A the racking wall is formed from the summation of wall panels that will make up the wall (i.e. the racking wall) whereas the PD method defines the racking wall as the summation of wall diaphragms along its length (see Figure 13.2). The racking wall is the length of wall being considered to resist the racking force in the orthogonal direction being considered. For both methods, racking walls must comply with the requirements of *9.2.4.1* in EC5, which require that there must be no failures under any loading combination, including failure against overturning and sliding.

13.4 LOADING

Characteristic values of wind actions are derived from the design rules in BS EN 1991-1-4:2005 [4], to be read with the UKNA to BS EN 1991-1-4:2005 [5] and values of vertical loads from the effects of permanent actions and variable imposed actions are derived from the rules in BS EN 1991-1-1:2002 [6], read with the UKNA to BS EN 1991-1-1:2002 [7]. Characteristic values of permanent actions for solid timber can also be obtained from the design data given in BS EN 338:2009 [8]. Variable actions for snow loading are obtained from BS EN 1991-1-3:2003 [9], read with the UKNA to BS EN 1991-1-3:2003 [10]. Design values of the above types of action are derived as described in Chapter 2.

When designing racking walls the following states have to be checked and the loading combinations that are relevant for each will be as stated:

(i) The racking strength of the wall:
The loading combination used for this analysis must simulate the design condition where the racking structure has to resist the largest stress resultants and this will occur when there is wind loading on the structure and no favourable contribution to racking strength from the effect of variable actions. This will involve the verification of static equilibrium in conjunction with member strength, using

(a)

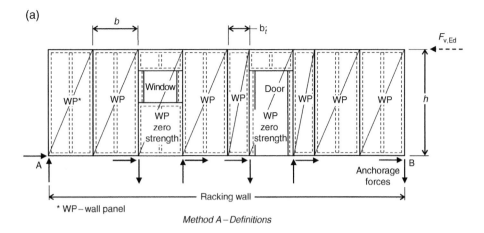

Method A – Definitions

(b)

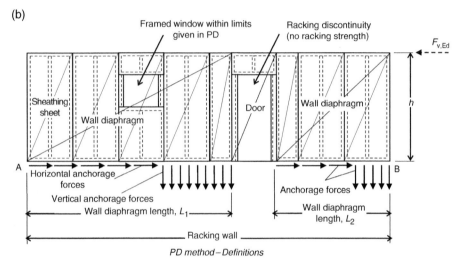

PD method – Definitions

Fig. 13.2. Alternative definitions used in Method A and PD Method for a racking wall.

the loading combination referred to in *Table NA.A1.2(A)* of the NA to BS EN 1990:2002 [11] based on *equation (6.10)* in BS EN 1990:2002 [12] and incorporating the following partial factors:

Permanent actions:
$\gamma_{Gj,sup} = 1.35$; $\gamma_{Gj,inf} = 1.15$ provided that applying $\gamma_{Gj,inf} = 1.00$ to both the favourable and unfavourable part of permanent actions does not give a more unfavourable effect. With racking calculations, $\gamma_{Gj,inf} = 1.00$ will invariably give the more unfavourable condition.

Wind actions:
$\gamma_Q = 1.5$ (unfavourable) or 0 (favourable).

(ii) The overturning stability of the wall:
The loading combination for this check will be the same as that required for a racking analysis and by designing the walls to comply with the requirements of

Method A or the PD Method, the static stability of the wall structure above ground floor level will be covered by the requirements of the racking strength calculation procedure. The only element that will have to be checked is the ability of the foundation structure to provide stability mass/strength at foundation level when the racking analysis shows there is an uplift force in the racking wall.

(iii) The stud and header/sole plate strength check:

The loading combination to be used for this design condition will be the one that maximises the forces on the studs and on bearing members. This condition will arise in leeward studs when the structure is subjected to the most unfavourable effect of vertical and lateral actions and will involve combined wind, permanent and variable actions. For internal studs, the critical condition is more likely to be due to the combined effect of permanent and variable actions with no wind present. The loading combination for either situation is referred to in *Table NA.A1.2(B)* in the NA to BS EN 1990 and in this book the combination associated with the use of *equation (6.10)* in BS EN 1990 is used. The partial factors to be used in the combination are:

Permanent actions:

$\gamma_{Gj,sup} = 1.35$; $\gamma_{Gj,inf} = 1.0$

Variable actions:

$\gamma_Q = 1.5$ (unfavourable) or 0 (favourable).

As stated above, for cases where the strength of internal stud/header/sole plates may be more critical, for this condition $\gamma_Q = 1.5$ (unfavourable) or 0 (favourable) and $= 0$ for wind actions.

(iv) The sliding resistance of the wall.

The loading combination for this check will be the same as that required for the racking strength, i.e. item (i) above.

13.5 BASIS OF METHOD A

13.5.1 General requirements

In this method the racking wall is formed from one or more wall panels, as shown in Figure 13.2a. Although not clearly stated in EC5, the sheathing to the wall panels should be wood-based panel products compliant with BS EN 13986:2004 [13] and, in the case of LVL sheathing, with BS EN 14279:2004 [14]. Softboards complying with the requirements of BS EN 622-4:2009 [15] can be used for wind bracing but, as required by EC5, *3.5*, the strength has to be verified by testing. It is not acceptable to use plasterboard as a sheathing material with this method.

A wall panel is formed by fixing a sheet of sheathing to one (or both) side(s) of stud wall framing and the minimum width, b, of sheathing able to be used to provide racking strength is $h/4$. For example, in the case of wall panel i in Figure 13.2a, h/b_i must be ≤ 4, which means if the wall is 2.4 m high the minimum width of wall sheathing (and hence wall panel) will be 600 mm. The sheathing fixings are normally nails or screws and their spacing must be uniform around the perimeter of the sheet and at intermediate studs the fixing spacing should be no greater than twice the perimeter spacing. Minimum spacing must comply with the requirements of EC5, *Chapter 8*, for the type of fastener being used and maximum spacing with the requirements of EC5, *10.8.2*.

(a)

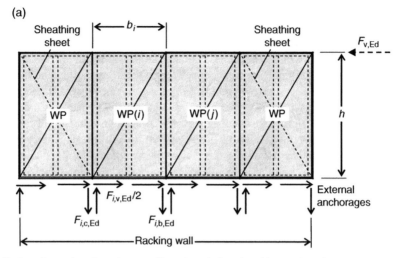

Each wall panel anchored externally at the windward and leeward studs

(b)

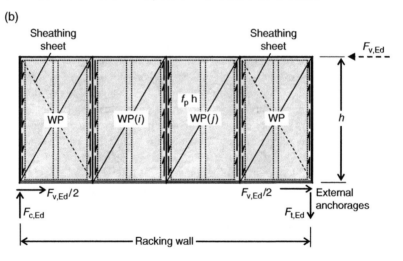

Each wall panel interconnected at the stud positions

Notes:

(1) $\sum_{i=1}^{n} F_{i,v,Ed} = F_{v,Ed}$

(2) f_p is the design value of the sheathing fastener/unit length around the perimeter of the sheathing in the wall panel.

(3) $f_p h$ is the inter wall panel shear strength.

Fig. 13.3. Alternative methods of anchoring wall panels in racking. walls.

When wall panels contain door or window openings they are not allowed to contribute to the load-carrying capacity of the racking wall and no guidance is given on minimum opening sizes below which a wall panel can be considered to be acceptable for racking purposes.

The method requires that each wall panel is fully anchored against uplift at its windward end and the EC5 requirement is that a tie-down is used to connect the windward end stud directly to the construction below by a suitable method, as shown diagramatically on Figure 13.3a. An alternative is that the windward stud is effectively anchored by vertical actions, or, if the vertical action is insufficient on its own, a combination of a limited strength anchorage plus vertical load is able to be used. When adjacent wall panels are connected to each other the racking strength of the racking wall will be the same as that obtained by using anchored wall panels. The advantage of connecting adjacent panels is that it reduces the number of base anchorages required, as shown on Figure 13.3b, and may be more cost effective and practical than fixing anchors. When connecting adjacent wall panels, to meet design requirements the design strength of the vertical connection/m height of the wall at abutting studs must not be less than the design strength/m of the fastener connecting the sheathing to the studs around the perimeter of the wall panel. Although not stated in EC5 it is recommended in this book that a head binder is used along the length of the racking wall as referred to in the Method B requirement.

13.5.2 Theoretical basis of the method

With this method, because each wall panel is anchored at its windward stud, by adopting a lower bound plastic analysis approach in accordance with the static theorem defined by B.G. Neal [16], and on the basis there will be no other form of brittle failure in the wall and its support elements, Kallsner et al. [17] have shown that when the fasteners are fixed at a uniform spacing around the perimeter of the panel and reach their failure strength, the internal force distribution shown in Figure 13.4 will comply with the requirements of the theorem and can be used to determine the racking strength. Ignoring any strength contribution from the wall panel fixings on the internal stud (which is a safe approximation), a pure shear force configuration will exist and the racking strength of the wall panel, $F_{i,v,Rk}$, and the leading stud anchor force, $F_{i,t,Rk}$, as shown on Figure 13.4 can be expressed in terms of the failure strength of the fastener/unit length of wall perimeter, f_k, as follows:

As the horizontal shear force through the height the wall panel will be uniform:

$$F_{i,v,Rk} = f_k \cdot b_i \qquad (13.1)$$

As the vertical shear force along the length of the wall panel will be uniform:

$$F_{i,t,Rk} = f_k \cdot h \qquad (13.2)$$

Also, considering the external forces on the wall panel:
$F_{i,c,Rk}=F_{i,t,Rk}=f_k h$ and the horizontal shear force to be resisted by the base fixings, assuming a fixing at each end, will be $F_{i,v,Rk}/2$.

If the wall panel is held in position by a vertical action at its windward stud, providing the value of the action exceeds $f_k \cdot h$ this will anchor the stud in position. Where there are actions along the top of the wall they can also be taken into account as as shown on Figure 13.4d. Where the action on the windward stud (which will also include any force transfer from vertical actions acting along the width of the wall panel, as shown in Figure 13.4d exceeds $f_k \cdot h$ there will be a resultant compression force$=((V2_{i,k}+(a_i \cdot V1_{i,k})/b_i)-f_k \cdot h)$ at the base of the stud. At the leeward side of the wall panel the reaction at the base of the stud will be $((b_i-a_i) \cdot V1_{i,k})/b_i+f_k \cdot h)$.

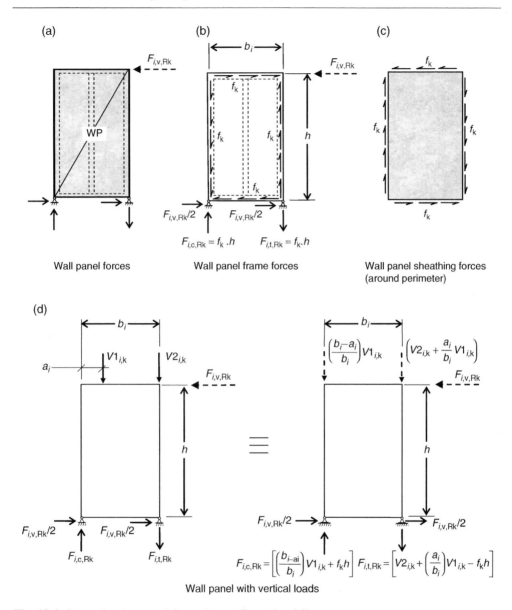

Fig. 13.4. Internal and external forces in a wall panel at failure.

When the racking wall is formed from n anchored wall panels and b_i is the length of wall panel i, the racking strength will be $\sum_{i=1}^{n} f_k \cdot b_i$. If the racking wall is made up from wall panels in which the windward stud of the windward wall panel is anchored and the remaining wall panels are interconnected at their abutting studs, the racking strength will still be $\sum_{i=1}^{n} f_k \cdot b_i$ and the required strength of the vertical connection per metre height of the racking wall at each abutting stud position will be f_k, as shown on Figure 13.5a.

Where there are vertical actions on the wall panel, as stated above these can be used to anchor the windward stud and this also applies where the racking wall is formed

(a)

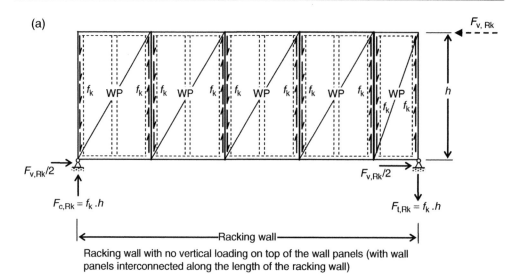

Racking wall with no vertical loading on top of the wall panels (with wall panels interconnected along the length of the racking wall)

(b)

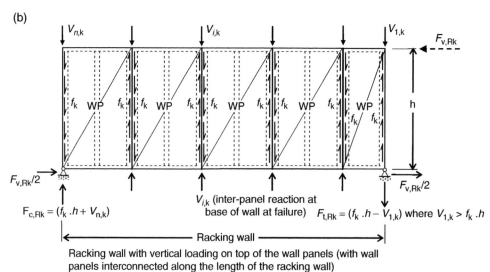

Racking wall with vertical loading on top of the wall panels (with wall panels interconnected along the length of the racking wall)

Fig. 13.5. Racking wall made up from interconnected wall panels.

using interconnected wall panels. Consider the racking wall shown in Figure 13.5b in which the vertical actions at the windward stud position of each wall panel have been derived by applying the procedure referred to in Figure 13.4d and the failure strength of the sheathing fastener/unit length of wall perimeter is f_{k}, then, at the failure condition the racking strength will be $\sum_{i=1}^{n} f_{\mathrm{k}} \cdot b_{i}$. The external vertical reactions at the base of each wall panel will be as shown on the Figure, and $V_{1,\mathrm{k}}$ must be greater than $f_{\mathrm{k}} \cdot h$.

13.5.3 The EC5 procedure

Based on the theoretical behaviour referred to in 13.5.2, the design procedure in EC5 for Method A is outlined in the following sections.

The failure strength of the perimeter fastener used in a wall panel, $F_{v,Rk}$, is derived in accordance with the relevant Johansen strength equations in EC5 and its design value, $F_{f,Rd}$, is taken to be:

$$F_{f,Rd} = \frac{k_{mod} \cdot \left(1.2 \cdot F_{v,Rk}\right)}{\gamma_M}$$ (13.3)

where

k_{mod} is the modification factor for load duration and service classes as given in *Table 3.1* in EC5 (and referred to in Chapter 2) for the design condition being considered.

$F_{v,Rk}$ is the characteristic lateral load carrying capacity of the sheathing fastener per shear plane, derived as described in Chapter 10.

γ_M is the partial coefficient for material properties, given in *Table NA 3* in the UKNA to EC5.

In equation (13.3), $F_{v,Rk}$ is multiplied by 1.2, which is a statistical factor commonly used to convert a characteristic strength value to a mean strength value. The EC5 argument is that where a significant number of fasteners are loaded in a line configuration (which is the failure model used in this method), the probability that all fasteners in the connection will only achieve the characteristic strength value is beyond the design basis and the code allows the mean strength value to be used. The edge distances of the fastener must comply with EC5 requirements and be determined on the understanding that all fastener loads act along the grain direction of the stud.

The design lateral load carrying capacity of wall panel i, $F_{i,v,Rd}$, (i.e. its design racking strength) as shown diagrammatically in Figure 13.6, is obtained from the following relationship:

$$F_{i,v,Rd} = \frac{F_{f,Rd}\, b_i\, c_i}{s}$$ (*EC5, equation (9.2)*)   (13.4)

where $F_{i,v,Rd}$ is as defined above; b_i is the length of the wall panel; s is the fastener spacing around the perimeter and $F_{f,Rd}/s$ is the design strength of the fastener/unit length of wall perimeter (i.e. the design value of the characteristic fastener strength per unit length of wall perimeter (f_k) referred to in equations (13.1) and (13.2)); c_i is a modification factor that will reduce the strength of the panel when its length is less than $h/2$, where h is the height of the wall panel.

The value of c_i is obtained from:

$$c_i = \begin{cases} 1 & for \quad b_i \geq \dfrac{h}{2} \\[2mm] \dfrac{2b_i}{h} & for \quad b_i < \dfrac{h}{2} \end{cases}$$ (based on *EC5, equation (9.22)*)   (13.5)

and has been introduced into the strength equation to prevent damage at the SLS condition.

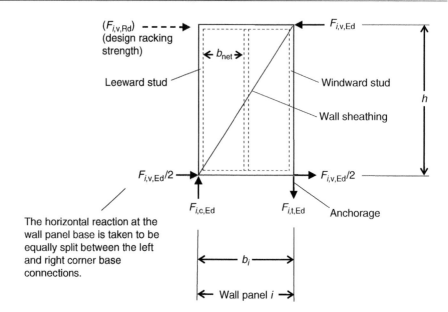

Fig. 13.6. A typical wall panel subjected to a racking force, $F_{i,v,Ed}$, and showing the external reaction forces on the panel.

The design racking strength $F_{i,v,Rd}$ of wall panel i, of width b_i and height h, is its lateral load withstand capability and can be considered to be equivalent to the maximum resistance force the panel will be capable of sustaining at its top level, as shown in Figure 13.6. The risk of the wall panel sheathing buckling under load can be ignored provided the following is complied with:

$$\frac{b_{net}}{t} \le 100 \qquad (13.6)$$

where b_{net} is the clear distance between the vertical members of the timber framing of the wall, and t is the thickness of the sheathing material.

Where wall panels are fitted with sheathing on both sides, the design racking strength of the panel will be:

(i)   where the sheathing material and the fasteners are of the same type but fastener spacing differs (i.e. s_1 one side and s_2 the other):

$$F_{i,v,Rd} = F_{f,Rd} b_i c_i \left(\frac{1}{s_1} + \frac{1}{s_2} \right) = \frac{F_{f,Rd} b_i c_i}{s1} \left(1 + \frac{s_1}{s_2} \right) \qquad (13.7)$$

(ii)   where different types of sheathing are used (sheathing 1 and 2), fixed with fasteners having similar slip moduli and the fastener strength in 1 ($F1_{f,Rd}$) is greater than the fastener strength in 2 ($F2_{f,Rd}$), unless some other value can be demonstrated to be valid:

$$F_{i,v,Rd} = b_i c_i \left(\frac{F1_{f,Rd}}{s_1} + \frac{0.75 \cdot F2_{f,Rd}}{s_2} \right)$$ (13.8)

(iii) where different types of sheathing are used (sheathing 1 and 2), fixed with fasteners having different slip moduli and the fastener strength in sheet 1 ($F1_{f,Rd}$) is greater than the fastener strength in sheet 2 ($F2_{f,Rd}$):

$$F_{i,v,Rd} = b_i c_i \left(\frac{F1_{f,Rd}}{s_1} + \frac{0.5 \cdot F2_{f,Rd}}{s_2} \right)$$ (13.9)

For a racking wall comprising several wall panels, as shown in Figure 13.7, the design racking strength, $F_{v,Rd}$, will be:

$$F_{v,Rd} = \sum_{i=1}^{n} F_{i,v,Rd}$$ (EC5, *equation (9.20)*) (13.10)

where n is the number of wall panels in the racking wall and the design racking strength of each wall panel, $F_{i,v,Rd}$, is calculated in accordance with equation (13.4), (13.7), (13.8) or (13.9), as appropriate. For the arrangement shown, where the width of each wall panel sheathing sheet $\geq h/4$ and the strength of wall panels with openings or doors (i.e. wall panels (3) and (5) on Figure 13.7) must be ignored, the design requirement will be to demonstrate that:

$$F_{v,Rd} \geq F_{v,Ed}$$

where $F_{v,Ed}$ is the design racking force on the racking wall.

For the condition $F_{v,Rd} = F_{v,Ed}$, the wall behaves in accordance with the theory developed by *Kallsner* et al., explained in 13.5.2. Where $F_{v,Rd} > F_{v,Ed}$ the wall panels will behave in an elastic or elasto-plastic manner, however in the analysis the design is undertaken assuming plastic behaviour exists.

Where a racking wall comprises wall panels, all fixed around their sheathing sheet perimeter by the same type and size of fastener and at a perimeter spacing s, equation (13.10) can be rearranged and by setting the design racking strength equal to $F_{v,Ed}$ the fastener spacing can be derived as follows:

$$s = \frac{\sum_{i=1}^{n} F_{f,Rd} b_i c_i}{F_{v,Ed}}$$ (13.11)

Where the racking walls have sheathing on both sides, by knowing the properties of the materials and the respective fixing sizes to be used and setting the fastener spacing associated with the weaker strength sheathing material (s_2) as a fraction of the fastener spacing in the stronger panel, (s_1), and $F2_{f,Rd}$ as a fraction $F1_{f,Rd}$, from equations (13.7),

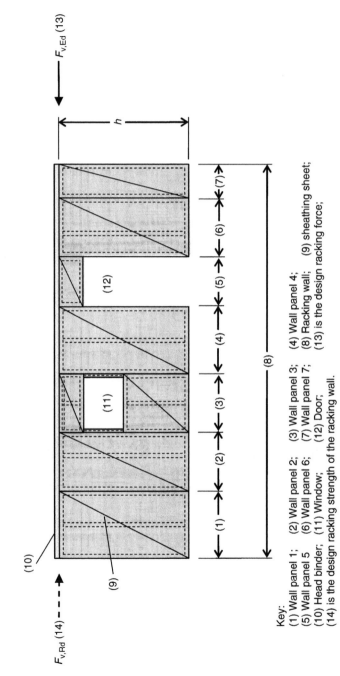

(1) Wall panel 1;   (2) Wall panel 2;   (3) Wall panel 3;   (4) Wall panel 4;
(5) Wall panel 5   (6) Wall panel 6;   (7) Wall panel 7;   (8) Racking wall;   (9) sheathing sheet;
(10) Head binder;   (11) Window;   (12) Door;   (13) is the design racking force;
(14) is the design racking strength of the racking wall.

Fig. 13.7. A wall assembly comprising several wall panels.

(13.8) or (13.9) an expression similar to that given in equation (13.11) can be set up in terms of s_1 for the relevant arrangement being considered. By adjusting the fastener spacing ratio, an acceptable spacing for s_1 and for s_2 can be determined.

The racking strength is also dependent on ensuring that the wall is anchored in accordance with the requirements of 13.5.2. Where the racking wall is formed from a single wall panel, as shown in Figure 13.6, the anchorage force at the leading edge ($F_{i,t,Ed}$) and the reaction force at the leeward edge ($F_{i,c,Ed}$) will be obtained from:

$$F_{i,c,Ed} = F_{i,t,Ed} = \frac{F_{i,v,Ed}h}{b_i} \qquad \text{(EC5, equation (9.23))} \quad (13.12)$$

Where the leading stud is loaded by the effect of downward actions, because of the plastic failure mechanism of the panel explained in 13.5.2 and shown in Figure 13.4, the value of the anchorage force will reduce and when the applied force from the actions on the stud exceed the value obtained from equation (13.12), the force at that anchorage position will be in compression and no tensile anchorage will be required. The value of $F_{i,c,Ed}$ will also be changed when the leeward stud is loaded by downward actions and for this condition the value will be obtained from the design value of the relationship shown on Figure 13.4d.

Where the racking wall comprises more than one wall panel, each wall panel should be anchored at its windward stud position and the anchorage forces in each wall panel, derived in accordance with the principles previously described will be as shown on Figure 13.8a. An alternative to this procedure is to anchor the windward stud of the leading wall panel in the racking wall and to interconnect the remaining wall panels at their abutting studs, in line with the procedure referred to in 13.5.2 and shown in Figure 13.3b. At the design condition the force configuration will be as shown on Figure 13.8b.

Both arrangements in Figure 13.8 assume the stiffness factor c_i for each wall panel in the racking wall is 1, however, where there is a wall panel for which the factor is less than 1, say wall panel i, to achieve internal wall equilibrium the maximum vertical force able to be taken in such a panel will be $\dfrac{F_{f,Rd}hc_i}{s}$ and for the condition covered in Figure 13.8b this force limit will also apply at the interconnection faces of adjacent wall panels. Theoretically this means that such panels cannot resist/transfer the vertical shear force from/to adjacent panels having a value $c=1$ and in such circumstances the panels at such locations need to be anchored. As an alternative approximate approach, as c_i is an empirical factor introduced for stiffness reasons to limit the maximum racking strength contribution from wall i, if this wall panel is detailed as a full strength panel (i.e. $c=1$), which, for production reasons, would normally be the case, the assumption can be made that it will be capable of withstanding a vertical shear force of $\dfrac{F_{f,Rd}h}{s}$ from interconnection with adjacent panels and an interconnected wall can be formed. The strength of wall panel i in the wall will still be derived using equation (13.4) with $c=c_i$ (not 1).

The racking force must also be resisted by horizontal reaction forces at the anchorage positions and for a racking wall comprising a single wall panel these forces will be as shown on Figure 13.6. As it is more common to provide an anchorage at each end of the wall rather than at one end (as has been adopted in EC5), the theory given in this book is based on the assumption the wall panel is restrained laterally at each end and

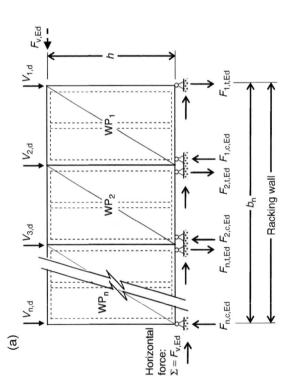

(a)

All wall panels anchored at their windward and leeward stud positions

Condition (a) – at the failure condition (and to comply with the anchorage requirements of Method A, there must be no tension forces at the leeward stud positions):

$$F_{1,t,Ed} = F_{v,Ed}\left(h/b_n\right) - V_{1,d} \qquad \text{(must be -ve)}$$

$$F_{1,c,Ed} = F_{v,Ed}\left(h/b_n\right) + V_{2,d}/2$$

$$F_{2,t,Ed} = F_{v,Ed}\left(h/b_n\right) - V_{2,d}/2 \qquad \text{(must be -ve)}$$

$$F_{2,c,Ed} = F_{v,Ed}\left(h/b_n\right) + V_{3,d}/2$$

$$F_{n,t,Ed} = F_{v,Ed}\left(h/b_n\right) + V_{n-1,d}/2 \qquad \text{(must be -ve)}$$

$$F_{n,c,Ed} = F_{v,Ed}\left(h/b_n\right) + V_{n,d}$$

Where $F_{v,Ed}/b_n \leq F_{f,Rd}/s$ and $F_{f,Rd}$ is the design load carrying capacity of the sheathing fastener and s is the fastener spacing around the perimeter of the sheathing.
The above equations have been derived for the failure condition i.e. $F_{v,Ed}/b_n = F_{f,Rd}/s$

Fig. 13.8. Anchorage forces in racking walls based on Method A.

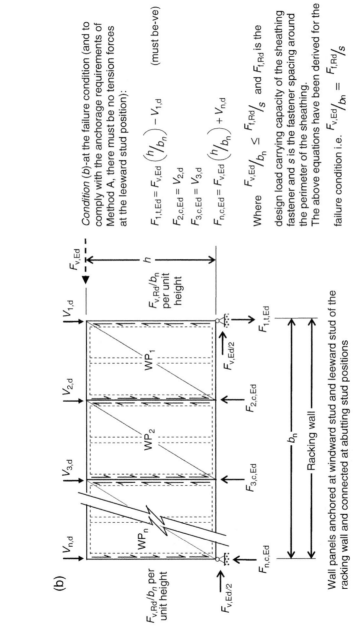

(b)

$F_{v,Rd}/b_n$ per unit height

$V_{n,d}$ $V_{3,d}$ $V_{2,d}$ $V_{1,d}$

$F_{v,Ed}$

h

$F_{v,Rd}/b_n$ per unit height

WP_n WP_2 WP_1

$F_{v,Ed/2}$

b_n

Racking wall

$F_{v,Ed/2}$

$F_{n,c,Ed}$ $F_{3,c,Ed}$ $F_{2,c,Ed}$ $F_{1,t,Ed}$

Wall panels anchored at windward stud and leeward stud of the racking wall and connected at abutting stud positions

Condition (b)-at the failure condition (and to comply with the anchorage requirements of Method A, there must be no tension forces at the leeward stud position):

$$F_{1,t,Ed} = F_{v,Ed}\left(\frac{h}{b_n}\right) - V_{1,d} \qquad \text{(must be -ve)}$$

$$F_{2,c,Ed} = V_{2,d}$$

$$F_{3,c,Ed} = V_{3,d}$$

$$F_{n,c,Ed} = F_{v,Ed}\left(\frac{h}{b_n}\right) + V_{n,d}$$

Where $\frac{F_{v,Ed}}{b_n} \leq \frac{F_{f,Rd}}{s}$ and $F_{f,Rd}$ is the design load carrying capacity of the sheathing fastener and s is the fastener spacing around the perimeter of the sheathing.

The above equations have been derived for the failure condition i.e. $\frac{F_{v,Ed}}{b_n} = \frac{F_{f,Rd}}{s}$

Fig. 13.8. (*continued*)

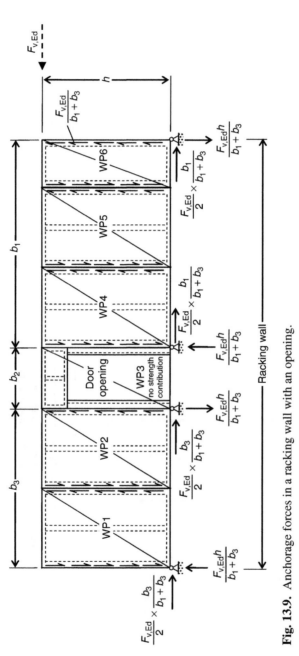

Fig. 13.9. Anchorage forces in a racking wall with an opening.

so the horizontal force to be taken at each position will be $F_{v,Ed}/2$. However, it will also be in accordance with the theoretical approach to use fixings through the base of the wall panel along its length, connecting the wall panel to its supporting structure, to resist this force. Where the racking wall is made up from a number of adjacent wall panels that are interconnected at abutting stud positions, the horizontal reaction forces will be at the ends of the racking wall as shown in Figure 13.8b. Where each wall panel in the racking wall is anchored at its ends, the designer can provide horizontal anchorage forces at the end of each wall panel as shown on Figure 13.8a and for this arrangement the horizontal force to be provided at the anchorage positions in wall panel i, of length b_i, and having a stiffness factor c_i, will be $\dfrac{F_{v,Ed} \cdot b_i \cdot c_i}{2}$. For the case where the horizontal anchorage force is provided by fixings through the base of the racking wall, the force to be resisted by these fixings per unit length of the racking wall will be $F_{v,Ed}/L$ where L is the length of the racking wall. These anchorage forces have been derived without taking into account the resistance provided by the effect of friction along the base of the racking wall. Although not stated in EC5 the friction force will contribute to the lateral resistance of the racking wall and where this is taken into account the value of the lateral force to be resisted at the anchorage positions will be reduced.

Where there is a window or door opening in a wall panel, although the panel is physically incorporated in the racking wall it will not contribute to the racking strength. In such situations the designer treats the wall panel as though there is a space in the racking wall and for the condition where interconnecting stud walls are being used, the wall is detailed as though the racking wall comprises two separate racking walls, as shown on Figure 13.9.

From the procedure described in this section the racking strength of racking walls is determined and as stated in 13.4(i) this will involve only the effects of permanent and wind actions. However, the wall studs and bearing members must also be validated against the maximum axial forces in the wall in accordance with the requirements of Chapter 5 and where they bear onto horizontal members, against the maximum compression stresses perpendicular to the grain, as described in Chapter 4. The maximum loadings on these members will be determined by analysing the wall as described in this sub-section but using the loading combination and partial factors referred to under 13.4(iii).

Example 13.8.1 demonstrates the application of the method to a three-storey structure.

13.6   BASIS OF THE RACKING METHOD IN PD6693-1

13.6.1   General requirements

In this method the racking wall is formed from one or more wall diaphragms, as shown in Figure 13.2b, with wood-based panel products compliant with BS EN 13986 forming the wall sheathing. Unlike Method A, plasterboard can be used as a sheathing material but the rules in PD 6693-1 do not permit it to be used together with wood-based panels in a racking wall. Racking walls that use only plasterboard are acceptable, but, with the exception of separating walls, their contribution to the racking strength of a building in any orthogonal direction should not exceed one third of the required racking strength. Where masonry is used to clad a timber framed building, because of the relative stiffnesses of the masonry and the timber structure, for buildings up to three storeys high, at corner areas the wind loading can be considered to be resisted by the masonry and rules are given in PD 6693-1 for taking this effect into account. On buildings with more than three storeys, no reduction in racking force from the masonry walls is permitted. The current rules lead to a very small reduction in the racking force on the timber frame, well below that previously given in relevant British Standards. No account is taken of the resistance provided by the interaction between the timber frame and the masonry arising from the masonry wall ties and this is an area where further work is required to achieve a more realistic assessment of the wind load to be taken by the timber frame.

A wall diaphragm is formed by fixing the sheathing to one (or both) side(s) of stud wall framing in which the stud spacing must not exceed 610 mm, the stud section size should not be less than 38 mm by 72 mm and must have a minimum strength class of C16, in accordance with the requirements of BS EN 338. The ends of the studs must be fixed to the top and bottom members of the frame by a minimum of two ring-shanked nails of diameter ≥ 3.1 mm that have a penetration into the stud ≥ 45 mm, or an equivalent detail. A wall diaphragm is formed from one or more wall panels, where a wall panel is formed by fixing the sheathing to a stud framework comprising only single width studs, as shown in Figure 13.10. Several sheets of sheathing can be used to form a wall panel and where wall panels abut in a diaphragm their abutting studs must be connected for the height of the wall diaphragm. Where there are no racking discontinuites within the length of the racking wall the length of the wall diaphragm and of the the racking wall will be the same. Racking discontinuites are formed when there are door openings in the racking wall or large openings (e.g. for a window) that exceed limits given in the PD. The strength contribution from areas of discontinuity is negligible and in the Method the racking strength obtained from the wall structure enveloping the length of a discontinuity is taken to be zero. A discontinuity also forms a boundary limit for a wall diaphragm and an example of this is shown on Figure 13.2b. Where there are openings within the limits defined for racking discontinuities and detailing and connection details comply with requirements also defined in the PD, the length of the wall diaphragm will not be affected but its racking strength will be reduced. For small openings in a diaphragm, compliant with limits and details that are also defined in the PD, there will be no reduction in racking strength.

Sheathing fixings are normally nails or screws, their diameter should be no greater than 0.09 times the stud thickness and their spacing must be uniform around the perimeter of the sheet and no greater than 150 mm c/c. At intermediate studs the fixing spacing should be no greater than twice the perimeter spacing and the minimum spacing must comply with the requirements of EC5, *Section 8* for the type of fastener being used. As in Method A, the forces in the perimeter sheathing fasteners are assumed to be loaded along the grain direction of the stud however minimum edge distances are less that the limits given in EC5 and are defined in the PD.

With Method A, racking walls are anchored to give full restraint against uplift at the windward stud position but in the PD Method each wall diaphragm is partially anchored along its base. The requirement is that in a wall diaphragm, where there is no net destabilising moment within the wall diaphragm, the fasteners securing the sheathing to the sole plate are designed to withstand the horizontal racking force taken by the diaphragm. Where there is a net destabilising moment, some of the fasteners will be required to resist the internal vertical forces arising from this effect with the

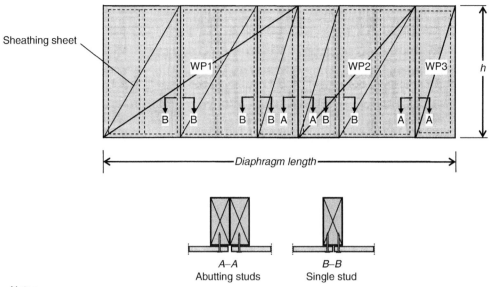

Sheathing sheet

WP1   WP2   WP3

B   B   B   B A   A B   B   A   A

h

←——————————Diaphragm length——————————→

A–A
Abutting studs

B–B
Single stud

Notes:
WP – denotes a wall in the diaphragm

Fig. 13.10. Diaphragm made up of wall panels.

remainder resisting the racking force, as shown in principle on Figure 13.2b. These internal forces must also be resisted by external forces provided by other fasteners or by holding down straps securing the wall diaphragm sole plate to the underlying structure supporting the diaphragm. Within each wall diaphragm, the shear strength of the sheathing to stud connection and, where the diaphragm includes abutting wall panels, of the abutting stud to stud connections of these panels, sets an upper limit for the vertical shear force able to be taken by the diaphragm. However, to simplify the design procedure, the implications of exceeding this limit have not been taken into account in the method, introducing an element of non-conservatism when such a condition arises.

Each wall diaphragm makes a contribution to the racking strength of the racking wall and to link the diaphragms in a racking wall a head binder, as referred to in Method B, should be used.

13.6.2   Theoretical basis of the method

The method is based on a semi-empirical approach that draws on the use of plastic theory to model the behaviour of diaphragm walls, adjusted by factors derived from the results of racking tests on wall diaphragms formed with and without openings.

Unlike the wall panels referred to in Method A, wall diaphragms are not fully anchored against uplift at their windward stud position. They are secured with fasteners along the length of the wall diaphragm base and these fixings must be capable of withstanding the internal uplift forces required to resist the overturning effect of the destabilising forces on the wall in addition to the induced internal shear and lateral forces.

Wall diaphragms can be categorised as being partially anchored walls and by adopting a lower bound plastic analysis approach in accordance with the static theorem defined by B.G. Neal, (referred to in 13.5.2) the theoretical behaviour of walls restrained by fixings along the base of the wall has also been investigated by Kallsner et al. [18, 19]. In publication [18], the destabilising (overturning) effect of the racking force is resisted by a length of the diaphragm base fixings providing resistance against uplift with the remainder providing lateral resistance to the racking force. Publication [19] is a further development of the approach and the methodology results in the wall exhibiting a greater racking strength. In both methods, overturning equilibrium is achieved by taking moments due to the external stabilising and destabilising forces about A (Figure 13.11), or, where the shear strength of the wall occurs within the wall length, about B (Figure 13.11), the position at which the vertical shear capacity of the wall is reached. An outline of the methodology used to determine the racking strength in these publications, based on the application of plastic theory and the principles of equilibrium, is summarised in Figure 13.11.

In a subsequent paper, based on development of the proposal in publication [17] and comparison with results of racking tests on panels, Kallsner et al. [20] proposed that overturning resistance could be derived by taking moments about point A when the shear capacity of the wall was exceeded, and, to comply with plastic theory behaviour, ignored any stabilising moment from loading over the length of the plastic zone of the diaphragm.

In the Kallsner methods the characteristic load carrying capacity/shear plane/unit length of the sheathing to timber fastener (f_k), is based on the mean value of the fastener strength (derived from the results of tests on panels) and for the same reason as given in 13.5.3, where the rules in EC5 are used to determine the characteristic value, in this explanation it is taken to be 1.2 times the characteristic value derived from the relevant EC5 strength rules.

Based on the approach in [20], the racking strength of a diaphragm of length L, height h, with no openings, subjected to say, a characteristic vertical uniformly distributed loading of q_k along the top of the diaphragm and within which the vertical shear capacity of the diaphragm is not reached, can be derived from:

$$F_{v,Rk} = f_k \ell_{eff} \tag{13.13}$$

where the functions are as defined on Figure 13.11a and:

$F_{v,Rk}$ is the characteristic racking strength;
f_k is 1.2 times the characteristic load-carrying capacity of the sheathing fastener/unit length, derived in accordance with the design rules in EC5;
ℓ_{eff} is the diaphragm length over which the racking force is resisted and is obtained from:

$$\ell_{eff} = -\frac{h}{\mu} + \left[\frac{h^2}{\mu^2} + L^2\left(1 + \frac{q_k}{\mu f_k}\right)\right]^{0.5} \tag{13.14}$$

The function μ ($\mu = f_{ax,Rk}/f_k$, where $f_{ax,Rk}$ is the characteristic withdrawal capacity of the fasteners securing the diaphragm base plate to the supporting structure, per unit length), enables the uplift capacity of the fasteners to be expressed in terms of f_k. The value of μ to be used in the equation will clearly be dependent on the withdrawal

Condition (a) – Analysis:

Definitions: f_k is the characteristic load carrying capacity of the perimeter sheathing fastener per unit length; μ is the following ratio: the characteristic withdrawal capacity of the bottom rail to floor connection per unit length/f_k.

(i) *Case where the vertical shear strength of the diaphragm is reached within its length L*: for length l, set up the moment equation about B due to the racking force in terms of the wall loading and the anchorage force μf_k; set up the vertical equilibrium equation for loading of the diaphragm along length l; set up the horizontal equilibrium condition $F_{v,Rk} = f_k(L - l + \ell_{eff,1})$. From these expressions $F_{v,Rk}$, l and ℓ_{eff} can be determined.

(ii) *Case where the vertical shear strength of the diaphragm is not reached within its length L*: for length L, set up the moment equation about A due to the racking force in terms of the wall loading and the anchorage force μf_k; set up the horizontal equilibrium condition $F_{v,Rk} = f_k \ell_{eff}$. From these expressions $F_{v,Rk}$ and ℓ_{eff} can be determined.

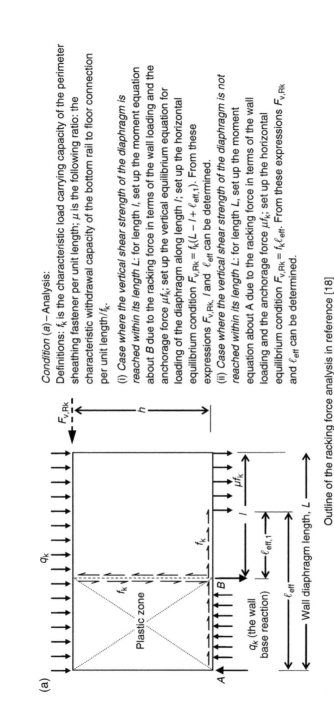

Outline of the racking force analysis in reference [18]

Fig. 13.11. Outline of the racking force analysis in references [18] and [19].

(b)

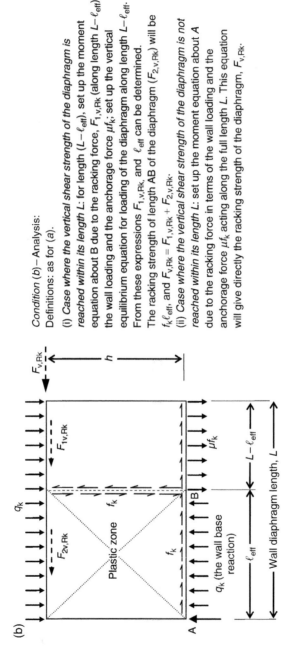

Condition (b) – Analysis:
Definitions: as for (a).

(i) *Case where the vertical shear strength of the diaphragm is reached within its length L:* for length $(L - \ell_{eff})$, set up the moment equation about B due to the racking force, $F_{1,v,Rk}$ (along length $L - \ell_{eff}$) the wall loading and the anchorage force μf_k; set up the vertical equilibrium equation for loading of the diaphragm along length $L - \ell_{eff}$. From these expressions $F_{1,v,Rk}$ and ℓ_{eff} can be determined. The racking strength of length AB of the diaphragm $(F_{2,v,Rk})$ will be $f_k \ell_{eff}$, and $F_{v,Rk} = F_{1,v,Rk} + F_{2,v,Rk}$.

(ii) *Case where the vertical shear strength of the diaphragm is not reached within its length L:* set up the moment equation about A due to the racking force in terms of the wall loading and the anchorage force μf_k acting along the full length L. This equation will give directly the racking strength of the diaphragm, $F_{v,Rk}$.

Outline of the racking force analysis in reference [19]

Fig. 13.11. (*Continued*)

strength of the diaphragm base fixings and will never be less than zero. By increasing the withdrawal capacity of the base fasteners the value of μ will increase but can never exceed 1 as at that value the strength of the sheathing fasteners will become the failure condition. The advantage of increasing the value of μ is that the racking strength of the diaphragm will be increased.

Equation (13.14), is derived from inequalities and will have the following validity requirements:

(i)   if $\ell_{eff} > L$ take $\ell_{eff} = L$
(ii)   if $\ell_{eff} < 0$ take $\ell_{eff} = 0$

Also, because a value for f_k will have been adopted in the analysis to enable equation (13.14) to be solved, situations can arise where $\ell_{eff} < L$ when $\mu = 0$. This means a length of the diaphragm base is not being used to contribute to the racking strength. In such situations, without changing the racking strength of the diaphragm, by reducing the value of f_k the value of ℓ_{eff} will be increased, with a minimum value being achieved for f_k when $\ell_{eff} = L$.

When the vertical shear capacity of the diaphragm is reached at a distance ℓ within the wall length, similar but more complex equations than (13.13) and (13.14) will be derived in terms of ℓ and $\ell_{eff,1}$.

Expressions can also be derived for cases where there are point loads along the head of the racking wall but require an iteration procedure to be used to achieve a solution.

In addition to satisfying the racking strength the lateral force resisted by the diaphragm must be able to be transferred to its supporting structure by friction or by a combination of friction and the lateral strength of the fixings connecting the diaphragm to the supporting structure. Also the tension forces over length $(L = \ell_{eff})$ have to be resisted by the supporting structure.

13.6.3   The PD6693-1 procedure

The design procedure used in the PD Method is based on the theoretical behaviour referred to in 13.6.2, and in particular the methodology used in the derivation of equations (13.13) and (13.14). There are however two areas where the Method has deviated from the theory and these relate to the approach to be used where the vertical shear capacity of a wall diaphragm is exceeded within the diaphragm length and the value to be used for the characteristic load-carrying capacity of the sheathing fastener per unit length.

Regarding the former, for those cases where the vertical shear capacity of the wall diaphragm will be reached within the diaphragm length the PD Method ignores the requirement to exclude the stabilising moment from any loading over the plastic zone of the diaphragm, introducing an element of un-conservatism for such cases. It is worth pointing out, however, that this approach has the advantage of enabling a solution to be derived for any loading configuration. The second change is that rather than use a fixed value of 1.2 times the fastener strength/shear plane/unit length in the strength derivation, from comparisons between the results of tests and from theory the fastener strength has been taken to be a function of nail spacing as well as basic strength and the 1.2 factor has been replaced by the function $(1.15 + s)$, where s is the perimeter fastener spacing in metres. For nail spacing at 50 mm the factor will equal 1.2, (i.e. the same as the value referred to in 13.6.2) and increases to a maximum of 1.3 when the spacing reaches the minimum spacing limit of 150 mm permitted in the PD Method.

Taking the above into account and using design values for relevant functions, the design value of the racking strength of a solid wall diaphragm i of height H_i, length L_i, where the sheathing is fixed around the perimeter of the diaphragm with fasteners having a design load carrying capacity in accordance with the rules in EC5 of $F_{v,Rd}$, fastener spacing of s, and subjected to a generalised loading along the top of the wall diaphragm, equations (13.13) and (13.14) become:

$$F_{i,v,Rd} = f_{pd} \ell_{i,eff} \tag{13.15}$$

$$\ell_{i,eff} = -\frac{H_i}{\mu_i} + \left[\frac{H_i^2}{\mu_i^2} + L_i^2 \left(1 + \frac{2M_{i,d}}{\mu f_{pd} L_i^2} \right) \right]^{0.5} \tag{13.16}$$

where $f_{pd} = \dfrac{(1.15 + s) F_{v,Rd}}{s}$; μ_i is the design value of μ as referred to against equation (13.14) (having the same numerical value as μ); $M_{i,d}$ for a single storey structure, is the value of the design moment at the leeward side of the diaphragm due to the vertical loading acting along the top at the wall diaphragm. The limits of validity of $\ell_{i,eff}$ remain as stated against equation (13.14).

In the PD Method, a function $K_{i,w}$ is used and for wall diaphragm i in a structure the function has been obtained from the relationship: $\ell_{i,eff} = K_{i,w} L_i$ i.e. $K_{i,w} = \ell_{i,eff} / L_i$, which, based on equation (13.16), equates to:

$$K_{i,w} = -\frac{H_i}{\mu_i L} + \left[1 + \frac{H_i^2}{\mu_i^2 L_i^2} + \frac{2M_{i,d}}{\mu_i f_{pd} L_i^2} \right]^{0.5} \tag{13.17}$$

and to comply with the limits of validity associated with $\ell_{i,eff}$, if $K_{i,w} > 1$ take $K_{i,w} = 1$ and if $K_{i,w}$ is negative, take $K_{i,w} = 0$.

Where there is more than one wall diaphragm contributing to the racking strength of the structure, the racking strength, $F_{v,Rd}$, is obtained from:

$$F_{v,Rd} = \sum_{i=1}^{i=n} F_{i,v,Rd} \tag{13.18}$$

To apply the method to multi-storey structures, the principles to be followed, and in particular the values to be used for $M_{i,d}$, are indicated on Figure 13.12 for a two storey structure and the application of the method to a three storey structure is given in Example 13.2.

As for design Method A in EC5, the risk of the wall panel sheathing buckling under load in the PD Method can be ignored provided the following is complied with:

$$\frac{b_{net}}{t} \leq 100 \tag{13.6}$$

the functions being as previously defined.

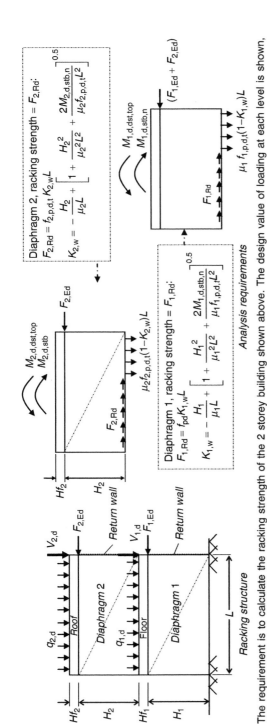

Diaphragm 2, racking strength = $F_{2,Rd}$:

$$F_{2,Rd} = f_{2,p,d,t}\, K_{2,w}L$$

$$K_{2,w} = -\frac{H_2}{\mu_2 L} + \left[1 + \frac{H_2^2}{\mu_2^2 L^2} + \frac{2M_{2,d,stb,n}}{\mu_2 f_{2,p,d,t} L^2}\right]^{-0.5}$$

$M_{2,d,dst,top}$
$M_{2,d,stb}$

$F_{2,Ed}$

$F_{2,Rd}$

$\mu_2 f_{2,p,d,t}(1-K_{2,w})L$

$V_{2,d}$ $q_{2,d}$ Roof   Diaphragm 2   Return wall $F_{2,Ed}$

$q_{1,d}$ Floor $V_{1,d}$ $F_{1,Ed}$ Return wall   Diaphragm 1

Racking structure

Hf_2 H_2 Hf_1 H_1 L

Diaphragm 1, racking strength = $F_{1,Rd}$:

$$F_{1,Rd} = f_{pd}\, K_{1,w}L$$

$$K_{1,w} = -\frac{H_1}{\mu_1 L} + \left[1 + \frac{H_1^2}{\mu_1^2 L^2} + \frac{2M_{1,d,stb,n}}{\mu_1 f_{1,p,d,t} L^2}\right]^{-0.5}$$

$M_{1,d,dst,top}$
$M_{1,d,stb,n}$

$F_{1,Rd}$

$(F_{1,Ed} + F_{2,Ed})$

$\mu_1 f_{1,p,d,t}(1-K_{1,w})L$

Analysis requirements

The requirement is to calculate the racking strength of the 2 storey building shown above. The design value of loading at each level is shown, derived from permanent actions and wind loading and the value to be used for the relevant partial factors will be as referred to in 13.4(i). The vertical loading on a diaphragm will be the design value of the load from the floor structure (or roof, as appropriate) it supports (i.e. in the example $q_{2,d}$ and $V_{2,d}$ will be the design load from loading at roof level and $q_{1,d}$ and $V_{1,d}$ will be the design load from the floor loading at that level). The respective loadings should also include the self weight of the diaphragm and any favourable permanent action from return walls that can be demonstrated by calculation to be able to be transferred to the diaphragm e.g. $V_{2,d}$ being the action from a return wall at first floor level and the strength of the fasteners connecting it to diaphragm 2 must be ≥ $V_{2,d}$. The horizontal loads, $F_{1,Ed}$ and $F_{2,Ed}$ are the wind loads from the floor and roof level diaphragms respectively. In the above equations the diaphragm heights and length are as shown; the depth of floor and roof structures are Hf_1 and Hf_2 respectively; μ is the ratio of the design withdrawal capacity of the bottom rail-to-floor connection/m length of the diaphragm/the design shear capacity of the sheathing fixings in the diaphragm per unit length ($f_{p,d,t}$), and: $M_{2,d,stb}$ —is the design stabilising moment at the leeward side of the diaphragm 2 due to the vertical loading acting along the top the diaphragm = $q_{2,d}L^2/2 + V_{2,d}L$; $M_{2,d,dst,\,top}$—is the design destabilising moment at the top level of diaphragm 2 = 0 for this example; $M_{1,d,stb}$—is the design stabilising moment at the leewardside of the diaphragm 1 due to the vertical loading acting along the top the diaphragm, plus $M_{2,d,stb} = M_{2,d,stb} + q_{1,d}L^2/2 + V_{1,d}L$; $M_{1,d,stb,n} = M_{1,d,stb} - M_{1,d,stb,\,top}$. To demonstrate racking compliance, it must be shown that $F_{2,Rd} \geq F_{2,Ed}$; $F_{1,Rd} \geq F_{2,Ed} + F_{1,Ed}$, and sheathing buckling as referred to in 13.5.3 must be satisfied. Sliding and overturning resistance at foundation level must also be checked.

Fig. 13.12. Procedure for calculating the racking strength of a multi-storey structure.

Wall diaphragms are made from one or more wall panels and as has been stated previously the abutting studs of adjacent wall panels must be connected by fasteners for the height of the diaphragm. The design load carrying capacity of the fasteners connecting the wall panels per unit height of the diaphragm must be ≥ the total design shear capacity per unit length of the perimeter sheathing fasteners contributing to the diaphragm strength ($f_{p,d,t}$). Where the wall panels in the wall diaphragm are faced with more than one sheathing sheet (e.g. sheathing on both faces or two layers on the same face etc.) the rules to be used to obtain $f_{p,d,t}$ are slightly different to those given for Method A and are;

(i)   where the sheathing material and the fasteners are of the same type with the same fastener of design strength $F_{f,Rd}$ and fastener spacing, s_1:

$$f_{p,d,t} = 1.75 F_{f,Rd} \left(\frac{1.15 + s_1}{s_1} \right) \qquad (13.19)$$

(ii)   where different types of sheathing are used on each side (sheathing 1 and 2), fixed with fasteners having different types and the fastener design strength in 1 ($F1_{f,Rd}$) is greater than the fastener design strength in 2 ($F2_{f,Rd}$):

$$f_{p,d,t} = \left(\frac{F1_{f,Rd} \left(1.15 + s_1 \right)}{s_1} + \frac{0.5 \cdot F2_{f,Rd} \left(1.15 + s_2 \right)}{s_2} \right) \qquad (13.20)$$

(iii)   where different types of sheathing are used (sheathing 1 and 2) on the same side, and the fastener strength in sheet 1 is $F1_{f,Rd}$ and in sheet 2 is $F2_{f,Rd}$:

$$f_{p,d,t} = \left(F1_{f,Rd} + 0.5 F2_{f,Rd} \right) \left(\frac{1.15 + s_1}{s_1} \right) \qquad (13.21)$$

In the PD Method the assumption is made that wall diaphragms will be fixed at their base level to the top of the supporting floor diaphragm structure (including rim beams where used) and, at their top level, to head binders (where used) and the underside of the floor/roof diaphragm structure. The floor structure between wall diaphragms, as well as the ground floor/foundation structure securing the ground floor diaphragm, must be able to resist and transfer the lateral and vertical diaphragm forces that arise at the respective levels. Through a combination of friction and the lateral resistance provided by the fixings securing the base of the diaphragm to its supporting structure over length $\ell_{i,\text{eff}}$ ($= K_{i,w} L_i$), as shown in Figure 13.13, the lateral force arising from the racking force at that level must be resisted. This force must also be able to be transferred through the floor structure in addition to the wind force in the supporting floor diaphragm. Although the lengths over which the lateral and tension forces at the head of a supporting diaphragm structure will be different to those at the base of the supported diaphragm, on the understanding that the floor depth being used does not exceed that associated with normal domestic construction, the respective lengths at the base of the supported diaphragm and at the head of its supporting diaphragm can be taken to be the same.

The lateral resistance force to be provided at the head of the supporting diaphragm will be the racking force at that level and for diaphragm 1 it will be $F_{1,Rd}$

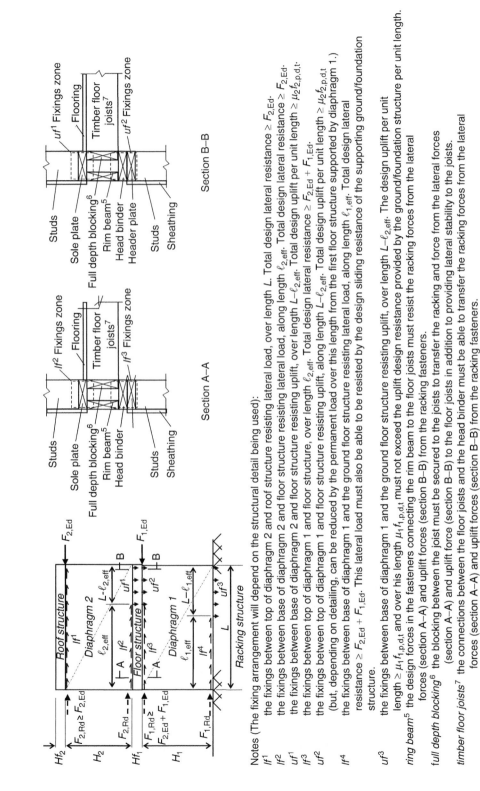

Notes (The fixing arrangement will depend on the structural detail being used):

lf^1 the fixings between top of diaphragm 2 and roof structure resisting lateral load, over length L. Total design lateral resistance $\geq F_{2,Ed}$.

lf^2 the fixings between base of diaphragm 2 and floor structure resisting lateral load, along length $\ell_{2,eff}$. Total design lateral resistance $\geq F_{2,Ed}$.

uf^1 the fixings between base of diaphragm 2 and floor structure resisting uplift, over length $L-\ell_{2,eff}$. Total design uplift per unit length $\geq \mu_2 f_{2,p,d,t}$.

lf^3 the fixings between top of diaphragm 1 and floor structure, over length $\ell_{2,eff}$. Total design lateral resistance $\geq F_{2,Ed} + F_{1,Ed}$.

uf^2 the fixings between top of diaphragm 1 and floor structure resisting uplift, along length $L-\ell_{2,eff}$. Total design uplift per unit length $\geq \mu_2 f_{2,p,d,t}$ (but, depending on detailing, can be reduced by the permanent load over this length from the first floor structure supported by diaphragm 1.)

lf^4 the fixings between base of diaphragm 1 and the ground floor structure resisting lateral load, along length $\ell_{1,eff}$. Total design lateral resistance $\geq F_{2,Ed} + F_{1,Ed}$. This lateral load must also be able to be resisted by the design sliding resistance of the supporting ground/foundation structure.

uf^3 the fixings between base of diaphragm 1 and the ground floor structure resisting uplift, over length $L-\ell_{1,eff}$. The design uplift per unit length $\geq \mu_1 f_{1,p,d,t}$ and over his length $\mu_1 f_{1,p,d,t}$ must not exceed the uplift design resistance provided by the ground/foundation structure per unit length.

ring beam[5]   the design forces in the fasteners connecting the rim beam to the floor joists must resist the racking forces from the lateral forces (section A–A) and uplift forces (section B–B) from the racking fasteners.

full depth blocking[6]   the blocking between the joist must be secured to the joists to transfer the racking and force from the lateral forces (section A–A) and uplift force (section B–B) to the floor joists in addition to providing lateral stability to the joists.

timber floor joists[7]   the connections between the floor joists and the head binder must be able to transfer the racking forces from the lateral forces (section A–A) and uplift forces (section B–B) from the racking fasteners.

Fig. 13.13. Diaphragm base and head fixing requirements in a multi-storey structure.

(i.e. $\geq F_{2,\mathrm{Ed}}+F_{1,\mathrm{Ed}}$), as shown on Figure 13.13. The friction resistance component at the design condition should be derived using a coefficient of friction of 0.4 and the remaining lateral force to be resisted over length $\ell_{i,\mathrm{eff}}$ can be taken by fixings securing the wall diaphragm base to its supporting structure.

When there is a tension force in the wall diaphragm it will occur over length $(L_i - \ell_{i,\mathrm{eff}})$ and the force to be resisted per unit length over this distance will be $\mu_i f i_{\mathrm{p,d,t}}$, as indicated on Figure 13.13. Where the lateral loading fixings referred to above are carried through the length of the tension zone, as they are not required to resist lateral loading in that zone they can be used together with additional fixings as required to resist the tension force. The tension force must be able to be resisted through the depth of the floor structure and, depending on the detailing within this depth, by the loading from the floor diaphragm structure resting on the supporting diaphragm. In Figure 13.13, the tension uplift force at the top of diaphragm 1 has been taken to equal the tension uplift at the base of diaphragm 2. At ground level the tension force must be resisted by the structure to which the diaphragm is fixed e.g. ground floor concrete slab and/or foundation structure.

For the compression forces in a wall diaphragm, where there will be a minimum of two studs within $0,1\,L$ of the leeward end of the diaphragm in a residential building of two or less storeys, no check on stud strength is required. Where strength checks have to be undertaken, an expression is given in the PD allowing the designer to determine the maximum compression load that should be designed for at the leeward end of the diaphragm, $(F_{\mathrm{c,d,leewd}})$, together with guidance on the wall zone over which this load can be considered to be supported by underlying structure within the diaphragm, including resistance offered by return walls. This enables the strength of the studs and bearing members at the most highly stressed positions to be checked under the worst loading condition, which will normally arise under the loading condition referred to in 13.4(iii). The requirement in the PD is:

$$F_{\mathrm{c,d,leewd}} \geq F_{\mathrm{cR,d}}$$
$$F_{\mathrm{c,d,leewd}} \geq 0.8W_{\mathrm{v,t,d}}\left[(M_{\mathrm{d,dst,base}}/M_{\mathrm{d,stb}})+(0.6/L)\right]$$

where

$F_{\mathrm{cR,d}}$ is the summation of the design compressive capacities of the studs in kN within $0,1\,L$ of the leeward end of the wall diaphragm. This will be the derived from the lesser of either the design buckling strength of the stud or the design bearing capacity on the horizontal framing members.

$W_{\mathrm{v,t,d}}$ is the design vertical load supported by the wall diaphragm, in kN;

$M_{\mathrm{d,stb}}$ is the design stabilising moment, in kN·m, about the leeward end of the wall diaphragm (calculated as shown in Figure 13.12)

$M_{\mathrm{d,dst,base}}$ is the design destabilizing moment, in kN·m, about the **base** of the wall diaphragm from design wind load;

L is the length of the wall diaphragm, in m.

Where there is a return wall at the leeward end of a diaphragm, provided that the construction within the depth of the floor zone supporting the diaphragm and the connection between the diaphragm and return wall can transfer the force from the return wall, up to $F_{c,d,lewd}/2$ may be redistributed to studs in the return wall that are no further than 1 m from the wall diaphragm.

The effect of small openings in wall diaphragms can be ignored and guidance is given in the PD on the sizes of openings that are permitted. Where openings exceed these limits, are framed, come within other greater limits and requirements that are also defined in the PD, the racking strength of the diaphragm must be reduced. This is achieved by the use of a modification factor $K_{opening}$, and is defined as follows:

$$K_{opening} = 1 - 1.9p \tag{13.22}$$

where $p = \dfrac{A}{HL}$ and A is the sum of 'valid' areas of the openings in the wall diaphragm.

Applying this factor, the racking strength of a wall diaphragm is obtained from:

$$F_{i,v,Rd} = K_{opening} f_{p.d.t} \ell_{i,eff} = K_{opening} K_{i,w} f_{p.d.t} L_i \tag{13.23}$$

where the functions are as previously defined.

When the opening factor is used, no guidance is given in the PD on the adjusted length over which the lateral resistance provided by the fasteners securing the wall diaphragm base can be taken to resist the lateral force on the diaphragm. In this book it is recommended a length of $K_{opening} \ell_{i,eff}$ (i.e. $K_{opening} K_{i,w} L$) is used for the design of the fastener, but the fasteners will be fitted along the full length $\ell_{i,eff}$.

In order to limit racking deflection, an empirical condition is given in the PD that must be satisfied and if exceeded the racking rules in the Method will not apply. The requirement is that the following condition should be applied for each floor diaphragm:

$$K_{i,w} f_{p,d,t} \leq 8(1 + K_{comb})(L/H)$$

where the functions are as previously defined and K_{comb} is defined in PD6693-1. For cases where there is sheathing on one face only,

$$K_{comb} = 0 \tag{13.24}$$

An example of the application of the PD Method is given in Example 13.8.2. With the approach as described in PD6693-1 the racking strength used is unable to be adjusted to equate to the racking force, resulting in a conservative solution when the racking force is below the racking strength. When Mathcad or equivalent software is being used the perimeter fastener spacing is readily able to be adjusted at minimum effort to better equate racking strength to racking force and optimise the design, and a procedure that can be used for such an analysis is outlined at the end of the example.

In timber frame construction in the UK, for designs in accordance with the rules in BS 5268-6 [21] it has been practice to use the racking strength of plasterboard faced walls and in PD6693-1 clauses are given for applying the PD Method to racking walls that are sheathed with plasterboard. The rules only apply to separating and internal plasterboard faced walls and the racking strength contribution from the latter must not exceed one third of the building racking resistance for the wind direction being considered. Values that are allowed to be used for $f_{p,d,t}$ with various specifications of plasterboard are given in the PD together with a list of the relevant rules associated with the PD Method that will apply.

See Example 13.8.2.

13.7   REFERENCES

1 PD6693-1-1:2012, Incorporating Corrigendum No1, *PUBLISHED DOCUMENT, Recommendations for the design of timber structures to Eurocode 5: Design of Timber Structures – Part 1–1: General – Common Rules and Rules for Buildings*, British Standards Institution.

2 NA to BS EN 1995-1-1:2004 + A1:2008, Incorporating National Amendment N0.2; *UK National Annex to Eurocode 5: Design of Timber Structures. Part 1-1: General – Common Rules and Rules for Buildings*, British Standards Institution.

3 Lancashire, R., Taylor, L. *Timber Frame Construction*, 5th edn, TRADA Technology Limited, High Wycombe, 2011

4 BS EN 1991-1-4:2005. *Eurocode 1: Actions on Structures – Part 1-4: General Actions – Wind Actions*, British Standards Institution.

5 NA to BS EN 1991-1-4:2005. *Actions on Structures Part 1-4: General Actions – Wind Actions*, British Standards Institution.

6 BS EN 1991-1-1:2002. *Eurocode 1: Actions on Structures. Part 1-1: General Actions – Densities, Self-Weight and Imposed Loads for Buildings*, British Standards Institution.

7 NA to BS EN 1991-1-1:2002. *UK National Annex to Eurocode 1: Actions on Structures – Part 1-1: General Actions – Densities, Self-weight, Imposed Loads for Buildings*, British Standards Institution.

8 BS EN 338:2009. *Structural Timber. Strength Classes*, British Standards Institution.

9 BS EN 1991-1-3:2003. *Eurocode 1 — Actions on Structures – Part 1-3: General Actions – Snow Loads*, British Standards Institution.

10 NA to BS EN 1991-1-3:2003. *Incorporating corrigendum no. 1, Eurocode 1 — Actions on Structures - Part 1-3: General Actions – Snow Loads*, British Standards Institution.

11 NA to BS EN 1990:2002 + A1:2005. *Incorporating National Amendment No. 1. Eurocode: Basis of Structural Design*, British Standards Institution.

12 BS EN 1990:2002 + A1:2005. *Incorporating corrigendum December 2008. Eurocode: Basis of Structural Design*, British Standards Institution.

13 BS EN 13986:2004. *Wood-based Panels for use in Construction. Characteristics, Evaluation of Conformity and Marking*, British Standards Institution.

14 BS EN 14279:2004. *Laminated Veneer Lumber (LVL) – Definitions, Classification and Specifications*, British Standards Institution.

15 BS EN 622-4:2009. *Fibreboards – Specifications, Part 4: Requirements for Softboards*, British Standards Institution.

16 Neal B. G. *The Plastic Methods of Structural Analysis*, 3rd edn, Chapman and Hall, London, 1977.

17 Källsner B., Girhammar U.A., Wu L. 'A Simplified Plastic Model for Design of Partially Anchored Wood-Framed Shear Walls', Proceedings CIB-W18 Meeting, Venice, Italy, 2001.

18 Källsner B., Girhammar U.A. 'A Plastic Lower Bound Method for Design of Wood- framed Shear Walls', Proceedings 8th World Conference on Timber Engineering, Lahti, Finland, June 14–17, 2004.

19 Källsner B., Girhammar U.A. 'Plastic design of partially anchored wood-framed wall diaphragms with and without openings', Proceedings CIB-W18 Meeting 38, Karlsruhe, Germany, 2005.

20 Källsner B., Girhammar U.A. 'Influence of framing joints on plastic capacity of partially anchored wood-framed shear walls', Proceedings CIB-W18 Meeting 37, Edinburgh, United Kingdom, 2004.

21 BS5268-6, (in 2 parts). *Structural Use of Timber – Part 6: Code of Practice for Timber Frame Walls*, (in 2 parts), British Standards Institution.

22 BS EN 12369-1:2001. *Wood-Based Panels – Characteristic Values for Structural Design. Part 1: OSB, Particleboards and Fibreboards*, British Standards Institution.

13.8 EXAMPLES

As stated in 4.3, in order to verify the ultimate and serviceability limit states, each design effect has to be checked and for each effect the largest value caused by the relevant combination of actions must be used.

However, to ensure attention is primarily focused on the EC5 design rules for the timber or wood product being used, only the design load case producing the largest design effect has generally been given or evaluated in the following examples.

Example 13.8.1 The lateral resistance of the gable wall of a three storey residential timber framed building is provided by timber stud walls faced on the outer face with 9 mm thick Oriented Strand Board (OSB) to BS EN 12369-1:2001 and on the inner face with plaster-board. The buildings is constructed using the platform method of construction, with the walls supporting a flat roof timber structure and timber floor structures as shown on Figure E13.8.1 and the racking resistance of the structure is to be designed to comply with the requirements of EC5, Method A. The wall panel height in each floor is 2.4 m and the floor structure zone is 350 mm. The floor structure forms a rigid support for wall panels, and is able to transfer vertical compression loads. The design values of floor/roof loading due to permanent actions (including permanent actions due to the self weight of the wall structure) plus the lateral wind loading at each level for the racking analysis and floor loading information arising from wind, permanent plus variable actions for checking the stud and bearing member strengths are given below the Figure. The environment in the buildings is equivalent to service class 2 conditions.

The studs are 38 mm wide by 140 mm deep, strength clas C16 timber to BS EN 338:2009 and the design lateral strength of the fastener (2.85 mm diameter nail, 50 mm long) used to connect 1.22 m wide OSB panels to the timber studding is 522 N, including for the 1.2 factor allowed in EC5, 9.2.4.2(5). All adjacent wall panels forming a racking wall are to be connected at abutting stud positions using fasteners (3.1mm diameter nails, 75 mm long) having a design capacity of 649 N and the leading studs in these walls are to be anchored to comply with the requirements of Method A. The plates at the top and bottom of the studs are 140 mm wide by 38 mm deep strength class C16 timber and are connected to supporting structure using fasteners with a design capacity of 649 N.

Based on analysis in accordance with the requirements of Method A in EC5, calculate:

(a) For each floor level, the perimeter fastener spacing required in wall panels to resist the design racking force at that level.

(b) The anchorage forces to be provided and the spacing required for the fixings connecting wall panels to supporting structure and at abutting stud positions at ground, first floor and second floor levels.

(c) The maximum axial force to be taken by a stud and bearing members at each floor level when subjected to wind loading.

The solution below applies to wind blowing from right to left of the structure shown on Figure E13.8.1. As wind can also blow from left to right the structural requirements will also apply to the mirror image of the structure when loaded in that direction.

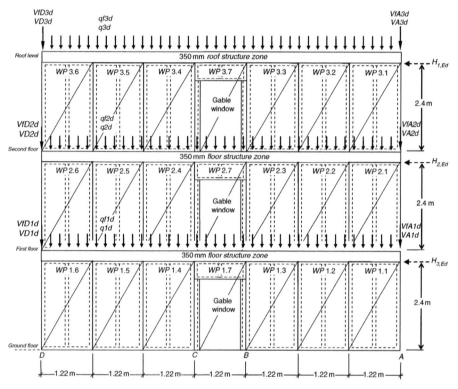

Fig. E13.8.1.

Design loading information:
Lateral wind loading: $H_{1,Ed}=6.55$ kN; $H_{2,Ed}=11.55$ kN; $H_{3,Ed}=11.55$ kN.
Floor loading for racking analysis:
    Roof level, $q3d=3.29$ kN/m, $VA3d=VD3d=1.8$ kN; Second floor level, $q2d=3.29$ kN/m, $VA2d=VD2d=1.8$ kN; First floor level, $q1d=3.29$ kN/m, $VA1d=VD1d=1.8$ kN
Floor loading for stud/bearing member analysis:
    Roof level, $qf3d=5.23$ kN/m, $VfA3d=VfD3d=2.43$ kN; Second floor level, $qf2d=6.28$ kN/m, $VfA2d=VfD2d=2.43$ kN
    First floor level, $qf1d=6.28$ kN/m, $VfA1d=VfD1d=2.43$ kN;

1. Geometric properties

Width of each stud, b	$b=38 \cdot mm$
Depth of each stud, h	$h=140 \cdot mm$
Maximum clear distance between studs, h	$b_{net}=\dfrac{1220 \cdot mm - 3 \cdot b}{2}$ $b_{net}=553\,mm$
Depth of floor structure, df	$df=350 \cdot mm$
Wall sheathing height, h_p	$h_p=2.4\,m$

Minimum sheathing width, ws $ws = 1220 \, \text{mm}$

Wall panels – all have same width, b_1 (will be the same value at each wall panel on each floor level)
    $b_1 = 1220 \, \text{mm}$ $b_2 = b_1$ $b_3 = b_1$ $b_4 = b_1$ $b_5 = b_1$ $b_6 = b_1$

Wall panel 7 – window panel, b_7 (will be the same value at each wall panel on each floor level)
    $b_7 = 1220 \, \text{mm}$

Wall panel 1 to 6 stiffness factor, c_1 (will be the same value at each wall panel on each floor level – so any adjustment when calculating the vertical shear strength at abutting wall panels will not apply)

$$c_1 = \begin{vmatrix} 1 & if & b_1 \geq \dfrac{h_p}{2} \\[2ex] \dfrac{2 \cdot b_1}{h_p} & otherwise \end{vmatrix} \qquad c_1 = 1$$

$$c_2 = c_1 \qquad c_3 = c_1 \qquad c_4 = c_1 \qquad c_5 = c_1 \qquad c_6 = c_1$$

Ratio of wall panel height/ wall length must be less than 4, r
9.2.4.2(2)
    $r = \dfrac{h_p}{b_1}$ $r = 1.97$

                                                    i.e. O.K

Length of wall 1 in racking wall at each floor level, L_1
    $L_1 = b_1 + b_2 + b_3$ $L_1 = 3.66 \, \text{m}$

Length of wall 2 in racking wall at each floor level, L_2
    $L_2 = b_4 + b_5 + b_6$ $L_2 = 3.66 \, \text{m}$

Thickness of OSB, $t_{o,s,b}$
    $t_{o.s.b} = 9.0 \cdot \text{mm}$

Fastener diameter, d_n
    $d_n = 3 \cdot \text{mm}$

Ratio of $b_{net}/t_{0,s,b}$
(Equation (13.6))
    $\dfrac{b_{net}}{t_{o.s.b}} = 61.44$ i.e. less than 100, OK.

2. Timber and nail properties, sheathing factor and other factors
Table 1.3 – C16 (BS EN 338:2009 *Table 1*):

Characteristic density of the timber frame in the wall panels, ρ_K
    $\rho_k = 310 \cdot \text{kg/m}^{-3}$

Design lateral capacity of fixing between the OSB sheathing and the timber frame (2.85mm nail diameter, 50mm long, including 1.2 factor), $Fs_{f.Rd}$
    $Fs_{f.Rd} = 522 \cdot \text{N}$

Design lateral capacity of fixing $F_{f.Rd}=649 \cdot N$
(3.1mm nail, 75mm long)
between the abutting studs in wall
panels (and also use between wall
base and a floor structure), $F_{f,Rd}$

Sheathing factor, k_n $k_n=1$ (sheathing only on one side of wall panels)

3. **Racking overturning/Sliding strength of racking wall (because of symmetry, the design requirements will apply to wind acting from right to left as well as from left to right. In this analysis the wind is taking to be blowing from right to left).**

3.1 Second floor level

Design value of the wind load at the wall head level, $H3_{v,Ed}$ $H3_{v.d}=6.55 \cdot kN$

Design value of the racking force at the wall head level, $F3_{v,Ed}$ $F3_{v.d}=H3_{v.d}$ $F3_{v.d}=6.55\times10^3 N$

Design value of the permanent uniformly distributed vertical load on wall head, $q3_d$ $q3_d=3.29 \cdot kN \cdot m^{-1}$

Design value of point load at A, $VA3_d$ $VA3_d=1.8 \cdot kN$

Design value of point load at D, $VD3_d$ $VD3_d=1.8 \cdot kN$

Fastener spacing in each wall panel at second storey level: s3 equation (13.11)

$$s3 = \frac{2 \cdot (b_1 + b_2 + b_3) \cdot Fs_{f.Rd}}{F3_{v.d}}$$ $s3=583.36\,mm$ (will use fasteners at 150 mm c/c – but retain calculated value in the analysis.)

Required anchorage force at the base of stud A3 (If the value is positive, it means the stud is in compression), $FA3b_{v,Ed}$ $FA3b_{v.d} = -\dfrac{F3_{v.d} \cdot h_p \cdot k_n}{L_1 +L_2} + VA3_d + q3_d \cdot \dfrac{b_1}{2}$

$FA3b_{v.d}=1.66\times 10^3$ N   i.e. no anchorage is required

Fastener spacing in abutting faces of wall panel 3.1/3.2, 3.2/3.3, 3.4/3.5 and 3.5/3.6 on second storey level: $s3_{12}$ Figure 13.3 $s3_{12} = \dfrac{F_{f.Rd} \cdot s3}{Fs_{f.Rd}}$ $s3_{12}=725.29\,mm$

will use fasteners at 300mm c/c

Downward reaction at the leeward end of wall panel 3,3:

$$FB3b_{v.d} = \frac{F3_{v.d} \cdot h_p \cdot k_n}{L_1 +L_2} + q3_d \cdot \frac{b_3 +b_7}{2}$$ $FB3b_{v.d}=6.16\times 10^3$ N

Required anchorage force at the base of stud C3 (If the value is positive, the stud is in compression), $FC3b_{v,Ed}$

$$FC3b_{v,d} = -\frac{F3_{v,d} \cdot h_p \cdot k_n}{L_1 + L_2} + q3_d \cdot \frac{b_4 + b_7}{2}$$

$FC3b_{v,d} = 1.87 \times 10^3$ N   i.e. no anchorage is required

Downward reaction at the leeward end of wall panel 3.6:

$$FD3b_{v,d} = \frac{F3_{v,d} \cdot h_p \cdot k_n}{L_1 + L_2} + q3_d \cdot \frac{b_6}{2} + VD3_d$$
$FD3b_{v,d} = 5.95 \times 10^3$ N

Sliding check at second storey level

Through the above calculation it is shown that the sheathing to stud fasteners at the top and bottom levels of the racking walls resist the racking force.

The condition to be checked is that the friction force plus sole plate to floor structure fixing strength can take the racking force. In this solution the lateral force will be checked ignoring friction and using only the fixings between the base of the wall panels and the floor structure:

Lateral force to be resisted/m run of wall:

$$F313_d = \frac{F3_{v,d}}{L_1 + L_2}$$
$F313_d = 894.81$ N/m

Maximum Fastener spacing required along wall base:

$$\frac{F_{f.Rd}}{F313_d} = 0.73\,\text{m}$$
use fasteners at 300 mm c/c.

3.2 First floor level

Design value of the wind load at the wall head level, $H2_{v,Ed}$ $H2_{v,d} = 11.55 \cdot$ kN

Design value of the racking force at the wall head level, $F2_{v,Ed}$ $F2_{v,d} = H3_{v,d} + H2_{v,d}$ $F2_{v,d} = 1.81 \times 10^4$ N

Design value of the permanent uniformly distributed vertical load on wall head, $q2_d$ $q2_d = 3.29 \cdot$ kN/m

Design value of point load at A, $VA2_d$ $VA2_d = 1.8 \cdot$ kN

Design value of point loat at D, $VD2_d$ $VD2_d = 1.8 \cdot$ kN

Fastener spacing in each wall panel at first floor level: s2 equation (13.11)

$$s2 = \frac{2 \cdot (b_1 + b_2 + b_3) \cdot Fs_{f.Rd}}{F2_{v,d}}$$
$s2 = 211.11$ mm
say use fasteners at 150 mm c/c.

Required anchorage force at the base of stud A2 (if the value is positive it means the stud is in compression), $FA2b_{v,Ed}$

$$FA2b_{v,d} = -\frac{F2_{v,d} \cdot h_p \cdot k_n}{L_1 + L_2} + VA2_d + q2_d \cdot \frac{b_1}{2} + FA3b_{v,d}$$

$$FA2b_{v.d} = -468.17 \text{ N}$$ i.e. the leading stud must be anchored to the leading wall stud on the first floor wall panel immediately below to resist this value of force

Fastener spacing in abutting faces of wall panel 2.1/2.2, 2.2/2.3, 2.4/2.5 and 2.5/2.6 on second floor level: $s2_{12}$

Figure 13.3

$$s2_{12} = \frac{F_{f.Rd} \cdot s2}{Fs_{f.Rd}}$$ $s2_{12} = 262.47 \text{ mm}$

Say 260 mm c/c

Downward reaction at the leeward end of wall panel 2.3:

$$FB2b_{v.d} = \frac{F2_{v.d} \cdot h_p \cdot k_n}{L_1 + L_2} + q2_d \cdot \frac{b_3 + b_7}{2} + FB3b_{v.d}$$ $FB2b_{v.d} = 1.61 \times 10^4 \text{ N}$

Required anchorage force at the base of stud C2 (if the value is positive it means the stud is in compression), $FC2b_{v.Ed}$

$$FC2b_{v.d} = -\frac{F2_{v.d} \cdot h_p \cdot k_n}{L_1 + L_2} + q2_d \cdot \frac{b_4 + b_7}{2} + FC3b_{v.d}$$

$$FC2b_{v.d} = -54.37 \text{ N}$$ i.e. the leading stud must be anchored to the leading wall stud on the first floor wall panel immediately below to resist this value of force

Downward reaction at the leeward end of wall panel 2.6:

$$FD2b_{v.d} = \frac{F2_{v.d} \cdot h_p \cdot k_n}{L_1 + L_2} + q2_d \cdot \frac{b_6}{2} + VD2_d + FD3b_{v.d}$$ $FD2b_{v.d} = 1.57 \times 10^4 \text{ N}$

Sliding check at first floor level

Through the above calculation it is shown that the sheathing to stud fasteners will at the top and bottom levels of the wall resist the racking force.

The condition to be checked is that the friction force plus sole plate to floor structure fixing strength can take the racking force. In this solution the lateral force will be checked ignoring friction and using only the fixings between the base of the wall panels and the floor structure:

Lateral force to be resisted/m run of wall:

$$F213_d = \frac{F2_{v.d}}{L_1 + L_2}$$ $F213_d = 2.47 \times 10^3 \text{ N/m}$

Maximum fastener spacing required along wall base:

$$\frac{F_{f.Rd}}{F213_d} = 0.26 \text{ m}$$

use fasteners at 260 mm c/c. Head of diaphragm also to be connected to soffit level of second floor structure by these fasteners at 260 mm c/c.

3.3 Ground floor level

Design value of the wind load at the wall head level, $H1_{\text{v,Ed}}$

$H1_{\text{v.d}} = 11.55 \cdot \text{kN}$

Design value of the racking force at the wall head level, $F1_{\text{v,Ed}}$

$F1_{\text{v.d}} = H3_{\text{v.d}} + H2_{\text{v.d}} + H1_{\text{v.d}}$ $F1_{\text{v.d}} = 2.96 \times 10^4 \text{ N}$

Design value of the permanent uniformly distributed vertical load on wall head, $q1_{\text{d}}$

$q1_{\text{d}} = 3.29 \cdot \text{kN/m}$

Design value of point load at A, $VA1_{\text{d}}$

$VA1_{\text{d}} = 1.8 \cdot \text{kN}$

Design value of point load at D, $VD1_{\text{d}}$

$VD1_{\text{d}} = 1.8 \cdot \text{kN}$

Fastener spacing in each wall panel at ground floor level: $s1$
equation (13.11)

$$s1 = \frac{2 \cdot (b_1 + b_2 + b_3) \cdot Fs_{\text{f.Rd}}}{F1_{\text{v.d}}}$$ $s1 = 128.87\,\text{mm}$ say use fasteners at 100 mm c/c.

Required anchorage force at the base of stud A1 (if the value is negative, it means the stud is in compression), $FA1b_{\text{v,Ed}}$

$$FA1b_{\text{v.d}} = -\frac{F1_{\text{v.d}} \cdot h_{\text{p}} \cdot k_{\text{n}}}{L_1 + L_2} + VA1_{\text{d}} + q1_{\text{d}} \cdot \frac{b_1}{2} + FA2b_{\text{v.d}}$$

$FA1b_{\text{v.d}} = -6.38 \times 10^3 \text{ N}$ i.e. anchorage is required

Fastener spacing in abutting faces of wall panel 1.1/1.2, 1.2/1.3, 1.4/1.5 and 1.5/1.6 on ground floor level: $s1_{12}$ Figure 13.3

$$s1_{12} = \frac{F_{\text{f.Rd}} \cdot s1}{Fs_{\text{f.Rd}}}$$ $s1_{12} = 160.23\,\text{mm}$

say 160 mm c/c

Downward reaction at the leeward end of wall panel 1.3:

$$FB1b_{\text{v.d}} = \frac{F1_{\text{v.d}} \cdot h_{\text{p}} \cdot k_{\text{n}}}{L_1 + L_2} + q1_{\text{d}} \cdot \frac{b_3 + b_7}{2} + FB2b_{\text{v.d}}$$ $FB1b_{\text{v.d}} = 2.98 \times 10^4 \text{ N}$

Required anchorage force at the base of stud C1 (If the value is positive it means the stud is in compression), $FC1_{\text{v,Ed}}$

$$FC1b_{\text{v.d}} = -\frac{F1_{\text{v.d}} \cdot h_{\text{p}} \cdot k_{\text{n}}}{L_1 + L_2} + q1_{\text{d}} \cdot \frac{b_4 + b_7}{2} + FC2b_{\text{v.d}}$$

$FC1b_{\text{v.d}} = -5.76 \times 10^3 \text{ N}$ i.e. anchorage is required

Downward reaction at the leeward end of wall panel 1,6:

$$FD1b_{\text{v.d}} = \frac{F1_{\text{v.d}} \cdot h_{\text{p}} \cdot k_{\text{n}}}{L_1 + L_2} + q1_{\text{d}} \cdot \frac{b_6}{2} + VD1_{\text{d}} + FD2b_{\text{v.d}}$$ $FD1b_{\text{v.d}} = 2.92 \times 10^4 \text{ N}$

Sliding check at ground floor level

Through the above calculation it is shown that the sheathing to stud fasteners will at the top and bottom levels of the wall resist the racking force.

The condition to be checked is that the friction force plus sole plate to floor structure fixing strength can take the racking force. In this solution the lateral force will be checked ignoring friction and using only the fixings between the base of the wall panels and the floor structure:

Lateral force to be resisted/m run of wall:

$$F113_d = \frac{F1_{v.d}}{L_1 + L_2} \qquad F113_d = 4.05 \times 10^3 \text{ N/m}$$

Maximum Fastener spacing required along wall base:

$$\frac{F_{f.Rd}}{F113_d} = 0.16 \text{ m}$$

use fasteners at 160 mm c/c. Head of diaphragm also to be connected to soffit level of second floor structure by these fasteners at 160 mm c/c.

Overturning

The racking structure comprises two walls, each 4.2 m long, and the leading stud must be anchored (or effectively held down by vertical loading on top of the leading wall panel) and by meeting this requirement overtuning stability will be assured. The critical requirement is to ensure the anchorage force at foundation level is able to be resisted by the mass of the foundation structure.

Foundation force/mass to be provided at position A to form anchorage.

$$FA1b_{v.d} = -6.38 \times 10^3 \text{ N}$$

i.e. $MassA = \dfrac{-FA1b_{v.d}}{g} \qquad MassA = 650.84$ kg

Foundation force/mass to be provided at position C to form anchorage.

$$FC1b_{v.d} = -5.76 \times 10^3 \text{ N}$$

i.e. $MassC = \dfrac{-FC1b_{v.d}}{g} \qquad MassC = 587.55$ kg

4 Forces on studs and wall plates

(The strength of the stud and of the bearing members has to be checked against the force on these members. If the stud force exceeds the stud strength the normal practice is to double up the stud (or use triple or more stud if required) connected along the height of the studs by fasteners. Also, where the bearing capacity of the header/footer wall plates is exceeded the use of double (or more abutting interconnected studs) at these positions, as required, will increase the bearing area to a value that exceeds the bearing force.)

The maximum stud (and plate) loading will be associated with load case 13.4. (iii) and the loadings for this condition are as follows.

4.1 Second floor level

Design value of the permanent uniformly distributed vertical load on wall head, $qf3_d$

$qf3_d = 5.23 \cdot \text{kN/m}^{-1}$

Design value of point load at A, $VfA3_d$

$VfA3_d = 2.43 \cdot \text{kN}$

Design value of point load to D, $VfD3_d$

$VfD3_d = 2.43 \cdot \text{kN}$

Downward reaction at the leeward end of wall panel 3.3:

$$FfB3b_{v.d} = \frac{F3_{v.d} \cdot h_p \cdot k_n}{L_1 + L_2} + qf3_d \cdot \frac{b_3 + b_7}{2} \qquad\qquad FfB3b_{v.d} = 8.53 \times 10^3 \text{ N}$$

Downward reaction at the leeward end of wall panel 3.6:

$$FfD3b_{v.d} = \frac{F3_{v.d} \cdot h_p \cdot k_n}{L_1 + L_2} + qf3_d \cdot \frac{b_6}{2} + VfD3_d \qquad\qquad FfD3b_{v.d} = 7.77 \times 10^3 \text{ N}$$

4.2 First Floor level

Design value of the permanent uniformly distributed vertical load on the wall head, $qf2_d$ $qf2_d = 6.28 \cdot \text{kN/m}$

Design value of point load at A, $VfA2_d$ $VfA2_d = 2.43 \cdot \text{kN}$

Design value of point load at D, $VfD2_d$ $VfD2_d = 2.43 \cdot \text{kN}$

Downward reaction at the leeward end of wall panel 2.3:

$$FfB2b_{v.d} = \frac{F2_{v.d} \cdot h_p \cdot k_n}{L_1 + L_2} + qf2_d \cdot \frac{b_3 + b_7}{2} + FfB3b_{v.d} \qquad FfB2b_{v.d} = 2.21 \times 10^4 \text{ N}$$

Downward reaction at the leeward end of wall panel 2.6:

$$FfD2b_{v.d} = \frac{F2_{v.d} \cdot h_p \cdot k_n}{L_1 + L_2} + qf2_d \cdot \frac{b_6}{2} + VfD2_d + FfD3b_{v.d} \qquad FfD2b_{v.d} = 2 \times 10^4 \text{ N}$$

4.3 Ground Floor level

Design value of the permanent uniformly distributed vertical load on wall head, $qf1_d$ $qf1_d = 6.28 \cdot \text{kN/m}^{-1}$

Design value of point load at A, $VfA1_d$ $VfA1_d = 2.43 \cdot \text{kN}$

Design value of point load at D, $VfD1_d$ $VfD1_d = 2.43 \cdot \text{kN}$

Downward reaction at the leeward end of wall panel 3:

$$FfB1b_{v.d} = \frac{F1_{v.d} \cdot h_p \cdot k_n}{L_1 + L_2} + qf1_d \cdot \frac{b_3 + b_7}{2} + FfB2b_{v.d} \qquad FfB1b_{v.d} = 3.95 \times 10^4 \text{ N}$$

Downward reaction at the leeward end of wall panel 6:

$$FfD1b_{v.d} = \frac{F1_{v.d} \cdot h_p \cdot k_n}{L_1 + L_2} + qf1_d \cdot \frac{b_6}{2} + VfD1_d + FfD2b_{v.d} \qquad FfD1b_{v.d} = 3.59 \times 10^4 \text{ N}$$

Applying the design requirements in chapter 5, the 38 mm by 140 studs will be acceptable generally but at the leeward end of wall panel 1,3 and wall panel 1,6, because of bearing requirements, three studs connected up their height using the stud to stud fastener at 160 mm c/c should be used.

Example 13.8.2. The lateral resistance of the gable wall of a three storey residential timber framed building is provided by timber stud walls faced on the outer face with 9 mm thick Oriented Strand Board (OSB) to BS EN 12369-1:2001 and on the inner face with plasterboard. The buildings is constructed using the platform method of construction and the wall supports a flat roof timber structure and timber floor structures as shown on Figure E13.8.2 and at each floor level the racking wall, which comprises two diaphragms, is anchored along the base in accordance with the requirements of the simplified method of analysis referred to in PD6693-1. The wall panel height in each floor is 2.4 m and the floor structure zone is 350 mm. The floor structure forms a rigid support for wall panels, and is able to transfer vertical compression loads. The design values of floor/roof loading due to permanent actions (including permanent actions due to the self weight of the wall structure) plus the lateral wind loading at each level for the racking analysis and floor loading information arising from wind, permanent plus variable actions for checking the stud and bearing member strengths are as given in Example E13.8.1 and the environment in the building is equivalent to service class 2 conditions. In addition to the gable window (also present in Example E.13.8.1) there is a 900 mm by 900 mm opening in one of the diaphragm walls on each floor level as shown in Figure E13.8.2 (i.e. within the requirements of 21.2.2, PD6693-1).

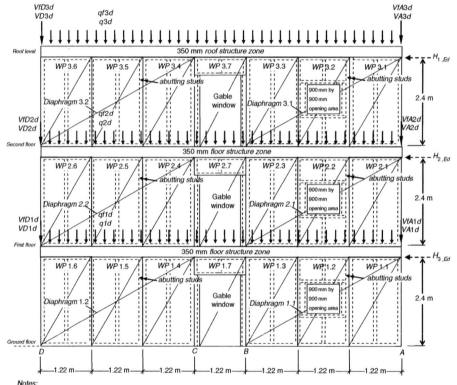

Notes:
Design loading information will be as given on Figure E13.8.1.
The connection between adjacent wall panels at each storey will be as shown on Figure 13.10, B–B unless defined as abutting studs, in which case Figure 13.10, A–A will apply.
The 900 mm by 900 mm openings in the wall diaphragms on each floor will comply with the framed opening limits given in PD6693-1 allowing the $K_{opening}$ factor referred to in equation (13.22) to be used.
Fastener spacing on intermediate studs will be 2 times the perimeter spacing (complying with the maximum spacing criteria in PD6693-1).

Fig. E13.8.2.

The studs are 38 mm wide by 140 mm deep, strength class C16 timber to BS EN 338:2009 and the design lateral strength of the fastener used to connect 1.22 m wide OSB panels to the timber studing is 435 N and complies with the maximum size criteria in PD6693-1. At abutting stud positions in wall panels in diaphragms, fasteners having a design capacity of 649 N shall be used. The plates at the top and bottom of the studs are 140 mm wide by 38 mm deep strength class C16 timber.

Based on the simplified analysis procedure referred to in PD6693-1, calculate:

(a) For each floor level, the perimeter fastener spacing required in the wall panels to resist the design racking force at that level.
(b) The anchorage forces to be provided and the spacing required for the fixings connecting wall panels at abutting stud positions at ground, first floor and second floor levels.
(c) The maximum axial force to be taken by studs and bearing members at ground floor level when subjected to permanent loading plus wind loading.

The solution below applies to wind blowing from right to left of the structure shown on Figure E13.8.2.

1. Geometric properties

Width of each stud, b	$b = 38 \cdot mm$	
Depth of each stud, h	$h = 140 \cdot mm$	
Maximum clear distance between studs, b_{net}	$b_{net} = \dfrac{1220 \cdot mm - 3 \cdot b}{2}$	$b_{net} = 0.553\,m$
Depth of floor structure, df	$df = 350 \cdot mm$	
Wall sheathing height, h_p	$h_p = 2.4\,m$	
Minimum sheathing width, ws	$ws = 1220\,mm$	
Wall panels – all have same width, b_1 (will be the same value at each wall panel on each floor level)	$b_1 = 1.22\,m \quad b_2 = b_1 \quad b_3 = b_1$ $b_4 = b_1 \quad b_5 = b_1 \quad b_6 = b_1$	
Wall panel 7 - window panel, b_7 (will be the same value at each wall panel b_7 on each floor level and acts as a discontinuity along the gable wall)	$b_7 = 1220\,mm$ All wall panels are 1220 mm long and so will comply with the rules in PD6693-1	
Length of diaphragm 1 in racking wall at each floor level (A-B), L_1	$L_1 = b_1 + b_2 + b_3$	$L_1 = 3.66\,m$
Length of diaphragm 2 in racking wall at each floor level (C-D), L_2	$L_2 = b_4 + b_5 + b_6$	$L_2 = 3.66\,m$

Thickness of OSB, $t_{o.s,b}$ $t_{o.s.b} = 9.0 \cdot \text{mm}$

Ratio of $b_{net}/t_{o,s,b}$
(equation (13.6)) $\dfrac{b_{net}}{t_{o.s.b}} = 61.444$ i.e. less than 100, O.K.

2. Timber and nail properties, and factors

Table 1.3 – C16 (BS EN 338:2009, *Table 1*):

Characteristic density of the timber frame in the wall panels, ρ_K $\rho_k = 310 \cdot \text{kg/m}^3$

Design lateral capacity of fixing between the OSB sheathing and the timber frame (2.85 mm nail, 50 mm long – from the design rules in Chapter 8, $Fs_{f.Rd}$ $Fs_{f.Rd} = 435 \cdot \text{N}$

Design lateral capacity of fixing to be used between the abutting studs in wall panels (and between diaphragm base and a floor structure) (3.1 mm nail, 75 mm long) – from the design rules in Chapter 8, $F_{f,Rd}$ $F_{f.Rd} = 649 \cdot \text{N}$

Design withdrawal capacity of the diaphragm base to floor structure connection from the design rules in Chapter 8, $F_{ax,Rd}$ $F_{ax.Rd} = 186 \cdot \text{N}$

Sheathing factor – will not apply as sheathing is only fixed to one side. Contribution from plasterboard on the other face is not permitted under the rules in PD6693-1.

Opening area in each diaphragm 1, A $A = 0.9 \cdot \text{m} \cdot 0.9 \cdot \text{m}$ $A = 0.81 \, \text{m}^2$

Opening area/diaphragm 1 area, p $p = \dfrac{A}{h_p \cdot L_1}$ $p = 0.092$

Opening factor for diaphragm 1, $K_{opening}$ (Equation 13.22) $K_{opening} = 1 - 1.9 \cdot p$ $K_{opening} = 0.825$

Coefficient of friction, μ $\mu = 0.4$

3. Racking/overturning/Sliding strength of racking wall (In this analysis the wind is taking to be blowing from right to left).

3.1 Second floor level

Design value of the wind load at the wall head level, $H3_{v,Ed}$ $H3_{v.d} = 6.55 \cdot \text{kN}$

Design value of the racking force at the wall head level, $F3_{v,Ed}$

$$F3_{v.d} = H3_{v.d}$$

$$F3_{v.d} = 6.55 \times 10^3 \text{ N}$$

Design value of the permanent uniformly distributed vertical load on wall head, $q3_d$

$$q3_d = 3.29 \cdot \text{kN/m}$$

Design value of the point load at A, $VA3_d$

$$VA3_d = 1.8 \cdot \text{kN}$$

Design value of point load at D, $VD3_d$

$$VD3_d = 1.8 \cdot \text{kN}$$

Design value of stabilizing moment on diaphragm 3.1 due to vertical loading, $M31_{d.stb}$

$$M31_{d.stb} = VA3_d \cdot L_1 + q3_d \cdot \frac{L_1^2}{2}$$

$$M31_{d.stb} = 28.624 \text{ kN} \cdot \text{m}$$

Design value of stabilizing moment on diaphragm 3.2 due to vertical loading $M32_{d.stb}$

$$M32_{d.stb} = \left[q3_d \cdot \left(\frac{b_7}{2} \right) \right] \cdot L_2 + q3_d \cdot \frac{L_2^2}{2}$$

$$M32_{d.stb} = 29.381 \text{ kN} \cdot \text{m}$$

Adopting minimum fastener spacing around perimeter of sheathing in diaphragm 3.1, $s31$

$$s31 = 0.15 \cdot \text{m}$$

Design shear capacity of perimeter fastener/m in diaphragm 3.1, $f31_{pdt}$ (equation (13.16))

$$f31_{pdt} = \frac{Fs_{f.Rd}}{s31} \cdot \left(1.15 + \frac{s31}{\text{m}} \right)$$

$$f31_{pdt} = 3.77 \text{ kN/m}$$

Initial estimate of base fastener spacing to floor structure for 3.1 and 3.2, $sb3$

$$sb3 = 300 \text{ mm}$$

Mu factor to be used for base fixings of diaphragm 3.1 and 3.2 to floor, $\mu31$

$$\mu31 = \min \left(1, \frac{\frac{F_{ax.Rd}}{sb3}}{f31_{pdt}} \right)$$

$$\mu31 = 0.164$$

Evaluation of $K_{i,w}$ factor for diaphragm 3.1, $K31_w$ (equation 13.17, and Figure 13.12)

$$K31_w = \min \left[1, \left[\left[1 + \left(\frac{h_p}{\mu31 \cdot L_1} \right)^2 + 2 \cdot \frac{M31_{d.stb}}{\mu31 \cdot f31_{pdt} \cdot L_1^2} \right]^{0.5} - \frac{h_p}{\mu31 \cdot L_1} \right] \right]$$

$$K31_w = 0.89$$

Racking strength of wall diaphragm 3.1, $F31_{v,Rd}$ (equation 13.23)

$$F31_{vRd} = K_{opening} \cdot K31_w \cdot f31_{pdt} \cdot L_1$$

$$F31_{vRd} = 10.133 \text{ kN}$$

Adopting the same value for the equivalent factors in diaphragm 3.2:

$$K32_w = \min\left[1, \left[\left[1 + \left(\frac{h_p}{\mu31 \cdot L_2} \right)^2 + 2 \cdot \frac{M32_{d.stb}}{\mu31 \cdot f31_{pdt} \cdot L_2^2} \right]^{0.5} - \frac{h_p}{\mu31 \cdot L_2} \right] \right]$$

$$k32_w = 0.909$$

Racking strength of wall diaphragm 3.2, $F32_{v.Rd}$

$$F32_{vRd} = K32_w \cdot f31_{pdt} \cdot L_2$$

$$F32_{vRd} = 12.543 \text{ kN}$$

Racking strength of racking wall, $F3_{v.Rd}$

$$F3_{vRd} = K31_{vRd} + F32_{vRd}$$

$$F3_{vRd} = 2.268 \times 10^4 \text{ N}$$

Racking force/racking strength on racking wall at level 3

$$\frac{F3_{v.d}}{F3_{vRd}} = 0.289$$

i.e. less than 1 – OK

(At the end of the example see the revised procedure that can be used to optimize on the design when Mathcad or equivalent software is being used.)

Spacing of fixings in abutting wall panel studs in both diaphragms, $sp3st$

$$sp3st = \frac{F_{f.Rd}}{f31_{pdt}}$$

$$sp3st = 0.172 \text{ m}$$
say 170 mm c/c

Length of 3.1 resisting uplift, $L31_{up}$

$$L31_{up} = (1 - K31_w) \cdot L_1$$

$$L31_{up} = 0.401 \text{ m}$$

Length of 3.1 resisting racking force

$$K_{opening} \cdot (L_1 - L31_{up}) = 2.688 \text{ m}$$

Length of 3.2 resisting uplift, $L32_{up}$

$$L32_{up} = (1 - K32_w) \cdot L_2$$

$$L32_{up} = 0.333 \text{ m}$$

Length of 3.2 resisting racking force

$$(L_2 - L32_{up}) = 3.327 \text{ m}$$

Determine lateral spacing of diaphragm base fixings to floor over racking force length $K31_w L_1$ and $K32_w L_2$, $sb3b$

$$sb3b = \left[\frac{F_{f.Rd}}{\frac{F31_{vRd}}{K_{opening} \cdot (L_1 - L31_{up})}} \right]$$

$$sb3b = 0.172 \text{ m}$$

i.e. use 150 mm c/c (by setting opening factor = 1, the spacing over length $K32_w L_2$ can be increased.)

Lateral spacing of diaphragm base fixings over uplift length of each diaphragm

$$sb3 = 0.3 \text{ m}$$

or continue with $sb3b$ in these lengths.

Check sliding condition at wall base to floor connection:

Frictional force available, $F3_{fric}$

$$F3_{fric} = \mu \cdot [2 \cdot VD3_d + q3_d \cdot (L_1 + L_2 + b_7)]$$

$$F3_{fric} = 12.679 \text{ kN}$$

$$\frac{F3_{fric}}{F3_{v.d}} = 1.936$$

i.e. greater than $F3_{v.d}$. O.K. without need to draw on base fixing strength.

Check compliance with racking deflection limit in PD6693-1:

$$\frac{K31_\text{w} \cdot f31_\text{pdt}}{8 \cdot (kN \cdot m^{-1}) \left(\dfrac{L_1}{h_\text{p}}\right)} = 0.275 \quad \text{and} \quad \frac{K32_\text{w} \cdot f31_\text{pdt}}{8 \cdot (kN \cdot m^{-1}) \left(\dfrac{L_2}{h_\text{p}}\right)} = 0.281 \qquad \begin{array}{l}\text{both less than 1}\\ \text{therefore OK}\end{array}$$

3.2 First floor level

Design value of the wind load at the wall head level, $H2_\text{v,Ed}$ $\qquad H2_\text{v.d} = 11.55 \cdot kN$

Design value of the racking force at the wall head level, $F2_\text{v,Ed}$ $\qquad F2_\text{v.d} = H2_\text{v.d} + H3_\text{v.d}$ $\qquad\qquad F2_\text{v.d} = 1.81 \times 10^4 \, N$

Design value of the permanent uniformly distributed vertical load on wall head, $q2_\text{d}$ $\qquad q2_\text{d} = 3.29 \cdot kN/m$

Design value of point load at A, $VA2_\text{d}$ $\qquad VA2_\text{d} = 1.8 \cdot kN$

Design value of point load at D, $VD2_\text{d}$ $\qquad VD2_\text{d} = 1.8 \cdot kN$

Design value of stabilizing moment on diaphragm 2.1 due to vertical loading, $M21_\text{d.stb}$

$$M21_\text{d.stb} = VA2_\text{d} \cdot L_1 + q2_\text{d} \cdot \frac{L_1^2}{2} + M31_\text{d.stb}$$

$$M21_\text{d.stb} = 57.248 \, kN \cdot m$$

Design value of destabilizing moment of diaphragm 2.1 at its top due to lateral load $H3_\text{v.d}$, $M21_\text{d.dst}$

$$M21_\text{d.dst} = \left(H3_\text{v.d} \cdot \frac{F31_\text{vRd}}{F31_\text{vRd} + F32_\text{vRd}} \right) \cdot \left(h_\text{p} + df \right)$$

$$M21_\text{d.dst} = 8.049 \, kN \cdot m$$

Net stabilizing moment at the base of diaphragm 2.1, $M21_\text{d.stb.n}$

$$M21_\text{d.stb.n} = M21_\text{d.stb} - M21_\text{d.dst}$$

$$M21_\text{d.stb.n} = 49.198 \, kN \cdot m$$

Design value of stabilising moment of diaphragm 2.2 due to vertical loading, $M22_\text{d.stb}$

$$M22_\text{d.stb} = \left[q2_\text{d} \cdot \left(\frac{b_7}{2} \right) \right] \cdot L_2 + q2_\text{d} \cdot \frac{L_2^2}{2} + M32_\text{d.stb}$$

$$M22_\text{d.stb} = 58.762 \, kN \cdot m$$

Design value of destabilising moment on diaphragm 2.2 at its top due to lateral load $H3_\text{v.d}$, $M22_\text{d.dst}$

$$M22_\text{d.dst} = \left(H3_\text{v.d} \cdot \frac{F32_\text{vRd}}{F31_\text{vRd} + F32_\text{vRd}} \right) \cdot \left(h_\text{p} + df \right)$$

$$M22_\text{d.dst} = 9.963 \, kN \cdot m$$

Net stabilising moment at the base of diaphragm 2.2, $M22_\text{d.stb.n}$ $\qquad M22_\text{d.stb.n} = M22_\text{d.stb} - M22_\text{d.dst} \qquad M22_\text{d.stb.n} = 48.799 \, kN \cdot m$

Initial estimate of fastener spacing around perimeter of sheathing in diaphragm 2.1, $s21$ $\qquad s21 = 0.15 \cdot m$

Design shear capacity of perimeter fastener/m in diaphragm 2.1, $f21_{pdt}$ (equation (13.16))	$f21_{pdt} = \dfrac{Fs_{f.Rd}}{s21} \cdot \left(1.15 + \dfrac{s21}{m}\right)$	
	$f21_{pdt} = 3.77 \text{ kN/m}^1$	

Initial estimate of base fastener spacing to floor structure for 2.1 and 2.2, $sb2$

$$sb2 = 300 \text{ mm}$$

Mu factor to be used for base fixings of diaphragm 2.1 and 2.2 to floor, $\mu21$	$\mu21 = \min\left(1, \dfrac{\dfrac{F_{ax.Rd}}{sb2}}{f21_{pdt}}\right)$	$\mu21 = 0.164$

Evaluation of $K_{i,w}$ factor for diaphragm 2.1, $K21_w$

$$K21_w = \min\left[1, \left[\left[1 + \left(\frac{h_p}{\mu21 \cdot L_1}\right)^2 + 2 \cdot \frac{M21_{d.stb.n}}{\mu21 \cdot f21_{pdt} \cdot L_1^2}\right]^{0.5} - \frac{h_p}{\mu21 \cdot L_1}\right]\right] \qquad K21_w = 1$$

Racking strength of wall diaphragm 2.1, $F21_{v,Rd}$

$$F21_{v,Rd} = K_{opening} \cdot K21_w \cdot f21_{pdt} \cdot L_1 \qquad\qquad F21_{v,Rd} = 11.381 \text{ kN}$$

Adopting the same value for the equivalent factors in diaphragm 2.2:

$$K22_w = \min\left[1, \left[\left[1 + \left(\frac{h_p}{\mu21 \cdot L_2}\right)^2 + 2 \cdot \frac{M22_{d.stb.n}}{\mu21 \cdot f21_{pdt} \cdot L_2^2}\right]^{0.5} - \frac{h_p}{\mu21 \cdot L_2}\right]\right] \qquad K22_w = 1$$

Racking strength of wall diaphragm 2.2, $F22_{v.Rd}$	$F22_{v,Rd} = K22_w \cdot f21_{pdt} \cdot L_2$	$F22_{v,Rd} = 13.798 \text{ kN}$
Racking strength of racking wall	$F2_{v,Rd} = F21_{v,Rd} + F22_{v,Rd}$	$F2_{v,Rd} = 2.518 \times 10^4 \text{ N}$
Racking force/racking strength on racking wall at level 2	$\dfrac{F2_{v.d}}{F2_{v,Rd}} = 0.719$	i.e. less than $1 - $ O.K., and as maximum spacing of 150 mm used, the spacing cannot be increased.
Spacing of fixings in abutting stud position in both diaphragms, $sp2st$	$sp2st = \dfrac{F_{f.Rd}}{f21_{pdt}}$	$sp2st = 0.172 \text{ m}$ say 170 mm c/c
Length of 2.1 resisting uplift, $L21_{up}$	$L21_{up} = (1 - K21_w) \cdot L_1$	$L21_{up} = 0 \text{ m}$
Length of 2.1 resisting racking force	$K_{opening} \cdot (L_1 - L21_{up}) = 3.019 \text{ m}$	
Length of 2.2 resisting uplift, $L22_{up}$	$L22_{up} = (1 - K22_w) \cdot L_2$	$L22_{up} = 0 \text{ m}$
Length of 2.2 resisting racking force	$(L_2 - L22_{up}) = 3.66 \text{ m}$	

Determine lateral spacing of diaphragm base fixings over racking force length $K22_{\mathrm{w}}L_2$, and $K21_{\mathrm{w}}L_1$, $sb2b$	$sb2b = \left[\dfrac{\dfrac{F_{\mathrm{f.Rd}}}{F21_{\mathrm{vRd}}}}{K_{\mathrm{opening}} \cdot \left(L_1 - L21_{\mathrm{up}}\right)} \right]$

$$sb2b = 0.172\,\mathrm{m}$$

i.e. use 150 mm c/c (by setting opening factor = 1, the spacing over length $K22_{\mathrm{w}}L_2$ can be increased.

Lateral spacing of diaphragm base fixings over uplift length of each diaphragm	$sb2 = 0.3\,\mathrm{m}$	as there is no tension uplift, these fixings are nominal.

Check sliding condition at wall base to floor connection:

Frictional force available , $F2_{\mathrm{fric}}$

$F2_{\mathrm{fric}} = \mu \cdot [2 \cdot VD2_{\mathrm{d}} + q2_{\mathrm{d}} \cdot (L_1 + L_2 + b_7)] + F3_{\mathrm{fric}}$	$F2_{\mathrm{fric}} = 25.357\,\mathrm{kN}$

$\dfrac{F2_{\mathrm{fric}}}{F2_{\mathrm{v.d}}} = 1.401$	i.e. greater than $F3_{\mathrm{v.d}}$ O.K. without need to draw on base fixing strength.

Check compliance with racking deflection limit in PD6693-1

$\dfrac{K21_{\mathrm{w}} \cdot f21_{\mathrm{pdt}}}{8 \cdot (\mathrm{kN \cdot m}^{-1}) \cdot \left(\dfrac{L_1}{h_{\mathrm{p}}}\right)} = 0.309$ and	$\dfrac{K22_{\mathrm{w}} \cdot f21_{\mathrm{pdt}}}{8 \cdot (\mathrm{kN \cdot m}^{-1}) \cdot \left(\dfrac{L_2}{h_{\mathrm{p}}}\right)} = 0.30$	both less than 1 therefore O.K.

3.3 Ground floor level

Design value of the wind load at the wall head level, $H1_{\mathrm{v.Ed}}$	$H1_{\mathrm{v.d}} = 11.55 \cdot \mathrm{kN}$

Design value of the racking force at the wall head level, $F1_{\mathrm{v.Ed}}$	$F1_{\mathrm{v.d}} = H1_{\mathrm{v.d}} + H2_{\mathrm{v.d}} + H3_{\mathrm{v.d}}$

$$F1_{\mathrm{v.d}} = 2.965 \times 10^4\,\mathrm{N}$$

Design value of the permanent uniformly distributed vertical load on wall head, $q1_{\mathrm{d}}$	$q1_{\mathrm{d}} = 3.29 \cdot \mathrm{kN/m}$

Design value of point load at A, $VA1_{\mathrm{d}}$	$VA1_{\mathrm{d}} = 1.8 \cdot \mathrm{kN}$

Design value of point load at D, $VD1_{\mathrm{d}}$	$VD1_{\mathrm{d}} = 1.8 \cdot \mathrm{kN}$

Design value of stabilising moment on diaphragm 1.1 due to vertical loading, $M11_{\mathrm{d.stb}}$

$M11_{\mathrm{d.stb}} = VA1_{\mathrm{d}} \cdot L_1 + q1_{\mathrm{d}} \cdot \dfrac{L_1^2}{2} + M21_{\mathrm{d.stb}} + M31_{\mathrm{d.stb}}$	$M11_{\mathrm{d.stb}} = 114.495\,\mathrm{kN \cdot m}$

Design value of destabilising moment on diaphragm 1.1 at its top due to lateral load $H3_{v.d}$, and $H2_{vd}$, $M11_{d.dst}$

$$M11_{d.dst} = \left(H3_{v.d} \cdot \frac{F31_{vRd}}{F31_{vRd} + F32_{vRd}} \right) \cdot \left(2 \cdot h_p + 2 \cdot df \right) + \left(H2_{v.d} \cdot \frac{F21_{vRd}}{F21_{vRd} + F22_{vRd}} \right) \cdot \left(h_p + df \right)$$

$$M11_{d.dst} = 30.455 \text{ kN} \cdot \text{m}$$

Net stabilising moment at the base of diaphragm 1.1, $M11_{d.stb.n}$

$$M11_{d.stb.n} = M11_{d.stb} - M11_{d.dst}$$

$$M11_{d.stb.n} = 84.04 \text{ kN} \cdot \text{m}$$

Design value of stabilising moment on diaphragm 1.2 due to vertical loading, $M12_{d.stb}$

$$M12_{d.stb} = \left[q1_d \cdot \left(\frac{b_7}{2} \right) \right] \cdot L_2 + q1_d \cdot \frac{L_2^2}{2} + \left(M22_{d.stb} + M32_{d.stb} \right)$$

$$M12_{d.stb} = 117.524 \text{ kN} \cdot \text{m}$$

Design value of destabilising moment on diaphragm 1.2 at its top due to lateral load $H3_{v.d}$, $M12_{d.dst}$

$$M12_{d.dst} = \left(H3_{v.d} \cdot \frac{F32_{vRd}}{F31_{vRd} + F32_{vRd}} \right) \cdot \left(2 \cdot h_p + 2 \cdot df \right) + \left(H2_{v.d} \cdot \frac{F22_{vRd}}{F21_{vRd} + F22_{vRd}} \right) \cdot \left(h_p + df \right)$$

$$M12_{d.dst} = 37.333 \text{ kN} \cdot \text{m}$$

Net stabilising moment at the base of diaphragm 1.2, $M12_{d.stb.n}$

$$M12_{d.stb.n} = M12_{d.stb} - M12_{d.dst}$$

$$M12_{d.stb.n} = 80.191 \text{ kN} \cdot \text{m}$$

Initial estimate of fastener spacing around perimeter of sheathing in diaphragm 1.1, $s11$

$$s11 = 0.1 \cdot \text{m}$$

Design shear capacity of perimeter fastener/m in diaphragm 1.1, $f11_{pdt}$ (equation (13.9))

$$f11_{pdt} = \frac{Fs_{f.Rd}}{s11} \cdot \left(1.15 + \frac{s11}{\text{m}} \right)$$

$$f11_{pdt} = 5.438 \text{ kN/m}$$

Initial estimate of base fastener spacing to floor structure for 1.1 and 1.2, $sb1$

$$sb1 = 300 \text{ mm}$$

Mu factor to be used for base fixings of diaphragm 1.1 to ground floor, $\mu11$

$$\mu11 = \min \left(1, \frac{\dfrac{F_{ax.Rd}}{sb1}}{f11_{pdt}} \right)$$

$$\mu11 = 0.114$$

Evaluation of $K_{i,w}$ factor for diaphragm 1.1 $K11_w$

$$K11_w = \min \left[1, \left[\left[1 + \left(\frac{h_p}{\mu11 \cdot L_1} \right)^2 + 2 \cdot \frac{M11_{d.stb.n}}{\mu11 \cdot f11_{pdt} \cdot L_1^2} \right]^{0.5} - \frac{h_p}{\mu11 \cdot L_1} \right] \right]$$

$$K11_w = 1$$

Racking strength of wall diaphragm 1.1, $F11_{v.Rd}$

$$F11_{vRd} = K_{opening} \cdot K11_w \cdot f11_{pdt} \cdot L_1 \qquad F11_{vRd} = 16.414\,kN$$

Adopting the same value for the equivalent factors in diaphragm 1.2:

$$K12_w = \min\left[1, \left[\left[1+\left(\frac{h_p}{\mu 11 \cdot L_2}\right)^2 + 2 \cdot \frac{M12_{d.stb.n}}{\mu 11 \cdot f11_{pdt} \cdot L_2^{\;2}}\right]^{0.5} - \frac{h_p}{\mu 11 \cdot L_2}\right]\right]$$

$$K12_w = 1$$

Racking strength of wall diaphragm 1.2, $F12_{v.Rd}$

	$F12_{vRd} = K12_w \cdot f11_{pdt} \cdot L_2$	$F12_{vRd} = 19.901\,kN$
Racking strength of racking wall	$F1_{vRd} = F11_{vRd} + F12_{vRd}$	$F1_{vRd} = 3.632 \times 10^4\,N$
Racking force/racking strength on racking wall at level 1	$\dfrac{F1_{v.d}}{F1_{vRd}} = 0.816$	i.e. less than $1-OK$
Spacing of fixings in abutting stud positions in both diaphragms, $sp1st$	$sp1st = \dfrac{F_{f.Rd}}{f11_{pdt}}$	$sp1st = 0.119\,m$ say 110mm c/c
Length of 1.1 resisting uplift, $L11_{up}$	$L11_{up} = (1 - K11_w) \cdot L_1$	$L11_{up} = 0\,m$
Length of 1.1 resisting racking force	$K_{opening} \cdot (L_1 - L11_{up}) = 3.019\ m$	
Length of 1.2 resisting uplift, $L12_{up}$	$L12_{up} = (1 - K12_w) \cdot L_2$	$L12_{up} = 0\,m$
Length of 1.2 resisting racking force	$(L_2 - L12_{up}) = 3.66\,m$	
Determine lateral spacing of diaphragm base fixings over racking force length $K12_w L_2$, and $K11_w L1$, $sb1b$	$sb1b = \left[\dfrac{F_{f.Rd}}{\dfrac{F11_{vRd}}{K_{opening} \cdot (L_1 - L11_{up})}}\right]$	

$sb1b = 0.119\,m$

i.e. use 110mm c/c (by setting opening factor = 1, the spacing over length $K12_w L_2$ can be increased.

Lateral spacing of diaphragm base fixings over uplift length of each diaphragm $sb1 = 0.3\,m$ or continue with $sb1b$ in these lengths.

Check sliding condition at wall base to floor connection:

Frictional force available, $F1_{fric}$

$$F1_{fric} = \mu \cdot [2 \cdot VD1_d + q1_d \cdot (L_1 + L_2 + b_7)] + F2_{fric} + F3_{fric} \qquad F1_{fric} = 50.715\,kN$$

$$\frac{F1_{fric}}{F1_{v.d}} = 1.71$$

i.e greater than $F1_{v.d}$ OK without need to draw on base fixing strength.

Check compliance with racking deflection limit in PD6693-1

$$\frac{K11_w \cdot f11_{pdt}}{8(kN \cdot m^{-1}) \cdot \left(\dfrac{L_1}{h_p}\right)} = 0.446 \text{ and } \frac{K12_w \cdot f11_{pdt}}{8(kN \cdot m^{-1}) \cdot \left(\dfrac{L_1}{h_p}\right)} = 0.446 \qquad \begin{array}{l}\text{both less than 1}\\ \text{therefore OK}\end{array}$$

Design compressive load on studs in kN on diaphragm studs within $0.1L$ from leeward end:

Diaphragm 1.1, total design load on studs within this length, $F11studs$

Vertical load in diaphragm 1.1

$$V11_d = VD3_d + VD2_d + VD1_d + \left(q3_d + q2_d + q1_d\right) \cdot \left(L_1 + \frac{b_7}{2}\right) \qquad V11_d = 47.545 \text{ kN}$$

Destabilising moment about diaphragm base, $M11_{d,dst,base}$

$$M11_{d.dst.base} = H3_{v.d} \cdot \frac{F31_{vRd}}{F31_{vRd} + F32_{vRd}} \cdot (3 \cdot h_p + 2 \cdot df) + H2_{v.d} \cdot \frac{F21_{vRd}}{F21_{vRd} + F22_{vRd}} \cdot (2h_p + df)$$

$$+ H1_{v.d} \cdot \frac{F11_{vRd}}{F11_{vRd} + F12_{vRd}} \cdot (h_p)$$

$$M11_{d.dst.base} = 62.538 \text{ kN} \cdot \text{m}$$

$$F11_{studs} = 0.8 \cdot \left[(V11_d) \cdot \left(\frac{M11_{d.dst.base}}{M11_{d.stb}}\right) + \frac{0.6}{\dfrac{L_1}{mm}} \right] \qquad F11_{studs} = 20.782 \text{ kN}$$

Diaphragm 1.2, total design load on studs within this length, $F12_{studs}$

Vertical load in diaphragm 1.2

$$V12_d = VD3_d + VD2_d + VD1_d + \left(q3_d + q2_d + q1_d\right) \cdot \left(L_2 + \frac{b_7}{2}\right) \qquad V12_d = 47.545 \text{ kN}$$

$$M12_{d.dst.base} = H3_{v.d} \cdot \frac{F32_{vRd}}{F31_{vRd} + F32_{vRd}} \cdot (3 \cdot h_p + 2 \cdot df) + H2_{v.d} \cdot \frac{F22_{vRd}}{F21_{vRd} + F22_{vRd}} \cdot (2h_p + df)$$

$$+ H1_{v.d} \cdot \frac{F12_{vRd}}{F11_{vRd} + F12_{vRd}} \cdot (h_p)$$

$$M12_{d.dst.base} = 76.41 \text{ kN} \cdot \text{m}$$

$$F12_{studs} = 0.8 \cdot \left[V12_d \cdot \left(\frac{M12_{d.dst.base}}{M12_{d.stb}} + \frac{0.6}{\dfrac{L_2}{mm}}\right) \right] \qquad F12_{studs} = 24.736 \text{ kN}$$

Larger axial loads within this $0.1\,L$ length may occur under combined permanent, variable and wind load and to obtain these the above procedure is repeated but using the loading: Roof level, $qf3d = 5.23\,\text{kN/m}$, $VfA3d = VfD3d = 2.43$ kN; Second floor level, $qf2d = 6.28$ kN/m, $VfA2d = VfD2d = 2.43$ kN; First floor level, $qf1d = 6.28$ kN/m, $VfA1d = VfD1d = 2.43$ kN;

Under this loading the maximum values will be: $F11_{\text{studs}} = 22.05$ kN; $F12_{\text{studs}} = 24.68$ kN

As stated under item 4.0 in Example 13.8.1, the strength of the stud and header/footer wall plates need to be checked at each floor and where required, additional abutting/ interconnected studs can be used to acheive the required strength. With these walls, where there is a return wall connected to it at the leeward end, it is also permitted to spread up to 50% of the load to studs in the return wall, up to a maximum of 1 metre from the diaphragm, providing the return wall is adequately connected to the diaphragm.

4.0 With the approach as described in PD6693-1 the racking strength used is unable to be adjusted to equate to the racking force, resulting in a conservative solution when the racking force is well below the racking strength. When Mathcad or equivalent software is being used the design procedure can be optimised to better equate strength to racking force and the procedure to be used for the level 3 analysis (given in section 3.1 above) is as described below. The same procedure can be applied to all floors to optimise the wall design.

Adjust the value used for the fastener spacing ($s31$) around perimeter of sheathing in diaphragm 3.1 until the racking strength equals the racking force (as shown against **Item A** below, $s3.1$

$$s31 = 0.576 \cdot m$$

In PD6693-1 the perimeter fastener strength/m length is a function of the fastener spacing, with a minimu permited spacing of 150 mm. In the optimising procedure spacings greater than 150 mm will be **assumed** in the **analysis** and for such values a maximum enhancement factor of 1.3 has been used in the analysis:

Design shear capacity of perimeter fastener/m in diaphragm 3.1, $f31_{\text{pdt}}$ (Equation (13.19))

$$f31_{\text{pdt}} = \min\left[\frac{1.3Fs_{\text{f.Rd}}}{s31}, \frac{Fs_{\text{f.Rd}}}{s31}\cdot\left(1.15+\frac{s31}{m}\right)\right]$$

$$f31_{\text{pdf}} = 0.982\ \text{kN/m}$$

Initial estimate of base fastener spacing to floor structure for 3.1 and 3.2 to the floor, $sb3$

$$sb3 = 300\,\text{mm}$$

In this approach the designer is allowed to use a smaller value for mu than the limit set in PD6693-1:

Estimate of mu factor to be used, $\mu31e$

$$\mu31e = 0.1$$

Mu factor to be used for diaphragm base fixings of diaphragm 3.1 and 3.2 to floor, $\mu31$

$$\mu31 = \min\left(1, \frac{\dfrac{F_{\text{ax.Rd}}}{sb3}}{f31_{\text{pdt}}}, \mu31e\right) \quad \mu31 = 0.1$$

Evaluation of $K_{i,w}$ factor for diaphragm 3.1, $K31_w$

$$K31_w = \min\left[1, \left[\left[1+\left(\frac{h_p}{\mu31\cdot L_1}\right)^2 + 2\cdot\frac{M31_{d.stb}}{\mu31\cdot f31_{pdt}\cdot L_1^2}\right]^{0.5} - \frac{h_p}{\mu31.L_1}\right]\right] \qquad K31_w = 1$$

Racking strength of wall diaphragm 3.1, $F31_{v.Rd}$

$F31_{vRd} = K_{opening}\cdot K31_w\cdot f31_{pdt}\cdot L_1$ $\qquad\qquad\qquad\qquad F31_{vRd} = 2.964$ kN

Adopting the same value for the equivalent factors in diaphragm 3.2:

$$K32_w = \min\left[1, \left[\left[1+\left(\frac{h_p}{\mu31\cdot L_2}\right)^2 + 2\cdot\frac{M32_{d.stb}}{\mu31\cdot f31_{pdt}\cdot L_2^2}\right]^{0.5} - \frac{h_p}{\mu31.L_2}\right]\right] \qquad K32_w = 1$$

Racking strength of wall diaphragm 3.2, $F32_{v.Rd}$

$F32_{vRd} = K32_w\cdot f31_{pdt}\cdot L_2$ $\qquad\qquad\qquad\qquad\qquad F32_{vRd} = 3.593$ kN

Racking strength of racking wall, $F3_{v.Rd}$

$F3_{vRd} = F31_{vRd} + F32_{vRd}$ $\qquad\qquad\qquad\qquad\qquad F3_{vRd} = 6.557\times10^3$ N

Item A: Racking force/racking strength on racking wall at level 3 (adjusted by changing the value of the fastener spacing ($s31$):

$$\frac{F3_{v.d}}{F3_{vRd}} = 0.999 \qquad\qquad\qquad \text{i.e. say–OK}$$

Spacing of fixings in abutting stud positions in both diaphragms, $sp3st$ $\qquad sp3st = \dfrac{F_{f.Rd}}{f31_{pdt}}$ $\qquad\qquad sp3st = 0.661$ m

$\qquad\qquad\qquad\qquad\qquad\qquad\qquad\qquad\qquad\qquad\qquad$ say 300 mm c/c

Length of 3.1 resisting uplift, $L31_{up}$ $\qquad L31_{up} = (1-K31_w)\cdot L_1$ $\qquad\qquad L31_{up} = 0$ m

Length of 3.1 resisting racking force $\qquad K_{opening}\cdot(L_1 - L31_{up}) = 3.019$ m

Length of 3.2 resisting uplift, $L32_{up}$ $\qquad L32_{up} = (1-K32_w)\cdot L_2$ $\qquad\qquad L32_{up} = 0$ m

Length of 3.2 resisting racking force $\qquad (L_2 - L32_{up}) = 3.66$ m

Determine lateral spacing of diaphragm base to floor fixings over racking force lengths $K32_w l_2$, and $K31_w L_1$, $sb3b$ $\qquad sb3b = \left[\dfrac{F_{f.Rd}}{\dfrac{F31_{vRd}}{K_{opening}\cdot\left(L_1 - L31_{up}\right)}}\right]$ $\qquad sb3b = 0.661$ m

$\qquad\qquad\qquad\qquad\qquad\qquad\qquad\qquad\qquad\qquad\qquad$ i.e. use 300 mm c/c

Lateral spacing of diaphragm base fixings over uplift length of each diaphragm $\qquad sb3 = 0.3$ m $\qquad\qquad$ no uplift in this zone so use nominal fixing.

Check sliding condition at wall base to floor connection:

Frictional force available, $F3_{fric}$

$$F3_{fric} = \mu \left[2 \cdot VD3_d + q3_d \cdot (L_1 + L_2 + b_7) \right]$$ $F3_{fric} = 12.679 \text{ kN}$

$$\frac{F3_{fric}}{F3_{v.d}} = 1.936$$ i.e. greater than $F3_{v.d}$ OK. without need to draw on base fixings.

Check compliance with racking deflection limit in PD6693-1

$$\frac{K31_w \cdot f31_{pdt}}{8 \cdot (\text{kN/m}) \cdot \left(\dfrac{L_1}{h_p} \right)} = 0.08 \quad \text{and} \quad \frac{K32_w \cdot f31_{pdt}}{8 \cdot (\text{kN/m}) \cdot \left(\dfrac{L_1}{h_p} \right)} = 0.08 \qquad \begin{array}{l} \text{both less than 1} \\ \text{therefore OK.} \end{array}$$

Continue the solution as in the above example and when analyzing level 2, which is covered in Section 3.2, use an equivalent insert to that given in 40 above at the appropriate position in the analysis. Also repeat this procedure for the ground floor, which is covered in Section 3.3.

Appendix A
Weights of Building Materials

The characteristic value of the loading, G_k, based on the mean value of material density imposed by typical building materials, is given in Table A1. The values are based on the content of BS 648:1964.

Table A1 Weights of building materials, (based on BS 648 : 1964)

Material	Unit mass
Asphalt	
Roofing 2 layers, 19 mm thick	42 kg/m²
Damp-proofing, 19 mm thick	41 kg/m²
Bitumen roofing felts	
Mineral surfaced bitumen per layer	44 kg/m²
Glass fibre	
Slab, per 25 mm thick	2.0–5.0 kg/m²
Gypsum panels and partitions	
Building panels 75 mm thick	44 kg/m²
Lead	
Sheet, 2.5 mm thick	30 kg/m²
Linoleom	
3 mm thick	6 kg/m²
Plaster	
Two coats gypsum, 13 mm thick	22 kg/m²
Plastic sheeting	
Corrugated	4.5 kg/m²
Plywood	
per mm thick	0.7 kg/m²
Rendering or screeding	
Cement:sand (1:3), 13 mm thick	30 kg/m²
Slate tiles	
(depending upon thickness & source)	24–78 kg/m²
	(continued)

Structural Timber Design to Eurocode 5, Second Edition. Jack Porteous and Abdy Kermani.
© Jack Porteous and Abdy Kermani 2013. Published 2013 by Blackwell Publishing Ltd.

Table A1 (*continued*)

Steel	
Solid (mild)	$7850\,\text{kg/m}^3$
Corrugated roofing sheet per mm thick	$10\,\text{kg/m}^2$
Tarmacadam	
25 mm thick	$60\,\text{kg/m}^2$
Tiling	
Clay, for roof	$70\,\text{kg/m}^2$
Timber	
Softwood	$590\,\text{kg/m}^3$
Hardwood	$1250\,\text{kg/m}^3$
Water	$1000\,\text{kg/m}^3$
Woodwool	
Slab, 25 mm thick	$15\,\text{kg/m}^2$

Appendix B

Related British Standards for Timber Engineering in Buildings

The following list of Eurocodes and their associated United Kingdom National annex:

Eurocode 0

BS EN 1990: Eurocode – Basis of structural design

Eurocode 1: Actions on structures

BS EN 1991-1-1: Densities, self-weight and imposed loads for buildings
BS EN 1991-1-2: Actions on structures exposed to fire
BS EN 1991-1-3: Snow loads
BS EN 1991-1-4: Wind actions
BS EN 1991-1-5: Thermal actions
BS EN 1991-1-6: Actions during execution
BS EN 1991-1-7: Accidental actions

Eurocode 5: Design of timber structures

BS EN 1995-1-1: General – Common rules and rules for buildings
BS EN 1995-1-2: General rules – Structural fire design
BS EN 1995-2: Bridges

BS EN standards

BS EN 300: Oriented Strand Board (OSB) – Definition, classification and specifications.
BS EN 301: Adhesives phenolic and aminoplastic, for load-bearing timber structures: Classification and performance requirements

---

Structural Timber Design to Eurocode 5, Second Edition. Jack Porteous and Abdy Kermani.
© Jack Porteous and Abdy Kermani 2013. Published 2013 by Blackwell Publishing Ltd.

BS EN 312: Particleboards – Specifications

BS EN 335-1: Durability of wood and wood-based products. Definitions of use classes. General.

BS EN 336: Structural timber – Sizes, permitted deviations

BS EN 338: Structural timber – Strength classes

BS EN 386: Glued laminated timber – Performance requirements and minimum production requirements

BS EN 338: Structural timber – Strength classes

BS EN 387: Glued laminated timber – Large finger joints – Performance requirements and minimum production requirements

BS EN 408: Timber structures – Structural timber and glued-laminated timber – Determination of some physical and mechanical properties

BS EN 636: Plywood – specifications

BS EN 789: Timber structures. Test methods. Determination of mechanical properties of wood based panels

BS EN 912: Timber fasteners – Specifications for connectors for timber

BS EN 1058: Wood-based panels. Determination of characteristic 5-percentile values and characteristic mean values

BS EN 1193: Timber structures – Structural timber and glued-laminated timber – Determination of shear strength and mechanical properties perpendicular to the grain

BS EN 1194: Timber structures – Glued-laminated timber – Strength classes and determination of characteristic values

BS EN 1313-1: Round and sawn timber – Permitted deviations and preferred sizes. Part 1: Softwood sawn timber

BS EN 12369-1: Wood-based panels – characteristic values for structural design. Part 1: OSB, particleboards and fibreboard

BS EN 12369-2: Wood-based panels – characteristic values for structural design – Part 2: Plywood

BS EN 14081-1: Timber structures – Strength graded structural timber with rectangular cross section – Part 1: General requirements

BS EN 14081-2: Timber structures – Strength graded structural timber with rectangular cross section – Part 2: Machine grading, additional requirements for initial type testing

BS EN 14272: Plywood. Calculation method for some mechanical properties

prEN 16351: Draft BS EN 16351. Timber structures – Cross laminated timber – Requirements.

British Standards

BS 4978: Specification for visual strength grading of softwood

BS 5268-2: Structural use of timber – Part 2: Code of practice for permissible stress design, materials and workmanship

BS 5756: Specification for visual strength grading of hardwood

BS 7359: Nomenclature of commercial timbers including sources of supply

BS 8417: Preservation of timber – Recommendations.

Appendix C

Possible Revisions to be Addressed in a Corrigendum to EN 1995-1-1:2004+A1:2008

At the time of publication of this second edition, a draft corrigendum is in the course of preparation for Amendment A2 to *Eurocode 5: Design of Structures. Part 1-1.* It has been under discussion for a considerable period but the content is still not fully agreed. Of the changes being considered, an outline of the possible changes, which, if approved, will be relevant to the content of this book, is given below. Where an amendment will result in a change to a design procedure described in the book, reference is made to the likely content of the draft proposal in the text.

The likely content of the draft amendments are given against the section number used in EC5 and those that, if approved, will result in a change to the relevant item in the book cover the following topics:

Section numbers 2.2.3(3), 2.2.3(4) and 2.3.2.2(1)

The code requirement in *2.2.3(4)* for the calculation of the final deformation in structures consisting of members having different creep behaviour is incorrect and to clarify what is required the content of *2.2.3(3)*, *2.2.3(4)* as well as *2.3.2.2(1)* are to be amended.

Change to 2.2.3(3): The final deformation u_{fin}, which remains as shown in EC5, *Figure 7.1*, should be obtained by adding the creep deformation u_{creep}, which is calculated using the quasi-permanent combination of actions, (as referred to in EN 1990, *6.5.3(2)(c)*) to the instantaneous deformation $u_{inst.}$ The instantaneous deformation is still calculated from the requirements of *2.2.3(2)* and the creep deformation is calculated using mean values of the appropriate elastic, shear and slip moduli and the relevant values of k_{def} give in *Table 3.2*.

Change to 2.2.3(4): Where the structure consists of members or components that have different creep behaviour, the time induced deformation arising from the quasi-permanent combination of actions is to be calculated using the final mean values of the appropriate elastic, shear and slip moduli according to *2.3.2.2(1)*. To obtain the final deformation, u_{fin}, the instantaneous deformation due to the quasi-permanent combination of actions is subtracted from the instantaneous deformation due to the

---

Structural Timber Design to Eurocode 5, Second Edition. Jack Porteous and Abdy Kermani.
© Jack Porteous and Abdy Kermani 2013. Published 2013 by Blackwell Publishing Ltd.

characteristic combination of actions, and the resulting value is added to the time-induced deformation referred to above.

Change to 2.3.2.2(1): The content will be amended to clarify that for the types of structure being referred to, the stiffness relationships given in *equations (2.7), (2.8)* and *(2.9)* will only relate to the calculation of the time induced deformation arising from the quasi-permanent combination of actions.

Section number *6.1.5(4)*

It is not clear from the presentation style used in EC5 that the content of *6.1.5(4)* is meant to cover for members that can also be subjected to uniformly distributed loading and the case given covers for situations where there are additional localised loads at a distance $\ell_1 \geq 2h$, as shown on Figure 6.2(b). The change to be formalised has already been incorporated (and clarified) in PD6693-1 and the code will incorporate a statement to also clarify the intent of this clause.

Section number *6.1.8*

It has been concluded that the value given for the shape factor, k_{shape}, for circular sections can be retained but those for rectangular sections are potentially unsafe. The values are to read:

$k_{shape} = 1.2$ for a circular cross-section

$k_{shape} = \min[(1 + 0.05h/b), 1.3]$ for a rectangular section.

Section number *6.2.3*

The condition where a member can fail by lateral torsional instability under major axis bending when subjected to combined bending and axial tension is not covered by the code and as a safe approach, it will be proposed that the method given in *6.3* will apply but the tensile stress should be taken to be 0, i.e. design the section taking into account instability requirements but ignore any stress due to tensile forces.

Section number *6.5.2(2)*

In *equation (6.60)* the function b should read b_{ef}, where b_{ef} is as defined in *6.1.7(2)*.

Section number *8.3.2(4)*

It is to be stated that when calculating t_{pen} in the pointside member, the length shall exclude the point length of the nail.

Section number *8.3.2(6)*

The solutions to *equations (8.25)* and *(8.26)* are in N/mm² units and this is to be stated in the proposed revision.

Section number *8.4(6)*

The current *equation (8.29)* is wrong and will be replaced by $M_{y,Rk} = 150d^3$.

Section number *8.4(7)*

In *8.4(7)* the value used for n_{ef} in a row of n staples is incorrect. For such a condition the value is to be taken to always equal n and the code is to be revised to state this.

Section number *8.5*

In *Table 8.5* there is an error in the minimum unloaded end distance, $a_{3,c}$, and the values to be used are to be:

$90° \leq \alpha \leq 150°$	$a_{3t} \lvert sin\alpha \rvert$
$150° \leq \alpha \leq 210°$	$max(3.5d; 40\,mm)$
$210° \leq \alpha \leq 270°$	$a_{3t} \lvert sin\alpha \rvert$

Section number *8.7.1*

Design requirements when using screws are to be clarified and the code revision will relate to the content of clauses *(1)*, *(4)*, *(5)* and *(6)*. They cover:

(1): When determining the load carrying capacity of a screw (e.g. when using the Johansen equations in EC5), the effect of the threaded part of the screw shall be taken into account by using an effective diameter, d_{ef}, when determining the yield capacity and the embedment strength. The value used for d_{ef} will be as defined in the existing clauses *8.7.1(2)* and *8.7.1(3)*. Also, it will be stated that when determining spacing, edge and end distances and the effective number of screws, the outer thread diameter, d, shall be used.
(4): When d_{ef} is greater than 6 mm, the rules in *8.5.1* will apply.
(5): When d_{ef} is less than or equal to 6 mm, the rules in *8.3.1* will apply.
(6): When the screw diameter is between 6 mm and 8 mm, relevant values for spacing, edge and end distances shall be obtained by linear interpolation between *Tables 8.2* and *8.4*.

The UK view is that the proposed change will not adequately clarify points of detail and a more detailed statement of requirement has been incorporated into PD6693-1.

Section number *8.7.2(4)*

In the first line it is to be stated that the content of *(4)* applies to softwood timber.

Section number *8.9*

In *Table 8.7*, the value '$1.5d_c$' in the last column in the row for $a_{3,t}$, is to be changed to read '$2.0d_c$'.

Section number *8.10*

In *Table 8.8*, the value '$2.0d_c$' in the last column in the row for $a_{3,t}$, is to be changed to read '$1.5d_c$'.

Annex A (Informative): Block shear and plug shear failure at multiple dowel-type steel-to-timber connections

The formula given against *equation (A.7) (d)(g)* is wrong and will be replaced by:

$$'t_{ef} = t_1 \left(\sqrt{2 + \frac{4M_{y,Rk}}{f_{h,k} d t_1^2}} - 1 \right),$$

Annex B.4 (Informative): Mechanical jointed beams

B4 Maximum shear stress

There is an error in *equation (B.9)*. Value h_2 in the *equation* is wrong and will be replaced by h^2.

Index

Structural Timber Design to Eurocode 5, Second Edition. Jack Porteous and Abdy Kermani.
© Jack Porteous and Abdy Kermani 2013. Published 2013 by Blackwell Publishing Ltd.

The Example Worksheets Order Form

All design examples given in this book are produced in the form of worksheets using Mathcad computer software, licensed by Parametric Technology Corporation (PTC), and are available on CD to run under Mathcad Version 11 or higher on any IBM/PC Personal Computer.

The worksheet files are labelled Example *l.m.n*.mcd where *l.m* refers to the chapter and section numbers and *n* to the example number in that chapter. For example, Example 6.7.2.mcd refers to Example 2 in Chapter 6, Section 7.

It is recommended that the user make a backup copy of the worksheets. This way he or she is free to experiment with the worksheets. When a worksheet is loaded it should be saved under a new file name so that the modified original disk file remains unaltered.

Although great care has been taken to ensure the accuracy of the example worksheets, it remains the responsibility of the user to check their results.

Please copy and complete the form below, enclosing a cheque for £30, which includes handling, postage and packing, made payable to A. Kermani, and send to:

A. Kermani, 4 Mid Steil, Glenlockhart, Edinburgh EH10 5XB, UK, email: a.kermani@napier.ac.uk.

Structural Timber Design to Eurocode 5 (2nd Edition) **Mathcad Worksheets**	
Name:	**Title:**
Position:	
Please indicate your profession	
Architect	☐
Building control officer	☐
Building engineer	☐
Civil/structural engineer	☐
Educator	☐
Student	☐
Address:	
Postcode: **Email**: **Date**:	

Note: There is no warranty with the worksheets.

Structural Timber Design to Eurocode 5, Second Edition. Jack Porteous and Abdy Kermani.
© Jack Porteous and Abdy Kermani 2013. Published 2013 by Blackwell Publishing Ltd.

Printed and bound by CPI Group (UK) Ltd, Croydon, CR0 4YY

27/10/2024

14580194-0001